W9-BYH-431
19767 $60.00

Virology

Virology

third edition

Jay A. Levy

Department of Medicine
Cancer Research Institute
University of California
School of Medicine
San Francisco, California

Heinz Fraenkel-Conrat

Department of Cell and Molecular Biology
and Virus Laboratory
University of California
Berkeley, California

Robert A. Owens

Molecular Plant Pathology Laboratory
Beltsville Agricultural Research Laboratory
U. S. Department of Agriculture
Beltsville, Maryland

PRENTICE HALL, *Englewood Cliffs, New Jersey 07632*

Library of Congress Cataloging-in-Publication Data

Levy, Jay A.
Virology / Jay A. Levy, Heinz Fraenkel-Conrat, Robert A. Owens—3rd ed.
p. cm.
Fraenkel-Conrat's name appears first on earlier editions.
Includes bibliographical references and index.
ISBN 0-13-953753-8
1. Virology. I. Fraenkel-Conrat, Heinz, II. Owens, Robert A., . III. Title.
[DNLM: 1. Virus Diseases. 2. Viruses. QW 160 L6684v 1994]
QR360.F714 1994
576'.64—dc20
DNLM/DLC 93-2450
for Library of Congress CIP

Acquistions editor: David Kendric Brake
Editorial assistant: Mary DeLuca
Copy editor: Barbara Liguori
Interior line art: Vantage Art, Inc.
Production coordinator: Trudy Pisciotti
Cover design: Design Solutions
Front cover: The structure of an icosahedral insect virus derived from cryo-electron microscopy and x-ray crystallography. The outer surface is the protein shell and the red contoured interior is RNA, both visualized by electron microscopy. The cylinders are helical regions of the protein subunits that interact with the RNA and the ordered duplex RNA (yellow space-filling model in center) is part of a molecular switch that controls the capsid architecture. The coordinates for the atoms in the cylinders and the ordered RNA were determined by x-ray crystallography. The figure was created by Holland Cheng and Timothy Baker who, with Norman Olson, did the cryo-electron microscopy and image reconstruction. The x-ray crystallography was done by Andrew Fisher, Vijay Reddy, and John Johnson (Fisher and Johnson, *Nature* **361**, 176–179, 1993).
Back cover: "Vase of Flowers," Jan Davidsz de Heem (1606-1683), Courtesy of the National Gallery of Art, Washington. Note the typical petal striping systems on the tulips caused by tulip breaking virus (see Chapter 1).

A Paramount Communications Company
Englewood Cliffs, New Jersey 07632

Printed in the United States of America
10 9 8 7 6 5 4 3 2 1

ISBN 0-13-953753-8

Prentice-Hall International (UK) Limited, *London*
Prentice-Hall of Australia Pty. Limited, *Sydney*
Prentice-Hall Canada Inc., *Toronto*
Prentice-Hall Hispanoamericana, S.A., *Mexico*
Prentice-Hall of India Private Limited, *New Delhi*
Prentice-Hall of Japan, Inc., *Tokyo*
Simon & Schuster Asia Pte. Ltd., *Singapore*
Editora Prentice-Hall do Brasil, Ltda., *Rio de Janeiro*

Contents

Preface

The concept of viruses as a natural phenomenon separate from other infectious organisms is less than 100 years old, and their nature began to be understood less than 50 years ago. The realization that many diseases of plants and animals as diverse as amoebas, insects, and people can be attributed to this newly recognized type of agent came even more recently. So did the recognition that similar agents kill bacteria. In more recent years, the appreciation that practically all living species may have viruses associated with them defines the widespread prevalence of these agents and their potential importance in nature. Although not all viruses are pathogenic in their host, several have raised great concern in this century because of the serious epidemic threats in humans (influenza in 1918, poliomyelitis in the 1930s, herpes in the 1960s, and AIDS in the 1980s). These diseases, possibly caused by changes to human societies and mutations of the viruses, have brought increased attention to this field of study.

Most texts do not attempt to cover the entire field of virology because of the breadth of this subject. We felt compelled to fill this gap because we were impressed by the fact that viruses, though immensely diverse, represent a definite entity with shared properties and concepts different from all other forms in nature. This fact has become clear largely through biochemical studies of the nature of viruses. Thus, no virology or biology text can now fail to focus on the physical, chemical, and biochemical aspects of its subject matter. We regard a general knowledge of biochemistry and biology, at least at the present high school level, a prerequisite to the use of this book.

We devote the first chapter to general discussions of the nature and history of viruses and virus diseases, certain methods of studying and testing viruses, and their classification. More details on these latter subjects are provided in the appendixes. In the next eight chapters we describe the properties of the major virus families or groups currently recognized, using the best-known members of each as examples. We focus on the biophysical and biochemical aspects of the viruses, that is, their size, shape, and composition, as well as their infection cycle. In discussing each virus family, we cover animal viruses first, then viruses of the lower kingdoms and of microorganisms. The latter by no means are less important. Work with bacteria, through the years, has uncovered important concepts in biology.

We discuss the virus families in a systematic manner, advancing from the smallest and simplest to the largest and most complex, with some evident exceptions. We start with RNA-containing viruses, because RNA may have been, in evolutionary terms, the first genetic material (see Postscript). Although one group of DNA viruses is smaller than many RNA viruses, all the large and structurally more complex viruses contain DNA genomes. Some of these approach living cells in the complexity of their structure. In connection with these virus families, we cite the major hosts and diseases. The general biological aspects of the responses of the host cells, the organisms, and populations to virus infection are dealt with in the later chapters. We end the book with our thoughts about the virologic challenges in the 1990s.

We have attempted to cover all of virology, and a discussion of the possible evolution of viruses in this book, and readers and teachers of courses concerned with only animal, plant, or bacterial viruses can find their subject matter in specific sections of chapters and the recommended references. We give the main facts that are currently available for each type of virus. The information and conclusions are usually derived from many experiments using different approaches and techniques. Each is subject to error and, often, to more than one interpretation. Instead of describing or discussing these numerous bits of evidence in detail we show a few typical examples of the primary experimental data, usually in the form of tables or figures from the original literature.

The student who wants to evaluate how hard or soft are the data leading to certain conclusions is referred to some original articles and selected reviews and books in Further Reading at the end of most sections. In a field where knowledge is still rapidly expanding and interpretations are often changing, references to information in current publications, which are not yet necessarily substantiated, should be more useful to the curious and interested student than a detailed description of possibly obsolete experimental data and conclusions.

We trust that this updated edition of *Virology* will be of value to both teachers and students following this constantly evolving and challenging field of science. This book can also serve as a useful reference for workers in the fields related to human and veterinary medicine and to agriculture.

We thank for their advice on this book: Amrik S. Dhaliwal, Loyola University; Jo-Ann C. Leong, Oregon State University; Philip I. Marcus, University of Connecticut; Edward K. Wagner, University of California, Irvine.

Jay A. Levy
Heinz Fraenkel-Conrat
Robert A. Owens

Virology

chapter

1

General Concepts of Virology

1.1 DISCOVERY OF VIRUSES AND A BRIEF HISTORY OF VIROLOGY

Although the field of virology is only about 100 years old, viruses have probably been present in living organisms almost since the origin of life. Whether or not viruses preceded or came after one-cell organisms is still a controversial subject. However, on the basis of the continual discovery of viruses in many different species, one could conclude that every species on this planet carries viruses. Recently, up to 100 million bacterial viruses (0.2 μm in size) per milliliter have been detected in water from natural habitats. While many viruses are harmful to their host, others are symbiotic and some virus infections may even give an advantage to the infected host. For example they could provide a virus gene that mediates a needed metabolic capability or drug resistance. In this sense, viruses could be among the most primitive means of **information transfer.** Virus infections are noticed most often because of deleterious effects on the host, including the lysing of bacterial cells, the symptoms of colds in humans, and worldwide epidemics. In other situations, the virus infection may remain unnoticed for months or years, or even for the life of the organism.

1.1.1 Viruses and Virologists

Viruses are of interest not only as agents of disease but also as tools for investigating the cellular and molecular biology of the host organism. Historically, the study of viruses has played a central role in understanding how cells function. Many important areas of research—the elucidation of the chemical nature of genes, as well as their transcription and control, DNA repair, translation of mRNA, splicing and other processing of RNA, protein modification and protein-protein interaction assembly, point and frameshift mutation, and malignancy—have relied extensively and in some cases almost completely on viruses as critical research tools. The intimate association of a virus with its host allows the study of viruses to advance the understanding of how cells function. In addition, viral genes and other virus-derived factors have become useful as "tools" in recent research on the genetic modification of organisms.

The main focus of clinically oriented, professional virologists is on virus transmission and the therapeutic control of virus infections. Frequently, their general aim is the eradication of at least the more dangerous viruses, an aim that seems to have been achieved for the **smallpox virus.** Other virologists are concerned with how viruses replicate, with the structures of virus particles, with fundamental questions about the effects of virus infections on cells, and with other questions that may eventually help in combating these pathogenic agents. Many of the advances in virology have been due to the efforts of biochemists, physicists, geneticists, and other newcomers in this field, as well as to scientists specifically trained in virology or microbiology. The resulting amalgamation of scientific approaches contributed greatly to the development of what came to be called **molecular biology.** Molecular biologists use viruses as models in the study of many intracellular biochemical functions. The interdisciplinary effort of molecular biology developed in large part because viruses turned out to be among the most useful subjects for studies of the chemical and physical bases of biological phenomena. Many such studies even took place in laboratories not directed toward understanding virological phenomena. Almost all the processes that in their totality we call life have been studied with the help of the viruses of lower and higher organisms. Virology thus has become an integral part of modern biology.

The concept of a biological disease agent, or **pathogen,** has its origins in the nineteenth century. Until about 1850, the intrinsic differences between various harmful agents, some of which were called by the Latin word for poison, *virus,* were poorly understood. Only with the realization of the nature of **bacteria** as potentially disease-carrying microorganisms did a distinction between bacteria on one hand, and poisons and toxins on the other, have become obvious. The **poisons** or **toxins** were then recognized to be harmful substances that, unlike bacteria, do not increase in the victim's tissue. Toxins or poisons may be elements such as mercury, or simple compounds such as carbon monoxide, but some are large and complex protein molecules.

Bacteria were recognized in the second half of the nineteenth century to be complete living organisms that metabolize and multiply in body fluids and cells. When methods were developed to culture bacteria in or on nutrient media *in vitro,* that is, in test tubes or on agar-agar gel in Petri dishes, their ability to multiply under such conditions became one of the most important criteria for their definition. Bacteria were found to be responsible for many of the infectious diseases and epidemics that were decimating animals and humans. Incidentally, many of the more virulent bacteria were found to produce highly toxic proteins, and in some instances such proteins were produced under the control of viruses that such bacteria carry (see Chapter 12).

1.1.2 Development of the Virus Concept

Whenever a branch of science has advanced to a new level of understanding, its practitioners tend to treat the new concept as if it were an indubitable and unalterable fact, a **dogma.** By so doing they often retard further advances. Because communicable diseases were proclaimed to be generally caused by microorganisms such as bacteria and fungi, failure to find and grow a microorganism and induce the disease with it was likely to be regarded as an indication of poor technique. Researchers under such circumstances become hesitant to report findings that do not agree with the accepted dogma for fear of being regarded as fools, if not heretics. And if the researchers do publish their results, such results tend to be overlooked or quickly forgotten.

In times of scientific or philosophical conformity, nonconformists are needed who have the courage to raise questions publicly concerning the dogma. Beijerinck, at Leiden in the Netherlands, was such a man. So were two German scientists, Loeffler and Frosch. At the end of the nineteenth century Beijerinck was working on a disease that disfigured tobacco leaves with a mosaic pattern of yellow or light green and dark green areas. Loeffler and Frosch were working on foot-and-mouth disease of cattle. By 1898, both diseases were known to be infectious, or transmissible, and thus presumably bacterial. However, the porcelain candles generally used to filter out bacteria were in both cases found incapable of "sterilizing" the infectious extracts or exudates.

What was Beijerinck's and Loeffler's and Frosch's explanation? In retrospect, it is quite obvious, although at that time it was heretical: Infectious agents can be small enough to pass through the pores of filters and thus are much smaller than the smallest bacteria known. They are "filterable viruses." Both Beijerinck and the German scientists grasped, surprisingly, that

filterable viruses were not only smaller than, but were inherently different from, microorganisms. Ivanowski in the Russian Crimea had made observations similar to those of Beijerinck but had interpreted them as being due to failure of the filters, even though controls indicated that the filters were fully functional. Even after he became aware of Beijerinck's results, Ivanowski refused to accept the Dutch author's explanation of their very similar observations.

TABLE 1.1
Some highlights in the history of virology

1798	Smallpox vaccine developed	E. Jenner. (Available in pamphlet, vol. 4232, Army Medical Library, Washington, D.C.)
1885	Rabies vaccine produced	L. Pasteur. *C. R. Acad. Sci.* 101:765.
1898	Viruses recognized as agents responsible for diseases of plants (tobacco mosaic) and animals (foot-and-mouth disease)	M. W. Beijerinck. *Zentralbl. Bakteriol. Parasitenkd.* Abt. II, 5:27; F. Loeffler and P. Frosch, *Zentralbl. Bakteriol. Parasitenkd.* Abt. I, 23:371.
1900	Discovery of the cause of yellow fever	W. Reed et al. *Phil. Med. J.* 6:790.
1904	A filterable agent causing equine anemia recognized (first lentivirus)	H. Vallee and H. Carre. *C. R. Acad. Sci.* 139:1239.
1908	A virus causing avian leukemia identified	V. Ellermann and O. Bang. *Zentralbl. Bakteriol. Parasitenkd.* 46:595.
1911	A virus causing avian sarcoma identified (RSV)	P. Rous. *J. Exp. Med.* 13:397.
1915, 1917	Discovery of bacteriophages	F. W. Twort. *Lancet* 189(2):1241; F. d'Hérelle. *C. R. Acad. Sci.* 165:373.
1933	A virus causing mammalian cancer identified (rabbit papilloma)	R. E. Shope. *J. Exp. Med.* 58:607.
	Human influenza virus isolated	W. Smith, C. H. Andrewes, and P. P. Laidlaw. *Lancet* 2:66.
1934	Bacteriophage purified to near homogeneity	M. Schlesinger. *Biochem. Z.* 237:326.
1935	Pure paracrystalline tobacco mosaic virus (TMV) isolated	W. M. Stanley. *Science* 81:1644.
1937	The nucleoprotein nature of plant viruses recognized	F. C. Bawden and N. W. Pirie. *Proc. R. Soc.* (London) 123:274.
1938	Comparative data on sizes of viruses	W. J. Elford. In *Handbuch der Virusfg.* Ed. R. Doerr and C. Hollauer. Springer-Verlag, Wien.
1939	A virus (TMV) visualized by the electron microscope	G. A. Kausche, E. Pfankuch, and E. Ruska. *Z. Naturfg.* 27:292.
1940	The replicative cycle of bacteriophage elucidated	M. Delbrück. *J. Gen. Physiol.* 23:643.
1942	A mammalian RNA tumor virus discovered (mouse mammary tumor virus)	J. J. Bittner. *Science* 95:462.
1949	Human cell cultures developed for the growth of poliovirus	J. F. Enders, T. H. Weller, and P. C. Robbins. *Science* 109:85.
1950	Induction of lysogenic phage demonstrated	A. Lwoff, L. Siminovich, and N. Kjelgaard. *C. R. Acad. Sci.* 231:190.
1951	A virus causing mammalian (mouse) leukemia identified	L. Gross. *Proc. Soc. Exp. Biol. Med.* 76:87.
1952	The DNA of a bacteriophage shown to be infectious	A. D. Hershey and M. Chase. *J. Gen. Physiol.* 36:39.
	A viral coat protein (TMV) chemically characterized	J. I. Harris and C. A. Knight. *Nature* 170:613.
	Transduction discovered	N. D. Zinder and J. Lederberg. *J. Bacteriol.* 64:679;
	Lysogenic phages discovered	E. Wollman. *Ann. Inst. Pasteur* 84:251; E. Lederberg and J. Lederberg. *Genetics* 38:51.

1953	Adenoviruses discovered	W. P. Rowe. *Proc. Soc. Exp. Biol. Med.* 84:570.
1955	Poliovirus crystallized	F. L. Schaffer and C. E. Schwerdt. *Proc. Natl. Acad. Sci. USA* 41:1020.
	Infectious TMV reconstituted from RNA and protein	H. Fraenkel-Conrat and R. C. Williams. *Proc. Natl. Acad. Sci. USA* 41:650.
1956	Infectivity of purified viral (TMV) RNA demonstrated	H. Fraenkel-Conrat. *J. Am. Chem. Soc.* 78:882; A. Gierer and G. Schramm. *Nature* 177:2702.
1957	Discovery of Interferon	A. Isaacs and J. Lindenmann. *Proc. Roy. Soc. London* 147:258.
1958	Chemical induction of viral (TMV) mutants demonstrated	A. Gierer and K. W. Mundry. *Nature* 182:1457.
1959	Parvoviruses discovered	L. Kilham and L. J. Olivier. *Virology* 7:428.
	Reversibility of denaturation of viral (TMV) coat protein demonstrated	F. A. Anderer. Z. *Naturfg.* 146:648.
1960	The amino acid sequence of a viral (TMV) coat protein established	A. Tsugita et al. *Proc. Natl. Acad. Sci. USA* 46:1463; F. A. Anderer et al. *Nature* 186:1922.
1962	The icosahedral structure of many isometric viruses established	D. L. D. Caspar and A. Klug. *Cold Spring Harbor Symp. Quant. Biol.* 27:1.
	In vitro translation of bacteriophage RNA demonstrated	D. Nathans, G. Notani, J. H. Schwartz, and N. D. Zinder. *Proc. Natl. Acad. Sci. USA* 48:1424.
1963	The occurrence of double-stranded RNA (in reovirus) discovered	P. J. Gomatos and I. Tamm. *Proc. Natl. Acad. Sci. USA* 50:878.
1965	*In vitro* replication of bacteriophage RNA demonstrated (Qβ)	S. Spiegelman et al. *Proc. Natl. Acad. Sci. USA* 54:919.
1967	*In vitro* replication of bacteriophage DNA reported (φX 174)	M. Goulian, A. Kornberg, and R. L. Sinsheimer. *Proc. Natl. Acad. Sci. USA* 58:2321.
	The nature of viroids elucidated	T. O. Diener and W. B. Raymer. *Science* 158:378.
1968	Epstein-Barr virus linked to infectious mononucleosis and possibly Burkitt's lymphoma	G. Henle, W. Henle, and V. Diel. *Proc. Natl. Acad. Sci. USA* 59:94.
	Multiple genomic RNAs of influenza virus demonstrated	P. H. Duesberg. *Proc. Natl. Acad. Sci. USA* 59:930.
1970	Presence of an oncogene in Rous sarcoma virus demonstrated	P. H. Duesberg and P. K. Vogt. *Proc. Natl. Acad. Sci. USA* 67:1673.
	Xenotropic viruses identified	J. A. Levy and T. Pincus. *Science* 170:326.
	Reverse transcriptase discovered	H. M. Temin and S. Mizutani. *Nature* 226:1211; D. Baltimore. *Nature* 226:1209.
1973	Evidence for stable transduction of "viral" information from rat cells to a mouse sarcoma virus	E. M. Scolnick et al. *J. Virol.* 12:408.
1976	A retrovirus (RSV) oncogene *(src)* shown to have counterpart in normal cells	D. Stehelin, H. E. Varmus, and J. M. Bishop. *Nature* 260:170.
	The ribonucleotide sequence of an RNA bacteriophage (MS2) determined	W. Fiers et al. *Nature* 260:500.
1977	The deoxyribonucleotide sequence of single-stranded bacteriophage DNA (φX 174) determined	F. Sanger et al. *Nature* 265:687.
	Messenger RNA splicing (in adenovirus transcription) discovered	L. T. Chow et al. *Cell* 12:1; S. M. Berget, C. Moore, and P. A. Sharp. *PNAS* 74:317.
1978	The deoxyribonucleotide sequence of double-stranded animal virus DNA (SV40) determined	W. Fiers et al. *Nature* 273:113; V. B. Reddy et al. *Science* 200:494.

	Recombinant DNA copies of a viral RNA (Qβ) shown to be infectious	T. Taniguchi, M. Palmieri, and C. Weissmann. *Nature* 274:223.
	The transforming gene *(src)* of an RNA tumorvirus shown to be a phosphokinase	M. S. Collett and R. L. Erikson. *Proc. Natl. Acad. Sci. USA* 75:2021.
1980	A human retrovirus associated with leukemia discovered	B. J. Poiesz et al. *Proc. Natl. Acad. Sci. USA* 77:7495.
	The eradication of a virulent human epidemic virus (smallpox) announced by the World Health Organization	*WHO Chronicle* 34:81.
1981	Viral vaccine (foot-and-mouth disease) made by recombinant DNA technology	D. G. Kleid et al. *Science* 214:1125.
1982	Reverse transcription reported to be involved in a DNA virus replication (hepatitis B)	J. Summers and W. S. Mason. *Cell* 29:403.
	Vaccinia virus used as a vector for antigens of other pathogens	D. Panicali and E. Paoletti. *Proc. Natl. Acad. Sci. USA* 79:4927; G. L. Smith, M. Hackett, and B. Moss. *Nature* 302:490.
1983	Isolation of a human retrovirus (lentivirus) associated with acquired immune deficiency syndrome (AIDS) reported	F. Barre-Sinoussi et al. *Science* 220:868.
1985	Retrovirus used as genetic vector to insert foreign genes into germ line of mice	H. v. d. Patten et al. *Proc. Natl. Acad. Sci. USA* 82:6148.
	Crystal structure of rhinovirus determined below 3 Å	M. G. Rossmann et al. *Nature* 321:145.
1986	Viroidlike structure of human hepatitis delta agent demonstrated	K. Wang et al. *Nature* 323:508; A. Kos et al. *Nature* 323:558; Chin et al. *Proc. Natl. Acad. Sci. USA* 83:8774.
	First evidence of ribozyme activity associated with viral RNA (satellite tobacco ringspot virus)	G. A. Prody et al. *Science* 231:1577.
	Use of transgenic plants to obtain resistance against viruses (TMV)	P. Powell-Abel et al. *Science* 232:738.
1989	Identification of hepatitis C virus by molecular techniques	Choo et al. *Science* 244:359.
1991	Cell-free *de novo* synthesis of complete infectious polioviruses	A. Molla et al. *Science* 254:1647.

At the beginning of the twentieth century, researchers in England showed that the human wart agent is filterable. As indicated in Table 1.1, there followed many other demonstrations of filterable viruses and the development of the concept of the virus as an agent entirely distinct from cellular disease agents such as bacteria and fungi. In particular, critical to the founding of the science of virology was the notion that virus particles have a relatively simple and regular chemical structure. Establishing the validity of this notion required the contributions of another researcher with an open mind and the courage to face and report his results with disregard to the prevailing "party line": Wendell M. Stanley was such a researcher.

By 1935 the active principle of the **tobacco mosaic disease** had been concentrated, and quantitative assays for infectivity had been developed. Other agents, now known to be viruses, also had been concentrated and assayed. However, although investigators had in hand what we now know were crude preparations of infectious virus particles, they were reluctant to regard what they had as anything other than cellular matter. Stanley's work occurred at a time when it had become generally accepted that all enzyme activities are due to proteins. The fact that each protein represented a chemically definable and uniform population of mol-

FIGURE 1.1

(a) A crystal of TMV in a hair cell of tobacco, as seen with the light microscope. (b) A crystalline deposit of TMV rods in a leaf cell of *Chenopodium amaranticolor* as revealed by electron microscopy. (Courtesy of R. G. Milne (1966) *Virology* 28:520)

(a)

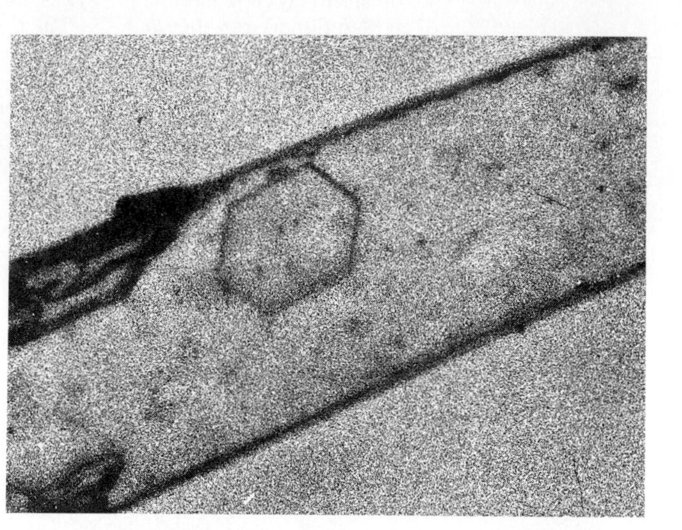

(b)

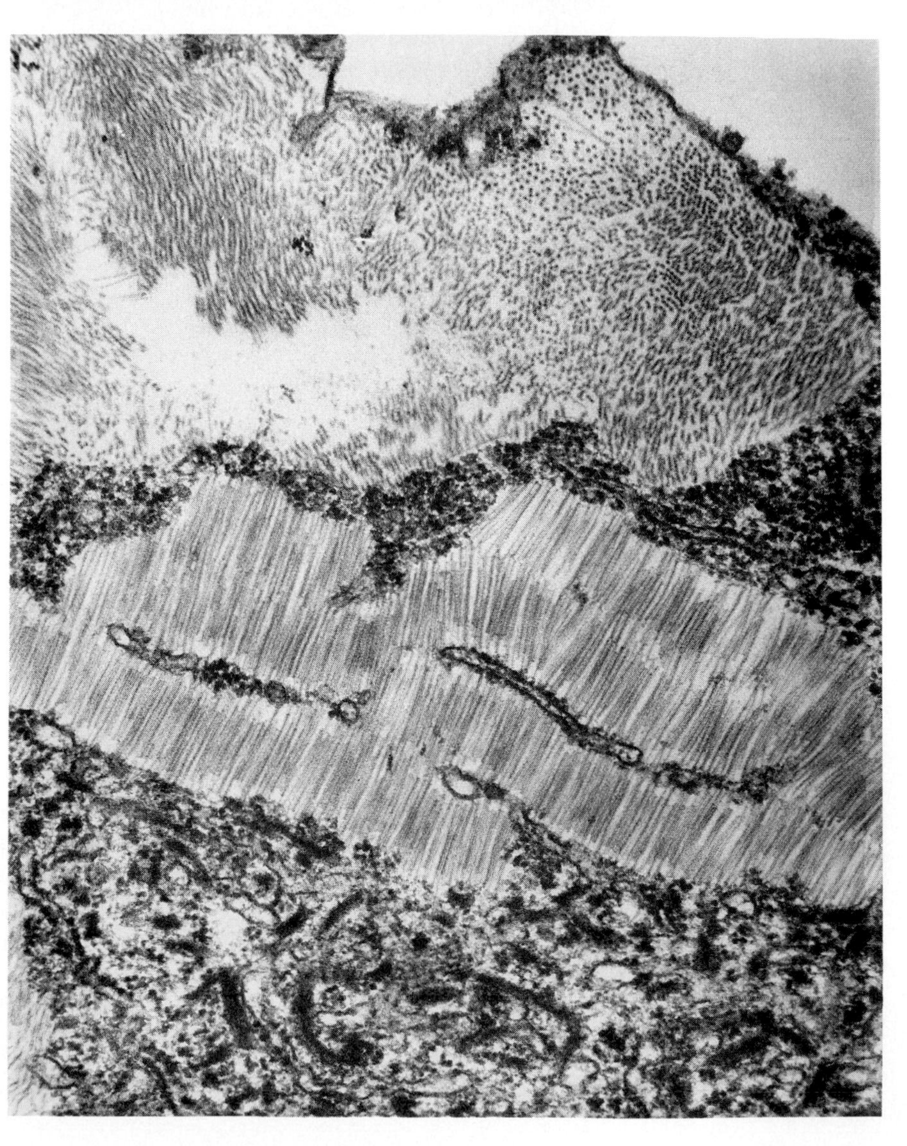

ecules was also a newly accepted concept. The great days of the crystallization of many of these enzymes, such as urease, pepsin, trypsin, and chymotrypsin, had just passed. Much of that progress had occurred in the department of John H. Northrop at the branch of the (then) Rockefeller Institute for Medical Research at Princeton, New Jersey, where Stanley, an organic chemist, started his work on viruses. The finding in that laboratory that an inert protein made in the pancreas, trypsinogen, could be transformed to the enzyme trypsin by the action of trypsin itself had naturally intrigued Stanley. Only a trace of trypsin in the test tube could produce large amounts of trypsin from an inert precursor protein, a process termed **autocatalysis.** Could autocatalysis resemble the process whereby a single virus particle infected a cell and gradually led to the production of millions of virus particles in the cell?

It is not surprising that Stanley tended to compare his virus, now known as tobacco mosaic virus, or TMV, with enzyme proteins in many respects. As far as crystallinity is concerned, solutions of purified TMV in water give a sheen that suggests crystals, and suspensions of carefully prepared precipitated TMV show long needles that are seen under the light microscope. We now call this a **paracrystalline** state because the particles arrange themselves in an orderly manner in only two dimensions, not three. Later, when the electron microscope was developed, actual three-dimensional crystals of TMV and other viruses were seen in sections of infected cells (Fig. 1.1). Thus, crystallizability, an indication of the relative purity of a substance, was demonstrated for viruses as it had been for enzymes.

Are viruses proteins? Stanley considered his observations and concluded that TMV is a protein. He only gradually accepted the viewpoint of the English scientists Bawden and Pirie that the small amount of phosphorus in TMV (0.5 percent) was not due to contaminations but actually represented evidence for the presence of a different type of macromolecule, nucleic acid. Because nucleic acids contain about 10 percent phosphorus, 0.5 percent phosphorus suggested the presence of about 5 percent nucleic acid in TMV. Other plant viruses, with **isometric** particles rather than the rod shape that is characteristic of TMV, were crystallized in England and were found to contain considerably greater amounts of phosphorus than does TMV. Ultimately, the characteristic presence of an amount of nucleic acid 10 times the amount of phosphorus became generally accepted as an intrinsic feature of typical viruses.

The idea that virus replication could be accounted for by an autocatalytic process resembling the trypsinogen-trypsin system was soon abandoned. It had then been known for 40 years that viruses increased only in living cells and are not formed in the test tube from a precursor. Virus replication does not represent a transition of some molecule from an inactive to an active state. Instead it reflects a genuine case of "increase and multiply," causing, in the case of TMV infection, a plant inoculated with much less than 1 μg of the virus to produce many million times as much virus in a few days.

FIGURE 1.2

Electron micrograph of T5 phage. (Courtesy of R. C. Williams)

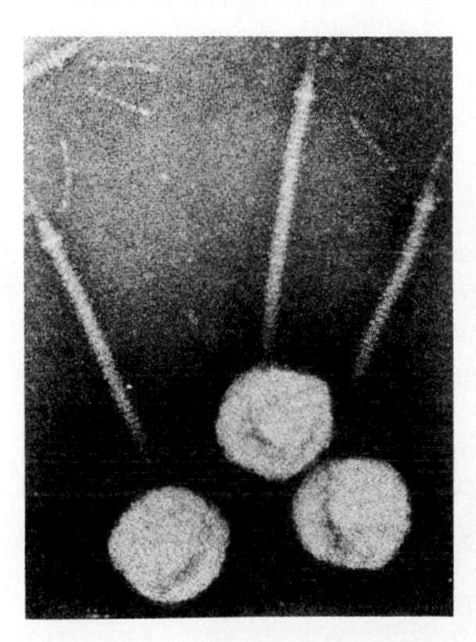

1.1.3 Bacteriophages and the Beginnings of Molecular Virology

Whereas the focus in the preceding paragraphs was on the viruses of plants and animals, a parallel development occurred with viruses of the bacteria and took center stage in the 1940s (Fig 1.2). Viruses of bacteria were termed **bacteriophages** when they were independently discovered in France and England because of their ability to eat holes (*phage* is derived from the Greek for "eating") on lawns of bacteria on agar-agar media in Petri plates. For a time, interest in these mysterious agents, which were then not generally thought of as viruses, focused on their bacteriocidal activity, and thus hoped-for medicinal usefulness (see Chapter 13). In Sinclair Lewis's book *Arrowsmith,* the researcher Dr. Arrowsmith was working along such lines. The connecting link between science and literature in this instance was Lewis's friend Paul de Kruif, then a scientist at the Rockefeller Institute in New York and later a famous science writer.

By the time that these **phages,** as they were beginning to be called, were found to be impractical for fighting bacterial diseases (see Chapter 12), they had been adopted by scientists who wanted to elucidate their nature and their mode of replication. These early studies of the "infection cycle" of phages in human's bacterial friend *Escherichia coli* (usually called *E. coli*) were carried out in part by former physicists. The phage researchers initiated the critical and quantitative study of virus replication. In fact, their experiments were among the first highly quantitative studies in biology. The understanding of the principles of bacteriophage replication was later found to be valid also for the viruses of higher organisms. It is primarily the time scale that is very different, for what takes minutes in bacteria takes hours in animals and days in plants.

The phage replication studies were conducted or directed not by chemists and biologists, but mostly by physicists who were remarkably uninterested in the chemical basis of what they were observing. In fact, a nearly pure bacteriophage preparation had been isolated in 1934 by Schlesinger, an early refugee from Hitler's Germany who was working in England. His findings, however, were generally disregarded. After his early death, they were forgotten. Bacteriophage research turned in a biochemical direction only about 20 years later, when it had become evident that the viruses of bacteria were not fundamentally different from those of higher organisms.

Molecular virology, the study of the biochemical and genetic nature of virus particles, virus replication, and virus interactions with cells, made its next great advance not from the study of phages but again from experiments with TMV. It was discovered in 1955 that under proper conditions TMV, and later other viruses, could be broken down into their components, protein and nucleic acid, and then reconstituted from those components into a fully infective form. In 1956, infectivity of the purified, protein-free nucleic acid of TMV was demonstrated. That is, the free nucleic acid of TMV and certain other viruses was itself able to produce the characteristic disease and engender new virus particles, although usually less efficiently than would be accomplished by the corresponding intact virus particle.

Gradually, as the science of virology developed and more viruses were studied, it became evident that the particles of a given virus contained *either RNA or DNA,* but not (an appreciable amount of) both; this virus RNA or DNA could be in either a single- or double-stranded form. Enzymes were detected in virus-infected cells and in some viruses that were able to replicate these nucleic acids. These enzymes made it possible to achieve *in vitro* synthesis of infectious, that is, disease-causing molecules—RNA in 1965 and DNA in 1967. Methods for the determination of amino acid sequences were applied to virus coat proteins in the 1960s. RNA from virus particles was incubated with cell extracts to generate authentic virus protein, thereby demonstrating the fidelity of *in vitro* protein synthesis. Later, methods for nucleotide sequence analysis were developed with viral nucleic acids as test molecules. The biochemical process by which certain viral nucleic acids became part of their host cells' genomes, and in some instances thereby **transformed** the cells into **cancer** cells, was first elucidated in 1970.

The virtual explosion of discoveries (and corresponding publications) since 1980 makes it difficult to summarize them comprehensively. Keeping recent entries in Table 1.1 from overflowing the entire text required the use of a selection process among numerous research claims, some of which have not yet stood the test of time.

1.1.4 Viruses in Human History

To understand the place of viruses in human history, it is useful to consider the prehistoric record. The earth is estimated to be about 4.5 billion years old. Fossil evidence for bacteria suggests that the planet has supported life for at least 3.5 billion years, and amoebas and protozoans probably emerged about 1.5 billion years ago. According to the current interpretations of fossils, primates considered to be *Homo sapiens* appeared on earth about 100,000 years ago. However, the most ancient evidence for virus diseases of humans is much more recent. **Rabies** and rabid dogs were well known in antiquity. A bas-relief of an Egyptian priest of 1500 B.C.E. shows a shriveled leg with an appearance that is consistent with the consequences of paralytic **polio** (Figure 1.3). Although evidence for what may have been smallpox occurs in Egyptian mummies and in ancient Chinese documents, nei-

FIGURE 1.3

A bas-relief showing the shriveled leg of a priest who has apparently recovered from paralytic poliomyelitis. Dated 1500 BC. (From the Ny Carlsburg, Glyptotchek, Copenhagen)

ther smallpox nor measles seems to have been known to Hippocrates (460–377 B.C.E). However, the occurrences of mumps and possibly influenza were recorded on the island of Thasos during that period. Apparently both **smallpox** and **measles** arrived in China in the period from 37 through 653 C.E. It has been suggested that the spread of smallpox and measles in the Roman empire contributed to its decline, and these diseases are considered to have been the principal cause of the sudden demise of the Aztec and Inca empires in South America after European conquest.

Possibly the earliest recorded attempts at controlling a virus disease comes from eleventh-century China. Itinerant wisemen inoculated children with extracts of the smallpox pustule. Although some children succumbed to smallpox infection after this treatment, many more were spared the most devastating effects of the disease. A more organized and successful approach was that of Lady Mary Worthy Montague, wife of the Ambassador from England to Turkey. She introduced what now would be termed vaccination for smallpox into London in 1721. The procedure had been well known among peasants in certain regions of Greece. Smallpox pustules were placed either under a fingernail or on the skin. Again, some recipients of this treatment died, but many more were saved. The practice became more widespread in the 1740s when children of the English royal family were immunized successfully. After the death of Louis XV from smallpox in 1774, treatment with smallpox pustules became well accepted in France, and George Washington is reported to have instituted a program of inoculation of Continental Army soldiers in 1776.

In 1798, a hundred years before evidence about the nature of viruses was to become available, the English country doctor Edward Jenner made an observation of enormous importance. He realized that milkmaids tended to catch a mild form of "the pox," presumably from cows, and were then protected against the typical disfiguring ravages of ordinary smallpox. This finding led to his developing a cowpox vaccination procedure (*vacca,* Latin for cow) using cow pustules now known to contain the **cowpox** virus. Recipients had only mild symptoms and were protected against the far more severe and dangerous smallpox disease (Fig. 1.4). This use of vaccines for protection against viral (and bacterial) infections is the major defense we now have for many of these agents in nature.

Other human virus diseases have occurred at various times and places and with various degrees of severity. Paralytic polio appears to have become endemic only at the beginning of the twentieth century. The most recent appearance of a devastating virus disease of humans is that of **acquired immune deficiency syndrome** (AIDS; see Chapter 14). How epidemics come about is discussed in Chapter 12.

FIGURE 1.4

A pustule formed on the skin following a smallpox vaccination. (Courtesy of E. Lennette)

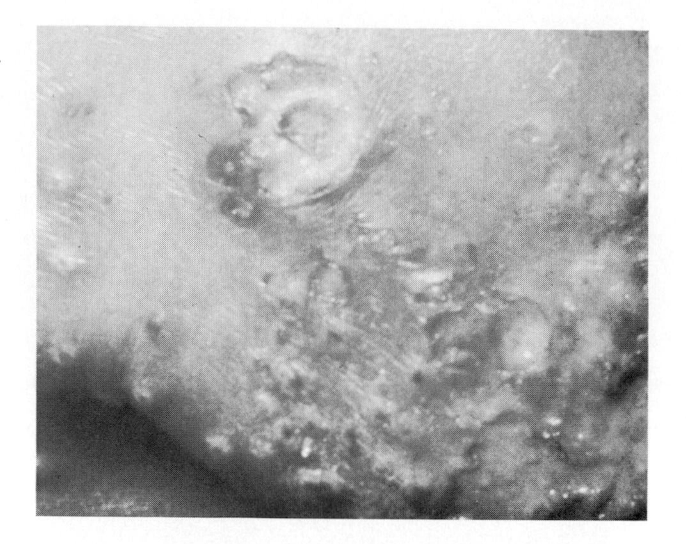

From time to time, viruses infecting plants have also had important, although sometimes indirect, effects upon human activities. The spectacular rippling color patterns in the flowers of tulips caused by viruses (see Fig. 1.5) was responsible for a period of unrestrained economic speculation known as "tulipomania" that threatened the financial stability of seventeenth-century Holland. At the height of tulipomania—before it was widely known that such plants could be produced by simple grafting techniques—a single infected bulb was exchanged for 4 tons of wheat, 8 tons of rye, 4 fat oxen, 8 fat pigs, 12 fat sheep, 2 hogsheads of wine, 4 barrels of beer, 2 barrels of butter, 1000 pounds of cheese, 1 bed with accessories, 1 full dress suit, and 1 silver goblet!

FIGURE 1.5

Vase of Flowers, a still-life painting by J. D. de Heem (1606–1683/84). The tulips show petal striping of the type induced by tulip breaking virus. (Courtesy of National Gallery of Art, Washington) *(See also backcover of book)*

1.2 NATURE OF VIRUSES: LIVING OR NONLIVING?

1.2.1 Host Cells of Viruses

Living organisms are in an extraordinarily intricate and dynamic state. A multitude of different macromolecules—mostly the proteinaceous catalysts, enzymes—cooperatively facilitate the production and utilization of chemical energy and the synthesis of macromolecules, such as nucleic acids and proteins, as well as subcellular particles and cell organelles. The result is a highly organized system, maintained in all active phases by continuous consumption of energy. The system decodes and acts on information carried in the structures of nucleic acid molecules to assure the functioning and continuity of the organism.

Organisms evolve. The processes that allowed organisms to be derived from presumably nonliving, primordial antecedents and to elaborate the variety of organisms in existence, including the sometimes self-destructive *Homo sapiens,* required the possibility of occasional random errors in information transfer and storage, that is, **mutations.** It is from such mutations within a population that the occasional advantageous **phenotype** appears, allowing natural selection to operate.

The features of viruses are best understood in the context of the characteristics of cellular organisms, which are of course the hosts for virus:

1. An organism has a limited lifespan, and populations are maintained by one or more processes of reproduction; thus, an organism has a life cycle with defined stages of development and a traceable pattern of lineage from ancestors.
2. The high state of organization of an organism is maintained by the consumption of chemical energy throughout its life cycle; organisms exhibit metabolism.
3. In all active phases of the life cycle, primary direction of the chemical reactions of metabolism, development, and reproduction resides in genetic material(s) that specify the structure of catalytic and other macromolecules; variations in the genetic material (mutations) result in variants of the organism (**mutants**); some mutations are inherited and appear in members of subsequent generations.

1.2.2 Virus Characteristics

In the early part of the twentieth century, viruses were considered almost entirely in negative terms: not visible in the light microscope, not retained by microbiological filters, and not cultivatable on any known nutrient medium. The first two of these properties are a consequence of viruses being generally smaller than even the smallest bacteria. However, it now is recognized that "elementary bodies" observed by light microscopy of smallpox-infected tissue are actually images of the smallpox virus particle (Chapter 8). Furthermore, most virus particles can be visualized by electron microscopy, and refinements in the preparation of membranes of controlled porosity allow virus particles to be retained and various viruses to be distinguished (Fig. 1.6). As early as the 1930s, Elford was able to differentiate virus particles on the basis of their size and shape using specially prepared filters. The third negative characteristic, the inability to "increase and multiply" on nutrient media, remains a diagnostic criterion for a viral agent. However, even this prototypical virus characteristic of obligate intracellular parasitism has been challenged by biochemical manipulations in which complex cell-free extracts have generated new, infectious virus particles under the direction of viral nucleic acid.

Understanding the nature of viruses requires distinguishing the "virus" from the **"virion."** The latter is the inert or nearly inert particle that is the extracellular phase of the virus infection cycle. The virion contains one or more molecules, either of DNA or of RNA, that is the genome of the virus. The term *virus* has a meaning that is much broader than that of virion. The virus consists of all virus-specified entities involved in all aspects of the infection cycle. "Virus" includes not only the virion but also virus-specified messenger RNAs, viral proteins that may not be incorporated into virions, and other features.

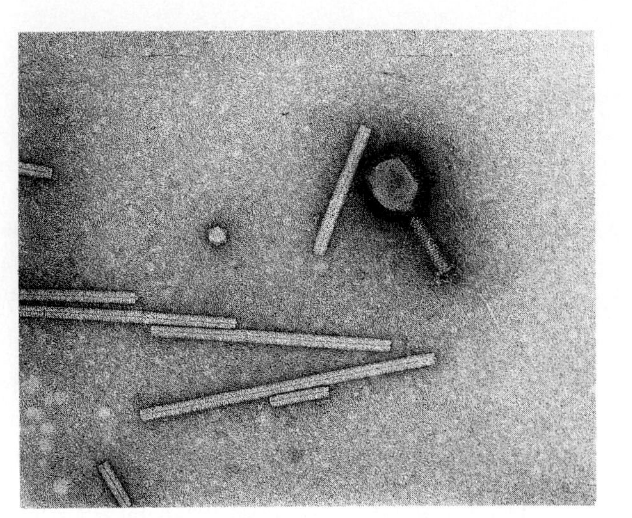

FIGURE 1.6

Illustration of the shapes of the three classes of viruses first studied by electron microscopy. At the right are typical bacteriophages, at the bottom are intact tobacco mosaic viruses, and in the center is the common icosahedral viral structure, as it occurs in animal, plant, and bacterial viruses.

To summarize:

1. A virus is an infectious agent and obligate intracellular parasite.
2. A virus has an infection cycle that includes a phase in which the agent consists of one or more inert particles (the virion or virions) composed of one or more molecules of nucleic acid, usually but not necessarily covered by a coat made up of one or more proteins and, in some instances, a membrane containing lipid and glycoprotein, as well as other substances.
3. A virus is the entity that is able to be transmitted to a suitable host cell, initiating another infection.
4. The information-bearing molecule(s) in the virion(s) is either RNA or DNA but not both; upon being exposed to the interior of a suitable cell, this **genomic nucleic acid** of the virus redirects the genetic and metabolic apparatus of the cell to produce new virions.

Whether viruses are living organisms is a philosophical question to which most informed biologists will answer no. Although a virus has the major features of cellular organisms that are presented in items 1, 2, and 3 of Section 1.2.1, viruses do not possess the machinery necessary for even such central aspects of metabolism as protein synthesis (see item 3 of Section 1.2.1). They cannot reproduce outside a host. Instead, virus genes are able to control the cell's metabolism and redirect it to produce virus-specified products.

Viruses differ from other agents: toxins, cellular obligate intracellular parasites, and plasmids. Toxins are not capable of multiplying. Although some toxins may be very potent, even a series of only a few transfers from a diseased organism to healthy organisms will result in dilution of the toxin to the point that it no longer induces disease. A biological disease agent will be expected to increase after each transfer to a healthy but susceptible host, preventing dilution to the point of failing to induce disease. As is indicated in Section 1.4, a virus infection cycle includes an "eclipse phase" in which no trace of the virion remains. Cellular obligate intracellular parasites do not have an eclipse phase, because the infection must be maintained by intact cells. **Plasmids,** which are DNA molecules capable of replicating in cells independently of the cells' genomic DNA, lack the protective structure that in the case of the virion prevents degradation of the virus genomic nucleic acid.

1.3 TERMS USED IN DESCRIBING THE SIZE, SHAPE, AND TOPOGRAPHY OF VIRUSES AND THEIR COMPONENTS

1.3.1 Terms for Mass, Dimensions, and Virus Growth Studies

The description and discussion of viruses frequently refer to their **size** or **mass** or to that of their components. Whenever we are dealing with aggregates of many molecules, usually of different nature, held together by secondary forces, such aggregates will be termed **particles** and described in terms of **particle weights.** Thus, viruses have particle weights rather than molecular weights, usually a number $\times 10^6$, since they are all in the millions (3 to 800 $\times 10^6$). Their component nucleic acids, proteins, and occasionally lipids and others are **molecules**, since they consist of covalently linked atoms, and their **molecular weights** will be given. Because the molecular weights of most viral nucleic acids are in the range of 100,000 to 100 million, we will also give them in terms of 10^6. Proteins usually have molecular weights of 10,000 to 200,000. For example, TMV coat protein has a molecular weight of about 17,500, or a mass of 17,500 daltons (Da), but will be abbreviated as the 17.5 kDa protein (k = 1000).

In discussing these frequently used numbers, we emphasize that they are usually not in a quantitative sense "hard" data. All particle and molecular weight analyses of macromolecules are subject to considerable methodological error. Thus, to say that a virus coat is made up of two proteins of 42 kDa and 22 kDa means only that the method employed revealed the presence of only two proteins and that one is about twice as large as the other, the actual molecular weight of either probably being within the range of ± 20 percent of the given value. The only way to obtain the exact molecular weight of macromolecules is by complete structural analysis, amino acid sequencing in the case of proteins, and (deoxy) ribonucleotide sequencing in the case of nucleic acids. The number of virus components that are known in such absolute terms has increased enormously in the recent years. Another means of characterizing particles and large molecules is by their S-value, derived from their sedimentation rate in the ultracentrifuge. These S-values are reproducible under specific conditions, but they are not directly proportional to the mass of the particle or molecule. Finally, when discussing **linear dimensions,** we use in this text only nanometers (nm, 10^{-9} m, formerly called mμ); we avoid the frequently used angstrom (Å), which equals 0.1 nm.

Two terms frequently used in virus studies are *in vivo* and *in vitro.* Because cell culture is generally done in glass dishes (*in vitro* in Latin), this term is often used for such studies. The term *in vitro* is also employed for biochemical studies of cellular components of disrupted cells. *In vivo* has, in general, been reserved for studies done in living hosts (e.g., animal or human subjects).

1.3.2 General Structural Principles and Terms

Most viruses are organized according to either of two structural principles, the **helical,** which gives them a rodlike or threadlike appearance, or the **isometric,** which gives them roughly spherical shapes. However, elongated viruses are also known that are not helical, and helical components may be embedded in isometric virus particles.

The minimal structural features for typical viruses consist of one molecule of nucleic acid integrated into a shell or coat made up of many identical protein molecules. The more complex viruses may contain several molecules of nucleic acid as well as several or many different proteins, internal bodies of definite shape, and complex envelopes with spikes that usually contain glycoproteins and lipids.

A group of terms has been proposed to identify the various structural components of viruses. The monomer of the protein that forms the viral coat is called the **structural subunit.** In isometric viruses a definite number of these subunits usually aggregate in a specific manner to form what is called a **morphological unit,** a body of a characteristic shape, generally discernible on electron micrographs; it is also called a **capsomer.** The orderly complex of many subunits or capsomers with viral nucleic acid is the **nucleocapsid,** as contrasted with the empty **capsid,** the aggregated protein shell lacking nucleic acid. The complete virus particle, which in addition to the nucleocapsid may contain additional structural proteins and/or envelopes, is called the **virion.** The lipid-rich outer coat that covers the nucleocapsid of many but not all viruses is called the **envelope.** The spikes protruding from the envelope may also be called **peplomers.** Most of these diverse structural features have been observed, at times in a very similar manner, in plant, animal, and bacterial viruses.

Viruses or virus components of helical structure are described in terms of length and diameter of the protein rod or fiber, diameter of the axial channel, pitch of the helix, number of subunits per turn, and so on.

In the isometric viruses the protein subunits are usually arranged in groups of equilateral triangles. The simplest and by far the most common shape of virions shows **icosahedral symmetry.** An icosahedron is a body with 20 equilateral triangular facets and 12 vertices. It appears that the icosahedral shell, the basic design of the geodesic dome, also represents a most efficient design for biological containers, one that requires minimal energy in assembly. This structure, in its simplest form consisting of 60 subunits (three per facet), is schematically presented in Figure 1.7. It is beautifully illustrated by comparison of the shadows cast by a model and by *Tipula* iridescent virus, a large insect virus (Figure 1.8).

Larger icosahedral virions are not formed simply by using larger protein subunits, presumably because this structure would be an inefficient use of the genetic coding capacity of a virus. Instead, the virion shell is built from more than 60 subunits. Clearly, the 20 12 vertices of the larger icosahedron could each still be composed of morphological units or capsomers containing five subunits (which are often called **pentamers,** or sometimes **pentons,** for short). However, some other arrangement must be employed to add additional subunits symmetrically onto the triangular facets, between the pentons. If these extra subunits are identical to those in the pentamers, as is often the case in larger viruses, they cannot be added in random configurations or numbers. In fact, Caspar and Klug realized that stable, *symmetrical icosahedra can be constructed from identical subunits only by using multiples of 60 subunits.* They showed that this can be done by dividing each triangular facet into smaller triangles defined by three subunits. In Figure 1.7(a), panel 3, and Figure 1.8(b), it is clear that subunits from the smaller triangles located on adjoining large triangular facets appear to be arranged in clusters of six (i.e., in capsomers called hexamers, or **hexons**). Caspar and Klug showed that identical subunits can be used to form the pentamers and hexamers of a larger icosahedron only if the subunit structure allows some relatively minor variation in the otherwise fairly precise intermolecular aggregation geometry of the capsomers. They introduced the concept of **quasi-equivalence** to describe the observation that identical subunits could form both pentamers and hexamers that, in turn,

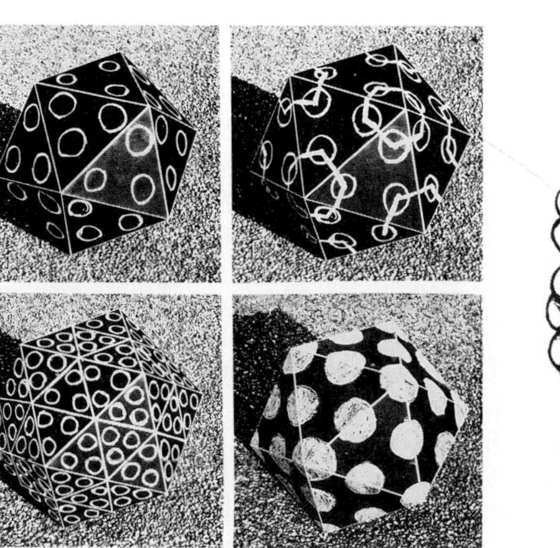
(a)

(b)

FIGURE 1.7

Structure of turnip yellow mosaic virus. (a) Panel 1 (top; left) shows the simplest arrangement of viral coat protein molecules on the surface of a dodecahedron, containing 60 capsomers that consist of identical protein molecules. In panel 2 (top; right), groups of five capsomers that form the vertices of this icosahedron have been connected by lines. Panels 3 and 4 show possible arrangements of structural capsomers in larger icosahedra with more capsomers. (b) Drawing of the icosahedron model in panel 3 illustrating the three-dimensional arrangement of capsomers. In both the model and drawing, note the groups of five capsomers (each forming vertices) on either side of each group of six capsomers (shaded). The clusters of six capsomers fill in the larger faces of this more complex structure.

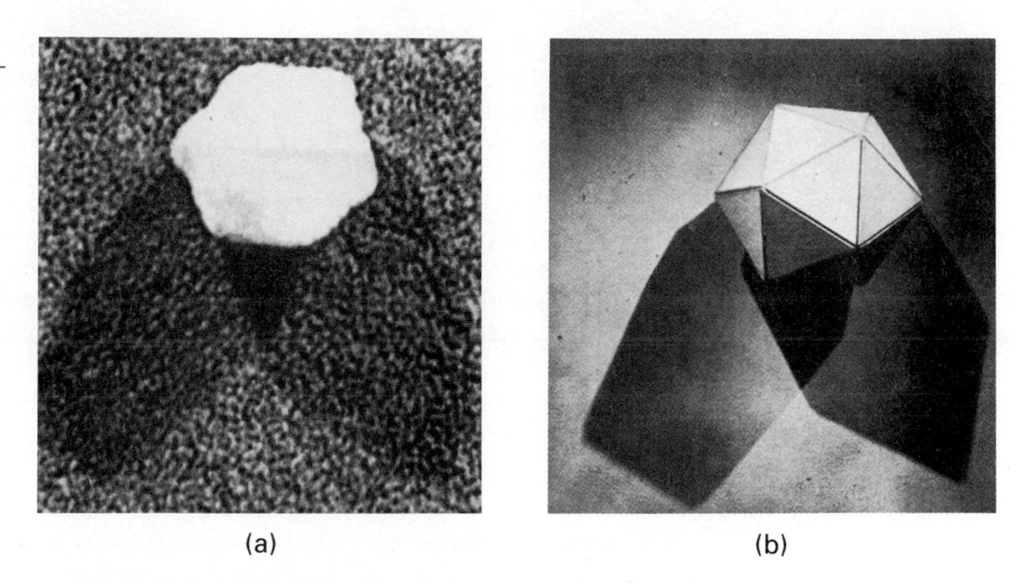

FIGURE 1.8

A particle of the large *Tipula* iridescent virus (a) shadowed for electron microscopy and placed next to an icosahedron (b) light-shadowed in the same manner from two directions. This figure clearly illustrates the icosahedral nature of the virus. (Courtesy of R. C. Williams)

could form stable icosahedral virions. In most plant and bacterial viruses identical coat protein subunits do actually make up both the pentamers and hexamers; but some animal viruses, such as adenoviruses, utilize different protein subunits in their pentons and hexons. Some virions may have a different, special protein in the center of each pentamer, exactly at the tip of each of the vertices.

It should also be noted that the quasi-equivalency principle allows for some flexibility and can thus account for the varying shapes among virus particle populations, including the oblong bodies and rods that have been observed among several principally isometric viruses, either normally or under certain conditions of viral infection. The total number of triangular facets needed to form large, geometrically stable bodies can be described mathematically by the so-called triangulation number, T, which is 3 (T = 3) for most of the many unenveloped icosahedral viruses of animals, plants, and bacteria. It is, however, 1 (T = 1) for Picornaviridae, and 4 (T = 4) for the one known member of Tetraviridae, the *Nudaurelia* β virus of insects.

1.4 THE VARIETY OF VIRUS INFECTIONS

Different virus–host cell combinations result in a variety of interactions. Although some viruses may barely perturb the cell metabolism, another virus–host cell combination may result in a very intimate interaction in which the viral genome is incorporated into the host cell's genome. Infections may result in cell death, or an infection may be persistent, even congenital. Some viruses are able to integrate their nucleic acid in a more or less reversible manner into the host's DNA; by such mechanisms some viruses become latent or **persistent** and others alter control mechanisms, possibly leading to cancer. Some aborted infections leave residual parts of the virus genome that are expressed in a fashion that may be more devastating to the host than a cytopathic infection. Viral genetic material may be passed to all the cells in the progeny of the host through infection of the germ cell (**vertical transmission**). In this case, the host for the virus is said to have been genetically transformed. In other situations, it can be passed to the fetus via the placenta or at birth (**horizontal transmission**). Some viruses have little or no pathogenicity in some host species, and these can represent useful tools in genetic and biological investigations. In the field of biological control of insect pests, viruses have been found to have several applications (see Chapter 13).

The obligatory **parasitic nature** and the more or less complete association of virus replication with normal cellular activities represent a major obstacle in the *chemotherapeutic control* of virus diseases (see Chapter 13). The relative simplicity of the virion, the **dormant phase** of the infection cycle, stands in contrast to the complexity of virus interactions with host cell components in the **vegetative or growth phase,** leading to the production of new virions.

Viruses differ greatly in their mode of infection of the host cell. For example, some bacteriophages of *E. coli,* as well as most animal viruses, seem able to enter the host cell at specific sites. Some bacteriophages have fairly sophisticated organs for piercing or by-passing the cell wall, whereas other viruses have affinities for specific components of their host cell membrane. Such components are known as **receptors** (see Sec. 11.2). No special "infection organs" are then required. In contrast, most plant viruses are dependent on mechanical damage to the cell wall, usually of the hair cells of the epidermis, before they can enter. Such damage results from physical forces (wind, abrasion), but more frequently the virus is introduced into the cell by a particular animal vector (insects, mites, nematodes, and so on) or by a soil-borne fungus through the roots. With certain viruses, infection of a plant can be achieved in the laboratory only by grafting with infected tissue. Infection, once started, spreads in plants largely through the abundant **plasmodesmata** connecting the hair cells and the lower epidermal cells. Animal cells, lacking cellulose walls, have more direct contact, but viruses also get carried through the body fluids (see Chapter 12). Bacterial viruses are generally transmitted through fluid media.

1.5 THE VIRUS INFECTION CYCLE

If all the susceptible cells in a uniform culture can be infected simultaneously with a uniform preparation of active virions, and if the culture can be maintained under controlled conditions, the events of the infection cycle may occur in each cell at nearly the same time. Results of analyses of aliquots removed from the culture could then be considered to reflect the results of events occurring in single cells. The degree of synchrony needed to achieve this goal has been obtained only with bacteriophages, using suspensions of bacterial cells *(E. coli)* in liquid medium. The result is a "one-step growth curve."

The most usual experimental procedure for studying virus replication is this **one-step growth curve** (certainly a misnomer, since virions do not "grow" by enlargement as do cells or multicellular organisms). The main idea is to synchronize infection of many cells to provide sufficient material for analyses of several replicate samples at different stages of a single cycle of the infection process. Typically, there are several such distinct stages of infection, which can be likened to developmental stages of higher organisms. The earliest stages are most generally called **adsorption** and **entry,** during which virions attach to the cell and the viral nucleic acid enters the cells, with or without other virion components, depending on the virus (see Chapter 11). During the next stage of infection, the main processes of virus replication occur: production of viral nucleic acids and proteins. Since no complete infectious particles are found during this stage, even if the cells are disrupted, early virologists called this stage the **eclipse period.** A more general term for this developmental stage is the **latent period;** the distinction between the two terms is that the eclipse period lasts only until virions are formed inside the cell (for applicable viruses), whereas the latent period lasts until newly assembled virions are released from the cell. The period of **assembly** of virions from their structural components and nucleic acids occurs late during infection, and the final steps of **maturation** of virions in infectious form may continue even during their release from the virion from the infected cell. The release of new virions is commonly called the **burst,** indicating the sudden disruption of the infected cell that occurs during many, but by no means all, viral infections.

A one-step growth curve can be useful, and a generalized form of such a curve is presented in Figure 1.9. to answer some of the questions about parameters of virus–host cell interactions posed above. Because the initial interaction of virion and cell occurs through random collision, starting the infection process with high concentrations of virus and cells facilitates rapid and nearly synchronous initiation of infection in all cells. In practice, not all infectious virions capable of adsorption do so in a reasonably short fraction of the whole replication cycle. It may therefore be desirable to take steps to eliminate unadsorbed virus, such as by washing the cells or adding trypsin or antiserum to eliminate the excess virus. Then at various times after initiation of infection, samples of culture medium or cells or both are tested for virus levels using an appropriate virus assay. At very early times, any infectious units observed will be due to residual unadsorbed virus. Soon thereafter, each infected cell acts as a single "infectious center" in the virus assay. Of course, if the cells are somehow disrupted, no infectious virus will be found inside until the end of the eclipse period.

FIGURE 1.9

The one-step growth curve. Schematic presentation of the virus reproductive cycle. The coordinates have no numbers because no specific type of cell or virus is envisaged.

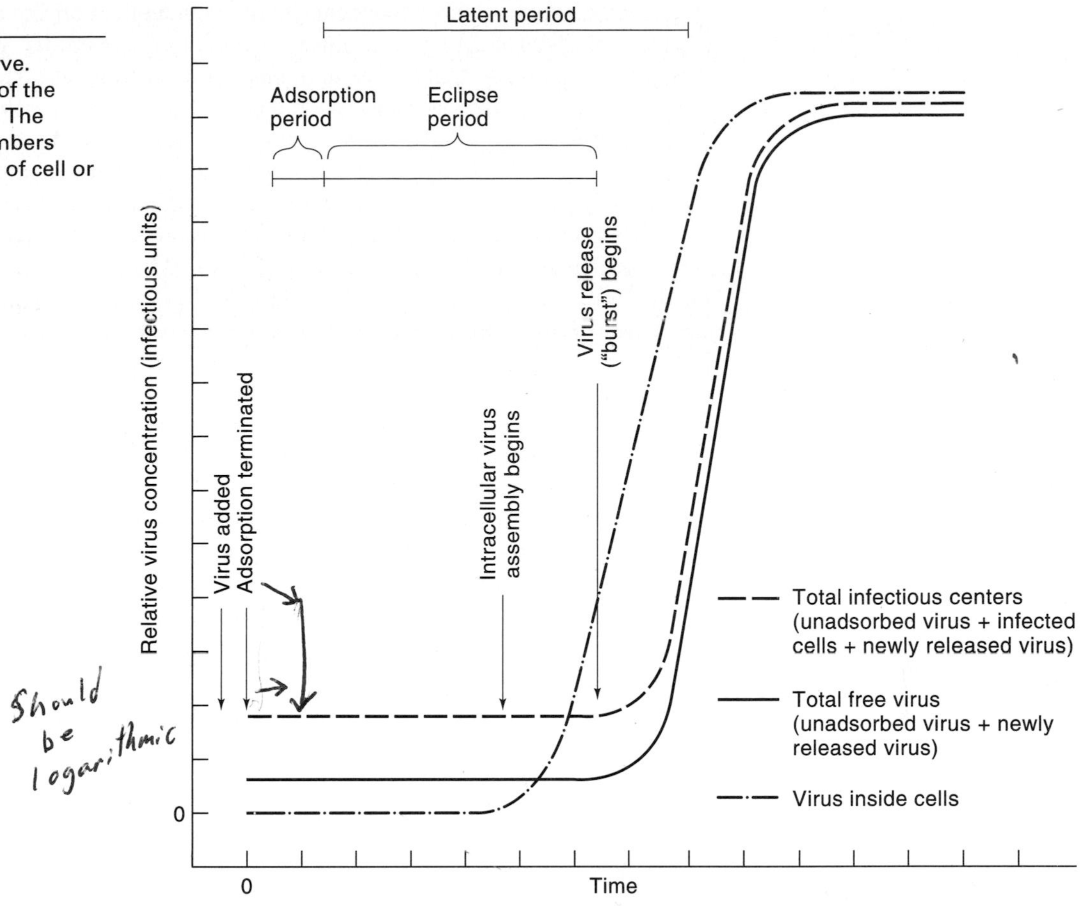

In the most elementary form of a one-step growth curve experiment, the only assay that is needed is to determine the number of **plaque-forming units (PFU)** in an aliquot taken from the culture at various times. In the experiment presented in the diagram (Fig. 1.9), one aliquot was of bulk culture. Another was of the supernatant obtained by centrifuging an aliquot of culture. The third aliquot was treated with chloroform, to disrupt bacterial cells, before the plaque assay. Several dilutions of each aliquot are prepared to accommodate the great range in number of PFU detected. That range is indicated by the use of the logarithmic scale for the ordinate in the diagram.

In the diagram, results are not directly presented as PFU but as PFU per cell in the culture just after adsorption (based on an assay for colony counts prior to infection or chemical analysis, e.g., for DNA). As expected, the value of PFU per cell is close to 1 early in the experiment. Subsequently the PFU per cell increases and reaches a plateau at several thousand PFU per cell. This gives the "burst size" for the infection. Note that chloroform treatment "exposes" the increase in infectivity (PFU) earlier than direct assay of the culture aliquots. This is evidence for the formation of fully competent viruses prior to lysis of the cells and release of virions.

During the "**eclipse period**" infectivity is not detected either in the supernatant or after the chloroform treatment. However, the PFU per cell remains at about 1. That is, the cells have the potential to produce viable bacteriophage particles, although no such particles are yet present. *The eclipse period is very characteristic of the virus infection cycle.*

The earliest time at which newly made virus is released from infected cells marks the end of the shortest possible latent period, a complete cycle for virus replication. This event may take a few minutes for bacterial viruses or several hours for **eukaryotic** viruses. The main reason for this difference in time scales seems to be the sheer enormity of animal and plant cells in comparison with bacterial cells—perhaps a 1000-fold difference in volume.

Because the various virus components involved in replication and assembly of virions can interact only after meeting each other via random collision, the greater volume through which virus components must move in eukaryotic cells greatly reduces the rate at which the inherently sluggish macromolecular components of the virus can interact by diffusion. As far as it is possible to determine, the actual rates of nucleic acid and protein synthesis are comparable in prokaryotes and eukaryotes, and in many cases the yields of virions per infected cell are comparable, despite the longer latent period in eukaryotes.

To determine the average number of virions produced per infected cell, or the average **burst size,** the final level of virus produced is divided by the number of infected cells producing the virus. In practice, the number of infected cells can be measured directly, by determining the number of **infectious centers** after the adsorption period, or the fraction of infected cells can be calculated from the known multiplicity of infection (m.o.i.), using the Poisson distribution (Appendix 1). Infectious centers can be quantitated by serologic procedures that detect viral proteins in individual cells. More commonly, the infected cells after virus adsorption are placed in suspension (washed or trypsinized) and added at different concentrations to nutrient agar (for **prokaryotes**) or fresh susceptible monolayer cells (for **eukaryotes**). Prior treatment of the virus-inoculated cells with X-rays or drugs (e.g., mitomycin C) to block cell replication is helpful in the assay. Each infected cell will then score for virus on the plated agar or target cells, and the percentage of cells initially infected can be calculated.

If we wish to learn the maximum or minimum amounts of viruses that are produced by different individual cells in an infected population or if we want to know whether two different virus particles can be replicated in a single cell, the conditions of infection must be varied somewhat from those of the basic one-step growth curve. Individual infected cells must be isolated so that the virus yield from a single cell, or a **single burst** as it is called, can be analyzed. Often the single-burst experiment involves great dilution of the infected culture and distribution of samples of the diluted cultures into tubes or dishes. Again the Poisson distribution is useful, not only to predict the fraction of cells that will receive one input virion, but also to predict the number of individual cultures that will actually receive just one cell. Because the cells are also of a particulate nature, the same sampling principles apply in both cases.

The simple one-step growth curve, which assays only for PFU, is a "black box" approach to the virus infection cycle that nevertheless is revealing. In summary,

The Steps Involved in the Virus Infection Cycle are:

1. Adsorption of virions to cell
2. Entry of virion nucleic acid into the interior of the cell, with or without other virion components
3. Exposure of genomic nucleic acid to cell's genetic machinery
4. Viral gene expression
5. Production of virion components, including the genomic nucleic acid
6. Virion assembly
7. Release of infectious viruses

Adsorption, entry, and exposure of the viral genomic nucleic acid to the genetic machinery of the cell are highly specific to a given group of viruses. Virion assembly and release similarly are specific to the virus–host cell combination. Gene expression and synthesis of virion components are so intimately connected to normal cell processes that the variation from virus to virus is much less apparent. Most of virology is concerned with what is happening during this eclipse period that is the time interval between viral nucleic acid entry into a cell and progeny virus production.

One-step growth curves can be attempted as well in cell culture where the monolayers have been synchronized through serum starvation or trypsinization. In this case, after virus inoculation, antibodies to the virus are placed in the culture medium to prevent second-round infection. These types of experiments have been helpful in duplicating the observations cited above with bacteriophage infection.

During the replication cycle, viruses generally produce identical copies of themselves but errors resulting from the action of their DNA or RNA **polymerases** and other causes, yield viruses that contain genetic mutations. The extent of such changes usually depends on the **error-prone nature** of the viral replicase. Some studies indicate differences in the mutation frequencies at different sites in the genome, but the exact mechanism for this specificity is not yet clear. In the case of the lentivirus (Chapter 14), the envelope and regulatory genes appear to be the most heterogeneous perhaps because of immune selection.

In general, RNA viruses are more error-prone in their replication than are the DNA viruses. A mutational frequency as high as 10^{-4} to 10^{-5} base pair (bp) substitutions per single base site have been observed. The DNA viruses generally mutate at a 10^{-7} to 10^{-11} frequency. These mutations give rise to a swarm of complex variant populations that have been termed "quasi-species" (see also Chapter 2). These variant populations, particularly with RNA genomes, can be selected by their adaptation to the environment (e.g., immune response, temperature, cell tropism) within just a few replicative cycles. These observations concerning the rapid mutational capability of RNA viruses have further supported the conclusions that the first nucleic acids formed were probably of the RNA species, which later gave rise to the more stable DNA species.

1.6 INTRODUCTION TO THE TAXONOMY OF VIRUSES

Figure 1.10 provides a broad classification of viruses based on the type of genomic nucleic acid and the types of molecules transcribed from the genomic nucleic acid as the virus directs the synthesis of messenger RNAs and new genomic nucleic acid. This figure presents some generalities about how viruses with particular types of genomic nucleic acids accomplish two critical tasks: synthesis of messenger RNA(s) for expression of virus genes and replication of the genomic nucleic acid. Appendix 2 lists some of the known virus families and some of their distinguishing properties.

If the genomic nucleic acid is single-stranded, there are several possible relationships between the genomic nucleic acid and messenger RNA specified by the genomic nucleic acid. To understand the possible relationships, it is necessary to define the **polarity** of a nucleic acid molecule. By definition, a single-stranded nucleic acid that has the same sequences as

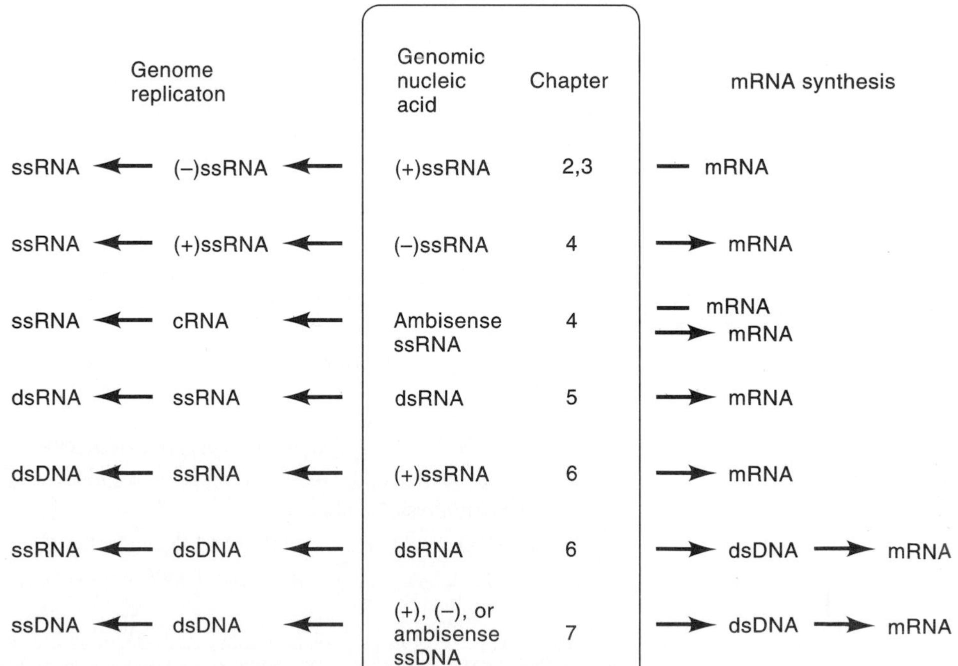

FIGURE 1.10

The arrows indicate transcription steps. The parallel lines alone indicate that the genomic RNA *is* mRNA. Because it is not possible to specify the polarity of the complement of an ambisense genomic RNA, such complementary RNA is specified as "cRNA." (Courtesy of G. Bruening)

Genome replicaton		Genomic nucleic acid	Chapter	mRNA synthesis
ssRNA ←	(−)ssRNA ←	(+)ssRNA	2,3	— mRNA
ssRNA ←	(+)ssRNA ←	(−)ssRNA	4	→ mRNA
ssRNA ←	cRNA ←	Ambisense ssRNA	4	— mRNA → mRNA
dsRNA ←	ssRNA ←	dsRNA	5	→ mRNA
dsDNA ←	ssRNA ←	(+)ssRNA	6	→ mRNA
ssRNA ←	dsDNA ←	dsRNA	6	→ dsDNA → mRNA
ssDNA ←	dsDNA ←	(+), (−), or ambisense ssDNA	7	→ dsDNA → mRNA
	dsDNA ←	dsDNA	8	→ mRNA

are found in messenger RNA (mRNA) is considered to be of the positive or (+) polarity. Often the (+) polarity can be recognized even without access to mRNA, if the nucleotide sequence is known. The nucleotide sequence is examined, taking three nucleotide residues at a time. Beginning at any AUG start codon, the sequence is examined (by computer) in the 5′ to 3′ direction until a stop codon (UAA, UAG, or UGA) is reached. The string of nucleotides beginning at the AUG and ending just before the stop codon is an **open reading frame (ORF).** An ORF is a *potential* polypeptide-encoding region of the molecule.

The presence of one or more ORFs indicates the mRNA polarity. Such long-ORF-possessing, single-stranded genomic nucleic acids, whether RNA or DNA, are said to be of the (+) polarity and are "positive sense."

If all the mRNA-specifying nucleotides of a single-stranded nucleic acid molecule are complementary to the mRNA sequences of the virus, the molecule is said to possess the (–) polarity and is negative sense.

A single-stranded genomic nucleic acid having mRNA-specifying sequences, some of which have mRNA sequences and others of which are complementary to mRNA sequences, is said to be "ambisense."

Accordingly, known viruses employ one of eight different **expression strategies** (see Fig. 1.10). The chapters in which these viruses are covered are indicated:

1. Genomic nucleic acid is single-stranded RNA of the positive polarity [(+)ssRNA] (Chapters 2 and 3)
2. Genomic nucleic acid is single-stranded RNA of the negative polarity [(–)ssRNA] or of both polarities (ambisense ssRNA) (Chapter 4)
3. Genomic nucleic acid is double-stranded RNA (dsRNA) of one or many segments (Chapter 5)
4. Genomic nucleic acid is single-stranded RNA or DNA of the positive polarity; transcription into dsDNA is essential for synthesis of new genomic RNA (Chapter 6)
5. Genomic nucleic acid is single-stranded DNA (ssDNA), which may have encoding sequences entirely of the (+) polarity, entirely of the (–) polarity, or both (ambisense) (Chapter 7)
6. Genomic nucleic acid is double-stranded DNA (Chapter 8)

Further details of the taxonomy of some viruses are given in Appendix 2. At this time, over 60 families have been identified. This taxonomy is based on many characteristics of the virus, one of which is the virion structure. To emphasize the great variety of forms that virions take, Figures 1.11–1.15 contain diagrams of most of the major groups of viruses that have received sufficient study to allow classification. These are presented according to host type, in the order used in most of the subsequent chapters of this book: viruses of vertebrates, viruses of invertebrates, viruses of plants and viruses of microorganisms.

FURTHER READING

McNeill, W. H. (1977) *Plagues and peoples.* Doubleday, New York.

Waterson, A. P., and L. Wilkinson. (1978) *An introduction to the history of virology.* Cambridge University Press, London.

White, D. O., and F. Fenner. (1986) *Medical virology.* Academic Press, Orlando, Fla.

Steinhauer, D. A., and J. J. Holland. (1987) Rapid evolution of RNA viruses. *Am. Rev. Microbiol.* 41:409.

Murphy, F. A. (1988) Virus taxonomy and nomenclature. In *Laboratory diagnosis of infectious disease, principles and practices,* ed. E. H. Lennette, P. Halonen, and F. A. Murphy. Vol 2. Springer-Verlag, New York.

Bergh, O., et al. (1989) High abundance of viruses found in aquatic environments. *Nature* 340:467.

FIELDS, B. N., and D. M. KNIPE. (1990) *Virology,* 2d ed. Raven Press, New York.

FRANCKI, R., et al. (1991) Classification and nomenclature of viruses. Fifth report of the International Committee on Taxonomy of Viruses. *Arch. Virol.* (suppl. 2). Springer-Verlag, Vienna.

JOKLIK, W. K., et al. (1992) *Zinsser's microbiology,* 20th ed. Appleton and Lange, Norwalk, Conn.

LUSTIG, A., and A. J. LEVINE. (1992) One hundred years of virology. *J. Virol.* 66:4629.

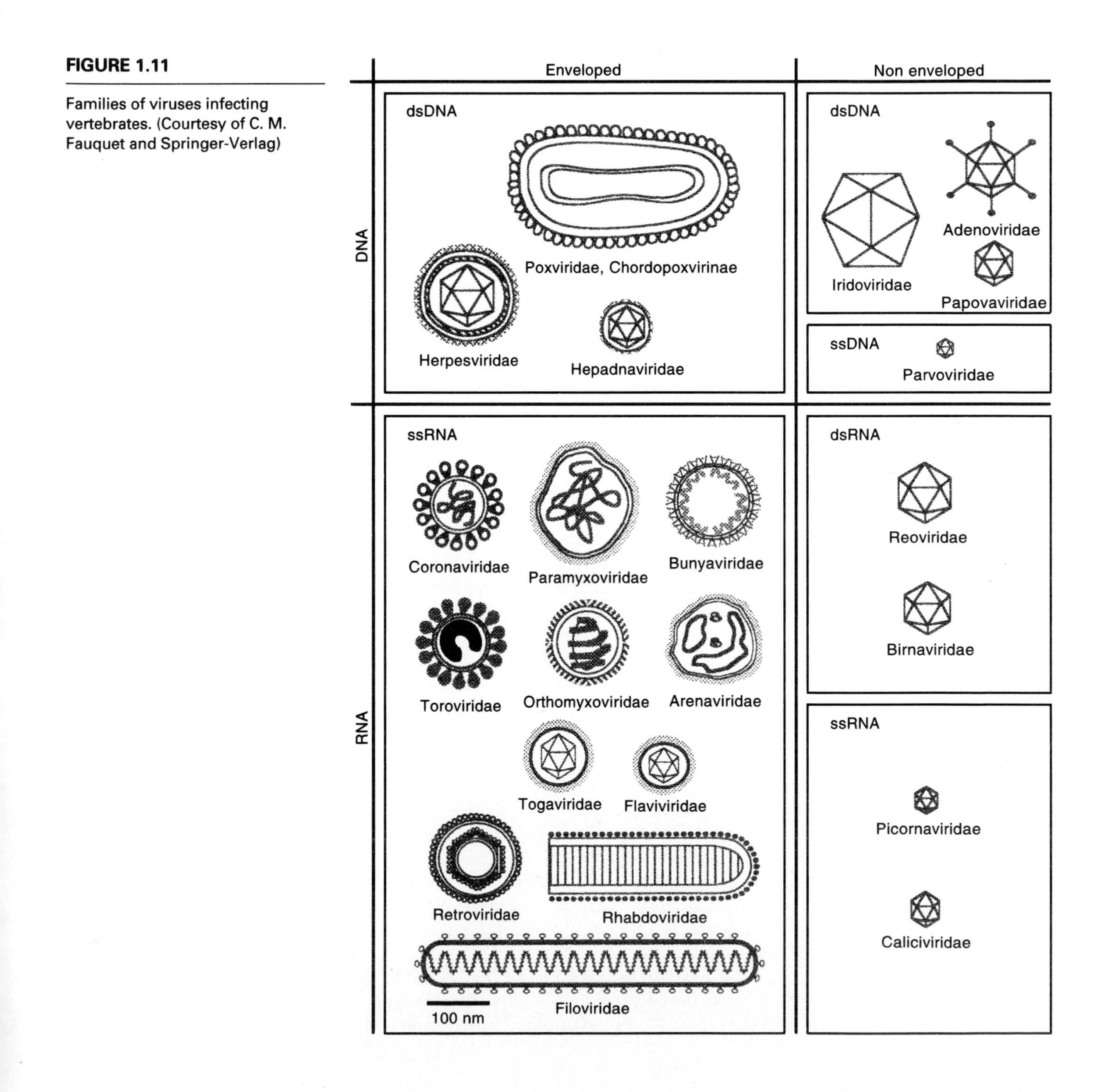

FIGURE 1.11

Families of viruses infecting vertebrates. (Courtesy of C. M. Fauquet and Springer-Verlag)

FIGURE 1.12

Families of viruses infecting invertebrates. (Courtesy of C. M. Fauquet and Springer-Verlag)

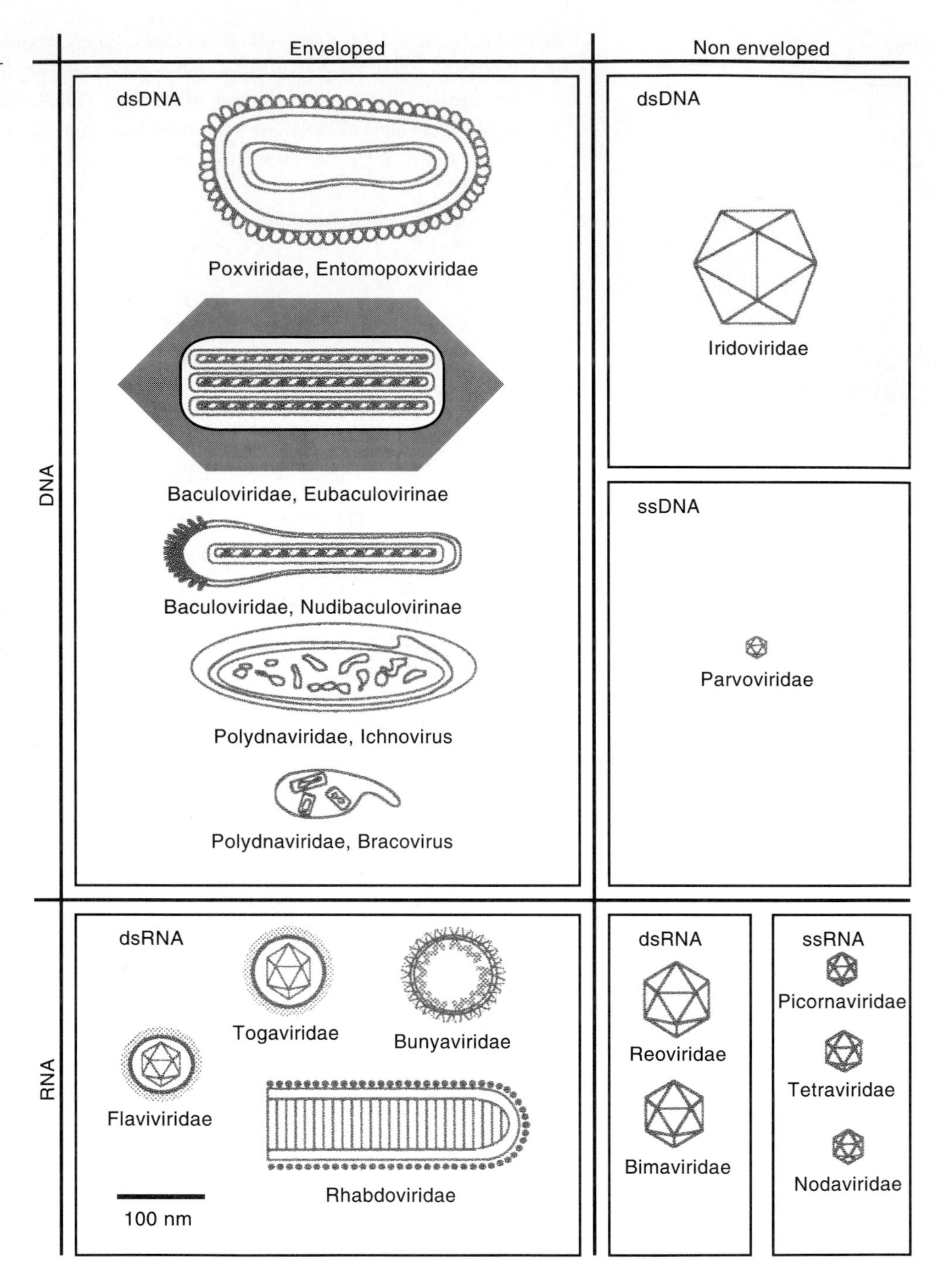

FIGURE 1.13

Families and groups of viruses infecting plants.
(Courtesy of C. M. Fauquet and Springer-Verlag)

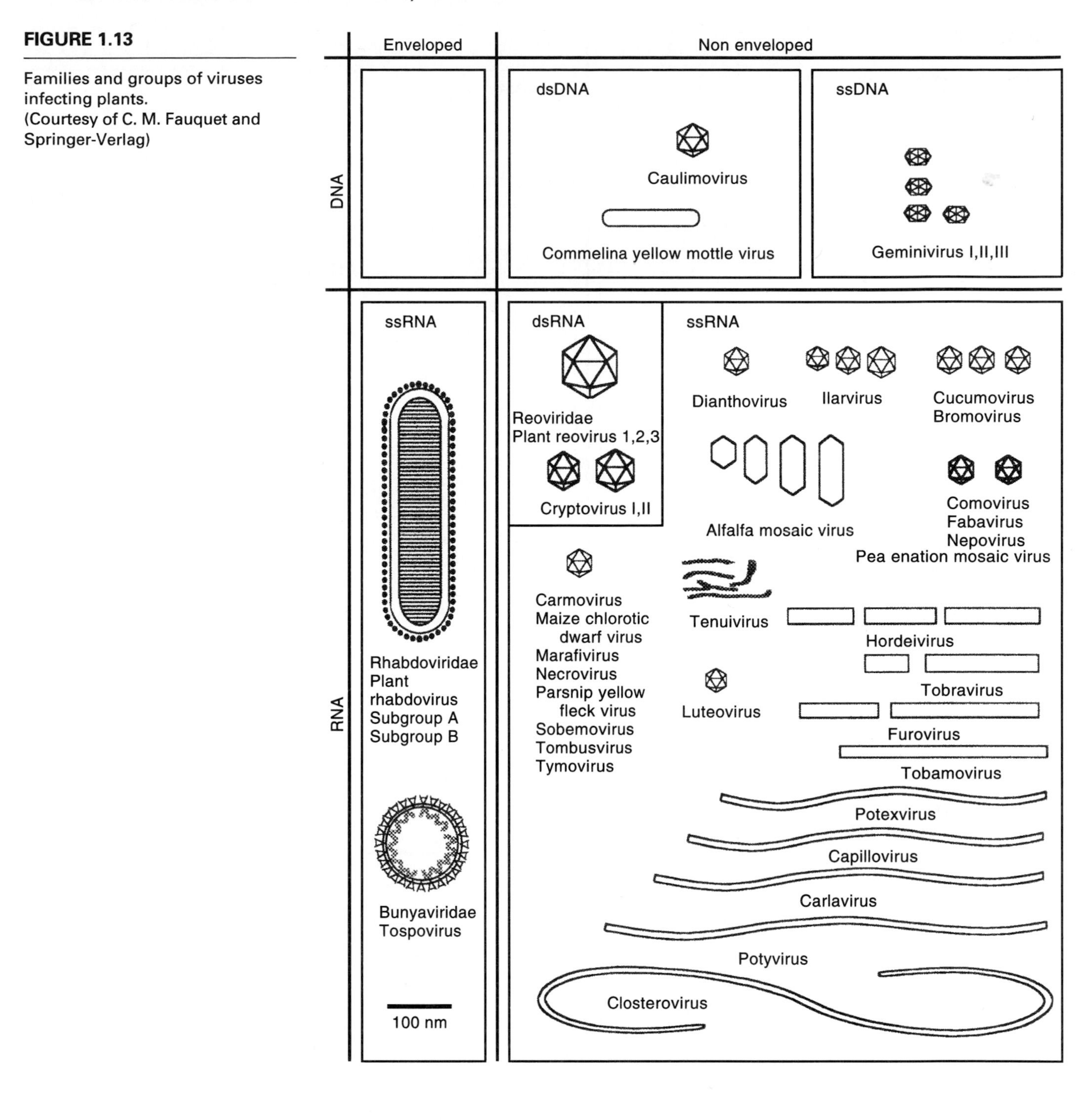

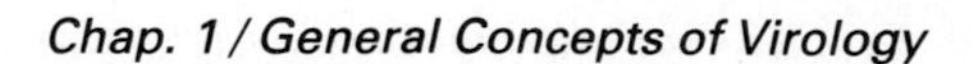

FIGURE 1.14

Families of viruses infecting algae, fungi, and protozoa.
(Courtesy of C. M. Fauquet and Springer-Verlag)

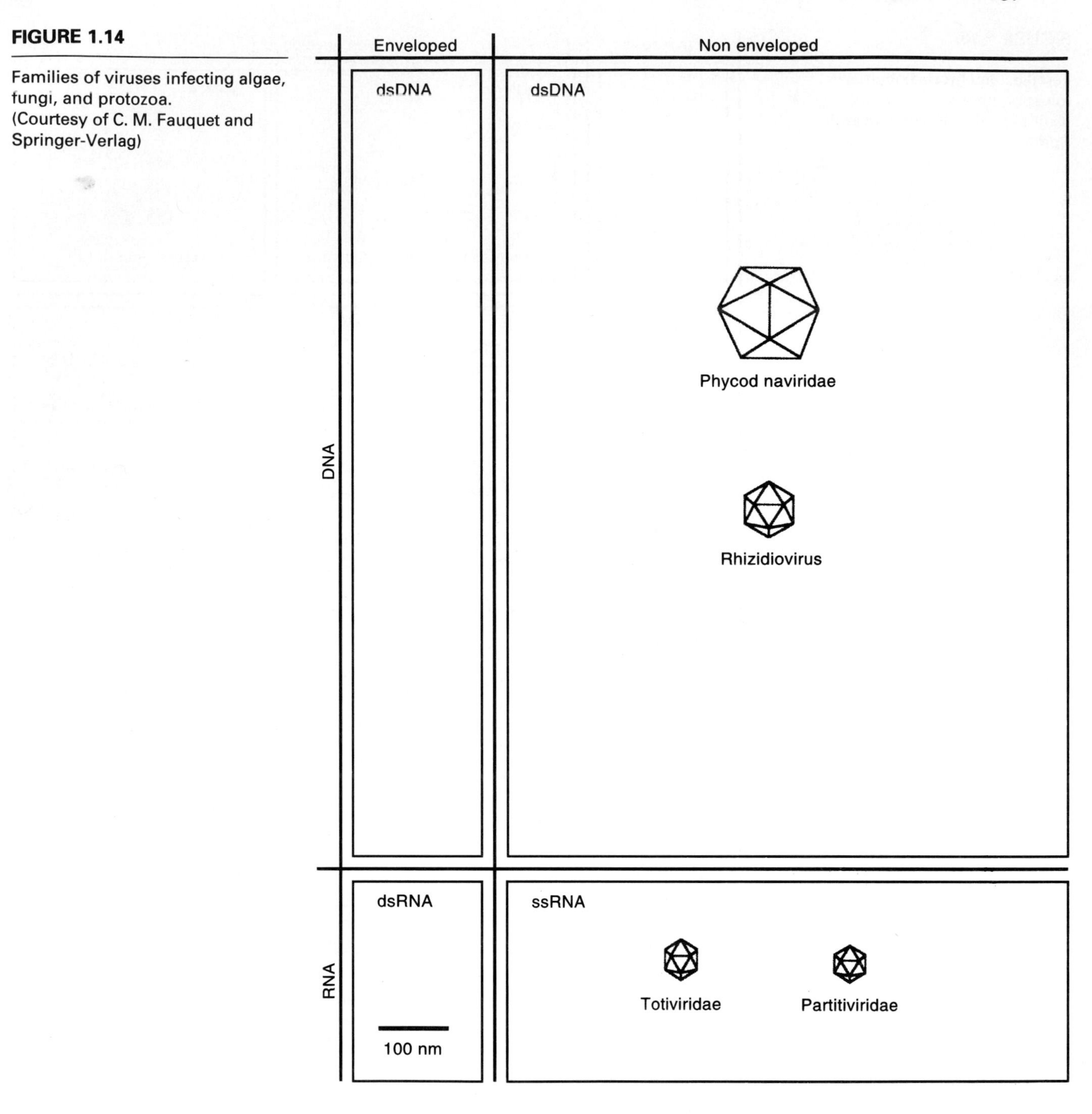

FIGURE 1.15

Families of viruses infecting bacteria. (Courtesy of C. M. Fauquet and Springer-Verlag)

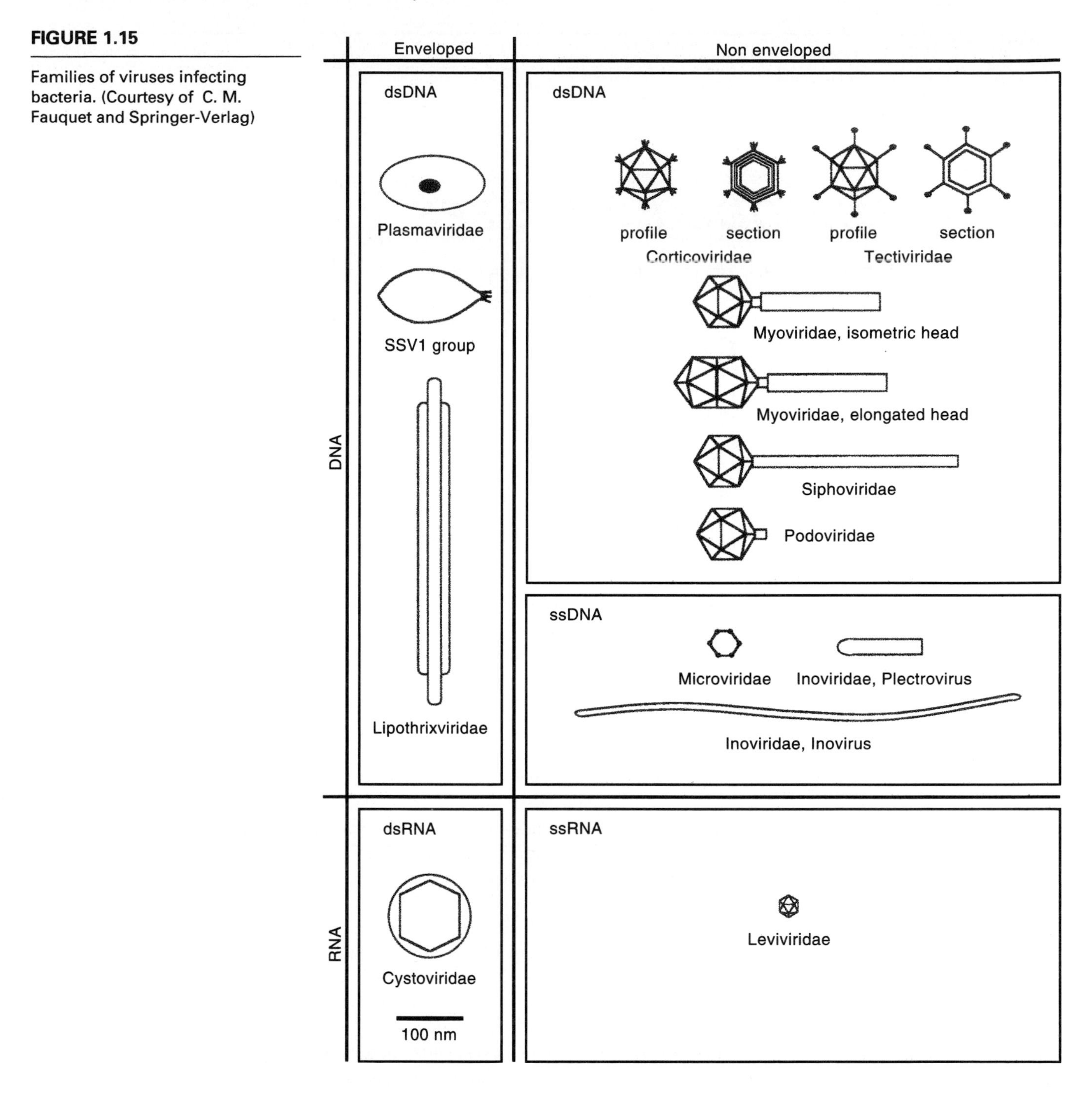

chapter

2

Nonenveloped Viruses with Positive-Sense RNA Genomes

This chapter deals with small nonenveloped viruses having positive-sense RNA genomes. These viruses are probably among the most primitive of all viruses—in biochemical as well as evolutionary terms. Such viruses occur in bacteria, where they have been termed **Leviviridae** (*levis,* "lightweight") and in animal cells, where the vast majority are termed **Picornaviridae** (*picorna,* "small RNA"). The greatest number and variety of viruses with single-stranded RNA genomes actually occur in plants, but these viruses have not yet been given family names.

2.1 INTRODUCTION

All these viruses share several properties: (1) They all carry positive-sense RNA genomes that contain 4000–10,000 nucleotides (1.2–3.2 × 10^6 Da) and, upon infection, act directly as mRNAs for the synthesis of viral proteins; (2) the bacterial as well as most of the plant and animal small RNA viruses have very similar shapes and dimensions and thus often look alike in electron micrographs; and (3) they all consist of only nucleic acid and protein and lack lipid-containing envelopes. Their protein coats consist of either multiple copies of a single protein (phages and most plant viruses) or as many as four proteins (animal picornaviruses). Although we will stress the similarities among all simple RNA viruses, there are distinct differences in the modes of infection and replication among viruses of pro- and eukaryotes that reflect the great differences in their hosts. In this chapter we first review the **replicative cycle** of this group of viruses. We then describe the properties of several individual viruses—first those infecting animals (e.g., picorna- and caliciviruses), next those infecting plants (e.g., tobacco mosaic virus), and, finally, those isolated from bacteria and other microorganisms (Qβ and related bacteriophages).

2.1.1 General Aspects of Viral RNA Translation

After attachment to the host cell, penetration, and uncoating, the ability of a virus to multiply depends on the **synthesis of viral proteins.** These viral proteins have three basic functions: (1) to ensure the replication of the viral genome; (2) to package the newly replicated viral genomes into virions; and (3) to alter the structure and/or function of the infected host cell. By definition, the genome of a positive-sense RNA virus is able to function as a viral mRNA. Although infectivity and mRNA synthesis do not depend on the presence of virion-associated enzyme(s), the viral genomic RNA may or may not be sufficient for synthesis of all the viral proteins. In many cases, expression of the genes of even small positive-sense RNA viruses depends on the synthesis of **subgenomic mRNAs.** It is also important to remember that viral proteins—especially those involved in genome replication—often only modify or assist the function of host cell proteins.

How do we know the number and nature of genes and corresponding proteins in any virus? Frequently, the existence of a gene product becomes evident from the study of **mutants** that lack a particular function. Two other strategies can also be used to directly study viral mRNA translation and thereby establish the number of virus-encoded polypeptides. Both the *in vivo* and *in vitro* strategies depend on the incorporation of radioactive amino acids into high-molecular-weight polypeptides, and each has certain advantages and disadvantages.

The *in vivo* approach is usually undertaken in cell culture. Virus-infected living cells are allowed to incorporate the amino acids for various time periods, and the proteins are then extracted and analyzed—usually by polyacrylamide gel electrophoresis (PAGE) (see Appendix 1). This technique has the advantage that protein synthesis is studied under normal biological conditions, but, because viral protein synthesis occurs concomitant with host protein synthesis, it may be difficult to distinguish between the two. Fortunately for the investigator (but not the cell), many viruses quickly diminish or "shut off" the synthesis of host nucleic acids and proteins after infection, and thus the only proteins synthesized to an appreciable extent are the viral proteins.

An alternative approach involves protein synthesis *in vitro,* and it is often the experimental strategy of choice for plant viruses and other viruses that do not substantially inhibit host protein synthesis. The first cell-free extracts were prepared from *E. coli,* and plant viral RNAs (e.g., TMV RNA) were among the initial available sources of mRNA. At present, extracts prepared from wheat germ and rabbit reticulocytes (immature red blood cells) are widely used. The advantages of the *in vitro* systems are that they are simpler and more subject to experimental control than is the living cell. Their disadvantage is that being simpler, they may be unable to carry out certain **posttranslational** modifications or **cleavages** of viral translation products.

A variety of immunological and biochemical tests can be used to prove identity between a known protein of the virion and a newly synthesized product (Appendix 1). Virus-specific translation products can be specifically precipitated by addition of antisera directed against either the intact virus or one of its protein components. The pattern of peptides resulting from digestion of the synthesized proteins by specific proteolytic enzymes (e.g., trypsin) can be compared with the pattern given by digests of the virion proteins, but definitive proof of the identity of two proteins requires determination of their complete amino acid sequences.

More recently, powerful new approaches to identify genes and gene products have evolved from the application of recombinant DNA technology and newer immunological approaches. Viruses often encode certain "minor" proteins whose existence is first suggested by nucleic acid sequencing studies but whose expression levels are too low for detection by conventional *in vivo* or *in vitro* labeling methods alone. In some such cases, such proteins have been detected by the use of antibodies made by immunizing animals with small synthetic peptide antigens containing 15 to 20 amino acids from a protein sequence predicted by computerized translation of nucleic acid sequences.

2.1.2 Replication of Viral RNAs

Although the replication of viral RNAs has been studied in many different biological systems for almost 30 years, the molecular events that occur during this process are still not known in detail for any viral RNA. When virus-specific RNAs are isolated from cells in which replication is occurring, two types of RNA differing from the viral genomic RNA are found. One type, having a distinctly lower sedimentation rate and pronounced resistance to ribonuclease, is double-stranded RNA; the other has an intermediate and variable sedimentation rate and is only partially **nuclease-resistant** (i.e., double-stranded). In analogy with the events in DNA replication (see Chapters 7–8), the first type of RNA has been called the **replicative form** (RF) and the second the **replicative intermediate** (RI).

It is evident from Figure 2.1 that nuclease treatment of RI under conditions specific for digestion of single-stranded RNA should produce material with some (but *not all*) of the properties of RF. Upon denaturation, genuine RF becomes two single strands of equal length (one positive- and one negative-sense), whereas nuclease-treated RI becomes one full-length and several shorter pieces of RNA. There is evidence that two or three phage RNA polymerase complexes can simultaneously copy one template RNA molecule, and this process accounts well for the properties of RI. This hypothetical mechanism requires that the copied (and thus double-stranded) 3′ end of the template RNA again interacts with an enzyme molecule, which proceeds to displace the new strand and to start a second copy, and so forth. In other words, the replication products are double-stranded until they are displaced by a "working" polymerase molecule. This process must be repeated to complete

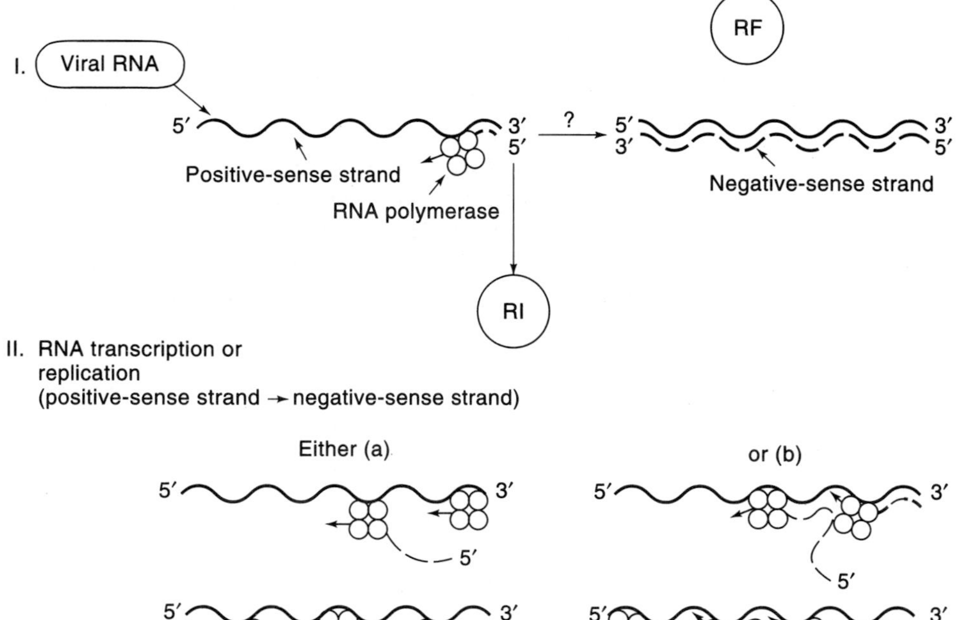

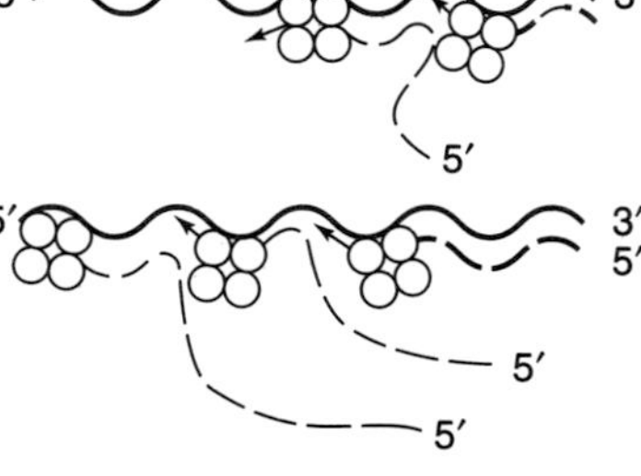

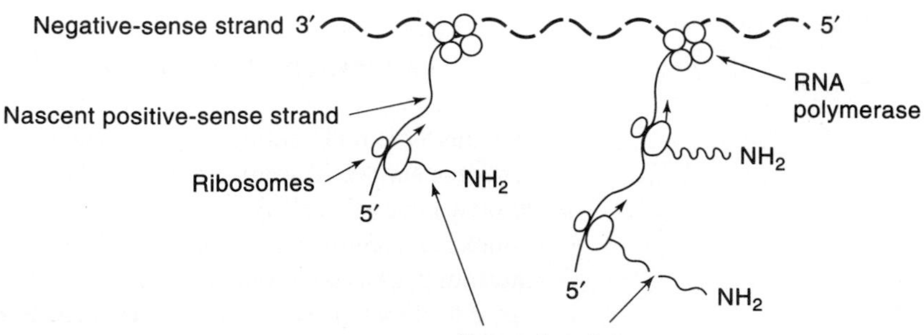

FIGURE 2.1

Modes of viral RNA replication. I. The RNA-dependent RNA-polymerase complex starts replication from the 3′ end of the positive-sense RNA template strand. Completion of that process yields a double-stranded product termed RF (replicative form). II. As shown in process (b), several enzyme molecules follow one another, pushing the primary transcripts aside as they proceed. The resulting replicative intermediate (RI) will have both double- and single-stranded character. In bacteria, however, it appears more probable that the newly made strand does not remain hydrogen-bonded to the template and that the RI is largely single-stranded *in situ,* as shown in process (a). III. During the second stage of RNA transcription, when the negative-sense strand is being transcribed, the nascent positive-sense strand will be available for translation only if the RI is of the type resulting from process (a).

the replication cycle. A single infecting positive-sense RNA first acts as a template for the production of several complementary negative-sense strands, and then these become templates for the synthesis of many positive-sense RNAs that serve both as mRNAs for large-scale virion protein production and as virion RNA for assembly of the progeny.

In this scheme, there is no real need for an RF, and RF may be a byproduct that only forms when replication slows down rather than a necessary intermediate in RNA replication. It has even been suggested that the double-stranded nature of RF and RI is largely an artifact; that is, it occurs only upon removal of associated protein(s) during the isolation of the viral RNA. According to this concept, template and product remain base-paired over only a short stretch (see Fig. 2.1). Although there is good evidence to support this view, the truth may well lie between the two extremes, since the presence of double-stranded RNA has definitely been demonstrated in eukaryotic cells infected with RNA viruses.

The RNA-dependent RNA polymerases that carry out the replication of positive-sense RNA virus are known as **RNA replicases.** This term denotes their ability to synthesize both positive- and negative-sense viral RNA. The RNA replicase of bacteriophage Qβ is by far the best characterized of such enzymes, and the complete Qβ replication apparatus contains five polypeptides. The "core replicase," containing a 65 kDa virus-encoded subunit responsible for phosphodiester bond formation plus host protein biosynthesis elongation factors EF-Tu and EF-Ts, is sufficient for all reactions except synthesis of the positive-sense (genomic) RNA. This reaction requires two additional polypeptides; that is, host ribosomal protein S1 plus six molecules of an RNA-binding protein known as "host factor." The 1965 demonstration by Spiegelman and collaborators that Qβ replication could be carried out *in vitro* was a landmark experiment in the development of molecular biology.

Purified preparations of cucumber mosaic virus RNA replicase contain two large (98 and 110 kDa) virus-encoded polypeptides encoded by RNAs 1 and 2, respectively, plus one major host-encoded polypeptide (50 kDa). RNA synthesis is **template-dependent** and produces a mixture of full-length double- and single-stranded products that includes both forms of the subgenomic RNA 4 (see Table 2.2 (Sec. 2.6). The single-stranded RNA produced is predominantly positive-sense.

Biochemical studies of poliovirus RNA replication have shown that viral protein 3D (see Fig. 2.4 in Sec. 2.2.2) is necessary and sufficient for chain elongation, but the mechanism of initiation of viral RNA synthesis has proven to be especially intractable. The role of a "host factor" present in uninfected cells in this process remains controversial, and the identity of the VPg donor (protein 3AB; see Fig. 2.4) has not been definitely established. Fortunately, the availability of poliovirus cDNA clones that produce infectious virus upon transfection of primate cells plus the development of a cell-free system that is able to synthesize infectious poliovirus *de novo* offer hope for more rapid progress in understanding picornaviral RNA replication in the near future.

2.1.3 RNA Population Genetics and the "Quasispecies" Concept

Mutation, a change in the nucleotide sequence of a genomic nucleic acid, is the all-important event that makes biological evolution possible. Single nucleotide replacements, usually termed **point mutations,** occur with much higher frequency in single-stranded than in double-stranded nucleic acids. Such changes also become immediately effective in the single-stranded genome, because it lacks the complementary (and unmutated) sequence. The same is true for **sequence reversion,** a process during which a mutation returns to the original or "wild-type" sequence. The high mutation and reversion rates of single-stranded nucleic acids are probably the reason why these molecules were rejected as genomes in the evolution of living cells and higher organisms.

Mutation of single-stranded genomes can result from either base alterations due to physical or chemical agents or the insertion or deletion of nucleotides due to errors during replication. Single-base modifications can occur spontaneously, a result of the inherent chemical properties of the bases themselves, but they can also be produced at much higher frequencies by reaction with certain chemical **mutagens.** Nitrous acid, the most useful reagent for this purpose, converts C to U and A to hypoxanthine (known as inosine in its nucleoside form), which can code like G. Its action on G is more complex. Reaction with hydroxylamine renders C ambiguous, allowing it to code as either C or U, and other mutagenic agents such as "nitrosoguanidine" (*N*-methyl-*N*-nitroso-*N'*-nitroguanidine) and many other nitroso-alkyl compounds or UV light act in a nonspecific manner. Subsequent to the primary mutational event, nature has developed several means of testing the advantages or disadvantages of a given mutation. Among these are **recombination** and **reassortment,** two processes that have played particularly prominent roles in the evolution of viruses.

Mutation rates for RNA genomes are much higher than those for DNA genomes—10^{-3} to 10^{-4} versus $\leq 10^{-7}$ changes per nucleotide per replication cycle. For thermodynamic reasons, all template-copying processes have limited accuracy, but RNA replicases appear to have little or no ability to remove misincorporated nucleotides via proofreading or other repair mechanisms. Thus, when one considers the average genome size of an RNA virus ($\leq$10,000 nucleotides) and number of progeny produced from a single infection (approximately 10^{12} particles), it is clear that even clonally derived virus preparations must contain a bewildering assortment of single, double, triple, and higher mutants. In the absence of selective pressure, very few (if any) RNA molecules will actually have a nucleotide sequence identical to the "consensus" or "master sequence" of the population. As described mathematically by Eigen and his collaborators (see Sec. 1.5), replicating RNA genomes are best considered **quasispecies** in which terms such as "wild-type," "mutant," and "revertant" must be defined in terms of **probabilities,** and biological selection assumes a new importance.

We will return to the population genetics of RNA viruses in Chapter 12, but, before turning our attention to the forces that stabilize their capsids, one example from the literature may help to make the concept of a quasispecies less abstract. In the early 1970s Charles Weissmann and his collaborators used T1 RNase fingerprinting techniques to compare the genomes of more than 160 clones of bacteriophage Qβ derived from a multiply passaged phage stock. If a random distribution of mutations among infectious molecules is assumed, the fact that 15 percent of these clones yielded T1 fingerprints different from that of the parental stock means that the sequence of each viable phage genome differs in 1 or 2 positions from the Qβ master sequence.

2.1.4 Virus Assembly

It has long been known that many virus infections, particularly those involving isometric viruses, produce not only infectious virions but also virus-like particles lacking most or all of their normal nucleic acid content. Because such non-infectious particles are necessarily less dense than complete virions, they are easily isolated by density-gradient centrifugation, where they are found near the top of the gradient. These empty particles are often, therefore, called "**top**" **components**. The coat proteins of such viruses evidently have a strong tendency to aggregate in a specific and surprisingly firm manner and can do so without the stabilizing support of the viral nucleic acid (i.e., purely through interprotein binding mediated by hydrogen bonds, ionic bonds, or hydrophobic interactions). Examples of viruses exhibiting this type of capsid assembly include turnip yellow mosaic and cowpea mosaic viruses.

Many other isometric viruses, however, do not produce top components, and their assembly *in vitro* (see below) occurs only after nucleic acid is added to the viral capsid protein. This assembly process is often quite promiscuous, leading to the formation of a virus-like shell around a definite amount of any nucleic acid—RNA, DNA, or even several molecules of a small RNA such as tRNA. Obviously, both protein-protein and protein-polynucleotide interactions are required for virion assembly in this case.

Grouping isometric viruses according to the degree to which their particles are more of one type or the other allows useful predictions to be made about their stability under certain biological conditions. For example, the infectivity of viruses that do not form empty shells, because they rely on the nucleic acid for their architecture, is quite sensitive to nucleases. Sensitive to salt concentration and slight changes in pH, such viruses are not very good viruses in terms of stability. Presumably they compensate for their structural deficiency by greater ease of infection, higher particle production, or some other feature. The plant virus, brome mosaic virus is one example of the class that requires RNA for virion formation, whereas the capsid proteins of tobacco mosaic or alfalfa mosaic virus can reassemble into, respectively, virus-like rod-shaped or isometric particles in the absence of RNA (Fig. 2.7, Sec. 2.6.1).

In 1955 Fraenkel-Conrat and Williams showed that infectious particles of TMV (a helical virus) could be reassembled *in vitro* from purified coat protein subunits and TMV RNA, particles that were indistinguishable from the native virus by a number of criteria. Several years later (in the late 1960s), Bancroft and his collaborators reported the first successful reconstitution of an isometric RNA virus, that of cowpea chlorotic mottle virus, a member of the bromovirus group. Only in the case of TMV, however, has the nature of the RNA-protein interaction responsible for the initiating reassembly been identified (see Sec. 2.2.2).

More recently, results from high-resolution X-ray crystallographic and molecular genetic analyses of several isometric RNA viruses [e.g., poliovirus and turnip crinkle virus (TCV)] have combined to provide increasingly detailed structural models for both their coat protein subunits and interactions between adjacent subunits. Thus, the coat protein of TCV has been shown to interact with its genomic RNA through sequence-specific as well as sequence-nonspecific interactions. The mechanisms for RNA condensation seem to differ among different icosahedral viruses, but in TCV, a plant virus that is primarily stabilized by RNA-protein interactions, co-condensation of RNA with polymerizing coat protein during capsid formation late in the replication cycle provides the most likely mechanism for RNA packing. Other examples of coat protein–RNA interactions that have profound regulatory effects during virus replication are discussed in the section on RNA bacteriophages (Secs. 2.7.3 and 2.7.4).

2.2 PICORNAVIRIDAE

Members of the Picornaviridae, are responsible for several serious and widespread diseases of humans and other animals. They have been intensively studied since the early days of molecular virology. Foot-and-mouth disease virus was discovered nearly 100 years ago when Loeffler and Frosch found that its causative agent could pass through filters used to retain all other known disease-causing microorganisms. **Poliomyelitis** was a very dangerous and disabling disease in the first half of this century, and the "war on polio" proclaimed by President Franklin Delano Roosevelt, a victim of that disease, in the early 1930s was the first instance of a large-scale and wide-ranging planned scientific attack by society against a disease. Although Landsteiner and Popper established the viral nature of poliomyelitis in 1909, the term *poliovirus* first appeared around 1955.

Poliovirus was the first animal virus to be purified and obtained in crystalline form and poliomyelitis was among the first human or animal diseases to be found transmissible by a pure RNA molecule. Poliovirions are isometric particles about 27 nm in diameter (see Fig. 2.2) and resemble those of RNA bacteriophages and many typical plant RNA viruses. Their three-dimensional structure, as well as those of several other picornaviruses, has been studied extensively and will be discussed in Section 2.2.1.

As shown in Table 2.1, members of the Picornaviridae have been grouped into five genera plus several as-yet-unassigned viruses. The Picornaviruses differ more or less markedly in their antigenic properties, pathogenicity, host range, and stability (e.g., sensitivity to acid). The acid-sensitive rhino- and aphthoviruses—agents that cause, respectively, the common cold and the foot-and-mouth disease of cattle—probably owe their localized pathogenicity to the fact that they are, in contrast with the enteroviruses, inactivated by the stomach's acidity. Many strains or serotypes are known for most of these viruses. This latter fact and the recognition that other viruses cause similar respiratory symptoms

TABLE 2.1
Taxonomy of picornaviridae

Genus	*No. of serotypes*	*Host range*	*Primary habitat*	*Examples*
Enterovirus	>70	Humans and other mammals (narrow–wide)	Gut	Human poliovirus 1–3; Human coxsackie virus; Theiler's murine encephalomyelitis; Human echoviruses
Rhinovirus[a]	>130	Humans and other mammals (narrow)	Upper respiratory tract	Human rhinoviruses; Bovine rhinoviruses
Aphthovirus[a]	7	Cloven-footed and other mammals (wide)	Generalized	Foot-and-mouth disease virus
Cardiovirus	1	Mice and other mammals (wide)	Heart, CNS	Encephalomyocarditis virus; Mengovirus
Hepatitis A	1	Humans and other primates (narrow)	Liver	Human hepatitis virus A
Unassigned	>3	Various		Equine rhinovirus 1 & 2; Cricket paralysis virus; *Drosophila* P virus

[a]Acid-sensitive.

explain why we do not develop immunity to the common cold and why useful vaccines have not been developed. Humans appear to gradually accumulate more of the necessary antibodies against these viruses as they age, so that older people are less susceptible to colds than are the young (see Chapter 12).

The **stability** of the enteroviruses in acid allows them to be ingested and to reach the intestinal tracts of animals and humans. Even in the absence of disease symptoms, a large number of enterovirus serotypes have been found in animal and human feces. These viruses appear to be ubiquitous, sometimes harmless, passengers in normal individuals. For this reason, some of these viruses were formerly given the name "echovirus," an acronym for *e*nteric *c*ytopathogenic *h*uman *o*rphan virus—meaning that, even though they may cause damage to certain cells in culture, their presence has no known relationship to disease. Several enteroviruses are associated with gastrointestinal disorders, meningitis/encephalitis, or respiratory illness, and human enterovirus 70 causes acute hemorrhagic conjunctivitis. As the name suggests, cardiovirus infections frequently tend to localize in the heart muscle, but the name of one of these viruses, mouse encephalomyocarditis virus (EMC), shows that it is equally likely to cause inflammations of the brain. One EMC variant can infect the pancreatic islet cells of mice, thus preventing synthesis of insulin. Coxsackie viruses have also been found associated with pancreatic disorders. These observations have led to suggestions that some forms of human diabetes may be caused by similar viruses during childhood infections (see Chapter 12).

2.2.1 Structure

The **isometric** virions of picornaviruses (160 S) are about 27 nm in diameter and consist of a single molecule of polyadenylated RNA containing 7209–8450 nucleotides (excluding the variable-length poly(A) tract) surrounded by an **icosahedral** protein shell. This protein coat is more complex than those of the small plant and bacterial viruses and consists of 60 molecules of each of four different proteins. These polypeptides, which were originally termed VP (*v*irion *p*rotein) 1–4 are now officially known as 1D, 1B, 1C, and 1A, respectively. Because these viruses, in contrast with most animal RNA viruses, have no lipid-containing envelope, they are not sensitive to ether and other water-immiscible solvents.

The structural details of polio- and rhinovirus virions have been elucidated at the level of individual amino acid side chains, and they are remarkably similar to those of isometrical plant viruses. As shown in Figure 2.2, each capsomer contains one copy of each of the three

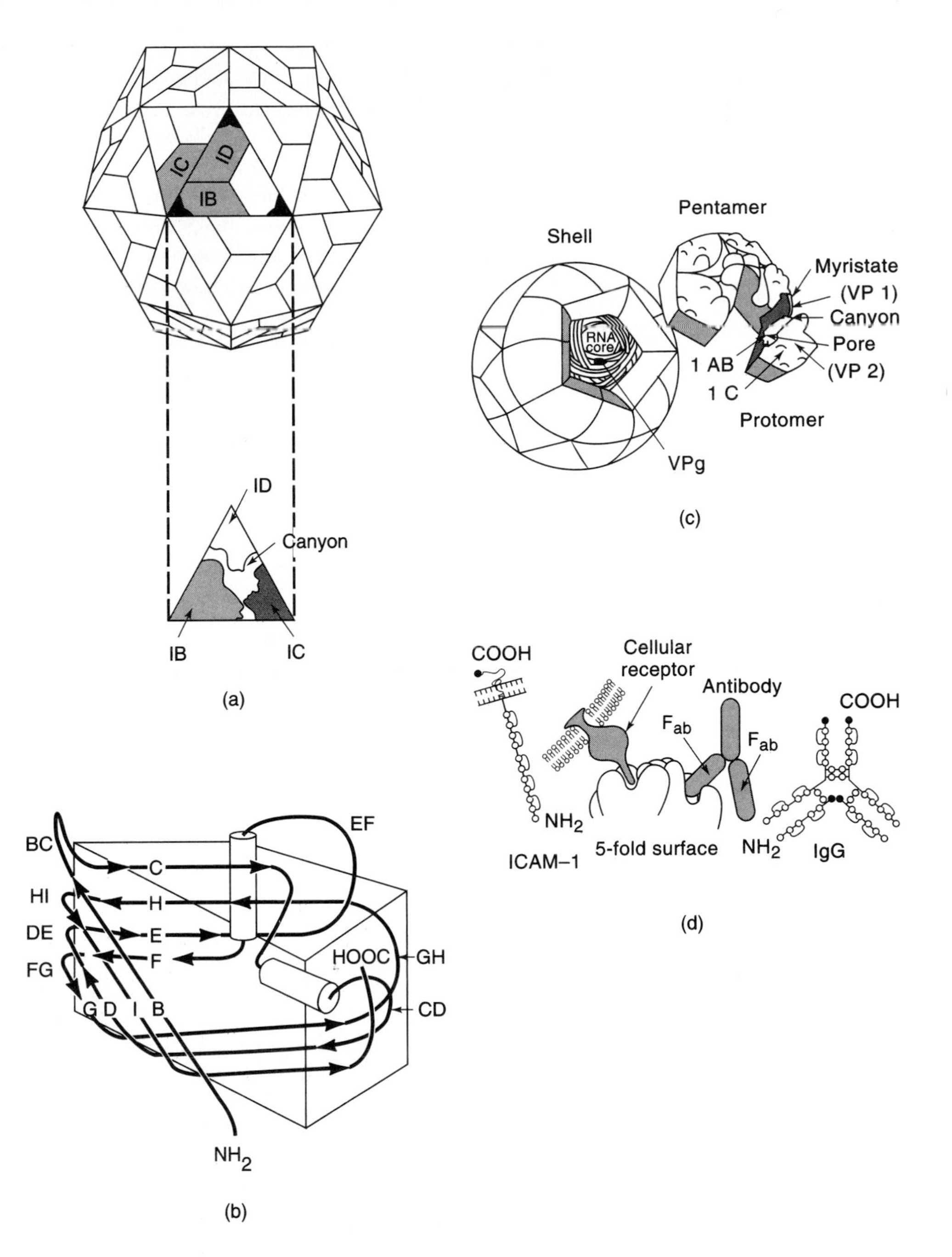

FIGURE 2.2

Key structural features of a typical picornavirus. (a) Pseudo T = 3 packing of 1D, 1B, and 1C in the picornaviral shell. The common folding pattern of the three proteins (known as an RVC, or RNA virus capsid domain) allows each to pack tightly at the fivefold or threefold axes of symmetry; VP4 is buried deep inside the virion at the base of the protomer. (b) The wedge-shaped topology of an idealized RVC domain. Each of the two apposed β sheets contains four antiparallel strands (denoted B–I). The sharp twist present in the sheet containing strands B, I, D, and G allows it to form both the inner surface and one side wall; strands C, H, E, and F form the other side wall. (c) Exploded diagram showing the internal location of the myristate residues on the N terminus of VP4. (d) Binding of the cellular receptor to the floor of the canyon. Note that the diameter of the virus-binding site on ICAM-1, the major rhinovirus receptor, is approximately half that of an IgG molecule. (**a, c, d** courtesy of R. R. Rueckert, **b** courtesy of S. C. Harrison.) (From chapters 3 and 20 of *Virology* 2d ed, eds. B. N. Fields and D. M. Knipe, 1990. Vol.1, Raven Press, New York)

distinct proteins 1D, 1C, and 1B. Protein 1A (VP4) is buried deep inside the particle in contact with the RNA and is not an integral component of the viral capsid. The three-dimensional structures of rhinoviruses and echoviruses are nearly identical, but it is not understood why the enteroviruses are so stable to acid. Furthermore, only the enteroviruses withstand disruption by strong anionic detergents such as sodium dodecyl sulfate (SDS). Precisely how the considerable differences in amino acid sequence among the various coat proteins lead to the differing stabilities of subunit interactions remains to be determined.

Besides the four capsid proteins, picornaviruses also contain an additional protein of unusual nature in the virion: Instead of the methylated G cap usually found at the 5′ terminus of cellular mRNAs (and many viral RNAs), a VPg protein containing 20 to 24 amino acids is attached to the 5′-terminal pUpU of the viral RNA via a phosphodiester linkage with a tyrosine-OH group. As discussed in Section 2.6.1, the RNAs of several plant virus groups also contain analogous 5′-terminal VPg moieties.

Like most eukaryotic mRNAs, picornavirus RNAs carry a string of adenylic acid residues at their 3′ terminus. The length of this poly(A) sequence is genetically determined and differs for different picornaviruses. Negative-sense RNAs, isolated from the replicative intermediate, contain poly(U) sequences of similar length and heterogeneity at their 5′ termini. Cardio- and aphthoviruses also contain a run of about 100 cytidylic acid within residues at a particular location within their 5′ untranslated region. Ranging in size from 624 to nearly 1200 nucleotides, the 5′ untranslated regions of picornaviral RNAs are unusually long when compared with those of cellular mRNAs, and their computer-predicted folding patterns are correlated with the virus groupings. The 3′ untranslated region is relatively short (47–126 nt) and follows a single long open reading frame containing 2178–2303 codons. The 5′ and 3′ untranslated regions are known to play important roles in viral replication. Thus, it is not surprising that changes of as few as one or two nucleotides may render the virus noninfectious or temperature-sensitive.

Isolated **picornaviral RNA is infectious;** indeed, the infectivity of aphthoviral RNA is nearly one infectious unit per molecule when microinjected into cells. RNA molecules with short poly(A) tracts have a lower specific infectivity, and infectivity is lost if the 3′-terminal poly(A) is removed. Although the 5′-terminal VPg is not required for infectivity, its removal from the RNA by a host-encoded "unlinking enzyme" prior to its translation appears to be an important regulatory signal that favors the flow of viral RNA to ribosomes early in the infection cycle. Double-stranded cDNA copies of certain picornaviral RNAs are infectious, provided that they are associated with a promoter sequence that can be recognized by the host cell DNA-directed RNA polymerase to generate a positive-sense RNA. Negative-sense RNA is not infectious.

2.2.2 Infection, Translation, and Posttranslational Cleavages

Our understanding of the process by which picornaviruses gain entry to the host cell is quite limited. Poliovirus infects only human and simian cells, but the naked viral RNA is not so exclusive—indicating that host specificity is often dictated by a chemical affinity between its coat (or envelope) proteins and a specific cell membrane site. Genes encoding the receptors for poliovirus as well as several other picornaviruses have been mapped to human chromosome 19, and ICAM-1 (*i*nter*c*ellular *a*dhesion *m*olecule-1, a member of the immunoglobulin gene superfamily) has been identified as the major **rhinovirus receptor** (see chapter 11). A number of antiviral chemicals neutralize the binding and/or subsequent uncoating of picornaviruses, and conformational changes in protein 1D induced by the so-called WIN compounds result in the loss of the ability of human rhinovirus 14 to bind to its cellular receptor (see Chapter 11). The binding affinity of HRV14 can also be modified by site-directed mutagenesis of amino acid residues in the floor of the canyon separating protein 1D from protein 1B and protein 1C in the capsid protomer (see Fig. 2.2 and 2.3).

Viral uncoating is believed to proceed via **receptor-mediated exocytosis**, a process in which virus-receptor complexes on the plasma membrane cluster at clathrin-coated pits and are subsequently internalized by invagination. Acidification of the resulting clathrin-coated vesicles then leads to release of 1AB, unfolding of hydrophobic regions of proteins 1B–D previously buried within the capsid, and formation of a pore through which the viral RNA can be transferred to the cytoplasm. The viral RNA may be released through the center of such an "exploded" pentamer, but until the exact location of VPg within the virion is established, we cannot know whether the release process involves one particular ("critical") pentamer. The presence of two uncleaved coat protein precursor molecules (protein 1AB; see below) may also play a role in releasing the RNA into the cytoplasm.

The first step in the replication of all positive-strand animal RNA viruses is the translation of the infecting RNA. In the case of all picornaviruses except the cardioviruses, this process is concomitant with a **marked decrease in translation of host m*RNAs*** and is a consequence of the proteolytic cleavage of the large (220 kDa) subunit of the cellular "cap-binding complex" (eIF-4). Rather than being mediated by one of the virus-encoded pro-

FIGURE 2.3

Organization and expression of the picornaviral genome. The position of a poly(C) tract found in the 5′ untranslated region of the cardio- and aphthovirus genomes is indicated by the three dots. Synthesis of the viral polyprotein is from left (N terminus) to right (C terminus). Growth functions (i.e., proteins needed for RNA synthesis and proteases required for polyprotein cleavage) are found downstream from the capsid protein: g^r, locus conferring resistance to guanidine, a drug believed to inhibit the initiation of RNA synthesis; h^r, a host-range determinant involved in RNA synthesis. Polyprotein cleavage is accomplished by three virus-encoded proteases: the M, or maturation, protease, the "early" 2A protease, and the 3C or 3CD protease. The *maturation cleavage* (1AB → 1A + 1B) occurs only after the viral RNA has been packaged in the protein shell. Proteins L and 2A carry out early cleavages of the polyprotein; all other cleavages are carried out by protease 3C or a precursor. The location of the sequence responsible for what is probably an *endoproteolytic* cleavage of 1AB → 1A + 1B has not been established. (R. R. Rueckert. (1990) in *Virology,* 2d ed., eds. B. N. Fields and D. M. Knipe. Vol. 1. Raven Press, New York)

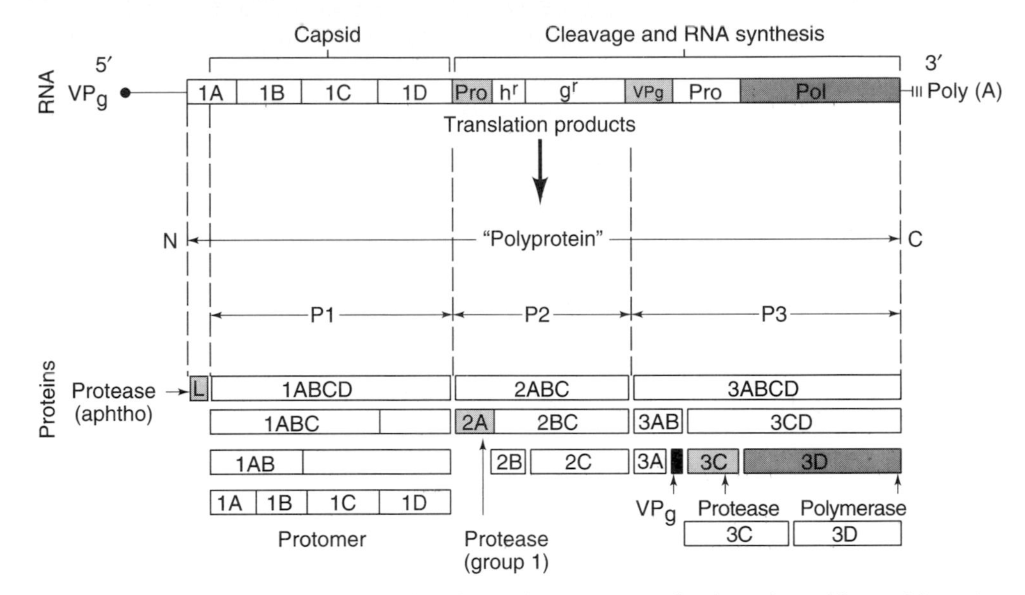

teases, this cleavage is carried out by a latent host protease that is activated by an 18–amino acid sequence present in viral 2A or L proteins (see Fig. 2.3). Because polio RNA lacks a cap, its translation is unaffected by this reaction. Instead, a series of stem-loop structures in the long (about 750 nt) 5′ noncoding region of picornavirus RNA constitutes an **internal ribosome entry site** that allows translation to begin without the need for ribosomal "scanning." Translation of a 2.4×10^6 Da mRNA would be expected to lead to unusually large polysomes, and polysomes containing up to 60 ribosomes on a single RNA molecule have been seen in picornavirus-infected cells (Fig. 2.4). Shutoff of host protein synthesis by cardioviruses is accomplished by a simple competition mechanism.

FIGURE 2.4

Electron micrograph of polysomes from poliovirus-infected HeLa cells. As many as 60 ribosomes can bind to one viral RNA molecule. (A. Rich et al. (1963) *Science* 142:1658)

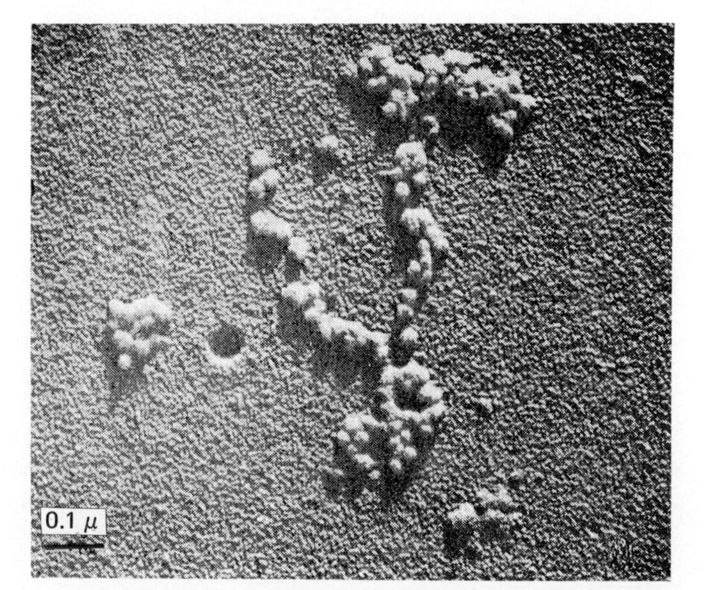

As shown in Figure 2.3, the large primary picornaviral translation product (or **polyprotein**) undergoes a complex series of proteolytic cleavages to yield a variety of intermediates and, ultimately, 11 to 12 final gene products among which are the capsid proteins cited above. Proteins required for RNA synthesis and proteolytic cleavage are contained within primary intermediates P2 and P3; that is, downstream from the four capsid proteins.

Construction of this genetic map was a difficult task that required nearly 20 years to complete. Confusing results were obtained when the pattern of proteins synthesized in polio-infected cells was first studied. The large primary **proteolytic cleavage products** (P1–3) are reasonably stable, but it was not possible to demonstrate the synthesis of proteins of ≥200 kDa unless processing was inhibited by the incorporation of certain amino acid analogues into the full-length polyprotein translation product. Another extremely useful experimental approach involved comparing the specific radioactivities of various viral proteins synthesized in the presence of pactamycin. At appropriately low concentrations, this drug blocks initiation of translation but allows chain elongation to continue. Thus, the relative amounts of the different gene products, determined by incorporation of radioactive amino acids, can be used as a measure of their distance from the initiation site; the specific radioactivity of viral proteins increases in the direction of translation ($5' \rightarrow 3'$). Parallel studies with different picornaviruses yielded fundamentally similar genetic maps whose validity has been confirmed by subsequent sequence analysis of viral cDNAs.

The production of viral proteins from a **single polyprotein precursor** is a most complex process, particularly as regards the P2-P3 region. The nature of the specific proteolytic cleavages is naturally of considerable interest, and it seems that these processing steps are very similar (but *not* identical) in all picornaviruses. Early cleavages that separate the P1 region from the remainder of the nascent polyprotein are catalyzed by proteins 2A and L (in those viruses having an L gene). Protein 3C, or its precursor, 3CD, appears to be responsible for all but one of the remaining cleavages, specifically cleaving 9 of the 13 glutamine-glycine bonds in the polyprotein. One of these cleavages separates capsid proteins 1C and 1D, but the "maturation cleavage" of all but two of the 1AB molecules into 1A + 1B occurs only after the RNA has been packaged into the viral capsid. Probably autolytic in nature, this cleavage may involve a serine protease. A serine residue from 1B is located near the 1A/1B cleavage site.

2.2.3 RNA Replication

The poliovirus infection cycle, shown schematically in Figure 2.5, lasts approximately 6 hours in cultured cells. RNA replication is thought to begin as soon as the first molecules of picornavirus polyprotein appear, that is, 10 to 15 minutes after the incoming genomic RNA has been uncoated. During the initial phases of infection, the rate of viral RNA synthesis doubles every 15 minutes, and many of the progeny positive-sense RNA strands are recruited as viral mRNA. When evaluated in laboratory studies, levels of RI appear to increase from 2 to 4 hours postinoculation, but RF (fully double-stranded RNA) accumulates more slowly and only to lower levels. Later, as RNA synthesis proceeds at a constant rate, synthesis of positive-sense RNA predominates, and approximately 50 percent of these molecules are packaged into virions. At this time, negative-sense RNA may constitute 5 percent to 10 percent of the total cellular RNA.

The 27 S replication complex is tightly associated with smooth membranes and contains at least two virus-encoded polypeptides (protein 3D and a VPg donor) plus a host-factor protein (67 kDa) in addition to the viral RNA template. Protein 3D acts as an RNA polymerase *in vitro,* using oligo U as a primer to copy poliovirus RNA. VPg is believed to act as the primer for RNA synthesis *in vivo.* Although genetic studies have implicated proteins 2B and 2C in RNA synthesis (see Fig. 2.4), their precise roles have not yet been determined.

The initial product of viral RNA replication sediments at 20 S and is presumed to be an RF molecule that, because of its more extended conformation, sediments more slowly than the more compact, single-stranded 35 S viral RNA. Later, many more RI molecules appear, each containing six to eight **nascent positive-sense strands.** As shown in Figure 2.1, the

FIGURE 2.5

Overview of picornavirus infection cycle. (R. R. Rueckert. (1990) In *Virology,* 2d ed., eds. B. N. Fields and D. M. Knipe. Vol. 1. Raven Press, New York)

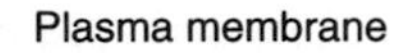

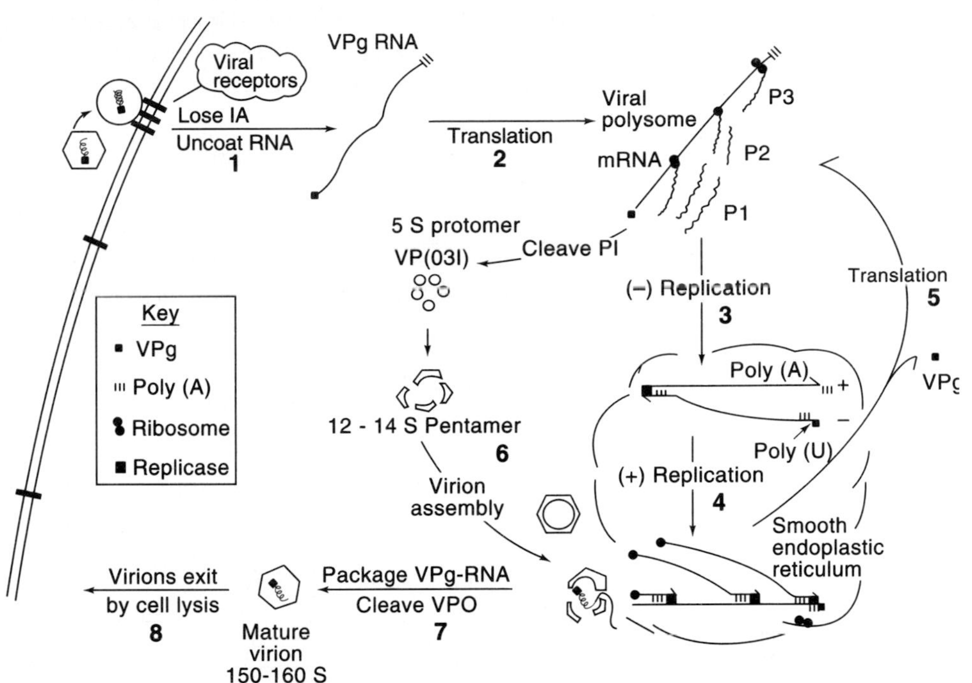

nascent positive-sense RNA progeny may have only minimal contact with the negative-sense "mother" strand. Synthesis of each RNA molecule requires approximately 45 seconds, and it is not known whether the same enzyme and/or cofactors are required for synthesis of both the negative and positive-sense RNA strands. Guanidine, a potent inhibitor of viral RNA replication that has been shown to interact with protein 2C (see Fig. 2.3), also inhibits the vacuolization and the proliferation of smooth membranes that normally accompany virus replication.

2.2.4 Assembly and Release of Virions

Intensive study of picornavirus morphogenesis suggests that the assembly process is driven by the orderly activation of certain **latent assembly domains** present in the coat protein protomer P1. Cleavage of two intersubunit bonds within polypeptide P1, catalyzed by protease 3C or 3CD and followed by substantial structural rearrangements, yields a 5S capsid protomer containing proteins 1C, 1D, and 1AB. Association of five such protomers to form a 14 S pentamer is followed by assembly into either a noninfectious RNA-containing provirion or a so-called NEC (or *n*aturally *e*mpty *c*apsid) particle.

Virion assembly appears to occur in association with the virus replication complex; that is, on the smooth rather than the rough endoplasmic reticulum. Furthermore, several lines of evidence, including the ability of guanidine to inhibit virion assembly as well as synthesis of polioviral RNA, point to a link between RNA synthesis and virion formation. The exact mechanism responsible for RNA packaging remains the central problem in picornavirus morphogenesis, however. Autolytic cleavage of most or all of the 60 molecules of protein 1AB present in the provirion completes the assembly process and is required for the generation of infectious particles.

Large portions of proteins 1B, 1C, and 1D are exposed on the surface of the mature virion, whereas protein 1A is located internally, close to the RNA. A most prominent feature of the virion surface, especially in the rhinoviruses, is a "canyon" that separates the exposed portions of protein 1D from those of 1B and IC. The ability of the virion architecture to accommodate great **sequence** (and thus antigenic) **variation** in the sequences surrounding the rim of the canyon without loss of infectivity provides a powerful mechanism for the virus to evade host defenses via mutation of these critical neutralization sites.

Very little is known about the mode of exit from the host cell for the mature picornavirus particle. Because these viruses are generally cytocidal, cell death may provide the means for release of the progeny virions. Nevertheless, cells in mitosis release virus without cytopathic effects.

2.2.5 Defective Interfering Particles

Many animal virus infections, particularly upon serial passage at high multiplicity, produce a progressively higher proportion of particles in which the viral genome is defective. Because such particles may interfere with the efficient replication of the normal or "helper" virions, probably by competion for the available replication machinery, they have been termed **defective interfering,** (or **DI), particles.** The biological properties of DI particles and the molecular mechanisms responsible for their generation are described in greater detail in Section 4.1.4.

In the case of poliovirus DI particles, the defective RNAs may be up to 30 percent shorter than that of the helper virus and are the result of deletions within the 5′ half of the wild-type genome. The reason for such preferential deletion of late viral functions (i.e., the capsid proteins) from the DI genome is presumed to involve the competitive disadvantage faced by mutants that lack early viral functions, especially RNA polymerase, in replicating in the same cell as the wild-type helper genome. In contrast to many other DI particle–helper virus combinations, the DI particles associated with poliovirus and other picornaviruses often interfere only slightly with the replication of their helper virus and undergo only slight enrichment during each successive round of replication. *Cis*-acting virus functions appear to mediate the interaction(s) between the DI particle and its helper virus, because (1) all naturally occurring poliovirus DI deletions preserve the original polyprotein reading frame and (2) secondary frame-shift mutations introduced downstream from such deletions result in a nonreplicating template that cannot be rescued by a helper virus.

2.3 CALICIVIRIDAE

Members of the Caliciviridae, a small group of nonenveloped viruses, share certain properties with members of both the Picornaviridae and the Togaviridae (see Chapter 3) and may thus be considered as intermediate between those two groups. The name **calicivirus** is derived from the cuplike appearance of the stained particle on electron micrographs. Although largely of veterinary importance, the Norwalk viruses, which cause severe gastroenteritis in humans, have also been assigned to this family. At least four different Norwalk virus serotypes have been described: Norwalk, Hawaii, Snow Mountain, and Taunton, named after the geographical locations in which they were first found.

The icosahedral particles of caliciviruses are 35 to 40 nm in diameter and, like those of picornaviruses, contain 60 coat protein trimers. In this case, only a single coat protein species (60–65 kDa) is present, however. The infectious, single-stranded genomic RNA contains approximately 8000 nt and is polyadenylated. Its 5′ terminus is not capped but is instead bound to a VPg-like protein.

Besides the presence of only a single coat protein, other differences between calici- and picornaviruses include the relative positions of structural and nonstructural genes within the genomes as well as the use of subgenomic RNAs and polyprotein processing to generate the mature virus proteins. Sequencing studies with feline calicivirus and rabbit hemorrhagic disease virus have shown that the coat protein ORF is located at the 3′ end of the genome and is probably expressed by translation of a prominent 2 kilobase (kb) polyadenylated subgenomic RNA. Nevertheless, the nonstructural proteins of these two viruses also contain (1) picornavirus 2C-like amino acid sequence motifs, (2) residues believed to be part of the 3C cysteine protease, as well as (3) the "GDD" motif conserved in many viral RNA-dependent RNA polymerases.

The recent determination of the complete 7696-nucleotide sequence of a Southampton (Snow Mountain–like) isolate of Norwalk virus has confirmed that these

FIGURE 2.6

Negatively-stained electron micrographs of (a) hepatitis virus E and (b) an astrovirus. (Courtesy of Fred Williams, Jr.)

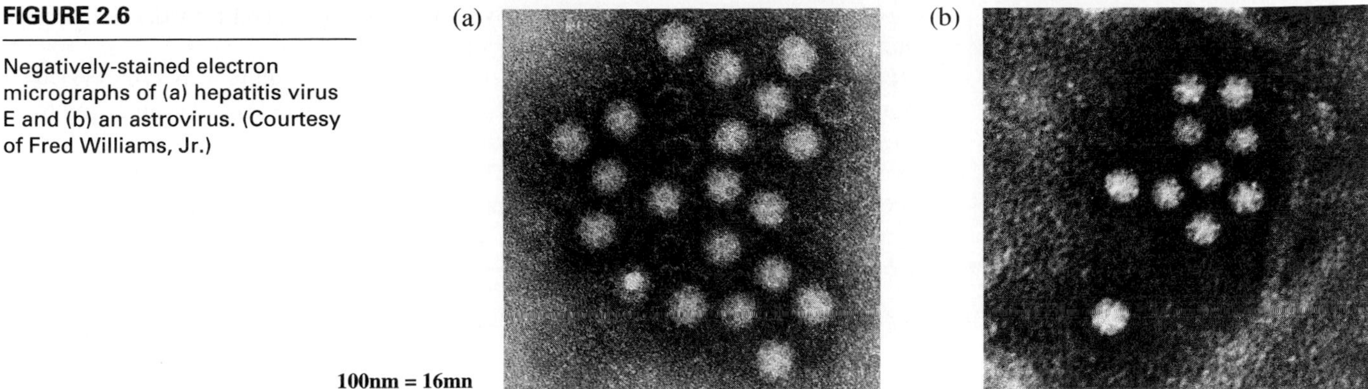

viruses, despite certain differences in the sizes of their ORFs and frameshift positions, can also be considered members of the Caliciviridae. The hepatitis E virus that is prominent in Southeast Asia and developing countries has also recently been found to be a probable member of the Caliciviridae (Fig. 2.6a). The viral genome (7.5 kb) has been cloned and sequenced, and development of a vaccine is underway. Additional sequence information and data concerning the genome organization and gene expression strategies of caliciviruses will be required before a definitive classification scheme can be proposed.

2.4 OTHER UNCLASSIFIED SINGLE-STRANDED RNA VIRUSES INFECTING HUMANS

Astroviruses, a group of as-yet-unclassified infectious agents associated with gastrointestinal disease, have been known since 1975. They were initially identified in fecal samples from young children suffering from nausea and diarrhea. Astroviruses are often associated with mild disease, and asymptomatic infections are not uncommon. They can be recognized by their small size (28 nm in diameter) and star-shaped surface (thus, their name) (Fig. 2.6b). Their single-stranded positive-sense RNA genomes are similar in size to those of picornaviruses, and astrovirus particles contain four structural proteins. Specific astrovirus serotypes are determined by immunofluorescent labeling techniques. Replication of human astrovirus serotype 2 in cultured cells leads to the synthesis of a 2.8 kb polyadenylated subgenomic mRNA. This RNA contains a single open reading frame encoding a putative 88 kDa capsid precursor. The deduced amino acid sequence of this polypeptide does not contain the conserved amino acid patterns reported for the capsid proteins of picorna- or caliciviruses. Recent analysis of RNA sequences has suggested the astro viruses should form a separate virus family.

2.5 INSECT VIRUSES (NODAVIRIDAE AND TETRAVIRIDAE)

2.5.1 Nodaviridae

The Nodaviridae family of insect viruses resembles the Picornaviridae in size, shape, and particle stability [29 nm isometric (T = 3) particles, stable to pH 3]. The virions of black beetle (BBV) and other nodaviruses, however, contain *two genomic RNAs,* both of which are required for infectivity. In this respect, nodaviruses resemble certain plant viruses with bipartite genomes (see Sec. 2.6). The ability of BBV to multiply efficiently in cultured

Drosophila melanogaster cells has made this virus a natural candidate for detailed molecular study. Nodaviruses are not usually host-specific.

BBV RNA 1 (3106 nucleotides) codes for two proteins. Protein A (105 kDa) is a component of the viral RNA polymerase, whereas protein B (10 kDa and synthesized from a subgenomic mRNA) is of unknown function. RNA 2 (1399 nucleotides) codes for the coat protein precursor (44 kDa). Both RNAs have a 5′-terminal cap, and their 3′ termini are also blocked (presumably by a protein). Both RNAs 1 and 2 are required for virion formation, but RNA 1 is capable of independent replication in cultured cells. The presence of RNA 2 strongly inhibits the synthesis of the subgenomic mRNA for protein B.

2.5.2 Tetraviridae

Nudaurelia β and a number of serologically related viruses were originally isolated from members of the *Lepidoptera,* principally from Saturniid, Limacodid, and Noctuid moths. Their icosahedral virions (35 nm diameter, stable at pH 3) have T = 4 symmetry (hence the name Tetraviridae; *tetra* = 4) and contain a single molecule of nonpolyadenylated positive-sense RNA (about 5500 nucleotides). Molecular masses of the single coat protein range between 60 and 70 kDa. Tetraviruses replicate primarily in the cytoplasm of gut cells, where they often form crystalline arrays within cytoplasmic vesicles. No infections of cultured invertebrate cells have been reported.

2.5.3 Other Viruses

Other small RNA viruses of invertebrates with properties common to the picornaviruses include the bee acute paralysis virus, bee virus X, *Drosophila* P and A viruses, and Queensland fruit fly virus.

SECTIONS 2.1 TO 2.5 FURTHER READING

Original Papers

Schaffer, R. L., and C. E. Schwerdt. (1955) Crystallization of purified MEF-1 poliomyelitis virus particles. *Proc. Natl. Acad. Sci. USA* 11:1020.

Summers D. F., J. V. Maizel, and J. E. Darnell. (1965) Evidence for virus-specific noncapsid proteins in poliovirus-infected cells. *Proc. Natl. Acad. Sci. USA* 54:505.

Bancroft, J. B., G. J. Hills, and R. Markham. (1967). A study of the self-assembly process in a small spherical virus. Formation of organized structures from protein subunits *in vitro. Virology* 31:354.

Jacobson, M. F., and D. Baltimore. (1968) Polypeptide cleavages in the formation of poliovirus proteins. *Proc. Natl. Acad. Sci. USA* 61:77.

Butterworth, B. E., and R. R. Ruc. (1972) Gene order of encephalomyocarditis virus as determined by studies with pactamycin. *J. Virol.* 9:823.

Cole, C. N., and D. Baltimore. (1973) Defective interfering particles of poliovirus. II. Nature of the defect. *J. Mol. Biol.* 76:325.

Domingo, E. et al. (1978) Nucleotide sequence heterogeneity of an RNA phage population. *Cell* 13:735.

Van Dyee, T. A., and J. B. Flanegan. (1980) Identification of poliovirus polypeptide P63 as a soluble RNA-dependent RNA polymerase. *J. Virol.* 35:732.

Racaniello, V. R., and D. Baltimore. (1981) Cloned poliovirus complementary DNA is infectious in mammalian cells. *Science* 214:916.

Hogle, J. M., et al. (1985) Three-dimensional structure of poliovirus at 2.9 Å resolution. *Science* 229:1358.

Rossmann. M. G., et al. (1985) Structure of a human common cold virus and functional relationship to other picornaviruses. *Nature* 317:145.

DASMAHAPATRA, B., et al. (1985) Structure of the black beetle virus genome and its functional implications. *J. Mol. Biol.* 182:183.

DASMAHAPATRA, B., et al. (1986) Infectious RNA derived by transcription form cloned cDNA copies of the genomic RNA of an insect virus. *Proc. Natl. Acad. Sci. USA* 83:63.

MOLLA, A., A. V. PAUL, and E. WIMMER. (1991) Cell-free, *de novo* synthesis of poliovirus. *Science* 254:1647.

TAM, A. W., et al. (1991) Hepatitis E virus (HEV): Molecular cloning and sequencing of the full-length viral genome. *Virology* 185:120.

HOLLAND, J. J., et al. (1991) Quantitation of relative fitness and great adaptability of clonal populations of RNA viruses. *J. Virol.* 65:2960.

MORI, K. -I., et al. (1992) Properties of a new virus belonging to Nodaviridae found in larval striped jack *(Pseudocaranx dentex)* with nervous necrosis. *Virology* 187:368.

BOROVEC, S. V., and D. A. ANDERSON. (1993) Synthesis and assembly of hepatitis A virus-specific proteins in BS-C-1 cells. *J. Virol.* 67:3095.

LAMBDEN, P. R., et al. (1993) Sequence and genome organization of a human small round-structured (Norwalk-like) virus. *Science* 259:516.

MONROE, S. S., B. JIANG, S. E. STINE, M. KOOPMANS, and R. I. GLASS. (1993) Subgenomic RNA sequences of human astrovirus supports classififcation of Astroviridae as a new family of RNA viruses. *J. Virol.* 67:3611.

ZHONG, W., and R. R. RUECKERT. (1993) Flock house virus: Down-regulation of subgenomic RNA 3 synthesis does not involve coat protein and is targeted to synthesis of its positive strand. *J. Virol.* 67:2716.

Books and Reviews

MOORE, N. F., B. REAVY, and L. A. KING. (1985) General characteristics, gene organization, and expression of small RNA viruses of insects. *J. Gen. Virol.* 66:647.

EIGEN, M., and C. K. BIEBRICHER. (1988) Sequence space and quasispecies distribution. In *RNA genetics,* ed. E. Domingo, J. J. Holland, and P. Ahlquist. Vol. 3, 211–245. CRC Press, Boca Raton, Fla.

CROUCH, R. B. (1990) Rhinoviruses. In *Virology,* 2d ed., ed. B. N. Fields. 607–629 Raven Press, New York.

MELNICK, J. L. (1990) Enteroviruses: Poliovirus, coxsackieviruses, echoviruses, and newer enteroviruses. In *Virology,* 2d ed., ed. B. N. Fields. 549–605 Raven Press, New York.

RUECKERT, R. R. (1990) Picornaviruses and their replication. In *Virology,* 2d ed., ed. B. N. Fields. 507–548 Raven Press, New York.

GREENBERG, H. B., and S. M. MATSUI. (1992) Astroviruses and caliciviruses: Emerging enteric pathogens. *Infect. Agents and Dis.* 1:71.

2.6 PLANT VIRUSES

The names of plant viruses are generally descriptive, reflecting the disease symptoms induced, the hosts in which they were discovered, or both; for example, tobacco mosaic virus (TMV), tobacco ringspot virus, or maize "rayado fino" virus. Although this terminology provides a convenient means to describe the economic impact of plant viruses, it is important to recognize that many viruses can infect species other than their original hosts—often producing quite different disease symptoms.

The small RNA viruses of plants are more varied and numerous than those of bacteria and animals. Most groups resemble their bacterial and animal counterparts in size and shape, but several have a helical architecture, giving them a rigid rodlike or flexible filamentous appearance. Others are variously elongated combinations of icosahedra with tubular arrangements of their proteinaceous building blocks or capsomers. Their capsids contain a single coat protein in all but one group, **the comoviruses,** which has equal numbers of two protein species. In contrast with the RNA phages (see Sec. 2.7), no adsorption proteins have been detected in plant viruses, and they do not necessarily lyse or kill their host cells.

By analogy with the RNA phages, the minimal number of genes in a plant RNA virus, would be two: a coat protein and an RNA replicase gene. Nevertheless, although the small-

TABLE 2.2
Properties of nonenveloped plant viruses containing single-stranded positive-strand RNA genomes

Group name[a]	No. of members	Example	Genes and gene products		Specific Features[d]
			RNAs[b]	Proteins[c]	
Isometric					
Bromovirus[e]	6	Brome mosaic (BMV)	3(1)	4(24)	Tripartite, cap, $tRNA^{tyr}$
Carmovirus	17	Carnation mottle (CarMV)	1(2*)	5(38)	Monopartite, cap
Comovirus	13	Cowpea mosaic (CPMV)	2	9(37,23)	Bipartite, VPg, poly(A), polyprotein two coat proteins
Cucumovirus[e]	4	Cucumber mosaic (CMV)	3(1)	4(24)	Tripartite, cap, $tRNA^{tyr}$
Dianthovirus	3	Red clover necrotic mosaic (RCNMV)	2(1*)	4(37)	Bipartite, cap
Fabavirus	3	Broad bean wilt virus (BBWV)	2	?(27,43)	Similar to comovirus?
Luteovirus	21	Barley yellow dwarf (BYDV)	1(2*)	6(22)	Monopartite, VPg
Machlovirus	3	Maize chlorotic dwarf (MCDV)	1(?)	?(18,30)	
Marafivirus	3	Maize rayado fino (MRFV)	1(?)	?(22,29)	Monopartite, replicates in insect vector
Necrovirus	4	Tobacco necrosis (TNV)	1(2*?)	?(30)	Monopartite
Nepovirus	36	Tomato black ring (TRSV)	2	7(53)	Bipartite, VPg, poly(A), polyprotein
Parsnip yellow fleck	3	Parsnip yellow fleck (PYFV)	1	?(31,26, 23)	Monopartite, VPg, polyprotein
Pea enation mosaic	1	Pea enation mosaic (PEMV)	2	9(21)	Bipartite, VPg
Sobemovirus	16	Southern bean mosaic (SBMV)	1(1*)	?(31)	Monopartite, VPg, polyprotein?
Tombusvirus	12	Tomato bushy stunt (TBSV)	1(2*)	5(41)	Monopartite
Tymovirus	19	Turnip yellow mosaic (TYMV)	1(1*)	4(20)	Monopartite, cap, $tRNA^{val}$, polyprotein
Quasi-isometric/bacilliform					
Alfamovirus[e]	1	Alfalfa mosaic (AMV)[e]	3(1)	4(24)	Tripartite, cap
Ilarvirus	20	Tobacco streak (TSV)	3(1)	?(29–30)	Tripartite, cap; similar to alfamoviruses
Rigid Rods					
Furovirus	11	Soil-borne wheat mosaic (SBWMV)	2(?*)	?(19)	Bipartite, cap, $tRNA^{val}$, fungus-transmitted
Hordeivirus	4	Barley stripe mosaic (BSMV)	3(2*?)	7?(22)	Tripartite, cap, $tRNA^{tyr}$
Tobamovirus	14	Tobacco mosaic (TMV)	1(3*)	4(17)	Monopartite, cap, $tRNA^{his}$
Tobravirus	3	Tobacco rattle (TRV)	2(3*)	5(25)	Bipartite, cap, nematode-transmitted
Filamentous					
Capillovirus	4	Apple stem grooving (ASGV)	1(?)	?(27)	Monopartite, poly(A), polyprotein
Carlavirus	56	Potato virus M (PVM)	1(2*?)	6(34)	Monopartite, poly(A), similar to potexviruses
Closterovirus	22	Apple chlorotic leafspot (ACLSV)	1(5*)	4(22)	Monopartite, poly(A) (some members), 5′ coterminal subgenomic mRNAs
Potexvirus	39	Potato virus X (PVX)	1(2*)	5?(21cp)	Monopartite, cap, poly(A), 3′ "Triple gene block"
Potyvirus	153	Tobacco etch (TEV)	1	7(30)	Monopartite, VPg, poly(A), polyprotein

[a]For shapes and sizes, see also Figure 1.13.
[b]Number of genomic and subgenomic (parentheses) RNAs;* nonencapsidated subgenomic RNAs.
[c]Number of ORFs or, where known, virus-specific polypeptides. Molecular weight of the viral coat protein(s) is shown in parentheses.
[d]Features, in order, include: *Genome organization* (mono-, bi-, or tripartite); *5′ terminus* (the presence of 5′-linked 7-Me-G or a covalently linked protein); *3′ terminus* (presence of poly(A) or an aminoacylatable tRNA-like structure); and the role of virus-encoded protease(s) in gene expression.
[e]For alfamo-, bromo-, cucumo-, and ilarviruses, the smallest encapsidated RNA species functions as a subgenomic coat protein mRNA. This RNA is *not* required for the infectivity of bromo- or cucumoviruses, but the infectivity of alfalfa mosaic or ilarvirus RNAs 1–3 depends on the presence of either RNA 4 or a low level of the coat protein itself.

est plant viral genomes are similar in size to those of the RNA phages (about 4000 nt), there is good evidence that the total number of their gene products is actually three to five. An interesting exception to this apparent uniformity in gene number and overall size is provided by the **plant viroids,** a group of subviral RNAs that are able to replicate autonomously despite their small size (250–400 nt) and the absence of a protective protein capsid (see Chapter 9).

Table 2.2 summarizes selected properties of the 27 currently accepted groups of nonenveloped positive-sense RNA plant viruses. Several characteristics underlying these groupings are discussed in the following sections: (1) the frequent presence of divided viral genomes, (2) additional strategies to facilitate gene expression from viral RNA containing more than one gene, (3) the helical architecture of many plant viruses (e.g., TMV), and (4) the presence of tRNA-like structures at the 3′ ends of many plant viral RNAs. Processes essential to the replication of all single-stranded positive-sense RNA viruses (e.g., RNA-directed RNA replication) were discussed at the beginning of this chapter. The sizes and shapes of all major plant virus groups were summarized schematically in Figure 1.13.

2.6.1 Viruses with Divided Genomes

A major feature that differentiates plant positive-sense RNA viruses from all RNA phages and animal picornaviruses (but *not* from their close relatives, the nodaviruses) is the frequent occurrence of **divided genomes.** Ten of the 27 virus groups included in Table 2.2 have divided genomes, and the virions of certain viruses such as brome mosaic and cucumber mosaic virus may contain more than a single species of genomic RNA. The multiple genomic RNAs of many more viruses are separately encapsidated, as illustrated by the four different types of alfalfa mosaic virus particles shown in Figure 2.7. Viral genomes consisting of two or three different nucleic acid components that are all required for infectivity are called **bipartite** and **tripartite,** respectively.

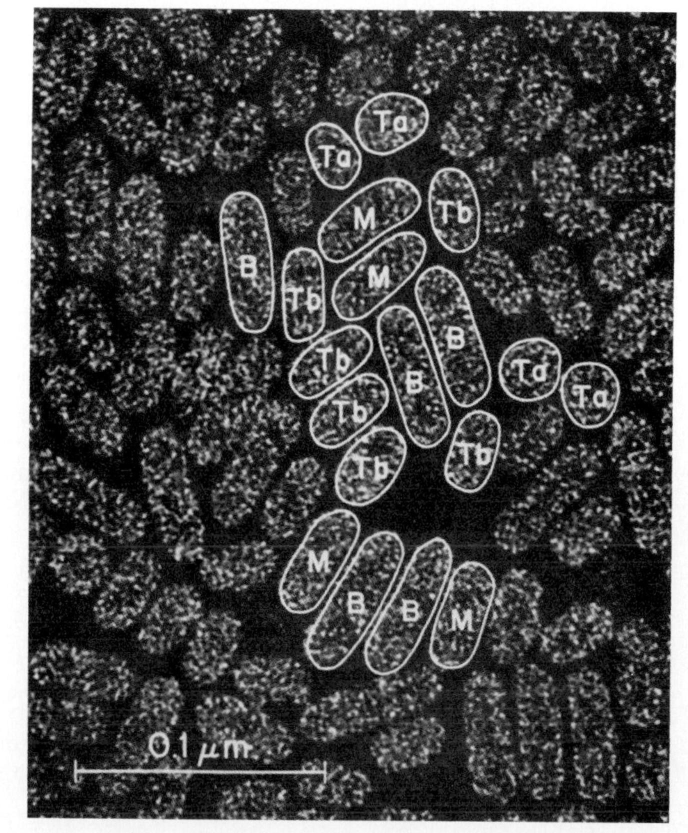

FIGURE 2.7

Electron micrograph of alfalfa mosaic virus. The largest B (or bottom) particles are 58 nm long and contain either RNA 1 or 2; the M (middle) particles contain RNA 3; and the Tb (top b) and Ta particles contain either RNA 4 or no RNA, respectively. (Courtesy of E. M. J. Jaspars)

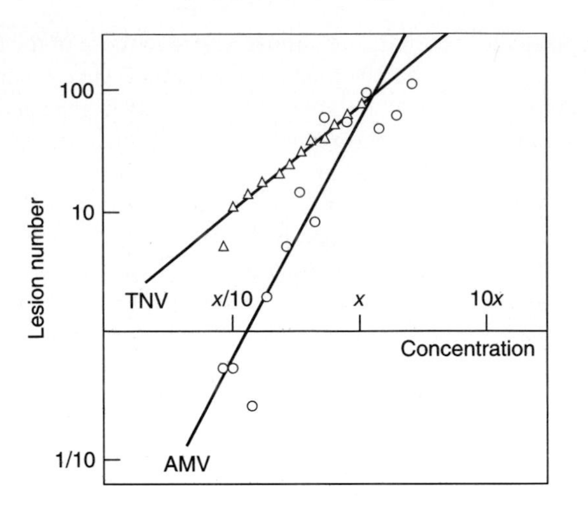

FIGURE 2.8

Comparison of infectivity dilution curves for single-particle (tobacco necrosis virus TNV) and multiple-particle (alfalfa mosaic virus AMV) viruses. Local lesion formation for both viruses was measured in *Phaseolus vulgaris.* (From L. van Vloten-Doting *Plant Virology,* 3d ed., Academic Press, New York)

Alfamo-, bromo-, and cucumoviruses have four RNAs—two large molecules containing about 3000 nucleotides plus two smaller ones containing either 2000 or <1000 nucleotides. Because viral infection and the production of progeny require only information present in the three largest RNA species (components 1 through 3), these are considered tripartite genomes. The smallest RNA is a monocistronic coat protein mRNA whose information is also present in the 3′ portion of component 3.

The genetic information for both the como- and nepoviruses is distributed between two RNA molecules, and no systemic infection results when plants are inoculated with only one RNA (i.e., the genome is bipartite). Tobraviruses also have two genomic RNAs, but in this case, the larger particle alone is infectious. Because the smaller genomic RNA contains the coat protein ORF, such infections result not in the formation of protein-coated virus particles but only in the replication of the larger genomic RNA.

The need for two (or more) particles to cooperate in establishing a successful infection suggests that **multipartite viruses** may operate at a considerable evolutionary disadvantage. As shown in Figure 2.8, the infectivity dilution curve for AMV (requiring B, M, and T_b particles for infectivity) is considerably steeper than that of tobacco necrosis virus (presumed to require only a single particle). Division of its RNA genome into two or more "minichromosomes" may have several important advantages for the virus, however, and these appear to outweigh any potential reduction in efficiency of virus transmission or infection.

Most eukaryotic mRNAs contain a single ORF, and 80 S ribosomes are adapted to translate only those ORFs immediately downstream from a short 5′ leader sequence. By placing several viral genes near the 5′ ends of their respective RNAs, multipartite genomes offer a variety of new regulatory possibilities. It has also been suggested that multipartite genomes may be an adaptative response to the high error rate associated with RNA-directed RNA replication. Virus preparations containing a high percentage of defective particles can give rise to wild-type virus via **multiplicity reactivation** or **genetic reassortment,** either during the replicative process itself or while the virus moves from cell to cell.

2.6.2 Additional Mechanisms Regulating Viral Gene Expression

In addition to the widespread occurrence of multipartite genomes, plant positive-sense RNA viruses have developed several other mechanisms to facilitate and/or regulate the expression of individual genes. Illustrated schematically in Figure 2.9, these mechanisms include the synthesis of subgenomic mRNAs as well as translational read-

FIGURE 2.9

Five strategies used by plant positive-sense RNA viruses to regulate gene expression. (Modified from R. E. F. Matthews. (1991) *Plant Virology,* 3d ed. Academic Press, New York)

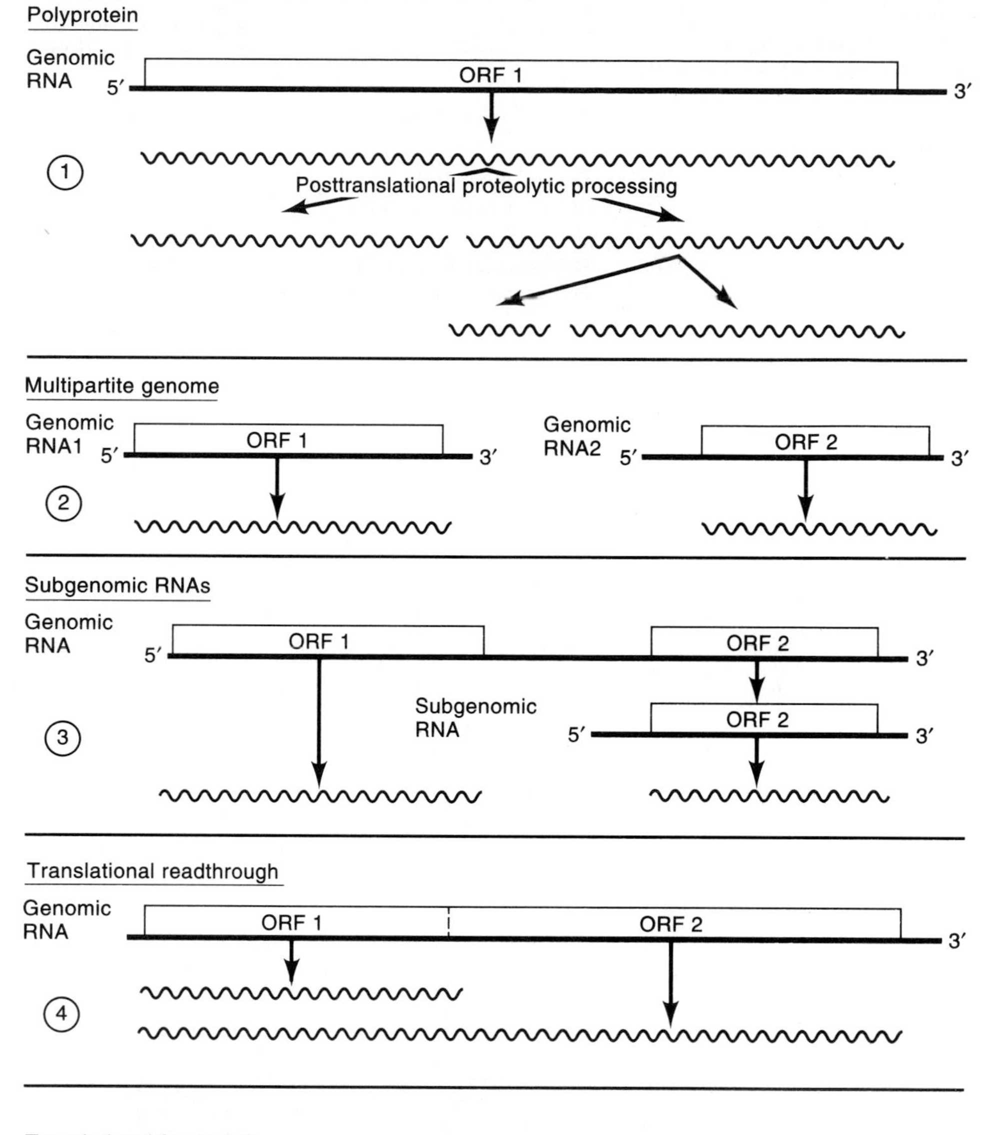

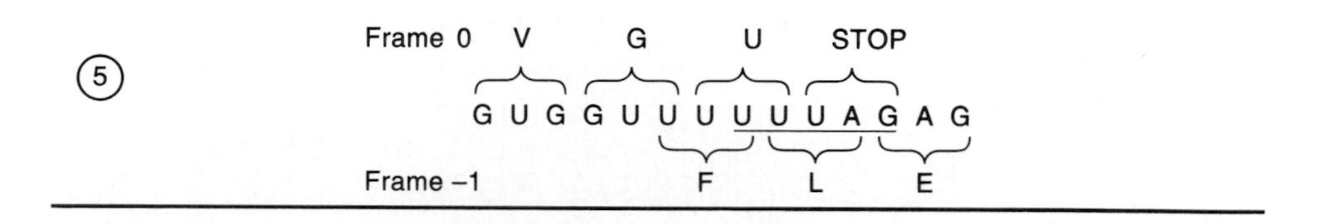

through and frame-shifting. Individual viruses often combine two or more of these mechanisms to achieve maximum regulatory flexibility. Even if translational readthrough and frame-shifting are considered to be variations on a single theme, as many as 15 different "**genome strategies**" are still available. As summarized in Table 2.3, the better characterized unenveloped viruses employ at least 7 distinct strategies. Other strategies are also possible (see Sec. 4.9 for a discussion of the genomic organization of tomato spotted wilt virus).

TABLE 2.3
Genome strategies adopted by positive-sense RNA plant viruses

Strategy	*Virus group(s) encoded*	*ORFs*	*Proteins*
Polyprotein	Potyvirus	1	8
Subgenomic RNAs	Potexvirus	5	4–5
	Tombusvirus	5	5
Subgenomic RNAs and Read-through/frame-shift	Tobamovirus	5	4–5
	Luteovirus	6	6–7
	Carmovirus	5	5–7
Subgenomic RNAs and Polyprotein	Tymovirus	3	3–5
	Sobemovirus	4	4–5
Multipartite genome and Polyprotein	Comovirus	2	9?
	Nepovirus	2	6?
Multipartite genome and Subgenomic RNAs	Bromovirus	4	4
	Cucumovirus	4	4
	Alfamovirus	4	4
	Ilarvirus	4	4
	Hordeivirus	7	7
Multipartite genome, Subgenomic RNAs, and Read-through/frame-shift	Tobravirus	5	5
	Furovirus	9	6–9
	Dianthovirus	4	4

Finally, a different and special situation is presented when one virus is dependent for its replication on the presence of a second, independently replicating virus. First described by Kassinis in 1962, such dependent viruses are known as **satellite viruses.** Satellite viruses are serologically unrelated to their "helpers," and the two viral genomes exhibit little, if any, sequence similarity. The term satellite has also been used in conjunction with small nongenomic RNAs found associated with tobacco ringspot, cucumber mosaic, and other viruses. Because these **satellite RNAs** have no coat protein of their own and are encapsulated with their respective helper viral RNAs, they represent a different type of subviral RNA. We will return to the plant satellite viruses and satellite RNAs in Chapter 9, where we compare their biological and biochemical properties with those of the similar-sized (but autonomously replicating) plant viroids.

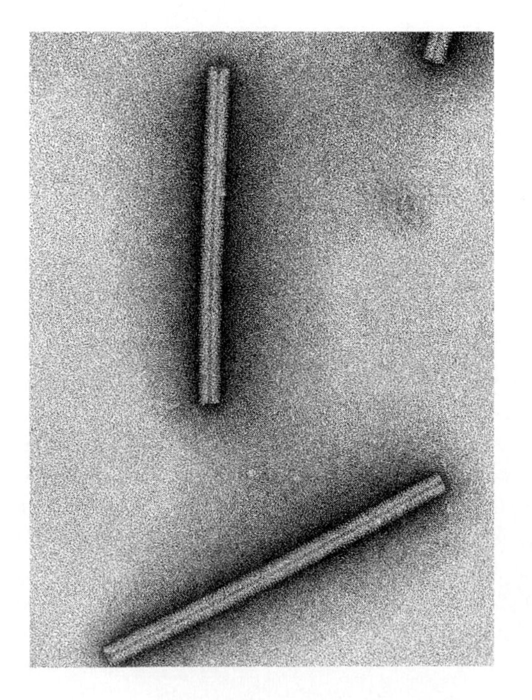

FIGURE 2.10

Electron micrograph of tobacco mosaic virus. The axial channel is clearly visible after negative staining. (Courtesy of R. C. Williams)

2.6.3 Tobacco Mosaic Virus: The Prototype Plant Virus

Having considered key features of two regulatory strategies used by many positive-sense RNA viruses of plants and animals (i.e., polyprotein processing and multipartite genomes), we now turn our attention to tobacco mosaic virus (TMV) (Fig. 2.10). Until quite recently, TMV occupied a unique position among plant viruses. Structural, biological, and genetic studies with TMV were well advanced even before the full impact of modern molecular genetic methods could be brought to bear. Results from these early studies helped establish subgenomic mRNAs and **translational readthrough** as important strategies for modulating viral gene expression. Studies with TMV have also made important contributions to our understanding of virus assembly mechanisms and the role of secondary and tertiary structure at the 3′ end of viral RNAs in regulating viral replication. First, however, some historical background about TMV.

The stability of the TMV virus particle probably accounts for its having been the first virus to be identified, purified to homogeneity, and then biochemically and biologically characterized (see Table 1.1). The helical aggregation of its relatively small coat protein (17.5 kDa) yields a narrow, rigid, rodlike particle (18 × 300 nm) that is quite characteristic of all tobamoviruses (Figs. 2.10, 2.11). The TMV rod contains an axial channel that is 4 nm wide, and the viral RNA lies within a groove in the surrounding protein helix. The pitch of this helix is 2.3 nm, and each turn contains 16 1/3 coat protein molecules. Thus, each full-

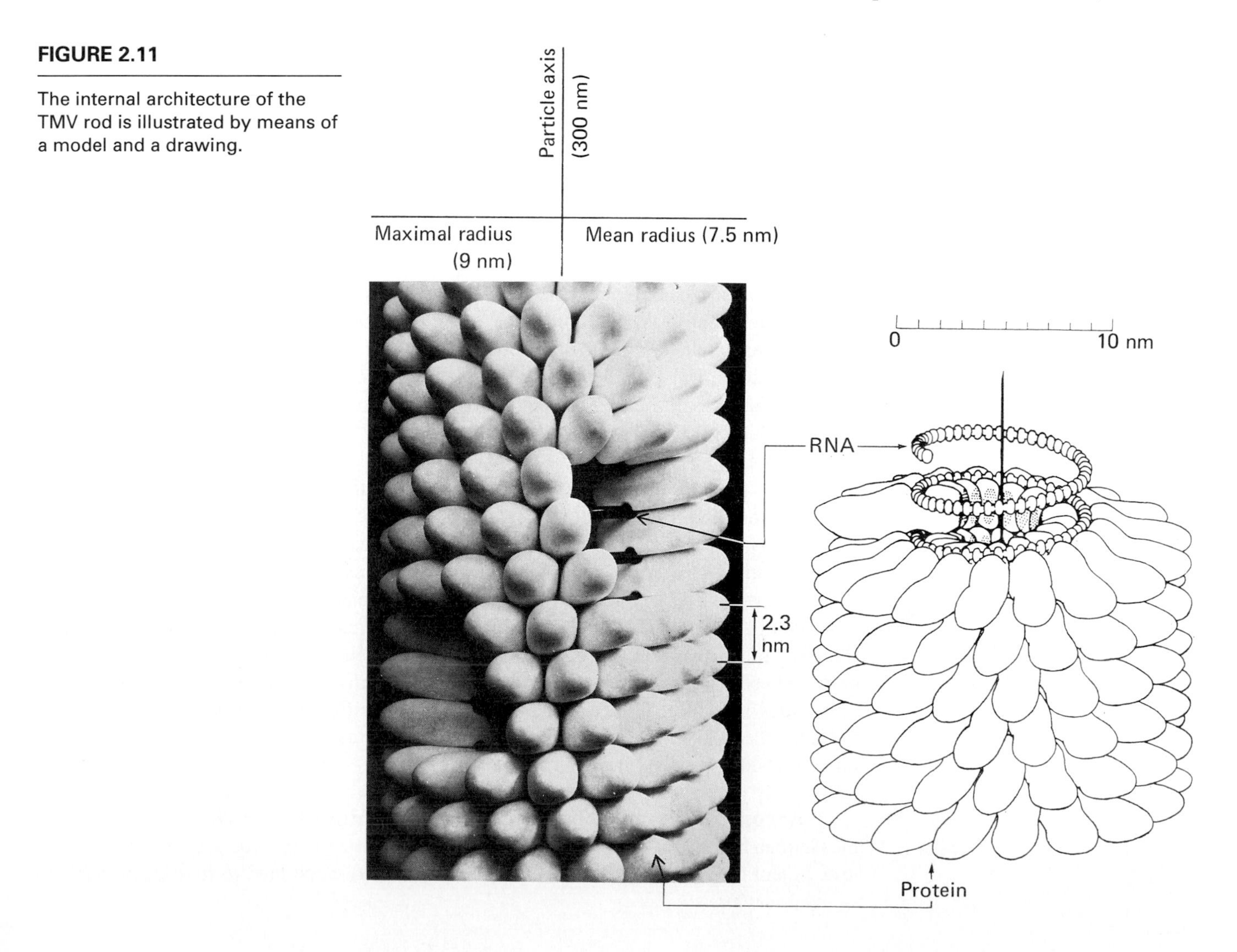

FIGURE 2.11

The internal architecture of the TMV rod is illustrated by means of a model and a drawing.

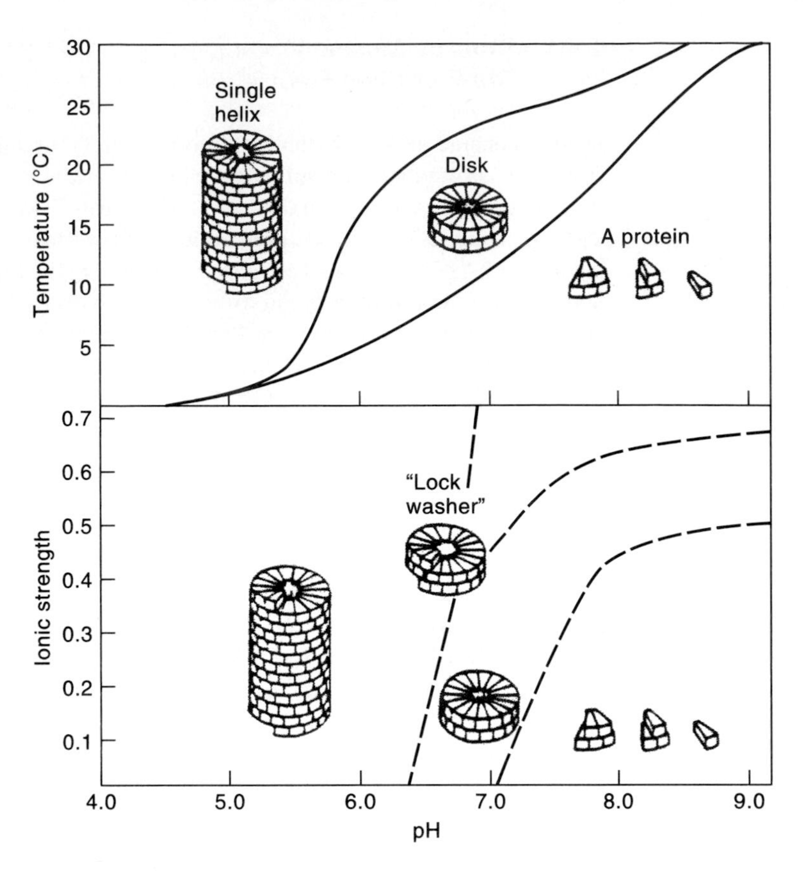

FIGURE 2.12

The effects of pH, ionic strength, and temperature on the aggregation states of TMV coat protein. The A (or acid-derived) protein, helical rod, and 20 S disk have been well characterized; the lock washer is the proposed intermediate in the initiation of TMV assembly (see Fig. 2.14). (From R. E. F. Matthews (1991) *Plant Virology,* 3d ed. Academic Press, New York)

length virion contains 130 helical turns. Other helical viruses or nucleocapsids, even if they have the same diameter, are more flexible, and their looser structure is reflected by a much lower degree of stability.

One consequence of this tightly packed architecture is a pronounced resistance to chemical and physical agents that inactivate many other viruses. That TMV RNA is protected against nucleases is neither unexpected nor unusual, but its coat protein is also remarkably resistant to typical proteolytic enzymes. The exact folded structure of this unusual protein molecule has been worked out in great detail by high-resolution X-ray crystallography. These studies showed that (1) both the carboxy terminus and the acetylated amino terminus of the coat protein are near the surface of the rod and (2) that the groove through which the RNA helix passes is about halfway between the interior channel of the rod and its exterior surface (see Fig. 2.11).

TMV particles, like almost all virus particles, fall apart in both alkaline and acid solutions. If the conditions (e.g., temperature, pH) are not too extreme, the viral protein molecules remain intact, and the denaturation is reversible. Removal of the denaturant allows the native structure of the viral protein to re-form, and near its isoelectric point (pH 4 to 6), TMV coat protein aggregates to form rod-shaped particles that look exactly like TMV virions (see Fig. 2.10 and 2.12). Thus, the shape of the virus particle is due to the amino acid sequence of its coat protein, which determines its tendency to fold into a particular oblong or oval shape. These molecules then tend to interact with one another, both side by side and up and down, to form the helical TMV rod. Such specific protein-protein interaction provides the mechanism for the formation of all helical virus structures, including the internal nucleocapsids of many animal viruses.

Reconstitution of TMV and the Nature of Virus Infectivity When, in 1955, the German biochemist G. Schramm first observed these virus-like rods of TMV protein and tested them for infectivity, he must have been greatly disappointed to find them unable to cause disease. Although we now regard such an experiment as naive, at that time no one knew that RNA was the genetic (and thus the essential) component of these viruses. In fact, no one even knew how to isolate large RNA molecules without damaging them.

Both acid and alkali are potentially harmful to RNA and DNA, but intact nucleic acids can be obtained by degrading viruses at neutral pH with either detergents (e.g., SDS) or 6 *M* urea or by extraction with phenol. When TMV RNA isolated by such methods is added to native TMV protein, rods form at neutrality (pH 7). Like natural TMV, these rods are stable from pH 3 to 9—in distinct contrast to rods containing only protein, which disassemble below pH 4 and above pH 6. Because these ribonucleoprotein rods are identical to TMV in all respects and can be fully infectious, Fraenkel-Conrat and Williams termed this process **virus reconstitution**.

In 1955, this result was particularly exciting, because it seemed that only the virus particles, and *not* the two separate components, were infectious. Somewhat later it became clear that the RNA alone was infectious, but only when inoculated at much higher concentrations. Why is almost 1000 times as much naked TMV RNA required to cause infection? Probably because RNA is easily degraded by intracellular nucleases, and the coat protein provides the protection needed for high infectivity. The efficiency of this protection varies greatly among different viruses, however, and some isometric and filamentous (threadlike) plant virus particles are no more infectious than are their naked RNAs.

How was the very low infectivity of TMV RNA proved to be a property of that molecule? Could the low infectivity of TMV RNA preparations have been due to the presence of a few contaminating virus particles? These questions were addressed by a series of experiments with TMV (and later with other viruses) that illustrated several important differences in the properties of viruses and their nucleic acids.

- **Sedimentation rate:** Because of their high particle weights, viruses sediment much more rapidly in a centrifugal field than do their nucleic acid genomes. Thus, if the infectivity does not sediment to the bottom of the centrifuge tube, where the virus is found, it is not due to virus particles.
- **Nuclease sensitivity:** The infectivity of viruses is more or less resistant to nucleases, whereas that of naked viral nucleic acid is very sensitive to such enzymes.
- **Inactivation by antibodies:** Viruses are antigenic. Provided that no nucleases are present, incubation with purified antisera will inactivate a virus but will have no effect on the infectivity of its isolated nucleic acid.
- **Effects of various denaturing and precipitating agents:** Detergents such as SDS, as well as urea and other chemicals, degrade viruses and thus cause loss of infectivity. They do not, however, inactivate nucleic acids. The use of precipitating agents such as ammonium sulfate as well as extraction with phenol can also discriminate between viruses and nucleic acids (see Appendix 1).

In the 1950s the conclusion that RNA alone could carry the genetic information necessary to cause disease was quite unexpected. Proof that the viral RNA was the sole determinant of tobacco mosaic disease was obtained by a **mixed reconstitution** of the type depicted in Figure 2.13. In this experiment the RNA of Holmes ribgrass mosaic virus (RMV)—a virus related to TMV that forms a characteristic type of local lesion on a certain variety of tobacco and whose coat protein contains histidine and methionine—was reconstituted with the usual 20-fold amount of coat protein from common (or wild-type) TMV. The common strain of TMV produces systemic disease in this tobacco plant variety and its coat protein lacks these amino acids. The mixed virus produced local symptoms typical of RMV infections, and the coat protein of the resulting progeny contained both amino acids absent from TMV common strain coat protein. Clearly, the viral RNA determines the nature of the disease and of the progeny virus—even when it is packaged in a different protein.

FIGURE 2.13

Schematic presentation of the result of reconstituting virus from the RNA and coat protein of two related viruses (*mixed reconstitution*). The progeny of the "mixed virus" contains the protein and exhibits all the genetic properties of the virus from which the RNA was isolated.

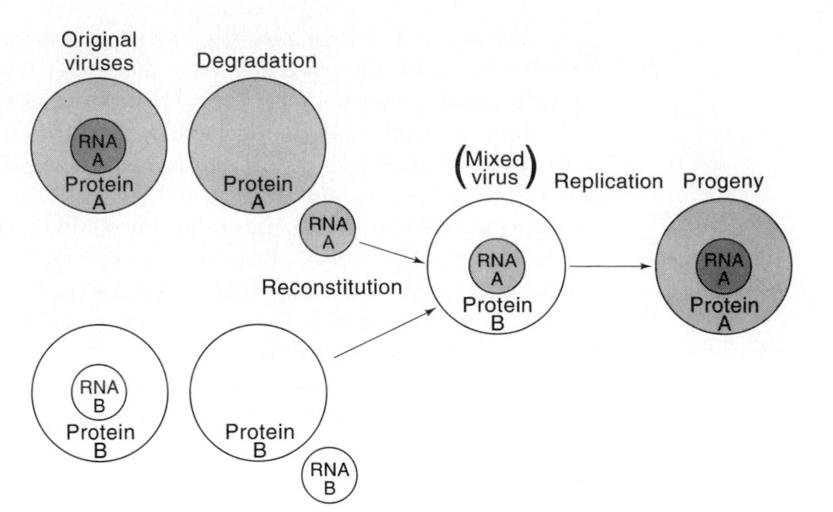

FIGURE 2.14

The mechanism of TMV assembly including initiation (a–c) and elongation (d–h). (a) The hairpin loop from the origin of assembly (OAS) sequence is inserted into the centrol hole of the 20 S disk. (b) The loop opens up as it intercalates between the two layers of coat protein subunits. (c) This RNA-protein interaction causes the disk to switch to the helical lock-washer (or protohelical) form, leaving both RNA tails to protrude from the same end of the protohelix. (d) A second double disk adds to the first on the side away from the RNA tails, switches to the helical form, and entraps two more turns of RNA. (e–h) Growth of the helical rod in the 5′ direction continues rapidly via addition of double disks. Extension of the rod in the 3′ direction appears to proceed by the addition of small A proteins aggregates. (Courtesy of P. J. G. Butler and reprinted from *Plant Virology,* (1991) 3d. ed. Academic Press, New York)

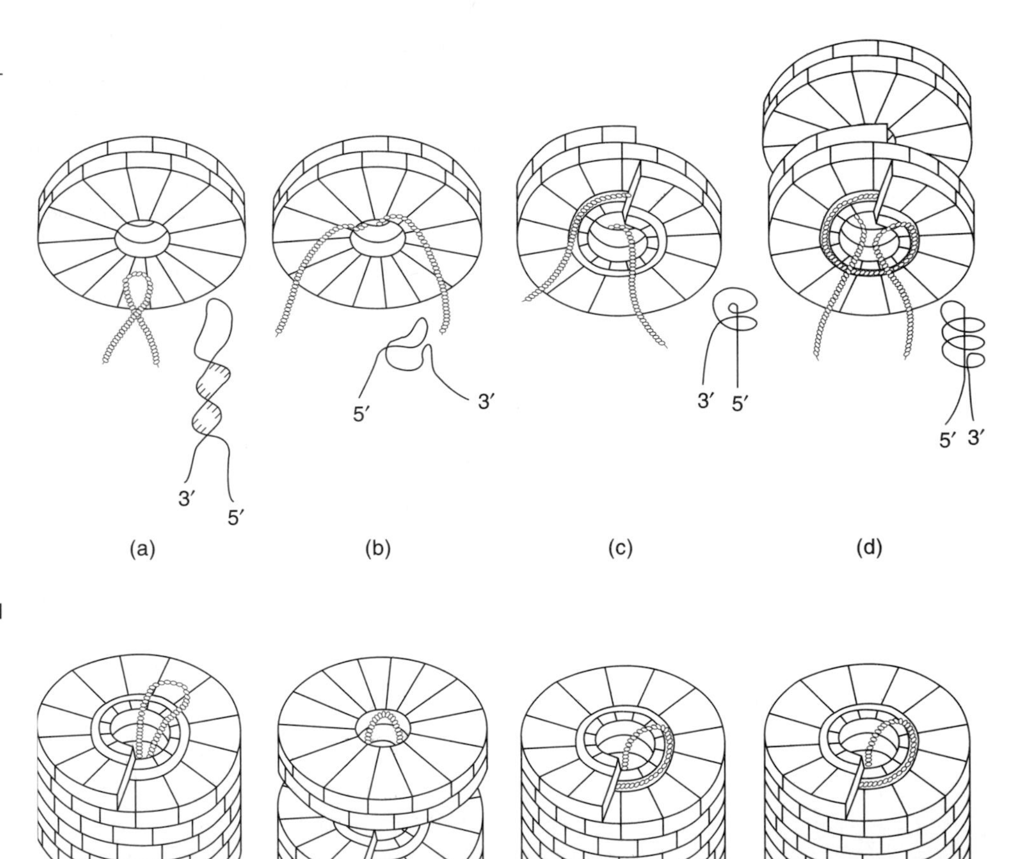

Assembly of Helical Viruses The actual mechanism of TMV virion assembly shown in Fig. 2.14 turns out to be quite different from what might have been expected. Thus, it is not TMV coat protein monomers but an aggregate of 33 such protein molecules (i.e., the "double disk" also shown in Fig. 2.12) that first combines with the viral RNA. Assembly does not begin at either RNA terminus, as first believed, nor is it a random process. Rather, it begins at a unique origin of assembly site (OAS) about 800 nucleotides from the 3′ terminus of TMV common strain RNA. Although downstream from the coat protein AUG initiation codon in the common strain of TMV, the location of the tobamovirus OAS is somewhat variable. Rod growth toward the 5′ terminus of the viral RNA is rapid, involving addition of double disks; encapsidation of the 3′ terminus proceeds more slowly, probably via addition of A protein monomers or small aggregates. Surprisingly, the RNA does not hang loose during this process but actually passes through the axial channel inside the growing rod and emerges as a loop at the site of rod growth in the 5′ direction. This mechanism facilitates the introduction of the RNA into the appropriate site of each protein molecule, where three specific basic amino acids bind three nucleotide phosphates.

It appears likely, but has not been proven, that similar mechanisms are involved in the assembly of other helical viruses. An important corollary is the biological means of liberating the RNA of the infecting virus particle. Recent studies strongly support a process termed **cotranslational disassembly,** in which coat protein subunits are displaced from the 5′ end of the RNA by ribosomes, and translation of the RNA proceeds concomitantly with the stripping off of the viral coat. This example is one of many instances of the simultaneous occurrence of events that we tend to regard as consecutive.

Structural Features of the TMV RNA 3′ Terminus Like many other plant viral RNAs (see Table 2.2), the 3′ end of TMV RNA ends with the sequence -C-C-C-A and can be charged with an amino acid (in this case, histidine). As shown in Figure 2.15, the 3′ noncoding region of TMV RNA can be folded into a tRNA-like structure preceded by a series of four pseudoknots. The ability to form these pseudoknots is conserved among various TMV strains, and many other nonpolyadenylated plant viral RNAs can also assume similar structures. The role of such structures in the virus life cycle has not been established, but at least four different possibilities have been suggested: (1) donating an amino acid during some stage of protein synthesis; (2) facilitating translation by disrupting base-pairing between the 3′ and 5′-terminal regions of the viral RNA; (3) acting as a recognition site for the viral replicase to initiate negative-strand synthesis; and (4) a molecular fossil from the original RNA world where tRNA-like structures tagged RNAs for replication and prevented the uncontrolled loss of nucleotides from their 3′ terminus.

The Roles of Subgenomic mRNAs and Translational Read-through in TMV Replication As described in Section 2.2, picornavirus gene expression is regulated via synthesis and proteolytic processing of a full-length **polyprotein.** Although the potyviruses (a large and plentiful group of viruses), depend on an identical regulatory strategy, many other groups of plant RNA viruses employ other regulatory strategies including the synthesis of **subgenomic mRNAs** as well as **read-through** or **frame-shift** proteins (see Table 2.3). TMV provided the model system in which many of the essential features of these strategies were first identified.

As summarized in Fig. 2.16 sequence analysis of the TMV genome revealed the presence of five open reading frames. The first two of these ORFs are separated by a "leaky" UAG termination codon, thereby allowing the synthesis of both a 126 kDa protein plus a larger 183 kDa read-through protein. Even before its complete nucleotide sequence had been established, *in vitro* translation experiments had shown that TMV RNA encodes two large polypeptides having the appropriate molecular weights and sharing a common amino terminus. The amount of 183 kDa protein produced has been shown to depend on the incubation temperature and the kind of $tRNA^{tyr}$ present in the cell-free extract, and both proteins have been detected in TMV-infected tobacco leaves and protoplasts. Site-directed mutage-

FIGURE 2.15

Proposed structure for the 3′ noncoding region of TMV RNA (vulgare strain). The tRNA-like structure is at the right; between this structure and the end of the coat protein gene is a stalk-like structure containing four pseudoknots. (a) Schematic representation; (b) three-dimensional artist's representation. (From Pleij et al. (1987) In *Positive-strand RNA viruses,* eds. M. A. Brinton and R. R. Rueckert. 299–316. Alan R. Liss, New York)

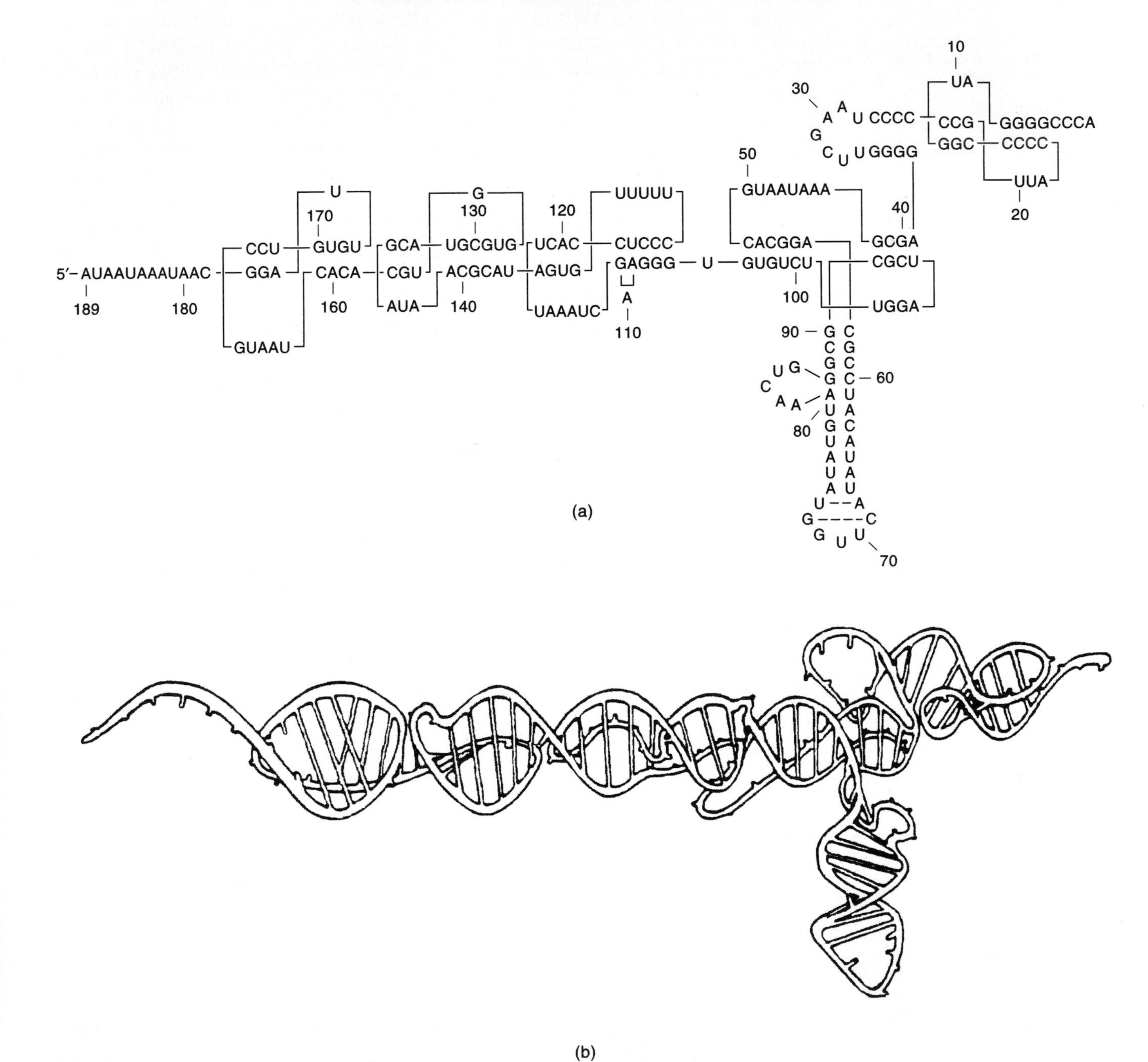

FIGURE 2.16

Organization and expression of the tobamovirus genome. (From *Plant Virology,* 3d ed. Academic Press, New York)

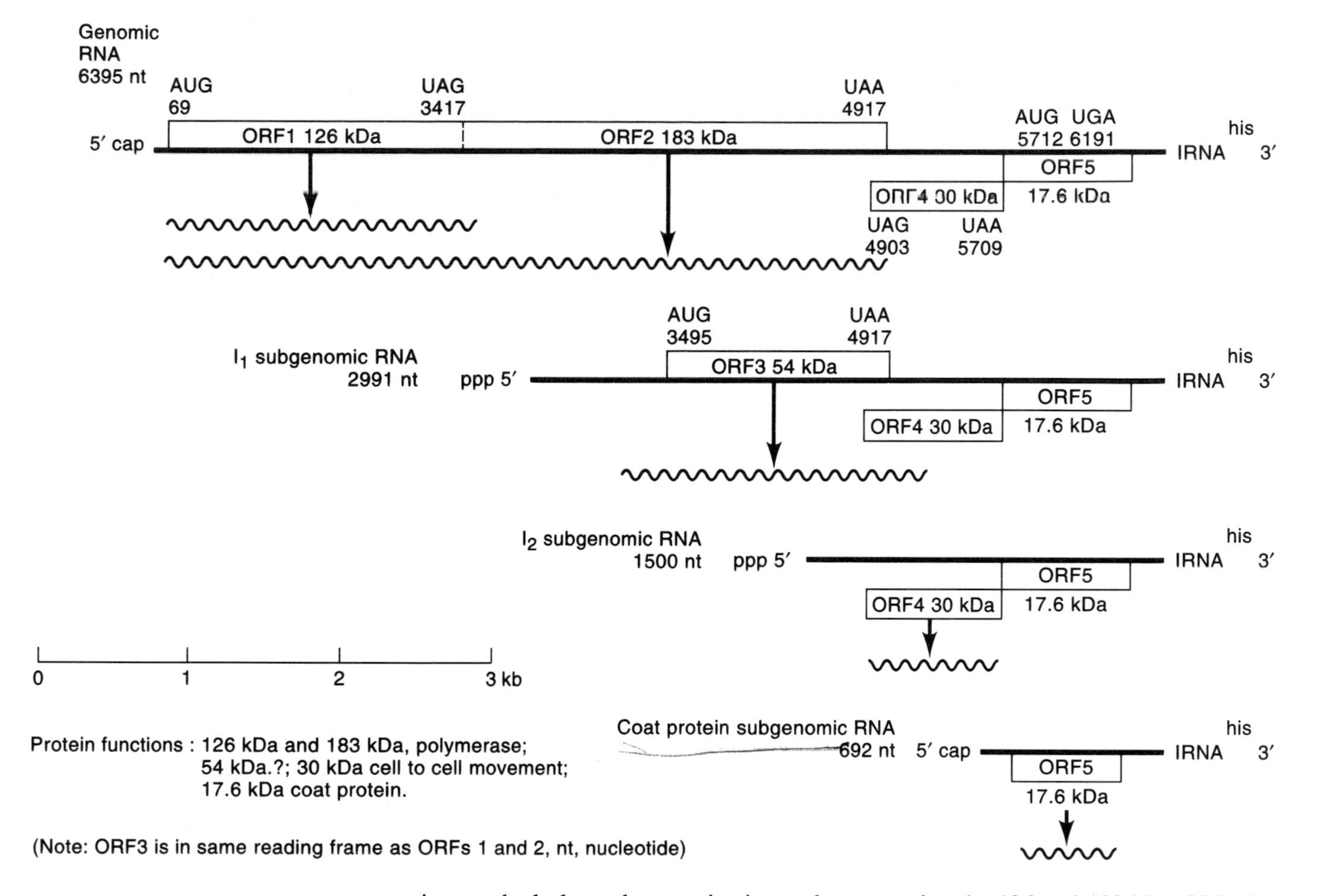

nesis near the leaky amber termination codon separating the 126 and 183 kDa ORFs has shown that TMV replication can proceed in the presence of the read-through protein alone.

Although TMV coat protein is by far the predominant virus-specific protein in infected cells, it is not found among the *in vitro* translation products of the full-length genomic RNA. Instead, it is translated from a small (692 nt) subgenomic RNA that can be isolated from virus-infected tissue. Two other subgenomic RNAs (the so-called I_1 and I_2 RNAs) have also been detected in TMV-infected tissue. The I_2 RNA encodes a 30 kDa protein that is essential for efficient cell-to-cell movement of TMV; the precise role (and even the *in vivo* existence) of the 54 kDa protein encoded by the I_1 RNA remains to be established. The use of this protein to create virus resistance in transgenic plants is discussed in Section 12.9.

Finally, to appreciate how *cis*-acting signals in viral RNAs control the synthesis of viral subgenomic mRNAs, let us turn our attention from TMV to brome mosaic virus (BMV) and the smallest of its three genomic RNAs (i.e., RNA3). In some of the earliest *in vitro* translation studies using plant viral RNAs, Kaesberg and his collaborators showed that BMV RNA4 (and only RNA4) could act as mRNA for BMV coat protein synthesis. Because the presence of RNA4 is not required for infectivity, it must be transcribed from one of the other three genomic RNAs. As shown in Fig. 2.17, the promoter for RNA4 synthesis is located between the ORFs encoding the cell-to-cell movement and coat proteins. It contains a core sequence of approximately 20 nucleotides flanked by an oligo(A) tract

FIGURE 2.17

Cis-acting signals in BMV RNA 3 that direct RNA replication and mRNA transcription. Coding regions for the 3a cell-to-cell movement and coat proteins are shown as open boxes, whereas filled boxes denote the various *cis*-acting regulatory sequences located within the noncoding regions. The sequence of the subgenomic RNA 4 promoter is that of the positive-sense RNA and terminates with AUG initiation codon for the BMV coat protein. Because negative-sense RNA 3 provides the templates for synthesis of RNA 4, the sequence of the promoter is actually complementary to that shown. (Courtesy of P. Ahlquist)

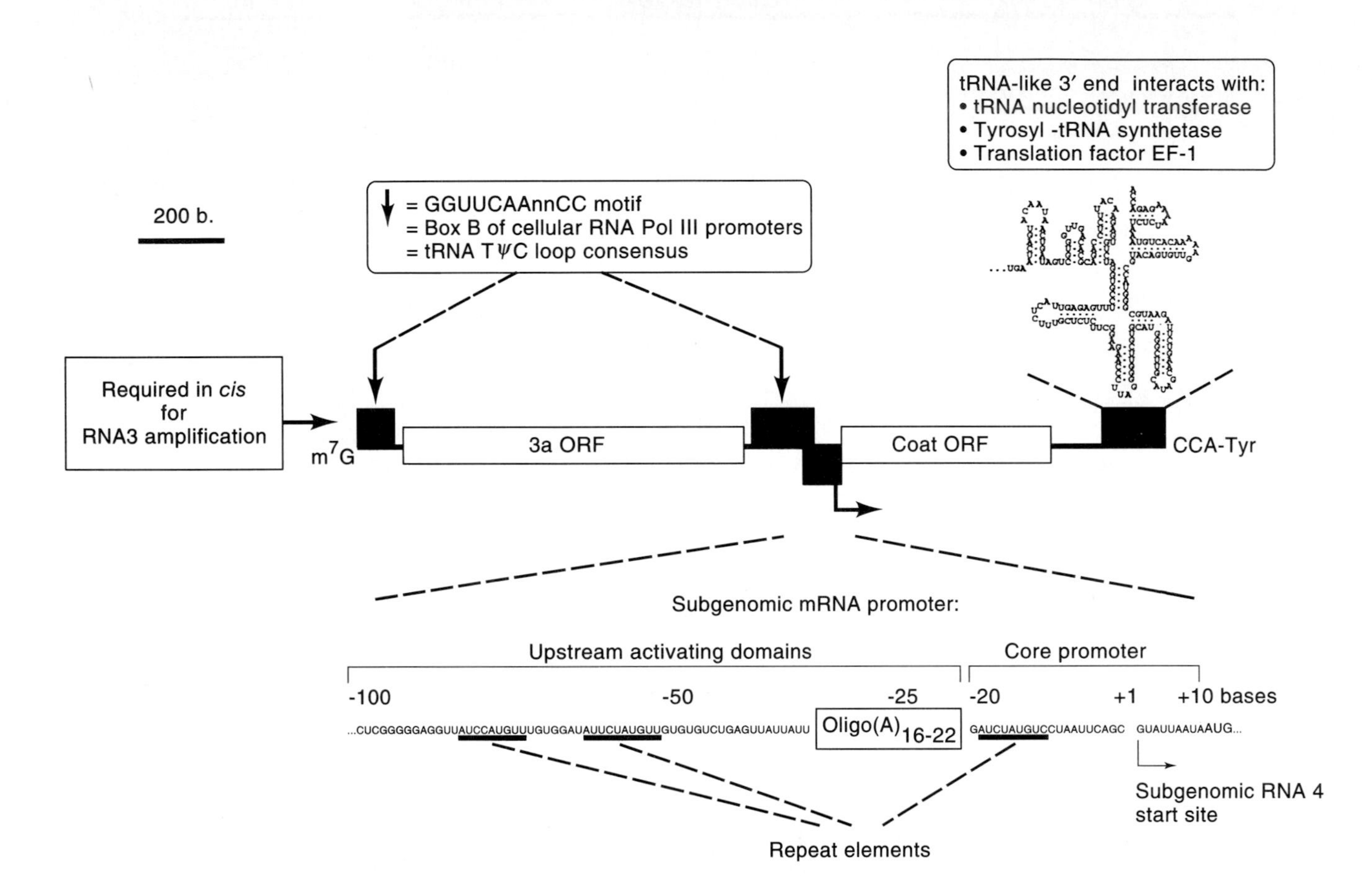

and a smaller domain that overlaps the 5′ untranslated portion of RNA4. This promoter bears no structural resemblance to the promoter at the 3′ termini of the three genomic RNAs, a tRNA-like structure that can be charged with tyrosine. Thus, it is quite likely that proteins other than the core polymerase are involved in the synthesis of the various virus-related positive-and negative-strand RNAs present in infected cells.

SECTION 2.6 FURTHER READING

Original Papers

FRAENKEL-CONRAT, H., and R. C. WILLIAMS. (1955) Reconstitution of active tobacco mosaic virus from its inactive protein and nucleic acid components. *Proc. Natl. Acad. Sci. USA* 41:690.

SCHRAMM, G., and W. ZILLIG. (1955) Uber die Struktur des Tabakmosaikvirus. IV. Die Reaggregation des Nucleinsaurefreien Proteins. *Z. Naturforsch.* 10b:493.

FRAENKEL-CONRAT, H. (1956) The role of the nucleic acid in the reconstitution of active tobacco mosaic virus. *J. Am. Chem. Soc.* 78:882.

FRAENKEL-CONRAT, H., and B. SINGER. (1957) Virus reconstitution. II. Combination of protein and nucleic acid from different strains. *Biochim. Biophys. Acta* 24:540.

SANGER, H. L. (1968) Characteristics of tobacco rattle virus. 1. Evidence that its two particles are functionally defective and mutually complementing. *Mol. Gen. Genet.* 101:346.

LEBEURIER, G., et al. (1977) Inside-out model for self-assembly of tobacco mosaic virus. *Proc. Natl. Acad. Sci. USA* 71:149.

BLOOMER, A. C., et al. (1978) Protein disk of tobacco mosaic virus at 2.8 Å resolution showing the interactions within and between subunits. *Nature* 276:362.

PELHAM, H. R. B. (1978) Leaky UAG termination codon in tobacco mosaic virus RNA. *Nature* 272:469.

AHLQUIST, P., et al. (1984) Multicomponent RNA plant virus infection derived from cloned viral cDNA. *Proc. Natl. Acad. Sci. USA* 81:7066.

BUJARSKI, J. J., and P. KAESBERG. (1986) Genetic recombination between RNA components of a multipartite plant virus. *Nature* 321:528.

SHAW, J. G., et al. (1986) Evidence that tobacco mosaic virus particles disassemble cotranslationally *in vivo. Virology* 118:326.

HAYES, R. J., and K. W. BUCK. (1990) Complete replication of a eukaryotic virus RNA *in vitro* by a purified RNA-dependent RNA polymerase. *Cell* 63:363.

JANDA, M., and P. AHLQUIST. (1993) RNA-dependent replication, transcription, and persistence of brome mosaic virus RNA replicons in *S. cerevisiae. Cell* 72:961.

DEMLER, S. A., D. G. RUCKER, and G. A. DE ZOETEN. (1993) The chimeric nature of pea enation mosaic virus: the independent replication of RNA 2. *J. Gen Virol.* 74:1.

Books and Reviews

GIBBS, A., and B. HARRISON. (1976) *Plant virology: The principles.* J. Wiley, New York.

BUTLER, P. J. G., and A. KLUG. (1978) The assembly of a virus. *Sci. Am.* 239:62.

BOS, L. (1983) *Introduction to plant virology.* Longman, New York.

FRANCKI, R. I. B., R. G. MILNE, and T. HATTA. (1985) *Atlas of plant viruses,* vols. 1 and 2. CRC Press, Boca Raton, Fla.

GOLDBACH, R. W. (1988) Evolution of plus-strand RNA viruses. *Intervirology* 29:260.

MATTHEWS, R. E. F. (1991) *Plant virology.* Academic Press, New York.

2.7 LEVIVIRIDAE (RNA BACTERIOPHAGES)

Although we will discuss only a few of the unenveloped positive-sense RNA bacteriophages that infect *E. coli,* many different bacterial species are known to harbor similar viruses. The Leviviridae family contains two genera—*Levivirus* and *Allolevivirus*—that show many similarities and may be evolutionarily related. RNA bacteriophages are rapidly replicated, causing lysis of the host cell in 30 to 40 minutes, and they yield very high levels of progeny particles (e.g., 10,000 per cell). The fact that these viruses are produced with very high efficiency *but only by male strains of E. coli* (and other bacteria) has made them a most useful material for the study of RNA virus structure, replication, genetics, and assembly. The best known of these phages are **MS2** plus the closely related R17 and f2 *(Levivirus)* and **Qβ** *(Allolevivirus).* During the 1960s, studies involving these phages were responsible for many of the advances in molecular biology.

2.7.1 Particle Structure

The RNA phages have particle weights and sedimentation rates of about 4×10^6 and 80 S, respectively. Their icosahedral virus particles are 26 to 27 nm in diameter and contain 180 molecules of a small (about 14 kDa) coat protein that are arranged in 12 groups of five at the vertices and 20 groups of six forming the facets plus one or a very few molecules of one or two larger proteins (41–44 kDa). This protein shell covers, but is also somewhat permeated by, a single molecule of RNA containing 3500 to 4700 nucleotides. Because the RNA is located close to the relatively thin protein shell, the phage particles have an internal cavity. They also contain about 1000 molecules of **spermidine,** a basic compound that neutralizes many of the negative charges on the RNA. The nature of the one (or, in the case of **Qβ**, two) additional proteins occurring in the virions is discussed below.

FIGURE 2.18

Amino acid sequences of two levivirus (MS2, fr) and one allolevivirus (Qβ) coat proteins. For fr, only residues that differ from MS2 are given. Residues shared by MS2 and Qβ after optimal alignment are enclosed in boxes.

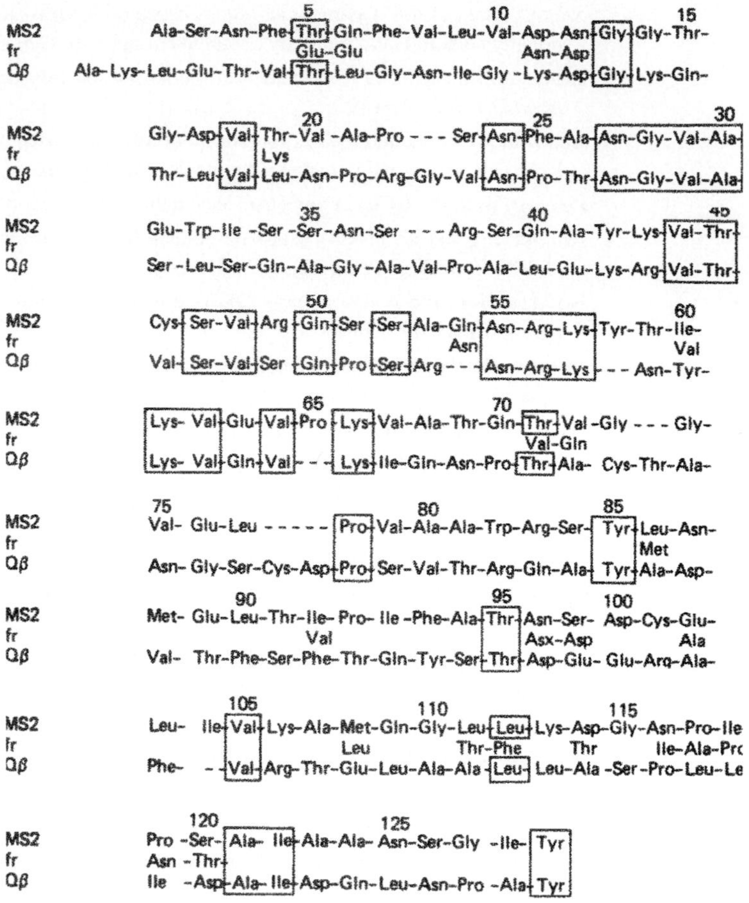

MS2 Ala-Ser-Asn-Phe-Thr-Gln-Phe-Val-Leu-Val-Asp-Asn-Gly-Gly-Thr- (5, 10, 15)
fr Glu-Glu Asn-Asp
Qβ Ala-Lys-Leu-Glu-Thr-Val-Thr-Leu-Gly-Asn-Ile-Gly -Lys-Asp-Gly-Lys-Gln-

MS2 Gly-Asp-Val-Thr-Val -Ala-Pro - - - Ser-Asn-Phe-Ala-Asn-Gly-Val-Ala- (20, 25, 30)
fr Lys
Qβ Thr-Leu-Val-Leu-Asn-Pro-Arg-Gly-Val-Asn-Pro-Thr-Asn-Gly-Val-Ala-

MS2 Glu-Trp-Ile -Ser -Ser-Asn-Ser - - - Arg-Ser-Gln-Ala-Tyr-Lys-Val-Thr- (35, 40, 45)
fr
Qβ Ser -Leu-Ser-Gln-Ala-Gly -Ala-Val-Pro-Ala-Leu-Glu-Lys-Arg-Val-Thr-

MS2 Cys-Ser-Val-Arg-Gln-Ser-Ser-Ala-Gln-Asn-Arg-Lys-Tyr-Thr-Ile- (50, 55, 60)
fr Asn Val
Qβ Val-Ser-Val-Ser-Gln-Pro-Ser-Arg - - -Asn-Arg-Lys - - Asn-Tyr-

MS2 Lys- Val-Glu-Val-Pro-Lys-Val-Ala-Thr-Gln-Thr-Val -Gly - - - Gly- (65, 70)
fr Val-Gln
Qβ Lys- Val-Gln-Val - - -Lys-Ile-Gln-Asn-Pro-Thr-Ala- Cys-Thr-Ala-

MS2 Val- Glu-Leu - - - - - Pro-Val-Ala-Ala-Trp-Arg-Ser- Tyr-Leu-Asn- (75, 80, 85)
fr Met
Qβ Asn- Gly-Ser-Cys-Asp-Pro-Ser-Val-Thr-Arg-Gln-Ala- Tyr-Ala-Asp-

MS2 Met- Glu-Leu-Thr-Ile- Pro- Ile -Phe-Ala-Thr-Asn-Ser- Asp-Cys-Glu- (90, 95, 100)
fr Val Asx-Asp Ala
Qβ Val- Thr-Phe-Ser-Phe-Thr-Gln-Tyr-Ser-Thr-Asp-Glu- Glu-Arg-Ala-

MS2 Leu- Ile-Val-Lys-Ala-Met-Gln-Gly-Leu-Leu-Lys-Asp-Gly-Asn-Pro-Ile (105, 110, 115)
fr Leu Thr-Phe Thr Ile-Ala-Pro
Qβ Phe- - -Val-Arg-Thr-Glu-Leu-Ala-Ala-Leu- Leu-Ala -Ser-Pro-Leu-Le

MS2 Pro -Ser- Ala- Ile-Ala-Ala- Asn-Ser-Gly -Ile- Tyr (120, 125)
fr Asn -Thr-
Qβ Ile -Asp-Ala- Ile-Asp-Gln-Leu-Asn-Pro -Ala- Tyr

The sequence of the MS2 coat protein, which is 129 amino acids long and lacks histidine, was determined soon after that of TMV. Several other phage coat proteins have since been partially or completely sequenced, and members of the MS2 group contain no more than two amino acid substitutions. Figure 2.18 compares the amino acid sequences of the MS2, fr, and **Q**β coat proteins. Note that two of the three additional amino acids in the **Q**β coat protein are located near its N terminus.

The complete 3569 nucleotide sequence of MS2 RNA as well as those of the R17, f2, and **Q**β genomes are now known. These studies started with the elucidation of the 5′- and 3′-terminal sequences, and the RNA of these viruses was the main material used in the development of methods for the direct sequencing of RNA—methodology that has now been largely supplanted by indirect methods using cDNA. The end-group analyses were followed by studies of the **ribosome-binding** (i.e., translation initiation) **sites** as well as other sites having specific affinities for certain proteins. It is for these historical reasons that we describe the methods used in the 1970s to elucidate the structure and function of various virus components.

These early studies showed that viral RNAs are not usually translated from beginning to end and that there are untranslated regions at both ends as well as between the individual genes of these **polycistronic** mRNAs. The first indications of how ribosome-binding sites, and thus the open reading frames for translation of genes to proteins, were determined on an mRNA were also obtained from study of phage RNAs. As they became known, the sequences of the ribosome-binding sites suggested the presence of extensive **base-pairing interactions** (see below), which were of great use in the sequencing studies as well as in helping to explain the regulation of the order and efficiency of translation of the genes of these RNAs. At 37°C, 74 percent of the bases are hydrogen-bonded in base-complementary fashion; at 0°C, ribonuclease T1 is able to attack the molecule at only a limited and reproducible number of G residues, yielding some relatively large primary cleavage products.

2.7.2 Genome Structure and mRNA Translation

In general, viruses tend to miniaturize their genetic requirements, presumably to maximize efficiency. The minimum number of genes required for RNA virus replication in a host that has no RNA-replicating mechanism would appear to be two: one for a coat protein and another to encode an enzyme to replicate the RNA—**an RNA-dependent RNA polymerase.** In fact, the action of RNA-dependent RNA polymerase is very complex and appears to require the association of several protein chains. In terms of information quantity, this obstacle has been overcome during RNA phage evolution by including within the viral genome only the information required for a single polypeptide (60.7 kDa in the case of MS2) that combines with three normal host proteins to form the RNA replicase.

A third protein termed the **A** (or **A2**) **protein** was found to be necessary for the phage to become adsorbed to the male-specific pili of *E. coli.* Also called the **maturation protein,** this 35–44 kDa adsorption protein apparently has several functions, although the evidence that it is needed for virion maturation is now less convincing. The existence of the **adsorption protein** in both MS2 and Qβ suggested the need for a third gene for this essential protein, and early genetic data also pointed to the existence of a fourth gene that provides a lysis function to disrupt the host cell and release newly made phage particles. The complete sequences corresponding genomes have now been determined, and their gene arrangements are known in detail (see Fig. 2.19). Surprisingly, although all RNA phages encode four genes, these genes were found to have different locations. Furthermore, seemingly analogous functions are supplied by different proteins in the MS2 and Qβ virus groups.

Elucidation of the structure of single-stranded RNA phage genes led to several profound revisions in our basic understanding of gene structure. The first novel finding involved the MS2 lysis gene. Initially, there was genetic evidence for this function but seemingly no room for it on the RNA. The viral genome appeared just long enough to encode the three proteins for which the sizes were already known. It is now clear that the additional protein, about 75 amino acids long, is translated from MS2 RNA in another reading frame—starting near the end of, the coat protein gene and ending 141 nucleotides within the polymerase gene (see Fig. 2.19). This unusual overlapping gene structure is not found in Qβ. Instead, a second form of **overlapping gene structure** encodes the Qβ A1 protein.

Determination of the means by which the Qβ A1 protein is synthesized was another milestone in our understanding of gene action. Here, it was found that two proteins of different function can result from translation of the same gene in the same reading frame by occasional read-through of a weak termination codon (UGA). Thus, the 14 kDa coat protein and 35 kDa A1 proteins of Qβ have the same N-terminal sequence of 132 amino acids, but the latter protein continues for an additional 200 amino acids. In some, as yet unexplained, way small amounts of A1 protein are required for Qβ particle assembly.

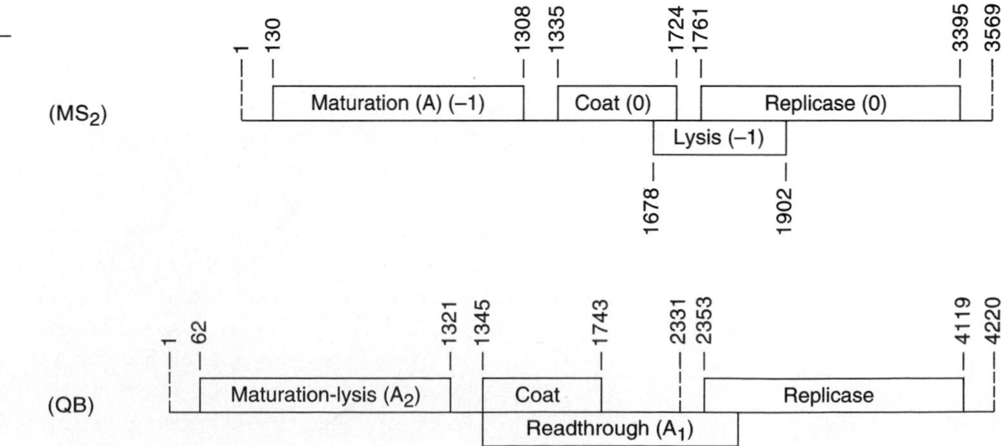

FIGURE 2.19

Genetic maps of MS2 and Qβ. Nucleotide numbers on the maps denote the boundaries of the individual genes as well as the RNA termini. (Redrawn from J. Van Duin (1988) In *The bacteriophages,* ed. R. Calender. Vol. 1,117–167. Plenum Press, New York)

This translation strategy represents yet another instance of **genetic miniaturization**—one nucleotide sequence coding for two structurally related proteins—but development of the need for an additional gene product was not gratis as far as Qβ is concerned. To allow for the additional information needed for the A1 protein, the nucleotide sequence between the genes for the coat protein and the RNA polymerase subunit is considerably longer in this phage than in the MS2 group. Thus, the genome of Qβ is 610 nucleotides longer than those of the other RNA phages.

2.7.3 Early Events in RNA Phage Replication

The infection cycle of the RNA phages lasts less than 1 hour: One or more (≤1000) virions become attached by their **adsorption protein** to the sides of the pilus of a male bacterium (see Fig. 2.20). In some way not yet understood (possibly through absorption or retraction of the pilus itself) a thin filament consisting of two protein fibers—the viral RNA of one or a few particles together with its A protein—enters the cytoplasm. There the cell's ribosomes attach to a binding site on the RNA. The only freely available site on the tightly folded viral RNA is the one at which translation of the coat protein is initiated (see Fig. 2.21). As the ribosome translates this sequence, the accompanying disruption of base-pairing interactions unveils an additional initiation site on the RNA molecule. This process leads to production of the 60 kDa replicase protein, a protein with a great affinity for three host proteins normally used for protein synthesis: elongation factors Tu and Ts (45 and 35 kDa) plus ribosomal protein S1 (70 kDa). The resulting active **RNA replicase** is a complex of four polypeptide chains that is generally quite unstable and has been isolated only from cells infected by Qβ or related phages. The availability of an active RNA replicase has led to some remarkably sophisticated studies concerning the mode of Qβ RNA replication, RNA sequencing, and site-specific mutagenesis (see Further Reading at the end of this section).

FIGURE 2.20

An F-pilus emerging from the *E. coli* cell (visible at the right bottom corner) covered with MS2 phage particles. At the end of the pilus, an fd phage (see Sec. 7.6) has attached itself. The thicker thread is a bacterial flagellum. (Courtesy of L. Caro)

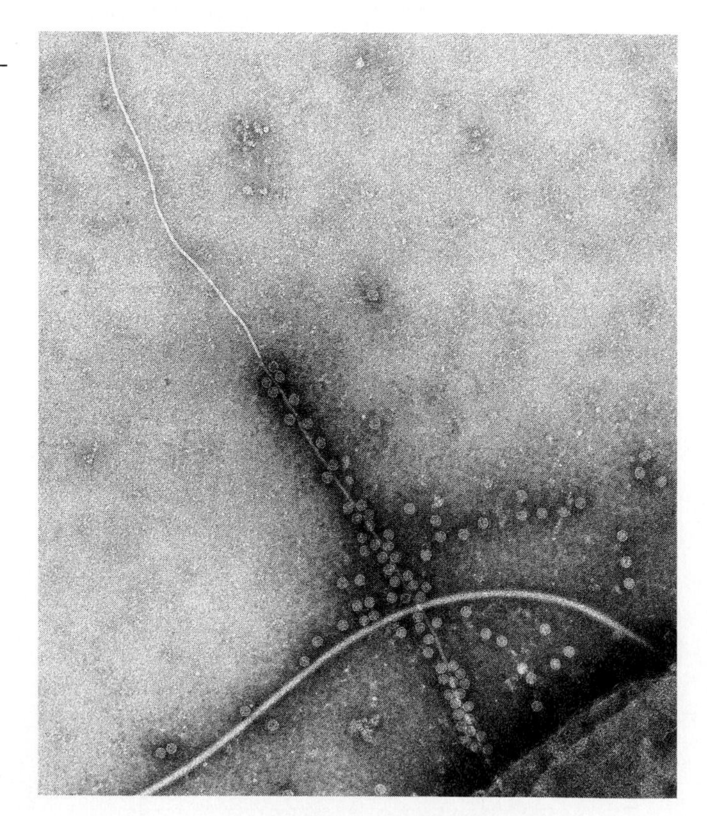

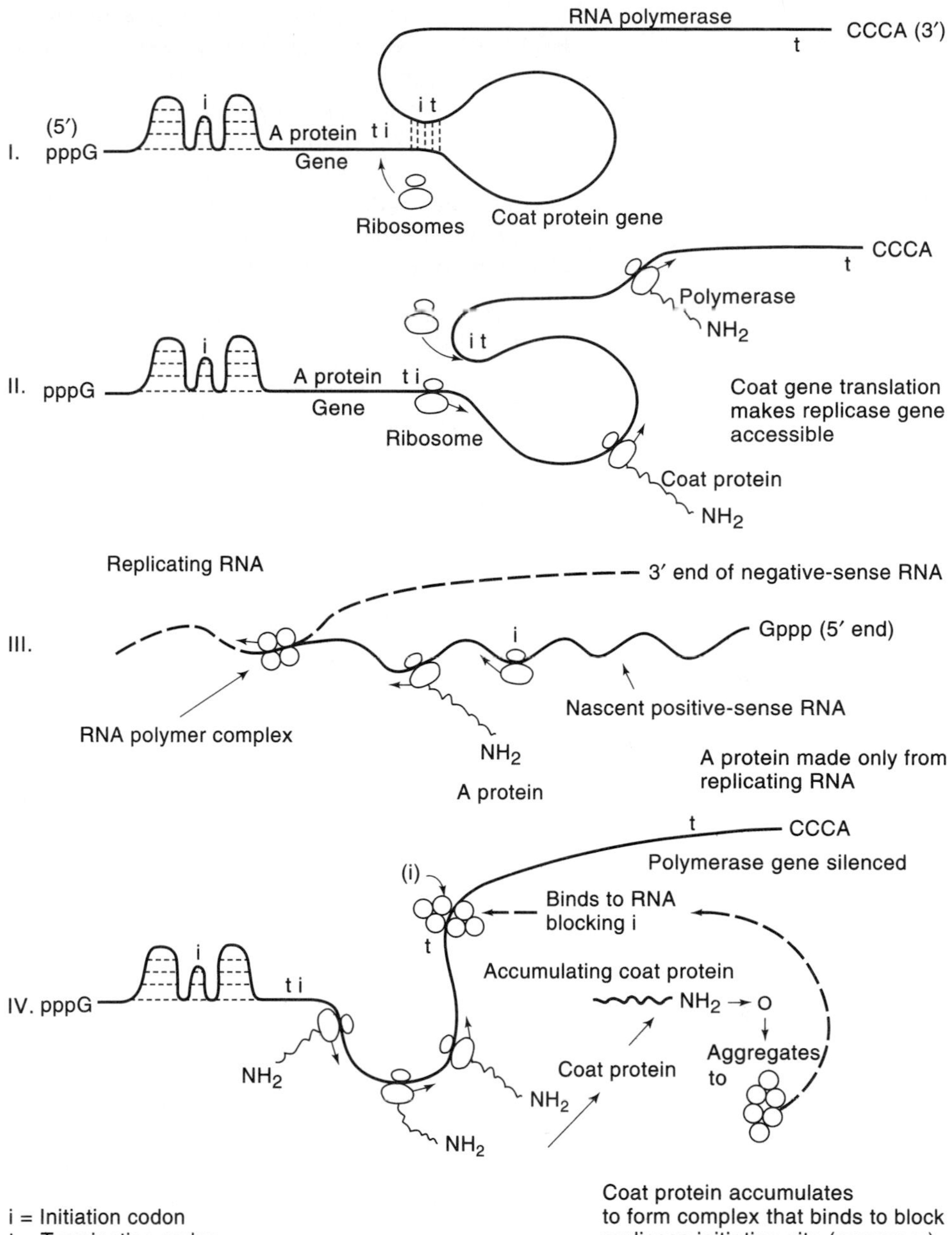

FIGURE 2.21

The mode of translation of an RNA phage. The genomic RNA is represented by a line starting with a 5′ guanosine triphosphate (pppG) and ending with CCCA-3′. Several hairpin foldings are schematically indicated to illustrate that the first and third initiation sites are not readily accessible to ribosomes during stage I. Upon translation of the coat protein gene, the initiation site for the third gene, the RNA polymerase, becomes available (stage II). Only upon replication of the RNA is the maturation gene sufficiently free of conformational constraint to become translated (stage III). As coat protein accumulates, it tends to aggregate and to bind to the RNA near the initiation site of the polymerase gene, thereby preventing its further translation (stage IV). The location and translation of the lysis gene of MS2 are not considered in this diagram.

The RNA replicase also has an affinity for the 3′ terminus of phage RNA and starts to replicate the RNA—starting with the penultimate C residue in the ... C-C-C-A terminal sequence and continuing transcription toward the left (i.e., in a 3′ → 5′ direction and leaving the 3′ terminal A uncopied). In the case of Qβ (but apparently *not* with other phages), an additional small (12 kDa) host protein known as **HF** and acting as a hexamer is needed for this primary transcription to produce the negative-sense RNA strand. Transcription of the resulting negative-sense RNA to produce positive-sense progeny requires neither S1 protein nor HF.

Because most mRNAs of eukaryotic cells and their viruses, in contrast with those of prokaryotes, contain a 3′-terminal poly(A), it was of interest to determine the effect of adding a similar poly(A) sequence to a prokaryotic mRNA. Qβ RNA containing such a poly(A) tail was not a suitable template for replication *in vitro,* but such RNA was infectious when tested in spheroplasts—which indicated that *E. coli* was able to remove the poly(A). The progeny Qβ RNA did not contain a 3′ poly(A) tail.

One aspect of Qβ replication that is not yet well understood is the accumulation of **6 S RNA,** a family of small RNAs containing 90 to 220 nucleotides. These are not mRNAs but serve as templates for the replicase complex. Synthesis of 6 S RNA is very susceptible to replication conditions, and its length changes in response to "environmental pressure." The largest of these variants, MDV-1 (positive sense) shows marked homology to the 3′ end of the Qβ negative-sense strand, and its formation may be related to the high error rate of Qβ replicase (10^{-3}–10^{-4} per nucleotide per replication). These small RNAs may represent a primitive equivalent to the defective interfering RNAs associated with certain larger plant and animal RNA viruses (see Sec. 4.1.4).

2.7.4 Later Events in RNA Phage Replication and Their Regulation

While its other end is still being replicated, the nascent (again positive-sense) RNA is immediately translated, starting at the first initiation site—63 or 130 nucleotides from the 5′ end of Qβ and MS2 RNA, respectively. This process leads to production of the A protein and provides a striking example of two biological processes—in this instance replication and translation of RNA—that occur simultaneously within the same molecule (see Fig. 2.21). Simultaneous transcription from the negative-sense strand and translation of the resultant positive-sense strand is possible, because mRNA synthesis proceeds in a 5′ → 3′ direction—the same direction as its translation and in the opposite direction to the use of the template. This combined process of replication and translation of phage RNA assures that a new A protein molecule is translated from each new molecule of RNA, a good regulatory strategy since these are needed in a 1:1 ratio for virion assembly.

Continuing translation of the completely replicated RNA produces the coat protein and the polymerase subunit. When the coat protein production reaches a certain level, several molecules bind to a specific site downstream and cover the third initiation site. This event blocks synthesis of the replicase subunit, which is no longer needed after the enzyme activity has reached a sufficient level (usually about 20 minutes after infection). The finding that a few molecules of phage coat protein tend to become bound at a specific site of the RNA and then are able to influence the translation strategy of that RNA represented the first instance of this type of a **feedback regulatory mechanism** in RNA translation. The site of binding of the coat protein at the polymerase initiation site accounts well for this activity. This process is analogous to the main method of control of gene expression from cellular DNA, where specific proteins, called **repressors,** are used to block the transcription of certain genes.

As a consequence of this event, most of the progeny RNA serves only as a messenger for coat protein, the protein whose initiation site in the free RNA is always the most available. Thus, production of coat protein, needed at a level of 180 molecules per virion, exceeds that of A protein. This latter protein is made primarily in synchrony with RNA replication, as well as the replicase, which is inhibited late in infection.

Regulation of lysis protein synthesis in MS2 to ensure that it is produced late in infection and in low quantities appears to result from a process called **translational interference.** Unlike the situation with the overlapping genes for the Qβ coat and A1 proteins, translation of the partially overlapping lysis cistron of MS2 RNA requires that ribosomes reading the upstream coat protein mRNA undergo a frame shift. Termination at either one of two nonsense codons just preceding the lysis cistron is then followed by initiation at the start codon of the lysis protein message. This process appears to be quite inefficient compared with initiation of the coat protein itself.

To summarize, the production rates of the phage coat, A, and replicase proteins are believed to be regulated in three different ways. The conformation of the potential mRNA makes certain ribosome-binding sites either available for initiation of translation or more or less inaccessible. Inaccessible or "silent" sites can become accessible and the mRNA information downstream expressed as a result of the act of translation itself. Such is the case when the ribosomes that synthesize coat protein disrupt the base-pairing between the coat cistron and the polymerase initiating site, thus favoring the reading of that message. Silent genes can also become active when, in the course of RNA replication, the inhibitory conformation has not yet been able to establish itself, as is the case with the nascent A protein gene. Finally, an initiation site can be secondarily silenced if a protein tends to bind to that part of the RNA and thus prevents ribosome binding. An example is the shut-off of polymerase production by coat protein late in infection. To begin to appreciate the true complexity of this regulatory scheme, remember that the viral RNA is actually folded in three (not two) dimensions.

2.7.5 Formation of Phage Particles

Virion assembly takes place throughout the infection cycle and finally reaches 10,000 to 20,000 particles per cell. Many of these particles, however, are either defective or represent empty protein shells, and the exact mode of *in vivo* assembly is not yet well understood. The A protein becomes associated with the RNA while it is still being translated and appears to lie near or at the surface of the mature union. Although still in contact with the RNA, the mature virion. The A protein is not firmly bound and can be removed from the phages by washing with salt solution. When this protein serves its adsorption function, it gets split into two fragments by a proteolytic enzyme, believed to be located in the bacterial cell wall. Cell lysis and **phage release** results from the action of the small (8 kDa) protein encoded by the fourth phage gene.

The importance of the A protein first became apparent during attempts to reconstitute RNA phages *in vitro* from their components following procedures for the reconstitution of plant viruses from purified coat protein and viral RNA (see Section 2.6.3). Assembly of phage-like particles containing RNA was observed, but these particles showed a somewhat lower sedimentation rate, contained less RNA, and were not infectious. Only when protein A, which is very insoluble and difficult to prepare, was also added were infectious particles obtained. These and similar observations suggested that protein A plays a role in particle maturation. It was later found that a complex of the RNA with protein A is sufficient to start the typical infection in male *E. coli* strains.

These results would seem to indicate that the A protein is required for phage infectivity. Such a conclusion is, however, erroneous, because phage RNA free of any protein had earlier been shown to be infectious. RNA is however, unable to enter the host cell in the usual way, that is through the pilus or directly through the cell wall. The latter has to be broken down by treating the bacteria with lysozymes or other agents. They turn them into spheroplasts, round cells lacking an intact cell wall and held together only by the cell membrane. Pure viral RNA can initiate an infection in **spheroplasts** prepared from female strains of bacteria. This result shows that the male specificity of these phages is really a reflection of the specificity of their A protein for the male pilus. In line with this observation, viral particles cannot infect spheroplasts and presumably neither can the RNA–protein A complex. Thus, protein A clearly represents a mechanism developed to attach the phage to the pilus and, subsequently, to bring the released RNA into the host cell.

2.7.6 RNA Phage Mutants

Many RNA phage mutants have been isolated, and the study of the effects of point mutations has proven to be particularly useful. **Point mutations** may lead to changes in the structure and character of a protein either through replacement of an amino acid (termed a **missense mutation**) or through blockage of protein synthesis by production of a termination codon (a **nonsense mutation**). Obviously, replacement of an amino acid by a very similar one is likely to be less damaging than when very different amino acids are involved. A point mutation can also be **conditionally lethal,** meaning that the altered or unfinished protein either retains its function or is not critically required under certain conditions. **Temperature-sensitive (ts) mutations,** where a protein becomes more easily denatured than the wild-type protein at higher (or lower) temperature, are the most common type of conditionally lethal mutation. They have proved to be particularly useful in viral genetic studies.

Surprisingly, the distribution of viable mutagenic events over the phage genome is quite uneven. The A protein is the most mutated gene in MS2 and related phages and the replicase gene is much less frequently mutated. This result is unlikely to be a simple reflection of the amount of variation that these proteins can tolerate and still retain their function, because in Qβ it is the replicase gene that shows the high mutation frequency. No satisfactory explanation exists for these observations.

Although nonsense mutations within mRNAs encoding essential proteins would appear to be lethal *a priori,* cells usually contain a few aberrant tRNA molecules that insert specific amino acids at termination signals. Particular host strains may be especially rich in one or another of these **suppressor tRNAs,** each of which specifically recognizes and suppresses the action of one of the three termination codons. Thus, a functional protein may be formed, and these mutants are generally termed *sus,* ***su**ppressor-**s**ensitive* **mutants,** to differentiate them from temperature-sensitive ones.

One frequently observed chain-terminating but suppressor-sensitive mutation in the coat protein gene of MS2 is the conversion of the CAG encoding its sixth amino acid, glutamine, into UAG. The so-called **polarity** of this mutation means that translation of the following gene (i.e., the replicase gene) is also depressed. Similar termination events at the 50th or greater amino acid of the coat protein have no such polarity effect. The reason for this polarity effect, and its limitation to the beginning of the coat protein gene, is now clear: Translation of the N-terminal portion of the coat protein cistron is necessary to break base-pairing interactions with the replicase initiation site and make that site accessible to ribosomes (see Fig. 2.21).

Changes in the noncoding parts of the genomic RNA or the untranslated portions of an mRNA are by no means harmless, but their results are often more difficult to evaluate. These parts of the phage genome probably function primarily to bind enzymes, ribosomes, and regulatory molecules. Indeed, their nucleotide sequences are more conserved within a virus group (i.e., less likely to remain functional after mutation) than are those carrying protein information.

SECTION 2.7 FURTHER READING

Original Papers

LANDSTEINER, K. AND E. POPPER. Ubertragung der Poliomyelitis acuta auf Affen. *Z Immunitatsforsch Orig* 1909:2:377.

LOEB, T., and N. D. ZINDER. (1961) A bacteriophage containing RNA. *Proc. Natl. Acad. Sci. USA* 47:282.

SPIEGELMAN, S., et al. (1965) The synthesis of a self-propagating and infectious nucleic acid with a purified enzyme. *Proc. Nat. Acad. Sci. USA* 51:919.

SUGIYAMA, T., and D. NAKANA. (1967) Control of translation of MS2 RNA cistrons by MS2 coat protein. *Proc. Nat. Acad. Sci. USA* 57:1744.

KAMEN, R. (1970) Characterization of the subunits of Qβ replicase. *Nature* 228:527.

KONDO, M., et al. (1970) Subunit structure of Qβ replicase. *Nature* 228:525.

SUGIYAMA, T., and D. NAKADA. (1970) Translational control of bacteriophage MS2 RNA cistrons by MS2 coat protein: Affinity and specificity of the interaction of MS2 coat protein with MS2 RNA. *J. Mol. Biol.* 18:349.

FLAVELL, R. A., et al. (1974) Site-directed mutagenesis: Generation of an extracistronic mutation in bacteriophage Qβ RNA. *J. Mol. Biol.* 89:255.

FIERS, W., et al. (1976) Complete nucleotide sequence of bacteriophage MS2 RNA: Primary and secondary structure of the replicase gene. *Nature* 260:500.

TANIGNCHI, T., et al. (1978) Qβ DNA containing hybrid plasmids giving rise to Qβ phage formation in the bacterial host. *Nature* 271:223.

KASTELEIN, R. A., et al. (1982) Lysis gene expression of RNA phage MS2 depends on a frameshift during translation of the overlapping coat protein gene. *Nature* 295:35.

MIELE, E. A., D. R. MILLS, and F. R. KRAMER. (1983) Autocatalytic replication of a recombinant RNA. *J. Mol. Biol.* 171:281.

BERKHOUT, B., et al. (1985) Translational interference at overlapping reading frames in prokaryotic messenger RNA. *Gene* 37:171.

BIEBRICHER, C. K., et al. (1986) Template-free RNA synthesis by Qβ replicase. *Nature* 321:89.

SHAKLEE, P. N. (1990) Negative-strand RNA replication by Qβ and MS2 positive-strand RNA phage replicases. *Virology* 178:340.

Books and Reviews

LOEFFLER, F. AND P. FROSCH. (1964) Report of the commission for research on foot-and-mouth disease. In: Hahon N., ed. *Selected papers on virology* 64–68. Prentice Hall, Englewood Cliffs, NJ .

WEISSMANN, C., et al. (1973) Structure and function of phage RNA. *Ann. Rev. Biochem.* 12:303.

ZINDER, N. D. (1975) *RNA phages.* Cold Spring Harbor Press, Cold Spring Harbor, New York.

VAN DUIN, J. (1988) Single-stranded RNA bacteriophages. In *The bacteriophages,* ed. R. Calender. Vol. 1, 117–167. Plenum Press, New York.

BIEBRICHER, C. K., and M. EIGEN. (1988) Kinetics of RNA replication by Qβ replicase. In *RNA genetics,* ed. E. Domingo, J. J. Holland, and P. Ahlquist. Vol. 1, 1–21. CRC Press, Boca Raton, Fla.

EIGEN, M., and C. K. BIEBRICHER. (1988) Sequence space and quasispecies distribution. In *RNA genetics,* ed. E. Domingo, J. J. Holland, and P. Ahlquist. Vol. 3, 211–245. CRC Press, Boca Raton, Fla.

ZINDER, N. D. (1989) Portraits of viruses: RNA phage. *Intervirology* 13:257.

chapter

3

Enveloped Viruses with Positive-Sense RNA Genomes

This chapter, like sections of Chapter 2, deals with small viruses containing positive-sense RNA genomes. The nucleocapsids of these viruses are surrounded, however, by a lipid-containing envelope. The many known members of this virus grouping are currently distributed among four families: the Togaviridae, Flaviviridae, Coronaviridae, and Toroviridae. None are known to infect plants. Several were formerly included among the **arboviruses**, because they are *ar*thropod-*bo*rne or transmitted and replicate in both their arthropod and vertebrate hosts. Because this terminology failed to consider either the chemistry of these viruses or their mode of replication, it has become obsolete. The abandonment of this group name made good sense because many quite unrelated viruses occur in and/or are transmitted by arthropods.

Many arboviruses have been placed in a single virus family, the Togaviridae. *Toga* (Latin, "cloak" or "mantle") indicates a group of viruses with lipid-rich envelopes, again not a particularly informative term, since the majority of the animal virus families are so enveloped. Members of all four of the current viral families are relatively small enveloped positive-sense RNA viruses whose monopartite genomes are encapsidated by a single species of protein. Differences in their genome structure and replication strategy led to the separation of the Flaviviridae from the Togaviridae. Members of the Coronaviridae have a much larger genome (27 to 30 versus about 12 kb) and a very distinctive particle morphology in negatively stained preparations. The sizes and principal structural features of these four families, as compared with all other animal virus families, are illustrated schematically in Figure 1.11.

3.1 TOGAVIRIDAE

As presently defined, the family Togaviridae contains three genera (see Table 3.1). Typical members of the *Alphavirus* genus include Sindbis, Semliki Forest, and equine encephalitis viruses; the first two of these viruses have been studied in great detail at the molecular level. Rubella virus, the sole member of the *Rubivirus* genus, is well known as the agent of German measles. This virus acquired great notoriety when it was realized that the virus is transmitted to embryos during early pregnancy and may cause serious consequences, including abortion and death to the newborn child. The 3'-terminal location of their coat protein gene plus the presence of a nested set of subgenomic mRNAs having a common leader sequence suggests, however, that members of the *Arterivirus* genus may be more closely related to the coronaviruses than to the togaviruses.

TABLE 3.1
Genera in the family Togaviridae

Genus	*Examples*
Alphavirus	Sindbis and Semliki Forest viruses Eastern equine encephalitis virus
Rubivirus	Rubella virus
Arterivirus	Equine arteritis virus Lactic dehydrogenase virus Porcine reproductive and respiratory syndrome virus Simian hemorrhagic fever virus

Most togaviruses have a broad vertebrate host range and grow well in both arthropods and arthropod cell lines. Alphaviruses are mosquito-transmitted, and certain of the equine alphaviruses are naturally maintained in bird populations, which serve as **reservoirs.** Transmission from bird to bird or, as an incidental infection, from bird to horse, is by mosquitoes. In horses the disease caused by equine encephalitis virus is so rapidly fatal that insufficient virus accumulates in the blood to allow horse-to-horse transmission via mosquitoes. In both horse and humans (when it occurs) the infection stops without further transmission. Rubella virus has no insect vectors and appears to be spread principally by aerosols.

Although all alphaviruses are serologically interrelated, six **antigenic complexes** have been identified. The clinical response to alphavirus infection is extremely varied, ranging from inapparent infection to fevers and rashes, arthritis, severe encephalitis, and death. Infection of their mosquito vectors is generally without major pathogenic effects. Alphaviruses produce extensive cytopathic changes in virtually all common cultured vertebrate cells, but most invertebrate cell lines respond with **persistent infections** without marked cytopathology.

3.1.1 Structure

Togaviruses are among the simplest of the enveloped animal viruses. Their single-stranded, positive-sense 12 kb RNA genome is polyadenylated at the 3′ end and capped with 7-methylguanosine at the 5′ end. This RNA is encapsidated by a single species of capsid protein, and the T = 3 icosahedral nucleocapsid is surrounded by a lipid bilayer derived from the host cell plasma membrane. Two viral-encoded glycoproteins (designated E1 and E2) project from the lipid bilayer and are embedded within it. Figure 3.1 illustrates the particle structure of Sindbis virus. Because of the high lipid content of their envelope, the togaviruses are very sensitive to ether and detergents; they also are inactivated by reducing agents, slightly elevated temperatures, and certain salts.

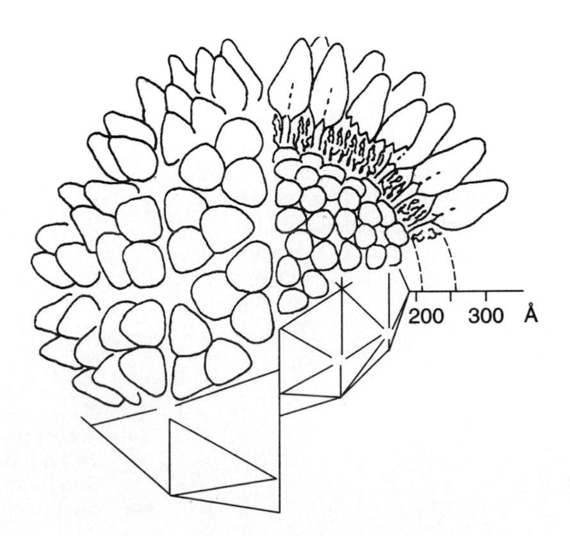

FIGURE 3.1

Organization of the Sindbis virus particle. The pear-shaped objects represent the external portions of glycoproteins E1 and E2, closely associated as heterodimers and further clustered into trimers in a T = 4 icosahedral lattice that describes the surface organization. There are 240 E1:E2 heterodimers in the complete particle, and hydrophobic transmembrane segments of E1 and E2 penetrate the lipid bilayer. A 33–amino acid internal domain on E1 is believed to mediate interaction with a nucleocapsid that contains 180 coat protein molecules packed in a T = 3 icosahedral structure. (From S. C. Harrison (1990) In *Virology,* 2d ed., ed. B. N. Fields et al. Vol. 1, 37–61. Raven Press, New York)

The envelopes of alpha viruses (and most if not all other enveloped viruses) share certain common features. Proteins of the viral envelope have regions that are rich in hydrophobic amino acid sequences (e.g. leucine, isoleucine, valine, phenylalanine), and these are always embedded in the plasma membrane. The hydrophilic regions form the exterior spikes. The lipids of the virion envelope are derived from host cell membranes and are therefore specified by host cell enzymes. Similarly, the addition of sugar residues to viral glycoproteins appears to involve only host cell enzymes.

Because viral envelopes generally lack host cell surface antigens (i.e., host cell proteins), the viral envelope proteins are thought to displace all host cell proteins from the regions of the plasma membrane region involved in virion maturation. Exceptions can be found with Retroviridae that can carry certain cellular MHC molecules on their surface (Sec. 6.1). It should also be noted that insertion of viral proteins into the cell membranes is responsible, in part, for a wide variety of biological effects on the host cell (see Chapter 10). Among these are the capability of most enveloped viruses to agglutinate red cells (or **hemagglutination**) which frequently leads to hemolysis. For the simplest enveloped viruses, the alphaviruses, this property is determined by glycoprotein E1. This fact was established by reconstituting particles lacking one or another viral protein.

3.1.2 Entry, Transcription, and Virus Production

After binding to the surface of susceptible host cells via the glycoprotein spikes that protrude from the virus membrane, alphaviruses enter the cytoplasm via **endocytosis.** The biochemical activities that release the genomic RNA from the endocytosed virion have not yet been completely defined, but, once released, the virion RNA functions as a normal cellular mRNA. Translation of Sindbis and other alphavirus RNAs resembles that of certain plant viruses, such as TMV.

Besides the 49 S genomic RNA, a 26 S (about 4.1 kb) subgenomic mRNA identical in sequence with the 3′-terminal one-third of the viral genome and carrying the information for the virion proteins (i.e., the capsid, and the three envelope proteins) is produced intracellularly early during infection. As with TMV (see Sec. 2.2.3), the smaller subgenomic

FIGURE 3.2

Organization and expression of the Sindbis genome. As the only portion of the genome to be directly translated, the nonstructural genes (nsP1–4) are denoted by a bold line. The four structural proteins are translated from a 26 S subgenomic mRNA. Both genomic and subgenomic RNAs have only a single site for initiation of protein synthesis, and the resulting polyproteins undergo proteolytic processing during translation. (From S. Schlesinger and M. J. Schlesinger. (1990) In *Virology,* 2d ed., eds. B. N. Fields and D. M. Knipe. Vol. 1, 697–711. Raven Press, New York)

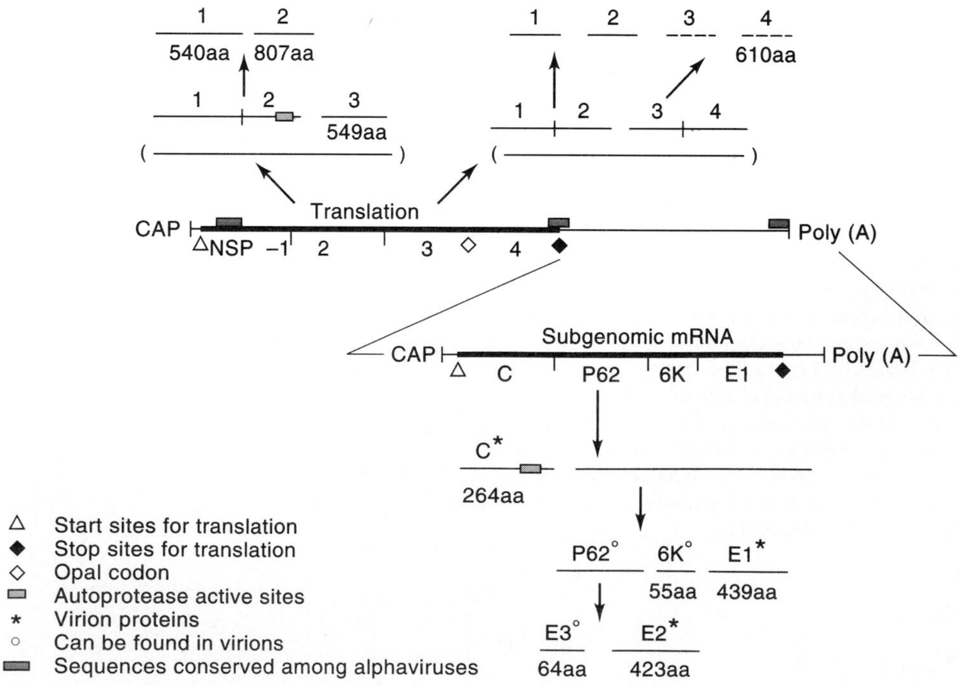

△ Start sites for translation
◆ Stop sites for translation
◇ Opal codon
▭ Autoprotease active sites
* Virion proteins
○ Can be found in virions
▬ Sequences conserved among alphaviruses

coat protein mRNA favors efficient production of those proteins that the virus needs in greatest amounts. The large genomic RNA supplies the smaller amounts of enzymatic products required. Determination of the complete nucleotide sequences of Sindbis and Semliki Forest viruses has provided detailed insight into their genomic organization and expression strategy (see Fig. 3.2).

Translation of the 49 S genomic RNA yields a polyprotein containing 3 to 4 nonstructural (ns) proteins. The presence of an opal termination codon several nucleotides upstream from the N-terminus of Sindbis nsP4 means that most translation products of the genomic RNA contain only nsP1–3. In Semliki Forest virus, there is no upstream termination codon, and translation continues to yield a polyprotein containing nsP1–4. As illustrated in Figure 3.2, an **autoproteolytic activity** present in nsP2 is responsible for some (and perhaps all) of the cleavages necessary to release all four nonstructural polypeptides from the polyprotein precursor(s).

These four nonstructural polypeptides catalyze replication of the genomic RNA as well as transcription of the 26 S subgenomic mRNA. RNA synthesis initiates at the 3′ terminus of the 49 S RNA, and this process presumably involves both the 19 conserved nucleotides adjacent to the poly(A) tract as well as another 51 conserved nucleotides near the 5′ terminus (see Fig. 3.2). The 26 S subgenomic RNA, whose 3′ terminus is identical with that of the genomic RNA, can*not* serve as template for negative-sense RNA synthesis. The resulting full-length negative-sense RNA serves as a template for the synthesis of two types of positive-sense RNAs. Whereas initiation at the 3′ end of the negative-sense strand yields full-length 49 S positive-strand progeny, initiation within the conserved **junction region** between the nonstructural and structural gene clusters leads to the formation of 26 S mRNA. Three of the four nonstructural proteins (nsP1, nsP2, and nsP4) contain domains that are closely related in sequence to domains in the RNA replicases of several RNA plant viruses. In addition to possible mechanistic similarities in their replication strategies, these sequence similarities also suggest that these diverse groups of viruses may have originated from a common ancestor (see Postscript).

Synthesis of both positive- and negative-sense RNA increases during the first several hours after infection. Negative-sense RNA synthesis then stops, whereas synthesis of 49 S and 26 S positive-sense RNAs continues at a nearly constant rate. Because most of the 49 S RNA is sequestered in nucleocapsids, the ratio of polysome-bound 26 S mRNA to 49 S RNA can be as much as 10:1. As a model for membrane biogenesis as well as for protein synthesis by most enveloped viruses, the translation of the 26 S subgenomic mRNA has been studied in great detail. It appears that cytoplasmic translation on free ribosomes occurs first and that the capsid protein (C) is released by a serine-type autoprotease activity (see Fig. 3.3).

Cleavage at the protease-sensitive site at the carboxy terminus of the capsid protein exposes a hydrophobic signal sequence that binds to **signal recognition particles** in the host cell that facilitate transport of the remaining nascent polypeptide across the endoplasmic reticulum (ER). Translation continues on the membrane surface with the product now

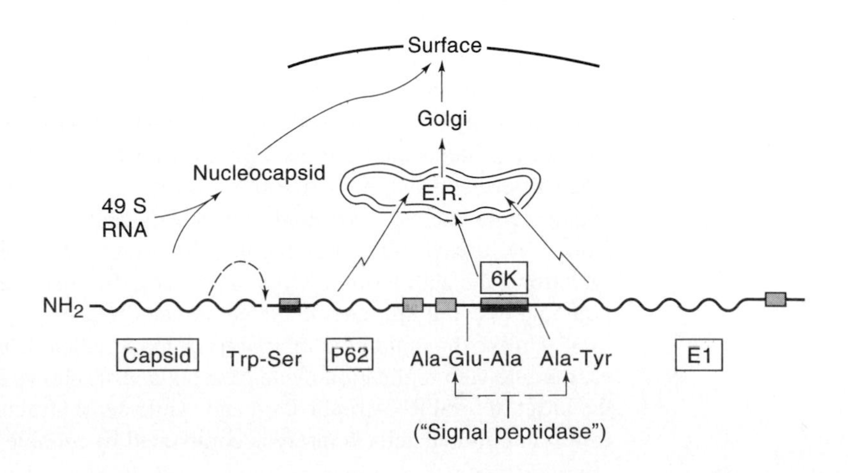

FIGURE 3.3

Proteolytic processing of alphavirus structural proteins during translation of membrane-associated 26 S mRNA. Capsid protein is released by autoproteolysis and forms nucleocapsids that ultimately combine at the cell surface with viral glycoproteins inserted into the endoplasmic reticulum (ER) and transported via the Golgi complex to the cell surface. Filled boxes, signal sequences; hatched boxes, hydrophobic sequences. (From S. Schlesinger and M. J. Schlesinger (1990) In *Virology*, 2d ed., eds. B. N. Fields and D. M. Knipe. Vol. 1, 697–711. Raven Press, New York) being released into the vesicles, where it becomes glycosylated and undergoes further proteolytic cleavage. Long-chain fatty acids are also added to cysteine residues as the proteins move from the ER through the Golgi apparatus to the plasma membrane. All these post-translational modifications are carried out by host cell enzymes.

The final stages of alphavirus replication occur at the plasma membrane, where the viral nucleocapsid, the glycoprotein spikes, and the lipid bilayer are brought together to form the mature virion. Nucleocapsids form in the cytoplasm when the newly synthesized capsid protein associates with 49 S positive-sense RNA. The amino terminus of the capsid protein is relatively rich in basic amino acids plus proline, and sequences near (but not at) the 5′ terminus of the 49 S RNA determine the specificity of the RNA–capsid protein association. **Virus budding** from the plasma membrane appears to be initiated by interactions between the nucleocapsid and a specific domain in membrane-bound E2 that is exposed to the cytoplasm. Conversion of p62 to E2 + E3 may promote this interaction.

3.1.3 Defective Interfering (DI) Particles

The **DI particles** that are frequently associated with togaviruses have been studied in some detail. As illustrated in Figure 3.2, the alphavirus genome contains four highly conserved regions: (1) 19 nucleotides at the 3′ terminus; (2) 21 nucleotides that span the junction between the nonstructural and structural genes and include the start of the 26 S subgenomic mRNA; (3) 51 nucleotides near the 5′ terminus; and (4) the 5′ terminus itself (more conserved in secondary structure than nucleotide sequence *per se).* All Sindbis or Semliki Forest virus DI RNAs contain the 3′-terminal conserved sequences, and most also contain the 51 conserved nucleotides near their 5′ terminus. **Deletion mapping** of DI cDNAs has shown that only sequences at the 3′ and 5′ termini are essential for amplification and packaging by the helper virus genome. Deletion of the conserved 51 nucleotides near the 5′ terminus of Sindbis DI RNAs did not destroy its biological activity. A quite unexpected finding was the discovery that the sequence of the 5′ terminus of many Sindbis DI RNAs is identical with nucleotides 10 to 75 of a cellular $tRNA^{asp}$. More information about the origin and replication of DI RNAs can be found in Section 4.1.4.

3.1.4 Rubella Virus

Like typical **alphaviruses,** rubella virus (genus *Rubivirus*) particles contain three structural polypeptides: a nonglycosylated capsid protein plus two glycoproteins (E1 and E2). Its small (60–70 nm), spherical virion contains a 30-nm electron-dense nucleocapsid "core" surrounded by a lipid envelope. Unlike other **togaviruses,** however, rubella virus has a distinctive electron-lucent zone visible between its core and envelope. Rubella virus also differs from most other togaviruses in that it has no known invertebrate host. Humans provide its only known natural reservoir.

Even though the structural proteins of rubella virus are serologically distinct and exhibit no major sequence homologies with the corresponding proteins from alphaviruses, the molecular biology of this virus appears to closely parallel that of the alphaviruses. Comparatively little is known about the properties of the rubella nonstructural proteins, but structural proteins are translated from a 24 S (about 3500 nt) polyadenylated subgenomic mRNA derived from the 3′ one-third of the 40 S genomic RNA. Rubella virus remains of considerable interest to virologists primarily because of several features that distinguish it from other togaviruses; these include the continued problems posed by congenital rubella infections, the ability of the virus to persist in the absence of detectable symptoms, and its potential to cause immune-mediated human disease.

Unlike the rapid lytic infections of mammalian cells typically caused by alphaviruses, the one-step replication cycle of rubella virus shows a 10-hour "eclipse" phase before the onset of viral RNA replication and synthesis of structural proteins. Rubella virus replication in cultured cells is rarely accompanied by reliable or distinctive cytopathic effects;

rather, the virus readily establishes **persistent infections** in which the growth rate of infected cells is slowed. The possible roles of DI particles and temperature-sensitive mutants in the establishment and maintenance of persistent infections remain unresolved.

Although rubella infections in early childhood or adult life are usually mild, infection of the developing fetus often has dire consequences—**abortion, miscarriage, stillbirth,** and a variety of fetal malformations. Up to 80 percent of children with congenital rubella show some type of neurological involvement. Fortunately, presently available live-virus vaccines have proved quite effective in reducing the incidence of congenital rubella in the United States. A serious concern in current vaccination programs is the risk to the developing fetus of mothers immunized during early pregnancy. However, inadvertent vaccination of a seronegative pregnant woman is not generally considered to be sufficient reason to routinely terminate the pregnancy. Reliable animal models exist for both symptomatic acquired and congenital rubella.

SECTION 3.1 FURTHER READING

Original Papers

SVREEVALSAN, T., and R. Z. LOCKART, Jr. (1966) Heterogeneous RNAs occurring during the replication of western equine encephalomyelitis virus. *Proc. Natl. Acad. Sci. USA* 55:974.

SCHLESINGER, M. J., and S. SCHLESINGER. (1973) Large-molecular-weight precursors of Sindbis virus proteins. *J. Virol.* 11:1013.

MONROE, S. S., and S. SCHLESINGER. (1983) RNAs from two independently isolated defective interfering particles of Sindbis virus contain a cellular rRNA at their 5′ end. *Proc. Natl. Acad. Sci. USA* 80:3279.

KAMER, G., and P. ARGOS. (1984) Primary structural comparison of RNA-dependent polymerases from plant, animal, and bacterial viruses. *Nucl. Acids Res.* 12:7269.

STRAUSS, E. G., C. M. RICE, and J. H. STRAUSS. (1984) Complete nucleotide sequence of the genomic RNA of Sindbis virus. *Virology* 133:92.

RICE, C. M., R. LEVIS, J. H. STRAUSS, and H. V. HUANG. (1987) Production of infectious RNA transcripts from Sindbis virus cDNA clones: Mapping of lethal mutations, rescue of a temperature-sensitive marker, and *in vitro* mutagenesis to generate defined mutants. *J. Virol.* 61:3809.

CONZELMANN, K.-K., et al. (1993) Molecular characterization of porcine reproductive and respiratory syndrome virus, a member of the arterivirus group. *Virology* 193:329.

SCHOEPP, R. J., and R. E. JOHNSTON. (1993) Directed mutagenesis of a Sindbis virus pathogenesis site. *Virology* 193:149.

Books and Reviews

SCHLESINGER, S., and M. J. SCHLESINGER. (1990) Replication of Togaviridae and Flaviviridae. In *Virology,* 2d ed., ed. B. N. Fields et al. Vol. 1, 697–711. Raven Press, New York.

WOLINSKY, J. S. (1990) Rubella. In *Virology,* 2d ed., ed. B. N. Fields et al. Vol. 1, 815–838. Raven Press, New York.

DOMS, R. W., et al. (1993) Folding and assembly of viral membrane proteins. *Virology* 193:545.

3.2 FLAVIVIRIDAE

The viruses in the family Flaviviridae, formerly regarded as a genus of the Togaviridae, resemble the alphaviruses in appearance, basic structure, **arthropod transmission,** and general pathogenesis. Their transcription processes, however, are distinct and resemble those of the picornaviruses. As presently defined (see Table 3.2), the family Flaviviridae contains three genera: In addition to the **flaviviruses** themselves, there are the **pestiviruses** such as bovine diarrhea and hog cholera viruses as well as human **hepatitis virus C,** a virus that is currently responsible for about one-third of all acute hepatitis in the United States (see Chapter 14).

Yellow fever virus, the prototype flavivirus, was among the first filterable agents shown to cause human disease as well as the first virus proven to have an arthropod vector. Other important flaviviruses, many of which are serologically interrelated, include dengue,

TABLE 3.2
Genera in the family Flaviviridae

Genus	*Examples*
Flavivirus	Yellow fever virus Dengue virus (types 1–4) Japanese encephalitis virus Tick-borne encephalitis virus
Pestivirus	Bovine diarrhea virus Hog cholera virus
Hepatitis virus C	Human hepatitis virus Simian hemorrhagic fever virus (possible)

West Nile, Kunjin, and Japanese encephalitis. Although the flaviviruses are of great interest to the molecular biologist, they are of even greater concern to the medical and veterinary professions because they cause serious (and at times lethal) diseases, such as **encephalitis** and **hemorrhagic** fever. Able to replicate without causing illness in their blood-sucking arthropod vectors, flaviviruses represent major epidemiological problems. Many of these diseases (e.g., **yellow fever** and **dengue)** are difficult to control for reasons other than the viruses' ability to persist in their insect vectors—in particular because of the frequent appearance of new, immunologically different strains. These properties are also characteristic of the Bunyaviridae (see Chapter 4).

The complex biology of flaviviruses is illustrated by Japanese encephalitis virus, which spreads from rats to mosquitoes to pigs to humans during the summer and persists in lizards over the winter. No virus is detectable in the lizards during their hibernation at 0°C, but virus reappears in the animals' tissues when the body temperature rises to summer levels. Yellow fever is, of course, the classic case of a mosquito-borne virus disease. Its epidemiology was studied by Walter Reed and his colleagues around the turn of the century (see Chapter 12). Some flaviviruses are transmitted by ticks rather than mosquitoes and cause fatal diseases in sheep in the United Kingdom and in monkey populations in India. Only rarely can some members of this family spread from animal to animal without requiring an insect vector.

3.2.1 Structure

Only 37–50 nm in diameter, flavivirus particles are smaller than those of togaviruses. The flavivirus particles consist of a nucleocapsid core containing a single molecule of genomic RNA that is complexed with a single 13 kDa C (capsid) polypeptide and the capsid is surrounded by a lipid bilayer that contains two virus-encoded proteins. The smaller of these two proteins (the 7 to 8 kDa M (membrane) protein) is synthesized as a larger glycosylated precursor molecule. The larger E protein ranges in size from 51 to 60 kDa and determines the serological specificity of the virus. Many (but not all) flavivirus E proteins are also glycosylated.

The genomic RNAs of flaviviruses contain a 5′ cap but lack the 3′-terminal poly(A) characteristic of togaviruses. As expected for a positive-sense RNA virus, the naked viral RNA is **infectious;** that is, when protein-free RNA is introduced into cells, progeny virions are produced. The complete 10,832 nucleotide sequence of yellow fever virus RNA has been determined, and Figure 3.4 illustrates the organization and expression of the viral genome.

3.2.2 Entry, Transcription, and Virus Production

A single, uninterrupted ORF covers most of the yellow fever viral genome. Translation initiates 118 nucleotides from the 5′ terminus and terminates at three closely spaced termination codons approximately 500 nucleotides upstream from the 3′ terminus. As in the case

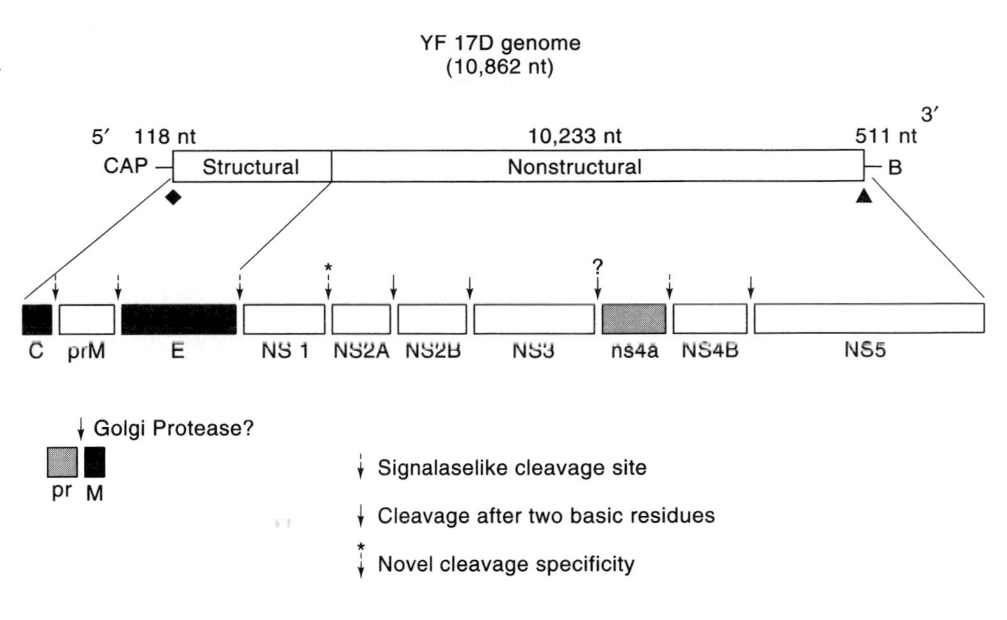

FIGURE 3.4

Organization and expression of the yellow fever 17D genome. In the top line, untranslated regions at the 5′ and 3′ termini are indicated as thin lines; translated regions are indicated as open boxes. A solid diamond represents the AUG codon where translation is initiated; the solid triangle is the UGA termination codon. In the bottom line, the solid boxes indicate structural proteins, open boxes are nonstructural proteins, and the hatched box indicates a protein that has not yet been isolated. (From S. Schlesinger and M. J. Schlesinger (1990) In *Virology,* 2d ed., ed. B. N. Fields et al. Vol. 1, 697–711. Raven Press, New York)

of the picornavirus genome, genes encoding the structural proteins are clustered at the 5′ end of the genome. Release of viral proteins by a combination of co- and posttranslational proteolytic processing appears to involve both viral and cellular proteases. There is no evidence for subgenomic viral RNAs.

Functions of the various nonstructural (NS) proteins of yellow fever virus are unclear, but amino acid sequences within NS5 are similar to those present in the RNA-dependent RNA polymerases of toga- and picornavirus. Several regions of the flavivirus genome are conserved in sequence or structure, including about 90 nucleotides at the 3′ terminus that appear to be organized into a stable **hairpin loop structure.** No extensive sequence conservation is present at the 5′ terminus, however. Flaviviruses mature on intracellular membranes rather than the plasma membrane; and the virions bud from the membranes of the ER and Golgi apparatus.

SECTION 3.2 FURTHER READING

Original Papers

Reed, W., et al. (1900) The etiology of yellow fever. A preliminary note. *Philadelphia Med. J.* 6:790.

Rice, C. M., et al. (1985) Nucleotide sequence of yellow fever virus: Implications for flavivirus gene expression and evolution. *Science* 229:726.

Chu, P. W. G., and E. G. Westaway. (1987) Characterization of Runjin virus RNA-dependent RNA polymerase: Reinitiation of synthesis *in vitro. Virology* 157:330.

Martell, M., et al. (1992) Hepatitis C virus (HCV) circulates as a population of different but closely related genomes: Quasispecies nature of HCV genome distribution. *J. Virol.* 66:3225.

Mandl, C. W., et al. (1993) Complete genomic sequence of powassan virus: Evaluation of genetic elements in tick-borne versus mosquito-borne flaviviruses. *Virology* 194:173.

Pryor, M. J., and P. J. Wright. (1993) The effects of site-directed mutagenesis on the dimerization and secretion of the NS1 protein specified by dengue virus. *Virology* 194:769.

Books and Reviews

Schlesinger, S., and M. J. Schlesinger. (1990) Replication of togaviridae and flaviviridae. In *Virology,* 2d ed., ed. B. N. Fields et al. Vol. 1, 697–711. Raven Press, New York.

Monath, T. P. (1990) Flaviviruses. In *Virology,* 2d ed., ed. B. N. Fields et al. Vol. 1, 763–814. Raven Press, New York.

3.3 CORONAVIRIDAE

Coronaviruses have been isolated from many mammals (including humans). The prototype of the virus family, a group of large, enveloped, positive-sense RNA viruses, is **avian infectious bronchitis virus.** Other important members of the Coronaviridae include mouse hepatitis and feline infectious peritonitis viruses. Viruses infecting poultry, cattle, and swine are responsible for substantial losses in food production.

Coronaviruses were first identified by their appearance in electron micrographs as enveloped particles 80 to 150 nm in diameter characterized by widely spaced, very large (12 to 24 nm) and club-shaped external spikes or **peplomers** (Fig. 3.5). The term *corona* refers to this crownlike appearance. These viruses are quite labile and easily degraded. They also appear to differ from many other virus families in that they bud from the ER and Golgi apparatus rather than the plasma cell membrane. As you recall, flaviviruses are also produced by this process (see Sec. 3.2).

FIGURE 3.5

Structure of coronaviruses. (a) Electron micrograph of a coronavirus. (Courtesy of R. C. Williams) (b) Model of coronavirus structure. The viral nucleocapsid is a long, flexible helix containing the 27 to 30 kb genomic RNA plus many molecules of a phosphorylated N (nucleocapsid) protein. The viral envelope includes a host-derived lipid bilayer and the viral glycoproteins E1 (20–30 kDa), E2 (180–200 kDa), and E3/HE (120–140 kDa). The large peplomers are composed of E2, which binds to cellular glycoprotein receptors and causes membrane fusion. Matrix glycoprotein E1 penetrates through the lipid bilayer and interacts with the nucleocapsid. Not found on all coronaviruses, E3 glycoprotein may form short peplomers. (From K. V. Holmes (1990) In *Virology,* 2d ed., ed. B. N. Fields et al. Vol. 1, 841–856. Raven Press, New York)

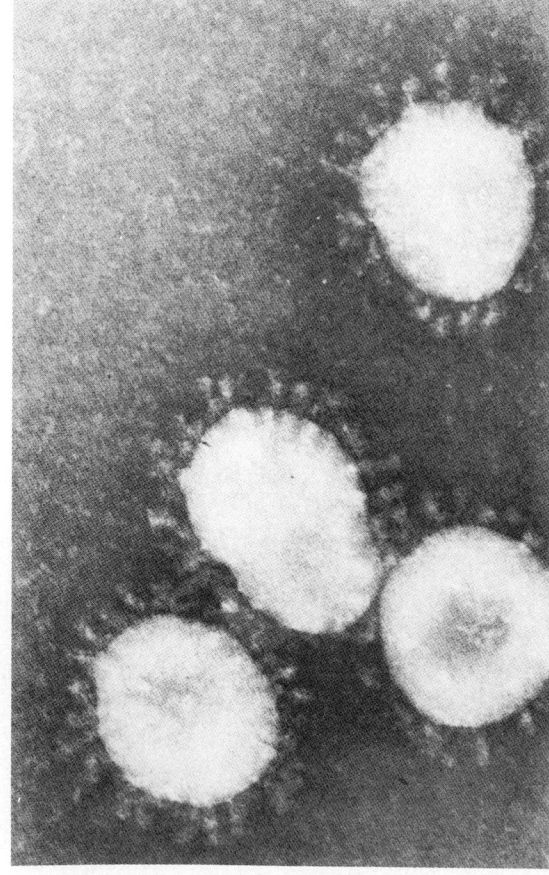

(b)

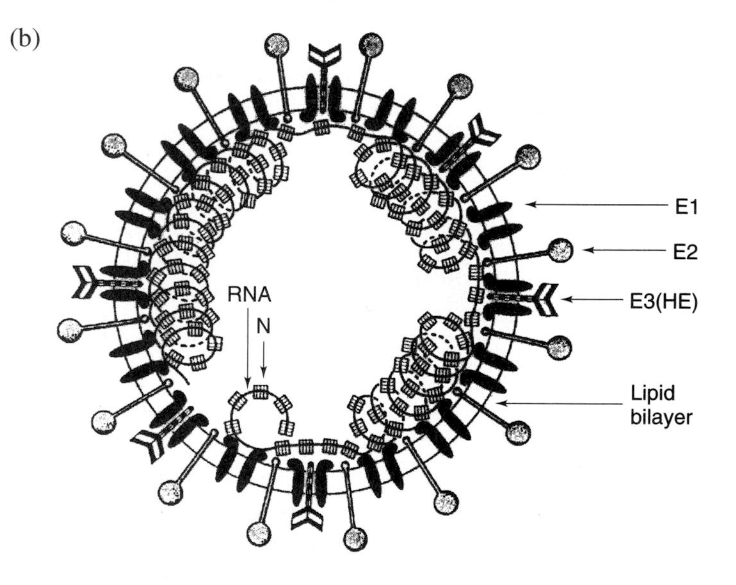

Depending on the virus strain and host cell type, coronaviruses may cause either cytocidal or **persistent infections** of cultured cells. Most coronaviruses show marked tissue tropism and will grow only in cells of their natural host species. Some require differentiated cells for complete virus replication. In immunologically competent hosts, coronavirus infections are usually acute and self-limited and do not lead to virus persistence. Human coronaviruses cause acute respiratory disease and, possibly, acute gastroenteritis. Mouse hepatitis virus is responsible not only for liver injury but also for serious immunological disorders, diarrhea, and wasting.

Feline infectious peritonitis (FIP) provides a particularly instructive example of the disease problems associated with coronavirus infections. In the past, FIP was believed to be a sporadic virus disease with a mortality rate of nearly 100 percent. There was no known cure or effective treatment for FIP, and no preventive vaccines were available. More recent epidemiological studies, however, suggest that 25 percent or more of all cats in the United States are infected with feline infectious peritonitis virus and that this virus may be involved in a variety of previously undiagnosed reproductive and neonatal disorders. Approximately one-quarter of all cats naturally exposed to the virus develop mild upper respiratory disease within 2 weeks after exposure, and most appear to recover uneventfully. The majority of these animals, however, will remain persistently infected, and only a very small number (≤5 percent) will eventually develop one of two fatal "secondary" disease syndromes. The classic symptoms of secondary disease—chronic weight loss, depression, fever, and lethargy—are believed to be immune-mediated.

3.3.1 Structure

Coronaviruses have the largest genome of any known RNA virus—a single infectious positive-sense RNA molecule containing 27,000 to 30,000 nucleotides. The 3′ terminus of the genomic RNA is polyadenylated, and its 5′ terminus is capped. Many molecules of a basic phosphoprotein (N, 50–60 kDa) encapsidate the genomic RNA to form a long, flexible nucleocapsid with **helical symmetry.** The envelopes of all coronaviruses are lipid bilayers that contain two viral glycoproteins, E1 (20–30 kDa) and E2 (180–200 kDa). A third envelope glycoprotein (E3/HE), having hemagglutinin-esterase activity, may also be present. Sequence similarities between E3 and the HA_1 subunit of influenza C (see Sec. 4.3) are probably the results of recombination between the genomic RNA of an ancestral coronavirus and the mRNA for influenza virus C glycoprotein.

3.3.2 Entry, Transcription, and Virus Production

A generalized scheme for coronavirus replication is presented in Figure 3.6. Coronaviruses attach to receptors on the membranes of target cells, and viruses that lack a hemagglutinin (glycoprotein E3) attach by means of E2. Replication occurs in the cytoplasm, and as for all positive-sense RNA viruses, the first event after penetration and uncoating is translation of the genomic RNA to yield an RNA-dependent RNA polymerase. For coronaviruses this stage of replication is not clearly understood. The gene at the 5′ end of the infectious bronchitis virus genomic RNA is very large (about 20 kb) and contains two overlapping ORFs. These ORFs encode 441 and 330 kDA polypeptides, and translation of the 42-nucleotide overlap region involves ribosomal frame-shifting. *In vitro* translation of the genomic RNA has been shown to produce several polypeptides larger than 200 kDA, but such an enormous polyprotein has not yet been detected.

Following synthesis of a full-length negative-sense RNA by the viral RNA-dependent RNA polymerase, this RNA acts as template for the synthesis of a variety of positive-sense RNAs. As shown in Figure 3.6, these positive-sense RNAs include a small (60 to 70

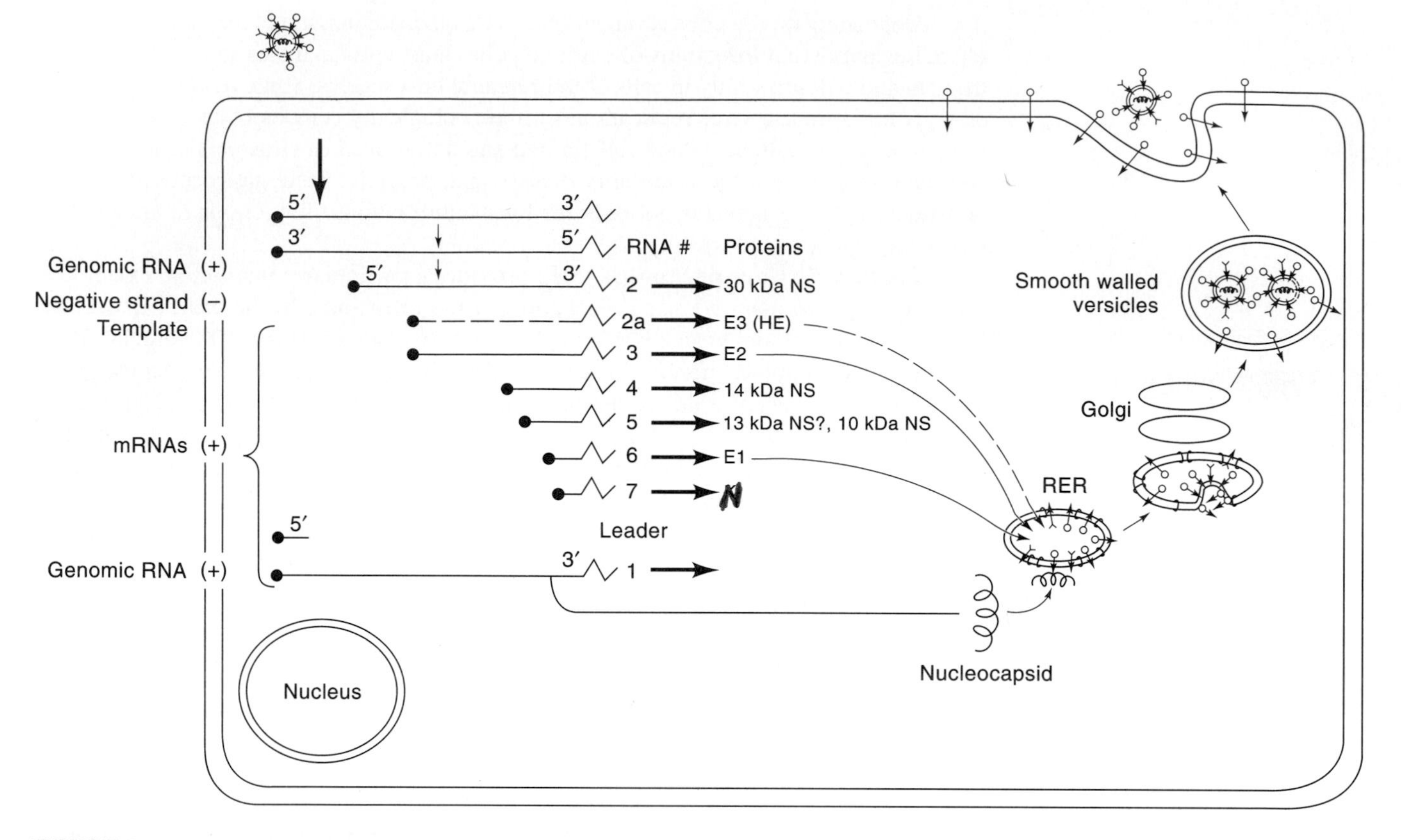

FIGURE 3.6

Replication of mouse hepatitis virus. The number of mRNAs as well as the sizes and location of nonstructural (NS) proteins vary for different coronaviruses. The genomic RNA acts as mRNA for synthesis of a large polyprotein that is processed and probably serves as an RNA-dependent RNA polymerase. This enzyme copies the genomic RNA to produce a full-length negative-sense template for the synthesis of (i) leader RNA, (ii) a nested set of overlapping, polyadenylated subgenomic mRNAs with common 3′ ends, and (iii) positive-sense genomic RNA progeny. All subgenomic mRNAs have, at their 5′ termini, the leader sequence encoded by the 3′ terminus of the negative-sense RNA. In general, each mRNA is translated to yield a single polypeptide encoded by its 5′ ORF. Translation of mRNA 7 into N protein occurs on free polysomes, and the subsequent assembly of nucleocapsids also occurs in the cytoplasm. E1–3 glycoproteins are translated on membrane-bound polysomes, and the proteins are modified during transport through the Golgi apparatus. Virions are formed by budding at membranes between the rough ER and Golgi apparatus, but *not* at the plasma membrane. (From K. V. Holmes (1990) In *Virology,* 2d ed., ed. B. N. Fields et al. Vol. 1, 841–856. Raven Press, New York)

nucleotide) **leader RNA** derived from the 3′ terminus of the negative-sense template (i.e., the 5′ terminus of the genomic RNA) as well as a set of seven "nested" mRNAs. Each of these mRNAs contain the leader sequence at its 5′ terminus and having identical 3′ termini. According to the currently accepted model for discontinuous, **leader-primed transcription** of coronavirus RNA, leader RNA becomes separated from its negative-sense template while remaining bound to the polymerase. The leader-polymerase complex may then reattach to the template farther downstream at one of the noncoding intergenic regions separating the various ORFs. Each intergenic region contains a short sequence that is complementary to a portion of the leader RNA and thus could serve to bind the polymerase-leader RNA complex. The ability of temperature-sensitive mutant viruses to synthesize leader RNA, but not mRNA, at nonpermissive temperatures indicates that leader RNA synthesis is indeed a process distinct from the transcription of mRNA.

The leader-primed replication and transcription of coronavirus RNA result in frequencies of mutation and **RNA recombination** that are unusually high for RNA viruses with a monopartite genome. When cell cultures were infected with a mixture of two different strains of mouse hepatitis virus containing leaders with different sequences as well as different temperature-sensitive mutations, virus could be recovered even when the cells were grown at a temperature that was nonpermissive for both mutant parents. RNA fingerprinting of its genomic RNA showed that the virus was a true recombinant. Its genome contained the leader and a large portion of the sequences from one parent and the 3′ end about 20 percent (3000 nucleotides) from the other parental virus. The various mRNA species synthesized by this recombinant virus contained the leader of the parent donating the 5′ end of the genome but the coding sequences of the other parental virus. This finding indicated that the leader of one parent was fused to the mRNA of the other during transcription of the recombinant virus. Discontinuous RNA transcription also favors the production of DI particles (see Sec. 4.1.4).

3.4 TOROVIRIDAE

In 1972 a virus was isolated during routine laboratory work on a horse that had died of a severe diarrhea in Berne, Switzerland. Serologic studies showed that it was unrelated to any known virus but that similar viruses were apparent in Swiss cattle and horse populations. Related viruses have since been isolated from animals in other countries in Europe and in the United States, and particles with the same unique morphology have also been observed in human materials. In 1987, these viruses were placed into a new family, the Toroviridae. This name derives from the shape of the nucleocapsid, which appears to be a hollow, tubular helix (diameter 23 nm) bent into an open torus (see Fig. 3.7). The nucleocapsid is surrounded by a tightly fitting membrane covered with projecting **peplomers.**

Studies of the genome organization and replication strategy of **Berne virus** (BEV) have revealed striking parallels with coronaviruses. The BEV genome consists of a single 25–30 kb positive-sense RNA, and virus-infected cells contain four 3′-coterminal mRNAs transcribed from the 3′ end of the genome. The corona- and toroviral genomes also exhibit the same basic gene order: 5′–polymerase–spike protein–membrane protein–nucleocapsid protein–3′. Although the 18.3 kDa N (nucleocapsid) protein of BEV seems to have little in common with its much larger (45–50 kDa) coronaviral counterpart, their membrane and peplomer proteins share many similar features.

Nucleotide sequence comparisons suggest that toroviruses and coronaviruses are related by divergence of their polymerase and envelope proteins from a common ancestor. These comparisons have also produced evidence for two independent nonhomologous RNA recombination events during BEV evolution. One of these recombinational events involved the hemagglutinin-esterase protein that is also found in certain coronavirus and influenza virus C.

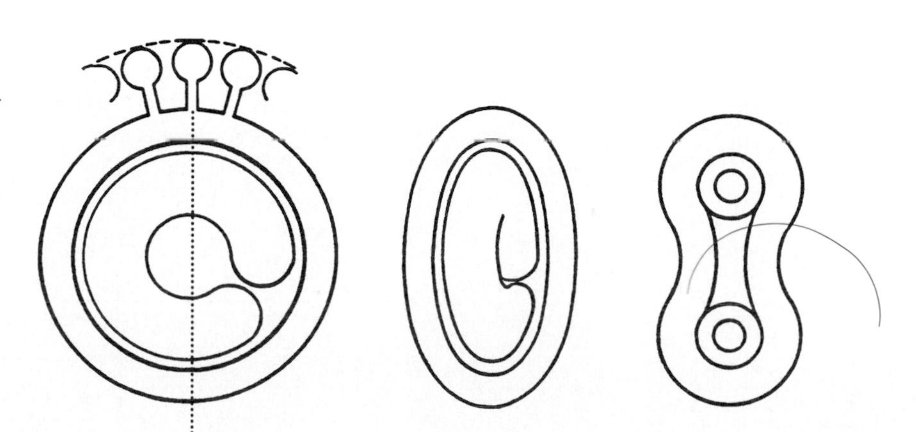

FIGURE 3.7

Architecture of the torovirion as compiled from electron micrographs of negatively stained and thin-sectioned particles. (Courtesy of M. C. Horzinek)

SECTIONS 3.3 AND 3.4 FURTHER READING

Original Papers

MAKINO, S., et al. (1986) High-frequency RNA recombination of murine coronaviruses. *J. Virol.* 57:729.

MAKINO, S., et al. (1986) Leader sequences of murine coronavirus mRNAs can be freely reassorted: Evidence for the role of free leader RNA in transcription. *Proc. Natl. Acad. Sci. USA* 83:4204.

BOURSNELL, M. E., et al. (1987) Completion of the sequence of the genome of the coronavirus avian infectious bronchitis virus. *J. Gen. Virol.* 68:57.

SNIJDER, E. J., et al. (1991) Comparison of the genome organization of toro- and coronaviruses: Evidence for two nonhomologous RNA recombination events during Berne virus evolution. *Virology* 180:448.

Books and Reviews

HORINEK, M. C., et al. (1987) A new family of vertebrate viruses: Toroviridae. *Intervirology* 27:17.

HOLMES, K. V. (1990) Coronaviridae and their replication. In *Virology,* 2d ed., ed. B. N. Fields et al. Vol. 1, 841–856. Raven Press, New York.

HRUBY, D. E., and C. A. FRANKE. (1993) Viral acylproteins: Greasing the wheels of assembly. *Trends in Microbiol.* 1:20.

chapter

4

Viruses with Negative or Ambisense Single-Stranded RNA Genomes

The negative-sense RNA viruses are agents that must transcribe their RNA genome into a positive-sense strand (i.e., mRNA) and then complete the infection cycle after producing the proteins required to yield infectious viruses. Most of the negative-sense RNA viruses use a similar process that involves an RNA replicase, but in many cases the exact protein responsible for this activity has not been clearly delineated. Instead, a replicase complex has been identified within the cell that gives rise to the positive-sense RNA.

4.1 RHABDOVIRIDAE

The Rhabdoviridae are a well-defined family of enveloped viruses numbering probably over 100 with a wide host range of vertebrates (including fish), arthropods (mostly insects but also crabs), amoebas, and plants. These viruses are primarily identified by their elongated (from the Greek *rhabdos,* "rod"), usually **bullet-shaped** appearance of 70 × 180 nm (Figs. 1.10 and 4.1) in size. They have lipid-rich envelopes, as indicated by their biochemical composition of 20 percent lipid, 13 percent carbohydrate, 3 percent RNA, and the remainder protein. Table 4.1 lists a few characteristic examples of this family, including their morphological, biochemical, and pathogenic properties. The viruses listed are not serologically interrelated. Two genera have been defined to classify members of the Rhabdoviridae infecting mammals: *Vesiculovirus,* a term derived

FIGURE 4.1

Electron micrograph of rhabdovirus particles (e.g. VSV). Most particles, permeated by the stain, reveal a helical internal structure. All virions show envelope and spikes clearly.

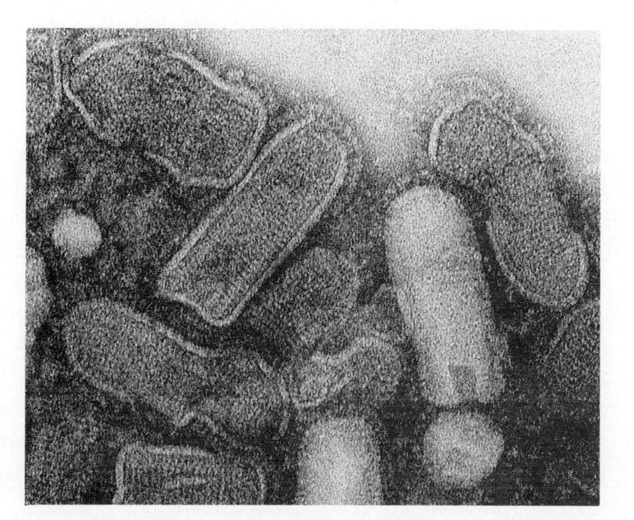

TABLE 4.1
Properties of a few typical Rhabdoviridae[a] of animals and plants

Name of virus	*Size (nm), mass, and shape*	*Hosts*	*Comments*
Vesicular stomatitis	170×70, 200×10^6 Da, bullet-shaped	Cattle, insects	Mild disease, cytocidal
Rabies	170×70, 200×10^6 Da, bullet-shaped	Vertebrates	Lethal, cytocidal
Infectious hematopoietic necrosis virus (IHNV)	170×70, $\sim 100 \times 10^6$ Da, bullet-shaped	Salmonid fish	Lethal, cytocidal
Lettuce necrotic yellows	230×70, 950 S, bacilliform	Sow thistle, lettuce aphids	Generally mild
Potato yellow dwarf	380×75, 1100×10^6 Da, bacilliform	Dicotyledons, leaf-hoppers	Generally lethal

[a]All Rhabdoviridae agglutinate goose red cells. Most of them (possibly all but rabies) are transmitted by insects.

from vesicular stomatitis virus (VSV), and *Lyssavirus,* composed of the rabies and rabies-like viruses. The Lyssaviruses have members whose infection cycles are limited to vertebrate species.

Most of the rhabdoviruses are transmitted by arthropods and are also replicated in these vectors. The diseases caused by the Rhabdoviridae range from mild to lethal. **Rabies** has a long history, recorded in documents from ancient times to the work of Louis Pasteur and the present development of a vaccine. Its unique mode of transmission to human beings through bites, frequently from domesticated animals, and its striking symptomatology and lethality makes it well known to virologists and physicians. A number of viruses closely related to rabies, such as the Makola virus, have been found in Africa, where this virus family may have originated. Additional rhabdovirus strains causing diseases of cattle, such as bovine ephemeral fever, continue to be found as virology focuses more intently on agricultural problems. VSV infection of humans can give rise to a variety of illnesses, ranging from a mild febrile reaction to headache, malaise, myalgia, pharyngitis, vomiting, and diarrhea. Some patients develop herpes-like vesicular lesions in the mouth and other areas on the face. Most clinical observations, however, suggest a high prevalence of asymptomatic infections.

4.1.1 Structure

The Rhabdoviridae are the first viruses to be discussed in this book in which the virion contains the RNA strand the sequence of which is complementary in its sequence to the mRNA carrying the genetic information of the virus. In biochemical terms, the rhabdoviruses are the simplest of these so-called negative-sense viruses. The **helical nucleocapsid** or ribonucleoprotein (RNP) core that gives them their characteristic shape is made up of (an N nucleocapsid) protein of about 50 kDa, plus the L (large) and NS (initially considered nonstructural) proteins. The rhabdovirus envelope contains an external glycoprotein (G) and internal matrix (M) proteins. The latter protein lines the inner surface of the viral membrane. Most effective antibody responses are directed against the G protein. It gives rise to type- and group-specific reactivity. In many species, control of rhabdovirus infection can be achieved by immunization with the G protein alone.

The RNP is covered, like that of all negative-sense RNA viruses, by a lipid-rich envelope with spikelike projections (Fig. 4.2). Also, as with all of these viruses, the rhabdovirions contain RNA-directed RNA polymerase activity that is capable of transcribing and replicating their RNA. The single-stranded RNA isolated from these, as from all viruses discussed in this chapter, is *not infectious,* since it lacks the enzyme that can replicate it and it cannot act as messenger until it has been replicated or at least transcribed to a positive strand. However, the nucleocapsid containing the RNA retains the proteins needed for RNA replication and is, therefore, infectious, though inefficiently so.

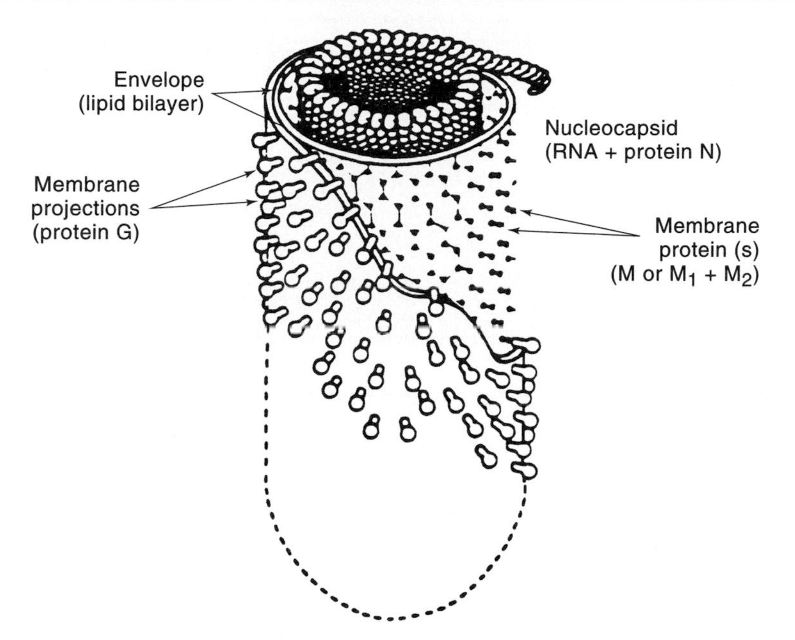

FIGURE 4.2

Model of a rhabdovirus particle cut through the middle to expose the internal structures. Bar = 10 nm. (After R. I. B. Francki and J. W. Randles. In *Rhabdoviruses,* ed. D. H. L. Bishop (1980) Vol. 3, 135 CRC Press,Boca Raton, Fla.)

An additional structural feature must be noted that appears to be shared by several quite unrelated virus families. **Inverted complementarity** of about 20 bases exists at the two ends of the rhabdoviral RNA:

$$\begin{array}{l}\text{—UUUUUUUGUCUUCGU}^{3'}\\ \text{—AAAAAAACAGAAGCAppp}^{5'}\end{array}$$

This terminal sequence is almost identical in the two serologically unrelated strains of VSV that were studied, called New Jersey and Indiana. However, in the latter strain the complementarity is slightly less perfect than that shown above for the New Jersey strain. This feature of **terminal-inverted complementarity,** which we will also encounter with many other viral nucleic acids, favors the formation of panhandlelike structures (Secs. 4.1.4, 7.1.1, Fig. 7.4).

4.1.2 Entry and Transcription

After attachment of the viral G protein to a receptor on a cell surface, the rhabdovirus penetrates by a temperature and low-pH-dependent process. Entry most likely involves **endocytosis** into coated pits and vesicles with subsequent fusion of the viral envelope and penetration of the genome into the cytoplasm. At this time the specific receptor is unknown but is probably a common constituent of a cell membrane, because the cellular host range of rhabdoviruses is so diverse.

In the most studied virus of this family, VSV, the single molecule of RNA of 4×10^6 Da (42 S) contains 11,162 nucleotides and is transcribed into five mRNAs coding for five proteins. These proteins are the glycoprotein (G, 65 kDa), the matrix protein (M, 26 kDa); the nucleocapsid protein (N, 50 kDa), L, the largest (160–190 kDa), and NS (40 kDa) recently called P. This letter is used because P is a structural protein that is highly *phosphorylated* by the host cell kinases and becomes complexed with L to form a functional enzyme. The M protein, also a phosphoprotein, is located beneath the envelope, which in turn consists of G and host cell plasma membrane lipids. M and N are close enough to one another to become cross-linked by small divalent chemicals (Figs. 4.2 and 4.3).

In contrast with free VSV RNA, the viral core, obtained by stripping off the envelope, including the G and M proteins, is, as noted above, slightly infectious. This finding indicates that the **transcriptase activity** is associated with one or all of the remaining internal proteins. Evidence now indicates that L and NS along with the capsid

FIGURE 4.3

PAGE pattern of the five proteins of a rhabdovirus (VSV). (R. R. Wagner. (1975) In *Comprehensive virology,* ed. H. Fraenkel-Conrat and R. R. Wagner. Vol. 4. Plenum Press, New York)

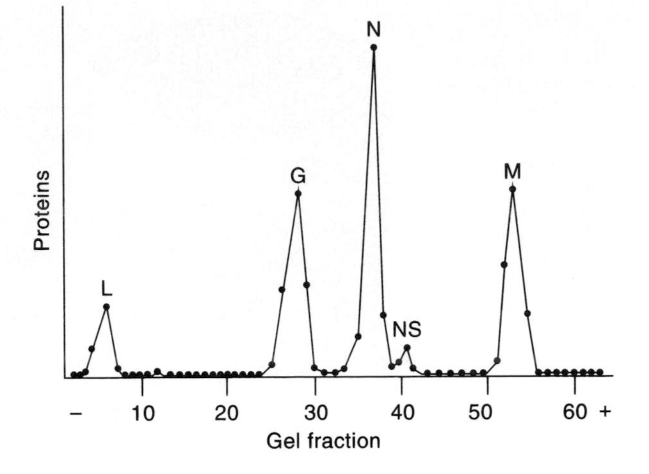

protein N are in a functional complex. However, the interrelationship among L, NS, and the nucleocapsid that accounts for the RNA polymerase activity of the core of the virion is not yet clearly defined.

Recent molecular cloning studies on the mRNA of another rhabdovirus, infectious hematopoietic necrosis virus (IHNV) of fish, revealed the presence of a sixth mRNA species. This mRNA encodes a non-structural virion rhabdovirus protein (about 12 kDa), designated NV, which is induced in infected cells but is not present in mature virions. R-loop mapping with cDNAs cloned into plasmids determined that the gene order on the IHNV genome is (3′) N-M1-M2-G-NV-L (5′), where two matrix proteins, M1 and M2 (about 30 kDa each), replace the NS and M proteins of VSV. The NV protein appears to be the first example reported of a non-structural viral protein discovered in a rhabdovirus, but its function is unclear. Whereas non-structural viral proteins of positive-sense RNA viruses (e.g., picornaviruses and togaviruses) are generally thought to be involved in transcription, IHNV, like other rhabdoviruses (see below), carries transcriptase in the virion, which can synthesize mRNA *in vitro* when purified virions are lysed with detergent. NV may be involved with viral replication infected cells or it could influence host cell processes. An additional small gene product (7 kDa) has also been detected in VSV.

Except for the NV protein, IHNV closely resembles VSV and rabies virus with respect to proteins, mRNA species, genome size, and genome organization. It is tantalizing that classic genetics with VSV mutants indicate the possible existence of six complementation groups, presumably genes, rather than five. However, cloning and sequencing studies on the entire genome of VSV have shown no indication of a sixth gene, and no additional mRNA or protein species have been reported in any work with other rhabdoviruses. Future studies on other piscine rhabdoviruses, such as spring viremia of carp virus, may reveal whether this extra gene is a common feature of fish rhabdoviruses.

Concerning the process of transcription in the viral core, the important question arises, Does it start separately for each of the five mRNAs, or does it proceed only from one end of the viral RNA and at first make one long positive-sense RNA? This question was approached by means of UV inactivation of the template RNA and other experiments. The UV experiments established the gene order of the negative-sense strand to be as shown in Figure 4.4. Transcription starts at a single site, requiring the 48-nucleotide leader sequence. The polymerase, starting at the 3′ end, reads stop-start signals and directly produces the five mRNA molecules that become capped and are polyadenylated. They differ, however, in the location of methyl groups near the cap. Transcription can be achieved by the purified nucleocapsid *in vitro* as well as by the viral core from

FIGURE 4.4

Schematic presentation of the rhabdovirus genome, its nontranscribed leader, and five consecutively transcribed mRNAs (black blocks). The letters refer to the proteins, and the numbers to the lengths of the mRNAs in terms of nucleotides. The gaps are only a few nucleotides long (two to four), except for the 5′-terminal nontranscribed sequence. (Modified from S. U. Emerson. (1985) In *Virology,* ed. B. N. Fields. 1119. Raven Press, New York)

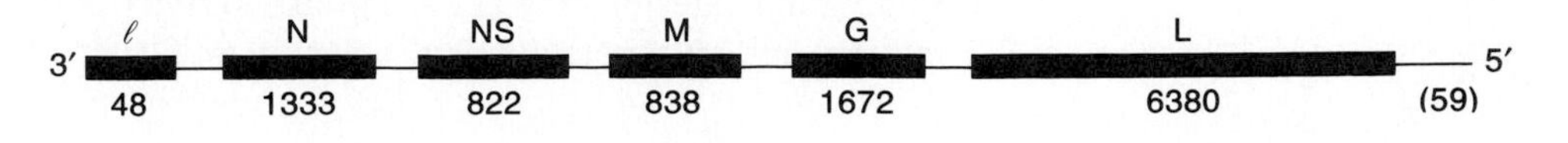

FIGURE 4.5

Steps in rhabdovirus infection:

1. Virus attachment to phosphatidyl serine receptor.
2. Virus penetrates the cell in an endosome.
3. Virus fuses with endosomal membrane and core enters the cytoplasm.
4. Uncoating of nucleocapsid occurs.
5. The viral negative-sense RNA is transcribed into positive-sense RNA.
6. Positive-sense RNA serves as a template for synthesis of the viral genome as well as the mRNA that gives rise to viral proteins. This type of replicative process occurs with many viruses that form positive RNA molecules.
7. The negative-sense RNA becomes incorporated into nucleocapsids (N).
8. These NC subsequently join the matrix protein (M) at the basal surface of the cell.
9. Budding of virus from the cell surface takes place. (Figure by J. Leong)

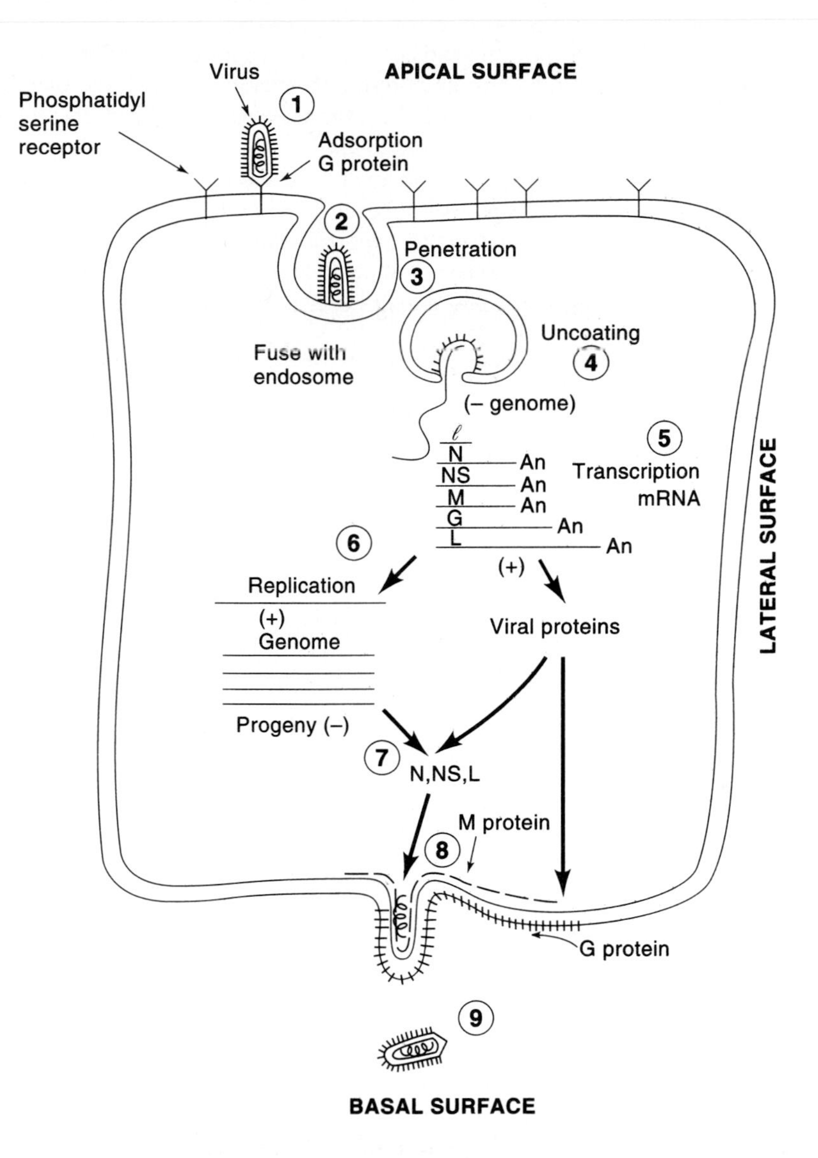

which the envelope has been removed by ether or detergent treatment. The replication of RNA is presumably achieved by the same viral enzyme that forms a positive-sense strand intermediate and the viral negative-sense RNA strand, but probably in particles that are nascent viral cores (Fig. 4.5). This remains the least well understood process in the virion's reproduction. Also, the control mechanism that leads some strands to become progeny viral RNA, and others to act as templates for the production of mRNAs, remains unknown.

Work however, with an *in vitro* system, involving defective interfering nucleocapsid templates and an mRNA-dependent reticulocyte lysate to support protein synthesis, has clarified the requirement for protein synthesis in rhabdovirus RNA replication. The results demonstrate that the major nucleocapsid protein (N) can by itself fulfill this requirement in RNA replication and allow complete replication, that is initiation and elongation, as well as encapsidation of genome-length progeny RNA.

For the purpose of becoming translated, the mRNA for G, the glycoprotein, associates with ribosomes, which are bound to membranes and the Golgi apparatus, all parts of the endoplasmic reticulum. This protein is made, like the Sindbis virus envelope proteins (see Sec. 3.1), by growing and becoming released from the ribosomes. The release occurs into, or through, the lipid bilayer of the endoplasmic reticulum or the microsomes to which these ribosomes are attached. The protein acquires its **oligosaccharides** in these membranes and ends up (like the nascent Sindbis envelope proteins) with most of the protein having its N-terminus sticking through the membrane into the lumen of the microsomes or vesicles. As these proteins fuse with the cell membrane, this part of the protein molecule becomes the protruding spikes. First the spikes appear on the cell surface where the virion is being assembled and later on the virion surface as it buds out of the cell membrane. By this time each spike molecule has acquired two or three branched oligosaccharide chains on its major N-terminal segment, while hydrophobic parts of the peptide chain pass through the bilayer. The C-terminal 30 amino acids are oriented toward the matrix protein, close enough so that these peptides can also be chemically cross-linked.

4.1.3 Assembly of the Virion

Assembly of the virion proceeds principally in the same manner as with all enveloped RNA viruses. As previously stated, certain sites of the cell's plasma membrane get fused with internal rough membrane material that carries viral glycoprotein molecules, oriented as discussed above. That area of the lipid bilayer membrane becomes a site from which most host proteins disappear by an as yet unknown mechanism. The M protein becomes internally attached, which in turn appears to attract **nucleocapsid particles** associated with L and NS protein molecules. These three proteins have by that time been synthesized by cytoplasmic ribosomes. The helical nucleocapsid or RNP then passes through the lipid bilayer, acquiring as a coat the M protein covered by the lipid bilayer with its associated G protein spikes. The membrane's sealing-off of the nucleocapsid releases the bullet-shaped virion into the intercellular medium or into **intracellular vacuoles** (see Fig. 4.5). Finally, budding from the cell surface occurs. It must be noted that small amounts of specific host plasma membrane proteins become incorporated in the viral envelope. This phenomenon has also been demonstrated with other viruses (e.g., retrovirus). As noted previously, the envelope carries small amounts of certain host cell proteins (e.g., MHC molecules).

4.1.4 Defective Interfering Particles

Most virus infections lead to the production of some defective (and usually noninfectious) particles (see Sec. 2.2.5). Packaged within the usual virus structural proteins, the genome of these **defective interfering** (DI) particles lack essential portions(s) of the wild-type genome and are unable to replicate in the absence of the homologous wild-type (or "helper") virus. In doubly infected cells, their replication occurs at the expense of the helper virus.

The first physical evidence for DI particles was provided by studies of influenza virus replication conducted by **von Magnus** more than 40 years ago. The usual method for growing influenza virus in embryonated eggs (i.e., to dilute the stock virus 1000-fold before inoculation) results in virus progeny that are highly infectious. Von Magnus showed that if undiluted stock virus was used instead as inoculum, the resulting progeny were both less infectious and more heterogeneous when examined by ultracentrifugation. This mixture of wild-type virus and DI particles showed one especially striking property—an ability to interfere with the replication of conventional high-titer, dilute-passage virus.

As first pointed out by Huang and Baltimore, most preparations, both of RNA and DNA animal viruses contain (DI) particles. Most commonly associated with such negative-sense RNA viruses as VSV and Sendai viruses DI particles and DI RNAs can also be detected in preparations of positive-sense RNA viruses—picornaviruses (see Sec. 2.2.5) as well as the plant carmo- and tombusviruses (see Sec. 9.2). This section focuses on the replication-related events responsible for generation of DI RNAs and only briefly mentions the possible role of DI RNAs in modulating viral disease *in vivo.* This topic is discussed in Chapter 13 in the context of potential antiviral strategies.

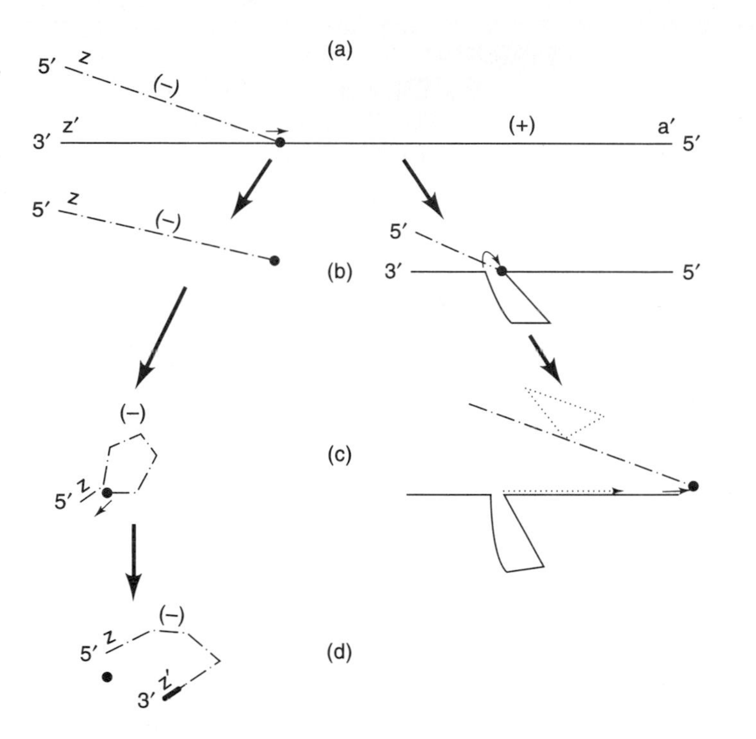

FIGURE 4.6

Generation of DI RNAs by a "copy choice" or "jumping polymerase" mechanism. (a) The viral replicase falls off or slides along its genomic RNA template during negative-sense strand synthesis. (b) Still carrying the nascent negative-sense strand, the replicase reinitiates synthesis—either on the original template or near the 5′ terminus of the nascent negative-sense strand. (c and d) Completion of the elongation reaction yields a shortened negative-sense strand containing either an internal deletion (right) or complementary termini (left). Slipping backward rather than jumping forward generates sequence repeats rather than deletions, whereas more complex rearrangements require repeated template switches. (From J. J. Holland. (1990) In *Virology*, 2d ed., eds. B. N. Fields and D. M. Knipe. Chapter 8. Raven Press, New York)

Structure and Origin Sequencing studies have shown that formation of most DI RNAs involves a series of **internal deletions** and **rearrangements** of sequences present in the helper genome. For negative-sense RNA viruses, four different types of DI RNAs have been described: (1) the internal deletion type, in which both the 3′- and 5′-viral termini are retained, but one or more internal segments are deleted; (2) the **stem** or **panhandle** type, in which the viral 5′ terminus is retained, but the 3′ terminus is lost and replaced by the sequences complementary to those at the 5′ terminus; (3) the **snapback** or **hairpin** type, which have termini similar to those of **panhandle-type** molecules but have lost all 3′-terminal noncomplementary sequences; and (4) the **mosaic** type, in which the termini may be of any type, and the internal sequences show evidence of complex and often repeated rearrangements. As shown in Figure 4.6, a "copy choice" or "jumping polymerase" mechanism similar to that believed responsible for RNA recombination (see Chapter 12) can account for the generation of all four types of DI RNAs.

Occasionally, the viral replicase may jump between two or more different templates before completing synthesis of a DI RNA. For example, the 3′ portion of the virulent satellite RNA C of turnip crinkle virus (TCV) is homologous to its helper virus, whereas its 5′ domain is homologous to the corresponding portions of two smaller TCV satellite RNAs (i.e., RNAs D and F; see Sec. 9.2 and Fig. 9.3). An even more dramatic example of **template switching** is provided by certain DI RNAs associated with Sindbis virus, where the sequence at the 5′ terminus is nearly identical with that of rat $tRNA^{asp}$. Evidently, RNA viruses are able to transduce genetic information encoded by the DNA genome of their hosts at an appreciable frequency (see also the discussion of oncogenes in Chapter 10).

Conclusive evidence has recently been presented for the *de novo* generation of several DI RNAs from the genomes of their helper viruses. DI RNAs appeared spontaneously in a substantial proportion of plants inoculated with RNA transcribed *in vitro* from cloned full-length turnip crinkle (TCV) or cucumber necrosis (CNV) virus cDNAs. The structure of one such DI RNA (i.e., TCV DI1 RNA) is shown in Figure 9.3. Mutations that abolish expression of the CNV 20 kDa nonstructural protein led to rapid *de novo* generation of DI RNAs without the need for serial passage. Whether this effect is caused by an increased frequency of transcriptional errors or the preferential accumulation of DI RNA arising at low levels during normal replication has not been determined.

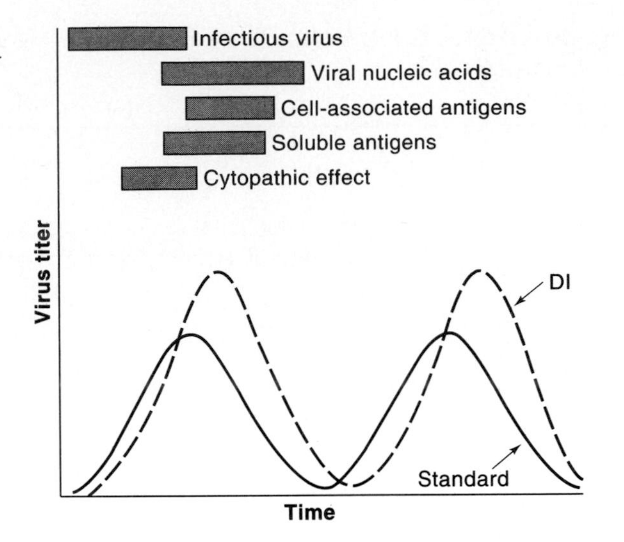

FIGURE 4.7

Cyclic fluctuations in the relative titers of helper virus and DI particles. (From A. S. Huang. (1988). In *RNA genetics,* ed. E. Domingo, J. J. Holland, P. Ahlquist. Vol. 3, 195–208. CRC Press, Boca Raton, Fla.)

Biological Effects DI particles were discovered on the basis of their ability to interfere with the replication of the parental (and now, "helper") virus. A likely mechanism for this interference is the ability of DI particles to outcompete their helper genomes for virus-encoded gene products required for replication or encapsidation. In addition to being shorter than the helper genome, many DI RNAs (especially those derived from negative-sense RNA viruses) have sustained other modifications that impair their ability to be transcribed into mRNA and thus favor their replication. Furthermore, DI transcription and translation products themselves may also be involved in interference, sometimes competing with wild-type virus for cell surface receptors.

Modulation of viral disease by DI particles is an extremely complex topic, especially if this phenomenon is considered at the whole animal or plant level. Fortunately, studies conducted in tissue culture are considerably easier to interpret, and Fig. 4.7 illustrates an important concept arising from such experiments. Here, serial undiluted passage of VSV preparations in Chinese hamster ovary cells leads to an alternating production and inhibition of helper virus, followed by a similar pattern for DI particles. Under such conditions, the recovery of *infectious* virus no longer shows its usual temporal relationship with the appearance of viral nucleic acids and proteins.

SECTION 4.1 FURTHER READING

Original Papers

Baltimore, D., A. S. Huang, and M. Stampfer. (1970) Ribonucleic acid synthesis of vesicular stomatitis virus. II. An RNA polymerase in the virion. *Proc. Natl. Acad. Sci. USA* 66:572.

Ball, L. A., and C. N. White. (1976) Order of transcription of genes of vesicular stomatitis virus. *Proc. Natl. Acad. Sci. USA* 73:442.

Katz, F. N., et al. (1977) Membrane assembly *in vitro*: Synthesis, glycosylation, and asymmetric insertion of a transmembrane protein. *Proc. Natl. Acad. Sci. USA* 74:3278.

Hagen, F. S., and A. S. Huang. (1981) Comparison of ribonucleotide sequences from the genome of VSV and two of its defective interfering particles. *J. Virol.* 37:363.

Emerson, S. U. (1982) Reconstitution studies detect a single polymerase entry site on the vesicular stomatitis virus genome. *Cell* 31:635.

Patton, J. T., et al. (1984) N protein alone satisfies the requirement for protein synthesis during RNA replication of vesicular stomatitis virus. *J. Virol.* 49:303.

Herman, R. C. (1986) Internal initiation of translation on the VSV phosphoprotein mRNA yields a second protein. *J. Virol.* 58:797.

Xu, L., et al. (1991) Epitope mapping and characterization of the infectious hematopoietic necrosis virus glycoprotein, using fusion proteins synthesized in *Escherichia coli*. *J. Virol.* 65:1611.

BANERJEE, M. E., and S. BARIK. (1992) Gene expression of vesicular stomatitis virus genome RNA. *Virology* 188:417.

HIRAMATSU, K., et al. (1992) Mapping of the antigenic determinants recognized by monoclonal antibodies against the M2 protein of rabies virus. *Virology* 187:472.

BOURHY, H., B. KISSI, and N. TORDO. (1993) Molecular diversity of the *Lyssavirus* genus. *Virology* 194:70.

SPIROPOULOU, C. F., and S. T. NICHOL. (1993) A small highly basic protein is encoded in overlapping frame within the P gene of vesicular stomatitis virus. *J. Virol.* 67:3103.

Books and Reviews

HUANG, A. S. (1973) Defective interfering viruses, *Ann. Rev. Microbiol.* 27:101.

NAYAK, D. P., et al. (1985) Defective-interfering (DI) RNAs of influenza viruses: origin, structure expression and interference. *Current Topics Microbiol.* Immunol. 114:103.

BANERJEE, A. K. (1987) Transcription and replication of rhabdovirus. *Microbiol. Rev.* 51:67.

WAGNER, R. R., ed. (1987) *The rhabdoviruses.* Plenum Press, New York.

BAER, G. M., et al. (1990) Rhabdoviruses. In *Virology,* 2d ed., eds. B. N. Fields and D. M. Knipe. 883–930 Raven Press, New York.

WAGNER, R. R. (1990) Rhabdoviridae and their replication. In *Virology,* 2d ed., eds. B. N. Fields and D. M. Knipe. 867–881 Raven Press, New York.

4.2 PARAMYXOVIRIDAE

Members of the Paramyxoviridae family are responsible for a number of serious diseases of humans and animals (mumps, measles, parainfluenza, pneumonia, Newcastle disease, distemper, rinderpest). Newcastle disease virus (NDV) is found in commercial avian flocks throughout the world. While some **avirulent** strains cause only mild or asymptomatic infections, other **virulent** strains are highly lethal to animals especially those in which they are able to replicate, and thus destroy, the host tissues needed for immunological response. Respiratory syncytial virus can cause life-threatening pneumonia, particularly in children. Many other paramyxoviruses are associated primarily with respiratory tract infections in humans and animals, although the canine distemper virus obviously has a tropism for neurological tissues.

The paramyxoviruses like rhabdoviruses have nonsegmented genomes and have been classified into three genera: *Paramyxovirus, Morbillivirus,* and *Pneumovirus.* In contrast to the rhabdoviruses, the paramyxoviruses have quite a narrow host range. Other distinctions include the fact that paramyxoviruses **agglutinate** mammalian and avian red cells and have **neuraminidase** activity; the morbilliviruses hemagglutinate but lack neuraminidase activity, and the pneumoviruses have neither of these activities. In addition, the pneumoviruses have a narrow nucleocapsid. The term **myxo** (Greek, *myxa:* "mucous") identifies the specific affinities of orthomyxoviruses and paramyxoviruses for mucopolysaccharides and glycoproteins—particularly **sialic acid**–containing receptors on the cell surfaces.

4.2.1 Structure

The virion's shape is quite irregular, ranging from approximate spheres to filaments, in diameters ranging from 150 to 300 nm; **pleomorphism** is the technical term for this variability. The large particles may contain no RNA molecules and are thus noninfectious. The lipoprotein envelope of paramyxoviruses is covered with short spikes, also called **peplomers** (see Fig. 4.8).

What is particularly characteristic about the paramyxoviruses, and differentiates them from the orthomyxoviruses and other viruses, is the nature of the RNA and the structure of their nucleocapsid. The nucleocapsid is a **helix,** 18 nm in diameter. It protects the RNA against ribonuclease attack. In this respect the nucleocapsid closely resembles the TMV rod, but is obviously very flexible and long (1000 nm) (Fig. 4.8). By comparing this length, with that of TMV (300 nm), the molecular weight of the RNA genome, can be es-

FIGURE 4.8

Electron micrograph of a paramyxovirus. The Newcastle disease virus particle is broken open and reveals the helical structure of the nucleocapsid. The spikes are clearly visible on the surface of the particles. Other particles are more or less permeated by the stain.

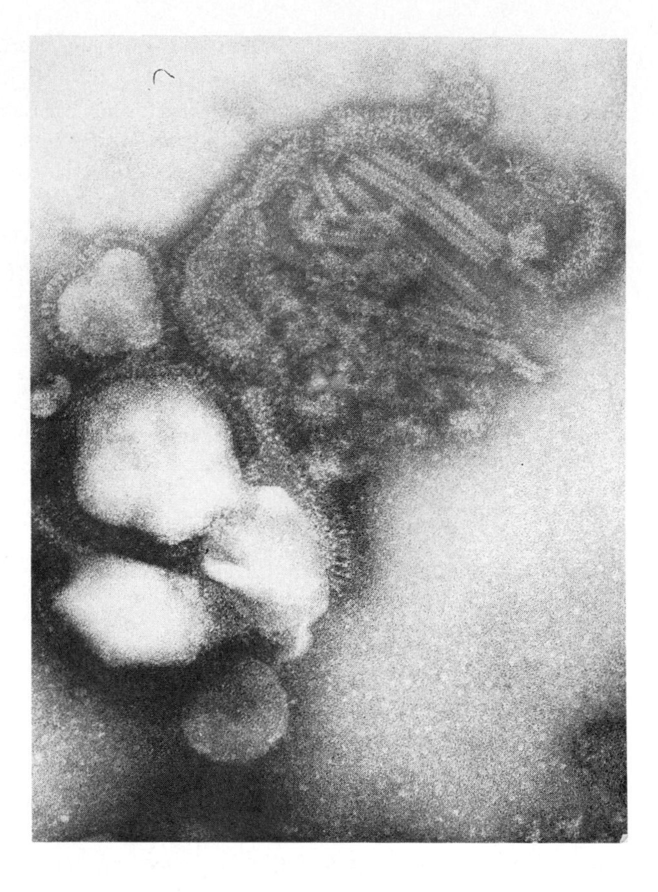

timated to be between 5 and 7×10^6 Da. This dimension is in agreement with the sedimentation rate of the isolated RNA (56 S), and with sequence data.

Seven virus-specific proteins have been detected, of which five play characteristic roles in the virion (Fig. 4.9). The largest, L, (200 kDa) is believed to represent the major component of the RNA transcriptase complex that is carried in the virion. In this, as in several other regards, the Paramyxoviridae are similar to, although slightly more complex than, the Rhabdoviridae. The enzyme activity is essential for virus replication, because this RNA is also a negative-sense strand: The enzyme activity always requires the core of the virion that comprises the nucleocapsid protein (NP, 60 kDa) as well as the membrane protein (M, 34 kDa) located in the inner surface of the virus envelope. The other associated proteins are P (polymerase-associated) (60 kDa) and C (22 kDa), which is encoded within the P gene but uses a different reading frame.

FIGURE 4.9

Schematic diagram of the probable arrangement of the structural components of a paramyxovirus. (Courtesy of R. Compans and CRC Press, Boca Raton, Fla.)

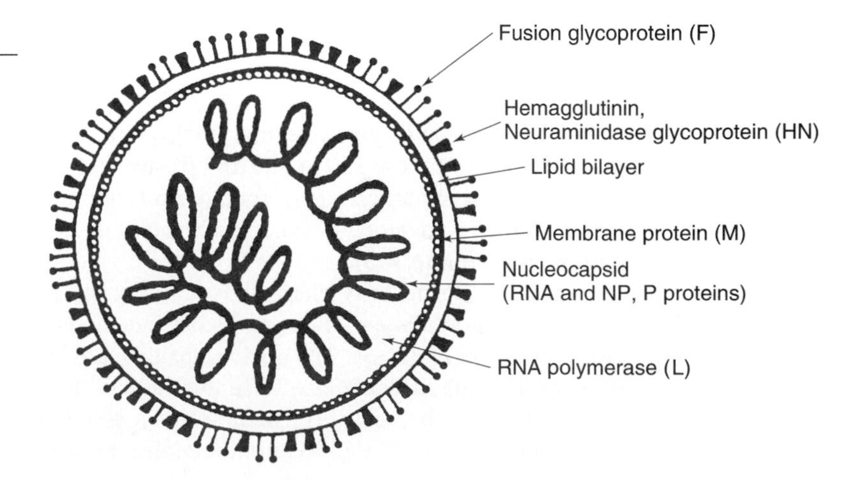

The lipid-rich envelope consists of two glycoproteins (HN and F, of about 80 and 65 kDa, respectively) (Fig. 4.9). The larger of these glycoproteins (HN) was found to form spikes that carry both hemagglutinating and, in most paramyxoviruses, also neuraminidase activity. This enzyme is able to split off the terminal sialic acid residues from the oligosaccharide chains on the host cell membrane. The envelope glycoproteins are needed for attachment to the cell surface receptor. The next step involves the fusion and delivery of the nucleocapsid into the cytoplasm. The F glycoprotein seems to be responsible for the biological activity characteristic of paramyxoviruses: **fusion**. It is involved in fusion of the virus lipoprotein membrane with the surface membrane of the host cell. It also causes fusion of cells and hemolysis. The F proteins are made in an inactive form and then activated by a **cleavage process** mediated by cellular trypsinlike enzymes. Moreover, unlike with influenza virus and the rhabdoviruses, fusion of the paramyxoviruses with the cell surface can occur at a neutral pH. In this regard, the lentivirus subfamily of retroviruses (e.g., HIV) follows a similar penetration pathway. Cleavage of the envelope by a cellular protease may be involved, and entry is pH-independent (Chapter 14).

This **cell-fusion activity** has proved to be a powerful tool in cell biology (see Sec. 11.2). UV-inactivated Sendai virus (a parainfluenza virus) can be used for this purpose because it is not pathogenic in humans. It appears that this fusion activity requires proteolytic cleavage of the smaller glycoprotein (F → F1, F2) and that viral infectivity also depends on this proteolytic cleavage mediated by cellular enzymes. The ability of these viruses to fuse cells is actually an expression of their infection mechanism, which relies on fusion of the viral envelope with the cell membrane. This activity is, however, dependent on the presence of the neuraminidase. The infection process of other enveloped viruses, particularly the orthomyxoviruses (discussed in the next section), and including several retroviruses is probably not fundamentally different, although fusion activity is most evident with the Paramyxoviridae (see Sec. 11.1).

Viruses within the pneumovirus genus (prototype, the **respiratory syncytial virus**) appear to be more complex than NDV and the parainfluenza viruses. A total of 13 virus-specific polypeptides have been identified in infected cells. Ten of the 13 proteins were shown to be unique and encoded by separate mRNAs. They are termed the L, G, F (F1, F2), N, P, M and the 24 kDa, 14 kDa, 11 kDa, and 9.5 kDa proteins. The seven largest proteins, L, G, F (F1, F2), N, P, M, and 24 kDa were clearly identified as virion structural proteins. The G (glycoprotein) protein is the counterpart to the HN or H (hemagglutination only) proteins of other Paramyxoviridae but lacks **hemagglutinating** and **neuraminidase** activities. The 24 kDa protein was characterized by detergent and salt dissociation studies as a fourth envelope-associated protein. Of the three smallest proteins, the 14 kDa and 11 kDa species were characterized as nonstructural proteins.

The amino acid sequence of a respiratory syncytial virus fusion protein (F) has been deduced from the sequence of a cDNA clone of its mRNA. The encoded protein of 574 amino acids is extremely hydrophobic and has a molecular weight of 63,371. The site of proteolytic cleavage within this protein was accurately mapped by determining a partial amino acid sequence of the N terminus of the purified larger subunit (F1). A sequence of Lys-Lys-Arg-Lys-Arg-Arg at the C terminus of the smaller N-terminal F2 subunit appears to represent the cleavage activation domain. Three extremely hydrophobic domains are present in the protein: (a) the N-terminal signal sequence, which is cleaved during insertion into the membrane; (b) the N terminus of the F1 subunit, which is analogous to the N terminus of the paramyxovirus F1 subunit and the HA2 subunit of influenza virus hemagglutinin; and (c) the putative membrane anchorage domain near the C terminus of F1. Thus, one subunit of the cleaved respiratory syncytial virus fusion protein and the parainfluenza and influenza hemagglutinins all seem to be anchored in their membranes by the same end, the amino terminus. In contrast the other respiratory syncytial virus fusion protein subunit may be anchored at the C terminus.

4.2.2 Entry and Virus Production

Following attachment and fusion with the cell membrane, the Paramyxoviridae replicate in a mode very similar to that of the Rhabdoviridae. The first step, transcription of the virion's RNA, occurs again within the core particle. In other words, no pure transcriptase has been isolated, and this process has so far been studied only with detergent-treated virions rather than at the molecular level. Quite possibly, as with rhabdoviruses, it is the structural interaction of the L and NP proteins that is required for enzyme action: a **transcription complex.** The transcripts are apparently six RNAs, capped, methylated, and polyadenylated. The largest mRNA, of 35 S, is translated to the L protein. The others, (about 18 S), some of which tend to aggregate to the larger (35 S) complex, code for the other proteins. However, the sufficiently efficient transcription of the small mRNAs requires the addition of cellular extracts, nonspecific in terms of source and unexplained in terms of their mode of action. From these studies it appears that the gene order (on the negative-sense strand) from 3′ to 5′ is l-NP-P(C)-M-F-HN-L and that the regulation of their translation frequency is determined by that order. The 50-nucleotide leader (l) sequence appears to be required for transcription of these viruses as with the Rhabdoviridae.

It is also noteworthy to note that this gene order resembles that of the rhabdoviruses, with the genes for the capsid protein at the 5′ terminus, the matrix or M protein in the middle, and the large enzymatic component at the 3′ end of the positive-sense strand. Yet clearly the Paramyxoviridae are more complex, since they code for at least seven rather than five or six genes and have a genome of 6×10^6 rather than 4×10^6 Da.

A previously unrecognized gene for a peculiar protein, SH (*s*mall *h*ydrophobic), has been identified on the virion RNA of the simian paramyxovirus SV5, between the genes for the fusion protein and the hemagglutinin-neuraminidase. An SH mRNA of 292 nucleotides plus poly (A) transcribed from the SH gene contains a single open reading frame that encodes a hydrophobic polypeptide of 44 amino acids with a molecular weight of 5012. This protein has been identified in SV5-infected cells, but its function is not yet known. Genes for similar proteins have yet to be identified in other paramyxovirus genomes.

Sequence data obtained on viral RNAs have further emphasized the similarities between rhabdovirus and paramyxovirus **transcription systems.** All the consensus intergenic and transcription initiation sequences of the genome of Sendai virus, a paramyxovirus, have been determined and compared with previously determined sequences of VSV a rhabdovirus. The Sendai virus intergenic sequence is similar to the analogous sequence of VSV, but there is no sequence homology between the mRNA start signals of the two viruses. Nevertheless, these mRNA start signals are organized in the same way, being 10 bases long and possessing two consensus regions divided by one (Sendai virus) or two (VSV) variable internal nucleotides. These findings extend the evidence that both families of negative-sense RNA viruses descended from a common ancestor and that an archetypal mechanism of transcriptional regulation has been conserved in their evolution.

Transcription and replication of respiratory syncytial viruses, which, as noted above, encode 10 different proteins, are not yet as well defined as they are for other paramyxoviruses. The mRNA sequences of two human respiratory syncytial viral nonstructural protein genes and of a gene for the 22 kDa protein have been obtained by cDNA cloning and DNA sequencing. The availability of a **bicistronic clone** confirmed the gene order for this portion of the genome. The translation of cloned viral sequences in the bicistronic and monocistronic clones produced two moderately hydrophobic proteins of 15.5 and 14.7 kDa, probably corresponding to the 14 kDa and 11 kDa proteins found in infected cells (see above). The 22 kDa (22,156 Da) proteins, containing 194 amino acids, corresponded to an unglycosylated 22 kDa protein seen in purified extracellular virus but not associated with detergent- and salt-resistant cores. A second open reading frame of 90 amino acids partially overlapping with the C terminus of the 22 kDa protein was also present within this sequence. The RSV matrix protein gene was previously shown to contain two overlapping reading frames. This mechanism is reminiscent of the overlapping P and C genes of NDV. Nevertheless, the finding of three additional viral transcripts encoding at least three identifiable proteins in human RSVs was a novel departure from the usual genetic organization of paramyxoviruses. Also, contrary to other unsegmented negative-sense RNA viruses, a 19-nucleotide intercistronic sequence was present between the NS1 and NS2 genes.

During virus infection, nucleocapsid formation occurs in the cytoplasm. The NC interacts with the M proteins lining the inner surface of the viral envelope glycoproteins located on the surface of cells. In this manner the virion particles subsequently bud from the cell membrane. Virus clumping appears to be prevented by the neuraminidase activity residing within the HN envelope protein.

Another feature of paramyxovirus infection can be "fusion from within." By this process, **trypsinlike proteases** on the cell surface can activate the F protein to cause fusion with nearby cell surfaces. Virus can then enter a new cell and escape possible interaction with antiviral antibodies. This intercell fusion activity leads to the large mononucleated cells **(syncytia)** characteristic of these and other fusion-inducing viruses (see Chapter 11).

SECTION 4.2 FURTHER READING

Original Papers

CHOPPIN, P. W., and W. STOECKENIUS. (1964) The morphology of SV5 virus. *Virology* 23:195.

DUESBERG, P. H., and W. S. ROBINSON. (1965) Isolation of the nucleic acid of Newcastle disease virus (NDV). *Proc. Natl. Acad. Sci. USA* 54:794.

SCHEID, A., and P. W. CHOPPIN. (1973) Isolation and purification of the envelope proteins of Newcastle disease virus. *J. Virol.* 11:263.

COLLINS, P. L., L. E. HIGHTOWER, and L. A. BALL. (1978) Transcriptional map of Newcastle disease virus mRNAs *in vitro*. *J. Virol.* 28:324.

GUPTA, K. C., and D. W. KINGSBURY. (1984) Complete sequences of the intergenic and mRNA start signals in the Sendai virus genome: Homologies with the genome of vesicular stomatitis virus. *Nucleic Acids Res.* 12:3829.

HIEBERT, S. W., R. G. PATERSON, and R. A. LAMB. (1985) Hemagglutinin-neuraminidase protein of the paramyxovirus simian virus 5: Nucleotide sequence of the mRNA predicts an N-terminal membrane anchor. *J. Virol.* 53:1.

ALKHATIB, G., and D. J. BRIEDIS. (1986) The predicted primary structure of the measles virus hemagglutinin. *Virology* 150:479.

PORTNER, A., R. A. SCROGGS, and C. W. NAEVE. (1987) The fusion glycoprotein of Sendai virus: Sequence analysis of an epitope involved in fusion and virus neutralization. *Virology* 157:556.

CROWLEY, J. C., et al. (1988) Sequence variability and function of measles virus 3′ and 5′ ends and intercistronic regions. *Virology* 164:498.

ELANGO, N., et al. (1988) Molecular cloning and characterization of six genes, determination of gene order and intergenic sequences and leader sequences of mumps virus. *J. Gen. Virol.* 69:2893.

MERSON, J. R., et al. (1988) Molecular cloning and sequence determination of the fusion protein gene of human parainfluenza virus type 1. *Virology* 167:97.

WAXHAM, M. N., et al. (1988) Sequence determination of the mumps virus HN gene. *Virology* 164:318.

ELANGO, N. (1989) The mumps virus nucleocapsid mRNA sequence and homology among the Paramyxoviridae proteins. *Virus Res.* 12:77.

TAKEUCHI, K., et al. (1991) Variations of nucleoside sequence and transcription of the SH gene among mumps strains. *Virology* 181:364.

COLMAN, P. M., P. A. HOYNE, and M. C. LAWRENCE. (1993) Sequence and structure alignment of paramyxovirus hemagglutinin-neuraminidase with influenza virus neuraminidase. *J. Virol.* 657:2972.

TANABAYASHI, K., et al. (1993) Identification of an amino acid that defines the fusogenicity of mumps virus. *J. Virol.* 67:2928.

Books and Reviews

VAINSCOPAA, R., et al. (1989) The paramyxoviruses: Aspects of molecular structure, pathogenesis, and immunity. *Adv. Virus. Res.* 37:211.

CHANOCK, R. M., and K. MCINTOSH, (1990) Parainfluenza viruses. In *Virology,* 2d ed., eds. B. N. Fields and D. M. Knipe. 963–988 Raven Press, New York.

NORRBY, E., and M. N. OXMAN. (1990) Measles virus. In *Virology,* 2d ed., eds. B. N. Fields and D. M. Knipe. 1013–1044 Raven Press, New York.

WOLINSKY, J. S., and M. N. WAXHAM. (1990) Mumps virus. In *Virology,* 2d ed., eds. B. N. Fields and D. M. Knipe. 989–1011 Raven Press, New York.

KINGSBURY, D. W., ed. (1991) The *Paramyxoviruses.* Plenum Press, New York.

4.3 ORTHOMYXOVIRIDAE

The Orthomyxoviridae family of animal RNA viruses, all called **influenza** or **flu viruses,** is small in number but large in unusual features, ranging from its epidemiology to its molecular biology and appearance. Type (equivalent to genus) A, the one responsible for most serious flu epidemics, one of which caused the death of more people in 1918–1919 than did World War I, is the most common. It infects humans and many species of mammals and birds. Type B and probably C infect only humans. They all show the same unique structural features of the group. Actually, type A has been studied much more than the others and you can assume from now on that we are referring to type A unless otherwise specified. How are the three types, A, B, and C, differentiated? By marked serological differences among their nucleocapsids that prove that these three types are not closely related. Within each type, particularly A, there exist many subtypes and strains, and these vary in their surface proteins; these differences are demonstrated most easily by serological tests. However, all members of one type show the same or very similar internal proteins, somewhat larger in type B than type A.

4.3.1 Distinct Characteristics of Influenza Virus

The flu virus is an enveloped RNA virus, meaning that it, like the Toga-, Rhabdo-, and Paramyxoviridae discussed previously, consists basically of an internal nucleocapsid and an envelope made up of an inner **matrix** protein, a **lipid bilayer,** and **external glycoproteins** (see Figs. 4.10 and 4.11). Yet here the similarity to other enveloped viruses ends, and the greater complexity of the flu virus becomes evident. We first list its main characteristic features and then discuss them in greater detail.

1. The **shape** of the virion is not uniform: besides more or less round particles of about 100 nm diameter (300×10^6 Da), there occur somewhat elongated, often bent, particles and also enormously long (at times several thousand nanometers) **filamentous particles** of the same diameter (Fig. 4.10).
2. The surface glycoprotein **spikes**, 10 to 12 nm long, each consisting of several peptide chains, represent two different shapes and chemical types and carry two distinct biological activities, **hemagglutination** and **neuraminidase activity.** As already stated, these surface proteins, particularly the hemagglutinin, determine the antigenic specificity of the subtypes.
3. The **RNA genome** of the flu virus of about 6×10^6 Da is divided. There are eight different molecules of RNA, each representing the gene for a protein, except that the two smallest RNAs carry two overlapping genes read in different phases and responsible in each case for two proteins. The multiplicity of RNA components in flu virions has only gradually become recognized, because of the tendency of the RNA to form an aggregate. As far as is known, no animal virus genes besides those of the Orthomyxoviridae and Reoviridae (Chapter 5) are physically segmented to such an extent (Fig. 4.12).

FIGURE 4.10

Electron micrograph of orthomyxovirus particles (flu virus, type A). Both typical (a) and filamentous particles (b) are shown. (From W. G. Laver. (1973) *Adv. Virus Res.* 18:65)

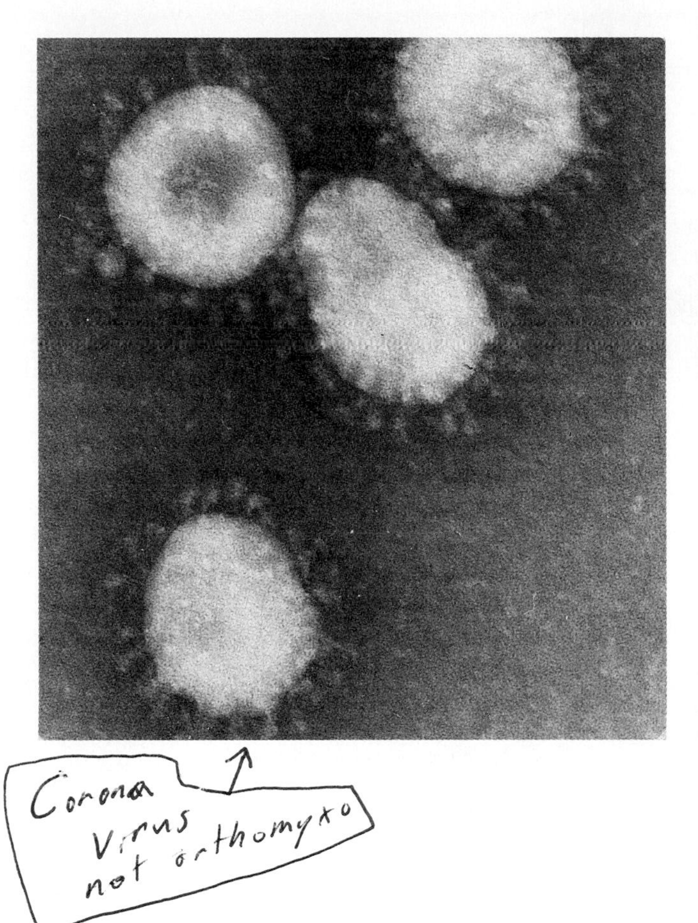

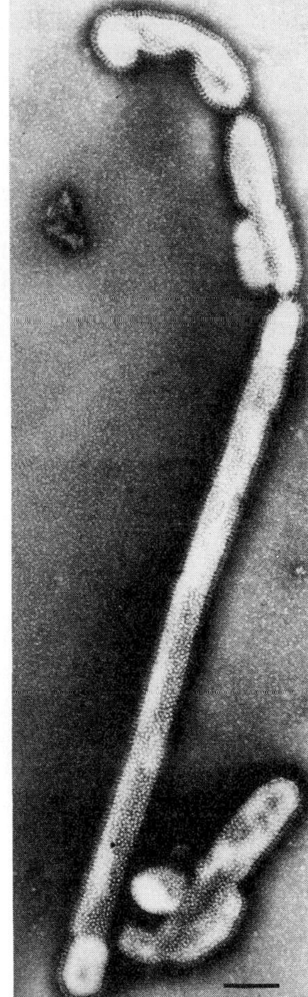

FIGURE 4.11

Schematic presentation of an influenza virus particle. (Modified from M. Girard and L. Hirth. (1980) *Virologic générale et moleculaire.* Doin Editeurs, Paris)

FIGURE 4.12

PAGE radioautogram of the influenza virus genome. The radioactive RNA segments of purified type A flu (PR8) and a new strain (HK = Hong Kong) are clearly evident and are identified by the names of the corresponding gene products. (Courtesy of P. Palese et al. (1977) *Virology* 76:114)

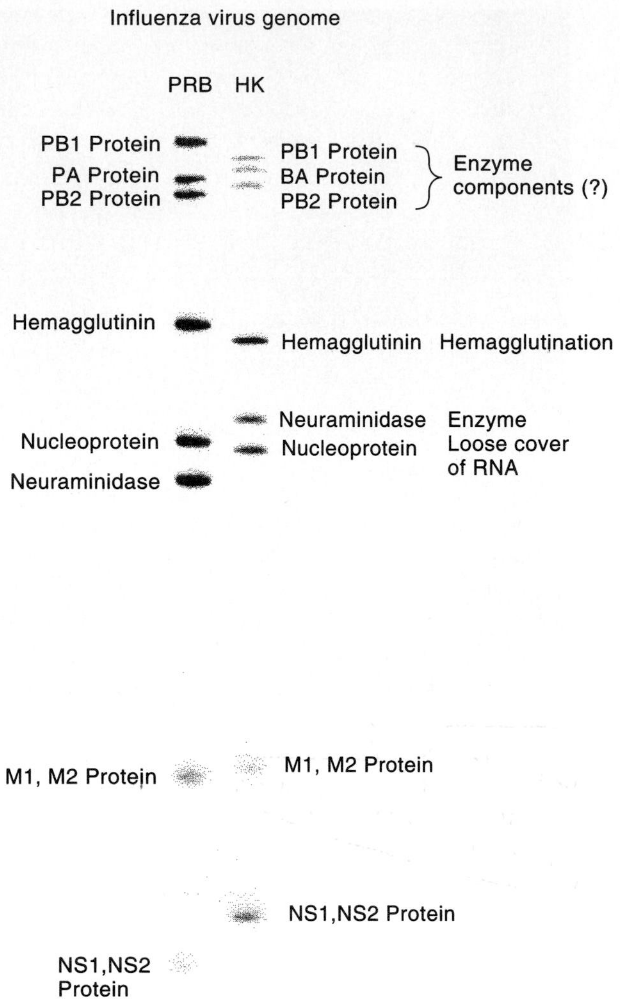

Influenza virus genome
PRB
HK
PB1 Protein
PA Protein
PB2 Protein
PB1 Protein
BA Protein
PB2 Protein
Enzyme components (?)
Hemagglutinin
Hemagglutinin
Hemagglutination
Neuraminidase
Enzyme
Nucleoprotein
Nucleoprotein
Loose cover of RNA
Neuraminidase
M1, M2 Protein
M1, M2 Protein
NS1,NS2 Protein
NS1,NS2 Protein

FIGURE 4.13

Antigenic drift strains of Hong Kong influenza virus illustrated by sequence data for the large subunit of the hemagglutinin observed in isolates from all over the world between 1968 and 1978. (Courtesy of G. W. Both and M. J. Sleigh)

Total base changes
Amino acid changes
Silent base changes
Number of changes in HA1
70
60
50
40
30
20
10
1968
1970
1972
1974
1976
1978

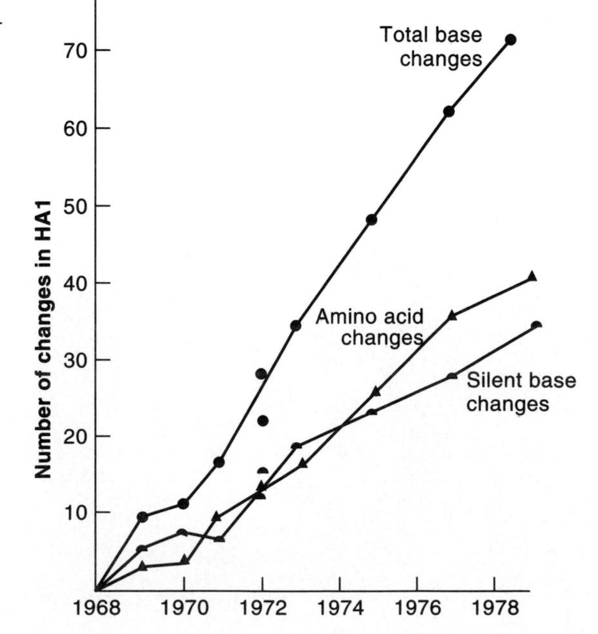

FIGURE 4.16

Electron micrograph of rose aggregates of neuraminidas molecules (5 × 8.5 nm) form upon removal of SDS. (Fron Laver. (1975) *Adv. Virus Res* 18:65)

4. As a consequence of the multiplicity of its genome components, the flu virus changes frequently and at times dramatically. **Antigenic shift** represents the abrupt appearance of new serotypes due to reassortment of RNA genome components in cells infected with two different strains. This phenomenon often leads to epidemics. **Antigenic drift** is the less marked result of spontaneous point mutations (Fig. 4.13).
5. The RNA molecules are weakly encapsidated as thin helical **filamentous** nucleocapsids, about 10 nm in diameter and of quite variable length, corresponding to the lengths of RNA they cover (average 50 to 60 nm). The RNA in the nucleocapsid, in contrast with that of paramyxoviruses, is sensitive to ribonuclease.
6. The transcription of the virion RNAs, particularly rich in U (35 percent), in contrast with that of most other RNA viruses, occurs in the nucleus.

4.3.2 The Envelope: Attachment and Entry

The first requirement in studying the properties of different parts of a virus is to find methods for separating them (see Appendix 1). Ether disrupts the flu virion, as it does all enveloped virions, by dissolving the lipid bilayer. Various gentle treatments with detergents, proteases (e.g., trypsin), or alkali release proteins from virions more selectively and often in biologically active forms. The main protein of the flu virus (M, 26 kDa) forms a shell around the nucleocapsid on which rests the envelope consisting of the lipid bilayer and the surface spikes. The neuraminidase and hemagglutinin activities of flu virus are located in these spikes.

The chemical and biological properties of the spikes (or **peplomers**) on the surface of flu virions have been studied longer and more intensely than those of most other animal virus problems. It was only gradually realized that two different protein microorganelles are responsible for the two well-known biological activities. We discuss first the chemical makeup of these spikes and then their biological activities.

The hemagglutinin (HA) is made up of rod-shaped protein molecules of 4 × 14 nm, dimensions that tend to form starlike aggregates, as seen in the electron microscope (Fig. 4.14). Such purified hemagglutinin preparations contain one of the large flu proteins, HA, a glycoprotein of about 75 kDa. Apparently this protein it is to a varying extent split by a specific peptide bond breakage into two chains (HA-1, HA-2) of roughly 50 and 25

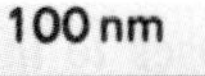

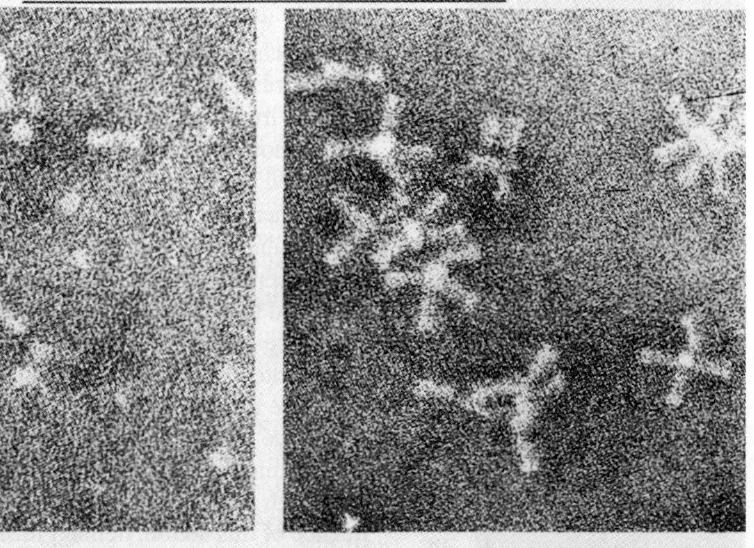

FIGURE 4.14

Electron micrograph of starlike clusters of hemagglutinin molecules (4 × 14 nm) forming upon removal of detergent. (From W. G. Laver. (1973) *Adv. Virus Res.* 18:65)

94

FIGURE 4.15

PAGE pattern of flu virus pro
The solid line indicates [^{14}C]-
labeled proteins, the dashed
[^{3}H]fucose-labeled
oligosaccharides. Protein NS
(25 kDa) comigrates with the
M protein. The hemagglutini
protein (HA) is partly split int
separate chains (HA-1, HA-2)
large proteins comigrating a
under those conditions can b
clearly separated into three
components (P1, P2, P3) by c
techniques. (P. W. Choppin a
W. Compans. (1975) In
Comprehensive virology, ed
Fraenkel-Conrat and R. R. W
Vol. 4. Plenum Press, New Y

4.3.3 Transcription and Virus Production

The orthomyxoviruses use the same kind of **transcription complex** (i.e., the nucleocapsid) described for the other negative-sense RNA viruses. In less than 20 minutes after infection, transcription of the viral RNA to mRNA molecules can be detected. By that time the virion has lost its envelope and the still intact nucleocapsid threads have entered the cell's nucleus. Recent evidence shows that cellular mRNAs are required and utilized as primers for the first step in viral replication, the transcription of its RNA. Actually, these host cell mRNAs are cannibalized by the flu virus, so that segments of 10 to 15 of their 5′-terminal nucleotides, including their cap, are transferred to the nascent flu mRNAs. Such splicing reactions, particularly of cap and leader sequences, have in recent years been found to occur frequently during the formation of eukaryotic host cell mRNAs, although usually involving only a single genome. It is the need for these nascent host cell mRNAs that orthomyxovirus replication depends on host cell **DNA–dependent RNA polymerase** (RNA polymerase II) function. Indications for such dependence were first obtained when flu replication was found to be inhibited by **actinomycin D** and **α-amanitin,** specific inhibitors of DNA-dependent, but not of RNA-dependent, polymerases. Flu transcription can be primed *in vitro* by the dinucleoside monophosphates ApG or GpG instead of the natural nuclear transcripts. This process is much less efficient and makes flu mRNAs that are 10 to 15 nucleotides shorter than the normal *in vivo* transcripts. They are also cap-less.

As already stated, this first transcription occurs within the virion's nucleoprotein core, which probably retains, besides the capsid protein (NP, 60 kDa), the three large polymerase-encoded proteins PB1, PA, and PB2 (about 90 kDa each). The three largest viral RNA segments encode these proteins. PB2 has an affinity for the mRNA cap, and the subsequent complex formation of the three proteins, and particularly PB1, performs the synthetic function. Although researchers generally use the term "enzyme" in discussing such transcription processes, they actually always have to use the particulate virus core complexes that remain intact upon gentle detergent treatment of the virus.

For 2 hours after infection, the production of viral mRNAs predominates in the cells, and translation products accumulate. These are, besides the structural proteins we have discussed in detail, the three large proteins (P on Fig. 4.15 = PA, PB1, PB2). These products, present in the virion, are made only in small amounts and are components of the RNA polymerase. There is also NS (25 kDa), a protein of unknown function, of which large quantities are found in the infected cell but not in the virion (Fig. 4.12). It remains unknown whether and how some of these proteins aggregate to make the transcriptase, which appears later to be independent of host cell mRNAs. Moreover, it is not clear whether the same enzyme acts ultimately (after 2 hours) to make the negative-sense RNAs for the progeny virus. As with the other negative-sense RNA viruses, control of RNA synthesis depends on the amount of nucleocapsid structure units.

The first products of this replication are complementary copies (mRNAs) of all of the influenza virus RNA segments. Subsequently, these RNA segments produce copies of the

TABLE 4.3

Relative amounts of RNA segments extracted from influenza virus of differing specific infectivity

*Specific infectivity (log$_{10}$)**	*RNA segment*						
	1 + 2	*3*	*4*	*5*	*6*	*7*	*8*
4.8	0.93	0.70	1.00	1.11	1.07	1.17	0.66
4.6	0.62	0.66	1.00	0.86	0.95	0.89	0.65
3.7	0.42	0.52	1.00	1.03	1.29	1.38	0.99
2.3	0.15	0.33	1.00	0.79	1.03	1.04	0.87
1.9	0.18	0.08	1.00	0.75	0.65	0.80	0.76
1.3	0.12	0.04	1.00	0.82	1.04	1.16	1.00
0.80	0	0	1.00	0.80	0.84	1.85	0.96

*Plaque-forming units hemaggutination activity

Source: W. M. Crumpton, et al. (1978) *Virology* 90:370.

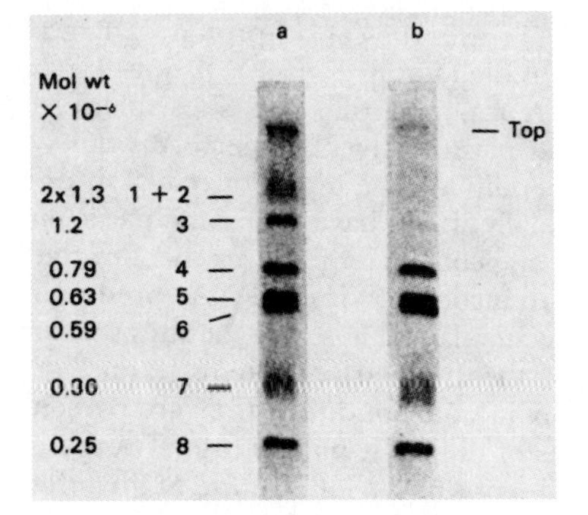

FIGURE 4.17

Comparison by PAGE radioautogram of genonomic RNAs extracted from standard (a) and defective (b) influenza virus stocks. Note the loss of RNA segments 1–3 from the defective virus stock. The molecular weights of the genome components are given. (Modified from N. J. Dimock et al. (1978) *Virology* 90:370)

genome segments (negative-sense). Apparently, some complementary mRNAs remain in the nucleus, where they can continue to produce genomic RNA segments. Other mRNAs and the negative-sense genomes rapidly move to the cytoplasm to complete progeny formation.

The events involved in infectious virus production, particularly the translational ones, proceed progressively more in the cytoplasm; the M protein accumulates in certain areas adjacent to the cell's plasma membrane at sites where the spike protein, the glycosylated hemagglutinin and neuraminidase, have accumulated. Host proteins become more or less excluded from these areas. The newly formed nucleocapsid complexes coated by the M protein then enter the membrane and acquire their lipid envelope traversed by the viral spike proteins. Subsequent budding from the cell surface occurs. These events probably take place by mechanisms similar to those recently elaborated for the Toga- and Rhabdoviridae. The selective incorporation of the negative-sense genomic segments into the nucleocapsid most likely depends on cellular proteins. A mechanism must also exist that places a complete set of RNA segments inside each virion particle.

4.3.4 Defective Interfering Particles

The phenomenon of the production of defective interfering (DI) virus particles was first observed with influenza virus. These **von Magnus particles,** which come to predominate upon repeated high-multiplicity passage, lack several, particularly the largest, of the typical flu RNA molecules (see Table 4.3 and Fig. 4.17). Their accumulation blocks replication of the normal influenza virions. Since the RNA polymerase function requires one or more of these large gene products, the reason for loss of infectivity is obvious. If the remaining smaller RNA components outcompete the large ones for the polymerase produced by the complete virions, the interfering aspect of the presence of these particles is also accounted for (see Sec. 4.1.4). Production of interferon can be a factor as well in the antiviral effects of DI (see Sec. 12.4).

SECTION 4.3 FURTHER READING

Original Papers

SMITH, W., et al. (1933) A virus obtained from influenza patients. *Lancet* i:66.

ANDREWES, C. H., et al. (1955) A short description of the myxovirus group (influenza and related viruses). *Virology* 1:176.

DUESBERG, P. H. (1969) Distinct subunits of the ribonucleoprotein of influenza virus. *J. Mol. Biol.* 42:485.

LAVER, W. G., and R. C. VALENTINE. (1969) Morphology of the isolated hemagglutinin and neuraminidase subunits of influenza virus. *Virology* 38:105.

DESSELBERGER, U. K., et al. (1978) Biochemical evidence that "new" influenza virus strains in nature may arise by recombination (reassortment). *Proc. Natl. Acad. Sci. USA* 75:3341.

INGLIS, S. C., et al. (1979) The smallest genome RNA segment of influenza virus contains two genes that may overlap. *Proc. Natl. Acad. Sci. USA* 76:3790.

HUANG, R. T. C., et al. (1980) The function of the neuraminidase in membrane fusion induced by myxoviruses. *Virology* 107:313.

MURPHY, B. R., et al. (1982) Reassortant virus derived from avian and human influenza virus A viruses is attenuated and immunogenic in monkeys. *Science* 218:1330.

VARGHESE, J. N., et al. (1983) Structure of the influenza virus glycoprotein antigen neuraminidase at 2.9 Å resolution. *Nature* 305:35.

AIR, G. M., et al. (1985) Gene and protein sequence of an influenza neuraminidase with hemagglutinin activity. *Virology* 145:117.

HERRLER, G., and H.-D. KLENK. (1987) The surface receptor is a major determinant in the cell tropism of influenza C virus. *Virology* 159:102.

PATTERSON, S., et al. (1988) The intracellular distribution of influenza virus matrix protein and nucleoprotein in infected cells and their relationship to haemagglutinin in the plasma membrane. *J. Gen. Virol.* 69:1859.

LAWSON, C. M., E. K. SUBBARAO, and B. R. MURPHY. (1992) Nucleotide sequence changes in the polymerase basic protein 2 gene of temperature-sensitive mutants of influenza A virus. *Virology* 191:506.

ROBERTS, P. C., W. GARTEN, and H. D. KLENK. (1993) Role of conserved glycosylation sites in maturation and transport of influenza A virus hemagglutinin. *J. Virol.* 67:3048.

Books and Reviews

WHO Memorandum. (1980) A revised system of nomenclature for influenza viruses. *Bull. WHO* 58:585.

COLRAN, P. M., and C. W. WARD. (1985) Structure and diversity of the influenza virus neuraminidase. *Current Topics Microbiol. Immunology* 114:177.

NAYAK, D. P., et al. (1985) Defective-interfering (DI) RNAs of influenza viruses: origin, structure, expression and interference. *Current Topics Microbiol. Immunology* 114:103.

WILEY, D. C., and J. J. SKEHEL. (1987) The structure and function of the hemagglutinin membrane glycoprotein of influenza virus. *Ann. Rev. Biochem.* 56:365.

KINGSBURY, D. W. (1990) Orthomyxoviridae and their replication. In *Virology,* 2d ed., eds. B. N. Fields and D. M. Knipe. 1075–1089 Raven Press, New York.

MURPHY, B. R., and R. G. WEBSTER. (1990) Orthomyxoviruses. In *Virology,* 2d ed., eds. B. N. Fields and D. M. Knipe. 1091–1152 Raven Press, New York.

4.4 FILOVIRIDAE

The single-stranded Marburg and Ebola RNA viruses were initially considered rhabdoviruses, but they have specific properties that warrant their placement in a separate family called *Filoviridae,* a term that describes their morphology (Fig. 4.18). They appear as long **filamentous forms** in a variety of shapes including U, 6-shaped, and circular. These viruses, with a diameter of 80 nm, have a helical nucleocapsid (50 nm) and are enveloped with surface projections. They contain an RNA of 4.2×10^6 Da and five major and two minor protein components, similar in size and presumably function to the rhabdovirus proteins. They can cause infections in monkeys, guinea pigs, hamsters, and mice. The replication of filoviruses resembles that of other negative-sense RNA viruses. By genome analysis they have some replication signals similar to those of the rhabdo- and paramyxoviruses. Infection of cells in culture leads to extensive cytopathic effects with **intracytoplasmic inclusion bodies** containing ribonucleocapsids. Since virion RNA is not readily detected in the cells, a fast replication process most likely occurs with negative-sense RNA being released rapidly within virions.

Marburg virus was first recognized in humans afflicted with severe and often fatal hemorrhagic fever in Marburg, Germany, in 1967. It infected researchers growing poliovirus in African green monkey kidney cells originally obtained from animals in Uganda. The recent demonstration that Marburg virus can infect cultured human endothial cells could explain in part its pathogenic nature. Ebola virus, first recognized in 1976,

(a)

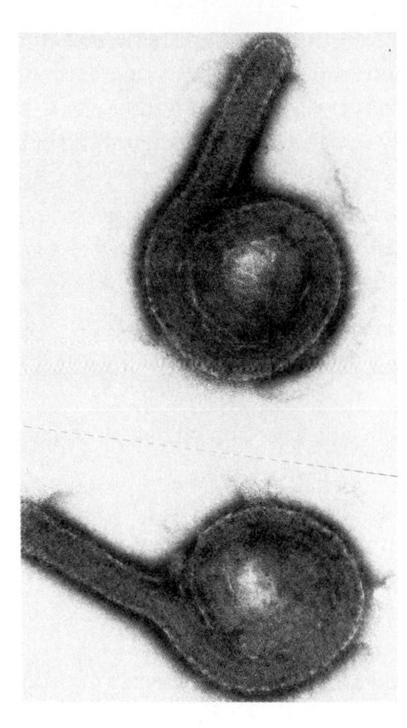

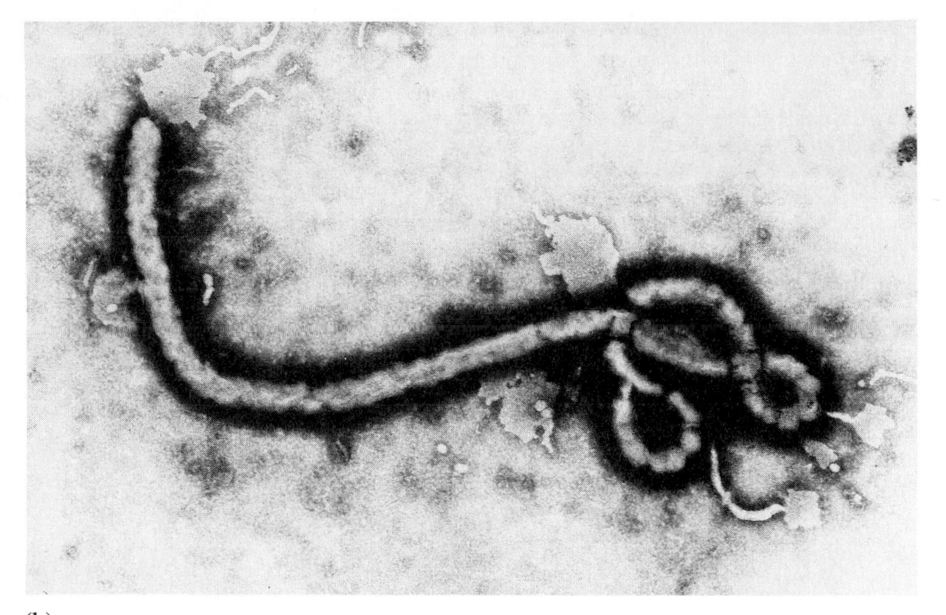

(b)

(c)

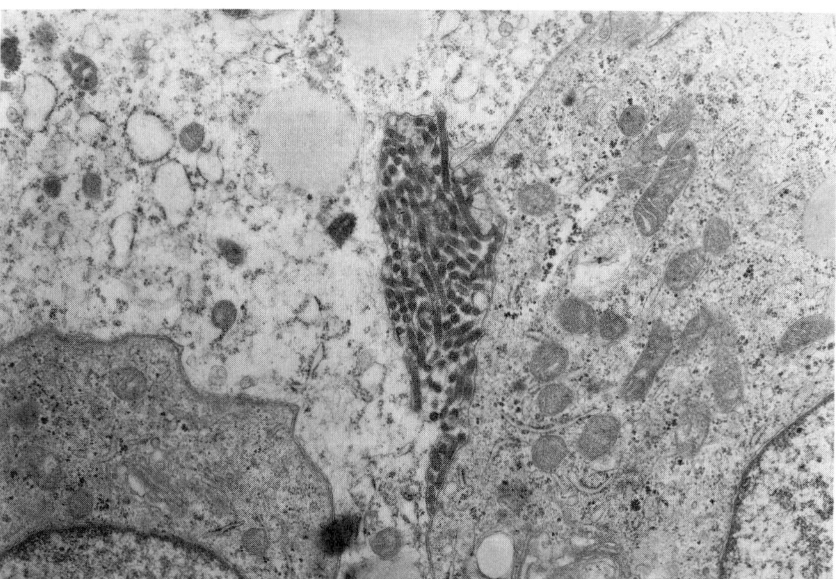

FIGURE 4.18

(a) Ebola virus passaged in monkey cells showing typical 6-shaped particles by negative stain (× 73,500). (b) Ebola virus isolated in 1976 on monkey Vero cells. Negative stain (× 70,000). This photo is actually the first Ebola virus ever visualized. (c) Marburg virus passaged from a human case (liver isolate) through monkey Vero cells. Virus particles are observed between the spaces of two cells (× 32,200). (Courtesy of Frederick A. Murphy)

represents a major problem in Africa, with a death rate from severe hemorrhagic fever of 50 percent to 90 percent in different countries due to at least two subtypes, Sudan and Zaire. The reservoir for these latter viruses in nature is still unknown. These filoviruses commonly give rise to **hemorrhagic disease** with major pathologic findings in the lungs, liver, and gastrointestinal tract. The infection can be contained by standard quarantine procedures, particularly avoidance of close contact with infected individuals and exposure to contaminated needles and syringes.

4.5 BORNA DISEASE VIRUS

Borna disease virus (BDV) is a recently recognized 8.5 kb RNA virus, which causes neurologic disorders in horses and sheep. Experimental infection has been achieved in a wide variety of animals including birds. It has been passed and studied in rats. BDV has not been visualized using the electron microscope, and is not yet classified. Its RNA has been found to have both positive and negative-sense polarity. Infected cells contain a large amount of negative-sense stranded BDV RNA, but they also have overlapping 3′ co-terminal viral mRNAs which are characteristic of the positive-sense coronavirus family (see Section 3.3). The role of this virus or a related one in human neurological diseases is under consideration. Some serologic evidence of human infections has been reported and needs to be confirmed. The replicative cycle of BDV most probably involves the nucleus of the cell.

SECTIONS 4.4 AND 4.5 FURTHER READING

Original Papers

SMITH, C. E. F., et al. (1967) Fatal human disease from vervet monkeys. *Lancet* ii:1119.

JOHNSON, K. M., et al. (1977) Isolation and painful characterization of a new virus causing acute haemorrhagic fever in Zaire. *Lancet* i:569.

COX, N. J., et al. (1983) Evidence for two subtypes of Ebola virus based on oligonucleotide mapping of RNA. *J. Infect. Dis.* 147:272.

ELLIOTT, L. H., et al. (1985) Descriptive analysis of Ebola virus proteins. *Virology* 147:169.

HAYASHI, T., and Y. MINOBE. (1985) Protein composition of rice transitory yellowing virus. *Microbiol. Immunol.* (Japan) 29:169.

KILEY, M. P., et al. (1986) Conservation of the 3′ terminal nucleotide sequences of Ebola and Marburg virus. *Virology* 149:251.

CARBONE, K. M., et al. (1991) Borna Disease: Association with a maturation defect in the cellular immune response. *J. Virol.* 65:6154.

FISCHER-HOCH, S. P., et al. (1992) Pathogeneic potential of filoviruses: Role of geographic origin of primate host and virus strain. *J. Infect. Dis.* 166:753.

PYPER, J. M., et al. (1993) Genomic organization of the structural proteins of borna disease virus revealed by a cDNA clone encoding the 38 kDa protein. *Virology* 195:229.

CARBONE, K. M., et al. (1993) Characterization of a glial cell line persistently infected with Borna disease virus (BDV): Influence of neurotrophic factors on BDV protein and RNA expression. *J. Virol* 67:1453.

SCHNITTLER, H. J., et al. (1993) Replication of Marburg virus in human endothelial cells: A possible mechanism for the development of viral hemorrhagic disease. *J. Clin. Invest.* 91:1301.

Books and Reviews

MURPHY, F. A., et al. (1990) Filoviridae. In *Virology,* 2d ed., eds. B. N. Fields and D. M. Knipe. 993–942 Raven Press, New York.

RICHT, J. A., et al. (1992) Infection with Borna disease virus: Molecular and immunobiological characterization of the agent. *Clin. Infect. Dis.* 14:1240.

4.6 BUNYAVIRIDAE

The Bunyaviridae, as well as the Arenaviridae, are often considered with negative-sense RNA viruses, but for some members, part of the replication cycle can involve an **ambisense strategy**. Within the cell, transcripts and proteins are found that suggest both positive- and negative-sense processes. The viral genome may have either a positive-sense RNA that directly acts as mRNA, or the positive-sense transcripts from the negative-sense genome make further negative-sense transcripts that code for other mRNAs and positive-sense viral genome species. The mechanism for this somewhat complicated replicative cycle, as it is now recognized, is not known. Negative and ambisense RNA viruses can also be found in certain plant virus families (see Secs. 4.8 and 4.9).

The Bunyaviridae represent a large group of about 350 mainly arthropod-transmitted viruses occurring most frequently in tropical countries. Although several members of the family are of medical and veterinary importance, causing severe disease such as encephalitis or hemorrhagic fever, most viruses never infect humans or domestic animals but cause lifelong persistent noncytopathic infections in their vectors. Five genera of animal Bunyaviridae are currently recognized: *Bunyavirus, Hantavirus, Phlebovirus, Nairovirus,* and *Uukuvirus,* as well as one plant virus family, the Tospoviridae to be discussed in Section 4.9. Most of these groups share some antigenic (serologic) properties, but the sizes of their RNAs and proteins differ.

Bunyaviruses, nairoviruses, and phleboviruses are transmitted to vertebrates by a wide variety of mosquito, biting-fly, and tick species. The hantaviruses are generally maintained as **persistent infections** of rodents; transmission to humans is via aerosolized excretions and bites. Most of these viruses replicate also in the insects that transmit them. Viruses that have been characterized at the molecular level include Bunyamwera, La Crosse, snowshoe hare, Hantaan, Rift Valley fever, Uukuniemi and the tomato spotted wilt viruses. Relatively little is known about the molecular features of other members of the family.

Most recently, a Hantaan virus was found as the cause of an acute respiratory disease often leading to death in Navajo Indians living in New Mexico. The cause of this illness was determined by epidemiological studies that suggested transmission by rodent droppings. Subsequent growth of the virus in cell culture led to its identification.

4.6.1 Structure

Viruses in the family Bunyaviridae share certain morphological and biochemical characteristics (Fig. 4.19). Virus particles are spherical, about 100 nm in diameter and enveloped, thus sensitive to lipid solvents. They contain three separately **encapsidated segments** of single-stranded RNA, designated L (large), M (medium), and S (small). The sizes of the RNA segments can vary widely: 0.3 to .9 kb for the S, 3.2 to 5 kb for the M, and 6.5 to 14.4 kb for the L, but the pattern of sizes is conserved within a genus. The RNA segments can cyclize because they have complementary, genus-specific, conserved terminal sequences at their 3′ and 5′ ends. The virions are composed of four structural proteins: two internal proteins, the L protein (RNA polymerase) encoded by the L RNA segment, the nucleocapsid protein (N) encoded by the S segments, and two surface glycoproteins (G1 and G2) that form spikes and are encoded by the M segment. The pattern of sizes of these proteins varies among the different genera. One or two additional nonstructural proteins (NS) are encoded by some viruses. The N and L proteins interact with each of the RNA segments to form **helical ribonucleoprotein complexes** termed nucleocapsids.

4.6.2 Entry, Transcription, and Virus Production

After being transmitted during an insect bite, the virus enters the cell by phagocytosis upon interaction of the viral envelope proteins with cell surface receptors. Multiplication, initiated by the viral transcriptase, occurs in the cytoplasm. The coding

FIGURE 4.19

Bunyavirion. (Courtesy of C. S. Schmaljohn)

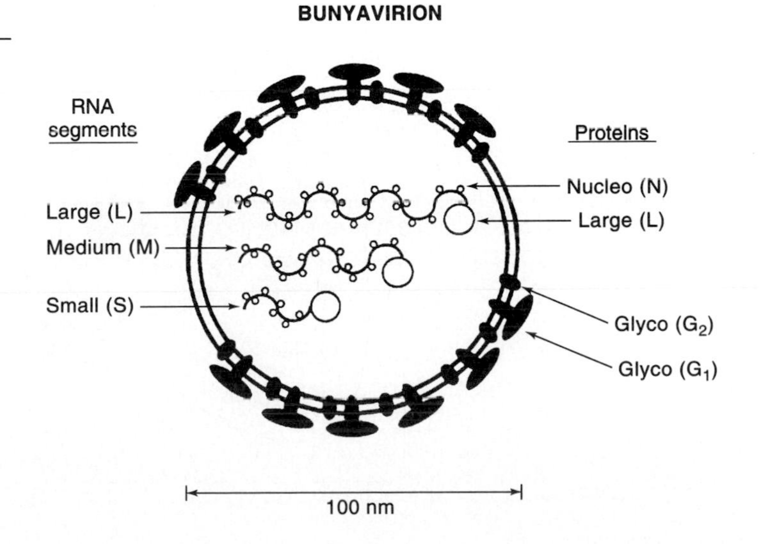

strategies of the individual RNAs are in large part negative-sense, but ambisense and positive-sense transcription has definitely been demonstrated to occur as well, particularly in the S and NS segments of phlebo- and kukuviruses, and the S RNA of their genera (Fig. 4.20). The viruses generally mature and are assembled within the Golgi apparatus (Fig. 4.20).

Many of the important biological properties of the viruses have been assigned to the M segment gene products (the glycoproteins), namely, attachment and fusion, hemagglutination, neutralization, and virulence. However the S, L and any NS segment gene products may have modulating effects on virulence. For the arthropod-transmitted

FIGURE 4.20

Summary of probable replication processes for viruses in the Bunyaviridae. Virion particles (inset on left) contain large (L), medium (M), and small (S) viral RNAs complexed with nucleocapsid proteins. The bilaminar, lipid envelope has integral virus-specified glycoproteins (G1 and G2) that interact across the membrane with the ribonucleoprotein structures. The precise location of the virion-associated polymerase is not known. Numbered events corresponding to those listed in the text include: (1) attachment; (2) entry and uncoating; (3) primary transcription to yield viral messenger RNAs (mRNA); (4) translation of L and S segment mRNA on free ribosomes: translation of M segment mRNA on membrane-bound ribosomes and primary glycosylation of the gene products (G1 and G2); (5) synthesis of antigenome templates; (6) genome replication; (7) secondary transcription; (8) further translation; (9) terminal glycosylation of G1 and G2 and assembly of virus particles by budding into Golgi vesicles; (10) transport of cytoplasmic vesicles to the cell surface, fusion, and release of mature virions. The stages of budding into smooth membrane vesicles (inset on right) are: (a) ribonucleoprotein structures accumulate on the cytoplasmic face of membranes which have G1 and G2 embedded into them and exposed on the luminal side; (b) involution of membranes; (c) completion of budding to yield a morphologically mature virion within a cytoplasmic vacuole. Abbreviations for cellular substructures are: N, nucleus; RER, rough endoplasmic reticulum. (From David H. L. Bishop (1990) *Virology* 2d ed. Vol. 2, eds. Fields and Knipe, 1183. Raven Press, New York.)

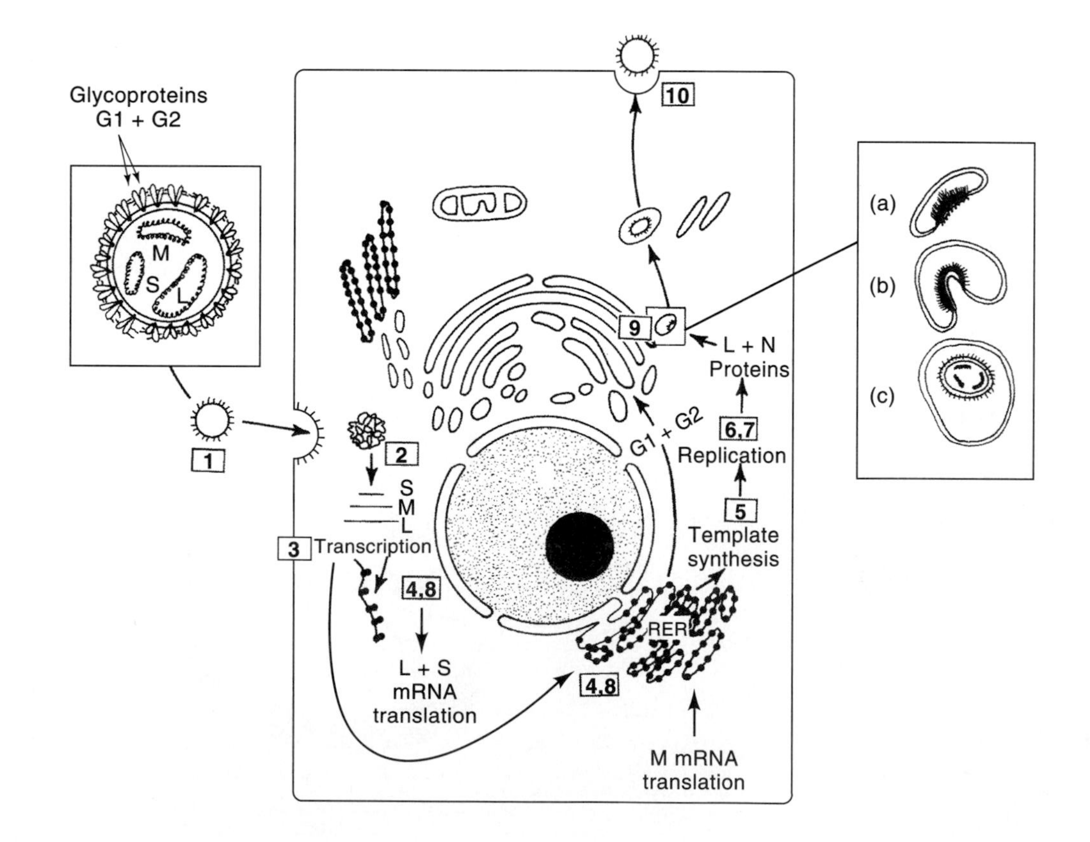

viruses, only a restricted number of vector species can maintain a particular virus. The M segment is the major viral determinant for the outcome of a particular virus-vector interaction. The segmented genome of the Bunyaviridae confers the capacity for sudden, dramatic antigenic shift by **genome segment reassortment,** comparable to that seen with influenza viruses. In view of their disease potential for humans, the Bunyaviridae warrant continual surveillance in nature. As with several other viruses that infect both insects and vertebrates, the infection of the insects is generally nonpathogenic and persistent.

SECTION 4.6 FURTHER READING

Original Papers

MURPHY, F. A., et al. (1973) Bunyaviridae: Morphologic and morphogenetic similarities of Bunyamwera serologic supergroup viruses and several other arthropod-borne viruses. *Intervirology* 1:297.

RANKI, M., and R. PATTERSSON. (1975) Uukuniemi virus contains an RNA polymerase. *Virology* 16:1420.

BEATTY, B. J., et al. (1981) Molecular basis of bunyavirus transmission by mosquitoes: Role of the middle-sized RNA segment *Science* 211:1433.

IHARA, T., H. AKASHI, and D. H. L. BISHOP. (1984) Novel coding strategy (ambisense genomic RNA) revealed by sequence analyses of Punta Toro phlebovirus S RNA. *Virology* 136:293.

JANSSEN, R. S., N. NATHANSON, M. J. ENDRES, and F. GONZALES-SCARANO. (1986) Virulence of La Crosse virus is under polygenic control. *J. Virol.* 170:505.

JIN, H., and R. M. ELLIOTT. (1991) Expression of functional Bunyamwera virus L protein by recombinant vaccinia viruses. *J. Virol.* 65:4182.

ROSSIER, C., R. RAJU, and D. KOLAKOFSKY. (1988) La Crosse virus gene expression in mammalian and mosquito cells. *Virology* 165:539.

RÖNHOLM, R. (1992) Localization of the Golgi complex of Uukuniemi virus glycoproteins G1 and G2 expressed from cloned cDNAs. *J. Virol.* 66:4525.

GOTT, P., et al. (1993) RNA binding of recombinant nucleocapsid proteins of Hantaviruses. *Virology* 194:332.

Books and Reviews

BISHOP, D. H. L., and R. E. SWOPE. (1979) Bunyaviridae. In *Comprehensive virology,* eds. H. Fraenkel-Conrat and R. R. Wagner. Vol. 14, 1–156. Plenum Press, New York.

ELLIOTT, R. M. (1990) Molecular biology of the Bunyaviridae. *J. Gen. Virol.* 71:501.

BOULOY, M. (1991) Bunyaviridae: Genome organization and replication strategies. *Adv. Virus Res.* 40:235.

BISHOP, D. H. L. (1990) Bunyaviridae. In *Virology*. 2d ed., eds. B. N. Fields and D. M. Knipe. 1155–1173 Raven Press, New York.

SCHMALJOHN, C. S. and J. L. PATTERSON. Replication of Bunyaviridae. In *Virology*. 2d ed., eds. B. N. Fields and D. M. Knipe. 1175–1194 Raven Press, New York.

4.7 ARENAVIRIDAE

Prototypes of the Arenaviridae family are the lymphocytic choriomeningitis virus (LCM) of mice and humans, the Tacaribe virus group (e.g., the Pichinde virus), and the virus responsible for Lassa fever. These and others usually cause mild and persistent diseases in the natural mammalian hosts, often a single species, but they can produce lethal hemorrhagic fevers when transmitted to primates. The animal reservoir includes rodents and in some cases fruit bats. LCM in human beings occurs mainly from infection of laboratory workers, but at least one outbreak associated with severe meningitis was traced to infected pet hamsters from a single infected colony.

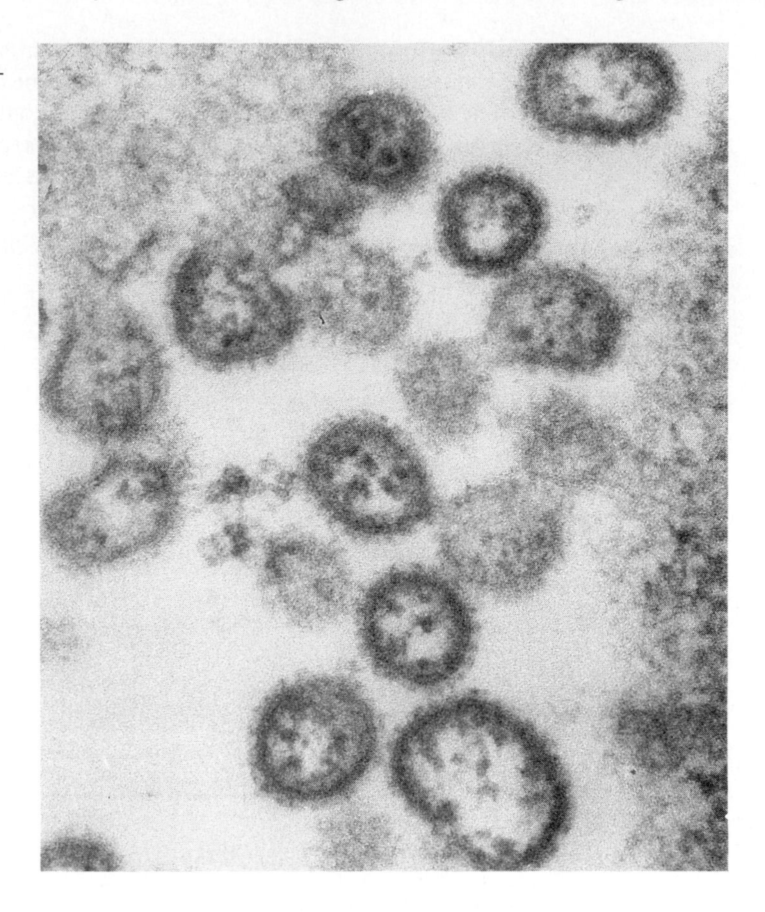

FIGURE 4.21

Electron micrograph of an arenavirus. (Courtesy of R. C. Williams)

4.7.1 Structure, Entry, and Replication

The arenaviruses are enveloped, **pleomorphic** (100–300 nm in diameter), and serologically interrelated. Their name is derived from the presence of particles in the virions that resemble sand (Latin, *arena*) on EM images (Fig. 4.21). These particles are likely to be host ribosomes that become incorporated during the assembly of the virions in areas of high ribosome concentration. The envelope proteins of these viruses form a small number of club-shaped spikes in which GP1 (see below) is external and serves in enabling the virion to attach to the host cell membrane. GP1 is dissociable by high salt from GP2, a glycoprotein that is imbedded in the membrane and can be cross-linked to the virion core. There is a much longer latent period from entry to appearance of progeny with arenaviruses than with bunya- and other negative-sense viruses.

The virion cores or nucleocapsids are beaded structures containing two RNA molecules and attached capsid proteins (N or NP). The larger (L) RNA ($2–3 \times 10^6$ Da) encodes the L protein (an RNA polymerase) and a small zinc-binding protein, (Z, or P11) the function of which has not yet been established. The S RNA (about 1.2×10^6 Da) encodes the nucleocapsid protein (63 kDa) and the two glycoproteins of the envelope GP1, GP2 (44 and 35 kDa respectively). These RNAs, lacking 5′ caps and 3′ poly (A), are termed "ambisense" since they carry the positive genomic information (though not translated) at the 5′ end. The actual genes (except for the Z protein) are negative-sense, i.e., in complementary form at the 3′ end. Since only the latter are probably operative and require transcription for the production of mRNAs, the Arenaviridae are functionally negative-sense viruses. The presence of mRNAs and/or an RNA transcriptase in the virion would be regarded as a requirement for all negative-sense viruses, since the polymerase becomes active only upon transcription and translation. LCM virus is symbiotic with its natural host, the mouse, and is excreted throughout the mouse's lifeline without producing detectable pathology.

SECTION 4.7 FURTHER READING

Original Papers

BUCHMEIER, M. J., and M. B. A. OLDSTONE. (1978) Virus-induced immune complex disease: Identification of specific viral antigens and antibodies deposited in complexes during chronic lymphocytic choriomeningitis virus infection. *J. Immunol.* 120:1297.

LEUNG, W. C., et al. (1981) Pichinde virus L and S RNAs contain unique sequences. *J. Virol.* 37:48.

RIVIERE, Y., et al. (1985) Genetic mapping of lymphocytic choriomeningitis virus pathogenicity: Virulence in guinea pigs is associated with the L RNA segment. *J. Virol.* 55:704.

MATSUURA, Y., et al. (1986) Expression of the S-coded genes of lymphocytic choriomeningitis arenavirus using a baculovirus vector. *J. Gen. Virol.* 67:1515.

IAPALUCCI, S. et al. (1989) The 5′ region of Tacaribe virus L RNA encodes a protein with a potential metal binding domain. *Virology* 173:357.

SALVATO, M. S., and E. M. SHIMOMAYE. (1989) The completed sequence of lymphocytic choriomeningitis virus reveals a unique RNA structure and a gene for a zinc finger protein. *Virology* 173:1.

BURNS, J. W., and M. J. BUCHMEIER. (1991) Protein-protein interactions in lymphocytic choriomeningitis virus. *Virology* 183:620.

LUKASHEVICH, I. S. (1992) Generation of reassortants between African arenaviruses. *Virology* 188:600.

XING, Z., and J. L. WHITTON. (1993) An anti-lymphocytic choriomeningitis virus ribozyme expressed in tissue culture cells diminishes viral RNA levels and leads to a reduction in infectious virus yield. *J. Virol.* 67:1840.

Books and Reviews

PEDERSON, J. R. (1979) Structural components and replication of arenaviruses. *Adv. Virus Res.* 24:277.

LEHMANN-GRUBE, F. (1984) The arenaviruses. *Intervirology* 22:121.

BISHOP, D. H. L. (1990) Arenaviridae and their replication. In *Virology,* 2d ed., eds. B. N. Fields and D. M. Knipe. 1231–1243 Raven Press, New York.

4.8 PLANT RHABDOVIRUSES

Plant rhabdoviruses are among the largest and most complex of all plant viruses, but precise measurements of particle size and morphology have been hampered by a variety of technical problems. Their particles contain approximately 70 percent protein, 25 percent lipid, 4 percent polysaccharide, and 1 percent RNA. Like animal rhabdoviruses, their bacilliform or (occasionally) bullet-shaped particles—about 50 to 95 nm in diameter and 200 to 500 nm in length—consist of an outer envelope that encloses a long, helically wound nucleoprotein strand. Virus particles contain at least four structural proteins plus a single molecule of single-stranded, negative-sense RNA approximately 11,000 to 13,000 nucleotides in length. As with the animal rhabdoviruses, the viral envelope is acquired from modified host cell membranes during particle maturation and contains a regular array of 6–10 nm projections (see Figs. 4.1 and 4.2).

A large number of plant rhabdoviruses have been described, based almost exclusively on the presence of rhabdovirus-like particles in plant cells or leaf-dip preparations from diseased plants. They infect both plants and insects, and a wide range of insects—aphids, leafhoppers, lacebugs, and mites—have been shown to act as viral vectors. Some viruses are **mechanically transmissible,** but none can be transmitted by more than a single insect vector. Plant rhabdoviruses are one of only two groups of plant viruses able to multiply in their insect vectors, and most are transmitted to only a limited number of plant species. Unfortunately, many of these viruses have been only poorly characterized, and taxonomic relationships among the plant Rhabdoviridae are unclear.

Their large size and distinctive morphology make plant rhabdoviruses particularly amenable to cytological studies of replication and particle maturation. They appear to fall into three groups: (1) Viruses like *Sonchus* yellow net virus (SYNV) that accumulate in the perinuclear space with only a few scattered particles in the cytoplasm. Structures resembling the inner nucleoprotein core are found within the nucleus, and the envelopes of some particles in the perinuclear space are continuous with the inner nuclear membrane. (2) Viruses like lettuce necrotic yellow virus that mature in association with the endoplasmic reticulum and accumulate in ER-associated vesicles. (3) Viruses like barley yellow striate mosaic virus where infection leads to the formation of large cytoplasmic viroplasms and replication appears not to involve the nucleus. SYNV is by far the best characterized plant rhabdovirus. Our consideration of the molecular properties will be limited to this one virus.

SYNV virions contain five structural proteins: a large (L) protein, a glycoprotein (G), a nucleoprotein (N), and two matrix proteins (M1 and M2). The N protein encapsidates the genomic RNA and is believed to form a transcription complex with the L (or polymerase) protein. The G protein forms the spikes that protrude from the surface of the virion, whereas the two matrix proteins have characteristics of either a phosphoprotein (M1) or envelope protein (M2). From the hybridization patterns observed when polyadenylated mRNA preparations from virus-infected tobacco were probed with cloned SYNV genomic cDNAs, the SYNV genome was shown to be transcribed into a short (144 nt) 3′-terminal "leader RNA" plus six mRNAs. In order of their appearance in the SYNV genome, the proteins encoded by these mRNAs are designated 3′-N-M2-sc4-M1-G-L-5′. Protein sc4 is *not* a structural polypeptide, but may be a **nonstructural polypeptide** that is synthesized at a specific stage of replication.

A virtually complete nucleotide sequence for SYNV is now available. As shown in Figure 4.22, the intergenic and flanking gene sequences of SYNV are highly conserved. They are similar to the corresponding sequences in vesicular stomatitis and rabies viruses and contain a GG dinucleotide spacer that does not appear in the viral mRNA. The 242 kDa SYNV L protein contains both putative polymerase and RNA binding domains. And, extended sequence alignment with other nonsegmented negative-strand RNA virus polymerases reveals 12 blocks of conserved sequence arranged sequentially along the proteins. The L protein of SYNV is more closely related to the polymerases of animal rhabdoviruses than to those of paramyxoviruses, but it is more distantly related to the animal rhabdovirus polymerases than they are to each other. Although plant and animal rhabdoviruses may have evolved from a common ancestor (perhaps via the use of insect vectors to cross taxa boundaries), **divergence** of the plant rhabdoviruses probably predates emergence of the different animal rhabdoviruses.

Finally, Figure 4.23 compares the genome map of SYNV with those of three animal rhabdoviruses. All are similar in having a leader RNA and the N gene near the 5′ termini and the G and L genes near the 3′ terminus. Note the variation in the number of genes as well as the location of the nonstructural genes. Although the VSV genome does not contain a separate gene encoding a nonstructural polypeptide, a sixth protein is synthesized by internal initiation of translation on the NS mRNA. The 417 nucleotide "rg" region between the rabies virus G and L genes is noncoding, but sequence comparisons suggest that this domain may represent a vestigial NV gene. Thus, rhabdoviruses have evolved different genome organizational strategies, but whether these differences are important in controlling gene expression and replication is not yet clear.

FIGURE 4.22

Comparison of the sequences at the SYNV gene junctions. Underlined nucleotides are 1, the putative polyadenylation signal at the 5′ terminus of each gene; 2, the untranslated intergenic sequence; and 3, the transcription initiation site at the 3′ terminus of each gene. (From Heaton et al. (1989) *Proc. Natl. Acad. Sci. USA* 86:8665)

leader	UUUCUUUUU	GG	UUGUA	N
N	AUUCUUUUU	GG	UUGUC	M2
M2	AUUCUUUUU	GG	UUGUC	sc4
sc4	AUUCUUUUU	GG	UUGAA	M1
M1	AUUCUUUUU	GG	UUGAA	G
G	AUUCUUUUU	GG	UUGUA	L
	1	2	3	

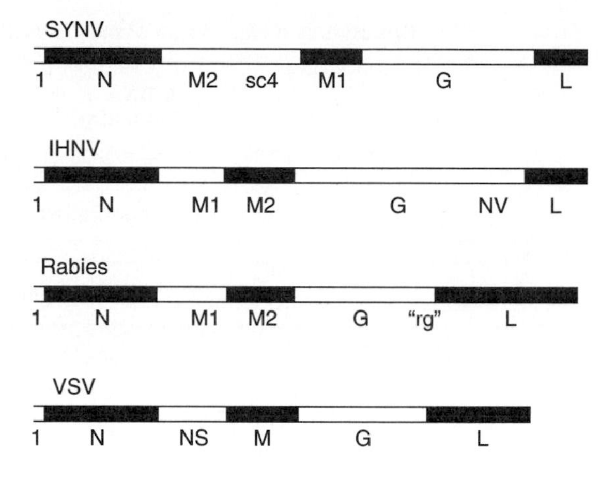

FIGURE 4.23

Comparison of the genome maps of SYNV and three animal rhabdoviruses (infectious hematopoietic necrosis, rabies, and vesicular stomatitis virus). Similar patterns in the maps correspond to genes that are thought to encode structural proteins having similar functions. There is, however, no evidence for similarity in the functions of sc4, NV, or "rg." (From Heaton et al. (1989) *Proc. Natl. Acad. Sci. USA* 86:8665)

4.9 PLANT BUNYAVIRUSES

Two groups of plant viruses—the **tospoviruses** (*to*mato *spo*tted wilt) and the **tenuiviruses** (Latin, *tenuis,* "thin, weak")—have been shown to utilize an ambisense genome expression strategy. The tospoviruses are considered to be a genus within the family Bunyaviridae, whereas the tenuiviruses (like most other plant viruses) are classified as a virus group.

4.9.1 Tospoviruses

Nearly 400 plant species in more than 50 families are susceptible to infection by tomato spotted wilt virus (TSWV). TSWV is transmitted by thrips, and the increased incidence of this viral infection in the Northern Hemisphere during the 1980s has been attributed to the sudden spread of the western flower thrip *Frankliniella occidentalis* (Pergande) throughout North America and Europe. This virus now causes serious disease losses in agricultural, horticultural, and ornamental crops in many tropical and subtropical regions.

Structure The roughly isometric particles of TSWV are enveloped, having a diameter of 70 to 110 nm (see Fig. 4.24). The virion contains four structural proteins: a nucleocapsid (N) protein associated with its three genomic RNA segments, two glycoproteins (G1 and G2) associated with the viral envelope, and a large (>200 kDa) protein believed to be the viral RNA polymerase. The complete nucleotide sequence of the tripartite TSWV genome has been determined. Its structure and expression strategy are diagrammed in Figure 4.25.

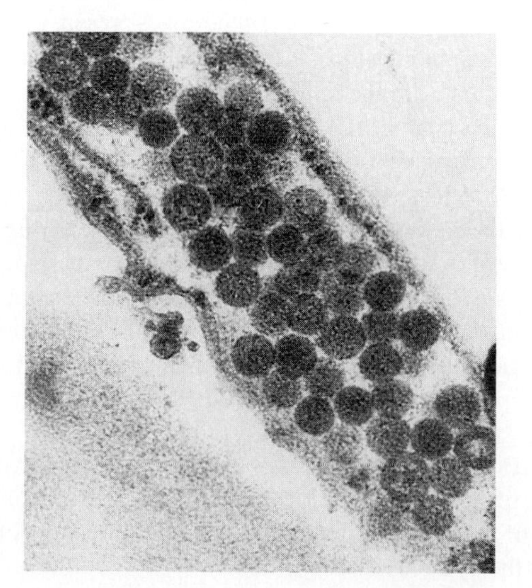

FIGURE 4.24

Impatiens necrotic ringspot virus in a thin section of tomato. (Courtesy of H-T. Hsu)

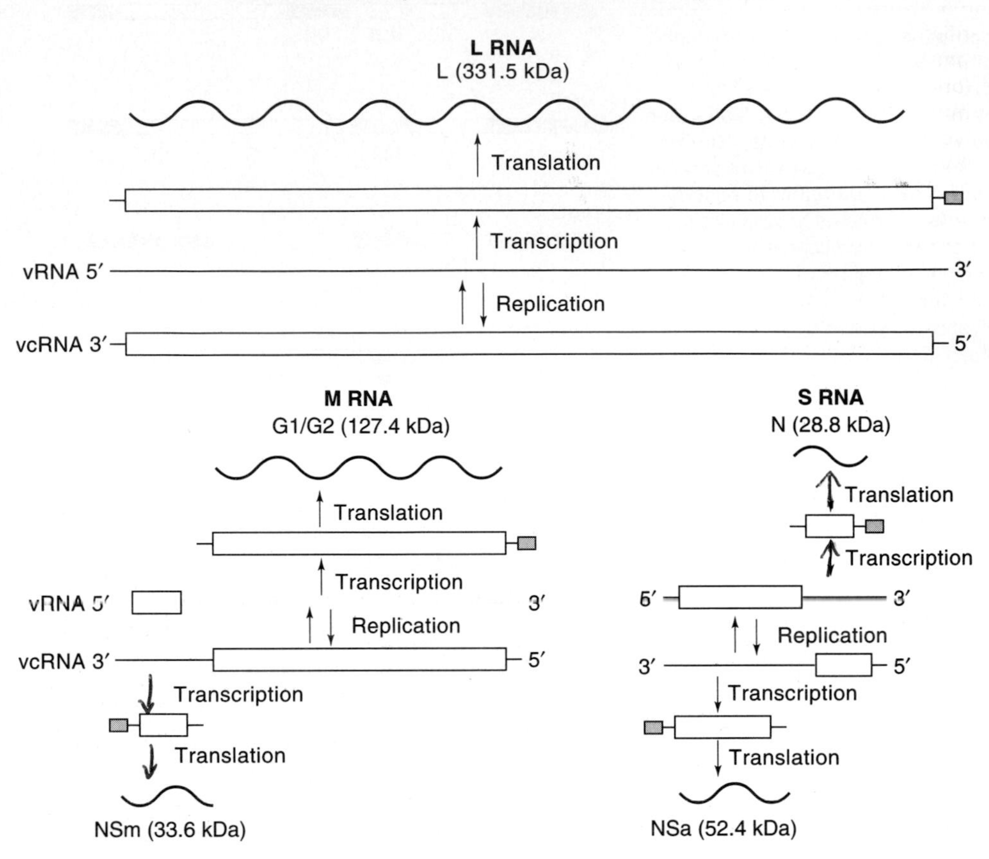

FIGURE 4.25

Genome structure and expression strategy of TSWV (isolate BR-01). Open reading frames are indicated as open bars, and the hatched boxes at the 5′ ends of mRNAs represent nonviral leader sequences. (Courtesy of R. W. Goldbach)

Transcription and Translation L RNA (8897 nucleotides) is a negative-sense RNA that encodes a single 331.5 kDa protein containing all the conserved sequence motifs characteristic of negative-sense RNA viral polymerases. Of all viruses compared, the L proteins of TSWV and Bunyamwera virus exhibit the greatest sequence similarity. The other two genomic RNAs (S and M) are ambisense RNAs. The M RNA (4821 nucleotides) contains two ORFs: one in the sequence order of the viral RNA encodes a 33.6 kDa nonstructural protein; the other in the sequence order complementary to the viral RNA encodes a 127.4 kDa precursor to the membrane glycoproteins G1 and G2. The S RNA (2916 nucleotides) encodes in the viral sense a 52.4 kDa nonstructural protein and in the viral complementary sense the 28.8 kDa nucleocapsid (N) protein. All four of the genes present on the M and S RNAs are expressed via synthesis of subgenomic mRNAs that terminate at A-U–rich **hairpin structures** located in the intergenic regions. Similar to the mRNAs of bunya- and orthomyxoviruses infecting animals, TSWV mRNAs contain 12 to 20 nonviral nucleotides at their 5′ termini. This observation indicates that transcription is initiated by a process known as "cap snatching" (see Section 4.6).

Based on its genome organization, TSWV has recently been classified as the sole member of the newly created genus **Tospovirus** within the family Bunyaviridae. Results from both serologic and nucleotide sequence comparisons involving their nucleocapsid (N) genes indicate that presently known TSWV isolates appear to represent four distinct viral species, namely, tomato spotted wilt, tomato chlorotic spot, groundnut ringspot, and *Impatiens* necrotic ringspot viruses. Their N gene amino acid sequences exhibit 100, 76, 78, and 55 percent homology, respectively.

Virus Production in the Vector The salivary gland of the thrip vector is the major site of TSWV replication in its vector. Virus particles are assembled in clusters in the cisternae of the endoplasmic reticulum. Dense masses of nucleocapsid material may accumulate in the cytoplasm.

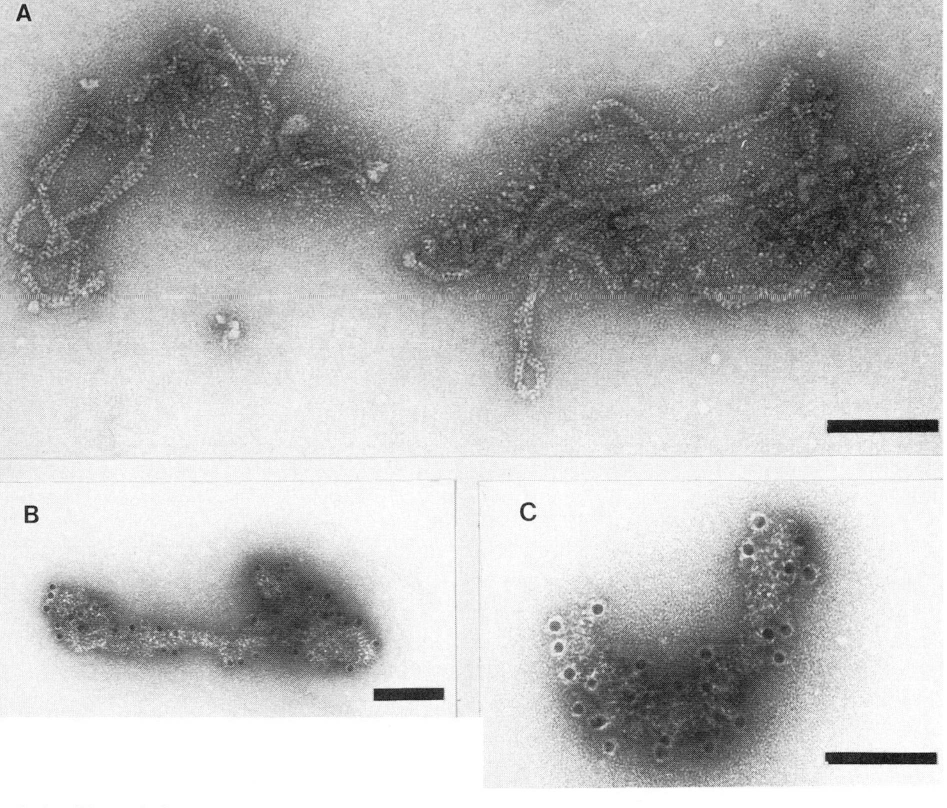

FIGURE 4.26

Purified rice hoja blanca virus ribonucleoprotein particles (RNP) before (a) and after (b and c) immunogold labeling using polyclonal antiserum against RHBV-RNP. Bar = 100 nm. (From A. M. Espinoza et al. (1992) *Virology* 191:619)

Original Papers

4.9.2 Tenuiviruses

Rice stripe, rice hoja blanca, and maize stripe viruses are three members of the tenuivirus group. As shown in Fig. 4.26, their filamentous particles (about 8 nm in diameter) are composed of a smaller, supercoiled ribonucleoprotein. The 16 to 18 kb tenuivirus genome is divided among four or five species of single-stranded RNA. This fact explains the variable particle length. Membrane-bound particles resembling those of the tospoviruses and other members of the Bunyaviridae have not been reported. Tenuiviruses are transmitted by leafhoppers in a persistent manner, but sap transmission is difficult. Their host range is restricted to members of the *Gramineae* (i.e., grasses).

Genome Structure The complete nucleotide sequence of the two smallest rice stripe and maize stripe virus RNAs has been determined. Each of these RNAs has an ambisense organization and encodes two proteins. One of the proteins encoded by RNA 3 is the N protein that, when associated with the viral RNA, makes up the supercoiled ribonucleoprotein fibers visible in Fig. 4.26. RNA 4 encodes a major noncapsid protein. Immunocytochemical studies with rice hoja blanca virus have shown viral ribonucleoproteins to be rather evenly distributed throughout the cytoplasm of infected cells, whereas the major noncapsid protein of this virus accumulates in **cytoplasmic inclusion bodies.**

SECTIONS 4.8 AND 4.9 FURTHER READING

HEATON, L. A., B. I. HILLMAN, B. G., HUNTER, D. ZUIDEMA, and A. O. JACKSON. (1989) Physical map of the genome of sonchus yellow net virus, a plant rhabdovirus with six gene conserved gene junction sequences. *Proc. Natl. Acad. Sci. USA* 86:8665.

KAKUTANI, T., et al. (1990) Ambisense segment 4 of rice stripe virus: Possible evolutionary relationship with phleboviruses and uukuviruses (Bunyaviridae). *J. Gen. Virol.* 71:1427.

HUIET, L., et al. (1991) Nucleotide sequence and RNA hybridization analyses reveal an ambisense coding strategy for maize stripe virus RNA 3. *Virology* 182:47.

ZHU, Y., et al. (1991) Complete nucleotide sequence of RNA 3 of rice stripe virus: An ambisense coding strategy. *J. Gen. Virol.* 72:763.

CHOI, T.-J., et al. (1992) Structure of the L (polymerase) protein gene of sonchus yellow-net virus. *Virology* 189:31.

de AVILA, A. C., et al. (1993) Classification of tospoviruses leased on phylogeny of nucleoprotein gene sequences. *J. Gen. Virol.* 74:153.

Books and Reviews

FRANCKI, R. I. B., R. G. MILNE, and T. HATTA. (1985) *Atlas of plant viruses.* Vol. 1, 73–100. CRC Press, Boca Raton, Fla.

JACKSON, A. O., R. I. B. FRANCKI, and D. ZUIDEMA. (1987) Biology, structure, and replication of plant rhabdoviruses. In *The rhabdoviruses,* ed. R. R. Wagner, 427–508 Plenum, New York.

BEATTY, B. J. and C. H. CALISHER. (1991) Bunyaviridae: Natural history. *Current topics in microbiology and immunology* 169:27.

chapter

5

Viruses with Double-Stranded RNA Genomes

Viruses with double-stranded RNA genomes have been found in all living species. Two distinct groups of virus families are recognized in terms of morphological structure and RNA content. The Reoviridae of animals, including insects and plants, have characteristic shapes and contain 10 to 12 RNA molecules. Members of the other group are **isometric** and contain only 1 to 3 RNA molecules. In this latter group, the Toti- and Partitiviridae occur seemingly only in molds and fungi. The Birnaviridae occur in only a few vertebrate species (but not in any mammalian species) and are found mostly in lower animals, including insects. The plant cryptoviruses show properties similar to those of this latter dsRNA virus group. Aside from the virus of Pseudomonas (φ6), all the double-stranded RNA viruses are nonenveloped.

5.1 REOVIRIDAE

The Reoviridae is a family of viruses of rather uniform properties. They are of considerable interest because of their unique structural and replicative peculiarities. It was in these viruses that double-stranded RNA was first detected as a stable natural product and then studied intensely.

Reoviruses with the same principal features occur in mammals, many other vertebrates, insects, and plants (see Table 5.1). The overall morphology of this family of viruses was considered in naming their different genera. **Rotaviruses** (*rota,* "wheel"), for example, have wheellike capsids in which spikes radiate from the inner capsid to the smooth viral outer capsid; **orbiviruses** *(orbis,* "ring") have ring-shaped capsomers on their inner shells.

TABLE 5.1
Physical and biological properties of some Reoviridae

Genus	*Prototype*	*Diameter of virion nm (S)*	*Number of RNA components and total genome mass (Da)*	*Hosts*
Orthoreovirus	Reovirus type 1	76 (630)	10, 15×10^6	Vertebrates
Rotavirus	Human rotavirus	70 (525)	11, 12×10^6	Mammals
Orbivirus	Blue-tongue virus	63 (550)	10, 12×10^6	Insects, mammals
Phytoreovirus	Rice dwarf virus	70 (510)	12, 17×10^6	Leafhoppers, grasses
	Wound tumor virus[a]	70 (514)	12, 15×10^6	Leafhoppers, dicotyledons
Fijivirus	Fiji disease virus	75	10	Sugar cane, insects
Cypovirus	Cytoplasmic polyhedrosis	55 (440)	10, 13×10^6	Insects

[a]Not oncogenic in natural host in the field.

Whereas the mammalian reoviruses (**orthoreoviruses**) and rotaviruses replicate only in vertebrates, the **orbiviruses** are primarily viruses of insects but also infect many mammals. Several of the **phytoreoviruses** and **fijiviruses** are replicated not only in plants but also in their insect vectors. The **cytoplasmic polyhedrosis** viruses (cypovirus) only grow in many species of insects.

Although the orthoreoviruses have been detected in several animals, including humans, they are not generally pathogenic in adult animals; hence the name **reoviruses** (from *r*espiratory-*e*nteric *o*rphan) was coined, with a rationale similar to that for echoviruses discussed in Section 2.2. However, Colorado tick fever is a mild febrile disease in humans caused, as the name implies, by a tick-borne reovirus. The orthoreoviruses have been divided into three subtypes as identified by neutralization and **hemagglutination inhibition** testing. For instance, only the reovirus type 3 agglutinates human red blood cells.

Mild or asymptomatic respiratory reovirus infections in many species may serve to facilitate initiation of severe bacterial respiratory disease. Avian reoviruses have been associated with arthritis in chickens, but no virions are observed in synovial cells of affected joints. A variety of diseases can be observed in reovirus-infected young mice, including severe neurological symptoms, pneumonia, myocarditis, encephalitis, and hepatitis; reoviruses may also be one of several families responsible for diarrhea in calves.

Six distinct groups of rotaviruses have been defined. Of these groups, A, B, and C are found in humans and animals. Generally, the viruses are species-specific. Within each group, the rotaviruses are further classified into serotypes and determined by neutralization assays. Probably over 30 different serotypes exist. Similar to the reoviruses and myxoviruses, the rotaviruses require **sialic acid** for binding to the cells. Rotaviruses are the cause of severe diarrhea in infants and children worldwide. They infect the intestinal mucosa and are activated for infectivity by **pancreatic trypsin** in the intestine. The rotaviruses were initially identified by electron microscopy of fecal material.

The orbivirus strains are numerous and give rise to diseases such as encephalitis in horses and African horse sickness. The illnesses are characterized by fever, reddening, and edema of the mucosal membranes, as well as congenital abnormalities. Sheep, white-tailed deer, and pronghorned antelope infected with blue-tongue virus may develop severe symptoms, whereas cattle have milder symptoms. There are at least 11 serotypes of blue-tongue virus. Since these viruses are transmitted by midges, they were originally considered arboviruses.

The phytoreoviruses cause severe diseases of rice, maize, and other crops. The wound tumor virus must be extensively replicated in its leafhopper vector to become transmitted. It then replicates indefinitely during the vegetative propagation of the infected plant. However, it gradually becomes unable to be vector-transmitted, due to the loss of part or all of four of its genomic components. This situation resembles the formation of DI particles of many other viruses (see Sec. 4.1.4). Finally, the cytoplasmic polyhedrosis virus can cause substantial damage in the silk industry by destroying the silkworms.

Most reoviruses are persistent and nonpathogenic in their insect vectors. Table 5.1 lists and describes the six recognized genera of reoviruses. These are not related to one another by serologic or nucleic acid hybridization features.

5.1.1 Structure

The virions of the mammalian reoviruses appear spherical, but are probably of icosahedral symmetry with double shells (see Fig. 5.1). The virions are about 70 nm in diameter, are not enveloped, and do not contain lipid. Three proteins make up the outer capsid and four to six, the core. The exact architecture of these shells is still uncertain, even though distinct protruding **capsomers** are clearly visible. Some studies suggest that the capsomers are either cylinders or truncated pyramids. Most observations suggest they exist as hexa-gonal or pentagonal units. The arrangement and number of capsomers constituting the core has not been determined. The outer shell can be digested away by proteolytic enzymes, leaving the 52 nm-diameter core intact. From the vertices of the icosahedral body emerge 12 short

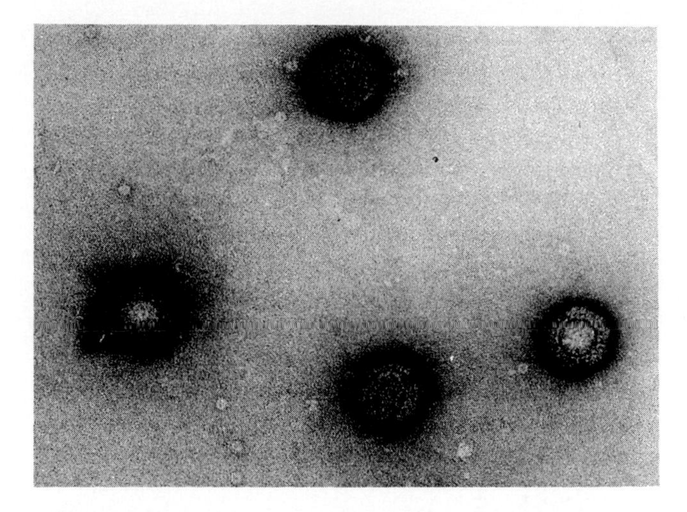

FIGURE 5.1

Electron micrograph of negative-stained reovirus. The capsomers and the double-shell structure can be clearly seen. (Courtesy of R. C. Williams)

(5 nm) stubby spikes with axial channels. Many species of orbiviruses seem to lack the sharply defined outer shell but also show these spikes and generally have a protein surrounding the core (see Figure 5.2).

The core contains—what is unique to this family—*10 to 12 molecules of double-stranded RNA,* a definite number for each reovirus genus or species. Also in the core is an RNA-dependent RNA polymerase. There are in mammalian reoviruses typically three large RNAs (L1, L2, L3, about 2.7×10^6 Da), three medium-sized RNAs (M1, M2, M3, about 1.3×10^6 Da), and four small RNAs (S1, S2, S3, S4 of 0.6 to 0.8×10^6 Da) (Fig. 5.3), a total double-stranded genome of about 15×10^6 Da. S 1 codes in different reading frames for two proteins, one of which is a **hemagglutinin.** All components carry methylated caps on the 5′ ends of the positive-sense strands and ppG- on the negative-sense strands, but there is almost as much small polynucleotide material, rich in adenylic and guanylic acid, in the core. The function of these oligonucleotides is unknown, but they are not required for productive infection.

5.1.2 Entry, Transcription, and Virus Production

Adsorption of reoviruses to the cell surface is rapid. They are subsequently taken up by phagocytic vacuoles within minutes and appear to enter "coated pits." Within an hour, most of the virus particles can be found in **lysosomes** following a fusion event with the vacuoles.

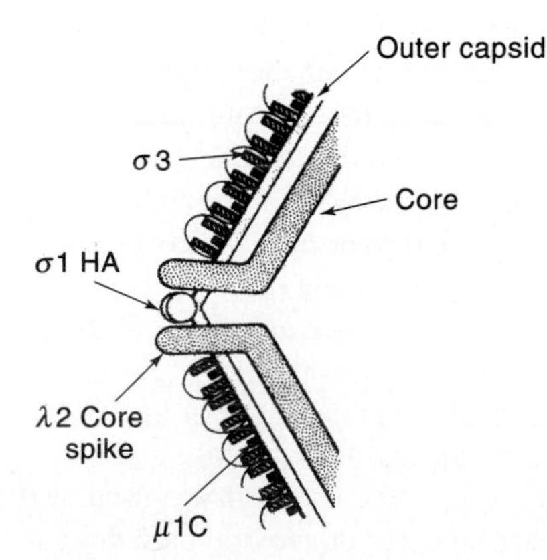

FIGURE 5.2

Surface proteins of reovirus outer capsid. (Courtesy of K. L. Tyler and B. N. Fields (1985). In *Virology,* ed. B. N. Fields, Raven Press, New York)

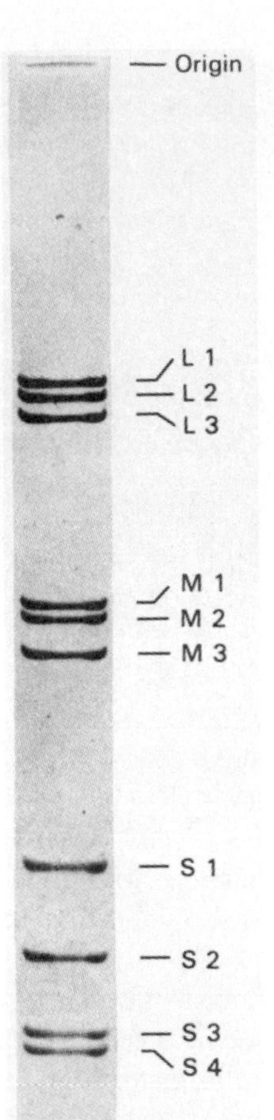

FIGURE 5.3

PAGE radioautogram of reoviral RNA (direction of electrophoresis from top). L1 is the largest and S4 the smallest component. (Courtesy of A. R. Schuerch)

A low intracellular pH is essential in the infection cycle. It leads to the digestion of the outer capsid with release of the inner core into the cytoplasm. Transcription then takes place most likely within the core.

Double-stranded RNA is inactive as mRNA. Therefore, the first step in reovirus replication is the transcription of one of the strands, by definition the negative-sense strand, to produce viral mRNA. Apparently, enzyme activity occurs at each RNA molecule, because **simultaneous transcription** of at least nine mRNAs has been observed by electron microscopy (Fig. 5.4). The same number of mRNAs can be detected in the infected cell as there are double-stranded molecules in the core. This process actually takes place within the intracellular equivalent of the core, called the **subviral particle,** which remains intact in the cell. Cores made by chymotrypsin treatment can also be used *in vitro,* and for long time periods, as RNA-producing machines. In contrast, the intact virion is not enzymatically active.

Not only are the core proteins very active as RNA polymerases, but they also have the needed enzyme activities to add the cap to the newly made mRNAs and methylate them at specific sites. The broad spike-like capsomers with axial holes are probably the channels through which the single-stranded RNAs are pushed out of the cores. The extruded mRNAs, which, like the viral RNA, have no 3′-terminal poly(A), are then translated, and 12 proteins of the expected molecular weights corresponding to the respective RNAs are produced. The proteins comprising the inner capsid of the virion [λ1, λ2, and σ2 (of 155, 90, and 38 kDa respectively)] then aggregate. Within this nascent core, back-transcription of the mRNA to the double-stranded RNA form occurs. **Pentamers** of protein λ2 (140 kDa) form the stubby core spikes. At least one (λ1) but probably three of these proteins (λ1, λ2, and σ2) are required for replicase activity, which may differ in composition or only in conformation from the transcriptase used at the beginning of the infection cycle (Fig. 5.5). The complete cores or subviral particles can serve either for a further transcription-translation cycle (Fig. 5.6), or, when enough outer capsid protein becomes available,

FIGURE 5.4

Electron micrograph of reovirus cores in the process of transcribing 12 mRNA molecules. (The bar is 200 nm.) (Courtesy of A. R. Bellamy)

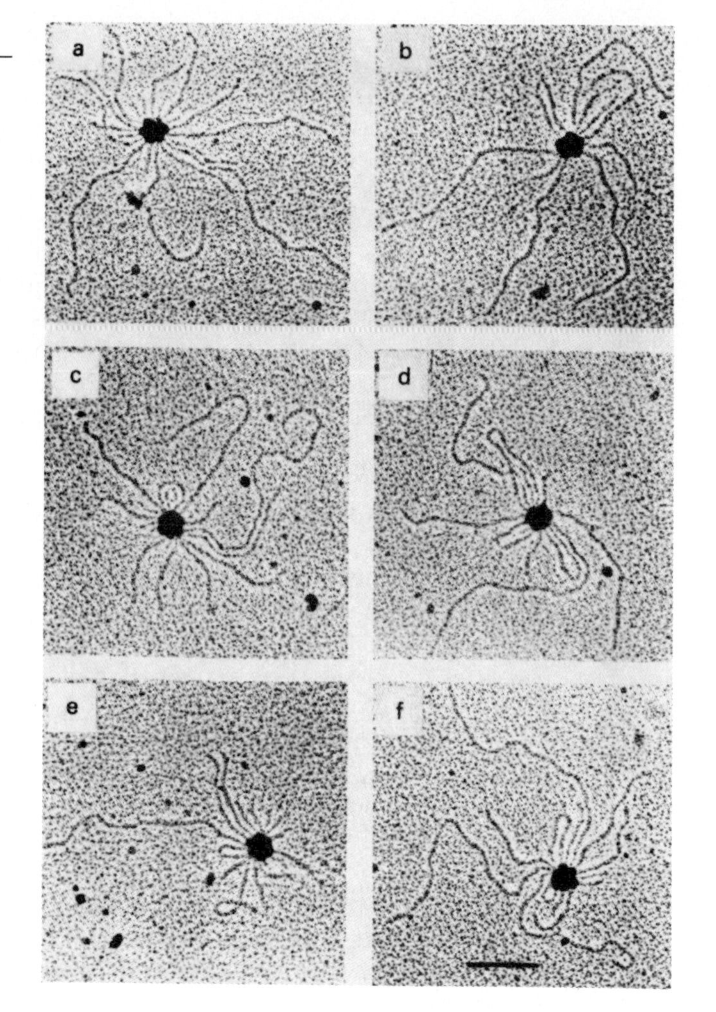

they can become virions. The proteins of the outer shell are μ1C (the C-terminal cleavage product of μ1), σ1, and σ3 (of 72, 46, and 34 kDa respectively)(Fig. 5.2). The last-mentioned protein is probably the last one to be added. This protein seals the virion so that it is no longer able to replicate its RNA. The σ1, the hemagglutinin, is responsible for attachment to the cell and for serotype specificity. Table 5.2 lists the relationship of the RNAs, their transcripts, locations, and probable functions.

TABLE 5.2
Reovirus RNAs and proteins

RNA species	*Protein species*	*Molecular weight (kD)*	*Approximate number of molecules per virus particle*	*Location*	*Function*
L3	λ1	155	120	Core	Replicase?
L2	λ2	140	60	Core spikes	Outer spikes
L1	λ3	135	12	Core	Structural
M2	μl	80	20	Outer shell	Structural
M2	μlC[a]	72	700	Outer shell	
M1	μ2	70	12	Core	Structural?
S1	σl[a]	42	48	Outer shell	Hemagglutinin
S1	σls	12		Nonstructural	Enzyme?
S2	σ2	38	180	Core	Replicase?
S4	σ3	34	720	Outer shell	
M3	μ3 (μNS)	75		Nonstructural	
S3	σ4 (σNS)	36		Nonstructural	

[a]Partly glycosylated.

FIGURE 5.5

PAGE radioautogram of reovirion proteins ranging from λ1, the largest, to σ3, the smallest.

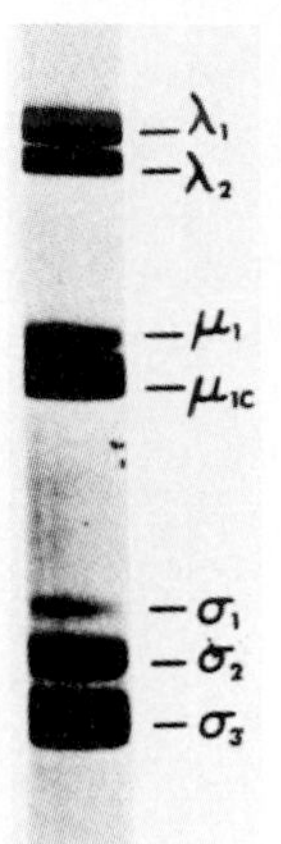

The S1 RNA that carries σ1 information has been found to be **bicistronic,** coding in a different overlapping reading frame for a second protein, σ1s, of about 12 kDa. The concept that eukaryotic mRNAs are always monocistronic has been proved wrong in an ever-increasing number of instances; dual in-phase translation has also been observed. Both types of dual translation were also found earlier in small phages and in plant viruses. (Sec. 2.6, 2.7).

The entire replication process of the Reoviridae occurs in the cytoplasm and transcription is unique in its conservative nature (see Fig. 5.6). The double-stranded RNA is *in vivo* never free of protein. Nevertheless the multiple nature of the reovirus genome, even though always protein bound, permits rearrangements to occur, similar to the processes that allow new strains of flu virus to appear. The mechanism by which the 10 unique segments of the viral RNA are placed within each particle is unknown, but probably occurs at the single-strand level and involves an RNA-binding protein. The formation of viral proteins occurs rapidly and can be found within 2 hours after infection. Mature virions are released by cell lysis.

The orbiviruses, whose prototype is the blue-tongue virus, are replicated in a manner similar to the orthoreoviruses. Their optimal temperature for transcription is 28°C, rather than 45°C as with the other animal reoviruses. The gene products of the orbiviruses are called vP1 to vP10. The proteins of the rotaviruses are similarly termed.

FIGURE 5.6

Schematic presentation of: (a) the "semiconservative" mode of transcription (displacement) of partitivirus and φ6 (see Secs. 5.5 and 5.6) as contrasted with (b) the "conservative" transcription of the Reoviridae. The solid lines represent parental RNA, the dashed lines, transcripts. Replication is always semiconservative.

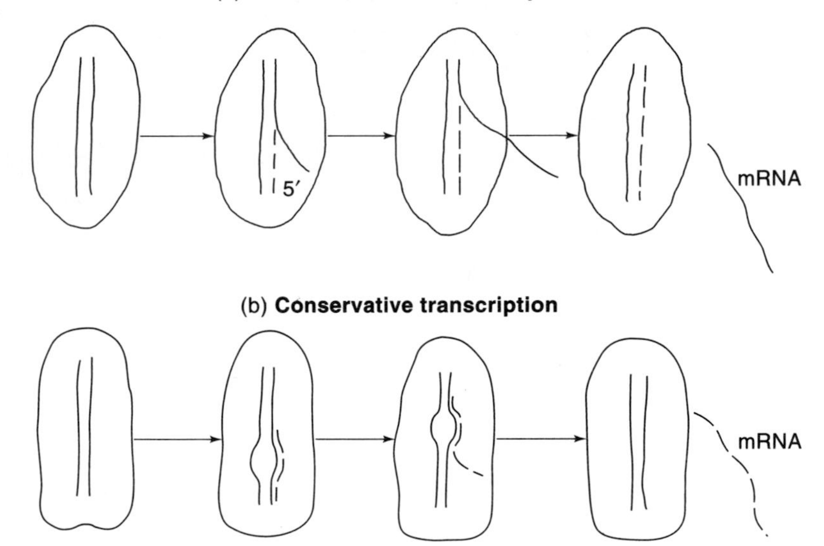

SECTION 5.1 FURTHER READING

Original Papers

GOMATOS, P. J., et al. (1962) Reovirus type 3: Physical characteristics and interactions with L cells. *Virology.* 17:441.

GOMATOS, P. J., and I. TAMM. (1963) The secondary structure of reovirus RNA. *Proc. Natl. Acad. Sci. USA* 49:707.

FURUICHI, Y., et al. (1975) 5′ Terminal m^7 G(5′)ppp(5′)G^mp *in vitro*: Identification in reovirus genome RNA. *Proc. Natl. Acad. Sci. USA* 72:742.

LEE, P. W. K., et al. (1981) Protein σ1 is the reovirus cell attachment protein. *Virology* 108:156.

HEINRICH, E., and A. J. SHATKIN. (1985) Reovirus hemagglutinin mRNA codes for two polypeptides in overlapping reading frames. *Proc. Nat. Acad. Sci. USA* 82:48.

TYLER, K. L., et al. (1986) Distinct pathways of viral spread in the host determined by reovirus S1 gene segment. *Science* 233:770.

PELLETIER, J., et al. (1987) Expression of reovirus type 3 (Dearing) σ1 and σs polypeptides in *Escherichia coli. J. Gen. Virol.* 68:135.

WINTON, J. R., et al. (1987) Morphological and biochemical properties of four members of a novel group of reoviruses isolated from aquatic animals. *J. Gen. Virol.* 68:353.

TYLER, K. L., et al. (1989) Antibody inhibits defined stages in the pathogenesis of reovirus serotype 3 infection of the central nervous system. *J. Exp. Med.* 170:887.

WIENER, J. R., and W. K. JOKLIK. (1988) Evolution of reovirus genes: A comparison of serotype 1, 2 and 3 *M2* genome segments, which encode the major structural capsid protein mu 1C. *Virology* 163:603.

DANIS, C., S. GARZON, and G. LEMAY. (1992) Further characterization of the ts453 mutant of mammalian orthoreovirus serotype 3 and nucleotide sequence of the mutated S4 gene. *Virology* 190:494.

NI, Y., R. F. RAMIG, and M. C. KEMP. (1993) Identification of proteins encoded by avian reoviruses and evidence for post-translational modification. *Virology* 193:466.

PONCET, D., C. APONTE, and J. COHEN. (1993) Rotavirus protein NSP3 (NS34) is bound to the 3′ end consensus sequence of viral mRNAs in infected cells. *J. Virol.* 67:3159.

TYLER, K. L., et al. (1993) Protective anti-reovirus monoclonal antibodies and their effects on viral pathogenesis. *J. Virol.* 67:3446.

Books and Reviews

SCHIFF, L. A., and B. N. FIELDS. (1990) Reoviruses and their replication. In *Virology,* 2d ed., ed. B. N. Fields and D. M. Knipe. 1275 Raven Press, New York.

TYLER, K. L., and B. N. FIELDS. (1990) Reoviruses. In *Virology,* 2d ed., ed. B. N. Fields and D. M. Knipe. 1307 Raven Press, New York.

JOKLIK, W. K. (1985) Recent progress in reovirus research. *Ann. Rev. Gen.* 19:537.

5.2 BIRNAVIRIDAE

The Birnaviridae family of viruses was named for its double-stranded RNA present in two (from the Greek, *bi,* "two") segments. The Birnaviridae therefore resemble the Reoviridae but have fewer gene segments. Members of this family include the **infectious pancreatic necrosis virus** (IPN) of fish (Figs. 5.7, 5.8) and the infectious bursal disease (IBD) virus of chickens, ducks, and turkeys. No mammalian counterparts are known. IBD virus gives rise to a highly contagious disease of chickens called **gumboro,** which involves immunodeficiency secondary to destruction of the bursa of Fabricius. The viruses are near-spherical of 70 nm diameter. Their RNAs are almost 3 kb and encode a 94 kDa protein, probably the DNA polymerase on segment B (Fig. 5.9). The mode of replication of birnaviruses is similar to that of other double-stranded RNA viruses but is less complicated because of the reduced number of segments. The **X virus** of *Drosophila* may be a birnavirus. Its virion of 60 nm diameter contains two double-stranded RNAs of

FIGURE 5.7

Electron micrograph of infectious pancreatic necrosis virus (IPNV) negatively stained with 1% phosphotunstic acid. The mean diameter of the virions in this hydrated state is 70 nm. (Courtesy of R. Hedrick, J. L. Fryer, and J. Leong)

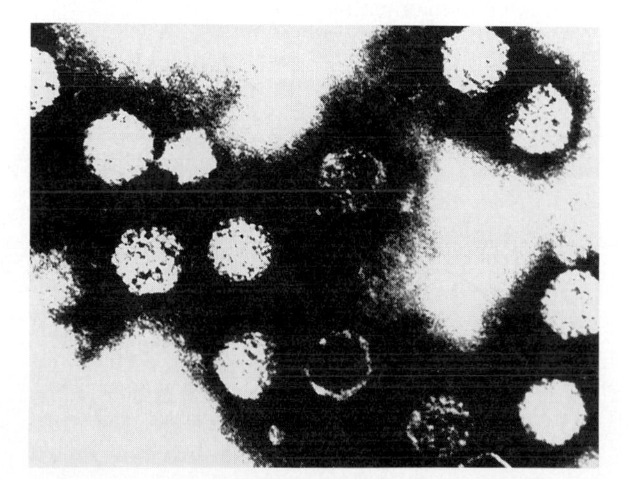

FIGURE 5.8

Fish inoculated with infectious pancreatic necrosis virus (IPNV), a birnavirus. The top fish is an uninfected, control fish. The three fish below show characteristic signs of IPNV infection, there is a prominent bump on the dorsal head, exophthalmia, petechial hemorrhages and an enlarged abdomen. Moreover, the diseased fish swim in a corkscrew pattern with fecal casts. The typical spinal curvature which results in the corkscrew swimming is shown in the third fish from the top. (Courtesy of J. Leong and L. Bootland)

about 3000 base pairs coding for a 104 kDa procapsid protein and a 44 kDa presumed RNA polymerase. A 94 kDa protein is linked to the ends of both RNAs. The procapsid protein is the source of two or three coat proteins. Identification of other members of this family may be made in the next few years.

Recently, **picobirnaviruses,** very small double-stranded RNA viruses, have been recovered from fecal material from humans, cows, guinea pigs, pigs, rats, and birds. Although, these viruses have been associated with diarrhea in animals, a pathological role in humans has not been established. The virus from guinea pigs and cows can be propagated in mammalian cells.

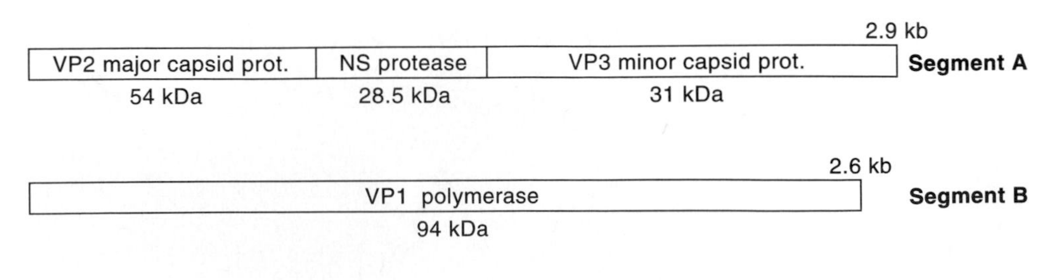

FIGURE 5.9

Genome structure of infectious pancreatic necrosis virus (IPNV), a fish birnavirus which contains two segments of double-stranded RNA in its nucleoid. Segment A mRNA is translated into a large polyprotein which is subsequently cleaved by the NS (non-structural) protein, an autocatalytic protease. Segment B mRNA encodes an RNA dependent RNA polymerase. (Courtesy of J. Leong)

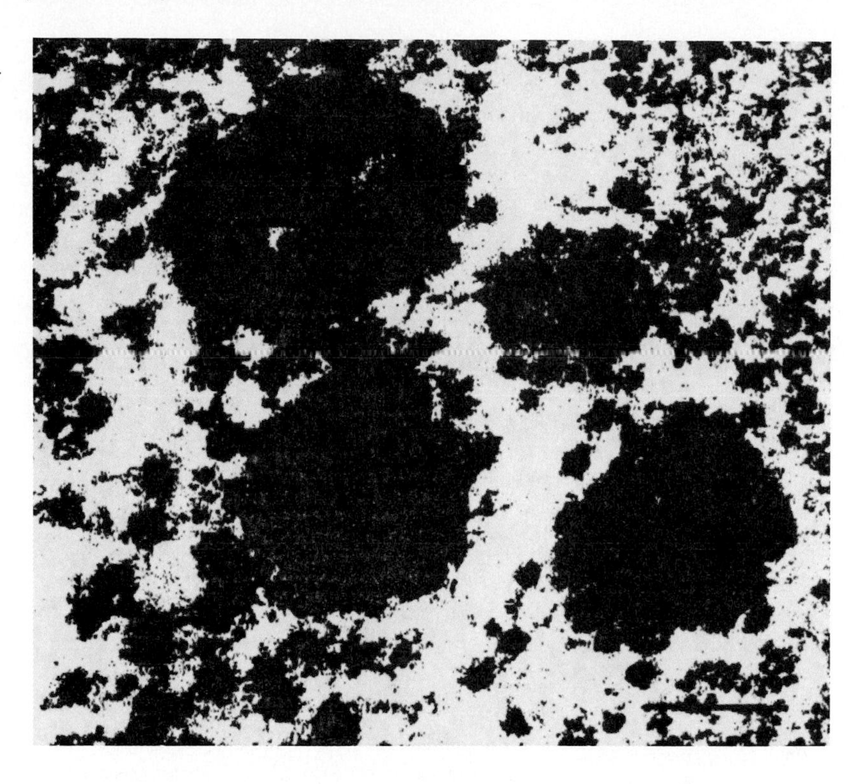

FIGURE 5.10

Developing polyhedral inclusion bodies of the *Bupalus piniarius*. Cytoplasmic polyhedrosis virus. Bar equals 250 nm.

5.3 CYPOVIRUSES (CYTOPLASMIC POLYHEDROSIS VIRUSES OF INSECTS)

Cytoplasmic polyhedrosis virions (members of the Reoviridae) are nonenveloped 60 nm particles (50×10^6 Da). Like other double-stranded RNA viruses, they are ether-resistant, thus lacking a lipid envelope. The viruses have 10 RNAs of $0.3–2.7 \times 10^6$ Da. They occur as polyhedra packaged in the cytoplasm in one major glycoprotein of 2 kDa of unknown origin (Fig. 5.10). They replicate like reoviruses in a wide variety of insects.

5.4 PHYTOREOVIRUSES AND CRYPTOVIRUSES

A wide variety of viruses containing dsRNA genomes have been isolated from plants. As shown in Table 5.3, these include both *phytoreoviruses* (two established plus one proposed genera) and two subgroups of *cryptoviruses*. Like the animal Reoviridae, plant reoviruses have double-shelled icosahedral particles that are 60 to 70 nm in diameter and contain 10 to 12 genome segments. They also replicate in cytoplasmic inclusion bodies. Particles of the many

TABLE 5.3
Plant viruses with double-stranded RNA genomes

Genus	*Genome segments*	*Vector*	*Examples (remarks)*
Phytoreoviruses			
Plant reovirus 1	12	Leafhopper	Wound tumor virus (Three serologically unrelated members)
Plant reovirus 2 [Fijivirus]	10	Plant hopper	Fiji disease and maize rough dwarf viruses (Three serologically unrelated virus clusters)
Plant reovirus 3	10	Plant hopper	Rice ragged stunt virus (Resemble *Cypoviruses* without matrix)
Cryptoviruses			
Subgroup I	2	None	White clover cryptic viruses 1 and 3 Beet cryptic virus 1–3 (RNAs = 1.20 and 0.97×10^6 Da)
Subgroup II	2	None	White clover cryptic virus 2 Carrot temperature virus 2 (RNAs = 1.49 and 1.38×10^6 Da)

known cryptic viruses are considerably smaller (30 to 35 nm in diameter) than reovirions and contain only two species of dsRNA. They thus resemble birnaviruses and several fungal viruses. Unlike conventional plant viruses, cryptic viruses are transmitted only *vertically* (i.e., through seed or pollen). Both plant reo- and cryptoviruses contain a particle-associated RNA transcriptase activity. In view of the obvious structural and functional similarities to animal reoviruses, our discussion of these two virus groups will focus on their biological properties.

5.4.1 Phytoreoviruses

All plant reoviruses are insect-transmitted, and the viruses multiply in their insect vectors (i.e., transmission is **propagative**). Following a variable acquisition period and a latency period of about 2 weeks during which the virus begins to replicate in the insect, the leafhopper or plant hopper vector remains able to transmit the virus for the rest of its life. The natural host range of **wound tumor virus** (WTV), the prototype member of the plant reovirus 1 subgroup, is unknown, but it can be experimentally transmitted to a wide range of dicotyledonous plants. Host ranges for other members of plant reovirus subgroups 1 through 3 are more restricted.

With the exception of rice dwarf virus, these viruses induce various types of neoplastic growths (e.g., tumors or enations) associated with the plant vascular system (see Fig. 5.11). WTV and rice gall dwarf virus appear to be confined to the neoplastic phloem-derived tissue. Both the host and viral genotypes are known to affect tumor size and frequency, but the molecular mechanism(s) responsible for tumor formation is unknown.

Many years of biological and molecular studies have made WTV by far the best-characterized plant reovirus. As shown in Table 5.4, the fully functional WTV genome

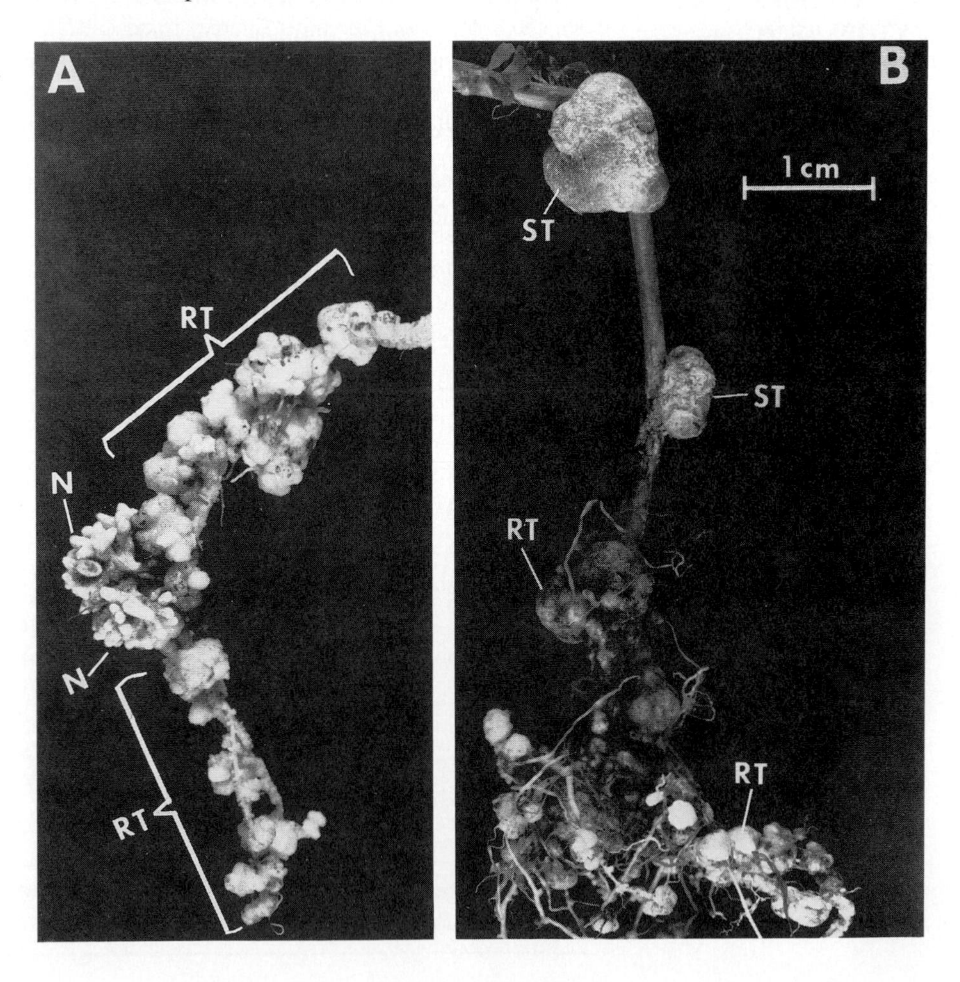

FIGURE 5.11

WTV-induced galls on the roots of clover. (Courtesy of L. M. Black and D. L. Nuss)

TABLE 5.4
Wound tumor virus: genome segments and gene products

Segment	Molecular weight[a] RNA (bp)	Protein (aa)	Nomenclature[b]	Location in virus
1	>3000	1550	P1	Nucleoprotein core
2	>3000	1300	P2	Outer protein coat
3	>3000	1080	P3	Nucleoprotein core
4	2565 bp	732	Pns4	—
5	2613	804	P5	Outer protein coat
6	1700	520	P6	Nucleoprotein core
7	1726	519	Pns7	—
8	1472	427	P8	Capsid
9	1182	345	P9	Capsid
10	1172	347	Pns10	—
11	1128	313	Pns11	—
12	851	178	Pns12	—

[a]Where known, sizes of individual genome segments and their protein products are expressed in base pairs (bp) and amino acids (aa), respectively

[b]P, structural protein present in the virion; Pns, nonstructural protein

contains 12 dsRNA segments. When virus isolates are maintained by vegetative propagation of infected plants rather than by insect transmission, certain genome segments suffer deletions and may even be lost entirely. Deletions in RNA segments 1, 4, and 10 impair the ability of WTV to be transmitted by its insect vector or to replicate in vector cell monolayers. Mutants that have completely lost segments 2 or 5 can still replicate in sweet clover and produce characteristic tumors.

Every WTV particle appears to contain a single copy of each genome segment. Equimolar amounts of each segment are present in RNA isolated from virus, and a single WTV particle is sufficient to initiate an infection. By determining the structure of a defective segment 5 RNA that was only one-fifth the normal length, yet able to compete with the parent molecule for packing, the signal responsible for packaging specificity was found to reside within 319 bp of the 5′ end or 205 bp of the 3′ end of the RNA. All 12 WTV RNAs contain fully conserved hexa- and tetranucleotide sequences at their 5′ and 3′ termini, respectively. As shown in Figure 5.12, these conserved sequences, in combination with adjacent inverted repeats, may represent the specific recognition sequences necessary to identify each individual genome segment.

5.4.2 Cryptoviruses

The term *cryptic virus* was first used by Kassanis to describe a small (30–35 nm in diameter) isometric viruslike particle isolated from sugar beet that caused no symptoms and was neither mechanically nor graft transmissible. This particle was highly **seed transmissible,** however. Since the initial report in the late 1970s similar particles have been isolated from a large number of plant species. Individual viruses appear to have narrow host ranges. Cryptic viruses are present in only low concentration, and efforts to transmit them by methods used for conventional plant viruses (i.e., mechanical inoculation, insect vectors, or even grafting) have been unsuccessful. Only radish yellow edge virus causes visible symptoms.

All the cryptic viruses examined to date contain at least two molecules of dsRNA. In the case of beet cryptic virus 1 and white clover cryptic virus 1, *in vitro* translation of the denatured genomic RNAs has been shown to yield two discrete polypeptides. The larger product (67–68 kDa) is derived from RNA 1 and may be involved in dsRNA replication. The smaller product (52–54 kDa) was precipitated by antiserum to virus particles, suggesting that it is the capsid protein. Other than because of their low concentrations, it is not known why these viruses cause only symptomless infections.

FIGURE 5.12

Organization of the wound tumor virus genome. The general organization of the genomic RNAs is illustrated at the top of the figure. Conserved hexa- (5′-terminal) and tetranucleotide (3′-terminal) sequences are indicated by large letters, the positions of the inverted terminal repeats (IR) by the open boxes, and the general position of the ORF by the solid rectangle. Terminal sequence domains from each of the 12 positive-sense strands of wound tumor virus are shown below. Segment-specific inverted repeats near the 5′ and 3′ termini have been oriented to indicate potential base-pairing interactions. The conserved hexa- and tetranucleotide sequences shared by all 12 genome segments are shown in white on black. (Modified from R. E. F. Matthews (1991) *Plant virology.* Academic Press, New York)

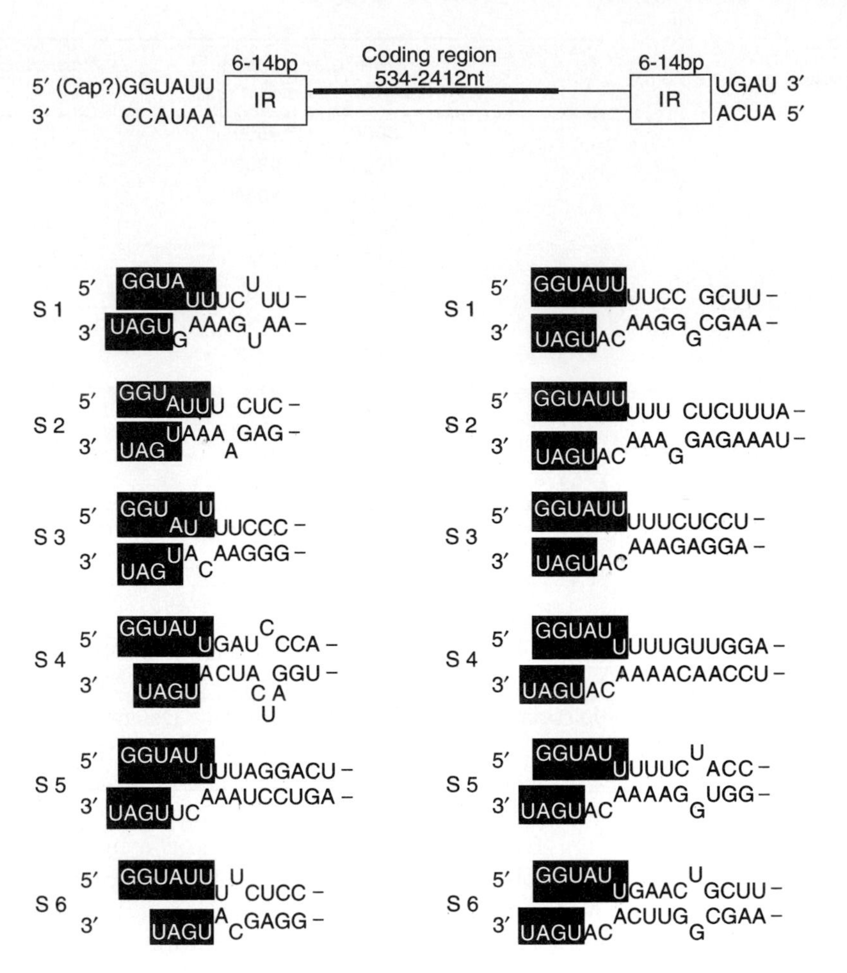

SECTIONS 5.2–5.4 FURTHER READING

Original Papers

Kassanis, B., et al. (1977) Beet cryptic virus. *Phytopath. Z.* 90:350.

MacDonald, R. D., and D. A. Gower. (1981) Geno- and phenotypic divergence among three serotypes of aquatic birnaviruses (infectious pancreatic necrosis virus). *Virology* 114:187.

Bocardo, G., and G. P. Accotto. (1988) RNA-dependent RNA polymerase activity in two morphologically different white clover cryptic viruses. *Virology* 163:413.

Pereira, H. G., et al. (1988) A virus with a bisegmented double-stranded RNA genome in rat *(Oryzonys nigripes)* intestines. *J. Gen Virol.* 69:2749–2754.

Duncan, R., et al. (1991) Sequence analysis of infectious pancreatic necrosis virus genome segment B and its encoded VPO1 protein: A putative RNA-dependent RNA polymerase lacking the Gly-Asp-Asp motif. *Virology* 181:541.

Grohmann, G. S., et al. (1993) Enteric viruses and diarrhea in HIV-infected patients. *N. Engl. J. Med.* 329:14–20.

Books and Reviews

Black, L. M. (1970) Wound tumor virus. CMI/AAB Descriptions of Plant Viruses, no. 34.

Dobos, P., and T. E. Roberts. (1982) The molecular biology of infectious pancreatic necrosis virus: A review. *Can. J. Microbiol.* 29:377.

Francki, R. I. B., R. G. Milne, and T. Hatta. (1985) Plant Reoviridae. In *Atlas of plant viruses.* Vol. 1, 47–72. CRC Press, Boca Raton, Fla.

Nuss, D. L., and D. J. Dall. (1990) Structural and functional properties of plant reovirus genomes. *Adv. Virus Res.* 38:249.

5.5 TOTIVIRIDAE AND PARTITIVIRIDAE

Members of the Totiviridae and Partitiviridae, double-stranded RNA viruses containing one to three unrelated RNA molecules, have been found in yeasts, molds, and protozoans. The term *mycovirus* is no longer favored for these viruses and they should not be confused with the DNA viruses from the family **Plasmaviridae** which infect mycoplasma (see Sec. 7.4). The totiviruses (one RNA) and the partitiviruses (two or three RNAs) resemble each other in their nonenveloped isometric structure and mode of replication. Most of these viruses are smaller than the animal birnaviruses (40 nm versus 75 nm diameter). They are generally transmitted by cell fusion, such as mating or sporogenesis, rather than by an extracellular route.

5.5.1 Totiviridae

The single double-stranded RNA molecule of the leishmania virus of *Entamoeba histolytica* has a molecular weight of about 4×10^6 (5284 nucleotide pairs). The RNA has three ORFs which encode the coat protein precursor, the RNA-dependent RNA polymerase, and a protein required to initiate translation. Many isolates contain particles with smaller RNAs, which may be satellite and/or DI RNAs (see Sec. 4.1.4 and 9.2). The best studied of these viruses is the L-A virus which infects yeast (*Saccharomyces cerevisiae*). The sequence of L-A RNA has been determined, and particles encoding "killer-proteins" have been identified. The LRVI virion of *Leishmania guyanensis* contains a partly single-stranded and partly double-stranded genome, suggesting a conservative and sequential mode of RNA replication (Fig. 5.6). This feature may also be the case for the L-A virus and most other double-stranded RNA virions, except the reoviruses. Other possible members of this family are *Giardia lamblia* virus and *Trichomonas vaginalis* virus.

5.5.2 Partitiviruses

These fungal viruses contain two (and at times three) separately encapsidated RNA molecules encoding the capsid protein and probably the RNA polymerase. Their hosts are *Penicillium, Aspergillus,* and other molds. That these viruses contain double-stranded RNAs was first noted when *Penicillium* culture filtrates, the source of penicillin, were found to have antiviral activity. Their inhibitory activity against influenza-, toga-, and picornaviruses was shown to be due to the presence of double-stranded RNA in these mycoviral filtrates. They induced the production of interferon (see Sec. 12.4). Later, it was realized that these double-stranded RNAs in mold culture fluids were actually viral genomes.

One virus species isolated from *Penicillium chrysogenum* (PcV) has 37 nm particles that contain a single capsid protein and three separately encapsidated RNAs of 1.9, 2.0, and 2.2×10^6 Da. The transcription of the double-stranded virion RNAs occurs in the same manner as that of φ6 (see Fig. 5.6). Viruses and viruslike particles (VLP) have been detected in many fungi, including the common mushroom, *Agaricus bisporus.* Green algae and other protozoans also contain many viruses and VLP of various dimensions and shapes, including TMV-like and filamentous structures. Few, however, have been characterized in terms of nucleic acid type and content.

5.6 CYSTOVIRIDAE

The Cystoviridae family includes one member, bacteriophage φ6 this bacterial virus of *Pseudomonas phaseolicola* is unique in several aspects. First, it is the only double-stranded RNA phage yet known. Second, as a consequence of the first point, it is one of the few phages with an intravirion transcriptase activity (see Chapter 8). Finally, it represents one of only a few phage groups known to have a lipid-containing envelope. Others, including the DNA phages of the PR and PMA families, are discussed in Section 7.6. Not all virion proteins enter the host, and apparently only one φ6 per cell is needed for infection. Entry is through the host pili. Presumably, as in the case of animal viruses, the envelope of φ6 is derived from the host cell membrane, but details of this process are unknown.

The genome of φ6 consists of three double-stranded RNA segments of 2.2, 3.2, and 5.0×10^6 Da. Nonionic detergents can remove the lipid envelope, leaving an icosahedral nucleocapsid consisting of at least five proteins and the double-stranded RNAs. The nucleocapsid shows RNA polymerase activity that produces three sizes of mRNA corresponding to the three double-stranded RNAs. The mechanism of RNA transcription is semiconservative (Fig. 5.6).

In vivo the nucleocapsid-associated polymerase produces the three mRNA species in about the same relative proportions. The identity of the enzyme responsible for replication of the double-stranded RNAs is not yet clear, but *in vitro* the virion polymerase does not appear to make double-stranded RNA.

SECTIONS 5.5 AND 5.6 FURTHER READING

Original Papers

VIDAVER, A. K., et al. (1973) Bacteriophage φ6: A lipid-containing virus of *Pseudomonas phaseolicola*. *J. Virol.* 11:799.

BUCK, K. W. (1978) Semiconservative replication of double-stranded RNA by a virion-associated RNA polymerase. *Biochem. Biophys. Res. Commun.* 84:639.

RATTI, G., and K. W. BUCK. (1980) Semiconservative transcription in particles of a double-stranded RNA mycovirus. *Nucl. Acid Res.* 5:3843.

VAN ETTEN, J. L., et al. (1980) Semiconservative synthesis of single-stranded RNA by bacteriophage φ6 RNA polymerase. *J. Virol.* 33:769.

ROMANCHUCK, M., and D. H. BAMFORD. (1985) Function of pili in bacteriophage φ6 penetration. *J. Gen. Virol.* 66:2461.

MCGRAW, T., et al. (1986) Nucleotide sequence of the small double-stranded RNA segment of bacteriophage φ6: Novel mechanism of natural translation control. *J. Virol.* 58:142.

STEELEY, H. T., JR., et al. (1986) Study of the circular dichroism of φ6 and φ6 nucleocapsid. *Biopolymers* 25:171.

BARBONE, F. P., et al. (1992) Yeast killer virus transcription initiation *in vitro*. *Virology* 187:333.

FRILANDER, M., et al. (1992) Dependence of minus-stranded synthesis on complete genomic packaging in the double-stranded RNA bacteriophage φ6. *J. Virol.* 66:5013.

KENNEY, J. M., et al. (1992) Bacteriophage φ6 envelope elucidated by chemical cross-linking, immunodetection, and cryoelectron microscopy. *Virology* 190:635.

RIBAS, J. C., and R. B. WICKNER. (1992) RNA-dependent RNA polymerase consensus sequence of the L-A double-stranded RNA virus: Definition of essential domains. *Proc. Natl. Acad. Sci. USA* 89:2185.

STUART K. D., et al. (1992) Molecular organization of the RNA of *Leishmania* virus. *Proc. Natl. Acad. Sci. USA* 89:8596.

WEEKS, R., R. F. ALINE, Jr., P. J. MYLER, and K. STUART. (1992) LRV1 Viral particles in *Leishmania guyanensis* contain double-stranded or single-stranded RNA. *J. Virol.* 66:1389.

OJALA, P. M., J. T. JUUTI, and D. H. BAMFORD. (1993) Protein P4 of double-stranded RNA bacteriophage φ6 is accessible on the nucleocapsid surface: Epitope mapping and orientation of the protein. *J. Virol* 67:2879.

Books and Reviews

HOLLINGS, M. (1978) Mycoviruses: Viruses that infect fungi. *Adv. Virus Res.* 22:1.

SAKSENA, K. N., and P. A. LEMKI. (1978) Viruses in fungi. In *Comprehensive virology,* ed. H. Fraenkel-Conrat and R. R. Wagner. Vol. 12. 103–143. Plenum Press, New York.

MINDICH, L., and D. M. BANFORD. (1988) Lipid-containing bacteriophages. In *The bacteriophages 2,* ed. R. Calendar. 475 Plenum Press, New York.

WICKNER, R. B. (1989) Yeast virology. *FASEB J.* 3:2258.

chapter

6

Viruses Using Reverse Transcription During Replication

Several single-stranded RNA viruses and a small group of DNA viruses use reverse transcription to complete their replication cycle. Through the function of an RNA-dependent DNA polymerase (**reverse transcriptase**), the RNA genomes are copied into a DNA. In many cases, this cDNA then becomes integrated into the chromosome of the cell as a proviral copy of the RNA virus. This chapter reviews the prominent family in which this reversal of the genetic code was first found, the Retroviridae, and other virus families that use this enzyme. Finally, we discuss several noninfectious reverse transcriptase–containing viruses that may replicate within the cell as well as retroviruslike sequences that are found in cellular genomes.

6.1 RETROVIRIDAE

Retroviridae is the taxonomic name for a large family of RNA-containing viruses that have a variety of pathogenic effects, including the ability to replicate and remain present in cells without any sign of infection. Several retroviruses cause malignant tumors or leukemias in animals. These viruses were formerly called RNA tumor viruses, leukoviruses, or oncornaviruses (from the Greek *onkos,* meaning "bulk," and RNA). The name retrovirus (from the Latin *retro,* "backward") is derived from the classifying criterion that these viruses contain reverse transcriptase. This RNA-dependent DNA polymerase copies DNA from the viral genome—a reversal of the usual genetic process. The retroviruses differ from all other RNA viruses in that their RNA genome, transcribed to DNA, can become incorporated into the host's cell genome. Depending on the genetic structure of a given virus, this event can lead to genetic transformation of the host cells, and it can result in malignant tumors or leukemias, although many of these viruses are not pathogenic, cytocidal, or transforming. **Rous sarcoma virus,** the prototype of this group, was identified as the causative agent of chicken malignancies early in the history of virology (see Sec. 1.1.2). The mouse mammary tumor virus (MMTV) and the Gross murine leukemia virus (MuLV) were the first mammalian oncogenic retroviruses discovered. Other retroviruses have frequently been isolated from birds and mice as well as from other species, including, recently, humans. Retroviridae is the only virus family known to be able to transform host cells and simultaneously produce virus.

6.1.1 Classification

Table 6.1 lists in taxonomic manner prototypes of the RNA viruses that on the basis of containing reverse transcriptase are now included in the family of Retroviridae. Three subfamilies were initially identified based on morphological, biological, and molecular features. Most recently the International Committee for Taxonomy of Viruses (ICTV) has subgrouped the retroviruses into seven classifications. They have maintained the Spumavirinae

TABLE 6.1

Comparison of "old" and "new" taxonomy of retroviruses

Taxon	"Old" Taxonomy	"New" Taxonomy
Family	Retroviridae	Retroviridae
Subfamily	Oncovirinae	[a]
Genus	Type C oncovirus	**MLV-related viruses**
Subgenus	Mammalian type C oncoviruses	Mammalian type C viruses
Species	MLV, FeLV, GALV	MLV, FeLV, GALV
Species	HTLV, BLV	[b]
Subgenus	Reptilian type C oncoviruses	Reptilian type C viruses
Species	CSRV, VRV	CSRV, VRV
Subgenus	Avian type C oncoviruses	Reticuloendotheliosis viruses
Species	SNV, REV	SNV, REV
Species	ALV, RSV	[b]
Genus	Type B oncovirus	**Mammalian type B viruses**
Species	MMTV	MMTV
Genus	Type D oncovirus	**Type D viruses**
Species	MPMV, SMRV	MPMV, SMRV
Genus	[b]	**ALV-related**
Species	[b]	ALV, RSV
Genus	[b]	**HTLV-BLV**
Species	[b]	HTLV-I, II, BLV
Subfamily	Lentivirinae	[a]
Genus	Lentivirus	**Lentivirus**
Subgenus	—	Ovine/caprine lentiviruses
Species	Visna, CAEV	Visna, CAEV
Subgenus	—	Equine lentiviruses
Species	EIAV	EIAV
Subgenus	—	Primate lentiviruses
Species	HIV, SIV	HIV, SIV
Subgenus	—	Feline lentiviruses
Species	FIV	FIV
Subgenus	—	Bovine lentiviruses
Species	BIV	BIV
Subfamily	Spumavirinae	[a]
Genus	Spumavirus	**Spumavirus**
Species	HSRV, SFV	HSRV, SFV

Source: Taken from J. Coffin. (1992) *The Retroviridae.* Vol. 1, ed. J. Levy, Chap. 2. Plenum Press, New York, with permission.

[a]The subfamily level of classification is no longer used.

[b]These groups are no longer considered to be sufficiently similar to the mammalian type C viruses to be classified with them.

Acronyms	*Full Names*
ALV	Avian leukosis (leukemia) virus
BIV	Bovine immunodeficiency virus
BLV	Bovine leukemia virus
CAEV	Caprine arthritis-encephalitis virus
CSRV	Corn snake retrovirus
EIAV	Equine infectious anemia virus
FeLV	Feline leukemia virus
FIV	Feline immunodeficiency virus
GALV	Gibbon ape leukemia virus
HIV	Human immunodeficiency virus
HSRV	Human spuma retrovirus
HTLV	Human T-cell leukemia (lymphotropic) virus
MLV (MuLV)	Murine (mouse) leukemia virus
MMTV	Mouse mammary tumor virus
MPMV	Mason-Pfizer monkey virus
REV	Reticuloendotheliosis virus
RSV	Rous sarcoma virus
SFV	Simian foamy virus
SIV	Simian immunodeficiency virus
SMRV	Squirrel monkey retrovirus
SNV	Spleen necrosis virus
VRV	Viper retrovirus

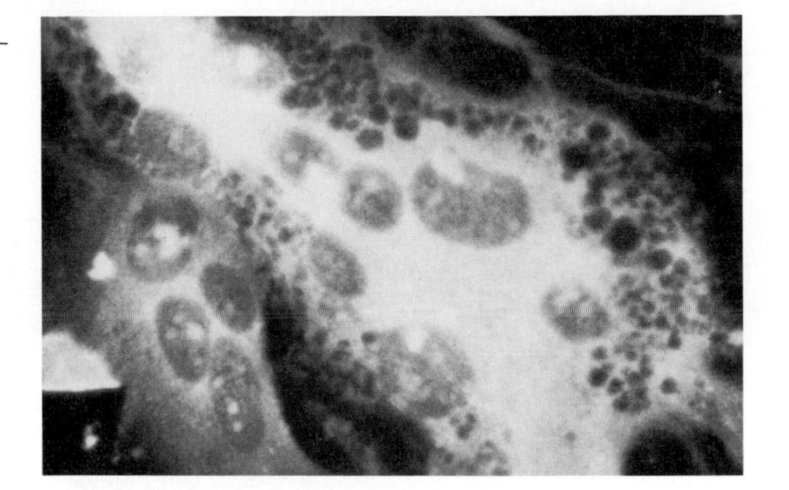

FIGURE 6.1

Syncytia formation induced in human epithelioid cells by human foamy virus. Note the multinucleated cells and cell degeneration. (From P. Loh, *The Retroviridae,* ed. J. Levy. Vol. 2 (1993) Plenum Press, New York)

and Lentivirinae as subgroups, but have divided the former Oncovirinae into five separate genera, as noted in the table. Many of these viruses are associated with leukemias and sarcomas in a wide variety of animal species (morphological types B, C, D). Others, however, do not cause malignancies. The Spuma-viruses (from the Greek *spuma,* "foam") derive their name from the vacuolar and soapsudslike features observed in cells infected with this genus of retroviruses (see Fig. 6.1). Although associated with cytopathic effects in cell culture, they have not been linked to any diseases. The Lentivirinae (from the Latin *lenti,* "slow") obtained their name from the long incubation period from infection to disease. Generally, this virus group is associated with chronic encephalopathies, pneumonitis, arthritis, **immune deficiencies,** and hemolytic anemia (in horses) (see Table 6.2). They have been identified in horses, goats, sheep, primates, cats, cows, and humans (see Chap. 14). Although these genera can be grouped according to their morphological and pathogenic features (Table 6.2), many overlapping biological properties can be found among them (Table 6.3) (see Sec. 10.3). The spuma- and lentiviruses are more often associated with **cytopathic changes** in infected cells and can be identified by distinct properties (Table 6.3). Nevertheless, all genera of the retroviruses can infect and replicate in cells without cytocidal activity.

The first five genera can be subclassified according to morphological criteria by electron microscopy (Table 6.1, Fig. 6.2). The A-type viruses (primarily found in mice) are located in the cytoplasm and are of two types: intracytoplasmic and intracisternal. The latter forms are found particularly in developing embryos and cross-react with a mouse type D virus. Their role in nature has not been defined. Recently a type A particle in a human T lymphocyte cell line was described. The intracytoplasmic forms of A particles appear to be

TABLE 6.2

Pathological conditions associated with animal retroviruses

Disease	*Animal Host*	*Virus Example*
Leukemia	Birds, mammals	Type C viruses
Sarcoma	Birds, mammals	Type C viruses: Rous sarcoma virus, Harvey sarcoma virus
Carcinoma	Mice, sheep	Type B (MMTV) and type C viruses
Immunodeficiency	Birds, mice, cats, monkeys, human	Type C and type D viruses and lentiviruses
Arthritis	Goats	Lentivirus (Caprine arthritis encephalitis virus)
Pneumonia	Sheep	Lentivirus (maedi)
Encephalopathy, neurological disease	Mice, sheep, goats, monkeys, human	Lentivirus (caprine arthritis encephalitis virus, visna); feral mouse type C virus
Autoimmunity (hemolytic anemia)	Horses	Lentivirus: equine infectious anemia virus

Source: Modified from J. A. Levy. (1986) *Cancer Res.* 46:5457. Courtesy of *Cancer Research.*

TABLE 6.3

Biological features of retroviruses[a]

Genus	Infects Nondividing Cells	Latent Infection	Syncytia Formation	Vacuole Formation	Cell Death	Transformation	Antigenic Variants[b]
MLV-related	–	++	++	+[c]	+[c]	+++	–
Spumavirus	–	++	+++	+++	++	–	–
Lentivirus	+	++	+++	+[c]	++	–	++

Source: Modified from J. A. Levy. (1986) *Cancer Res.* 46:5457. Courtesy of *Cancer Research.*

[a]Comparable degree of expression by genus is represented by plus signs.
[b]Refers to generation of envelope antigenic variants that escape host neutralizing antibody.
[c]Limited to certain subtypes.

FIGURE 6.2

Comparisons by electron microscopy of morphologies of A-, B-, C-, and D-type retrovirus particles. (a) Intracisternal A particles from a rhabdomyosarcoma-derived cell line of the mouse. Note budding from the endoplasmic reticulum into cisternal spaces; no extracellular forms. (b) A cluster of intracytoplasmic A particles from a C_3H mammary tumor. (c) Mouse mammary tumor virus (MMTV) type B particles budding into a lumen of a C_3H mammary tumor. (d) Extracellular mature MMTV; C_3H tumor. (e) Type C virions budding from mouse L cells. Note crescent nucleoid forming concomitantly with bud. (f) Cluster of mature type C virions in an L-cell culture. (g) Small cluster of squirrel monkey retrovirus (SMRV) intracytoplasmic A particles (in canine fetal thymus cells). (h) Budding SMRV D-type virion; typically like MMTV with preformed nucleoids and distinct fringe of spikes. (i) Extracellular mature type D SMRV virions. (Courtesy of John E. Dahlberg)

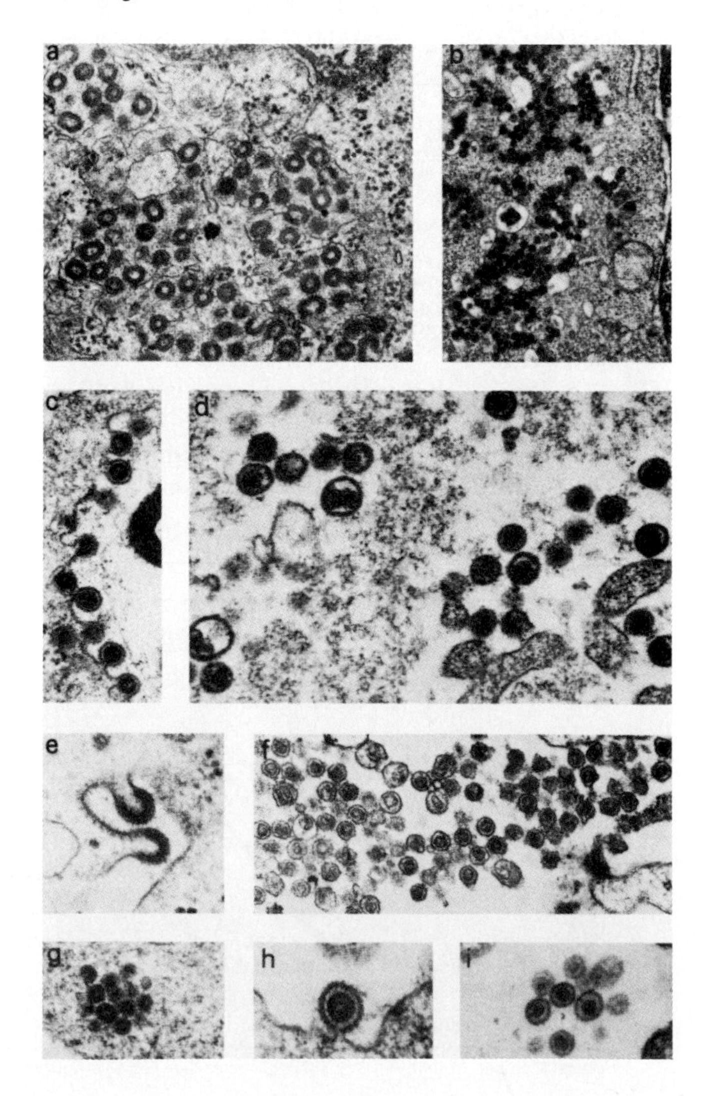

precursors of type B particles and are inserted into the viral envelope when the virion buds off the plasma membrane. In contrast, nucleoids of type C particles form during the budding process. The type B particles can be distinguished by large spikes and the **eccentric nucleoid** in a mature particle, rather than the central nucleoid of the type C virus. Finally, the type D particles appear to form a complete nucleoid before budding.

The shape of the viral core can also distinguish some retroviruses. For instance, the type D and the lentiviruses have cylindrical and cone-shaped forms, respectively, in contrast with the cuboidal shape of the other retroviruses. Recently, one amino acid change alone in the core protein of a type D retrovirus has been found to modify the shape of the nucleoid from a cylindrical to a cuboidal form.

The reverse transcriptase of most of the Retroviridae prefers Mg^{2+} to Mn^{2+} as a cation for its activity. Only the MLV-related and type D viruses differ in their preference for Mn^{2+}. The HTLV-BLV and ALV-related genera, in contrast with the conventional animal retroviruses with type C morphology, use Mg^{2+} rather than Mn^{2+} as the cation during reverse transcription. Based on biological, serological, and molecular studies, it appears that all these retroviruses were evolutionarily related.

The **lentiviruses** are particularly noteworthy for their heterogeneity. Genetic differences in the envelope region as well as the accessory genes in these viruses can be observed among many isolates. These variations can help in epidemiologic and transmission studies in which particular viruses are being traced. The increased mutation rate among the lentiviruses is most likely related to their **error-prone** reverse transcriptase. In some ways, the enzyme permits survival of the virus through its ability to alter the virus sufficiently to escape immune surveillance. The general properties of the retrovirus genera are summarized in Table 6.4.

6.1.2 Structure of Retroviruses and Components

Due to the intense interest in the relationship of retroviruses to cancer, encephalopathies, and, more recently, to AIDS, the amount of material published on these viruses far exceeds

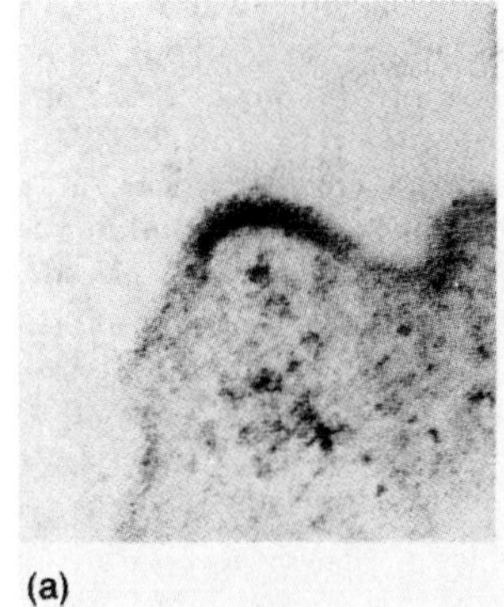

(a)

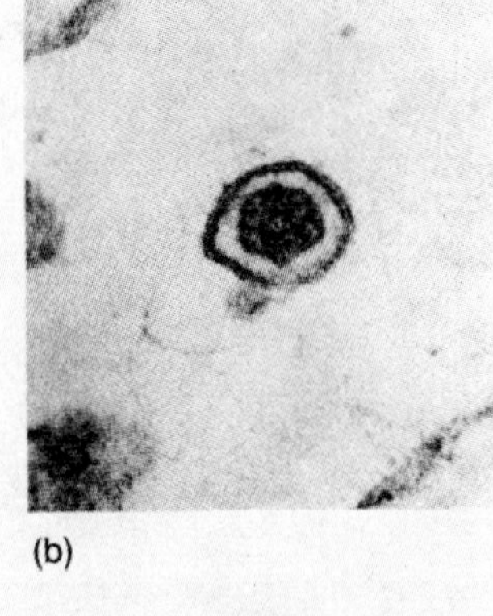

(b)

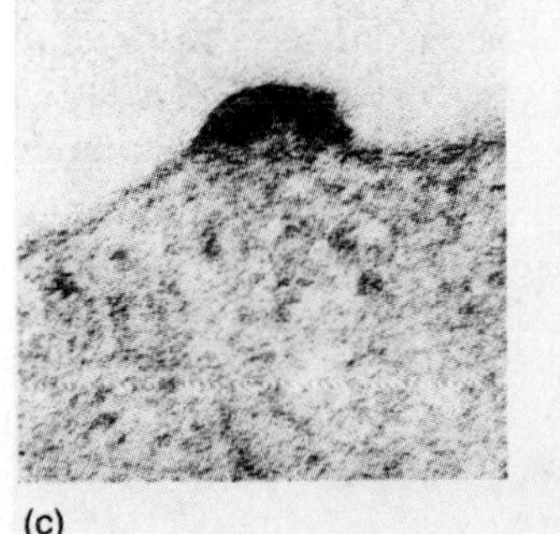

(c)

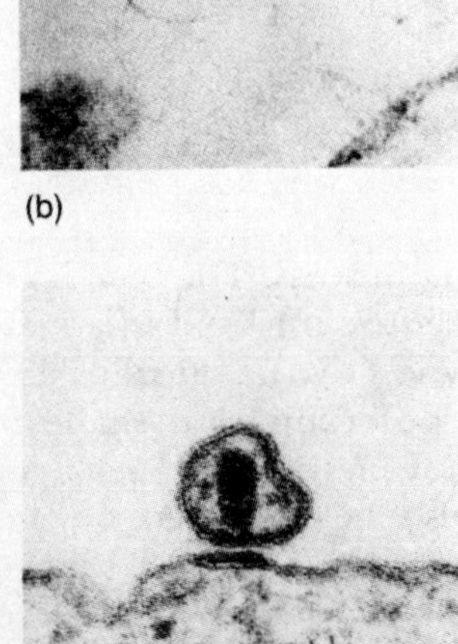

(d)

FIGURE 6.3

Electron micrograph examination of the human T-cell leukemia virus. type I (HTLV-I) and the human immunodeficiency virus (HIV). (a) Budding HTLV-I. (b) Mature HTLV-I. (c) Budding HIV. (d) Mature HIV. Note the lack of a translucent area between the nucleoid and the cell membrane during the budding process of HIV, as well as this virus's cone-shaped core. Marker = 100 nm. (Photos provided by L. S. Oshiro, courtesy of Little, Brown and Co.)

TABLE 6.4
Characteristics of the retrovirus genera

Genus Subgenus	*Examples*	*EM Type*[a]	*Genome Size (kb)*[b]	*"Extra" Genes*	*Endogenous Members?*	*Members with Oncogenes?*	*Pathogenesis*
1 **MLV-related**							
Mammalian C type	Murine leukemia virus	C	8.3	None	Yes	Yes	Various: malignancies, immunodeficiencies, and neurological diseases
Reticuloendotheliosis	Avian reticuloendotheliosis virus						
Reptilian C type	Viper retrovirus						
2 **Mammalian B type**	Mouse mammary tumor virus (MMTV)	B	10	1 *(orf)*	Yes	No	Mammary carcinoma; T-cell lymphoma
3 **Type D**							
Primate type D	Mason-Pfizer monkey virus, simian retrovirus (SRV)	D	8.0	None	Yes	No	Immunodeficiencies
Ovine type D[b]	Jaagsiekte retrovirus						(Malignancy)
4 **ALV-related**	Avian leukosis and sarcoma viruses	C	7.2	None	Yes	Yes	Malignancies, osteopetrosis, immunological disease
5 **HTLV-BLV**	Human T lymphotrophic virus (HTLV), bovine leukemia virus	C	8.3	2 *(tax, rex)*	No	No	T- or B-cell lymphoma, neurological disease
6 **Lentivirus**							
Primate immunodeficiency viruses (human, simian)	HIV, SIV	L	9.2	~6 *(tat, rev, nef, vif, vpu, vpr)*	No	No	Immunodeficiencies, autoimmune degenerative, neurological, and other diseases
Ovine-caprine lentiviruses	Visna/Maedi virus						
Equine lentiviruses	Equine infectious anemia virus (EIAV)						
Feline lentiviruses	Feline immunodeficiency virus (FIV)						
Bovine lentiviruses	Bovine immunodeficiency virus (BIV)						
7 **Spumavirus**	Human, simian, feline foamy virus	S	11	~4(bel 1, 2, 3, bet)	No	No	None known

Source: Modified from J. Coffin. (1992) *The Retroviridae.* Vol.1, ed. J. Levy. Chap. 2. Plenum Press, New York, with permission.

N*ote:* For most retroviruses, the reverse transcriptase uses Mg^{2+} as its cation. For the MLV-related and type D virus genera, Mn^{2+} is used.

[a]Classification by electron microscopy.
[b]Approximate.
[c]Tentative.

that on any other virus family. We focus in this chapter on the molecular biology of one of the first of these agents studied, an avian retrovirus, the Rous sarcoma virus. Most of the major features of retroviruses from other species resemble those of this avian prototype. HIV is discussed in Chapter 14. The biological, serological, and oncogenic properties of the retroviruses, their host specificities, modes of transmission, and relationship to host genes are also considered in detail in Sections 10.3, 12.1, and 12.7.

The Rous sarcoma virus (RSV) resembles most enveloped RNA viruses in that its presumed **icosahedral nucleocapsid** is covered by an envelope consisting of a **lipid bilayer** and spikes formed by transmembrane (TM) and surface (SU) glycoproteins. On the external region of the glycoprotein are found both group- and type-specific antigens. The type-specific antigen serves as the identification target for neutralizing antibodies and for serological subtyping. The four internal structural proteins of RSV, about 65 percent of the total, are identical within the RSV group and are termed **group-specific antigens** (*gag*). In other retroviruses these proteins can also have some type-specific epitopes. In most retroviruses, the *gag* precursor protein is cleaved into three to five capsid proteins. Among these are those making up the matrix (MA), the capsid (CA), and the nucleic acid binding protein (NC). The MA is located just under the virion envelope and serves as a stabilizing structure for the virus.

The nucleocapsid of RSV consists of two phosphoproteins of 19 and 12 kDa that bind to the RNA, and a component of 15 kDa; a shell around the capsid contains the major *gag* protein (27k) that corresponds to the M protein of other enveloped families (see Fig. 6.4). Inside the core is also the **reverse transcriptase** (62 and 96 kDa) or polymerase (*pol*), closely associated with the viral RNA.

The **envelope** (*env*) glycoproteins of RSV are about 85 and 37 kDa and represent split products of a larger unglycosylated precursor. Two or three other small proteins of unknown role are also often detected in small amounts. As in other virus families, the envelope lipids (about 30 percent) are derived from the plasma membrane, and the polysaccharide components of the glycoproteins are also host cell–derived, though virus-specific.

Retroviruses differ from all other viruses in having a **diploid genome,** which means that their two RNA molecules are identical. The reason for this feature is not clearly understood. It appears probable that the interaction of the two molecules serves a regulatory function in reverse transcription. The diploid nature of the viral RNA may also permit a higher rate of recombination, particularly if **heterozygotes** form in which each RNA molecule within a virion core comes from a different virus. In the 70 S RNA, the strands are, through base pairing, hydrogen-bonded to one another near their 5′ ends, and there is other evidence for a specific conformation (Fig. 6.5).

The genome of RSV (70 S) is dissociated by heating into its two molecules of single-stranded RNA of about 38 S (3.3×10^6 Da). The RNA is a positive-sense strand and has the features of typical eukaryotic mRNAs: a methylated 5′-terminal cap and 3′-terminal poly(A), about 200 residues long.

In addition to that large RNA, retroviruses contain a mixture of small RNAs of 4 to 10 S, which appear to be degradation products and cellular tRNAs. However, during the denaturing treatment of the purified 70 S RNA, which is necessary to separate the 38 S molecules, a small amount of a specific *4* S RNA is released that is predominantly a *single tRNA* and an *essential primer* for RNA transcription (see Sec. 6.1.3).

The molecular organization of a retroviral RNA strand is (5′)R-U5-*gag,pol,env*-U3-R(3′). The R region at both ends of the RNA represents short repeated sequences (16 to 65 nucleotides). This region is required for dimerization of the two RNA strands (Fig. 6.5).

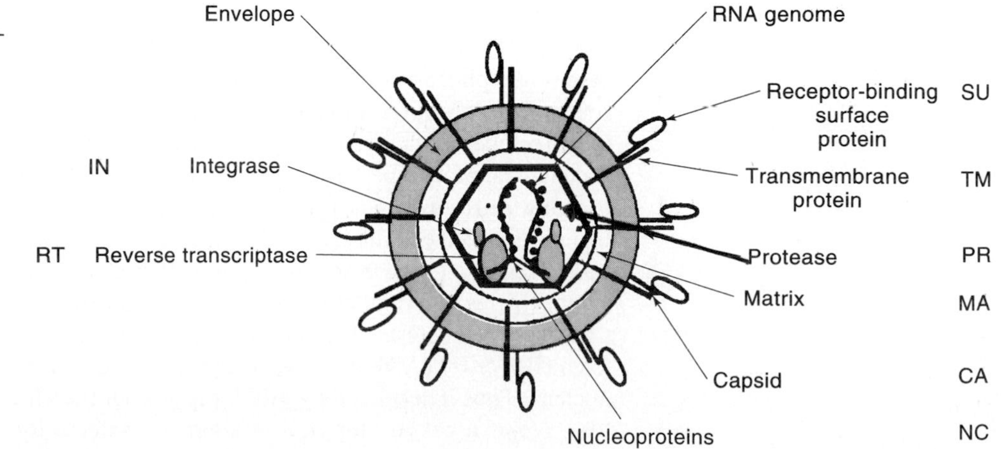

FIGURE 6.4

The retrovirus virion. The figure is schematic, depicting the current ideas of the locations and relationships of the various components. (Modified from J. Coffin *The Retroviridae,* ed. J. Levy, Vol. 2 (1993) Plenum Press, New York)

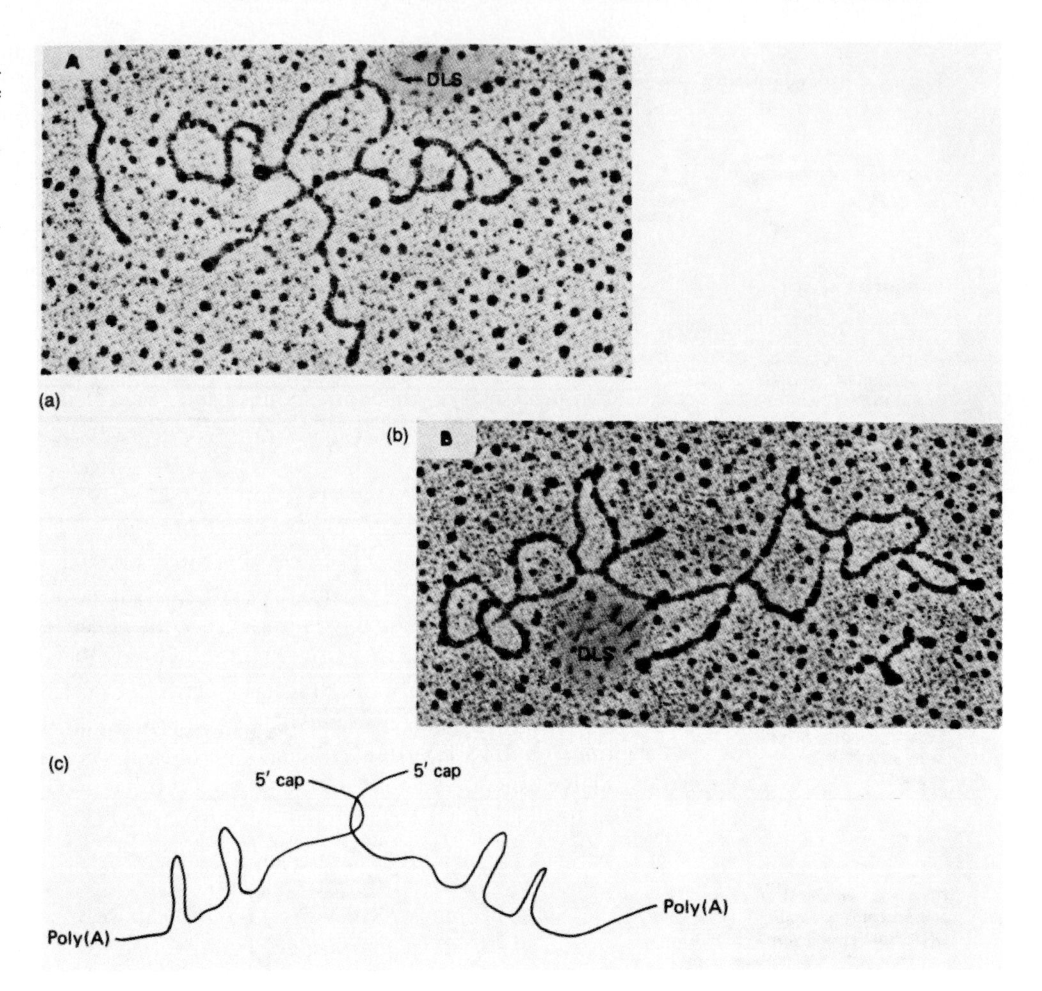

FIGURE 6.5

(a) and (b) Electron micrographs of the dimeric Rous sarcoma virus RNA (two molecules of 38 S RNA with characteristic intertwining at D (dimer)); conformation of long and short hairpins are indicated by L and S (c) Schematic presentation. (Unpublished, courtesy of N. Davison)

U5 and U3 represent sequences unique, respectively, to the 5′ and 3′ ends of the RNA genome. These three sets of end sequences help form the **long terminal repeats** (LTR) in the viral genome (see Fig. 6.6). They contain regions that are important for promotion, enhancement, and regulation of virus replication. The internal regions of the viral RNA code for the structural proteins (noted above) of fully infectious retroviruses: gag (core proteins), pol (polymerase), and env (envelope) glycoproteins. In addition to these three structural genes, the **oncogenic viruses** may have an additional gene, called an **oncogene** (*onc*), associated with cellular transformation; it is not required for replication. In some viruses, such as RSV, this gene, called *src,* is an additional region in the viral genome (Figs. 6.8, 6.9). With other oncogenic viruses, parts of the structural genes may be deleted or partly substituted by an **oncogene.** These viruses cannot replicate and require the help of a replication-competent virus to provide the protein products lacking for their assembly or replication. In general, the usual protein missing in these oncogenic viruses is the envelope. Thus, these replication-defective viruses, when produced, carry the outer protein coat of a helper virus (see Sec. 6.1.5). Further discussion on oncogenes and transforming retroviruses is found in Section 10.3. Certain retroviruses, particularly the lentiviruses, have additional coding regions in their genome that can affect virus replication. Common to these so-called **complex retroviruses** (HTLV, BLV, lentiviruses, and spumaviruses) are regulatory genes that can up-regulate virus replication through transactivating proteins (Tax, Tat, and Bel). Moreover, another regulatory protein (Rev, or Rex from HIV or HTLV respectively) appears to prevent splicing of small mRNAs within the nucleus. Thus it permits long mRNA species of the virus to enter the cytoplasm and code for proteins necessary for viral structure and infectivity.

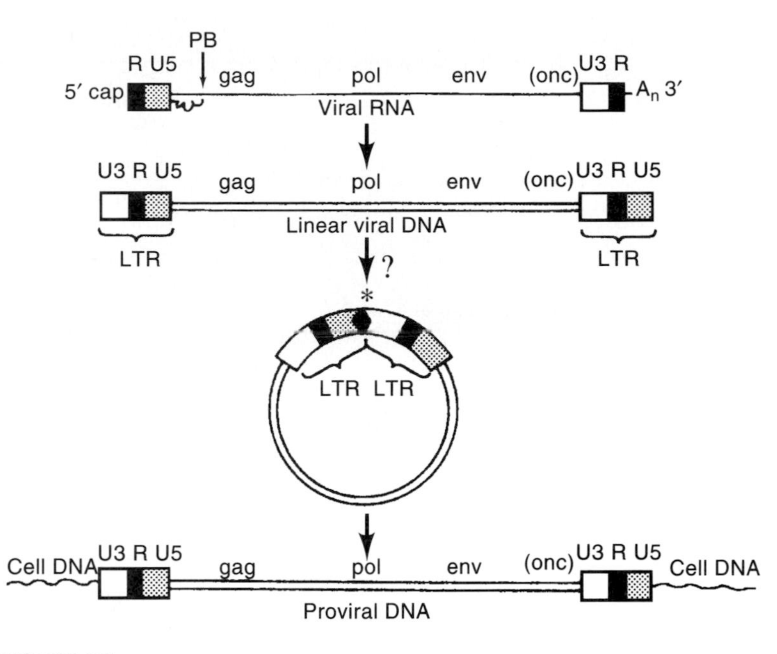

FIGURE 6.6

Major features of the retroviral RNA and its DNA provirus structure. The viral genome is *diploid,* containing two copies of the viral RNA, each approximately 3 x 10^6 Da in size. These are linked together in the mature virion at the 5′ termini by complementary base pairing. (*Top*) structural features of a retroviral RNA: the capped nucleotide (5′-cap) at the 5′ terminus, the polyadenylate tract at the 3′ terminus, and the site of binding (PB) of the cellular tRNA used to prime DNA synthesis via the reverse transcriptase. The short sequence R (for repeat sequence) is present at both ends of the viral RNA; *U5,* unique sequence found at the 5′ end of viral RNA; *U3,* sequence found uniquely at the 3′ end of the viral RNA. During replication, the viral RNA is transcribed into a double-stranded linear DNA molecule with flanking LTRs. *(Middle)* organization of these sequences as flanking LTR units in the linear viral DNA which forms during viral replication. The linear viral DNA is probably converted to a non-convalently bound circular DNA molecule before integration into the host cellular DNA *(bottom)* where the LTRs flank the proviral DNA and are specifically linked to the cellular DNA. *, circular junction or attachment site for integration of the proviral form into cellular DNA. (Adapted by J. Leong from R. Weiss et al. (1982) RNA tumor viruses. Cold Spring Harbor Press. J. A. Levy. (1986). *Cancer Res.* 46:5457. Courtesy of *Cancer Research*)

Another regulatory protein of HIV is coded by the *nef* gene and appears to be pleiotropic. In some cells *nef* suppresses HIV replication (thus its name, for *negative factor*). However, recently some *nef* gene products have enhanced HIV production in certain cell types. Thus, its true function is not yet known. Other accessory viral proteins (Vp) (e.g., Vif, Vpr, Vpu) appear to help in virus infectivity and replication.

6.1.3 Entry and Reverse Transcription of Retroviridae

Retroviruses begin the infection process by attaching to specific receptors on the cell surface. In recent years the identity of these receptors has been defined for some avian and murine retroviruses and for the human immunodeficiency virus (e.g., CD4 molecule). The receptor for the Moloney murine leukemia virus is a permease that resembles a membrane pore or channel protein. After attachment, a fusion event takes place either at the cell surface, as is suggested for lentiviruses, or through endocytotic pits for other retroviruses (see Sec. 11.2). The RNA is introduced into the cytoplasm associated with the core proteins. Reverse transcription takes place in the cytoplasm and is completed in the nucleus, where circular non-covalently bound DNA forms (called "copy" DNA or cDNA) are made that eventually integrate into the cell chromosome. The positive-sense RNAs of other animal

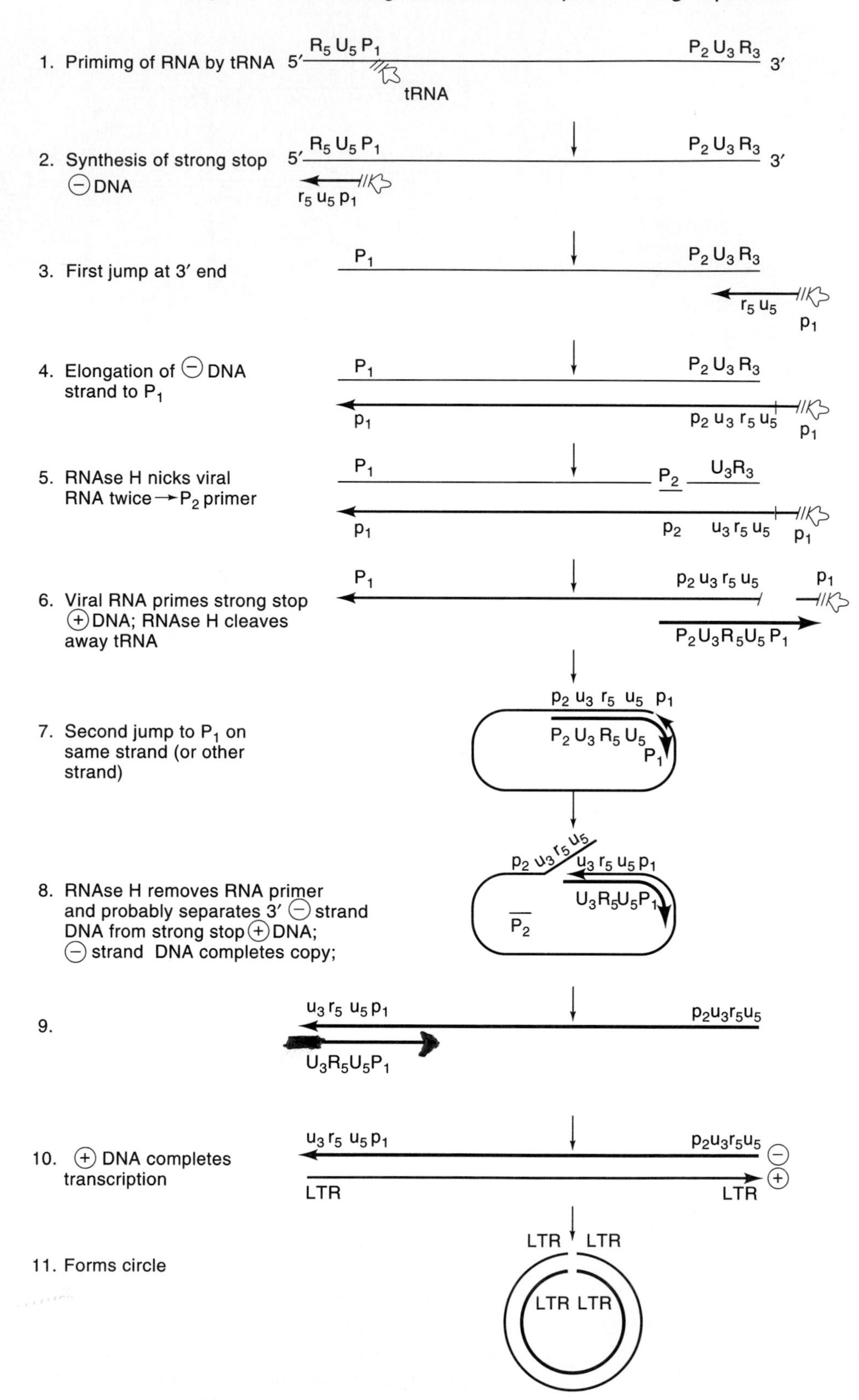
1. Primimg of RNA by tRNA
5′
R5 U5 P1
P2 U3 R3
3′
tRNA
2. Synthesis of strong stop
⊖ DNA
r5 u5 p1
3. First jump at 3′ end
P1
r5 u5
p1
4. Elongation of ⊖ DNA strand to P1
p1
p2 u3 r5 u5
5. RNAse H nicks viral RNA twice → P2 primer
P2
U3R3
p2
u3 r5 u5
6. Viral RNA primes strong stop ⊕ DNA; RNAse H cleaves away tRNA
p2 u3 r5 u5
P2U3R5U5 P1
7. Second jump to P1 on same strand (or other strand)
p2 u3 r5 u5 p1
P2 U3 R5 U5
8. RNAse H removes RNA primer and probably separates 3′ ⊖ strand DNA from strong stop ⊕ DNA; ⊖ strand DNA completes copy;
u3 r5 u5 p1
U3R5U5P1
9.
p2u3r5u5
10. ⊕ DNA completes transcription
LTR
11. Forms circle
LTR LTR

FIGURE 6.7

For initial transcription of the viral genome a primary binding site is used (P_1 on figure) located at the 3′ end of the U_5 region of the viral RNA. The first transcription product, negative-sense DNA, is begun at this site by addition of nucleotides to the 3′ end of the tRNA primer (step 1). Transcription then proceeds leftward to make a DNA copy of the nucleotides out to the 5′ end (step 2). This first replication product has been termed "strong-stop" negative-sense DNA strand because the polymerase tends to stop after its synthesis (at least *in vitro*) when the enzyme reaches the 5′ end of the RNA and runs out of template. At this point the RNAse H function of the enzyme may remove the 5′ end of the RNA and the methylated cap. In this way the short-stop negative-sense DNA strand has a copy of the R region that is free to anneal with the R sequences at the 3′ end of the viral RNA subunit (step 3). This RNA subunit may be either the original template or the other RNA in the virus genomic dimer. The relocation of transcription from the 5′ region of the RNA template to the 3′ end is known as the "first jump." After this jump, negative-sense DNA strand transcription spans almost the entire genome but stops at the binding site region (P_1) (step 4). Next, it appears that RNAse H nicks the viral genomic RNA twice, adjacent to the 5′ end of the U_3 region, to generate an 11–16 bp viral primer fragment (step 5). From this initiation point (termed P_2 in the figure) a "short-stop" positive-sense DNA strand is made that terminates in the tRNA primer binding site (P_1) (step 6). During this step, RNAse H has presumably cleaved away other portions of the genomic RNA and the tRNA.

To complete the negative-sense DNA strand, a second LTR ($U_3R_5U_5$) must be copied. The P_1 region of the short-stop positive-sense DNA strand then anneals to the P_1 region of the negative-sense DNA strand, permitting a second transcriptional jump to a new template (steps 7, 8). At this point the RNA primer (P_2) is believed to be removed by RNAse H to facilitate base pairing between the two P_1 sequences. After this second jump, negative-sense DNA strand synthesis is completed (step 8). The short-stop positive-sense DNA strand then uses the full-length negative-sense DNA strand as a template to complete its synthesis (steps 9, 10). How the negative-sense DNA strand separates from the positive-sense strand short-stop fragment is not clear, but it may involve RNAse H or another enzyme function associated with the polymerase product. The final result (Step 11) is a double-stranded viral DNA larger than the initial RNA template because of the addition of the U_3 and U_5 regions to either end, forming the long terminal repeats (LTR). (The preferred preintegration DNA form (see Figure 6.6) is a very closely associated but not covalently bound circular DNA.) (Revised from Gilboa et al. (1979) *Cell* 18:93, with help from W. Yong.)

viruses are first translated, because their RNA replicase protein must be produced before replication can begin. The retrovirus particles, however, carry many molecules of the active enzyme needed for their replication. This aforementioned enzyme, the reverse transcriptase, is a comparatively small protein, a single chain of 60 to 70 kDa. In some viruses, such as the avian retroviruses, it exists as a dimer. This transcriptase has at least three different activities: It makes a DNA complementary to the viral RNA; it has a nuclease activity (called **ribonuclease** H) that digests only the RNA strand of this DNA/RNA hybrid; and it makes the remaining DNA strand double-stranded. In some retroviruses an **endonuclease** (integrase) is also coded by the *pol* gene.

The transcription of viral DNA from RNA involves two "jumps" that are mediated by the viral reverse transcriptase and permits the transfer of the growing chain of DNA to similar sequences on the same, or the other, RNA strand. Eventually a double-stranded DNA copy (cDNA) of the viral RNA genome is produced. From the viral double-stranded DNA, termed **provirus,** a viral positive-sense RNA becomes transcribed by the host's DNA-dependent RNA polymerase II, although probably only after the viral DNA has become integrated into the cell's genome.

The mechanism of reverse transcription of retrovirus RNA is unique. This enzyme, like all typical DNA polymerases, needs a primer. The primer in RSV is a molecule of *tryptophan-tRNA* bound to the 38 S RSV RNA by base complementarity after the 101st nucleotide from the 5′ ends. In other Retroviridae this binding is 10 to 40 nucleotides more up-or downstream, and another tRNA may be used. The enzyme also has a particular binding affinity for the specific tRNA. The steps in the transcription process leading to **double-stranded DNA** in circular form are shown in Figures 6.6 and 6.7. The viral DNA then becomes supercoiled, enters the nucleus, and can in this form be integrated at probably a

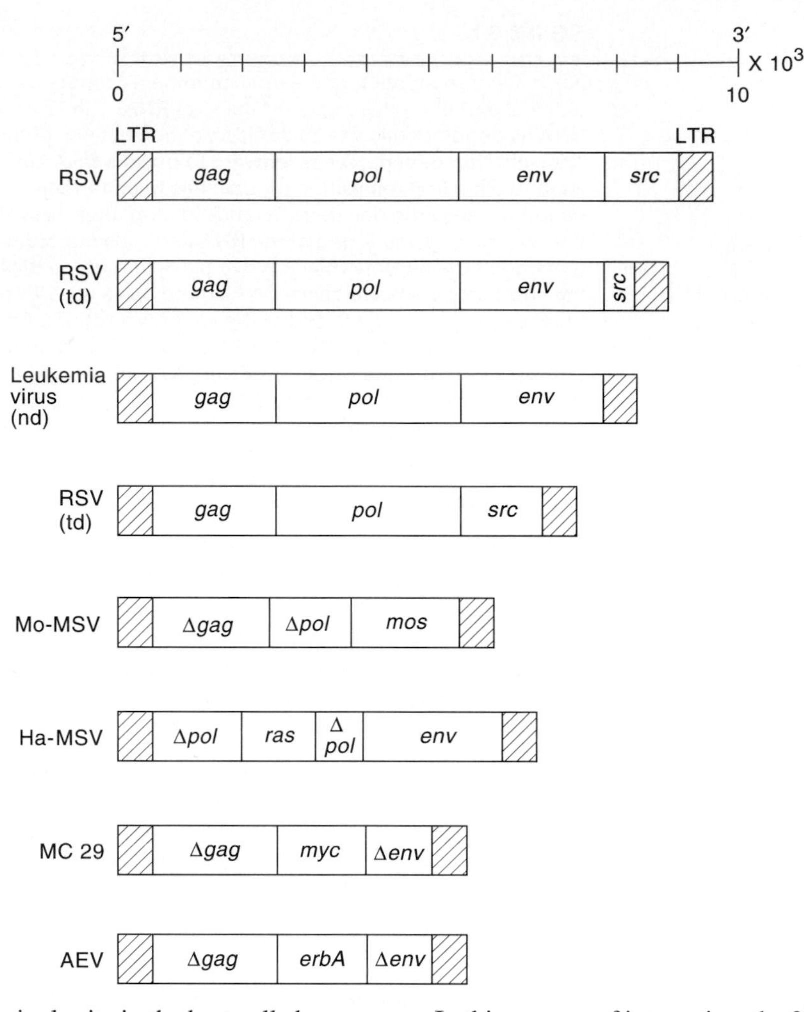

FIGURE 6.8

Genetic maps of oncogenic retroviruses. The shaded segments are the parts of the genome responsible for malignancies: td, transformation-defective; nd, nondefective; rd, replication-defective; RSV, Rous sarcoma virus; Mo-MSV, Ha-MSV, Moloney and Harvey mouse sarcoma viruses; MC-29, avian myelocytoma virus; AEV, avian erythroblastosis virus. The Δ indicates defective genes. (Modified from P. Duesberg and K. Bister. (1980) Proc. 3d Internat. Feline Leukemia Virus Meeting)

single site in the host cell chromosome. In this process of integration, the 3′ ends of the viral DNA are joined to the 5′ ends of cellular DNA, generating four base pair (bp) repeats in the host cell target sequence, while two base pairs from the viral DNA are eliminated. Within 5 hours an infectious double-stranded DNA can be detected in infected cells, and within 9 hours progeny virus is produced.

Recently, **integration** was found to be mediated by the association of the integrase protein with the ends of the viral DNA. This complex first cleaves a region in the chromosomal DNA and then inserts the viral DNA. Integration may occur randomly in sites of DNA that are transcriptionally active (i.e. having "open" chromatin structure). This conclusion could explain how nondividing macrophages and T cells arrested in division can still be infected by retroviruses and undergo some virus integration.

The terminal redundancy of the DNA of retroviruses suggests an integration mechanism similar to that of certain transposable genetic elements of bacteria, called **transposons** (see Sec. 6.4). The extent of similarity between transposons and retroviral RNA as well as the presence of reverse transcriptase in a wide variety of animal and plant viruses (see Sec. 6.3) has been recognized. The possible existence of this enzyme in normal cells is also actively being investigated.

6.1.4 Transcription, Translation, and Assembly of Retroviridae

Transcription of retrovirus genes from proviral DNA is influenced by the LTRs (see Fig. 6.7). As noted previously, these sequences contain several distinct elements, including **gene promoters,** ribosome binding sites, cellular protein binding sites, transcription termination signals, and the customary signal for addition of the 3′ poly(A). In addition, LTRs contain elements similar to those recently found in many other eukaryotic genes, called **enhancers** they serve to increase transcription of nearby genes. Enhancers stimulate initiation of transcription at a promoter and must be located on the same DNA molecule (i.e., *cis* position) as the stimulated gene. However, enhancer activity, unlike promoter activity, can be independent of the polarity of the enhancer-gene arrangement, i.e., the enhancer can be located upstream (5′) or downstream of the stimulated gene, and in either orientation.

Molecular clones of the LTR region of a number of avian retroviruses have been analyzed by DNA sequencing and found to form a family of interrelated, but distinct, LTRs. Results of gene expression tests indicated that the large differences in observed gene expression correlated directly with LTR enhancer activity. Since the more oncogenic avian viruses tested contained LTRs producing greater gene expression, these results suggested that LTR enhancer activity may be the primary determinant of avian retroviral LTR transcriptional activity and, hence, oncogenic potential. Moreover, these regions of the genome influence the **cell tropism** of retroviruses. Nevertheless, studies with murine retroviruses show important contributions by other viral genes as well (e.g., *gag*).

The gene stimulation action of enhancers can be affected by cellular factors, generally thought to be gene-regulating proteins. For example, the proviral DNA of mouse mammary tumor virus (MMTV) contains a regulatory region closely associated with its promoter that subjects **transcription** to the control of glucocorticoid hormones. Sequence analysis of a chimeric MMTV LTR–thymidine kinase gene (LTR–tk) has shown that the hormonal regulation sequence is confined to 202 nucleotides preceding the LTR-specific RNA initiation site. This fragment confers hormonal inducibility onto attached genes over distances of 0.4 or 1.1 kb. This **hormonal response region** functions when it is placed at either 5′ or 3′ of the regulated gene in both of the possible orientations and is thus reminiscent of an enhancer sequence.

Although the infecting 70 S RNA of RSV is the template for reverse transcription, it is, as previously stated, not active as mRNA, which accounts for its not being infectious. In contrast, the 38 S virion RNA, and this same RNA as transcribed from free or integrated provirus DNA, is an active mRNA. In addition, two RNAs of about 28 S and 21 S that have mRNA properties have been found in infected cells. Each of these mRNAs carries the 5′-terminal leader sequence, including the cap of the large RNA, added to them by a mechanism now known to occur in the expression of many eukaryotic viral as well as cellular genes, **splicing.** This gene control mechanism actually involves cleavage and rejoining of single-stranded RNA molecules at highly specific recognition sequences of nucleotides. The process is generally carried out with the aid of proteins that may recognize the sequence and provide catalytic activities needed for cutting and rejoining the RNA strands. However, the occurrence of **autocatalytic splicing** of RNA molecules has also been detected. We return to the subjects of splicing and of enzymatic action of RNA in subsequent chapters.

The translation of the large (38 S) RNA of RSV mainly produces the primary *gag* gene protein of 76 kDa, which is then proteolytically split into the four capsid proteins; the C-terminal one is the 15 kDa protease needed for these cleavages (Fig. 6.8). However, smaller amounts of a 180 kDa protein are also observed. This product includes the translation of the *pol* gene that is not separately initiated. It first appeared probable that its product, the reverse transcriptase, resulted from occasional read-through followed by specific cleavage. This mode of translation would explain why much less enzyme than capsid protein is made. However, DNA sequence analyses of cloned Rous sarcoma virus mRNA have shown that the *pol* gene is positioned downstream of the *gag* gene in a different, briefly overlapping reading frame. Thus, the results now strongly favors a **frame-shifting** mechanism on the ribosome, rather than RNA splicing, to explain this case of differential gene expression. Similar sequence-specific frame-shifting mechanisms are probably involved generally in mRNA production and expression of proteins by retroviruses (e.g., HIV) and perhaps also by cellular genes.

The first-formed enzyme produced by proteolytic cleavage of the *gag-pol* fusion product of RSV is called β (96 kDa), and this polymerase product is proteolytically trimmed further to 63 kDa (α). The biochemistry of the reverse transcriptase of RSV is quite remarkable. Both the α and β chains alone are active in all three activities required for this enzyme, the β chain forming a dimer. From the latter is derived the αβ complex, made by proteolytic processing, and this complex of 140 kDa appears to be the active form used in the cell. An additional translation product, a protein of 32 kDa, has been found that shares peptides with β. This protein binds to nucleic acids and acts as the endonuclease on circular double-stranded DNA.

The middle-sized mRNA of RSV (28 S) is translated to envelope proteins, analogous to the translation of other viral envelope glycoproteins (see Sec. 3.1). The smallest RNA (21 S), like the largest, is translated in the cytoplasm and yields the *src* protein of 60 kDa (see Fig. 6.8).

When the proteins have accumulated sufficiently, some of the 38 S RNA and the group-specific (internal) viral proteins move toward the cell membrane, where they encounter the glycoproteins. Only upon maturation and budding of the progeny virus through the plasma membrane are the dimers of the 70 S RNA again formed. As noted above, this description of the molecular events involved in RSV replication is a prototype for all retroviruses although some differences can be noted among members of the seven genera (see Secs. 10.3, 14.1, 14.2).

Because of their replication process, some retroviruses (particularly the amphotropic murine virus) have served in recent years as excellent vectors for introduction of foreign genomic material, into cells since they can infect such a wide variety of cells. Moreover, the vector genomes can be manipulated sufficiently so that replication of virus does not take place after insertion of the foreign gene (see Chap. 13).

6.1.5 Defective Retroviruses

Except for certain strains of avian sarcoma viruses (e.g., Rous sarcoma virus), the oncogenic retroviruses transform but do not replicate in cells. They lack part or all of one of the structural genes needed for replication (Figs. 6.7, 6.9, and 6.10). They therefore de-

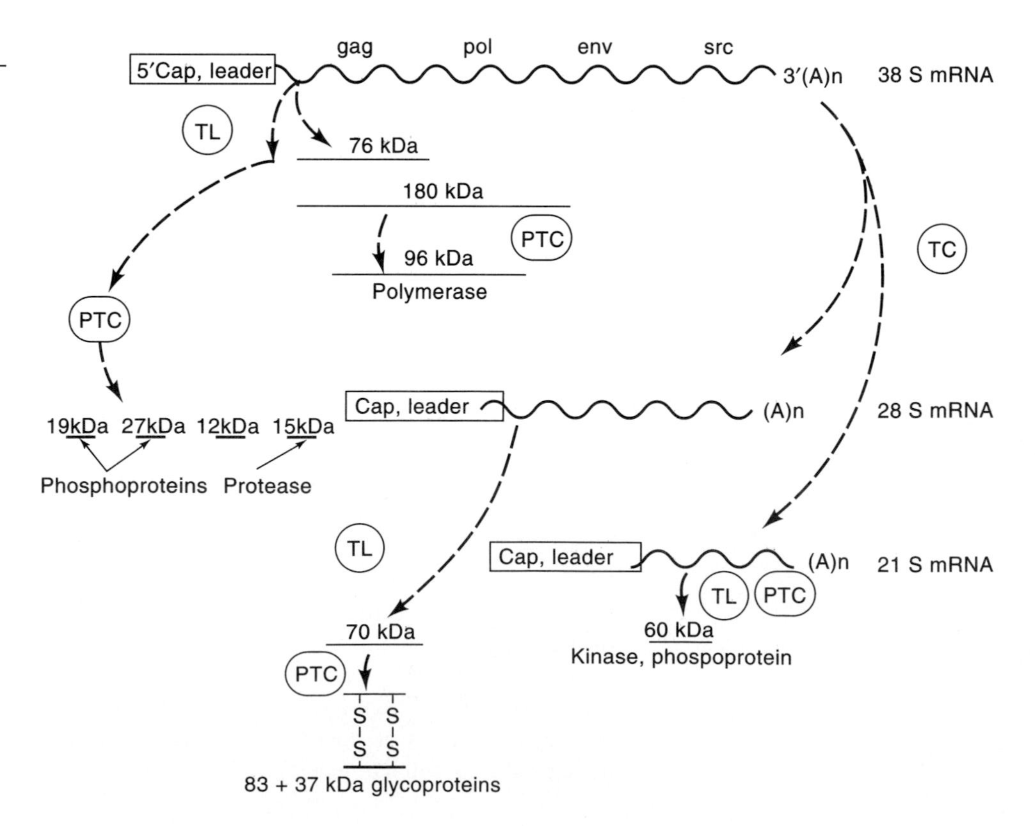

FIGURE 6.9

Gene map and the transcription (TC), translation (TL), and posttranslational cleavage (PTC) and processing strategies of Rous sarcoma virus.

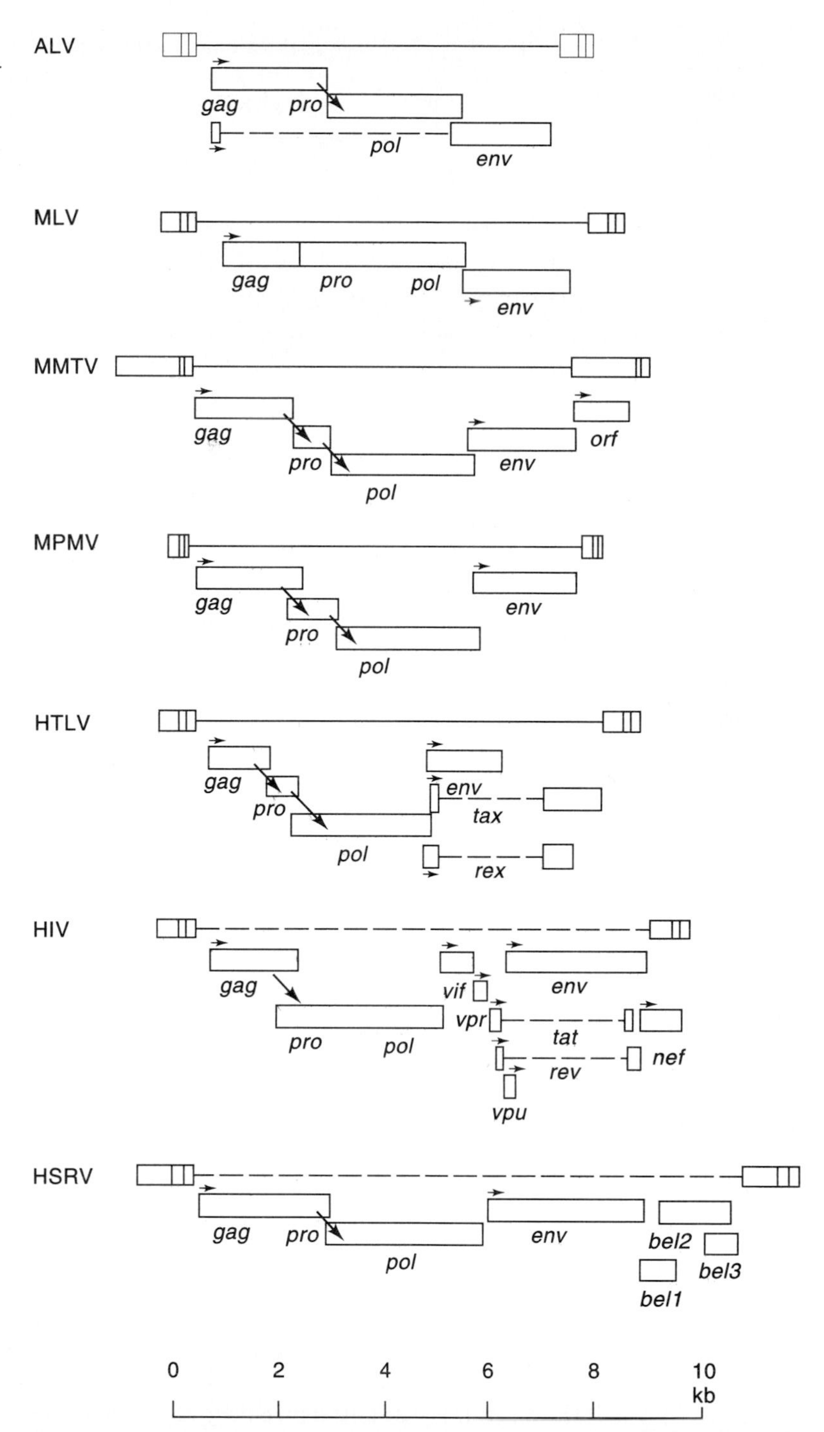

FIGURE 6.10

Coding regions of retrovirus genomes. The top lines depict the proviral DNA with the LTRs boxed. Under them are the coding regions, with each box corresponding to a separate reading frame. Horizontal arrows indicate points of translational initiation; vertical lines indicate terminators that may be partially suppressed during translation; diagonal arrows indicate frame-shift sites. Dashed lines show reading frames joined by splicing events. ALV, avian leukosis virus; MLV, murine leukemia virus; MMTV, mouse mammary tumor virus; MPMV, Mason-Pfizer monkey virus; HTLV, human T lymphotropic virus; HIV, human immunodeficiency virus; HSRV, human spuma retrovirus. (From J. Coffin *The Retroviridae,* ed. J. Levy, Vol. 1 (1992) Plenum Press, New York)

pend on an accompanying nondefective (nd) "helper" virus for their infectious progeny production (see third diagram in Figure 6.9). These replication-competent viruses are frequently transmitted through the germ line and thus exist as **endogenous** (inherited) viruses (see also Chapters 10 and 12). They lack *src* or other recognizable oncogenes but may by other processes cause the slow development of leukemias and, rarely, sarcomas (see Chapter 10). Other endogenes viruses are never oncogenic and may play a role in normal development; however, their replicating ability enables them to serve as vectors for the nonreplicating transformimg viruses (Fig. 6.11).

The **defective nature** of most sarcoma viruses is reflected by the diminished size of their RNAs, which can be as low as 1.5×10^6 Da, compared with the unusual 3.3×10^6 Da

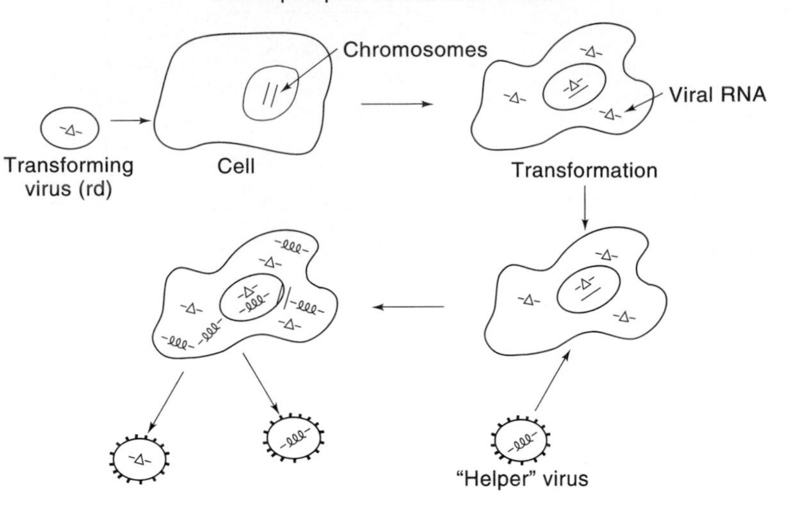

FIGURE 6.11

Rescue of transforming gene. A replication-defective transforming virus enters the cell, integrates, and causes transformation. Subsequent infection of the transformed cell with a replication-competent retrovirus leads to the incorporation of the transforming viral RNA into the envelope of the "rescuing" or "helper" retrovirus. Subsequent production of both the superinfecting and transforming viruses results—with both carrying the outer envelope of the replicating virus. The pseudotype sarcoma virus formed is now capable of infecting and transforming those cells expressing receptors that bind to the envelope of the helper virus.

size of the RNA of RSV. As illustrated in Figure 6.8, RSV has not only all the viral structural genes but also the additional *src* gene; it is therefore replication-competent. The larger size of RSV led to the initial recognition of the *src* gene. This observation paved the way toward the eventual identification of oncogenes in other retroviruses and **proto-oncogenes** in cells (Sec. 10.3).

Through the simultaneous infection of the host cell by both a replication-competent virus and a replication-defective transforming virus, the transforming viral genome becomes incorporated into the envelope of the replication-competent helper virus. The resulting oncogenic viruses are called **pseudotypes** because their envelope is not coded for by their own genome. A similar biological phenomenon is observed with other viruses (see Secs. 9.2 and 9.4). By this process, these transforming viruses can then complete their replicative cycle and have the host range of the helper retroviruses. They infect cells containing the receptor for the envelope of the helper virus. This exchange of envelope coats among viruses, called **phenotypic mixing,** is discussed further in Section 11.7.1.

During the replication of the competent avian sarcoma viruses, some viruses with deletions in part or all of the *src* genes have emerged; these are transformation-defective (td) (Fig. 6.9). These viruses act like conventional helper retroviruses and can in some cases cause chronic leukemias. The mechanism for oncogenesis is the same as that of other nonacutely transforming retroviruses (see Sec. 10.3.6).

In some situations the defective viral genome carried within the envelope of a helper virus infects a cell in the absence of an accompanying replicating retrovirus. Integration and transformation will then occur if viral reverse transcriptase is present. However, subsequent infection of this transformed cell with a replicating retrovirus is needed to rescue the transformed genome from the cell and bring about progeny virus production (Fig. 6.11). This event is a second example of the interaction of a **replication-defective transforming** virus and a competent helper virus. Production of pseudotype sarcoma viruses can occur both by co-infection and superinfection. The process permits the transfer of oncogenes among a variety of cells and potentially into other hosts.

Since cell transformation in the absence of virus production has been observed in some animal systems, researchers have attempted to uncover a human sarcoma gene by infecting certain cultured human cancers with retroviruses. This approach has not led to the demonstration of a human transforming virus, but it has uncovered a rat sarcoma virus from transformed rat cells. The latter observation was a dramatic demonstration of the possibility, suggested by oncogene research, that genes with transforming potential may be present in all cells (i.e., **proto-oncogenes;** see Sec. 10.3.7 and Postscript). Under certain circumstances these cellular genes may be activated directly by retroviruses or get incorporated into their genomes. This latter process is analogous to transduction in bacteria (see Sec. 11.11).

Recently, an immune deficiency in mice has been linked to a defective virus associated with a murine leukemia virus. In this murine AIDS model (MAIDS), the defective virus has a genome size of 4.9 kb. Expression of its internal *gag* viral protein on the cell surface is associated with a **polyclonal activation** of B lymphocytes, and results in immunodeficiency (presumably caused by cytokine production). This observation led several groups to look for a defective retrovirus in human AIDS, but none has been detected.

6.1.6 Human Retroviruses

The retrovirus family has members of human origin. The first discovered was the human foamy virus, recovered in 1971 from cultured cells of a human cancer. However, its relationship to any human disease has not been established. The second was a virus found associated with T-cell leukemias in the southern islands of Japan, the southeastern United States, and the Caribbean. This virus, called the **human T-cell leukemia virus** (HTLV-I) or adult T-cell leukemia virus, preferentially infects certain cells, particularly of the helper T-cell subset, and can immortalize them into established cell lines (see Chapter 14). These HTLV viruses can also infect fibroblasts in culture. Moreover, HTLV-I has been found to infect rats and rabbits. A subtype of HTLV-I, called HTLV-II, was recovered from a human leukemia cell line, but its association with disease has not been documented. HTLV-II appears to preferentially infect human CD8+ cells. Recently, antibodies to HTLV-I and II have been found with increased frequency in human populations outside Japan. In particular, HTLV-II infection appears to be more common in intravenous drug users.

The most recent human retroviruses identified are lentiviruses that are associated with the **acquired immunodeficiency syndrome** (AIDS) (see Sec. 14.1). These human immunodeficiency viruses (HIV-1 and HIV-2) are readily distinguished from the other known human pathogenic retroviruses (HTLV-I, HTLV-II) by morphology (Fig. 6.3) and structure (see Figure 14.4). HIV-2, which differs in genome structure by 55% from the other HIV-1 subtype, primarily in its envelope genes, was first discovered in western Africa. It has not spread as widely as HIV-1 and has on only relatively limited occasions been detected in individuals outside Africa. Primates have an HTLV-I-like virus associated with leukemia called simian T-cell leukemia virus (STLV). In addition, as reviewed in Section 6.1, simian immunodeficiency viruses (SIV), lentiviruses, have been isolated from a variety of non-human primate species. They cause AIDS in certain strains of monkeys, but not in others. This latter observation has led to intense research aimed at determining what factors are responsible for suppression of disease development in a lentivirus-infected host.

SECTION 6.1 FURTHER READING

Original Papers

Rous, P. (1911) Transmission of a malignant new growth by means of a cell-free filtrate. *J. Am. Med. Assoc.* 56:198.

Bittner, J. J. (1942) The milk-influence of breast tumors in mice. *Science* 95:462.

Gross, L. (1951) "Spontaneous" leukemia developing in C3H mice following inoculation, in infancy, with AK-leukemia extracts or AK-embryos. *Proc. Soc. Exp. Biol. Med.* 76:27.

Bernhard, W. (1960) The detection and study of tumor viruses with the electron microscope. *Cancer Res.* 20:712.

Rubin, H. A. (1960) Virus in chick embryos which induces resistance *in vitro* to infection with Rous sarcoma virus. *Proc. Natl. Acad. Sci. USA* 46:1105.

Hanafusa, H., T. Hanafusa, and H. Rubin. (1963) The defectiveness of Rous sarcoma virus. *Proc. Natl. Acad. Sci. USA* 49:572.

Huebner, R. J., et al. (1966) Rescue of the defective genome of Moloney sarcoma virus from a noninfectious hamster tumor and the production of pseudotype sarcoma viruses with various murine leukemia viruses. *Proc. Natl. Acad. Sci. USA* 56:1164.

Bentvelzen, P. A., et al. (1968) Genetic transmission of mammary tumor inciting viruses in mice: Possible implications for murine leukemia. *Bibl. Haematol.* 31:101.

Huebner, R. J., and G. J. Todaro. (1969) Oncogenes of RNA tumor viruses as determinants of cancer. *Proc. Natl. Acad. Sci. USA* 64:1087.

BALTIMORE, D. (1970) Viral RNA-dependent DNA polymerase. *Nature* 226:1209.

DUESBERG, P. H., and P. K. VOGT. (1970) Differences between the ribonucleic acids of transforming avian tumor viruses. *Proc. Natl. Acad. Sci. USA* 67:1673.

PARKMAN, R., J. A. LEVY, and R. C. TING. (1970) Murine sarcoma virus: The question of defectiveness. *Science* 168:387.

TEMIN, H. M., and S. MIZUTANI. (1970) RNA-dependent DNA polymerase in virions of Rous sarcoma virus. *Nature* 226:1211.

ACHONG, B. G., et al. (1971) An unusual virus in cultures from a human nasopharyngeal carcinoma. *J. Natl. Cancer Inst.* 46:299.

ROWE, W. P., et al. (1971) Noninfectious AKR mouse embryo cell lines in which each cell has the capacity to be activated to produce infectious murine leukemia virus. *Virology* 46:866.

KEITH, J., and H. FRAENKEL-CONRAT. (1975) Identification of the 5′ end of Rous sarcoma virus RNA. *Proc. Natl. Acad. Sci. USA* 72:3347.

GILBOA, E., et al. (1979) A detailed model of reverse transcription and tests of crucial aspects. *Cell* 18:93.

POIESZ, B. J., et al. (1980) Detection and isolation of type C retrovirus particles from fresh and cultured lymphocytes of a patient with cutaneous T-cell lymphoma. *Proc. Natl. Acad. Sci. USA* 77:7415.

SHINICK, T. M., R. A. LERNER, and J. G. SUTCLIFFE. (1981) Nucleotide sequence of Moloney murine leukemia virus. *Nature* 293:543.

BARRE-SINOUSSI, F., et al. (1983) Isolation of a T-lymphotropic retrovirus from a patient at risk for acquired immune deficiency syndrome. *Science* 220:868.

DESGROSEILLERS, L., E. RASSART, and P. JOLICOEUR. (1983) Thymotropism of murine leukemia virus is conferred by its long terminal repeat. *Proc. Natl. Acad. Sci. USA* 80:4203.

PANGANIBAN, A. T., and H. M. TEMIN. (1984) Circles with two tandem LTRs are precursors to integrated retrovirus DNA. *Cell* 36:673.

BATTLES, J. K., et al. (1992) Immunological characterization of the *gag* gene products of bovine immunodeficiency virus. *J. Virol.* 66:6868.

DEAN, G. A., et al. (1992) Hematopoietic target cells of anemogenic subgroup C versus nonanemogenic subgroup A feline leukemia virus. *J. Virol.* 66:5561.

MARTINEAU, D., et al. (1992) Molecular characterization of a unique retrovirus associated with a fish tumor. *J. Virol.* 66:596.

MOK, E., T. V. GOLOVKINA, and S. R. ROSS. (1992) A mouse mammary tumor virus mammary gland enhancer confers tissue-specific but not lactation-dependent expression in transgenic mice. *J. Virol.* 66:7529.

GELMAN, I. H., and H. HANAFUSA. (1993) *src*-specific immune regression of Rous sarcoma virus-induced tumors. *Cancer Res.* 53:915.

GIRON, M. L., et al. (1993) Human foamy virus polypeptides: Identification of *env* and *bel* gene products. *J. Virol.* 67:3596.

GRANDGENETT, D. P., et al. (1993) Comparison of DNA binding and integration half-site selection by avian myeloblastosis virus integrase. *J. Virol.* 67:2628.

GRAY, K. D., and M. J. ROTH. (1993) Mutational analysis of the envelope gene of Moloney murine leukemia virus. *J. Virol.* 67:3489.

LYNCH, W. P., and J. L. PORTIS. (1993). Murine retrovirus-induced spongiform encephalopathy: Disease expression is dependent on postnatal development of the central nervous system. *J. Virol.* 67:2601.

YOUNG, J. A. T., P. BATES, and H. E. VARMUS. (1993) Isolation of a chicken gene that confers susceptibility to infection by subgroup A avian leukosis and sarcoma viruses. *J. Virol.* 67:1811.

Books and Reviews

TEMIN, J. M. (1980) Origin of retroviruses from cellular movable genetic elements. *Cell* 21:599.

WEISS, R., et al. (1982) *RNA tumor viruses.* Cold Spring Harbor Press, Cold Spring Harbor, New York.

GROSS, L. (1983) *Oncogenic viruses.* Pergamon Press, Elmsford, New York.

BALTIMORE, D. (1985) Retroviruses and retrotransposons: The role of reverse transcription in shaping the eukaryotic genome. *Cell* 40:481.

HAASE, A. T. (1986) Pathogenesis of lentivirus infections. *Nature* 322:130.

LEVY, J. A. (1986) The multifaceted retrovirus. *Cancer Res.* 46:5457.

VARMUS, H. (1988) Retroviruses. *Science* 240:1427.

COFFIN, J. M. (1992) Structure and classification of retroviruses. In *The Retroviridae,* ed. J. A. Levy. Vol. 1, 19. Plenum Press, New York.

KOZAK, C. A., and S. RUSCETTI. (1992) Retroviruses in rodents. In *The Retroviridae,* ed. J. A. Levy, Vol. 1, 405. Plenum Press, New York.

LUCIW, P. A., and N. J. LEUNG. (1992) Mechanisms of retrovirus replication. In *The Retroviridae,* ed. J. A. Levy, Vol. 1, 159. Plenum Press, New York.

PAYNE, L. N. (1992) Biology of avian retroviruses. In *The Retroviridae,* ed. J. A. Levy. Vol. 1, 299. Plenum Press, New York.

TEMIN, H. M. (1992) Origin and general nature of retroviruses. In *The Retroviridae,* ed. J. A. Levy. Vol. 1, 1. Plenum Press, New York.

LOH, P. C. (1993) Spumaviruses. In *The Retroviridae,* ed. J. A. Levy. Vol. 2. 361. Plenum Press, New York.

SUGAMURA, K., and Y. HINUMA. (1993) Human retroviruses: HTLV-1, HTLV-2. In *The Retroviridae,* ed. J. A. Levy, Vol. 2. 399. Plenum Press, New York.

6.2 HEPADNAVIRIDAE

Hepatitis B virus (HBV), formerly called serum hepatitis virus, was first identified as an infectious agent in the late 1960s (Fig. 6.12). Its biology indicates that it has an extremely limited host range, being infectious only for liver cells of humans and higher primates. Viral hepatitis B has an incubation period from 45 to 120 days and is generally passed by serum or blood. Although the virus is present in saliva, this body fluid as a source of infection is rare, unless delivered through biting. Mother-to-child transmission has also been recorded. HBV is a prototype for other animal viruses recently detected, including agents from the Beechey ground squirrel (ground squirrel hepatitis virus, GSHV), Pekin duck (duck hepatitis B virus, DHBV), herons, and woodchuck (**woodchuck hepatitis virus,** WHV). Isolates from marsupials and tree squirrels have also been identified.

The avian and mammalian hepadnaviruses appear to represent two separate genera because of differences in their genomic nucleotide sequences, gene numbers, and organization. Moreover, the duck virus has a core covered with spikelike projections that are not present on those of the mammalian viruses. Oddly enough, one plant virus family, the **caulimoviruses** (Sec. 6.3), shows properties quite similar to those of the hepatitis viruses. All the animal hepadnaviruses have common morphology, characteristic production of excess surface antigen particles, typical genome size, relative hepatotropism, and a particular replication process. The use of **RNA-dependent DNA polymerase** during replication of the DNA genome merits their inclusion in this chapter. These viruses have been classified as a new family named for their DNA genome and tropism for liver cells.

Differences among the hepadnaviruses can be found, particularly in host range, pathogenic spectrum, and routes of transmission. The duck virus, for instance, can replicate in liver, pancreas, kidney, and spleen and is transmitted almost exclusively by the **vertical** route—from viremic adults to hatchlings via the egg. In contrast, the mammalian hepadnaviruses are passed primarily by **horizontal spread** and infect only the liver (Sec. 12.1.1). Moreover, hepatic neoplasms are common in chronically infected woodchucks, but not ground squirrels. Their association with hepatic cancers in humans has now been well established. This fact may reflect the profound inflammatory consequence of infection in the host. No viral oncogene has been found associated with this family of viruses; their induction of malignancy must be by a more indirect mode (see Sec. 10.2).

6.2.1 Structure

The presence of HBV was first detected as a noninfectious serum antigen, a 22 nm particle first termed Australian antigen and now known as HB_SAg (hepatitis B *surface* antigen). This protein has become a useful diagnostic marker of HBV infection. This antigen is actually a large complex of viral envelope and other proteins. Later, a rarer Dane particle was detected in sera and found to be the actual HB virion. This is a 42 nm *enveloped* particle

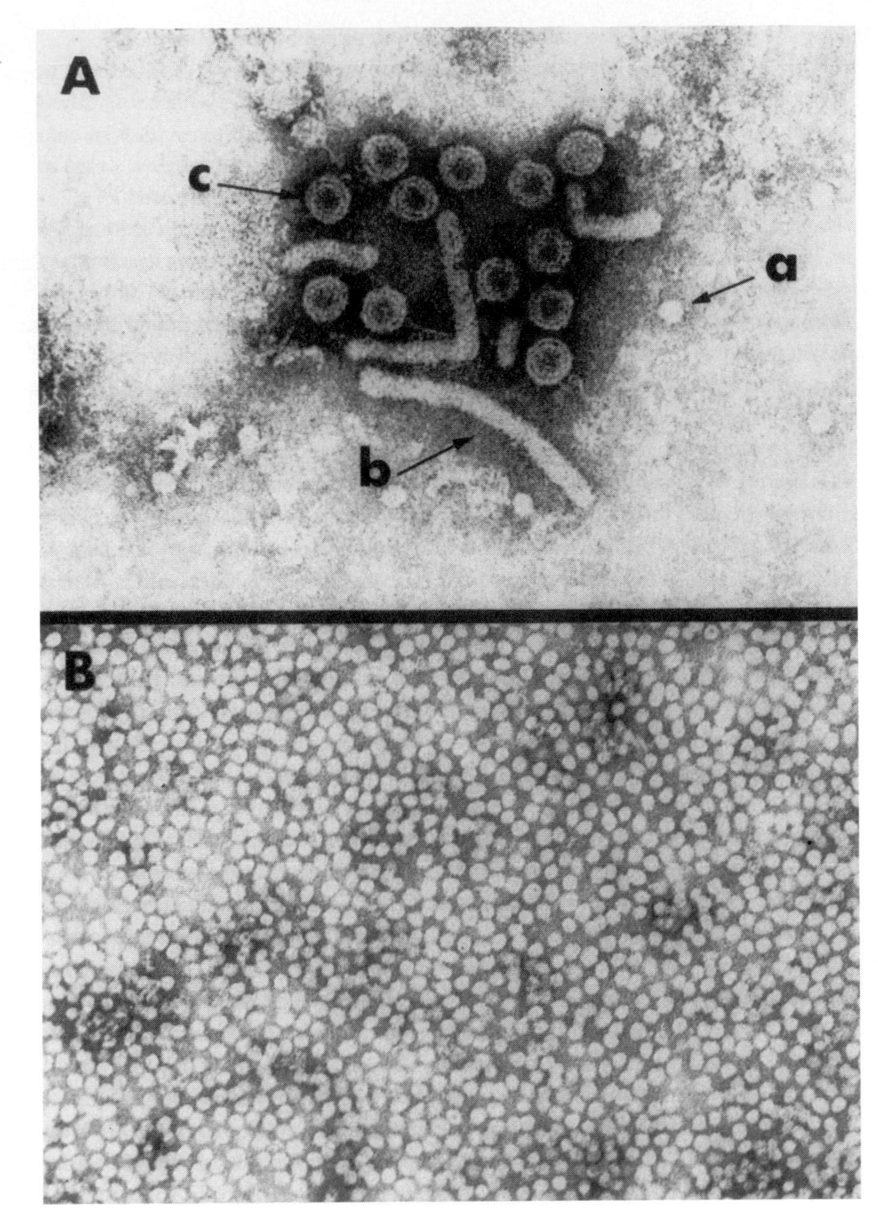

FIGURE 6.12

(a) Electron micrograph of serum showing the presence of three distinct morphological entities: a, 20-nm pleomorphic spherical particles; b, tubular or filamentous forms with a diameter of 20 nm; and c, 42-nm spherical particles now considered to be HBV (Dane particle). ×132,000. (b) Electron micrograph of a purified preparation of 17- to 25-nm pleomorphic spherical particles containing HBsAg, ×77,000. Similar preparations currently are being used in human hepatitis B vaccines. (From F. B. Hollinger, J. L. Dienstra, and P. D. Swenson, Hepatitis viruses. In *Manual of Clinical Microbiology,* ed. A. Balons. 1985 American Society of Microbiology)

with a 28 nm electron-dense *core* containing, in addition to viral DNA, the HB_cAg and HB_eAg proteins, DNA polymerase, and protein kinase activity. Its buoyant density of about 1.18 g/ml reflects the **lipid-containing** envelope of the hepadnavirions (similar to retroviruses). Particles containing lipid, protein, and carbohydrate without virion cores have also been found in high amounts in serum of infected hosts. The HB_S antigen has both group-specific determinants and subtype determinants (d,y and w,r), which are mutually exclusive. The **antigenic heterogeneity** of this virus family could involve other determinants as well. These differences have been helpful in epidemiological studies monitoring the spread of virus in human populations.

The DNA of the hepadnaviruses has unusual properties shared only in certain respects with the DNA plant viruses of the caulimo group (cauliflower mosaic virus). The complete genome of HBV is a small negative-sense strand DNA of 3200 nucleotides, a circular molecule with a break. A complementary positive-sense strand of 1700 to 2800 nucleotides is hydrogen-bonded to this negative-sense strand with a 5′ end in different molecules, always about 300 nucleotides beyond the break in the other chain (see Fig. 6.13). The negative-sense strand is blocked at the 5′ end by a covalently attached protein that may act as a primer in replication.

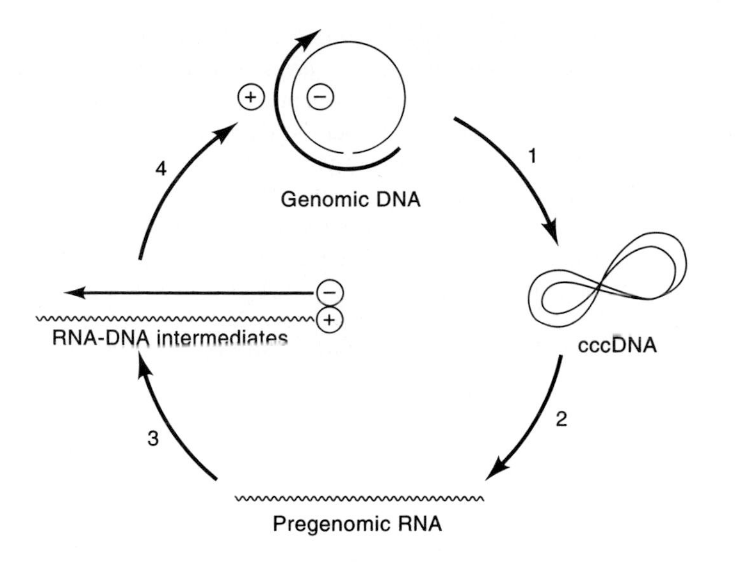

FIGURE 6.13

Replication of hepatitis B virus genome by reverse transcription. (From D. Ganem and H. Varmus (1987). *Ann. Rev. Biochem.* 56:651)

Molecular studies have suggested the presence of at least four promoter elements in the viral DNA. The viral genome has three major regions: the *S* gene that codes for the glycosylated envelope protein (226 amino acids) and the core protein (183 amino acids), the *P* gene that appears to encode a protein containing the DNA polymerase (95 kDa) or reverse transcriptase activity and the protein primer for DNA negative-sense strand synthesis, and finally the *X* gene that is associated with **transactivating** properties. Overlapping reading frames in the genome permit coding of many more proteins than would have been possible with a genome of the same size.

6.2.2 Entry, Transcription, and Replication

Virus entry has not been well defined, because cellular receptors or attachment sites on the virions cannot be studied without appropriate tissue culture assays. With the duck hepadnavirus, infection of primary duck liver cells was found not to be dependent on pH. Thus, entry appears to involve fusion at neutral pH, as it does with certain enveloped viruses, such as the herpesviruses and several retroviruses.

The presence in the hepadnavirus of both double- and single-stranded DNA as well as DNA/RNA hybrid molecules and RNA led to the eventual recognition that replication utilizes an **RNA intermediate** template. The viral DNA polymerase acts as a reverse transcriptase. The common finding of this enzyme associated with the hepadnaviruses suggests a past divergence from the retroviruses. The hepadnaviruses differ from retroviruses in certain characteristics: besides packaging viral DNA in the virions, they use a protein as a primer for the negative (first strand) DNA synthesis, whereas retroviruses use tRNA. Moreover, the hepadnaviruses have no major regulatory region such as the retroviral LTR.

The hepadnaviral DNA is replicated in a complex fashion that is unique to this group of agents (Fig. 6.13). The first step in replication is the conversion of the asymmetric DNA to covalently closed circle DNA (cccDNA). This process, which occurs within the nucleus of infected liver cells, involves completion of the DNA positive-strand synthesis and ligation of the DNA ends. In the second step, the cccDNA is transcribed by host transcriptase to generate a 3.5 kb RNA template (the **pregenome**). This pregenome is complexed with protein in the viral core. The third step involves the synthesis of the first negative-sense DNA strand by copying the pregenomic RNA by reverse transcriptase using a protein primer. Finally, the synthesis of the second positive-sense DNA strand occurs by copying the first DNA strand, using an oligomer of viral DNA as primer (Fig. 6.13). The pregenome also transcribes mRNA for the major structural core proteins.

As noted above, explanted hepatocytes from ducklings have been placed in primary culture and have been shown to be susceptible to exogenous DHBV infection. Although infection is transient, the cells can show productive infection with all the viral DNA and

RNA forms observed *in vivo*, and infectious progeny virions are produced. Moreover, well-differentiated human hepatoma cell lines can produce viral replicative intermediates and Dane particle–like structures following transfection with cloned HBV DNA. Progeny genomes can be seen, but there are some differences in the replication process using this approach. Nevertheless, with the development of cell culture systems, research with the hepadnaviruses will be greatly facilitated.

It is known that the virions released from the cell have either immature or mature forms of viral DNA. The particles appear first within the endoplasmic reticulum; whether budding occurs is not yet known.

6.2.3 HBV and Human Disease

HBV, which frequently appears together with other hepatitis-causing agents, represents a severe medical problem. Furthermore, HBV viral DNA in double-stranded form may on occasion integrate into the host genome by mechanisms still unknown. Hepatocellular cancer, which is frequently detected after HBV infection, may be dependent upon such integration. HBV and its related animal viruses are very stable and persist in latent form in hepatic cells for long periods of time. Its antibody complex is infectious and circulates in symptomless carriers for years, this condition has frequently caused transmission of hepatitis during immunization, blood transfusion, and other medical procedures. Persistence of infection can lead to chronic hepatitis, characterized by autoimmune sequelae and progression to cirrhosis.

An additional virulent hepatitis-causing agent, **hepatitis delta,** has been detected in association with infections with HBV. This circular RNA that resembles a viroid is described in Sec. 9.4. This virus can be transmitted only in conjunction with hepatitis B virus, which supplies the outer coat. Its replication, however, can occur without any "help" from HBV. For further discussion of the relation of HBV to liver cancer see Sec. 10.2.

SECTION 6.2 FURTHER READING

Original Papers

GALIBERT, F. E., et al. (1979) Nucleotide sequence of the hepatitis B virus genome (subtype ayw) cloned in *E. coli. Nature* 281:646.

TWIST, E. M., et al. (1981) Integration pattern of hepatitis B virus DNA sequences in human hepatoma cell lines. *J. Virol.* 37:229.

BEASLEY, R. P., et al. (1981) Hepatocellular carcinoma and hepatitis B virus: A prospective study of 22,707 men in Taiwan. *Lancet* ii:1129.

SIDDIQUI, A., et al. (1981) Ground squirrel hepatitis virus in DNA: Molecular cloning and comparison with hepatitis B virus DNA. *J. Virol.* 38:393.

SUMMERS, J., and W. S. MASON. (1982) Replication of the genome of a hepatitis B–like virus by reverse transcription of an RNA intermediate. *Cell* 29:403.

VENTO, S., et al. (1987) Prospective study of cellular immunity to hepatitis B virus antigens from the early incubation phase of acute hepatitis B. *Lancet* ii:119.

WILL, H., et al. (1987) Replication strategy of human hepatitis B virus. *J. Virol.* 61:904.

CHEN, H. S., et al. (1993) The woodchuck hepatitis virus X gene is important for establishment of virus infection in woodchucks. *J. Virol.* 67:1218.

GUSTINE, K., et al. (1993) Characterization of the role of individual protein binding motifs within the hepatitis B virus enhancer 1 on X promoter activity using linker scanning mutagenesis. *Virology* 193:653.

POLLACK, J. R., and D. GANEM. (1993) An RNA stem-loop structure directs hepatitis B virus genomic RNA encapsidation. *J. Virol.* 67:3254.

Books and Reviews

GANEM, D., and H. VARMUS. (1987) The molecular biology of the hepatitis B viruses. *Ann. Rev. Biochem.* 56:651.

HOLLINGER, F. B. (1990) Hepatitis B virus. In *Virology,* 2d ed., ed. B. N. Fields and D. M. Knipe. 2171. Raven Press, New York..

6.3 PLANT CAULIMO- AND BADNAVIRUSES

Plant viruses containing dsDNA genomes have been divided into two groups, the **caulimoviruses** (type member, cauliflower mosaic virus (CaMV)) and **badnaviruses** (type member, *Commelina* yellow mottle virus). Whereas the isometric particles of caulimoviruses are about 50 nm in diameter, badnaviruses have approximately 130 × 30 nm bacilliform particles (hence the name, *ba*cilliform *DNA*-containing *virus*). The size and structure of the caulimo- and badnavirus genomes are very similar, namely, a single 7.5–8 kb dsDNA molecule that contains single-strand discontinuities at specific sites. As with the hepadnaviruses (see Sec. 6.2), replication of caulimo- and badnaviruses involves a virus-encoded reverse transcriptase activity, but *not* integration into the host DNA. Thus, all three groups of viruses have sometimes been termed *pararetroviruses.*

Both caulimo- and badnaviruses have narrow host ranges and are transmitted by insects in a semipersistent manner—CaMV and related viruses by aphids, and badnaviruses by mealybugs. Caulimoviruses are also mechanically transmissible. Rice tungro bacilliform virus is transmissible by leafhoppers only from plants that are co-infected by rice tungro spherical virus.

CaMV was the first plant virus to have its genome completely sequenced, and its biological and molecular properties have been intensively investigated for nearly 15 years. Its so-called **35 S promoter,** responsible for synthesis of a greater-than-full-length RNA transcript of the entire CaMV genome, has been widely used to drive gene expression in transgenic plants (see Chapter 13).

6.3.1 Caulimoviruses (CaMV)

Examination of unstained, frozen-hydrated samples of CaMV by low-irradiation, cryo-electron microscopy and three-dimensional image reconstruction procedures has shown that the virus particles have a multilayered structure. As shown in Fig. 6.14, the spherical particles are composed of three concentric layers surrounding a large, solvent-filled cavity. The outermost layer contains 72 capsomeric morphological units constructed from 420 molecules of coat protein (37–42 kDa) arranged with T = 7 icosahedral symmetry. The double-stranded DNA genome is distributed in layers II and III along with a portion of the viral coat protein.

CaMV and the animal polyomaviruses (see Sec. 7.2.4) are similar-sized, isometric viruses, each with $T = 7$ icosahedral symmetry and a single major capsid protein of similar mass (about 40 kDa). Whereas the CaMV genome appears to be in close contact with virion-coded proteins, those of polyoma viruses are complexed with cellular histones in a central “core.” Although CaMV particles are quite resistant to extremes of pH, temperature, and chemical environment, they are more readily distorted by electron microscopy procedures than are those of most other viruses.

The organization of the 8 kb caulimovirus genome as well as an overview of the replication cycle is shown in Figure 6.15. ORFs I and II encode proteins required for cell-to-cell movement and aphid transmission, respectively. ORF III encodes a non-sequence-specific DNA binding protein, and the product of ORF IV is the 57 kDa precursor of the coat protein. The 79 kDa ORF V product is present in viral replication complexes and has been shown to have reverse transcriptase activity. It also contains protease and RNAse H domains. The product of ORF VI is the major protein found in CaMV inclusions (or viroplasms) and also functions in *trans* to activate translation of the five other viral genes (see below). All eight ORFs begin with AUG initiation codons, but because they are not present in all caulimoviruses, the importance of ORFs VII and VIII is unclear.

Key steps in the transcription and replication of the CaMV genome are illustrated in Figure 6.15. Replication involves reverse transcription of a greater-than-full-length 35 S RNA via a pathway very similar to that used by hepadnaviruses (see Sec. 6.2). The gene VI protein is the only CaMV polypeptide translated from its own mRNA (i.e., the 19 S RNA). The 5′ region of the 35 S RNA contains several small ORFs and appears to fold into an extensive secondary structure. Gene VI protein has been shown to promote translation of ORFs I to V from the CaMV 35 S RNA in *trans.* Translation of the 19 S gene VI mRNA, in turn, is regulated by *cis*-acting sequences the negative effects of which are relieved by the presence of gene VI protein (i.e., **autoregulation**).

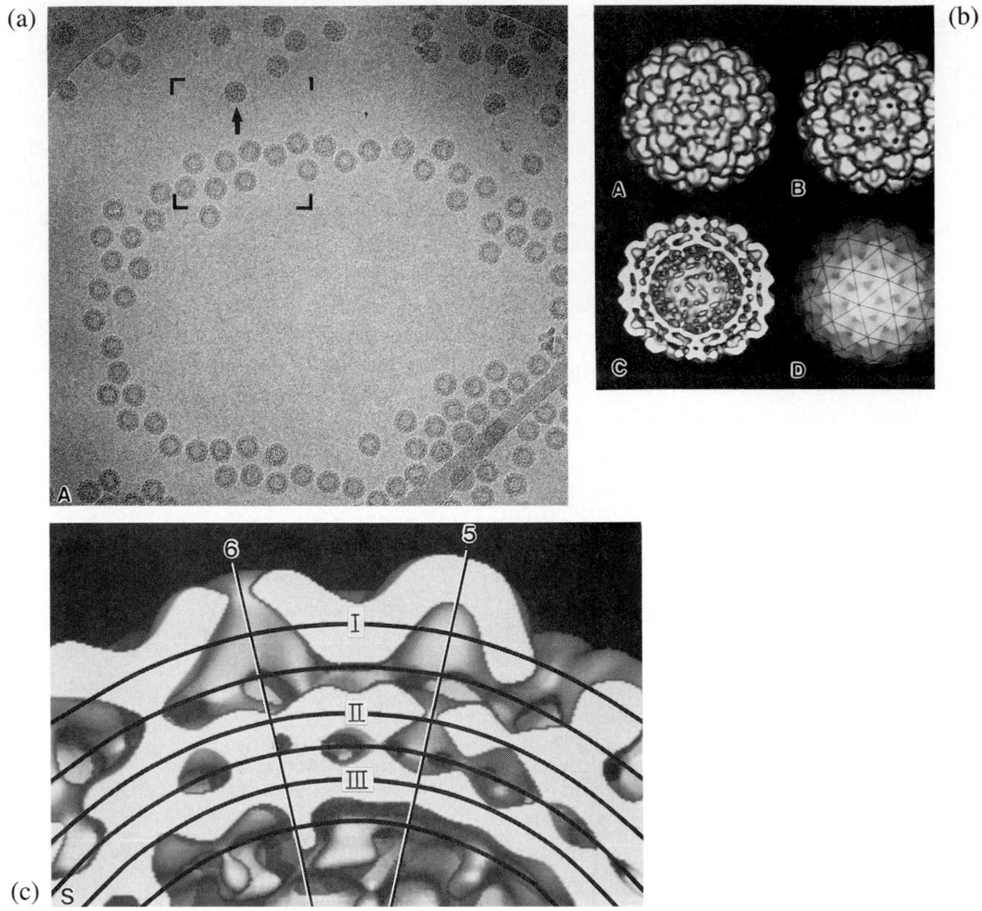

FIGURE 6.14

Structure of the CaMV virion. (a) Electron micrograph of a vitrified sample of CaMV containing polyomavirus (arrow) as an internal calibration standard; (b) Image reconstructions derived from (a). Left, reconstruction of the Cabb-B isolate computed from 21 independent particle images and viewed along an icosahedral, twofold axis of symmetry; middle, depth-cued representation of Cabb-B overlayed with a T = 7/(left-handed) icosahedral lattice net whose points of intersection identify the positions of 72 capsomers (bright regions appear closer to viewer); right, cutaway view revealing the multilayered Internal structure and a large central cavity; (c) internal features of the multilayer CaMV capsid. Circular arcs represent peaks and troughs seen in a radial density plot. (Modified from R. H. Cheng et al. (1992) *Virology* 286:655)

6.3.2 Badnaviruses

The dsDNA genomes of two badnaviruses, *Commelina* yellow mottle (CoYMV) and rice tungro bacilliform virus (RTBV), have been completely sequenced. The organization of the RTBV genome, approximately 0.5 kb larger than that of CoYMV, is illustrated in Fig. 6.16. Although it exhibits certain similarities to caulimoviruses, RTBV is much more similar to CoYMV in its genome organization than it is to CaMV.

The ORFs for RTBV P12 and P24 are very similar in size and location to the two smallest ORFs of CoYMV, but CoYMV lacks an ORF analogous to P46. The absence of a P46 ORF accounts for most of the difference in the size of the two viral genomes. No homologies between RTBV P12 or P46 and the proteins encoded by CoYMV have been reported. In both viruses, the largest ORF appears to encode a polyprotein containing the viral coat protein as well as the protease, reverse transcriptase, and RNase H domain of the viral polymerase. The *pol* gene of badnaviruses, like those of retroviruses and retroelements, appears to be expressed as a *gag-pol* fusion protein. In contrast, expression of the CaMV *pol* gene (i.e., gene V) appears to be independent from that of the *gag* gene (i.e., gene IV = coat protein).

6.4 RETROELEMENTS

Reverse transcriptase genes have also been found associated with retroelements in plants, animals, fungi, protozoa, and even in prokaryotes. The observations indicate that these elements replicate through an RNA intermediate. Among the retroelements are **retrotransposons** and **retroposons.** In contrast with retroviruses, these elements are not infectious; they do not have an envelope and thus undergo only an intracellular replicative cycle. Nevertheless, they can be transmitted horizontally through cell fusion or the rare uptake of viruslike particles. In this sense they appear to be primitive or vestigial viruses.

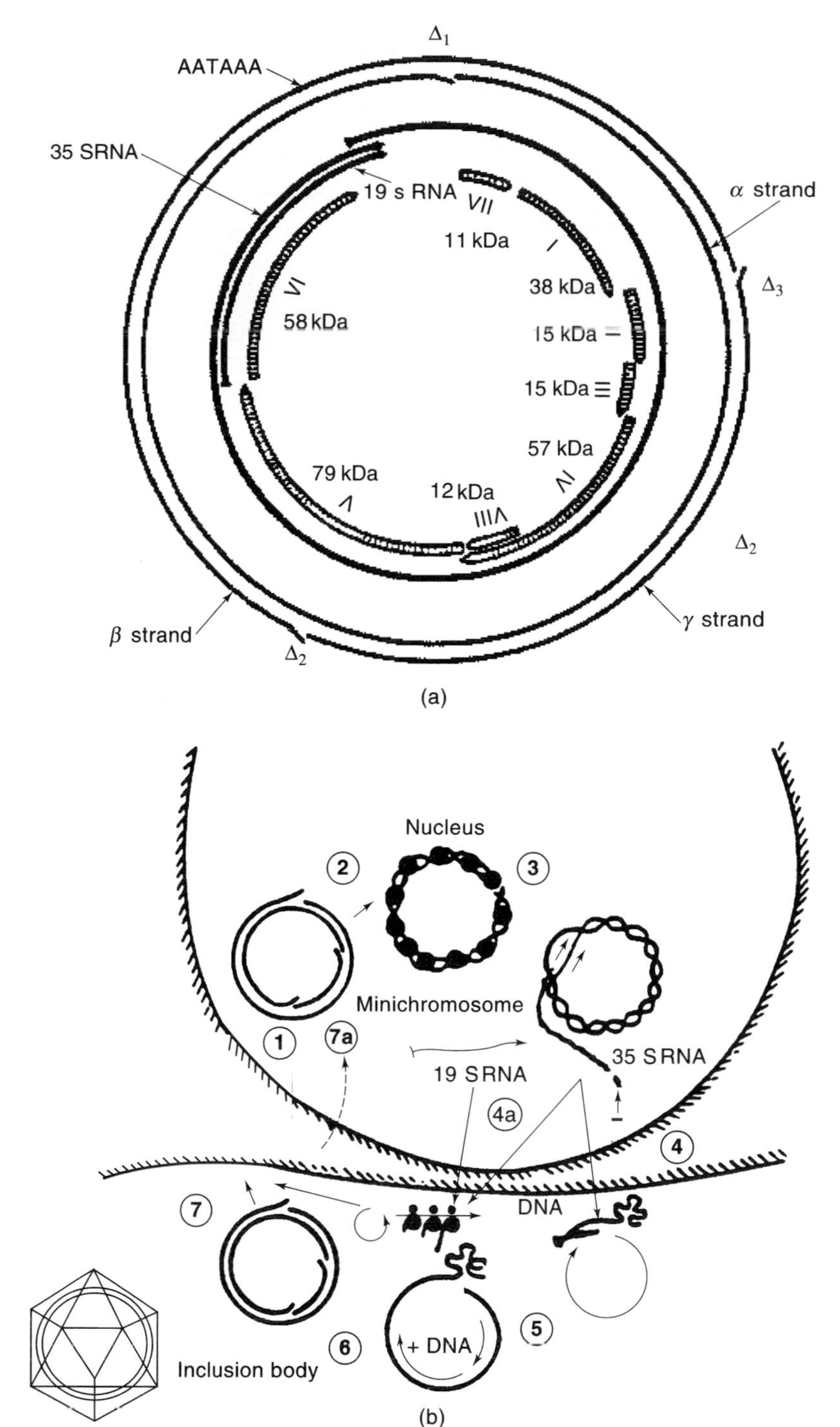

FIGURE 6.15

Organization and expression of CaMV genome. (a) The 8 kb viral DNA contains three single-stranded interruptions, one (1) in the α (or coding) strand and two (2 and 3) in the noncoding strand, that define the β and γ DNA species. The capped and polyadenylated 19 S and 35 S RNAs have different promoters but share the same 3′ termini. As shown below, the two mRNAs are transcribed from a fully double-stranded supercoiled form of the DNA and *not* from the gapped form shown here. (b) Replication cycle of CaMV. Upon infection, DNA released from the incoming virions (1) undergoes repair and ligation in the nucleus, where it associates with histones to form a transcriptionally active minichromosome (2). The 19 S and 35 S RNA transcripts are exported to the cytoplasm for translation (4a). Some of the 35 S RNA enters a viroplasm, where it associates with a molecule of met-tRNA (4) before being reverse-transcribed into DNA. After digestion of the RNA template and synthesis of the second DNA strand (5), the viral DNA (6) is packaged into virions (7). Two replicative pathways may coexist: The 35 S RNA may be encapsidated as early as (4) and DNA synthesis may occur in preparticles; alternatively, replication complexes may generate free viral DNA that is directed back into the nucleus for amplification. (Modified from R. E. F. Matthews. (1991) *Plant virology.* Academic Press, New York)

All these reverse transcriptase (RT)-containing elements have many shared characteristics (Table 6.5). For example, retrotransposons contain *gag* and *pol* genes that are near long terminal repeats (LTR). Retroposons can have the *gag* and *pol* genes, but not an LTR; they have another repeated segment at the beginning of the element which is required for replication and integration. It is just recently that RT-containing elements, called **retrons,** were discovered in bacteria. They have genetic cross-reactivity with retroelements, but other characteristics are still not well defined. They are **episomal** in nature (i.e., remain unintegrated in the cell), are highly heterogeneous, and use reverse transcriptase to increase in number. Retons usually have a single-stranded DNA that is covalently attached to an RNA.

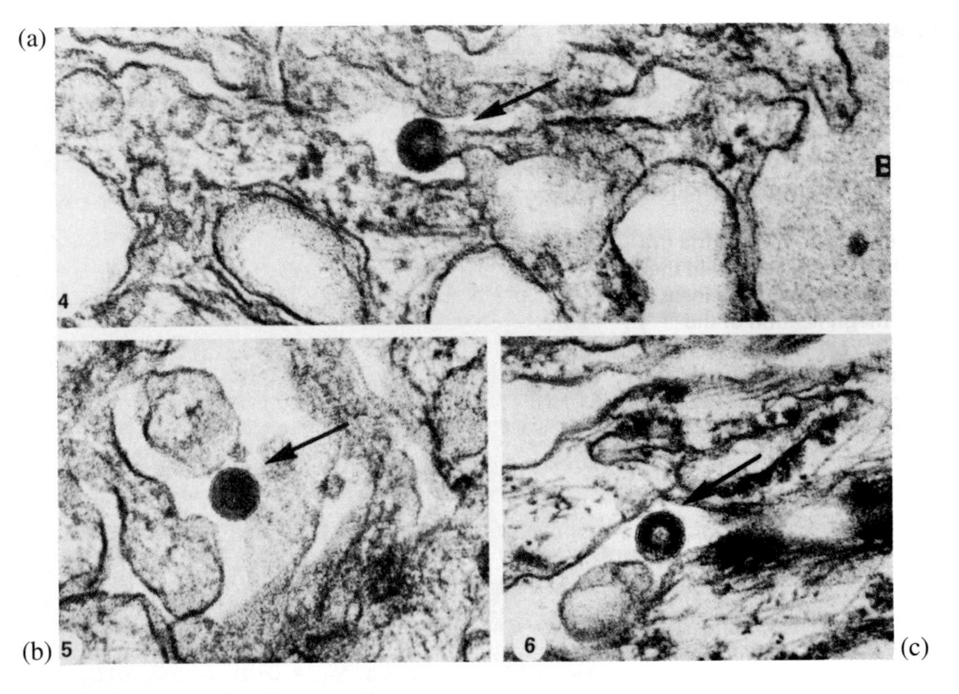

FIGURE 6.19

Retrovirus-like particles in human placenta. (a) A budding viruslike particle (arrow) near the basal lamina (B) of a syncytiotrophoblast cell. Note that the membrane coating the particle is closely applied to the nucleocapsid. Uranyl acetate and lead citrate. × 100,000 (b) A free viruslike particle (arrow). Uranyl acetate and lead citrate. × 100,000 (c) Viruslike particle (arrow) near demosome of syncytiotrophoblast cell. Uranyl acetate and lead citrate. × 100,000 (From E.R. Dirksen and J.A. Levy (1977) *J. Natl. Cancer Inst.* 59:1187)

SECTIONS 6.3 AND 6.4 FURTHER READING

Original Papers

KALTER, S. S. et al. (1973) C-type particles in normal human placentas. *J. Natl. Cancer Inst.* 50:1081.

DIRKSEN, E. R., and J. A. LEVY. (1977) Virus-like particles in placentas from normal individuals and patients with systemic lupus erythematosus. *J. Natl. Cancer Inst.* 59:1187.

FRANK, A., et al. (1980) Nucleotide sequence of cauliflower mosaic virus DNA. *Cell* 21:285.

GARDNER, R. C., et al. (1981) The complete nucleotide sequence of an infectious clone of cauliflower mosaic virus by M13mp7 shotgun sequencing. *Nucl. Acids Res.* 9:2871.

PAULSON, K. E., et al. (1985) A transposon-like element in human DNA. *Nature* 316:359.

LIEB-MOSCH, C., et al. (1990) Endogenous retroviral elements in human DNA. *Cancer Res.* 50:5636s.

MEDBERRY, S. L., et al. (1990) Properties of *Commelina* yellow mottle virus's complete DNA sequence genomic discontinuities and transcript suggest that it is a pararetrovirus. *Nucl. Acids Res.* 18:5505.

HAY, J. M., et al. (1991) An analysis of the sequence of an infectious clone of rice tungro bacilliform virus, a plant pararetrovirus. *Nucl. Acids Res.* 19:2615.

CHENG, R. H., et al. (1992) Cauliflower mosaic virus: A 420 subunit ($T = 7$), multilayer structure. *Virology* 286:655.

SCHOLTHOF, H. B., et al. (1992). Regulation of caulimovirus gene expression and the involvement of *cis*-acting elements on both viral transcripts. *Virology* 190:403.

HIROSE, Y. et al. (1993) Presence of *env* genes in members of the RTLV-H family of human endogenous retrovirus-like elements. *Virology* 192:52.

WILKINSON, D. A., et al. (1993) Evidence for a functional subclass of the RTLV-H family of human endogenous retroviruslike sequences. *J. Virol.* 67:2981.

BOUHIDA, M., B. E. L. LOCKHARDT, and N. E. O. OLSZEWSKI. (1993) An analysis of the complete sequence of a sugarcane bacilliform virus genome infectious to banana and rice. J. Gen. Virol. 74:15.

Books and Reviews

MASON, W. S., J. M. TAYLOR, and R. HULL. (1987) Retroid virus genome replication. *Adv. Virus Res.* 32:35.

PFEIFFER, P., et al. (1987) The life cycle of cauliflower mosaic virus. In *Plant molecular biology.* ed. D. von Wettstein and N.-H. Chua. 433. Plenum, New York.

XIONG, Y., and T. H. EICKBUSH. (1988) Similarity of reverse transcriptase-like sequences of viruses, transposable elements, and mitochondrial introns. *Mol. Biol. Evol.* 5:675.

BROSINS, J. (1991) Retroposons: Seeds of evolution. *Science* 251:753.

GARFINKEL, D. J. (1992) Retroelements in microorganisms. In *The Retroviridae,* ed. J. A. Levy. Vol. 1, 107. Plenum Press, New York.

chapter

7

Viruses with Small DNA Genomes

A considerable number of unrelated virus families that exist share the property of having DNA genomes of only about 1.5 to 3 $\times$ 10^6 Da. Among these, only members of the Papovaviridae are **double-stranded.** The small DNA viruses, as many other viruses, are faced with the problem of packing into a small particle sufficient nucleic acid to carry all the information needed for the viruses to enter the cells of their hosts, interact with their metabolic processes in various ways, become replicated, and again leave the cell dead, alive, or transformed. In contrast to positive-sense RNA viruses of prokaryotes and eukaryotes, these DNA viruses encounter a cellular machinery capable of producing all their components. Nevertheless, the smallest DNA viruses have four to nine genes.

Among the virus families that have small DNA genomes are a group of **icosahedral** animal viruses carrying **single-stranded** DNA, the Parvoviridae. Members of another family, the Papovaviridae, include the polyoma and SV40 viruses that contain about twice as much double-stranded (ds) DNA. Genomes of viruses in the second genus of that family, the papillomaviruses, are somewhat larger, containing 5 $\times$ 10^6 Da of double-stranded DNA. They show some properties similar to the polyoma-type viruses, and are included in this discussion. Only a few of the parvo- and papovaviruses represent serious dangers to animals or humans. In contrast, hepatitis B virus, which is also a small DNA virus, is causing major medical concern. It was discussed in Chapter 6 because of its unusual replication properties.

Plant viruses containing single-stranded DNA, the gemini viruses, have also been discovered. These are discussed in this chapter along with circoviruses. The double-stranded caulimoviruses which show resemblances to Retroviridae and the Hepadnaviridae were dealt with in Chapter 6. Finally, two types of bacteriophages carrying single-stranded (ss) DNA in their virions, one icosahedral and the other fibrous or threadlike are considered. These families are called Microviridae and Inoviridae, respectively. A schematic drawing of the DNA structures of the small DNA viruses compared with those of the large DNA viruses discussed in Chapter 8 is given in Figure 7.1.

7.1 PARVOVIRIDAE

The Parvoviridae animal virus family consists of the smallest and simplest virus particles—hence the name *parvo*viruses, from the Latin word for "small." They are icosahedral, about 20 to 25 nm in diameter, and have a particle weight of about 6 $\times$ 10^6 Da (Fig. 7.2). The parvoviruses have one molecule of single-stranded DNA of about 1.5 $\times$ 10^6 Da and a coat composed of possibly 32 capsomers made of three comparatively large (60–80 kDa) proteins. Particles lacking DNA are of 65 S, compared with 110 S for the infectious virus.

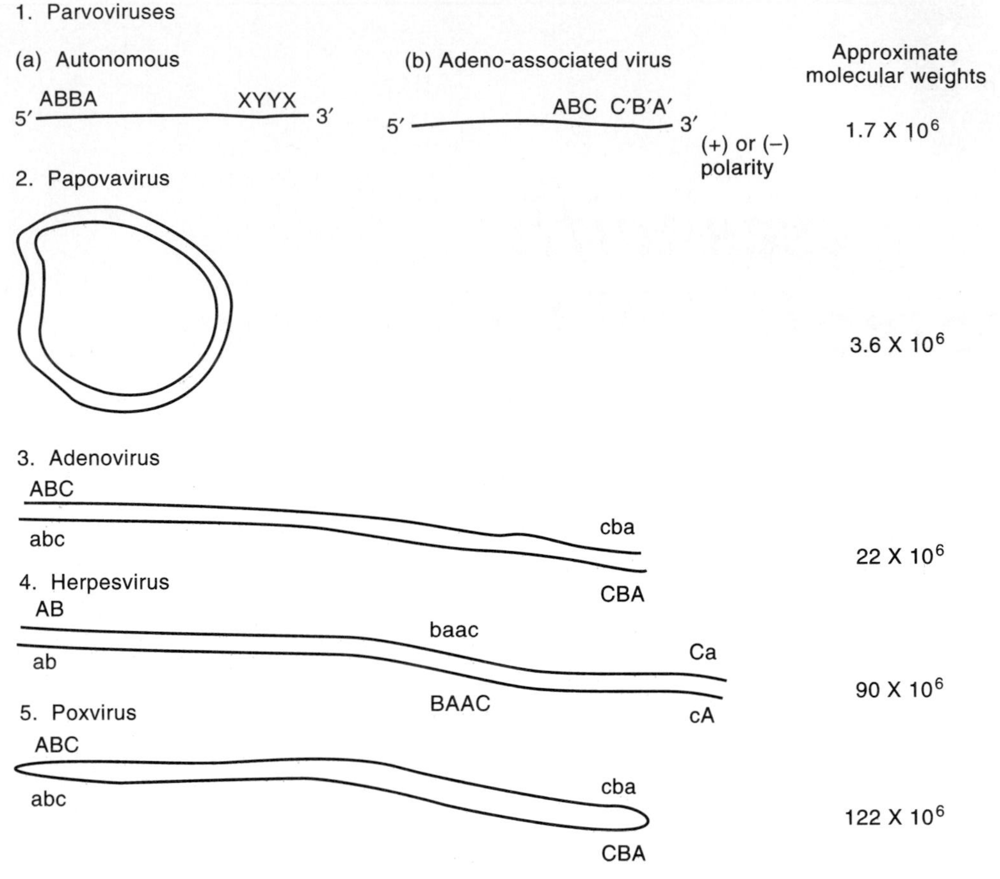

FIGURE 7.1

Schematic comparative drawing of the structures of the DNA of major DNA virus groups.

There are three genera of Parvoviridae: the **parvoviruses** present in rodents and other mammals; the **densoviruses** (or *densonucleosis* viruses) found in invertebrate species; and the **dependoviruses** (adeno-associated viruses [AAV]) present mostly in mammals, including humans. The first two groups are autonomous, nondefective viruses. Generally, their cellular host range is species-specific. The adeno-associated viruses are, as their name suggests, dependent for their replication on the presence of adenoviruses or herpesviruses of the same host species (see Secs. 8.1, 8.2).

The parvoviruses, unlike the papovaviruses, cannot induce DNA synthesis in host cells and require cell division for their replication. Because of this latter characteristic, their pathological effects associated with developing fetuses, intestinal epithelium, and the hematopoietic systems are not surprising. The extent of pathology may also be related to a particular stage of cell differentiation. The spread of parvoviruses may be by feces, urine, saliva, nasal secretion, and perhaps contact with genital fluids.

The viruses **hemagglutinate** erythrocytes of certain animal species and can be quantified by this process (Sec. 11.2). They are never produced in great amounts in culture and often cause no detectable cytopathology; they are also generally not pathogenic in most adult hosts. In the young of certain species, however, parvoviruses can take a drastic toll. For example, infectious feline and mink enteritis syndromes are caused by this family of viruses. Parvovirus inoculation into the brain of hamsters gives rise to severe neurological consequences.

The mink parvovirus causes Aleutian disease, which is characterized by increased plasma cells in the blood with high production of immunoglobulins. These antibodies, mostly to the virus, lead to chronic renal failure, resulting from immune complex glomerulonephritis. Thus, parvoviruses can cause **autoimmune disease.** A worldwide epizootic disease of dogs, first seen in 1978, appears to be due to this mink virus, or a mutant of it. In cats the disease is also called human panleukopenia (a deficiency of all types of white

FIGURE 7.2

The B19 human parvoviruses (× 160,000). (Courtesy of Y. Cossart)

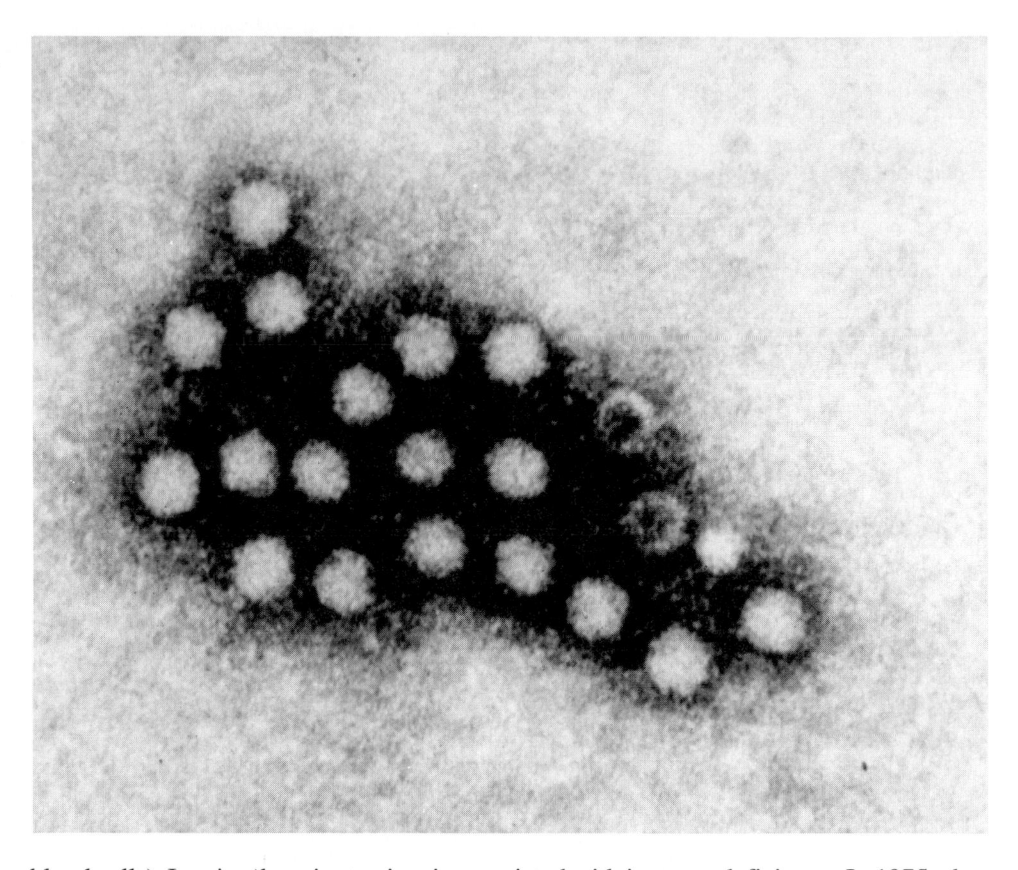

blood cells). In mice the minute virus is associated with immune deficiency. In 1975 a human parvovirus (B19) (Fig. 7.2) was discovered in the sera of normal blood donors and subsequently identified as the causative agent of (a) transient aplastic crisis of hemolytic disease; (b) **fifth disease,** a common childhood rash called **erythema infectiosum;** and (c) in normal adults, a polyarthralgic syndrome. The symptoms associated with B19 virus infection share features with rubella and, like that virus, pose a danger to the fetus. Because of the minimal effects on cultured cells, the detection and isolation of parvoviruses in biochemically useful amounts is often difficult. They are more heat-resistant (56° for 60 min) than most other viruses, and they can also withstand various treatments with proteases, nucleases, detergents, lipid solvents, and weak acids (pH 3.0). For this reason, the viruses may persist for long periods of time (months) in the environment and be capable of infection of new susceptible hosts. Thus, for example, houses and grounds previously inhabited by a parvovirus-infected dog need to be decontaminated before introducing a new dog into the area.

7.1.1 Structure

The virion contains no viral or cellular enzymes. The DNA of autonomous parvoviruses is usually a negative-sense strand and lacks terminal base complementarity or repeated sequences. Thus, no panhandle or other types of rings can form. However, these DNAs also carry complementary **palindromic sequences** favoring **Y-shaped hairpins** at both ends (Figs. 7.3 and 7.4). These structures are highly conserved in the evolutionary sense. Four apparently unrelated parvoviruses show very similar sequences, although each differs at the two ends and none of them resemble the termini of the adeno-associated viruses.

The B19 human parvovirus has some distinguishing characteristics. It is a single-stranded DNA of about 1.8×10^6 Da, which encodes two capsid proteins of about 84 and 58 kDa. The virus replicates in the nucleus as high-molecular-weight intermediate forms linked through terminal hairpin structures, as have been described for other parvoviruses. Unlike the conventional parvoviruses, however, similar quantities of both positive- and

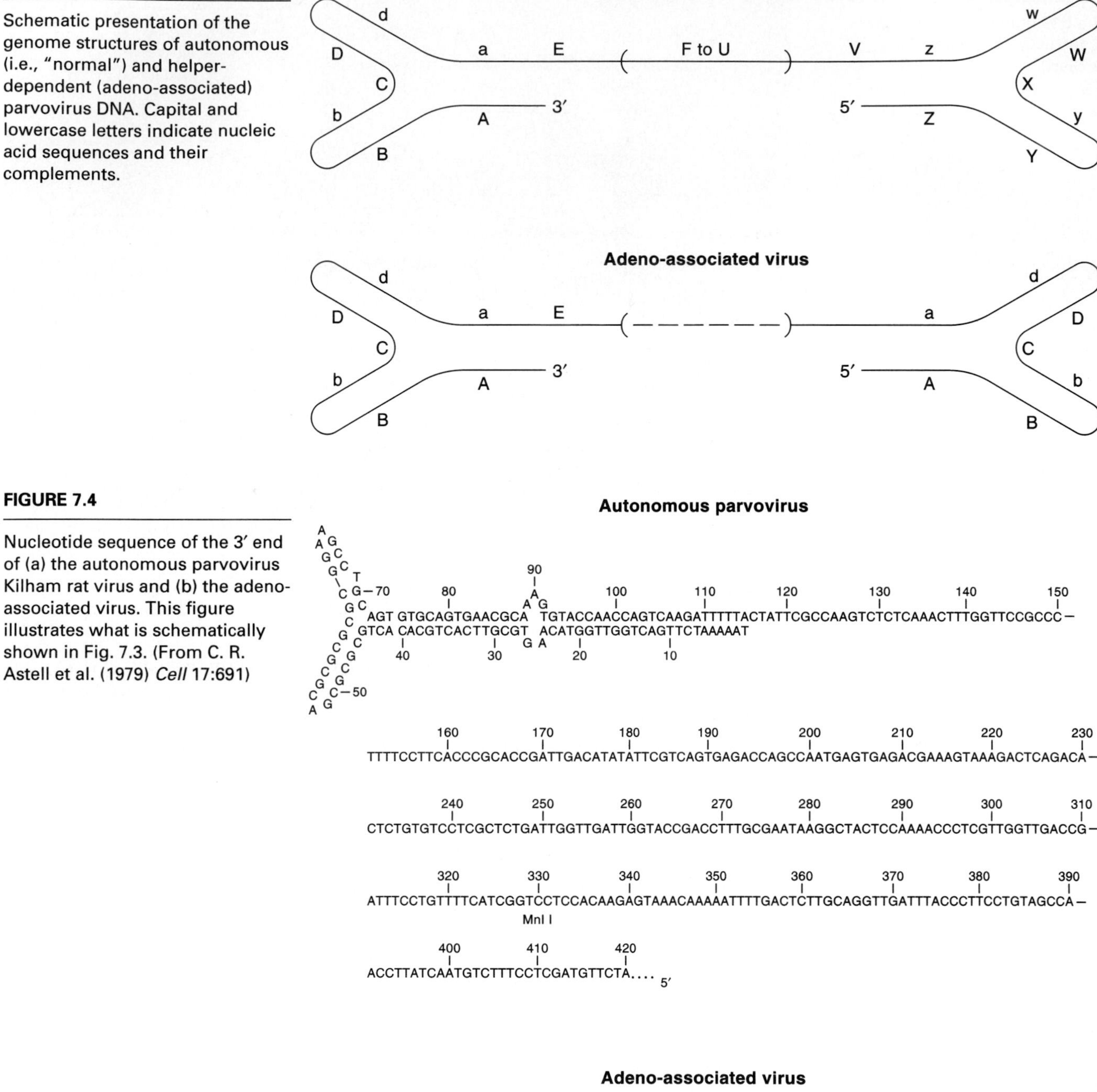

FIGURE 7.3

Schematic presentation of the genome structures of autonomous (i.e., "normal") and helper-dependent (adeno-associated) parvovirus DNA. Capital and lowercase letters indicate nucleic acid sequences and their complements.

FIGURE 7.4

Nucleotide sequence of the 3′ end of (a) the autonomous parvovirus Kilham rat virus and (b) the adeno-associated virus. This figure illustrates what is schematically shown in Fig. 7.3. (From C. R. Astell et al. (1979) *Cell* 17:691)

negative-sense DNA strands are found separately encapsidated, as they are in the other parvovirus genera (see below). The target cell for B19 appears to be an immature cell in the erythroid lineage, that is responsive to **erythropoietin.**

The *dependoviruses* and *densoviruses* share a biochemical peculiarity that was first discovered with these virus groups, but may be unique only in quantitative terms: Single strands of *both polarities* (positive and negative) are produced in similar amounts, and both become encapsidated. Thus, when AAV DNA is isolated, the two strands liberated from different virions tend to anneal, and molecules of *double-stranded* DNA of about 3×10^6 Da are obtained, much larger than the composition of these small virions permits. This finding caused considerable confusion until the mystery was solved: It was recognized that the double-stranded state was an artifact occurring during isolation of the DNA.

Another feature of AAV DNA structure was later found to be characteristic also of adeno- and many other viruses. As noted above, *inverted complementary sequences* are present at the two ends of the molecule. This structure enables the molecule to form single-stranded circles, held together by the base-paired **panhandle-shaped terminal** sequences (Fig. 7.5). This feature may also play a role in the capability of these viruses to become integrated into the host cell genome, even in the absence of a helper virus, and thus remain in a latent, nonexpressed state. AAV can subsequently be rescued and become replicated by infection with a helper virus (e.g. an adenovirus or herpes virus). Whether integration is involved in a latent state for the autonomous parvoviruses is not known.

Recently the defectiveness of AAV has been challenged by the observation that pretreatment of some cell lines with a variety of agents (e.g., UV light, cyclohexamide, chemical carcinogens) can lead to AAV replication. Whether these agents would have an effect *in vivo* is unknown; the replication of AAV is primarily associated with helper viruses. The results suggest that all AAV-enhancing properties are associated with increasing expression of certain cellular proteins.

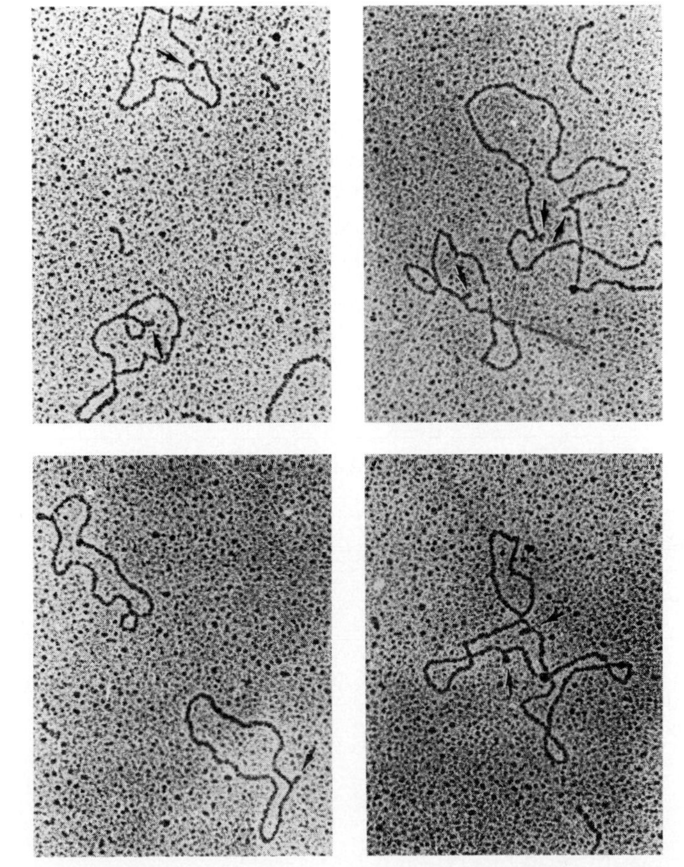

FIGURE 7.5

Electron micrograph of the DNA of adeno-associated virus. The "panhandles" giving the DNA its circular structure are evident. They are about 100 base pairs long. Double-length molecules can also be seen. (From A. K. Berns and T. J. Kelly, Jr. (1974) *J. Mol. Biol.* 82:267)

Noninverted repetitions have also been observed at each end. A further structural feature of the parvoviruses is internally complementary sequences at the termini, which allow the terminus of a DNA to fold back on itself, creating **terminal hairpins** showing the typical Y-shaped structures of the parvoviruses. As we shall see, all these features play critical roles in the transcription and replication of the DNA of these viruses. The respective possible structures of the two types of parvoviral DNAs are represented schematically in Fig. 7.3.

Sequencing of several parvovirus genomes has shown three ORFs in the positive-sense strand, the long left-most one coding for the nonstructural proteins (NS1, NS2) and the long right-most one for the three coat proteins. Studies of the silkworm densovirus shows a third open reading frame on the complementary strand. Through the use of overlapping and nested mRNAs, as well as splicing, much more information is carried by these, as by so many other viral genomes, than their length would suggest.

The single-stranded linear DNA of these viruses is, like that of most single-stranded phages, rich in **thymine** (about 30 percent). The DNA of the autonomous parvoviruses has not been found to be infectious; only the double-stranded form of AAV DNA has been found to be infectious when helper virus is present.

7.1.2 Entry, Transcription, and Virus Production

The receptor needed for attachment of a parvovirus to the cell has not yet been determined, but treating cells with neuraminidase prevents binding, which suggests the need for sialic acid, as in myxovirus infection. The replication of the DNA of the parvoviruses takes place in the nucleus without integration. It depends on the host cell DNA replicating machinery, particularly DNA polymerase γ; NS1 may also be involved. It occurs almost only during the S and G2 phase of cellular DNA synthesis. In fact, these DNAs cannot be produced in cells that are not replicating their own DNA, since the DNA-synthesizing enzymes are active in animal cells only during the S phase. The products of the autonomous and defective parvoviruses appear principally similar. Replication seems to proceed without cyclization and *without the need for RNA primers.* This fact reflects the fold-back of the **self-complementary sequence** at the 3′ end of molecules (Fig. 7.4). The structure produces the primer—OH that is required for DNA polymerase action. That part of the molecule, being rich in A/T and next to a G/C-rich sequence, has all the earmarks of a DNA replication origin. Mechanisms have been proposed to account for the replication of that terminal hairpin and the entire molecule; they are illustrated in Fig. 7.6. This event requires a specific nicking of the parental strand opposite the replication initiation site. Double-length replicative intermediates have

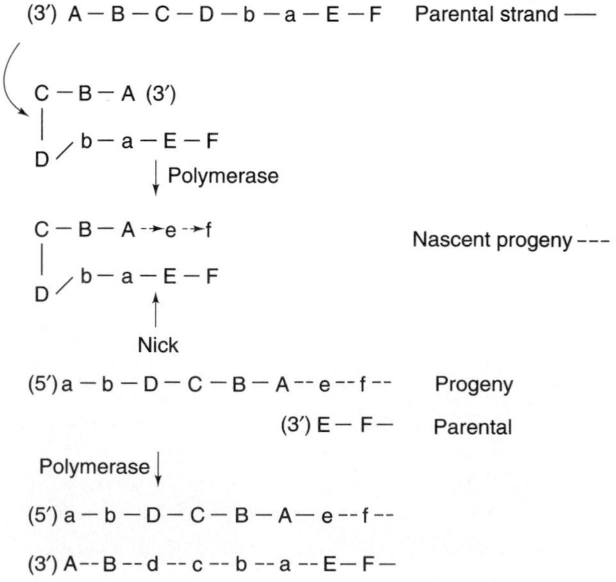

FIGURE 7.6

Model for formation of a duplex replicative intermediate for AAV DNA. The nucleotide sequence of the terminal repetition is represented as *ABCDba E, F* where *AB* are nucleotides 1–41 that are complementary to *ab* (nucleotides 125–85), *C* and *D* are two shorter self-complementary hairpin sequences (nucleotides 42–62 and 64–84, respectively), and *E, F* represent the continuing DNA sequence. (See Fig. 7.4 (b).) (Modified from W. W. Hauswirth and A. K. Berns. (1979) *Virology* 93:57)

also been observed. The fact that the autonomous mammalian parvoviruses have different terminal sequences must in some way be related to their replicating predominantly negative-sense strands. This end result is in contrast with the AAVs, which have terminal inverted redundancy and produce similar amounts of strands of both polarities.

The transcription of the parvovirus DNAs by host cell RNA polymerase requires the ds replicative form. It begins at a promoter site near the 3′-terminal hairpin segment. The main product is an mRNA of about 0.9×10^6 Da, which corresponds to the molecular weight of the largest protein component of these viruses (about 90 kDa). A larger transcript almost equaling the molecular weight of the DNA occurs in the nucleus. Splicing appears to be involved in generating the ultimate mRNAs. Three major mRNA species are formed: Two encode the nonstructural proteins, NS1 and NS2. The third mRNA gives rise to a species that is spliced and gives rise to the coat proteins, VP1 and VP2. VP2 is then cleaved into VP3; the ratio depends on the time course of the infection. Recent evidence has shown that parvoviruses produce a nonstructural protein early upon infection that up-regulates transcription of their late genes. This **transactivation** resembles that of other DNA viruses and human retroviruses. The NS proteins may also complex with cellular proteins that are involved in the replication of the virus. Synthesis of the coat proteins occurs after that of the NS proteins. Progeny virus is observed after about 24 hours, with virion assembly taking place in the nucleus. Virus replication leads to death of the cells.

SECTION 7.1 FURTHER READING

Original Papers

CRAWFORD, L. B. (1966) A minute virus of mice. *Virology* 29:605.

TATTERSAL, P., and D. C. WARD. (1976) Rolling hairpin model for replication of parvovirus and linear chromosomal DNA. *Nature* 263:106.

HAUSWIRTH, W. W., and K. I. BERNS. (1979) Adeno-associated DNA replication: Non-unit-length molecules. *Virology* 93:57.

MORTIMER, P. P., et al. (1983) A human parvovirus-like virus inhibits hematopoietic colony formation *in vitro*. *Nature* 302:426.

SUMMERS, S., et al. (1983) Characterization of the genome of the agent of erythrocyte aplasia permits its classification as a human parvovirus. *J. Gen. Virol.* 64:2527.

CHEN, K. C., et al. (1986) Complete nucleotide sequence and genome organization of bovine parvovirus. *J. Virol.* 60:1085.

COTMORE, S. F., et al. (1986) Identification of the major structural and nonstructural proteins encoded by human parvovirus B19 and mapping of their genes by procaryotic expression of isolated genomic fragments. *J. Virol.* 60:548.

OZAWA, K., et al. (1986) Replication of the B19 parvovirus in human bone marrow cell cultures. *Science* 233:883.

BANDO, H., et al. (1987) Organization and nucleotide sequence of a densovirus genome imply a host-dependent evolution of the parvoviruses. *J. Virol.* 61:553.

RHODE, S. L. III, and S. M. RICHARD (1987) Characterisation of the trans-activation responsive element of the parvovirus H-1 P38 promoter. *J. Virol.* 61:2807.

YACOBSON, B., et al. (1987) Replication of adeno-associated virus in synchronized cells without the addition of a helper virus. *J. Virol.* 61:972.

BLOOM, M. E., et al. (1988) Molecular comparisons of *in vivo*- and *in vitro*-derived strains of Aleutian disease of mink parvovirus. *J. Virol.* 62:132.

CLEMENTS, K. E., and D. J. PINTEL. (1988) The two transcription units of the autonomous parvovirus minute virus of mice are transcribed in a temporal order. *J. Virol.* 62:1448.

CHAPMAN, M. S., and M. G. ROSSMAN. (1993) Structure, sequence, and function correlations among parvoviruses. *Virology* 194:491.

PATOU, G., et al. (1993) Characterization of a nested polymerase chain reaction assay for detection of parvovirus B19. *J. Clin. Microbiol.* 31:540.

TAM, P., and C. R. ASTELL. (1993) Replication of minute virus of mice minigenomes: Novel replication elements required for MVM DNA replication. *Virology* 193:812.

Books and Reviews

BERNS, K. I., ed. (1984) *The parvoviruses.* Plenum Press, New York.

BERNS, K. I. (1990) Parvoviridae and their replication. In *Virology,* 2d ed., ed. B. N. Fields and D. M. Knipe. 1743 Raven Press, New York.

7.2 PAPOVAVIRIDAE

This nonenveloped Papovaviridae family has come under close scientific scrutiny for two main reasons. First, their **circular double-stranded** DNAs represent miniature models for eukaryotic cell DNA. The study of these DNAs has therefore greatly increased our understanding of cellular DNA replication and transcription. These viruses have actually become the eukaryotic equivalents of phage φX in prokaryotes. Second, many of the papovaviruses can under certain conditions transform cells in a manner resembling malignancy. Thus, they can also serve as miniature models in the field of cancer research. Some members of this virus family are listed in Table 7.1.

The name Papovaviridae, is derived from three characteristics of the viruses: They can cause small neoplastic tumors of the skin (*pa*pillomas) and multifocal expression of tumors (*po*lyomas), and they can induce *va*cuoles in infected cells. This latter feature was particularly noted in the simian virus 40 (SV40) but has also been recognized in other members of the papovavirus family.

Recent evidence indicates that, among the members of this family, the papillomaviruses can clearly be distinguished from the mouse polyoma virus, the simian virus, SV40, and others, based on their molecular structure and replicative cycle. The Papovaviridae have thus been subclassified into two genera. Of these, the **papillomaviruses** are much larger (50 to 55 nm) than the **polyoma viruses** (44 nm). The papillomaviruses carry all their genetic information on one of the two circular supercoiled DNA strands, of about 5.2×10^6 Da, whereas the polyomaviruses carry this information on both strands of a 4×10^6 Da DNA. Papillomavirus DNA is rarely, if ever, integrated into the host genome but remains as a **circular plasmid** in the nucleus of the transformed cell. Moreover, papillomaviruses, in contrast with polyomaviruses, are limited in their cell tropism (epithelial cells) and are difficult to grow in tissue culture. Finally, distinct differences in the mode of transcription and the processing signals of the two virus genera can be noted.

TABLE 7.1
Hosts and pathogenicity of papovaviruses

Virus	*Host*	*Tumor (Host) or Disease*
Shope papilloma	Rabbit	Papillomas, carcinomas (rabbits)
Bovine papilloma	Cow	Papillomas (cow) Lymphomas (hamster)
Wart-papilloma	Human	Warts, laryngeal papillomas, and probably cervical and carcinomas
Polyoma	Mouse	Multiple types (rodents)
Simian virus 40 (SV40)	Monkey	Lymphomas, sarcomas (hamsters, rats)
Lymphotropic papovaviruses	Monkey	Sarcomas (hamster)
JC	Human	Associated with progressive multifocal leukoencephalopathy (PML)
BK	Human	Associated with Wiscott Aldrich syndrome, immunosuppression
K	Mice	—
Rabbit kidney vacuolating virus	Rabbit	—

7.2.1 Papillomaviruses

Papillomas, keratinized skin tumors or warts, are widespread in the animal kingdom. The responsible papillomaviruses are generally species- and tissue-specific and usually not oncogenic in the host species. However, when transferred to another animal species, they can often cause **malignancies.** The first virus linked to a mammalian malignancy was actually the papillomavirus of rabbits discovered by Shope (Table 1.1). This agent, as is observed with other papillomaviruses, gives rise to benign skin tumors in these animals, but under some conditions it can induce carcinomas.

Papillomaviruses have been detected in warts from rabbits, hamsters, mice, sheep, goats, deer, cattle, horses, dogs, and monkeys. Nevertheless, aside from the Shope rabbit papillomavirus, only the bovine and human viruses have been studied in any detail.

Up to 70 different types of human papillomaviruses (HPV) have been identified associated with warts and epidermal and epithelial lesions at different sites. Distinct isolates are responsible for specific papillomas such as the common wart, flat wart, and genital warts, and as noted below, with some carcinomas. The HPV can be classified as cutaneous or mucosal types.

7.2.2 Structure

The papillomavirus particle has a spherical protein coat of approximately 72 capsomers. As with the parvoviruses, the major component of the virion is the **capsid protein** (55 kDa). It makes up about 80 percent of the total viral protein. A minor protein (70 kDa) is also present. Characteristically, the papillomaviruses can be visualized by electron microscopy only in epithelial cells undergoing differentiation (Fig. 7.7). It is surmised that their genome is present in plasmid form in the basal cell layer of the skin but does not ex-

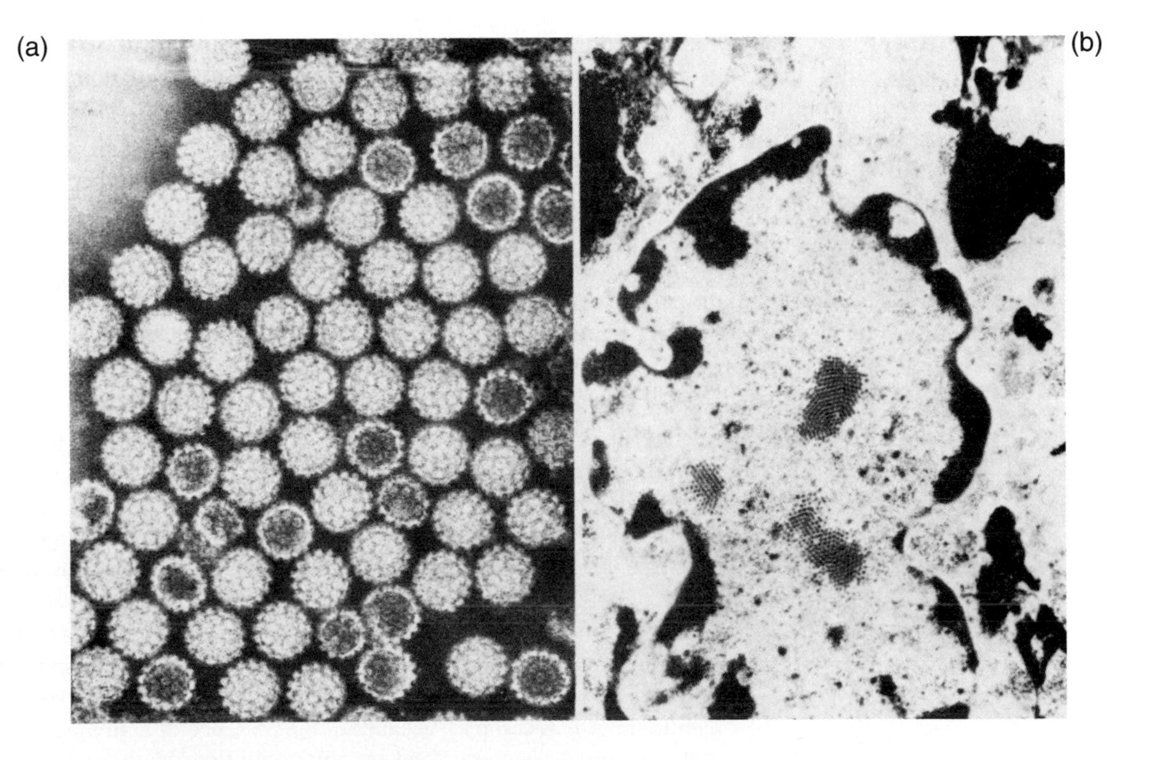

FIGURE 7.7

Electron micrograph of human papillomavirus. (a) The diameter of the particles is 55 nm. The capsomers can clearly be seen. Stain-filled particles probably lack the DNA. (Courtesy of R. C. Williams) (b) An intranuclear factory of papovavirus DNA in the epithelial cells of a human wart.

press its full viral protein complement until differentiation takes place. For this reason, viral forms are noted only in the top differentiated levels of common warts. This nonpermissive state of less-differentiated cells may contribute to the potential malignant properties that have become recognized with papillomaviruses. At this time, some of the human papillomaviruses (particularly HPV types 5, 6, 8, 16, and 18) have been found associated with carcinomas of the skin, cervix, anus, larynx, and lung. Their detection has been achieved only through procedures involving DNA hybridization and recently polymerase chain reaction (PCR) (see Appendix 1) using very small tissue samples and appropriate viral genomic probes. These approaches have become possible through advances in genetic engineering techniques. The molecular cloning of the first papillomavirus genome in the late 1970s paved the way for detailed analyses of this virus group. Thus far, all HPVs cross-hybridize with each other under low stringency conditions. Thus, screening of tumor material for HPV-related sequences is facilitated. For a papillomavirus to be considered a new type, its genome must have less than a 50 percent homology with that of other known types.

7.2.3 Entry and Production of Papillomaviruses

Although the human wart virus [see Fig. 7.7(a)] was one of the first viruses recognized at the turn of the century, the papillomaviruses up until recently have not been studied as intensely as members of the other genus, the polyomavirus family. This fact reflects the restrictive cell tropism of the papillomaviruses and the difficulty in growing them in tissue culture. The bovine papillomavirus type 1 (BPV-1) recently has been a helpful prototype for studies of the molecular biology and **transforming capability** of papillomaviruses in general. Initial reports suggested that papillomaviruses could be replicated in culture under selected temperature and environmental conditions, but until now, no reproducible procedure for growth of these agents *in vitro* has been achieved.

How papillomaviruses enter a cell is not known. Their route of transmission is not evident but appears to be by skin. Their tissue specificity suggests that specific receptors exist on the cell surface. How viruses emerge from the cells is also not clear, particularly since the particles are not found in the cytoplasm. Papillomaviruses are probably distributed throughout the body with certain subtypes "homing" out in some regions (e.g., HPV 16, 18 in genital areas).

Double-stranded DNAs (7900 bp), (5.2×10^6 Da) of certain papillomaviruses have been cloned using recombinant DNA techniques. The gene map resembles, in greatly simplified form, that of the adenoviruses. The genome has an **early region** coding for proteins needed for transformation, a **late region** that encodes the capsid proteins, and a **regulatory region** that contains the origin of replication and many of the control elements for transcription and replication. Some promising work has come from the introduction of these cloned viral DNAs into cells treated by a calcium/phosphate procedure that favors the uptake of viral DNA by the cells.

Based on recent studies a **two-stage model** for papillomavirus replication has been proposed. Initially, the viral genome undergoes amplification; subsequently, the nonencapsulated viral plasmid replicates on average about once every cell cycle (maintenance stage). Finally, there is the productive stage of infectious virus replication in differentiated epithelial cells. The replication of the viral genome depends on the expression of early viral genes and in this way resembles the process with **polyomaviruses.** Transcription of the papillomavirus is complex because of the presence of many promotors within the genome, multiple splicing patterns, and the differential production of mRNA species in different cells. All but 10 ORFs are on the same papillomavirus DNA strand and give rise to either early or late viral proteins. Close to 20 different mRNA species have been identified in some papillomavirus-transformed cells. Moreover, within the papilloma DNA is a long control region (LCR) that appears responsive to various *cis* elements that can regulate virus transcription. As with other DNA viruses with transforming capabilities, the early gene products are associated with cellular transformation besides having **transactivating** activity via the viral LCR. The late (L) gene products are expressed only in productively infected cells. It is estimated that the incubation period between virus infection and papilloma development ranges between 3 and 18 months. Cellular immunity appears to be the major mechanism for control of HPV expression.

7.2.4 Polyomaviruses

The mouse polyomavirus was first detected in 1953 when Ludwik Gross was experimenting with the mouse leukemia virus (see Sec. 6.1). Extracts of mouse tissues containing this retrovirus were contaminated with the polyoma agent and gave rise to salivary gland tumors when inoculated into newborn mice. The polyomavirus has been found in many populations of wild and laboratory mice but does not seem to produce any naturally occurring disease. However, when large amounts of virus are injected into neonatal mice, they either may show no symptoms or die of infections of kidney, salivary gland, thyroid, and other organs. Some, as noted above, develop tumors in various organs, particularly salivary gland, kidney, and thymus. This ability to induce tumors in several organs, which contrasts with the papillomaviruses and most of the RNA tumor viruses, earned this DNA virus the name polyoma. It is noteworthy that inoculation of mouse polyomavirus into a variety of rodents (e.g., hamsters, rats) can produce tumors in which virus replication does not take place. A similar observation has been made with some papillomaviruses (e.g., **bovine**). This characteristic of oncogenic DNA viruses—transformation without replication—is another important distinction between this transforming virus group and the RNA tumor viruses (see Sec. 10.3.1).

SV40, a polyomalike virus, was first discovered in 1960 in cultured monkey kidney cells that were found to contain several simian viruses. This 40th virus isolate (SV40) was later detected as a contaminant in polio and adenovirus vaccines when these latter viruses were grown in large amounts in monkey kidney cells (see Chapter 8). Another papovavirus with unusual characteristics has also been isolated from African green monkeys. This **lymphotropic papovavirus** (LPV) can be propagated only in B lymphocytes and transforms hamster fibroblasts. Other polyomaviruses now recognized include those of hamster and budgerigar. The latter virus is the first nonmammalian polyomavirus described. The hamster papillomavirus has been associated with leukemia and lymphomas in newborn hamsters.

Two types of **human polyomalike viruses** have been isolated from subjects suffering from either neurological disease (the JC agent) or immunological dysfunction (BK virus). They cross-react serologically with some proteins of polyoma and SV40 but are distinct human viruses. Infection by these agents usually occurs in childhood, and they are then carried for life. Generally, by age 40 to 50 over 80 percent of the human population show serum antibodies against these viruses. Neither of these human papovaviruses is pathogenic or naturally oncogenic. In immunocompromised individuals however, the JC agent can give rise to the rare human neurological disease, **progressive multifocal leukoencephalopathy** (PML) (see Sec. 12.2). These human viruses, like other papovaviruses, can induce tumors in newborn hamsters and can transform animal and human fibroblasts in culture. Mouse and the human polyomaviruses attach to and agglutinate red cells. As with other viruses, a hemagglutination assay can be used to quantify the physical particles of the virus (see Sec. 11.2).

Structure of Polyomavirus and SV40 Viruses The most commonly studied members of the Papovaviridae are the polyomavirus and SV40 virus. They have the **icosahedral** symmetry of papovavirons with clearly visible **capsomers** (see Fig. 7.7(a)). Although these two virus groups are not closely related, they do have many similarities. Their virion has a diameter of about 41 nm. It contains 12 percent double-stranded circular DNA of 3.2×10^6 Da, which represents a similarly small

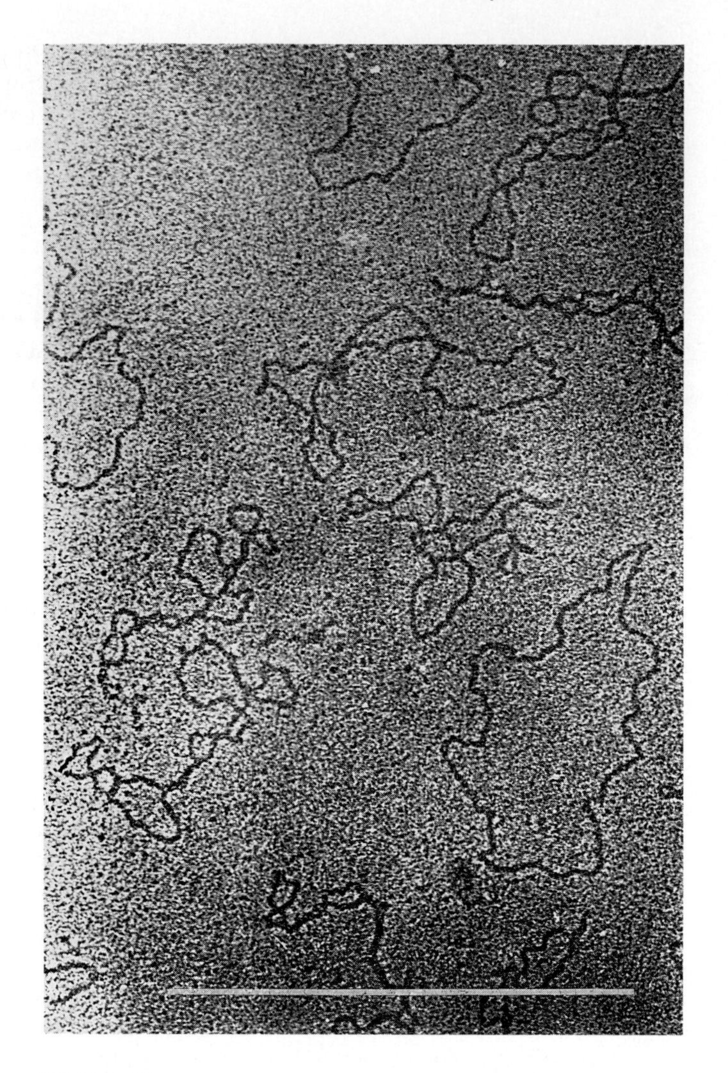

FIGURE 7.8

Polyomavirus DNA electron micrographic image obtained by the Kleinschmidt spreading technique. Supercoiled and open circles as well as a few linear molecules can be seen.

amount of genetic information as that present in the parvovirions. However, in contrast, as discussed below, the polyomaviral DNA carries part of the genetic information on each strand, and these can become integrated into the host cell genome. As usual, supercoiled and relaxed circular forms as well as linear DNA molecules can be seen (Fig. 7.8). The DNA by itself is infectious. Three coat proteins make up the virion (VP1, VP2, VP3). As with the papillomavirus, there are several **cellular histones**, the small basic proteins occurring in chromosomes that become associated with the viral DNA. These histones actually interact specifically with the viral DNA, forming beads on a string resembling the **nucleosomes** of chromatin. A fourth protein, termed agnoprotein, of 61 predominantly basic amino acids, is coded in the late leader region. This protein may play a role in virion assembly.

The peptide maps of the virus-specific coat proteins show evidence of overlap. Now that the complete nucleotide sequences of SV40 and polyomavirus are known, two of the three proteins have been found encoded in the same reading frame on the same part of the genome (see Fig. 7.9). VP1, representing 75 percent of the total virion protein, is read in a different reading frame; it is modified to yield six electrophoretically different species, which play various important roles in viral development.

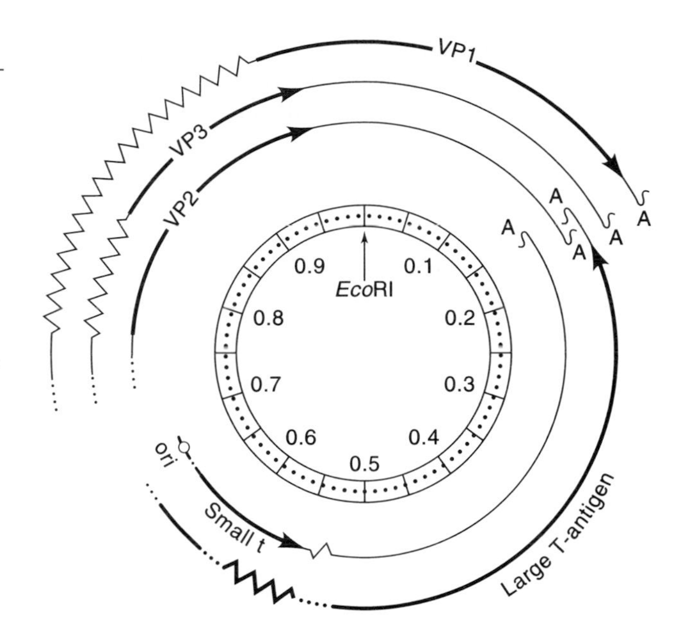

FIGURE 7.9

Genome map of SV40 DNA. The single cleavage site by the restriction endonuclease *Eco* RI is termed 0 (or 12 o'clock). The DNA replication origin is at 0.663, and the translation origin is in the same general area. The five virus-coded proteins are indicated by open arrows. The coat proteins (VP 1, 2, 3), translated clockwise, overlap, as do the t and T proteins translated anticlockwise. Zigzag lines represent spliced-out genome components ("introns"). Untranslated nucleic acid segments are indicated by single (solid) lines. (From W. Fiers et al. (1978) *Nature* 273:115)

7.2.5 Entry and DNA Replication

The polyomavirus replicative cycle begins with adsorption of the virus to the cell surface and its entry by **endocytosis.** Because infection can be blocked with **sialidase,** adsorption may involve sialic acid residues. Similar receptors could also be used by the human JC virus. VP1 is a major viral capsid protein that mediates this interaction with the cell membrane. Specific binding must occur, since entry by alternative nonspecific means leads to a degradation in lysosomes.

The site of entry of the human polyomalike viruses is not known but may be the respiratory tract. After infection, the polyomavirus is transported to the cell nucleus, where it fuses with the nuclear membrane to enter. Uncoding occurs and transcription of early mRNAs takes place with production of the T antigens. Subsequent virus DNA replication occurs via the late mRNAs. Translation of the late proteins gives rise to the progeny particles.

The DNA of these viruses closely resembles the RF of the single-stranded DNA phages (see Sec. 7.5) as well as typical plasmids. Its replication requires prior transcription of the early genes and formation of the T antigen (see below) that binds to the "origin" site on the parental DNA. Replication is by the **theta mode,** bidirectionally from a single replication origin. During this process the parental molecule remains partially supercoiled, the relaxed state apparently being generated only locally and temporarily by enzyme action (Fig. 7.10). The process is presumed to require the usual steps of DNA replication: RNA primer syntheses, helix unwinding, formation of DNA nascent strands (Okazaki fragments), primer degradation, gap filling, ligase activity, and the action of **topoisomerase II,** the enzyme that supercoils DNA. The entire replication process, which occurs in the nucleus of the cell, takes about 5 minutes. The location of the origin

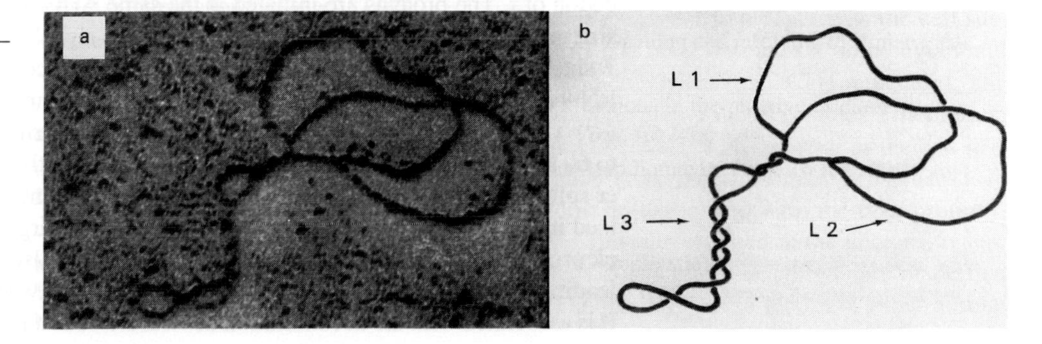

FIGURE 7.10

(a) Electron microscopic and (b) schematic presentation of a replicating SV40 DNA molecule. The only partially supercoiled nature of the molecule is evident. L1 and L2 indicate the relaxed parental and progeny parts of the molecule, L3, the still-supercoiled part. (Courtesy of N. Salzman)

LIVINGSTON, D. M., and M. K. BRADLEY. (1987) Review of the simian virus 40 large T antigen: A lot packed into a little. *Mol. Biol. Med.* 4:63.

LAMBERT, P. F., et al. (1989) The genetics of bovine papillomavirus type 1. *Ann. Rev. Gen.* 22:235.

ZUR HAUSEN, H. (1989) Papillomaviruses as carcinomaviruses. In *Advances in viral oncology.* Vol. 8, 1. Raven Press, New York.

ECKHART, W. (1990) Polyomavirinae and their replication. In *Virology,* 2d ed., ed. B. N. Fields and D. M. Knipe. 1593 Raven Press, New York.

HOWLEY, P. M. (1990) Papillomavirinae and their replication. In *Virology,* 2d ed., ed. B. N. Fields and D. M. Knipe. 1625 Raven Press, New York.

PIPAS, J. M. (1992) Common and unique features of T antigens encoded by the polyomavirus group. *J. Virol.* 66:3979.

7.3 PLANT VIRUSES WITH SINGLE-STRANDED DNA GENOMES

For years, plant pathologists suspected that a number of whitefly- or leafhopper-transmitted diseases were of viral etiology but were unable to identify causative agents. In the mid 1970s, **isometric** particles 18 to 20 nm in diameter and usually occurring in pairs were isolated from maize streak–diseased plants. Soon thereafter, the genome of bean golden mosaic virus, a virus with similar properties, was shown to consist of two small DNA molecules. These discoveries coincided with the beginnings of plant molecular biology, and a full array of molecular biological techniques was rapidly applied to these and other plant viruses containing ssDNA genomes. At present, at least two types of viruses with small ssDNA genomes are known: **geminiviruses,** which have the characteristic geminate or twinned (*gemini,* "twins") particle morphology described above, plus a small number of viruses resembling the porcine circoviruses in particle morphology.

7.3.1 Geminiviruses

Figure 7.11 illustrates the unique, twin-shaped particle morphology of geminiviruses. Their geminate particles, about 18 × 30 nm, consist of two incomplete icosahedra with $T = 1$ symmetry and contain a total of 22 capsomers. Each particle contains only one molecule of single-stranded circular DNA (≤ 3000 nt).

As shown in Table 7.2, differences in genome structure, host range, and vector specificity have led to the creation of three subgroups within the *Geminivirus* group. Except for

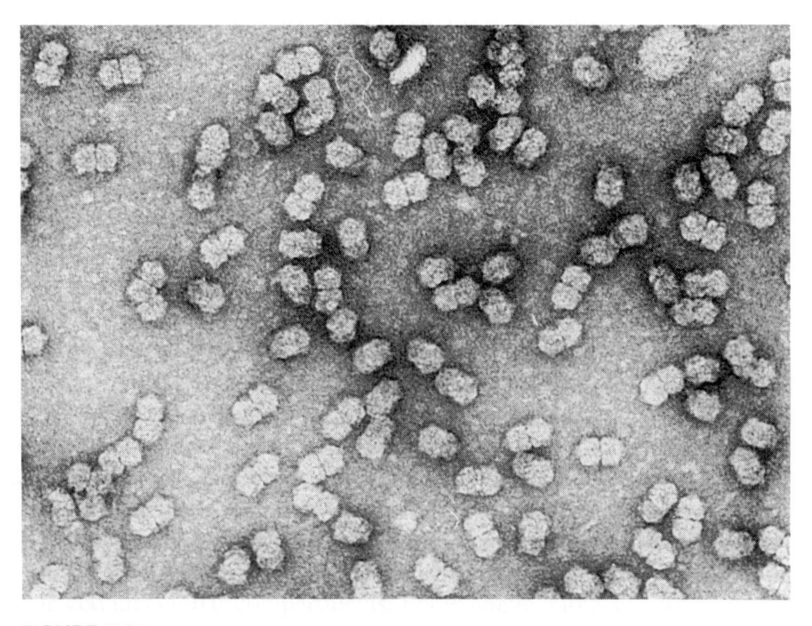

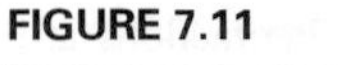

FIGURE 7.11

Purified preparation of maize streak viruses. (Courtesy of M. I. Boulton)

TABLE 7.2
Biological and molecular properties of plant geminiviruses

Subgroup	*Type Member*	*Hosts*	*Vectors*	*Genome Structure*
I	Maize streak virus (MSV)	Monocots	Leafhoppers	Monopartite
II	Beet curly top virus (BCTV)	Dicots	Leafhoppers	Monopartite
III	Bean golden mosaic (BGMV)	Dicots	Whiteflies	Bipartite

beet curly top virus, most individual geminiviruses have narrow host ranges. Geminiviruses are transmitted by insects in nature, subgroups I–II by leafhoppers and subgroup III by whitefly. Members of subgroup III have a bipartite genome.

For the many geminiviruses that are not mechanically transmissible, the discovery that they can be experimentally transmitted by **agroinoculation** has greatly facilitated their molecular characterization. In this technique, a greater-than-full-length clone of the viral DNA(s) is inserted into the Ti plasmid of *Agrobacterium tumefaciens, (*a plant pathogenic bacteria that is capable of transforming plant cells (see Sec. 13.5). Upon entry into the host cell, a full-length, covalently closed, circular viral dsDNA escapes from the Ti plasmid DNA via homologous recombination and initiates virus replication.

As diagrammed in Figure 7.12, all geminivirus DNAs exhibit several common organizational features. All are slightly less than 3 kb in size, and transcription is **bidirectional**; that is, discrete polyadenylated viral mRNAs complementary to both the viral and complementary DNA strands can be isolated from infected cells. Furthermore, the

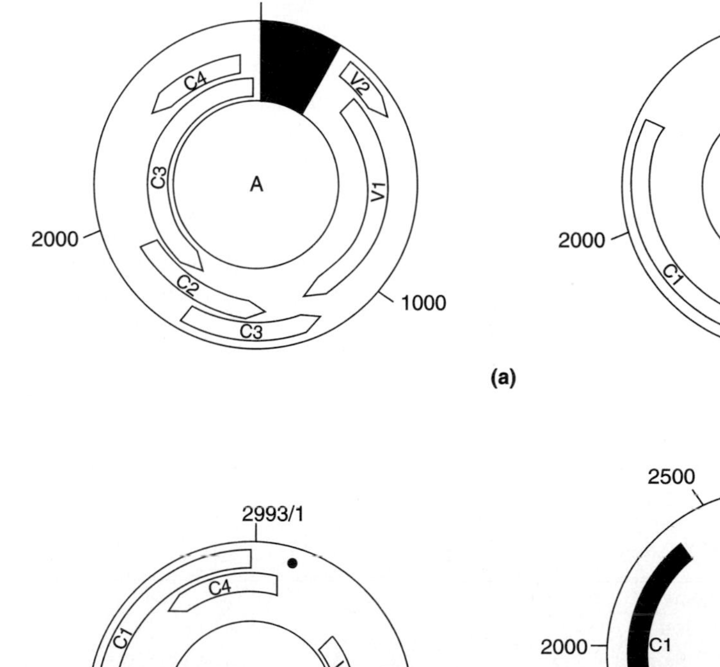

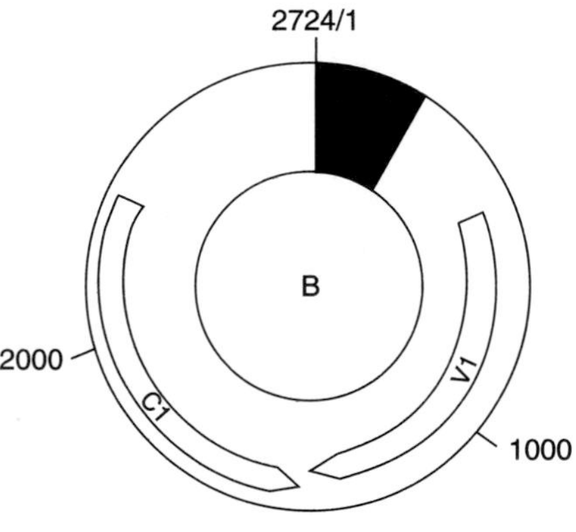

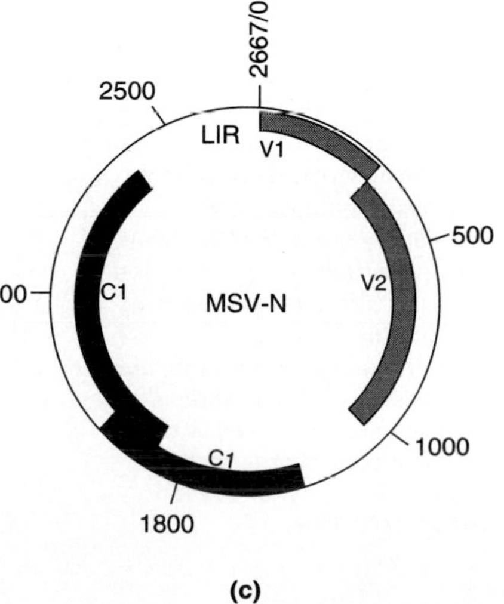

FIGURE 7.12

Genome organization of representative geminiviruses. Positions of known or potential ORFs and the direction of mRNA transcription are indicated by arrows. The conserved intergenic or "common" region is located between twelve and one o'clock [i.e. shaded area in (a)]. (a) African cassava mosaic virus (subgroup III); (b) beet curly top virus (subgroup II); (c) maize streak virus–Nigerian isolate (subgroup I). (Courtesy of J. Stamley and M. I. Boulton)

approximately 200 nucleotide **intergenic region** between twelve and one o'clock on each genome map is able to form a hairpin loop. This loop contains a **conserved sequence** that is present in all geminiviruses. Because it is so highly conserved, this region (or structure) is considered to be essential for both DNA replication and RNA transcription. In geminiviruses with bipartite genomes (e.g., African cassava mosaic virus (ACMV)), this is the only portion of the genomic DNAs that is nearly identical in sequence in each genome.

As mentioned above, geminiviruses have been subjected to intensive molecular genetic analysis. For the bipartite viruses, such as ACMV, the coat protein gene has been shown to reside on DNA A (i.e., ORF V1 in Fig. 7.12) and to determine vector specificity. Although synthesis of functional coat protein is not essential for infectivity, systemic spread, or symptom induction by subgroup III viruses does require coat protein. Because ACMV DNA A can replicate independently in protoplasts, it must contain all the genes required for DNA replication. The failure of DNA A to infect whole plants means that protein(s) encoded by DNA B must be necessary for cell-to-cell movement.

7.3.2 Other Plant Viruses with Single-stranded DNA Genomes

In addition to the geminiviruses, there is also a second, as yet officially unnamed group of plant viruses containing ssDNA genomes. Members of this group include coconut foliar decay, banana bunchy top, and subterranean clover stunt viruses, and their 20 nm icosahedral particles resemble half-geminate particles of *Chloris* striate mosaic virus (a subgroup I geminivirus). In the case of coconut foliar decay virus (CFDV), however, there is no detectable sequence homology with either caulimoviruses or *Digitaria* streak virus, a geminivirus known to infect a grass growing in association with the diseased coconut palms.

The organization of the CFDV genome is illustrated in Figure 7.13. Its 1291 nucleotide ssDNA genome contains six ORFs encoding proteins larger than 5 kDa. The largest open reading frame (ORF1) codes for a leucine-rich 33.4 kDa protein with a nucleoside triphosphate-binding motif that may be involved in viral replication. The small (6.4 kDa), arginine-rich polypeptide encoded by ORF3 may be the virus coat protein. Like the genomes of all known geminiviruses, single-stranded CFDV DNA can assume a **stem-loop structure** (see Fig. 7.13). In the virus complementary (or negative-sense) orientation, the sequence of the loop is highly homologous to the conserved geminivirus TAATATTAC motif.

Both banana bunchy top (BBTV) and subterranean clover stunt (SCSV) viruses appear to have multipartite genomes. In the case of BBTV, there may be as many as six small (1000 to 1200 nucleotides) ssDNAs, each containing a single ORF. The genome of SCSV may be distributed among as many as seven similar-size ssDNAs. Cloning and sequence analysis of both the BBTV and SCSV genomes is underway, and much more information about possible taxonomic relationships with other viruses should soon be available.

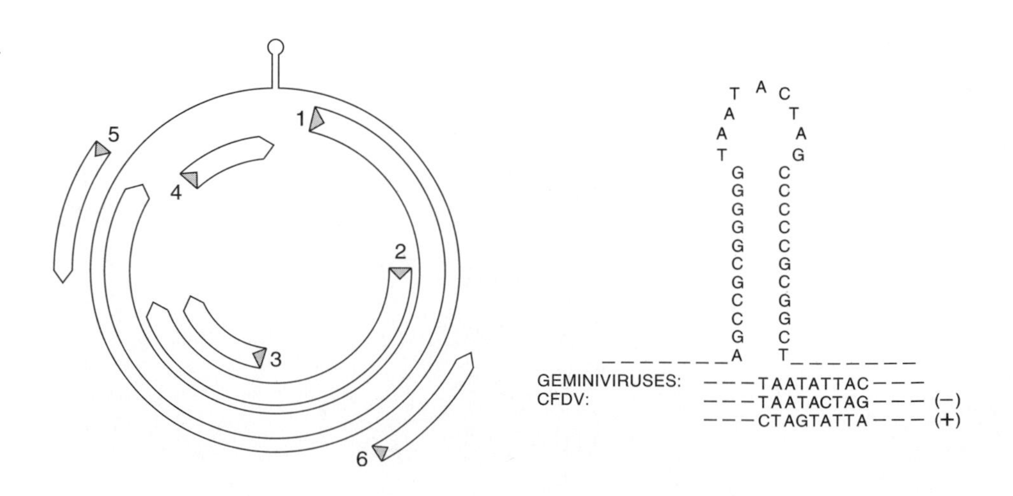

FIGURE 7.13

Genomic organization of coconut foliar decay virus. (Left) relative locations of putative ORFs (denoted by numbered arrows) and a potentially stable stem-loop structure; (right) stem-loop structure for the virion-complementary DNA strand. Loop sequences for both strands of the viral DNA are compared with the canonical geminivirus motif. (From W. Rohde et al. (1990) *Virology* 176:648)

SECTION 7.3 FURTHER READING

Original Papers

GOODMAN, R. M. (1977) Infectious DNA from a whitefly-transmitted virus of *Phaseolus vulgaris*. *Nature* (London) 266:54.

STANLEY, J., and M. GAY. (1983) Nucleotide sequence of cassava latent virus DNA. *Nature* (London) 301:260.

ROGERS, S. L., et al. (1986) Tomato golden mosaic virus: A component DNA replicates autonomously in transgenic plants. *Cell* 45:593.

GRIMSLEY, N., et al. (1987) *Agrobacterium*-mediated delivery of infectious maize streak virus into maize plants. *Nature* (London) 325:177.

CHU, P. W. G., and K. HELMS. (1988) Novel viruslike particles containing circular single-stranded DNAs associated with subterranean clover stunt disease. *Virology* 167:38.

ROHDE, W., et al. (1990) Nucleotide sequence of a circular single-stranded DNA associated with coconut foliar decay virus. *Virology* 176:648.

STANLEY, J., et al. (1992) Mutational analysis of the monopartite geminivirus beet curly top virus. *Virology* 191:396.

GILBERTSON, R. L., et al. (1993). Pseudorecombination between infectious cloned DNA components of tomato mottle and bean dwarf mosaic geminiviruses. *J. Gen. Virol.* 74:23.

HARDING, R. M., et al. (1993). Nucleotide sequence of one component of the banana bunchy top virus genome contains a putative replicase gene. *J. Gen. Virol.* 74:323.

Books and Reviews

LAZAROWITZ, S. G. Geminivirus: Genome structure and gene function. In *CRC critical reviews in plant sciences,* ed. B. V. Conger. Vol. 11. CRC Press, Boca Raton, Fla. In press.

DAVIES, J. W., et al. (1987) Structure and replication of geminivirus genomes. *J. Cell Sci., Suppl.* 7:95.

7.4 MICROVIRIDAE (ISOMETRIC SINGLE-STRANDED DNA BACTERIOPHAGES)

7.4.1 Structure

The prototype of the Microviridae is φX**174**. This virus, which is now frequently called simply φX, has been studied since about 1960 and was the first virus found to have a *single-stranded circular DNA* genome. The virion consists of 60 copies of a 50 kDa protein that makes up the 20 facets of the icosahedral capsid. The spikes at the 12 vertices (see Fig. 7.14), contain five molecules of a 20 kDa and one molecule of a 37 kDa protein.

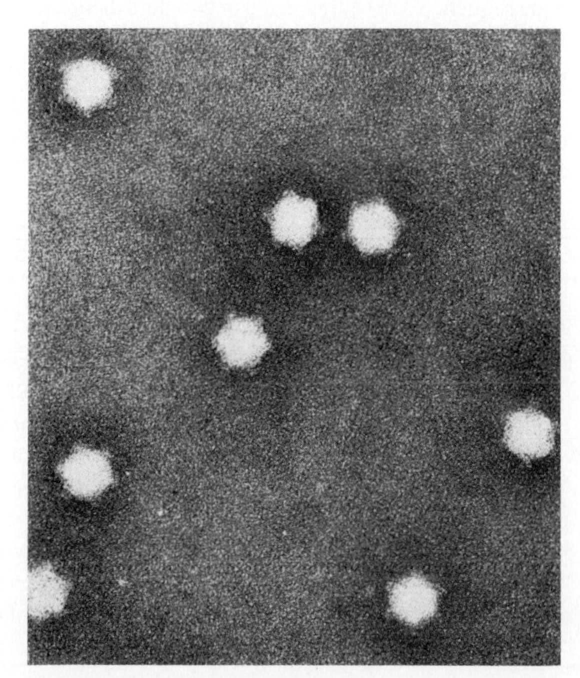

FIGURE 7.14

Electron micrograph of the DNA phage φX174. The diameter of the particles is 25 to 35 nm. The spikes can clearly be seen. (Courtesy of R. C. Williams)

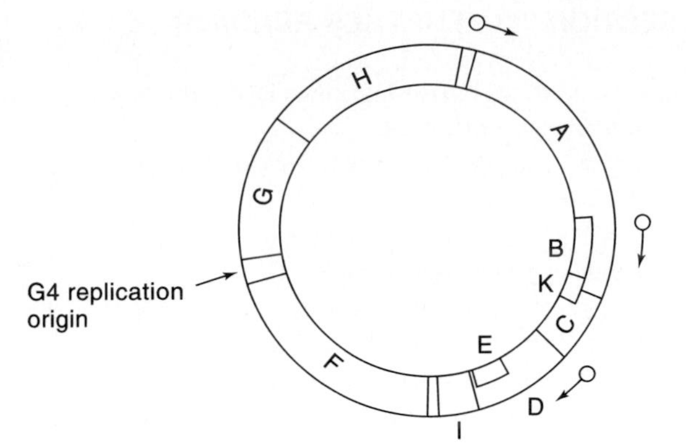

FIGURE 7.15

Generalized gene map of an isometric DNA bacteriophage. Positions of the origin of G4 DNA replication (↓) and the promoter sites for φX174 mRNA transcription (♂) are marked. Note the overlapping genes A-B-K-C-D-E; each uses a different reading frame.

The complete 5386 nucleotide sequence of φX174 DNA was published by F. Sanger and his colleagues in 1977. This was the first viral genome to be completely sequenced, and the sequencing strategy developed during these studies (i.e., the dideoxy chain termination method) has subsequently become one of the standard methods for DNA sequence analysis. In 1980, Dr. Sanger received a Nobel Prize (his second) for this work.

The sum of the molecular weights of these proteins and of the genomic DNA comes to a particle weight of 6.35×10^6 Da, and the crystal structure of φX174 has been determined. The particles have diameters of 25 to 35 nm, depending on whether they are measured from the tip of the spikes or from the valleys. The 37 kDa protein at the tip of each spike has been compared with the A protein of the RNA phages, in that it appears to be necessary for infection of intact cells. This protein has an affinity for specific lipoproteins on the outer membrane of the host cell. It then seems to enter the cell membrane together with the viral DNA and has been termed a pilot protein. Like the A protein of the RNA phages, this **pilot protein** also appears to play a role in the maturation of the phage. It is not essential for viral infectivity, however, because spheroplasts can be infected with the pure phage DNA.

Initial estimates of the number of genes in ϑX and related phages (e.g. G4) came from the study of mutants. These studies indicated the presence of eight genes, called **complementation groups** when established by genetic methodology. Several of these could be identified with specific proteins made *in vitro*. Quite unexpected, however, was the finding, suggested by genetic evidence and proved in the course of a sequencing of ∂X DNA, that the same nucleotide sequence can be translated in two (and theoretically three) reading frames and can thus code for two (or three) proteins. As shown in Fig. 7.15, with these phages gene E lies within gene D, and gene B within gene A. Thus, a short nucleotide sequence within gene A thus actually codes for—in addition to protein A—protein B, yielding a different amino acid sequence, and protein A*, yielding the same amino acid sequence (see Fig. 7.16). Translation in different phases was later discovered in RNA phages (see Sec. 2.7.2) and in many animal viruses. In Qβ

TABLE 7.3

Genes and proteins of isometric phages

Gene	*Protein*	*Function and Comments*
A	60 kDa	Site-specific nickase and ligase (RFI→RFII→ss DNA)
*A** (internal initiation within *A*)	37 kDa	Possible role in turn-off of host DNA synthesis
B (overlapped by *A*)	30 kDa	Morphogenesis
C	5.8 kDa	Maturation (RF→ss DNA)
D	14 kDa	Scaffolding protein (Morphogenesis)
E (within *D*)	10 kDa	Lysis (hydrophobic N terminus)
J	4 kDa	Internal protein
F	50 kDa	Major capsid component (60 per virion)
G	20 kDa	Major spike component (60 per virion)
H	37 kDa	Spike component, pilot protein (12 per virion)
K (overlapped by *A* and *C*)	8 kDa	Membrane-related function (?) (hydrophobic at N terminus)

FIGURE 7.16

Portion of the φX174 genome encoding overlapping proteins A, B, K, and C. Nucleotide number 1 lies within the unique Pst I recognition site. Since the viral DNA is single positive-sense, its translation products can be read directly, even though it must first become double stranded and then transcribed, before the mRNA is translated. (From F. Sanger et al. (1978) *J. Mol. Biol.* 125:225)

```
A  Gly Leu Gly Ala Lys Glu Trp Asn Asn Ser Leu Lys Thr Lys Leu Ser Leu Leu Pro Lys
B                       Met Glu Gln Leu Thr Lys Asn Gln Ala Val Ala Thr Ser Gln G
GGTCTAGGAGCTAAAGAATGGAACAACTCACTAAAAACCAAGCTGTCGCTACTTCCCAAG
   5067      5077      5087      5097      5107      5117

A  Lys Leu Phe Arg Ile Arg Met Ser Arg Asn Phe Gly Met Lys Met Leu Thr Met Thr Asn
B   Ala Val Gln Asn Gln Asn Glu Pro Gln Leu Arg Asp Glu Asn Ala His Asn Asp Lys Ser
AAGCTGTTCAGAATCAGAATGAGCCGCAACTTCGGGATGAAAATGCTCACAATGACAAAT
   5127      5137      5147      5157      5167      5177

A  Leu Ser Thr Glu Cys Leu Ile Gln Leu Thr Lys Leu Gly Tyr Asp Ala Thr Pro Phe Asn
B   Val His Gly Val Leu Asn Pro Thr Tyr Gln Ala Gly Leu Arg Arg Asp Ala Val Gln Pro
CTGTCCACGGAGTGCTTAATCCAACTTACCAAGCTGGGTTACGACGCGACGCCGTTCAAC
   5187      5197      5207      5217      5227      5237

A  Gln Ile Leu Lys Gln Asn Ala Lys Arg Glu Met Arg Leu Arg Leu Gly Lys Val Thr Val
B   Asp Ile Glu Ala Glu Arg Lys Lys Arg Asp Glu Ile Glu Ala Gly Lys Ser Tyr Cys Ser
CAGATATTGAAGCAGAACGCAAAAAGAGAGATGAGATTGAGGCTGGGAAAAGTTACTGTA
   5247      5257      5267      5277      5287      5297

A  Ala Asp Val Leu Ala Ala Gln Pro Val Thr Thr Asn Leu Leu Lys Phe Met Arg Ala Ser
B   Arg Arg Phe Gly Gly Ala Thr Cys Asp Asp Lys Ser Ala Gln Ile Tyr Ala Arg Phe Asp
GCCGACGTTTTGGCGGCGCAACCTGTGACGACAAATCTGCTCAAATTTATGCGCGCTTCG
   5307      5317      5327      5337      5347      5357

A  Ile Lys Met Ile Gly Val Ser Asn Leu Gln Ser Phe Ile Ala Ser Met Thr Gln Lys Leu
B   Lys Asn Asp Trp Arg Ile Gln Pro Ala Glu Phe Tyr Arg Phe His Asp Ala Glu Val Asn
ATAAAAATGATTGGCGTATCCAACCTGCAGAGTTTTATCGCTTCCATGACGCAGAAGTTA
   5367      5377                 11        21        31

A  Thr Leu Ser Asp Ile Ser Asp Glu Ser Lys Asn Tyr Leu Asp Lys Ala Gly Ile Thr Thr
B                       Met Ser Arg Lys Ile Ile Leu Ile Lys Gln Glu Leu Leu Leu
K    Thr Phe Gly Tyr Phe ***
ACACTTTCGGATATTTCTGATGAGTCGAAAAATTATCTTGATAAAGCAGGAATTACTACT

A  Ala Cys Leu Arg Ile Lys Ser Lys Trp Thr Ala Gly Gly Lys ***
K   Leu Val Tyr Glu Leu Asn Arg Ser Gly Leu Leu Ala Glu Asn Gluee Lys Ile Arg Pro Ile
C                                          Met Arg Lys Phe Asp Leu Ser
GCTTGTTTACGAATTAAATCGAAGTGGACTGCTGGCGGAAAATGAGAAAATTCGACCTAT
   101       111       121       131       141       151

K  Leu Ala Gln Leu Glu Lys Leu Leu Leu Cys Asp Leu Ser Pro Ser Thr Asn Asp Ser Val
C   Leu Arg Ser Ser Arg Ser Ser Tyr Phe Ala Thr Phe Arg His Gln Leu Thr Ile Leu Ser
CCTTGCGCAGCTCGAGAAGCTCTTACTTTGCGACCTTTCGCCATCAAATAACGATTCTGT
   161       171       181       191       201       211

K  LYS Asn
C   Lys Thr Asp Ala Leu Asp Glu Glu Lys Trp Leu Asn Met Leu Gly Thr Phe Val Lys Asp
CAAAAACTGACGCGTTGGATGAGGAGAAGTGGCTTAATATGCTTGGCACGTTCGTCAAGG
   221       231       241       251       261       271

C  Trp Phe Arg Tyr Glu Ser His Phe Val His Gly Arg Asp Ser Leu Val Asp Ile Leu Lys
ACTGGTTTAGATATGAGTCACATTTTGTTCATGGTAGAGATTCTCTTGTTGACATTTTAA
   281       291  mRNA start  301      311       321       331

                                                                       Met
D  Glu Arg Gly Leu Leu Ser Glu Ser Asp Ala Val Gln Pro Leu Ile Gly Lys Lys Ser
C
AAGAGCGTGGATTACTATCTGAGTCCGATGCTGTTCAACCACTAATAGGTAAGAAATCAT
   341       351       361       371       381       391

D  Ser Gln Val Thr Glu Gln Ser Val Arg Phe Gln Thr Ala Leu Ala Ser Ile Lys Leu Ile
GAGTCAAGTTACTGAACAATCCGTACGTTTCCAGACCGCTTTGGCCTCTATTAAGCTCAT
   401       411       421       431       441       451

D  Gln Ala Ser Ala Val Leu Asp Leu Thr Glu Asp Asp Phe Asp Phe Leu Thr Ser Asn Lys
TCAGGCTTCTGCCGTTTTGGATTTAACCGAAGATGATTTCGATTTTCTGACGAGTAACAA
   461       471       481       491       501       511

D  Val Trp Ile Ala Thr Asp Arg Ser Arg Ala Arg Arg Cys Val Glu Ala Cys Val Tyr Gly
E                                                               Met Val
AGTTTGGATTGCTACTGACCGCTCTCGTGCTCGTCGCTGCGTTGAGGCTTGCGTTTATGG
   521       531       541       551       561       571
```

RNA, one gene is translated into two proteins, one an extension of the other, resulting from read-through of a weak termination signal (sec. 2.7.2). A similar situation exists in the case of φX gene *A*, which codes for the A protein and the A* protein of 59 and 37 kDa, respectively. The latter results from a **second internal initiation** site on the A gene (Table 7.3).

The 10 to 12 proteins of ϑX174 that have definitely been identified are listed in Table 7.3. Their functions are discussed when we describe the replication cycle of these phages. The ability of viruses (and other organisms) to utilize the same nucleotide sequence more than once is possible only because of the **degeneracy** of the genetic code, which allows two to six different triplets for all but two rather rare amino acids. It represents a remarkable achievements toward packaging a maximum amount of genetic information into minimal chemical matter and thus space.

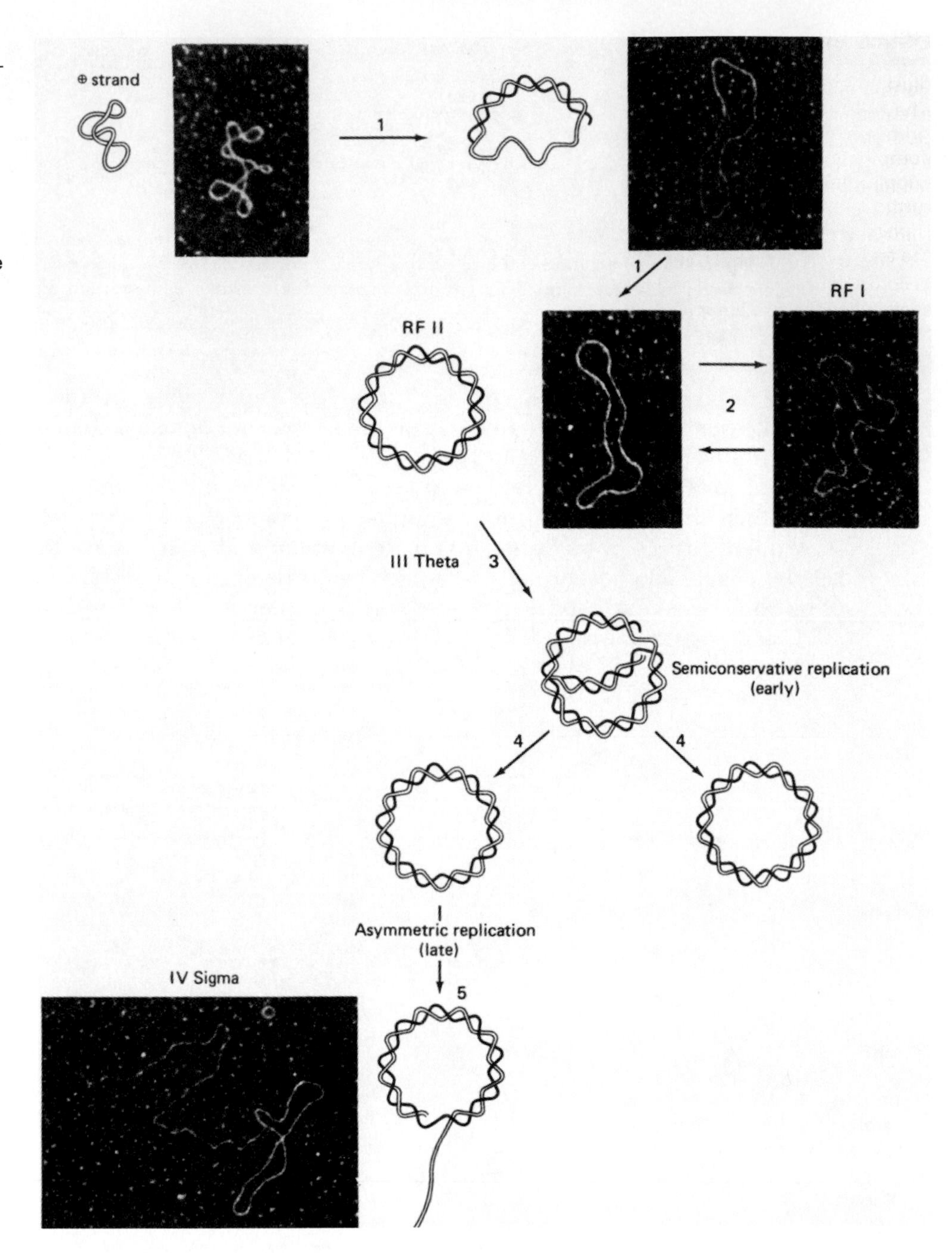

FIGURE 7.17

Schematic presentation of the replication of single-stranded circular DNA, such as that of the Micro- and Inoviridae. Step 1 is making the molecule double-stranded. It may then become supercoiled (step 2). Early in infection both strands then become copied in opposite directions to give theta-formed intermediates (step 3) and double-stranded daughter circles (step 4). These become copied with only one strand of the circle serving as the template, resulting in circles with a single-stranded tail (sigma σ) on step 5. [Courtesy of D. Dressler. (1978) *Proc. Nat. Acad. Sci. USA* 75:605]

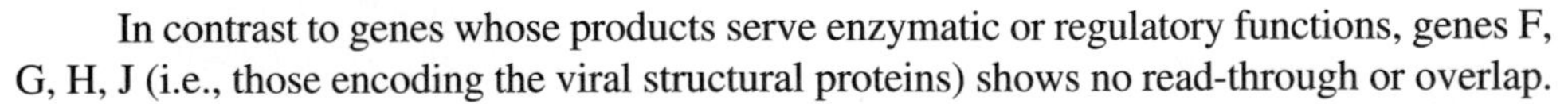

In contrast to genes whose products serve enzymatic or regulatory functions, genes F, G, H, J (i.e., those encoding the viral structural proteins) shows no read-through or overlap.

7.4.2 Replication

Formation of Double-stranded Circles (ssDNA→RF) Figure 7.17 presents a broad outline of the replication of the phage genome. The first step in the replication of these small single-stranded DNA molecules is the formation of double-stranded circles (RF I), and this reaction requires not only a DNA *polymerase* and a DNA *template* strand, but also a *priming 3′—OH group* (Figs. 7.16, 7.17). In contrast with DNA polymerases, RNA polymerases can initiate synthesis *de novo,* and it is now well established that most DNA synthesis starts at the 3′ end of oligoribonucleotides. The function of obligoribonucleotide primers is illustrated in Fig. 7.18, and, surprisingly, the requirements for the production of such **primers** differ for the three small phage DNAs that have been studied in detail.

Primer: oligoribonucleotide

(5′) A U C U G A^{OH} ← pppC, then ppG, etc.

3′– C A T A G A C T G C. . . etc. (leads to release of pyrophosphate (pp))

(Template) DNA

FIGURE 7.18

Illustration of the action of a primer. The AUCUGA ribonucleotide primer, hydrogen bonded to a complementary DNA sequence, supplies a 3′—OH group to which further (deoxyribo-) nucleoside 5′-triphosphates complementary to the template can be added.

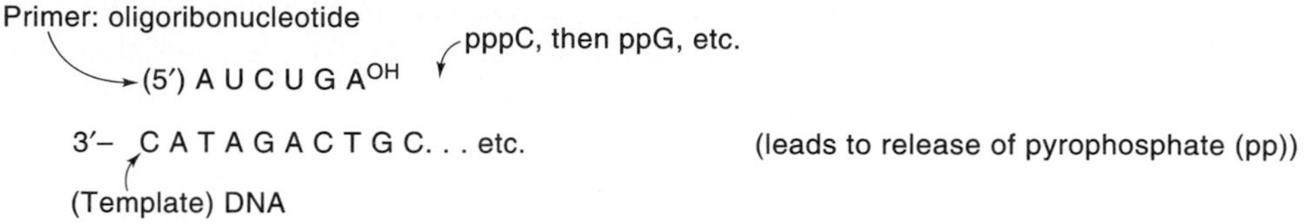

For the isometric phages ∅X and G4 as well as the flamentous phages (e.g. M13) discussed in Section 7.5, the infecting circular molecule is first covered by *the single-stranded DNA binding protein* present in the host cell. In the case of φX, but probably not G4, this complex must then interact with several additional *E. coli* gene products. Subsequently in the case of both these phages and in contrast with M13, *E. coli* primase, (or the *dna* G gene product), rather than the classical DNA-dependent RNA polymerase, enters into a complex termed a **primosome.** The latter remains attached to the replication origin of the DNA throughout several cycles of replication. It also makes the necessary primer oligoribonucleotide, complementary to that specific site on the phage DNA (Fig. 7.19). In the case of G4, this primer is usually 29 nucleotides long, but shorter primers containing both ribo- and deoxyribonucleotides are also active. Priming and complementary strand synthesis begins at a unique site between genes F and G on G4 DNA (see Fig. 7.17). φX174 DNA replication can apparently start at several sites, although it seems probable that the same origin as in G4 is preferred.

The complete process of RF formation requires the following steps: First the bacterial **DNA polymerase III** holoenzyme works its way around the circle; next **DNA polymerase I** (the original Kornberg enzyme, which has exonuclease activity) removes the priming ribonucleotides and fills in the resultant gap with the corresponding deoxynucleotides; then the **DNA ligase** forms the ring-closing phosphodiester bond; and finally an enzyme termed topoisomerase II or **DNA gyrase** causes the supercoiling of the circular molecule (see Fig. 7.17). Supercoiling is not necessary as long as the primosome remains attached to the DNA, however. All these steps that are required for RF formation also rely on at least 13 host enzyme functions.

Transcription and Translation (RF→mRNA→protein) Transcription and translation must occur next, because at least one phage-coded protein is required for replication of this RF molecule (see the next section). The viral DNA can be considered

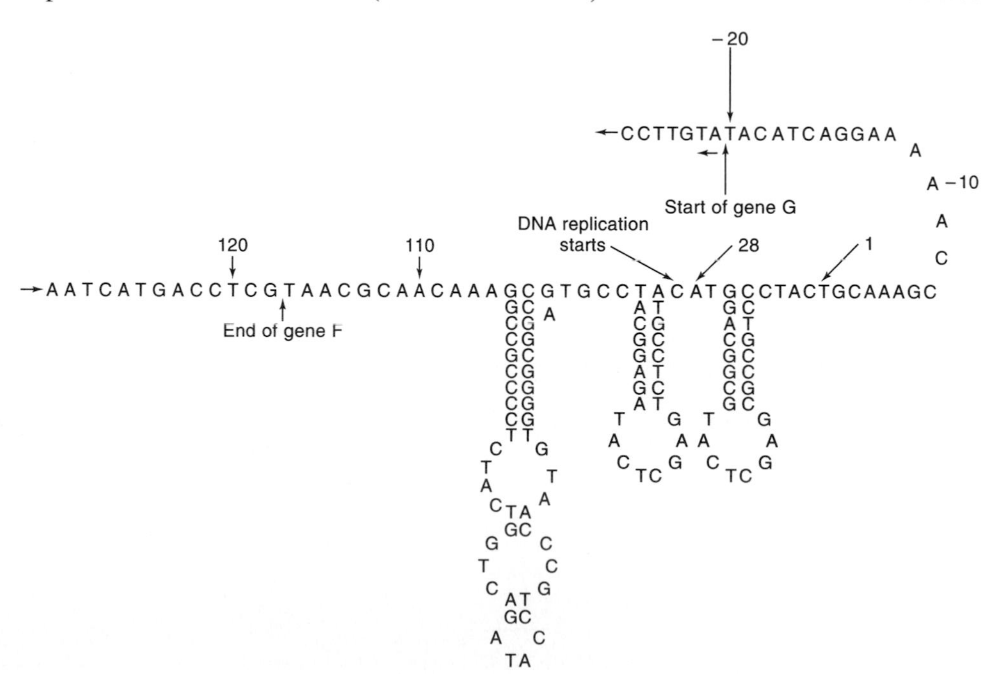

FIGURE 7.19

Primary and secondary structure of G4 phage positive-strand DNA template between genes F and G. A complementary RNA primer, starting with pppA-GUAG, hylindizes with nucleotides in the right-most loop starting at nucleotide number 1 and supplies the primer that permits DNA replication to begin at nucleotide 29. That is stable loop also probably acts as a recognition site for the priming protein.

positive sense, because its complement in the RF has been shown to be the template from which the mRNAs are transcribed. Transcription starts with the binding of the bacterial DNA-dependent RNA polymerase at specific promoter sites. These sites, indentified by the nucleotide sequence, were beautifully confirmed by electron microscopy (Fig. 7.20). The promoter sites are not equally active, and the resulting difference in transcription rate appears to control the ultimate amounts of the various gene products. Since the total number of gene products exceeds the number of transcriptionally active genome segments, several of the mRNAs must be polycistronic. They appear to be arranged in such a manner that each promoter activates the transcription of a functionally related group of genes (see Fig. 7.15 and Table 7.3).

Replication of RF and Formation of Single-stranded Viral Progeny DNA

The one feature of φX174 and other small DNA viruses that has made them so exciting to molecular biologists is that the intracellular replication of their RF represents a most useful system for studying DNA replication in general. We discuss DNA replication here, and not in later sections dealing with larger, double-stranded DNA viruses, because the properties of φX RF are not different from those of most types of circular double-stranded DNA (viral or cellular) known to exist in cells.

These molecules generally occur in two forms, differing in sedimentation rate and other respects: a **superhelical form** (RFI), characterized by both strands being covalently closed and the **relaxed form** (RFII), resulting from at least one break in one of the strands, usually due to nuclease action (Figs. 7.15, 7.21). A break in the other strand, opposite or near the first, results in linear molecules (RFIII), but this is not a normal event in the replication of these phages. We will repeatedly encounter these various forms of viral DNA when discussing the DNA of other viruses.

If two such breaks occur a few nucleotides apart in a circular double-stranded DNA, as they frequently do from restriction endonuclease action (see the next section), short single-stranded ends are formed that are complementary to each other and therefore cohesive or "sticky." When such sticky ends through base pairing reestablish the circular state, DNA ligases can reform the missing phosphodiester bond and thereby covalently close the circle. Thus, the transition from superhelical to relaxed circular to linear forms is reversible by means of cleavage and resynthesis of one or two phosphodiester bonds plus the twisting effect of gyrase. Simply joining the two ends of a linear double-stranded DNA molecule does not create a superhelix—the ends must be rotated relative to each other, resulting in a twisting force in the molecule once both strands are made circular. DNA gyrase is but one of a class of enzymes known as **topoisomerases.** Its remarkable mechanism of action actually involves cutting both strands of the double-stranded relaxed circle, holding the two ends apart without rotation, passing a distant region of the circle through the cut, and resealing the ends.

FIGURE 7.20

Visualization of the specific binding of *E. coli* RNA polymerase to φX174 DNA. The phage DNA RF was cleaved by the restriction enzyme *PstI* after binding the enzyme. The sites are at 6, 74, and 91 percent of the length of the cleaved genome. The bar represents 100 nm. (Courtesy of R. C. Williams)

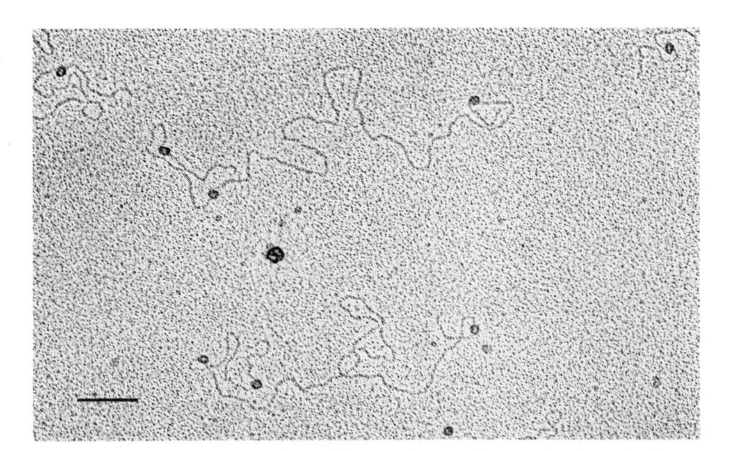

FIGURE 7.21

The φX174 replication cycle. The arrow point between one and two o'clock denotes the origin of replication and later site of attachment of the DNA nicking and closing enzyme (the product of gene *A*) that remains associated with the replicating DNA. Rolling-circle replication leads to the synthesis of single-stranded DNA In contrast bacteriophage λ (see Sec. 8.5), synthesis of the θ form illustrated in Fig. 7.17 is rare in this phage and is not shown. (Courtesy of D. Dressler)

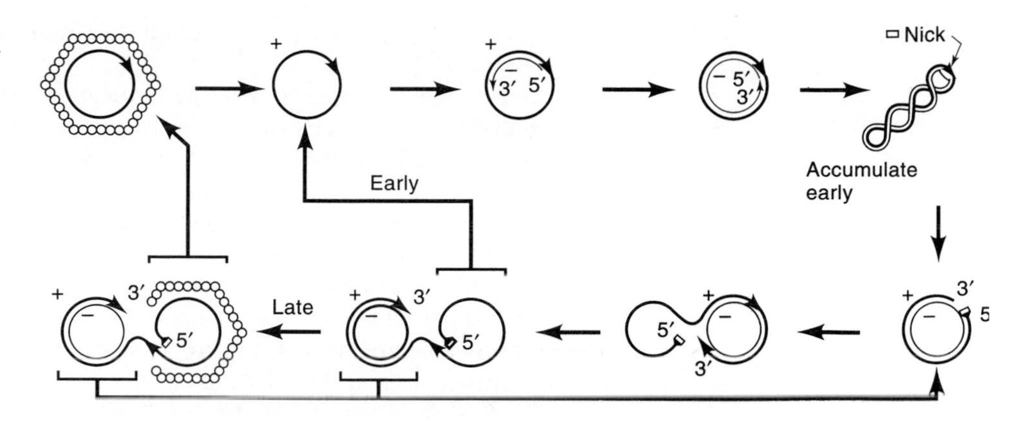

The replication of RFI of these phages faces the same problem as do all DNA replications. The first step is a break in one chain, caused by a site-specific "nicking" enzyme, the gene A product. This may be the only phage-coded protein required for RF→RF replication. This protein binds to the phage DNA at an A/T-rich region near the beginning of the *A* gene in φX. It then remains covalently attached at the 5′ end of the break it causes, ready to fulfill further functions (see Fig. 7.21).

The relaxed RFII is now able to serve as template for DNA polymerase III, starting at a destabilized part of the helix which binds two *E. coli* proteins, the so-called helix-unwinding and DNA-binding proteins (see Fig. 7.22). First, a primer oligonucleotide must again be made by the *E. coli* primase or RNA polymerase, as discussed above. One strand of the parental DNA, termed the **leading strand,** can then be copied continuously in the direction of the progressing **replication fork.** However, the other strand cannot be replicated in the same manner, since it is of opposite polarity. It is instead copied in short segments, the so-called **Okasaki fragments** (see Fig. 7.22), also in the 5′-to-3′ direction, and thus, away from the fork. Each of these short segments must also be initiated at the 3′—OH group of an RNA primer. These fragments then become connected by the action of **DNA polymerase I,** (which degrades the primer and fills the gap) and DNA ligase. It now appears probable that both strands of DNA may actually be transcribed in such a discontinuous manner in sections of a few thousand nucleotides. This form of **semiconservative DNA replication** involving a ϑ form is characteristic for most (if not all) cellular circular DNAs, but appears to occur only in the early stages of replication of these phages.

FIGURE 7.22

Schematic presentation of a replication fork showing the roles of the helix-unwinding and DNA-binding proteins, the DNA polymerase, the RNA primer, the Okasaki fragments, and the DNA ligase in DNA replication.

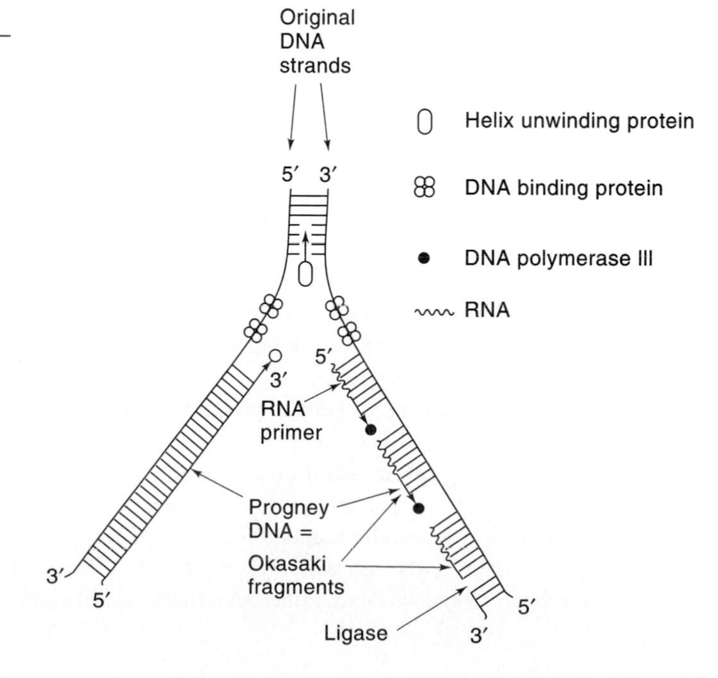

During the later stages of phage replication, an asymmetric process called the **rolling-circle** (or σ) mechanism predominates (Figs. 7.17 and 7.21) Here a 3′ end released by a break in the parental DNA can serve as primer, and uninterrupted synthesis of a DNA strand is possible. The rolling-circle mechanism of producing multiple copies of a single template strand achieves the same objective as the replicative intermediate (RI) in the replication of RNA described in Sec. 2.7. However, with RNA the multiple copies have the same molecular weight as the template, whereas the rolling circle mechanism yields multiple DNA copies *in tandem* (i.e., attached to one another end to end, so-called **concatemers**). These then need to be cut to proper molecular lengths before they can be circularized by appropriate enzymes. In the case of the small DNA phages, the circle rolls only one turn. The nicking enzyme, which remains attached to the 5′ end of the parental strand (the gene *A* product in the case of φX) cuts off the newly made strand and probably also circularizes it. There is some similarity in the cutting and sealing action of this type of enzyme and the topoisomerases mentioned above.

As we will see later, Rolling-circle replication can also serve for production of double-stranded DNA but it is of particular advantage if single-stranded DNA is the required end product. At early stages of infection the resultant single-stranded circular DNA molecules re-enter the DNA replication-RNA transcription process

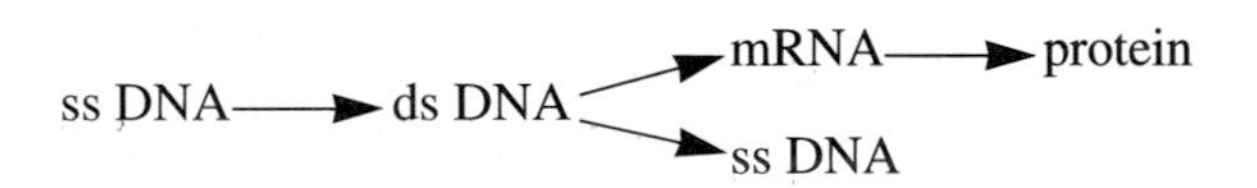

In this manner both progeny DNA, mRNA, and proteins are made. When enough of the phage coat proteins have accumulated (gene *F* possibly being the limiting factor), the single-stranded DNA circles start to bind these proteins and are thus prevented from serving as template for further replication. This leads to the process of virion maturation (see below). Thus, primase, DNA polymerases III and I, in some instances DNA-dependent RNA polymerase, and ligase action are all required for replication of the smallest single-stranded phage DNA in the prokaryotic cell. To this structure must be added the DNA helix-unwinding and binding proteins and several other proteins, some host- and some phage-coded, that control the shape of the DNA to make it available or unavailable to the various enzymes, as the need arises. These functions, in the case of the small phages, assist in its conversion to a single-stranded molecule.

Morphogenesis and Lysis Assembly of the 114 S Microviridae particles is triggered by the accumulation of single-stranded DNA and requires all the genes that favor this process. Particles of 6 S, 9 S, and 12 S that are rich in the products of gene *G* and/or *F* have been observed in infected cells and are precursors in this maturation process. A large particle of 108 S that also contains *H* and *D* gene products (the latter is not present in mature phage) is also seen under certain circumstances. The gene *H* product, with its very hydrophobic N-terminal region, has particular affinity for the cell membrane, and it is there that assembly is accomplished. The first DNA-containing phage precursor particle consists of these same proteins, as well as additional ones—some from the host. The role of the various nonstructural proteins remains uncertain. Until phage assembly can be achieved by means of purified phage components *in vitro*, the exact mechanism of the formation of these particles will not be known.

The mechanism of lysis of phage-filled cells is also unclear, but the hydrophobic N-terminal domain of the protein coded by gene *E* may play a role in that function. Lysis can be prevented by **UV irradiation** and **mitomycin C,** possibly because of the failure of sufficient amounts of that protein to be made under these conditions. Active lysis is not an obligatory step in single-stranded phage propagation, and considerably more phage actually accumulates in the cell when lysis is retarded. The entire replication cycle of the φX 174 and related phages is summarized diagrammatically in Figure 7.21.

SECTIONS 7.4 FURTHER READING

Original Papers

SINSHEIMER, R. L. (1959) Purification and properties of bacteriophage φX174. *J. Mol. Biol.* 1:37.

FIERS, W., and R. L. SINSHEIMER. (1962) The structure of the DNA bacteriophage φX174. III. Ultracentrifugal evidence for a ring structure. *J. Mol. Biol.* 5:424.

FREIFELDER, D., et al. (1964) Electron microscopy of single-stranded DNA: Circularity of DNA of bacteriophage φX174. *Science* 146:254.

BENBOW, R. M., et al. (1971) Genetic map of bacteriophage φX174. *J. Virol.* 7:549.

BARRELL, B. G., et al. (1976) Overlapping genes in bacteriophage φX174. *Nature* 264:34.

SANGER, F., et al. (1978) The nucleotide sequence of bacteriophage φX174. *J. Mol. Biol.* 125:225.

EISENBERG, S., and A. KORNBERG. (1979) Purification and characterization of φX174 gene *A* protein. *J. Biol. Chem.* 254:5328.

KOTH, K., and D. DRESSLER. (1980) The rolling circles capsid complex of an intermediate in φXDNA replication and viral assembly. *J. Biol. Chem.* 255:4328.

LOW, R. L., K. I. ARAI, and A. KORNBERG. (1981) Conservation of the primosome in successive stages of φX174 DNA replication. *Proc. Natl. Acad. Sci. USA* 78:3.

AOYAMA, A., et al. (1983) *In vitro* synthesis of phage φX174 by purified components. *Proc. Natl. Acad. Sci. USA* 80:4195.

LUBITZ, W., et al. (1984) Requirement for a functional host cell autolytic system for lysis of φX174. *J. Bact.* 159:385.

HAMATAKE, R. K., et al. (1985) The J gene of bacteriophage φX174: *In vitro* analysis of J protein function. *J. Virol.* 54:345.

AOYAMA, A., and M. HAYASHI. (1986) Synthesis of φX174 *In vitro*. *Cell* 47:99.

Books and Reviews

DENHARDT, D. T. (1977) The isometric single-stranded DNA phages. In *Comprehensive virology.* Vol. 7, ed. H. Fraenkel-Conrat and R. R. Wagner. 1–104. Plenum Press, New York.

GODSON, G. N., et al. (1978) Comparative DNA sequence analysis of the G4 and φX174 genomes. In *The single-stranded DNA phages.* ed. D. T. Denhardt, D. Dressler, and D. S. Ray. 51–86 Cold Spring Harbor Laboratory, Cold Spring Harbor, New York.

MCMACKEN, R., et al. (1978) Priming of DNA synthesis on viral single-stranded DNA *in vitro.* In *The single-stranded DNA phages.* ed. D. T. Denhardt, D. Dressler, and D. S. Ray. 273–85 Cold Spring Harbor Laboratory, Cold Spring Harbor, New York.

COZZARELLI, N. R. (1980) DNA gyrase and the supercoiling of DNA. *Science* 207:953.

KORNBERG, A. (1983) Mechanism of replication of the *Escherichia coli* chromosome. *Eur. J. Biochem.* 137:377.

HAYASHI, M. et al. (1988) Biology of the bacteriophage φX174 In *The bacteriophages 1.* ed. R. Calendar. Plenum Press, New York.

7.5 INOVIRIDAE (FILAMENTOUS SINGLE-STRANDED DNA BACTERIOPHAGES)

When compared with the **isometric** phages, the **filamentous** phages fd, f′, and M13 are paradoxically both similar and different in appearance and behavior. Using the prototype **fd** and its close relative **M13** as examples. The following comparisons can be made:

1. Phage fd is a long, thin fiber (see Fig. 7.23), 900 nm long and 6 nm in diameter, that nevertheless accommodates a single molecule of circular single-stranded DNA, containing 6408 nucleotides.
2. The virion of fd consists of about 2000 copies of a single small coat protein (5.2 kDa) plus a few molecules of a single attachment or pilot protein (42 kDa) at one end of the fiber.
3. These DNA phages infect only male bacteria by first attaching themselves to the end of a pilus. In contrast, φX is not male-specific. The RNA phages that are male-specific attach themselves to the sides of the bacterial pilus (see Fig. 2.20).
4. The fd phages do not lyse or kill their host cell but are continuously released.

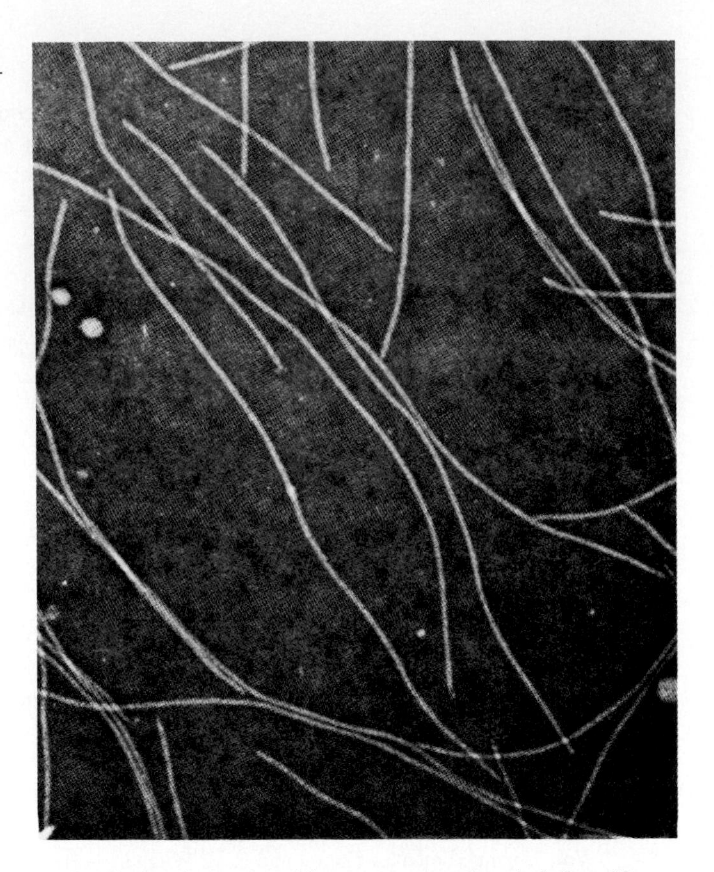

FIGURE 7.23

Electron micrograph of phage fd, 900 nm long and 6 nm in diameter (see also Fig. 2.20). (Courtesy of R. C. Williams)

5. These phages are able to accommodate their 9 or 10 genes within 6408 nucleotides with only one overlap and without the need to utilize more than one reading frame, probably because of their simpler capsid composition and their nonvirulence (Fig. 7.24).

7.5.1 Structure and Virus Entry

Quite possibly the most paradoxical feature of the filamentous phage virions, (6 nm in diameter and 900 to 1900 nm in length) is that they contain a **circular DNA** (12 percent). This DNA represents 12 percent of the total particle weight and is not self-complementary. In the prototype, fd, the DNA runs up and down the particle, with loops at the ends, with little if any base-pairing interaction. Like this virus, φX174, this virus contains a considerable excess of one base (thymine, 34 percent). In some of the extra-long phages of this family, the presence of some complementary DNA sequences may lead to partial double-strandedness.

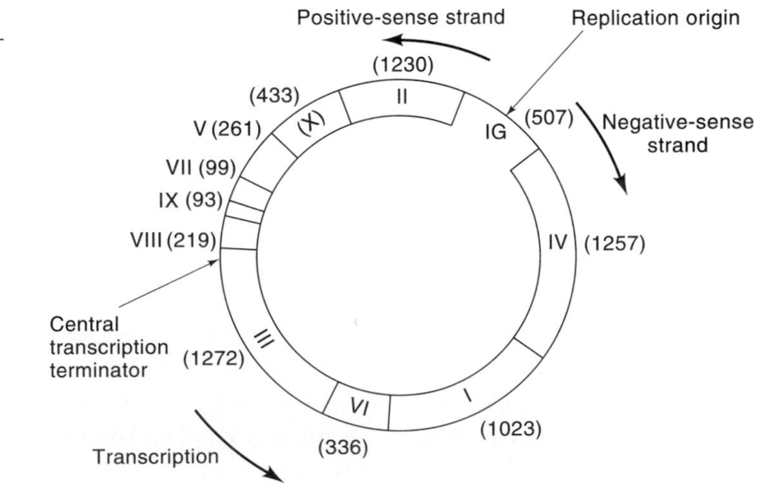

FIGURE 7.24

Generalized genome map of a filamentous DNA phage. The replication origin (bidirectional) is marked. Transcription is unidirectional, initiating at several sites and terminating at central transcription terminator. IG denotes the noncoding "intergenic" region. The number of nucleotides in each genome segment is also shown.

FIGURE 7.25

Amino acid sequence of the major capsid protein (gene VIII product) fd. The presence of acidic groups (underlined) among the 10 C-terminal residues, basic groups (overlined) among the 10 N-terminal residues, and the many hydrophobic residues between residues 21 and 39 are noted.

Acidic: 1 H–Ala–Glu–Gly–Asp–Asp–Pro–Ala–Lys–Ala–Ala–Phe–Asp–Ser–Leu–Gln–Ala–Ser–Ala–Thr–Glu 20

Hydrophobic: 39 Ily–Gly–Ile–Thr–Ala–Gly–Val–Ile–Val–Val–Val–Met–Ala–Trp–Ala–Tyr–Gly–Ile–Tyr 21

Basic: 40 Lys–Leu–Phe–Lys–Lys–Phe–Thr–Ser–Lys–Ala–Ser–OH 50

The **coat protein** of the filamentous phages is a very small protein (50 amino acids for fd) which is arranged as a helix around the DNA thread. The pitch of the helix is 1.5 nm, with nine coat protein molecules in two turns. The walls of this helical tube are 3.5 nm thick.

The coat protein has many basic (and thus positively charged) amino acids in its C-terminal third, which is located toward the inside of the particle and the negatively charged DNA. The protein's hydrophobic middle segment presumably causes interprotein aggregation, and the acidic N-terminal third is oriented toward the outside of the thread (Fig. 7.25).

Another protein of interest is the 42 kDa **adsorption protein**, probably only one or a few molecules of which are located at one end of the virion. Entry of nucleic acid into a bacterium, along or possibly through the pilus appears to facilitate the infection process, therefore phages using this route, be they RNA- or DNA-containing, need only one type of protein molecule as an infection organ. In contrast, infection through the cell wall of most bacteria requires a structure at least as complex as the spike of φX, which contains six protein molecules (see Sec. 7.4), and most of the larger phages have developed sophisticated tail structures for this purpose. Probably only the fd DNA and adsorption protein (also called **pilot protein**), enter the cell, possibly by retraction of the pilus. Both the initiation of the replication cycle and phage maturation appear to occur in the cell membrane and are probably controlled by the hydrophobic segments of the coat protein and its precursor.

It should be noted that the existence and the mode of action of a **restriction endonuclease** was first discovered and studied with phage fd. These DNases occur in many bacteria and cleave invading foreign DNAs at specific base sequences, that are absent or protected by methyl groups in the DNA of the respective bacterial strain. The discovery of these enzymes with their remarkable and varied specificities, together with the resolving power of *polyacrylamide gels* for the separation of the DNA fragments generated by these nucleases, has made the enormous advances in DNA sequencing technology possible. These three techniques have, in turn, opened up the field of **recombinant DNA research** and **"genetic engineering."** M13 is now a favored vector for cloning genes because at least 40,000 nucleotides (six times the length of the phage DNA) of "foreign" DNA can be inserted into the phage genome without affecting the infectivity. The virion merely becomes longer to accommodate the excess DNA.

7.5.2 Intracellular Events and Phage Replication

The replication cycle of the filamentous phages does not differ in principle from that of the isometric phages, or any DNA. The first event is the *formation of a double-stranded circle* (ss DNA→RF). This occurs in the membrane and is in some way directed by the pilot protein. Enzymatic requirements are the host cell DNA binding protein, DNA-dependent RNA polymerase, DNA polymerase III holoenzyme, DNA polymerase I, and DNA ligase. The origin of replication has been identified (Fig. 7.26). **Transcription** of this RF (RF→mRNA) is the next event, starting at several promoter sites, some stronger (i.e., more active) than others. All transcriptions proceed in the same direction and end at the same site, the *"central terminator."* In this manner different amounts of the resultant mRNAs are made, and the location of the promoter sites plays an important role in regulating the amounts of the different messages that are being made. This step is followed by translation of the mRNAs into 9 or 10 proteins (termed I to IX) (Fig. 7.24), whose molecular weight and function (as

replication. One of these, of 160 kDa, binds specifically to the virus replication origin, possibly representing a type I topoisomerase. Either 3′ end can serve as template for uninterrupted 5′-to-3′ replication, whereas the other parental strand that carries the 55 kDa protein becomes displaced as a single strand. Nevertheless, that protein is essential for the replication process to begin. We have stressed the need for DNA polymerases to start at a primer. The 55 kDa protein bound to the 5′ ends of the viral DNA (Fig. 8.3), or possibly its 80 kDa precursor, probably plays a key role as a primer.

In contrast to the parvoviruses, initiation at folded-back 3′ termini appears not to be the mechanism used by the double-stranded adenoviruses, although the presence of inverted complementary sequences at the ends, about 100 nucleotides long (C,B,A–a,b,c in Fig. 8.3), has been noted. Thus, when this DNA is denatured and allowed to reanneal, single-stranded circular molecules can form that are held together by a **panhandlelike** structure. A similar structure is also observed with adeno-associated viral DNA, although the adenovirus DNA is more than 10-fold longer.

Most investigators now believe that adenovirus DNA replication starts at either end but not at both ends (Fig. 8.3). All agree that single-stranded DNA is an intermediate product, probably the displaced parental strand upon the first round of replication. Frequently, replicating molecules with two single-stranded tails are seen, as well as partly double- and partly single-stranded ones (Fig. 8.4). The preferred model for adenovirus DNA replication and the role of the terminal protein(s) in this process is illustrated in Fig. 8.5.

FIGURE 8.4

Electron micrographs of replicating adenovirus DNA molecules. (a) A double-stranded molecule with a single-stranded tail starting at a replication fork can be seen, as well as (b) a molecule with two forks, and (c) a partially double-stranded strand. All are of adenovirus DNA length. (From T. J. Kelly, Jr. (1980) *Proc. Natl. Acad. Sci. USA* 77:5105)

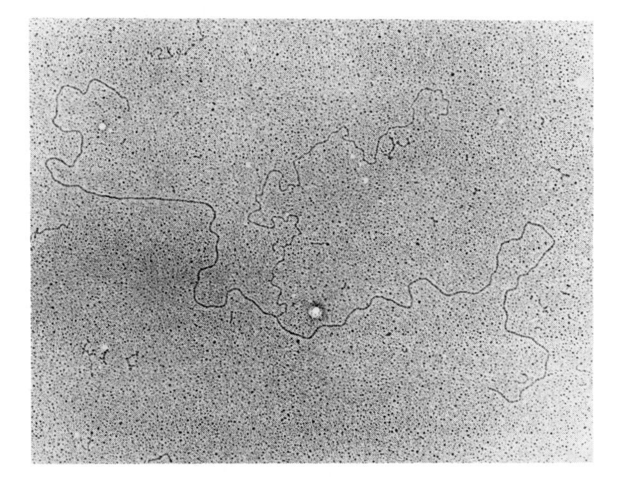

(a)

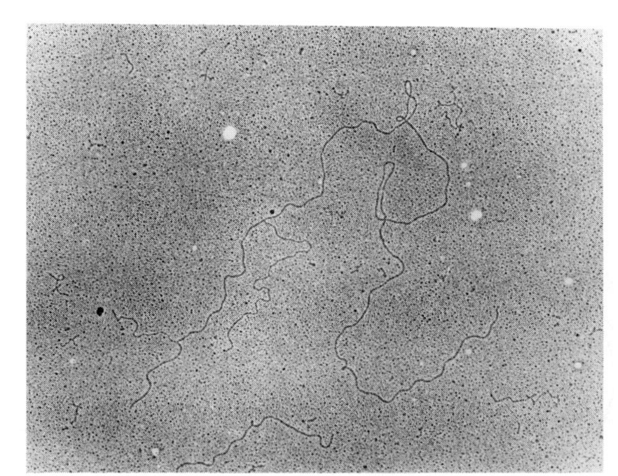

(b)

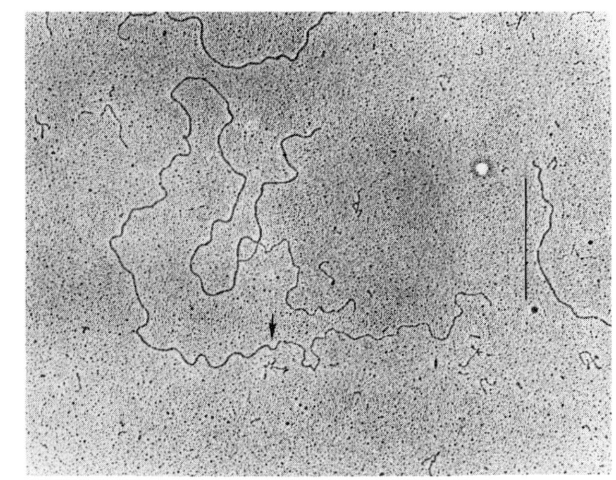

(c)

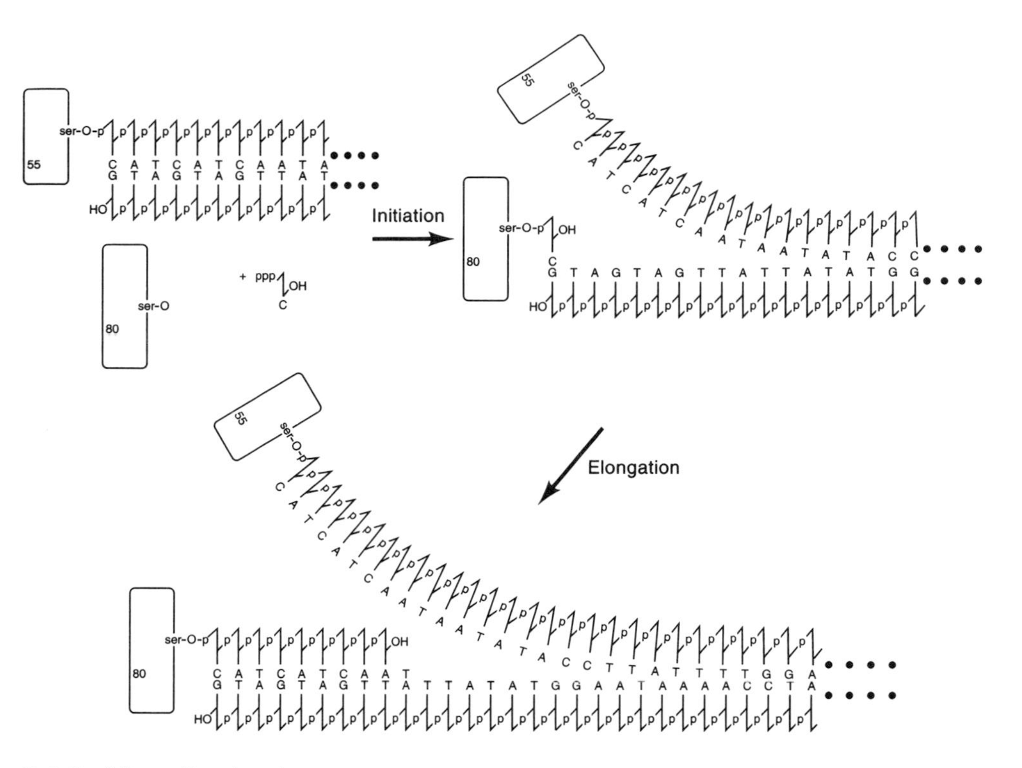

FIGURE 8.5

Model for the initiation of adenovirus DNA replication. The boxes labeled 55 and 80 refer to the 55 kDa 5′-terminal protein and its 80 kDa precursor. (From M. D. Challberg, et al. (1980) *Proc. Natl. Acad. Sci. USA* 77:5105)

8.1.3 Virus Production

The whole adenovirus replication cycle is 24 to 30 hours for different serotypes; the early mRNAs appear during the first 8 to 10 hours, but they continue to be made throughout the infection cycle (Fig. 8.6). The left-most gene on the r-strand (EIA) (for early gene product) plays multiple important roles. Located in the nucleus, it has three gene products, two of which are identical at the N and C termini but differ by an inserted peptide. One of these induces DNA synthesis, another activates and/or enhances transcription from the other viral promoters; a third interacts with cellular oncogenes (e.g., c-*ras*) to produce fully transformed phenotypes.

Finally, adenoviruses, as do polyomaviruses, give rise to tumor (T) -associated antigens (17–58 kDa) that are detected by antibodies from animals carrying these malignancies. These T antigens are products of the early transcribed regions of the adenovirus genome (EIA, EIB) [Fig. 8.6(a)]. EIA resembles functionally, and to a certain extent structurally, the SV40-T, papilloma E2, and herpes IE (see Secs. 7.2 and 8.2) genes. They all share the need for early production of active proteins.

E2A codes for a 22 kDa DNA-binding phosphoprotein. It also gives rise to several very short (4 S) RNAs, but most of the transcription products are large and polycistronic and are processed after transcription by extensive splicing. It was actually in these virus studies that splicing of transcripts to ultimate mRNAs was first observed. E2B from the left strand codes for a 72 kDa protein, a single-stranded DNA-binding protein that is required for the initiation of viral DNA synthesis.

Through the use of many **temperature-sensitive mutants,** and by the technique of heteroduplex mapping, most adenovirus genes have been localized and identified with specific mRNAs, but the function of several of the proteins made upon translation remains uncertain. This is because host cell DNA synthesis and transcription continues during productive adenovirus replication. Also, host cell protein synthesis is not inhibited during the early phase of virus infection. Later, in the course of increasing virus transcription and translation, host cell protein synthesis and, secondarily, DNA synthesis are arrested. Translation takes place in the cytoplasm in a manner indistinguishable from cell host protein synthesis.

The *late* transcripts, made after onset of DNA replication, are almost all from the r-strand, their tripartite leaders starting near map unit 16.5. After capping and polyadeny-

pesviruses (Table 8.2) appear to vary greatly in nucleotide composition. The DNA contains terminal and repeat sequences as shown by heteroduplex mapping and other techniques. The intervening segments code for a number of structural and regulatory proteins as well as enzymes required for replication. Different herpesvirus groups show different numbers and lengths of internal and terminal repeats, some inverted. This variability in the number of sequence reiterations affects the size of the individual genome; their magnitude may reflect an evolutionary order.

Herpes virus DNA contains nicks, so that alkali denaturation renders it smaller and heterogeneous. Generally, the pattern of **restriction endonuclease sensitivity** is stable for a particular HSV strain coming from specific regions of the world; however, changes in the repeated sequences distributed throughout the genome can take place over time within the host. By these parameters, herpesvirus transmission can be monitored. Herpes simplex DNA shows transcriptional peculiarities suggesting that it consists of two, covalently linked, segments. Circularization through complementary regions has also been observed.

The isolated DNA was found to show some infectivity, but with a genome of that size the possibility of the presence of associated protein molecules is almost impossible to rule out. The nucleocapsids lacking the envelope are considerably more infectious, but still much less than the complete virus particles, of which as few as four have been reported to initiate an infection in cultured cells. The nucleocapsids are the predominant intracellular particles. The tegument and envelope are acquired by the nucleocapsids during their exit passage through the nuclear and/or plasma membrane. They also greatly facilitate the entry of the virion through the plasma membrane upon infection.

8.2.3 Entry, Transcription, and Virus Production

Virus entry involves attachment of the herpes virus to a cellular receptor and fusion of the viral envelope with the cell membrane. This process is pH-independent. For HHV-1 and 2, heparin sulfate proteoglycans appears to be a binding site for the virus. For EBV, it is part of the **complement** receptor (CR21). The viral capsid is then transferred from the cytoplasm through nuclear pores into the nucleus, where the DNA is released and the transcription process begins.

Recent evidence suggests that a retrograde transfer of HSV from peripheral sensory nerves to the ganglia occurs via **microtubules;** drugs blocking microtubule action prevent this transport in experimental animals. The onset of herpesvirus infection does not, in contrast to the situation with other DNA viruses, require or stimulate host cell DNA synthesis and/or mitosis. In permissive cells (that is, cells with appropriate receptors and intracellular milieu), and for alpha-herpesvirinae that includes almost all cells, replication proceeds quite rapidly, with viral DNA beginning to appear in the nucleus within 3 hours after infection. It is completed within 12 to 14 hours. As with other DNA viruses, transcription precedes this event, relying on host cell RNA polymerase II interacting with certain viral proteins. It seems to proceed *in toto,* with less clearly separated early and late phases than observed with most other DNA viruses. Two groups of more or less early transcripts, α and β, reach high concentrations consecutively prior to appreciable DNA synthesis. These map in noncontiguous parts of the genome. The third group of genes, γ, are transcribed only after DNA synthesis has become maximal, 6 hours post-infection. Yet even then some early transcripts continue to be made, although at much lower rates. It seems that the rate of processing of the mRNAs in the nucleus, which involves capping and polyadenylation, represents an additional control mechanism for herpesvirus protein synthesis.

Replication of the DNA, presumably by the virus-specific DNA polymerase and by a rolling-circle mechanism, requires ribonucleotide primers and involves the joining of short segments, apparently for both strands. Long concatemeric molecules accumulate and appear to become processed to viral DNA monomers only in the course of virion assembly into a capsid in a manner similar to phage DNA processing (see Sec. 8.7).

The herpesvirus genome carries the information for many enzymes performing functions involved in DNA synthesis. These include a DNA polymerase (140 kDa), a DNA binding protein (124 kDa), a protein that binds to the origins *(ori)* for viral DNA replication (94 kDa), a protein that binds to double-stranded DNA (62 kDa), and three additional proteins that form a complex that has primase and helicase functions. Other coded proteins are responsible for the processing and packaging of the genomic viral DNA into capsids.

Viral mRNAs are processed by conventional means with relatively little splicing. All viral proteins are made in the cytoplasm. HSV-1, as an example, produces at least 75 peptides that are extensively modified through processes that include cleavage, phosphorylation, glycosylation, sulfation, and poly(AdP)-ribosylation. Some of the **glycoproteins** are associated with cell fusion.

The viral envelope is added to the DNA-containing capsids at the nuclear membrane; thus, full virions of HSV can be seen by electron microscopy within the cytoplasm of cells. Some enveloping may also occur at the plasma membrane. Virus particles, or capsids, appear to be transferred by microtubules through the cytoplasm to the external surface of the cell. Eventually, a rupture of the cell due to the inhibition of normal cellular metabolism releases infectious virus. An important part of the replicative pathway of EBV and some other herpesviruses is the formation of circular DNA forms that appear as **plasmids** in the cell. They can accumulate in very large numbers and transcribe viral proteins without integration. Nevertheless, it appears that at least one viral genome integrates into the cell chromosome.

Most herpesviruses under varying conditions tend to become persistent and/or cryptic; part or all of the DNA of others becomes integrated into the host genome without any of the typical phenotypic expressions of transformation. These phenomena usually involve specific cells, such as B lymphocytes or ganglia. The mechanism for this latency is not yet clear. With persistent infection of B cells, often only the nuclear binding protein may be detected. In fact, in studying the association of Epstein-Barr virus with **Burkitt's lymphoma** and **nasopharyngeal carcinoma** (see Sec. 10.2.5), the viral proteins coded by the genome can be important in diagnosis and prognosis. For example, the early antigens (EA) can be divided into the restricted (R) and diffuse (D). Antibodies to EA-R are observed more commonly in Burkitt's lymphoma, whereas EA-D antibodies are found in nasopharyngeal carcinoma. The latter are increased when the tumor spreads. In addition, antibodies to the nuclear binding antigen, EBNA, reflect an appropriate T-cell response to the virus and generally appear 6 months after infection (see Fig. 13.1). Thus, the presence of this antibody indicates prior infection with EBV. Finally, since the early viral antigens are made only during active replication, the presence of antibodies to these proteins can indicate primary infection or reactivation of EBV.

SECTION 8.2 FURTHER READING

Original Papers

COLE, R., and A. G. KUTTNER. (1926) A filterable virus present in the submaxillary glands of guinea pigs. *J. Exp. Med.* 44:855.

ROWE, W. P., et al. (1956) Cytopathogenic agents resembling human salivary gland virus recovered from tissue cultures of human adenoids. *Proc. Soc. Exp. Biol.* (NY) 92:418.

SMITH, M. G. (1956) Propagation of the salivary gland virus of the mouse in tissue cultures. *Proc. Soc. Exp. Biol.* (NY) 86:435.

WELLER, T. H., et al. (1957) Isolation of intranuclear inclusion producing agents from infants' illnesses resembling cytomegalic inclusion disease. *Proc. Soc. Exp. Biol.* (NY) 94:4.

EPSTEIN, M. A., et al. (1964) Virus particles in cultured lymphoblasts from Burkitt's lymphoma. *Lancet* i:702.

HENLE, G., W. HENLE, and V. DIEHL. (1968) Relation of Burkitt's tumor-associated herpes-type virus to infectious mononucleosis. *Proc. Natl. Acad. Sci. USA* 59:94.

WOLF, K., and R. W. DARLINGTON. (1971) Channel catfish virus: A new herpesvirus of actalurid fish. *J. Virol.* 8:525.

teins are secondarily processed. There is also some evidence for a most unusual mode of regulation of gene expression in poxviruses. Early enzymes are not made at late times, but the early mRNAs are still transcribed. Thus, presumably, translation of early mRNAs is specifically inhibited, possibly by a "late" viral function. The entire growth cycle lasts about 24 hours, with pox DNA appearing after 8 hours.

Assembly of the poxviruses occurs in electron-dense regions in the cytoplasm that become evident after 3 to 4 hours and have been called "factories." These are proportional in number to the multiplicity of infection. The entire virion is here built up, including the lipid-containing membrane, which in these viral coats, in contrast with all other virus envelopes, differs for various poxviruses in composition and amount (34 percent of the particle weight for fowlpox). It is always quite different in composition and structure from the plasma membrane. The viral membrane is generally particularly rich in phospholipids. As stated, the membrane is usually covered by an external protein coat, which accounts for the fact that many poxviruses are ether-resistant. The mode of reproduction of the poxviruses is summarized schematically in Fig. 8.12. As with other DNA viruses, progeny production leads to death of the infected cell.

8.4 AFRICAN SWINE FEVER VIRUS

The African swine fever virus (ASF) has a double-stranded DNA genome of about 110_R10^6Da (170 kb). Its large genome induces synthesis of about 100 proteins, and the genes transcribed late in infection map to the central part of the genome. The virion (123–300 nm) consists of a lipid envelope surrounding an icosahedral nucleocapsid. Apparently, viral DNA replication begins in the nucleus and continues in the cytoplasm after budding through the nuclear membrane. The classification of ASF has not been conclusively established. It shares several properties with poxviruses: Its genome contains hairpin loops and terminal-inverted repeats, and its production requires early RNA synthesis. However, the ASF replicative cycle and particularly the morphology resemble those of iridoviruses (see Sec. 8.5).

SECTIONS 8.3 AND 8.4 FURTHER READING

Original Papers

JOKLIK, W. K., and Y. BECKER. (1964) The replication and costing of vaccinia DNA. *J. Mol. Biol.* 10:432.

GESHELIN, P., and K. I. BYRNS. (1974) Characterization and localization of the naturally occurring cross-links in vaccinia virus DNA. *J. Mol. Biol.* 88:785.

GARON, C. F., et al. (1978) Visualization of an inverted terminal repetition in vaccinia virus DNA. *Proc. Natl. Acad. Sci. USA* 75:4863.

VARICH, N. L., et al. (1979) Transcription of both DNA strands of vaccinia virus genome *in vivo*. *Virology* 96:412.

POCO, B. G. T. (1980) Terminal crosslinking of vaccinia DNA strands by an *in vitro* system. *Virology* 100:339.

RICE, A. P., and B. E. ROBERTS. (1983) Vaccinia virus induces cellular mRNA degradation. *J. Virol.* 47:529.

SMITH, G. L., et al. (1985) Infectious vaccinia virus recombinants that express hepatitis B virus surface antigen. *Nature* 302:490.

DELANGE, A. M., et al. (1986) Replication and resolution of cloned poxvirus telomeres *in vivo* generates linear minichromosomes with intact viral hairpin termini. *J. Virol.* 39:249.

KITAMOTO, N., et al. (1987) Monoclonal antibodies to cowpox virus: Polypeptide analysis of several major antigens. *J. Gen. Virol.* 68:239.

WRIGHT, C. F., and B. MOSS. (1987) *In vitro* synthesis of vaccinia virus late mRNA containing a 5′ poly(A) leader sequence. *Proc. Natl. Acad. Sci. USA* 84:8883.

GERSHON, P. D., et al. (1989) Poxvirus genetic recombination during virus transmission. *J. Gen. Virol.* 70:485.

GOEBEL, S. J., et al. (1990) The complete DNA sequences of vaccinia virus. *Virology* 179:249.

UPTON, C., et al. (1990) Myxoma virus and malignant rabbit fibroma virus encode a serapinlike protein important for virus virulence. *Virology* 179:618.

GARCIA-BEATO, R., et al. (1992) Role of the host cell nucleus in the replication of African swine fever virus DNA. *Virology* 188:637.

GIERMAN, T. M., et al. (1992) The eukaryotic translation initiation factor 4E is not modified during the course of vaccinia virus replication. *Virology* 188:934.

BALDICK, C. J., Jr., and B. MOSS. (1993) Characterization and temporal regulation of mRNAs encoded by vaccinia virus intermediate-stage genes. *J. Virol.* 67:3515.

YANEZ, R. J., and E. VINUELA. (1993) African swine fever virus encodes a DNA ligase. *Virology* 193:531.

Books and Reviews

VIÑUDA, E. (1985) African swine fever virus. In *Iridoviridae,* ed. D. B. Willis. 151 Springer-Verlag, Berlin.

MCGEOCH, D. J., et al. (1987) Some highlights of animal virus research in 1986. *J. Gen. Virol.* 68:1501.

MOSS, B. (1990) Poxviridae and their replication. In *Virology,* 2d ed., ed. B. N. Fields and D. M. Knipe. 2079 Raven Press, New York.

8.5 INSECT DNA VIRUSES

Many vertebrate and plant viruses are transmitted by insects, and many of these insects not only carry such viruses but also actually support their replication. Insect-borne Rhabdo-, Bunya-, Toga-, or Reoviridae, have been described in other chapters. Some members of the Reo-, Noda-, and Poxviridae, also have insects as their prime or only host, but these viruses also do not require separate discussion. There exist, however, certain insect viruses that have distinct characteristics warranting a separate family classification. These are the Baculoviridae and Iridoviridae. The latter family now also includes some animal viruses.

8.5.1 Baculoviridae

The Baculoviridae family name was coined to cover two similar genera of viruses that infect only invertebrates, particularly many insects. The two subgroups are the **nuclear polyhedrosis viruses** (NPV) and the **granulosis viruses** (GV). The term *polyhedrosis viruses* refers to the fact that these viruses are usually embedded in polyhedral inclusion bodies. This is also the case for both the cytoplasmic polyhedrosis viruses, insect members of the Reoviridae (Sec. 5.1), and insect poxviruses (entomopox). As the names indicate, the NPVs, in contrast with the cytoplasmic polyhedrosis viruses, are located in *nuclear* polyhedral bodies. The GVs are also embedded in nuclear inclusion bodies, which appear less polyhedral (Sec. 11.4).

As indicated by the name, the baculoviruses are rod-shaped and thus of helical symmetry. Individual NPV particles are of about 40×300 nm, and GV virions are 65×280 nm (see Fig. 8.13). The Baculoviridae contain double-stranded, circular, and supercoiled DNA genomes of $90–230 \times 10^6$ Da. Replication takes place mostly in the nucleus. The prototype host is *Autographa californica NPV.*

The virions are structurally complex, containing 12–30 structural polypeptides and have a lipid-rich envelope. The enveloped nucleocapsids (Fig. 8.13a) are embedded in a crystalline occlusion body composed of a 25–33 kDa **polyhedrin** or **granulin** protein. The polyhedra **serve** to protect the virions, so that they remain viable for many years in soil. The GV virions are usually embedded singly rather than in bundles [Fig. 8.13(b)]. In contrast with the lipid-enveloped virions, the inclusion bodies are very stable and resistant to bacteria, heat, and many chemical agents, including low pH. They are soluble in sodium carbonate and become degraded by a protease carried in the virion at pH10. This is near the pH in the host's midgut.

The Baculoviridae, like the cytoplasmic polyhedrosis viruses and the Iridoviridae, primarily infect insects as larvae that feed on polyhedron-contaminated leaves. As the viruses replicate, they come to fill the entire body of the larva. Baculoviruses are thus of great concern wherever insects are used commercially, the silk and honey industries for examples.

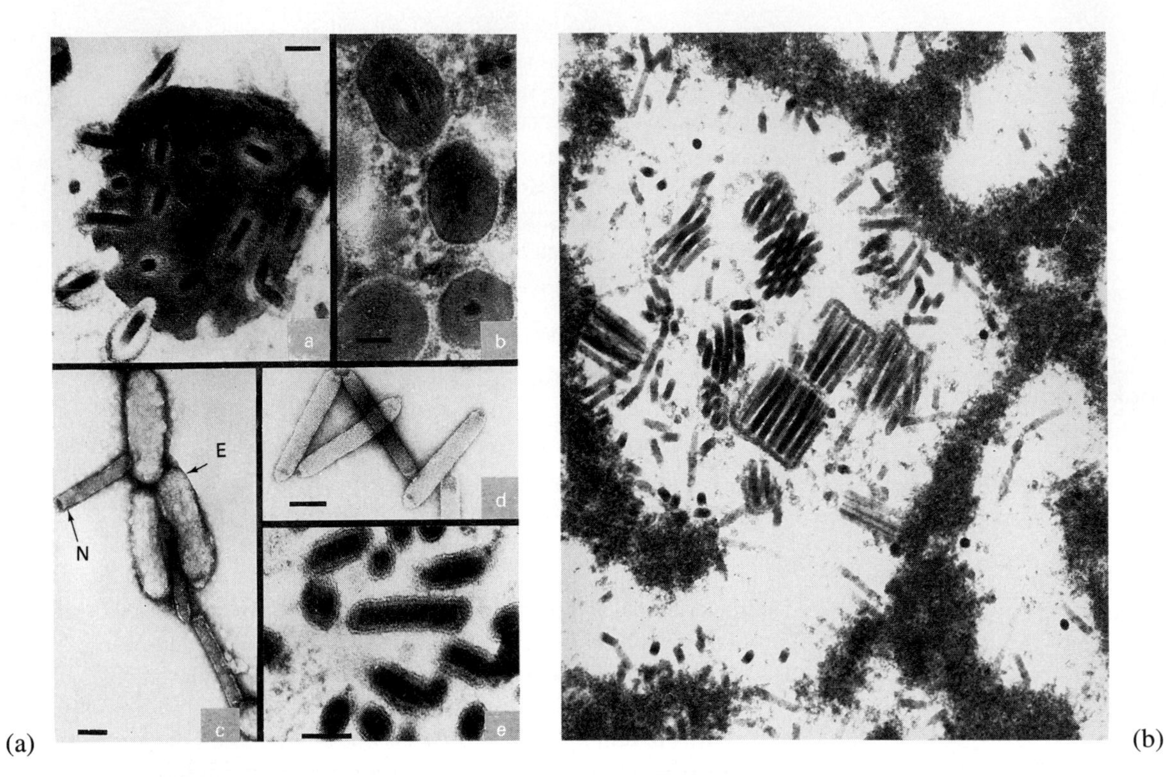

FIGURE 8.13

Electron micrographs of nuclear polyhedrosis and granulosis viruses. Left photo: (a) Immature baculovirus nuclear polyhedron in section. (b) Granulosis virus "capsules" in section. (c) Nuclear polyhedrosis virus particles after alkali-release from polyhedra (E = enveloped nucleocapsid, N = nucleocapsid). (d and e) Nuclear polyhedrosis virus particles in section and stained (bar = 100 nm). Right photo: Bundles of nuclear polyhedrosis virus particles, with envelopes partly visible. (Courtesy of K. A. Harrap)

During infection, two forms of virions—single nucleocapsids or extracellular virions (ECV) and occluded virions—are produced. Cell-to-cell spread is presumed to involve the ECV, while the occluded form is important for **horizontal transmission** (see Sec. 12.1.2).

The use of Baculoviruses as vectors for the expression of foreign genes as well as their uses to control certain insect pests is discussed in Sec. 12.4

8.5.2 Iridoviridae

The unusual Iridoviridae virus family includes several genera whose members infect many insects as well as amphibians and fish. The name of its prototype, *Tipula iridescent virus* (TIP) (see Fig. 1.4), is derived from the blue-green iridescent appearance of heavily infected tissue.

The icosahedral particles of various iridoviruses are 125–300 nm in diameter and contain 5–9% lipid (predominantly phospholipid) as an integral part of their shells. Some members also contain an additional plasma membrane-derived envelope that enhances (but it is not required for) infectivity. Virions contain 13–35 structural polypeptides, one (possibly two) molecules of linear circularly permuted as DNA (MW = 100–250 $\times 10^6$ Da), and several virion-associated enzymes.

Iridoviruses enter the cell by pinocytosis and are uncoated in phagocytic vaculoes. Transcription of the viral DNA appears limited to the nucleus, but DNA replication occurs in both the nucleus and cytoplasm. Paracrystalline virus inclusions are formed in the cytoplasm. Although virions may be released from infected cells, most virus remains cell associated. Many iridoviruses appear to have restricted host ranges, but certain exceptions are known. **Frog virus** 3 grows in a wide variety of cultured cells, and Tipula iridescentvirus infects a wide range of insects.

SECTION 8.5 FURTHER READING

Original Papers

XEROS, N. (1954) A second virus disease of the leatherjacket, *Tipula paludosa*. *Nature* 174:562.

KNUDSON, D. L., and T. W. TINSLEY. (1978) Replication of a nuclear polyhedrosis virus in a continuous cell line of *Spodoptera frugiperda:* Partial characterization of the viral DNA, comparative DNA-DNA hybridization and patterns of DNA synthesis. *Virology* 87:42.

MCCARTHY, W. J., et al. (1978) Characterization of the DNA from four *Heliothis* nuclear polyhedrosis virus isolates. *Virology* 90:374.

SUMMERS, M. D., and G. E. SMITH. (1978) Baculovirus structural polypeptides. *Virology* 84:390.

ADANG, M. J., and L. K. MILLER. (1982) Molecular cloning of DNA complementary to mRNA of the baculovirus *Autographa californica* nuclear polyedrosis virus: Location and gene products of RNA transcripts found late in infection. *J. Virol.* 44:782.

HOOFT VAN IDDEKINGE, B. J. L., et al. (1983) Nucleotide sequence of the polyhedrin gene of *Autographa californica* nuclear polyhedrosis virus. *Virology* 131:561.

WARD, V. K., and J. KALMAKOFF. (1987) Physical mapping of the DNA genome of insect iridescent virus type 9 from Wiseana spp. larvae. *Virology* 160:507.

CAMERON, J. R. (1990) Identification and characterisation of the gene encoding the major structural protein of insect iridescent virus type 22. *Virology* 178:36.

PASSARELLI, A. L., and L. K. MILLER. (1993) Identification and characterization of *lef*-1, a baculovirus gene involved in late and very late gene expression. *J. Virol.* 67:3481.

Books and Reviews

HARRAP, K. A., and C. C. PAYNE. (1979) The structural properties and identification of insect viruses. *Adv. Vir. Res.* 25:273.

TINSLEY, T. W. (1979) The potential of insect pathogenic viruses as pesticidal agents. *Ann. Rev. Entomol.* 24:63.

MILLER, L. K. (1984) Exploring the gene organization of a baculovirus. *Methods Virol.* 7:227–258.

DEVANCHELLE, G., et al. (1985) Comparative ultrastructure of iridoviridae. In *Iridoviridae,* ed. D. B. Willis. 1 Springer-Verlag, Berlin.

MURTI, K. G., et al. (1985) An unusual replication strategy of an animal iridovirus. *Adv. Vir. Res.* 30:1.

WILLIS, D. B., ed. (1985) *Iridoviridae.* Springer-Verlag, Berlin.

WILLIS, D. B. (1985) Macromolecular synthesis in cells infected by frog virus 3. *Current Topics Microbiol. Immunol.* 116:77.

FAULKNER, P., and E. B. CARSTENS. (1986) An overview of the structure and replication of baculoviruses. In *The molecular biology of baculoviruses,* ed. W. Doerfler and P. Böhin. Springer-Verlag, Berlin.

ROHRMANN, G. F. (1986) Polyhedrin structure. *J. Gen. Virol.* 67:1499.

8.6 VIRUSES OF GREEN ALGAE AND FUNGI

A large number of nonenveloped, polyhedral viruses infecting green algae have been isolated based on their ability to form plaques on one of three strains of *Chlorella* that were originally isolated as endosymbionts of either *Paramecium bursaria* or *Hydra viridis.* Assigned to a separate taxonomic family (the **Phycodnaviridae**), the particles of these viruses (100–200 nm in diameter) contain 20 to 50 structural proteins plus 5 to 10% lipid as an integral part of the polyhedral shell. The viral genome, a single molecule of linear nonpermuted double-stranded DNA (250–350 kb) with cross-linked **hairpin termini,** contains variable amounts of 5-methyldeoxycytine plus, in some cases, N^6-methyldeoxyadenosine. Virion attachment to the cell wall occurs in a host- and site-specific manner, and DNA replication and particle assembly occur in the cytoplasm. Progeny release occurs via lysis of the host cell.

As shown in Figure 8.14, sequence analysis has demonstrated that the termini of the *Paramecium burscaria-chlorella* virus (PBCV-1) genome contain identical 2185 nucleotide inverted repeats. The termini of two poxviruses, vaccinia and African swine

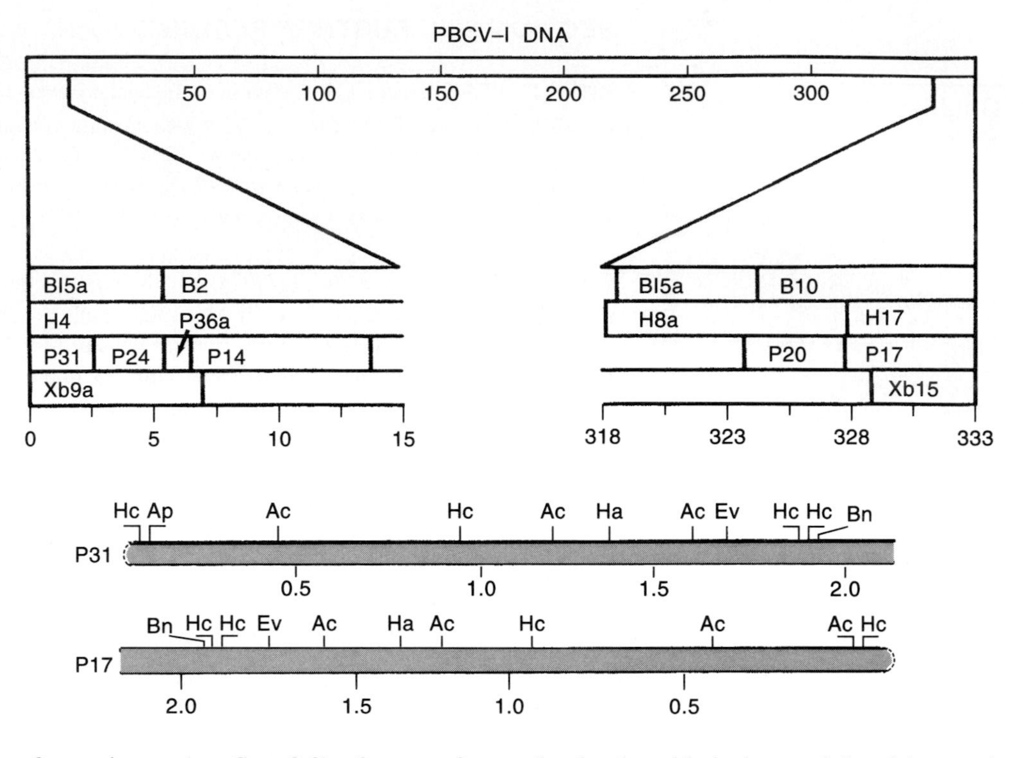

FIGURE 8.14

Restriction site maps of the PBCV-1 DNA termini. *(Top)* The entire 333 kb genome of PBCV-1. *(Middle)* Relative locations of various restriction fragments near the "left" and "right" ends of the genome. Fragments P17 and P31 (shaded) contain the covalently closed (i.e., hairpin) genome termini. *(Bottom)* Restriction maps of the inverted terminal repeats located within fragments P17 and P31. (From Strasser et al. (1991) *Virology* 180:763–769)

fever viruses (see Sec. 8.3), also contain covalently closed hairpin termini and inverted terminal repeats, but their terminal repetitions are much larger (about 10 kb vs. 2.2 kb) and contain many identical tandem direct repeats (50–125 nucleotides). The termini of PBCV-1 contain several direct repeats, but the degree of repetition is much less than with Poxviridae.

An isometric virus containing a linear approximately 25 kb double-stranded DNA genome has been isolated from *Rhizidiomyces,* an aquatic fungus. The 60 nm particles of this virus contain at least 14 polypeptides (26–84.5 kDa) and are first visible in the nuclei of infected cells. The virus appears to be transmitted in a latent form in **zoospores** of the host, and virus activation by stress conditions (e.g., heat or nutrient stress) results in cell lysis. It has been placed in the genus *Rhizidiovirus* within the family **Adenoviridae.**

Many large DNA viruses resembling bacteriophages have been isolated from the prokaryotic blue-green algae. In general, these **cyanophages** are nonenveloped, icosahedral viruses—members of the Myo-, Sipho-, and Podoviridae. Some of these viruses (e.g., LPP-1) have been studied in great detail.

SECTION 8.6 FURTHER READING

REISSER, W., et al. (1988) A comparison of viruses infecting two different *Chlorella*-like green algae. *Virology* 167:143–149.

STRASSER, P., et al. (1991) The termini of *Chlorella* virus PBCV-1 genome are identical 2.2-kbp inverted repeats. *Virology* 180:763–769.

SHERMAN, L. A., and R. M. BROWN. (1978) Cyanophages and viruses of eukaryotic algae. In *Comprehensive virology,* ed. H. Fraenkel-Conrat and R. R. Wagner. Vol. 12, 145–223. Plenum Press, New York.

8.7 BACTERIOPHAGES

In the early days of systematic phage research M. Delbrück, the father of this field, named a group of *E. coli* phages that shared the ability to lyse the host cell, T1 to T7. These **lytic** phages were thus differentiated from the possibly more numerous **lysogenic** phages, (originally termed **temperate**) that can exist in a latent suppressed state in the bacterium (see Sec. 11.11). The differentiation between these two groups of phages has proved less definitive than first believed, however.

Most of the typical DNA phages contain a single molecule of double-stranded DNA that, with the exception of the lipid-containing phages of the PR and PM2 types (see Sec. 8.7.3), range in molecular weight from 20 to 150×10^6 Da and generally represent about 50 percent of the particle weight. They vary greatly in size, shape, and nature of tail (if any) as illustrated schematically in Figure 1.15. This figure also shows their family names and compares them with the single-stranded RNA (Leviviridae) and small single-stranded DNA–containing phages (Micro-, Ino-, Cortico-, and Tectiviridae) discussed in Sections 2.7 and 7.4–7.6.

The classic studies on the biology of phages carried out by the research groups of Delbrück, Luria, Hershey and others in the 1940s and 1950s dealt largely with the particular series of the dissimilar T1 to T7 phages of *E. coli*. These are now recognized as representing examples of many other phages in several other hosts. Physical and chemical investigations starting in the 1960s also largely utilized the T phages, particularly the closely related T-even phages, T2 and T4. In contrast with these virulent lytic phages, the discovery of lysogeny focused attention on a different *E. coli* phage, lambda (λ). We begin our discussion with the smaller and structurally simpler T-odd phages before turning to the large phages exemplified by the T-even group.

8.7.1 Podoviridae

The Podoviridae family consists of the T3 and T7 phages. We will discuss T7 as an example.

Structure These closely related phages have isometric particles with very short tails (Fig. 8.15) a feature from which they derive their name (Gr., *podus,* foot or short tail). Their particle weights are about 45×10^6Da, half of which is a double-stranded DNA molecule. The major head protein is about 40 kDa, and three proteins are associated with the short tail structure. Determination of the complete 39,936 nucleotide sequence of T7 DNA revealed the presence of 55 possible genes. Proteins encoded by 44

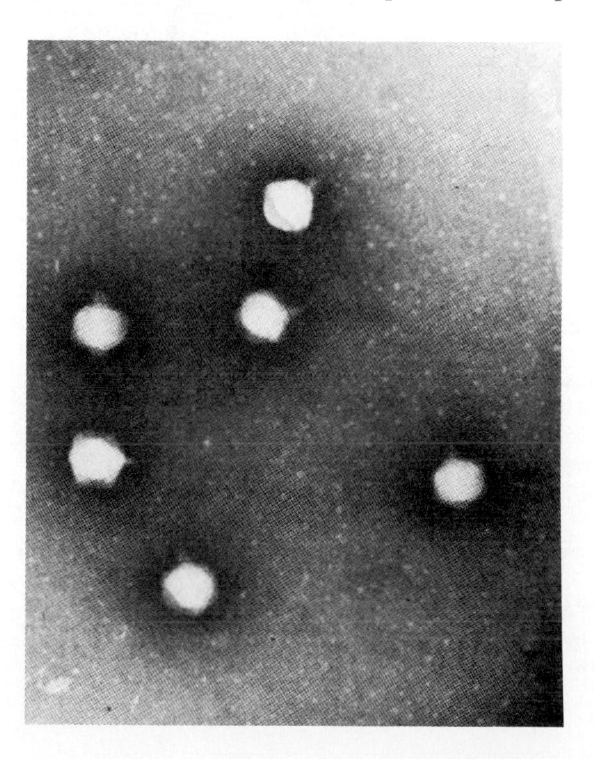

FIGURE 8.15

Electron micrograph of T3 phage. (Courtesy of R. C. Williams)

of these genes have been identified, and the functions of approximately 30 genes are known (see Fig. 8.16).

Transcription It is obvious from Fig. 8.16 that the genes for proteins involved in early transcription and translation are located near one end of the DNA, whereas those involved in DNA metabolism and phage structure/assembly are farther downstream. Most complex DNA bacteriophages modify the host RNA polymerase in order to ensure the efficient and timely expression of the various classes of phage genes. For example, the proteins encoded by genes 33 and 55 of phage T4 replace the host σ factor, thereby limiting transcription to the late genes. T7 gene I, however, encodes an entirely new **RNA polymerase** that is highly specific for T7 late gene promoters. Transcription of the T7 early genes (including gene I) is carried out by *E. coli* RNA polymerase.

The Replication Cycle The replication of the phage DNA results from production of a **specific DNA polymerase,** the product of gene *5*. The origin of replication is between genes *2* and *3*, and replication proceeds in both directions (i.e., it is bidirectional). Since the path to the left end is much shorter than that to the right, Y-shaped molecules are frequently seen. T7 DNA is **terminally redundant,** 260 base pairs of the genome being almost completely identical in sequences at both ends. Circularization does not seem to occur with these phages. Instead, **linear concatemers** form through recombination at the redundant terminal sequences. This process is illustrated schematically in Figure 8.17. It appears probable that **recombination** is a necessary step in the replication of linear DNAs to overcome the lack of priming sites at the 5′ end. Genes for specific nucleases and other proteins playing a role in the recombination process (see below) occur in all these phages. However, the host cell is also rich in the enzymes needed for the cleavage and recombination of DNA

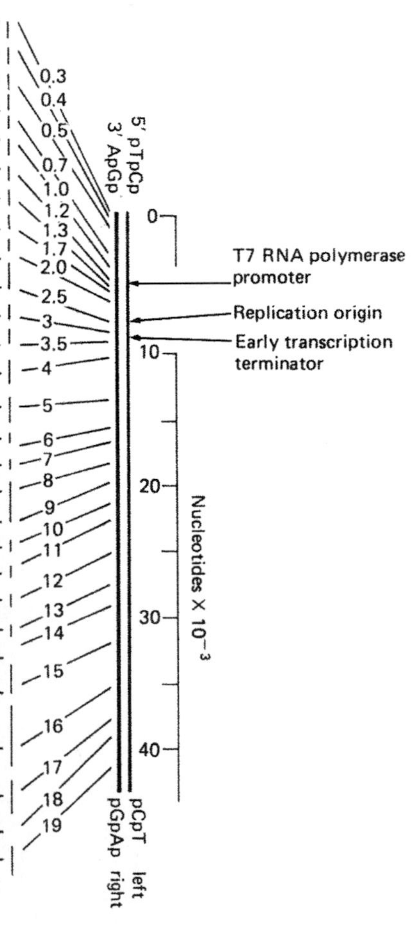

FIGURE 8.16

The genetic map of phage T7 as refined by sequencing. Along the double-stranded DNA molecule are indicated the approximate positions of genes and other sites, and transcription proceeds from top to bottom. The molecular weight of each gene product formed, and the location on the nucleotide sequence are listed. Minor products, reading frame changes, as well as some minor corrections, have not been included. (Modified from Dunn and Studier (1983) *J. Mol. Biol.* 166:147)

	Protein (daltons)	Nucleotides (number)
Abolishes host restriction	(13,678)	925-1276
		1278-1431
		1496-1637
Protein kinase	(41,124)	2021-3098
RNA polymerase	(98,092)	3171-5820
DNA replicase	(10,059)	6007-6133
DNA ligase	(41,133)	6475-7552
?	(22,053)	8166-8754
Inactivates host transcription	(7,043)	8898-9090
SS DNA binding protein	(25,562)	9158-9854
Endonuclease I	(17,040)	10257-10704
Lysozyme	(16,806)	10706-11159
Single strand DNA binding	(62,656; 55,743)	11565-13263
DNA polymerase	(79,692)	14353-16465
Exonuclease	(39,995)	17359-18403
Host range protein	(15,303)	19129-19528
Head-tail protein	(58,989)	20239-21847
Head assembly	(33,766)	21949-22870
Head proteins (A,B)	(36,414)	22966-24159
Tail protein	(22,289)	24227-24815
Tail protein	(89,265)	24841-27223
Internal virion protein	(15,852)	27306-27720
Internal virion protein	(20,836)	27727-28315
Internal virion protein	(84,210)	28324-30565
Internal virion protein	(143,840)	30594-34548
Tail fiber protein	(61,441)	34623-36282
DNA maturation	(10,145)	36552-37280
DNA maturation	(66,130)	37369-39127

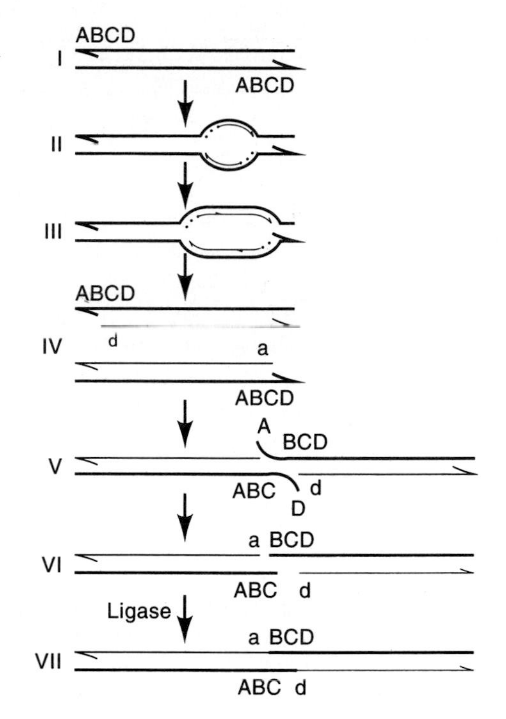

FIGURE 8.17

Suggested role of recombination in the replication of linear DNA. (I =) linear duplex DNA; (II =) formation of a bidirectional replicating loop (a single origin is shown for simplicity); — = parental DNA; — = newly replicated DNA; ⟶ = 3′ terminus; ...= DNA binding protein; (III =) further replicated DNA, showing discontinuous synthesis; (IV =) two replicated DNA molecules; (V =) recombination at the incompletely replicated ends; (VI =) presumed trimming of single-stranded ends by products; (VII =) covalently sealed recombinant molecule. (C. K. Mathews. (1977) In *Comprehensive virology.* Ed. H. Fraenkel-Conrat and R. R. Wagner. Vol. 7. Plenum Press, New York)

molecules. Compared to the 30 proteins required for replication of the host cell genomic DNA, a more limited number of proteins (both phage- and host- encoded) are required for T7 DNA replication. As noted above, T7 DNA polymerase is encoded by gene 5, but its activity requires formation of a complex with thioredoxin, a host protein. Primase and helicase activities are supplied by the product of gene 4, and gene 6 encodes a nuclease that removes the RNA primers. Gene 2.5 encodes a ssDNA binding protein.

Processing that involves a specific nuclease and gap filling produces the terminal redundancy of phage DNA molecules. This event probably takes place at the time of virion assembly in the cell membrane. The T3 and T7 phages, in contrast with the phages discussed in Sec. 8.7.2, show all the properties of *phage virulence.* They shut off host transcription, degrade the host's DNA, and finally lyse the cell.

8.7.2 Siphoviridae

Structure The Siphoviridae family contains among many members the lambda and T5 phages of *E. Coli.* Their family name is derived from their shape (Gr., *siphon,* tube). We will discuss lambda (λ) as an example of this virus group. λ, the prototype of the **lysogenic** (or temperate) phages has probably been studied longer and in greater detail than any other virus.

The λ virion consists of an icosahedral head 55 nm in diameter and a thin noncontractile tail of 150 × 8 nm with short terminal and subterminal tail fibers.

Its genome is a 48,502 nucleotide linear molecule with single-stranded **cohesive ends** (12 nucleotides long), at each end. These enable the DNA to quickly citularize upon infection:

5′GGGCGGCGACCT→
←CCCGCCGCTGGA3′

In the linear DNA, genes encoding head proteins or playing a role in its assembly are located in the first 18 percent, the tail-related functions in the next 18 percent, and genes responsible for lysogenic functions and their regulation occupy the next 50 percent of the genome. Genes for DNA replication and cell lysis complete the genome map (see Fig. 8.18).

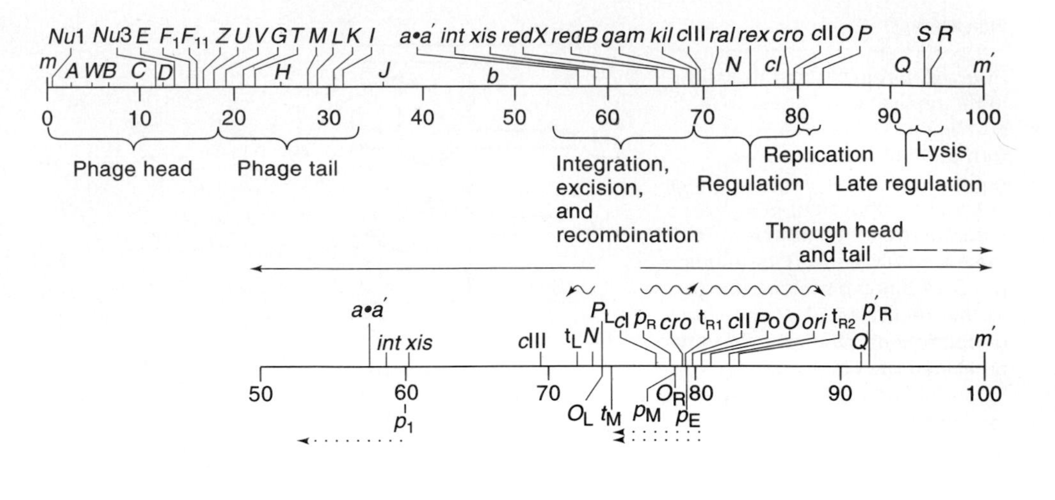

FIGURE 8.18

The genetic map of phage λ and its early mode of transcription. The genes that code for proteins of defined function are shown in the upper part of the figure. The central region is silent in terms of defined viral functions, although it does code for several proteins. The regulatory sites and their function are indicated in the lower part of the figure. The different stages of transcriptional activity and the DNA region involved are indicated by the arrows; the actual length of DNA transcribed can be determined from the intersection of the promoter (p) and terminator (t) lines with the horizontal calibrated line representing the λ genome. (From H. Echols and H. Murialdo. (1978) *Microbiol. Rev.* 42:577)

Replication The first event upon infection by λ phage is the **covalent circularization** of the molecule via joining of the cohesive single-stranded ends by host cell DNA ligase. This process allows **supercoiling,** which is a requirement for integration (Sec. 7.2.4) and transcription. Transcription by host cell RNA polymerase begins when the enzyme binds to the O_L and O_R sites and transcribes leftward and rightward, respectively (see Fig. 8.18). The first genes transcribed encode regulatory proteins, *N* and *cIII* on the leftward and *cro* and *cII* on the rightward transcript. The N product favors continuation of these particular transcription processes. When the *cII* product is transcribed, it activates transcription of c_I and **int.** Rightward transcription leads to products O and P, which initiate replication of the phage DNA starting at *ori.* Cro protein represses transcription of the early genes that are no longer needed in the productive replication cycle.

The replication of λ DNA follows the usual steps for DNA replication. Early in infection, bidirectional replication produces more circular molecules. Later in productive λ infections there is a switch to a unidirectional rolling circle mechanism similar to that used by ⦰X174 (see Sec. 7.5) this process yields σ-type molecules with very long concatemeric linear tails. These forms, in contrast with those observed with the single-stranded DNA phages, are double-stranded. As they enter the phage assembly process, they are cut into typical 47,000 base-pair-length molecules with cohesive ends. This process involved *site-specific cleavage* of the phage DNA, as opposed to *length-specif-*

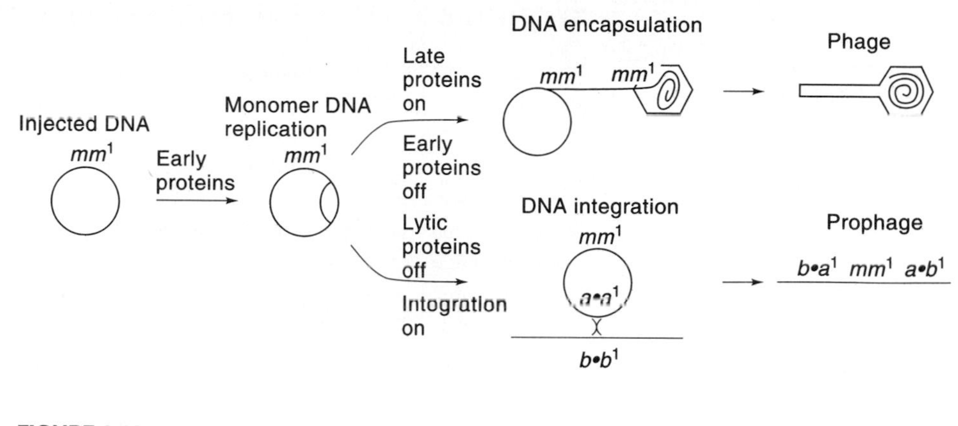

FIGURE 8.19

The two developmental paths available to lysogenic phages. The injected DNA forms a covalently closed circle through pairing of the cohesive ends, m and m′, followed by ligation. After an early symmetrical replication phase common to both pathways, viral development may follow the productive or lysogenic pathways. In the productive pathway, synthesis of encapsulation and lysis proteins is turned on, synthesis of early proteins is turned off, and replication switches to a rolling-circle mode that generates the multimeric DNA used as a substrate for encapsulation of linear DNA with its cohesive ends. In the lysogenic pathway, synthesis of lytic proteins is turned off, and the circular viral DNA is inserted into the host genome by a specific recombination event. (From H. Echols and H. Murialdo. (1978) *Microbiol. Rev.* 42:577)

ic or *"headful"* mechanisms that operate in T-even phages. Thus, deletion mutants of λ carry smaller DNA molecules in the virion, whereas such mutants of T-even phage carry DNA molecules of the same physical size as the wild type, but with larger terminal redundancies.

The assembly of the phage head involves at least 10 phage gene products and one host cell protein. Five proteins form a precursor particle, termed **prohead** or petit λ. Some of these proteins are proteolytically processed in the course of assembly, and two are joined during assembly. One protein, Nu 3, plays a role similar to the **scaffolding proteins** that are prominent in T7 assembly. This protein is processed upon entering the prohead, and exits prior to the entry of the DNA. The prohead, like the final capsid, contains 420 molecules of gene product E. The tails are assembled separately, and the finished tail consists of 135 molecules of gene product V, plus a few molecules of other proteins. The tails are able to attach themselves to the heads *in vitro*. The mechanisms of the processes of integration, repression, and excision, and their control, which involve proteins coded for by half of the phage's genome, are discussed in Section 11.11. The nature of the two paths that lysogenic phages such as λ can take are summarized schematically in Fig. 8.19.

There exist a great number of lysogenic phages that are more or less similar to λ. One that has been intensively studied is **phage Mu**. This phage is of particular interest because its integration properties differ from those of λ in a manner that makes one realize how tenuous is the dividing line between virology and general biology. Mu behaves like **transposon,** inserting itself anywhere in the host's genome—causing mutations and suppressing expression of genes rather than at one or a small number of sites like λ.

SECTION 8.7.1 TO 8.7.3 FURTHER READING

Original Papers

STUDIER, F. W. (1972) Bacteriophage T7. *Science* 176:367.

KAISER, D., et al. (1975) DNA packaging steps in bacteriophage lambda head assembly. *J. Mol. Biol.* 91:175.

SAUER, R. T., et al. (1979) Regulatory functions of the λ repressor reside in the amino-terminal domain. *Nature* 279:396.

SHAW, J. E., and H. MURIALDO. (1980) Morphogenetic genes *C* and *Nu3* overlap in bacteriophage λ. *Nature* 283:30.

DUNN, J. J., and F. W. STUDIER. (1983) Complete nucleotide sequence of bacteriophage T7 DNA and the locations of T7 genetic elements. *J. Mol. Biol.* 166:477.

HARSHEY, R. M., et al. (1985) Primary structure of phage Mu transposase: Homology to Mu repressor. *Proc. Natl. Acad. Sci. USA* 82:7676.

DREXLER, H., and J. R. CHRISTENSEN. (1986) T1 *pip:* A mutant which affects packaging initiation and processing packaging of T1 DNA. *Virology* 150:373.

SCHMUCKER, R., et al. (1986) DNA inversion in bacteriophage Mu. *J. Gen. Virol.* 67:1123.

KARSKA-WYSOCKI, B., et al. (1987) Characterization of morphogenetic intermediates and progeny of normal and alkylated bacteriophage T7. *Virology* 157:285.

TEMPLE, L. M., et al. (1991) Nucleotide sequence of the genes encoding the major tail sheath and tail tube proteins of bacteriophage P2. *Virology* 181:358.

TYAGARAJAN, K., et al. (1991) RNA folding during transcription by T7 RNA polymerase analysed using the self-cleaving transcript assay. *Biochem.* 30:10920.

HENDRIX, R. W., and R. L. DUDA. (1992) Bacteriophage λ PaPa: Not the mother of all λ phages. *Science* 258:1145.

KIM, S., and A. LANDY. (1992) Lambda int protein bridges between higher order complexes at two distant chromosomal loci *att*L and *att*R. *Science* 256:198.

MÉNDEZ, I., et al. (1992) Initiation of φ29 DNA replication occurs as the second 3′ nucleotide of the linear template: A sliding back mechanism for protein-primed DNA replication. *Proc. Natl. Acad. Sci. USA* 87:9579.

SON, M., and P. SERWER. (1992) Role of exonuclease in the specificity of bacteriophage T7 DNA packaging. *Virology* 190:824.

CALDENTEY, J., C. LUO, and D. H. BAMFORD. (1993) Dissociation of the lipid-containing bacteriophage PRD1: Effects of heat, pH, and sodium dodecyl sulfate. *Virology* 194:557.

CHAUDHURI, B., et al. (1993) Isolations, characterization, and mapping of temperature-sensitive mutations of the genes essential for lysogenic and lytic growth of the mycobacteriophage L1. *Virology* 194:166.

MORITA, M., M. TASAKA, and H. FUJISAWA. (1993) DNA packaging ATPase of bacteriophage T3. *Virology* 193:748.

Books and Reviews

RABUSSAY, D., and E. P. GEIDUSCHEK. (1977) Regulation of gene action in the development of lytic bacteriophages. In *Comprehensive virology.* ed. H. Fraenkel-Conrat and R. R. Wagner. Vol. 8, 1–196. Plenum Press, New York.

ECHOLS, H., and H. MURIALDO. (1978) Genetic map of bacteriophage lambda. *Microbiol. Rev.* 42:577.

HAYES, S. (1980) Bacteriophage λ. *Intervirology* 13:133.

STUDIER, F. W., and J. J. DUNN. (1983) Organization and expression of bacteriophage T7 DNA. *Cold Spring Harbor Symp. Quart. Biol.* 47:999.

SALAS, M. (1988) Phages with protein attached to the DNA ends. In *The bacteriophages I.* ed. R. Calendar. 169 Plenum Press, New York.

MCCORQUODALE, D. J., and H. B. WARNER. (1988) Bacteriophage T5 and related phages. In *The bacteriophages I.* ed. R. Calendar. 439 Plenum Press, New York.

BERTAND, E., E. W. SIX, and R. CALENDAR. (1988) P2-like phages and their parasite P4, The *T7 group.* 73 Plenum Press, New York.

HAUSMANN, R. (1988) The T7 group. In: *The bacteriophages,* ed. R. Calendar, 259 Plenum Press, New York.

HENDRIX, R. W. et al. (1983) Lambda II. Cold Spring Harbor Laboratory. Cold Spring Harbor, NY.

8.7.4 Myoviridae

The most intensely studied phages, with the possible exception of λ, are the T-even phages of *E. coli.* Their family name is derived from their contractile tail (Gr., *myos,* muscle). Apart from having a genome four times as large as λ (about 120×10^6 Da), these phages show two features of DNA structure first observed with these viruses. One feature, **circular permutation**, is not restricted to the DNA of these phages but has also been observed for several other smaller phages as well as with certain animal viruses (see Chap. 8). The other unusual feature of the T-even phage DNAs is the presence of a **modified base, 5-hydroxymethylcytosine (HMC),** instead of cytosine.

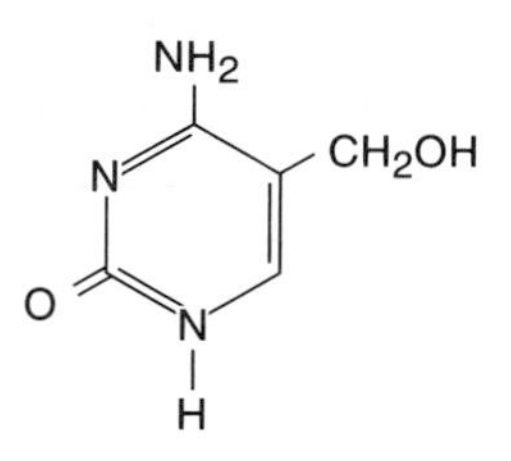

Except for the occurrence of methyl groups on the 5′-terminal guanosine residues and 2-position of certain ribose residues in mRNAs of cells and viruses as well as in the genomes of small DNA phages (Secs. 7.5 and 7.6), modified bases have not been mentioned in this book. However, their occurrence in biological nucleic acids is by no means rare and tRNAs, contain many specifically **modified** or **hypermodified bases.** However, in no nucleic acids except the DNAs of some bacteriophages is one of the four typical bases quantitatively replaced by a modified one.

Because the particular component of the T-even phages, HMC, has been found only in the T-even phages, it can be regarded as the result of a unique event in their evolution. However, events of this general type appear to have occurred more than once, for we know of other phage families having other unusual bases. Thus, a group of large interrelated *B. subtilis* phages, the prototype of which is called SP01, contain *5-hydroxymethyluracil* instead of thymine, a change of a —CH_3 to a —CH_2 OH group. Another *B. subtilis* virus, PBS1, carries *uracil* instead of thymine, unusual in terms of DNA. S-2L, a cyanophage of *Synecocochus,* has *2-amino-adenine* instead of adenine. Even hypermodified bases are found in specific phages, such as *dihydroxypentyl-U* which contains an added five-carbon dialcoholic chain. Some other phages show less-than-complete replacement of the normal by the modified base. None of these base modifications affect their base-pairing properties, even though their one or more glucose molecules may be attached to some of the 5 —CH_2 OH groups. The function of these unusual bases may be to protect the phage DNA from attack by a phage-specified enzyme that degrades unmodified DNA.

Structure The head of T2 and T4 is an icosahedron, elongated by one or two extra bands of hexamers, of 85 × 110 nm (Fig. 8.20). Its shell consists of 5-nm-diameter capsomers composed of three major proteins, as well as lesser amounts of several other proteins. Each capsomer may consist of six molecules each of gp* 23 and gp soc, and one of gp hoc. The contractile tail (25 × 110 nm) is an extraordinarily complex organ, composed of a tube, a sheath, a connecting neck with a collar and whiskers, and a complex base plate with pins and carrying long jointed fibers (see Fig. 8.21). The neck consists of six molecules each of four proteins and 18 of another, all about 33 kDa. Each whisker contains 38 molecules of a 53 kDa protein. The tail tube is made of 144 molecules of a 19 kDa protein, and the sheath of the same number of molecules of a 70 kDa glycoprotein.

*gp stands for gene product. The names of the genes are usually in italics, but not the protein gene products.

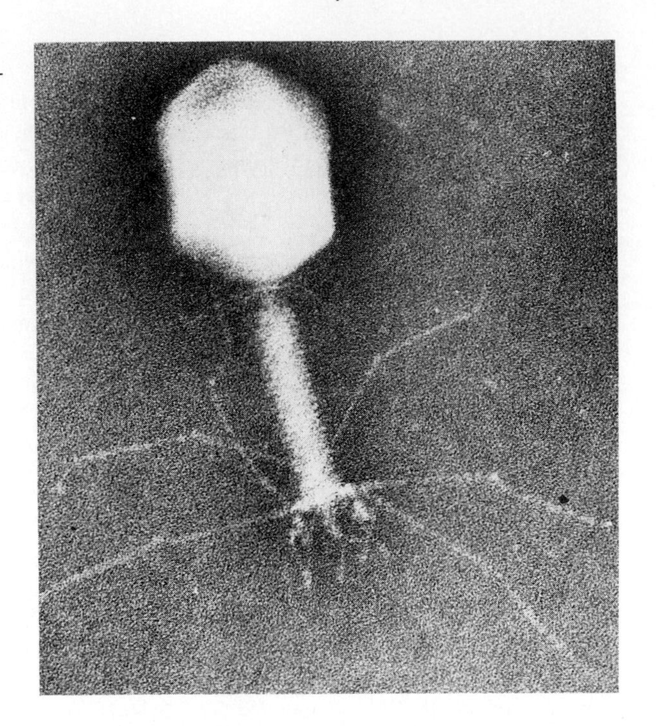

FIGURE 8.20

Electron micrograph of T2 phages. (Courtesy of R. C. Williams)

The sheath protein molecules are arranged in 24 rings; as a consequence of a conformational change of that protein, the rings become larger and their number is reduced to 12, leading to the shortened, "contracted" form. This event plays a critical role in the infection mechanism. The base plate plus spikes requires 14 proteins, ranging from 15 to 140 kDa, most of which are present in six copies (a few are multiples of six). The six fibers consist of two proteins of 145 and 115 kDa forming dimers and two small proteins, one dimeric and the other single. The structural proteins of the virion thus number to over 30 proteins.

The DNA of the T-even phages is linear, 52 nm long (i.e., 650 times the diameter of the phage; see Fig. 8.22.). It occupies about half of the head's capacity. It is associated with one major protein (18 kDa) and lesser amounts of several smaller proteins and

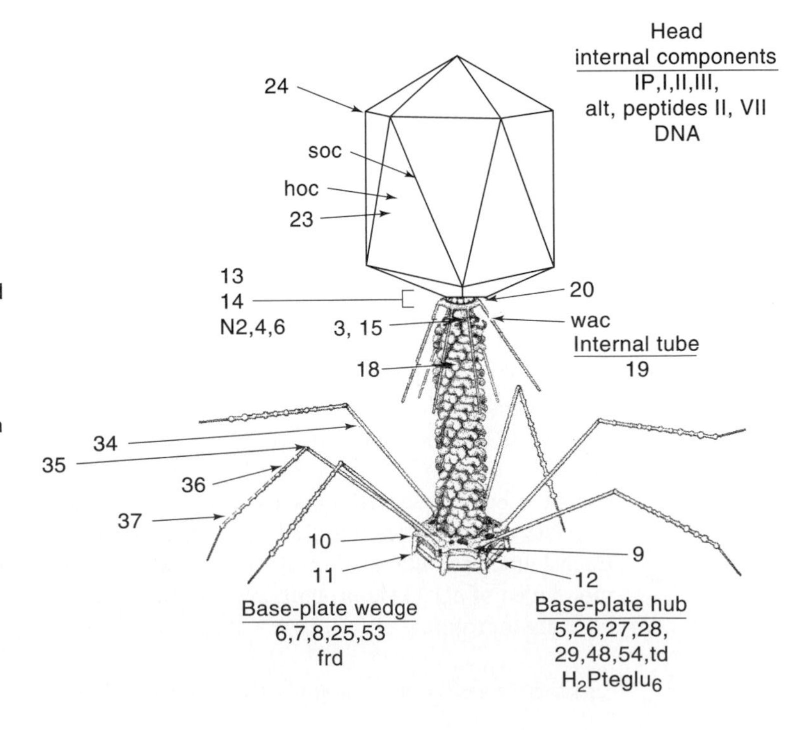

FIGURE 8.21

Structure of bacteriophage T4, based on chemical crosslinking and on electron microscopic structure analysis to a resolution of about 2 to 3 nm. Near the head and tail are shown the locations of the known viral proteins. The icosahedral vertices of the head are made of cleaved gp24. The gene 20 protein is located at the connector vertex, bound to the upper collar of the neck structure and the lower part of the head-tail connector is made of gp13 and 14. The six-whiskers and the collar structure appear to be made of gp13 and 14. The six whiskers and the collar structure appear to be made of a single protein species, gp*wac.* The gp18 sheath subunits fits into holes in the baseplate, and the gp12 short tail fibers are shown in a stored position. The central tube made of gp19 is filled with gp29 and terminated by gp3. The baseplate is assembled from a central hub and six wedges. The positions of most of the baseplate wedge proteins are known, but less is known about the hub proteins. The proximal half of the tail fibers is made of gp34, the distal half of the fiber from three proteins which taper at the end that binds to the bacterial cell surface. (Courtesy of F. Eiserling)

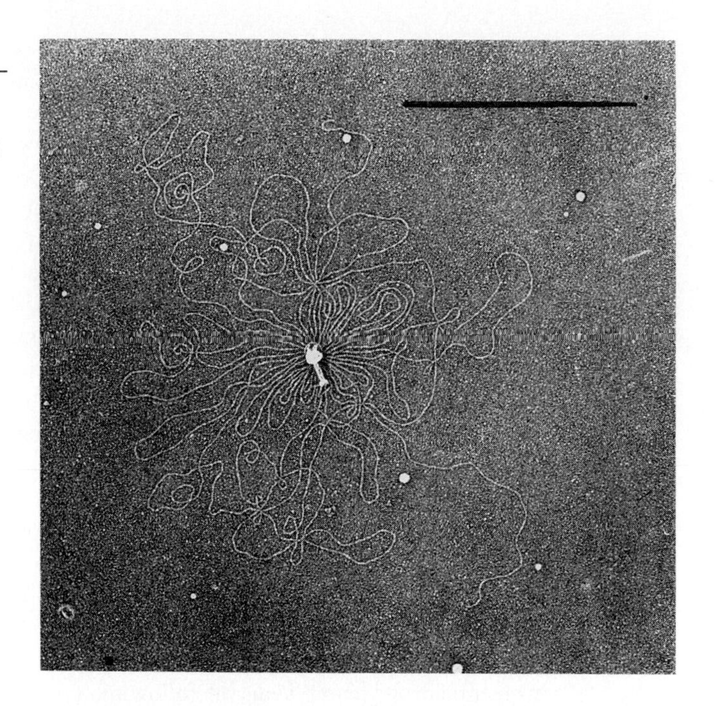

FIGURE 8.22

Electron micrograph of a burst T-even phage particle revealing the entire linear DNA (ends at top and bottom). (The bar represents 1 μm.)

polypeptides, as well as considerable amounts of the basic compounds spermidine and putrescine acting as counterions to the phosphate groups.

Like most, if not all, linear phage DNAs and many animal virus DNAs, the T-even DNAs are **terminally redundant,** with a sequence of 3000 bases occurring twice. They are also **circular permutated.** The mystery of 35 years ago that the T-even phage genomes, although clearly linear (see Fig. 8.22), behaved genetically as if genes from the two ends were neighbors is clearly explained by this phenomenon. As we have already seen, certain insect viruses share this property (Sec. 8.5.2).

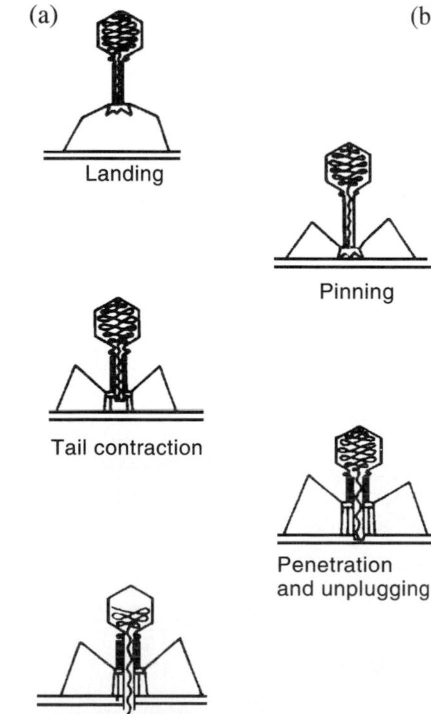

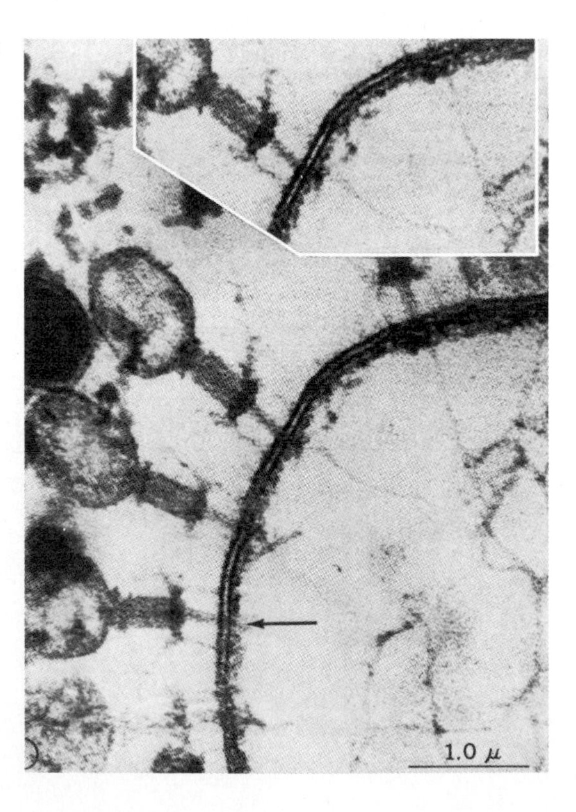

FIGURE 8.23

Stages of T4 phage attachment and DNA injection. (a) Schematic presentation. (b) Thin-section electron microscopic image that shows the entry of the tail tube into the bacterium (at arrow) and the thin threads of injected DNA (particularly in insert). (From C. K. Mathews. (1977) In *Comprehensive virology,* ed. H. Fraenkel-Conrat and R. R. Wagner. Vol. 7. Plenum Press, New York)

Virus Entry The physical appearance of the tail with its appendages, as well as the number of different proteins that it contains indicate that the tail must be an organ of considerable functional sophistication. The tail fibers are normally held near the phage head by the whiskers and their distal ends represent the site of chemical affinity to cell wall lipoproteins in the outer membrane of the bacterium. Contact by any one fiber may suffice for primary attachment, but at this stage, environmental changes can cause the phage to release and abort infection. Subsequently, the other tail fibers neatly set the phage in a perpendicular manner on the surface, and a change in the angle of the two parts of the fiber brings the base plate into contact with the cell surface. At that stage the infection process can no longer be aborted. Attachment leads to considerable conformational changes of the proteins of both the sheath and the base plate and causes the tail to contract. The changes in the base plate enlarge a central hole so that the tail tube can pass through it and pierce the cell wall. The end result is injection of the head's contents, mostly DNA, into the cell. Figure 8.23 illustrates some of these processes.

Replication Cycle The DNA of the T-even phages may contain as many as 160 genes, of which over 120 have been identified. These are usually presented on a circular map, because the permuted state of the DNA makes it functionally circular, even though it is physically linear. Fig. 8.24 illustrates the high level of evolution that these phages have reached and the amount of work done to elucidate their genome structure.

Inspection of the map reveals the following: (1) Genes of similar numbers and related in functions are located near one another on the map. This phenomenon has already been encountered with simpler viruses. (2) Almost half of the genome is needed for nonstructural proteins, such as enzymes, enzyme modifying agents, and regulatory proteins.

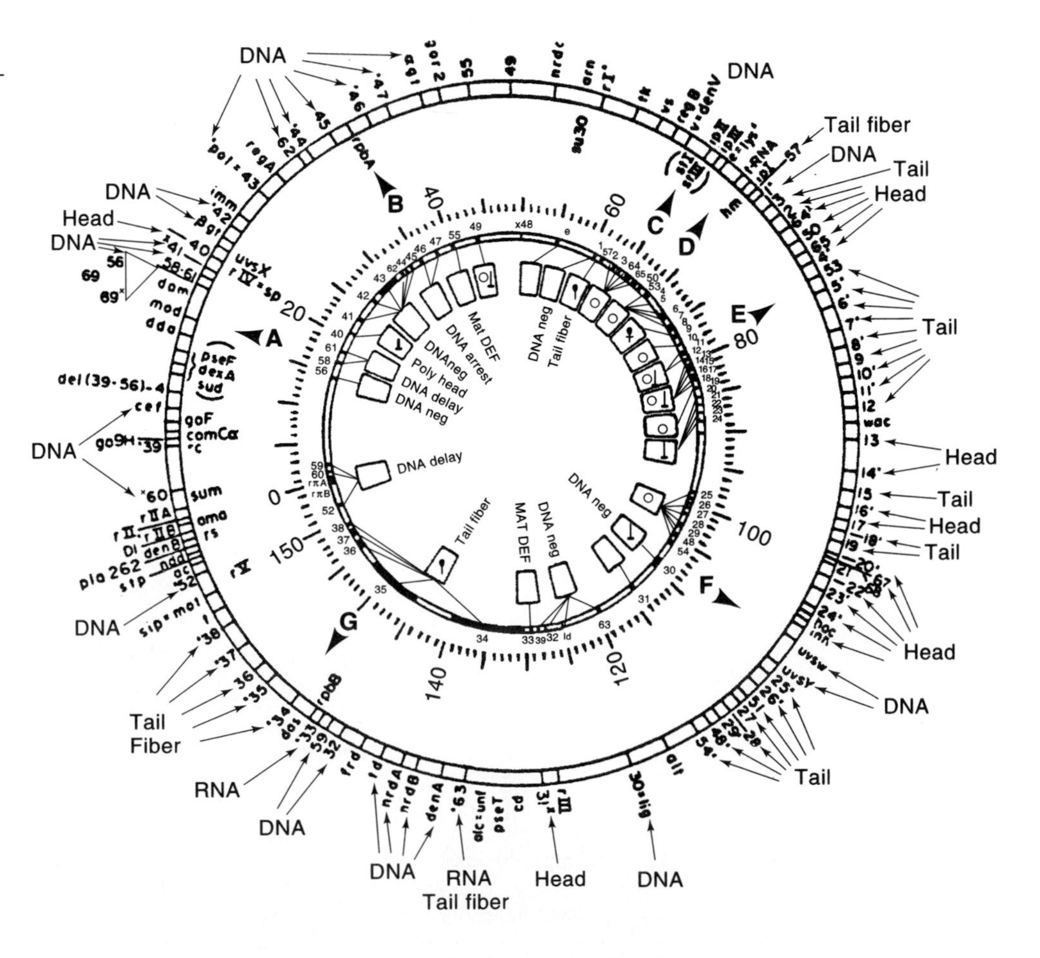

FIGURE 8.24

A map of the known T4 genes (Table I). Distances are based on recombination frequencies (inner circle; Edgar and Wood, 1966), on electron microscopy and restriction enzyme analysis (middle circle; Kutter et al., 1987), and on the probability of packaging cuts (outer circle; Mosig, 1968). The origins of DNA replication are indicated by large capital letters. (Courtesy of G. Moskik)

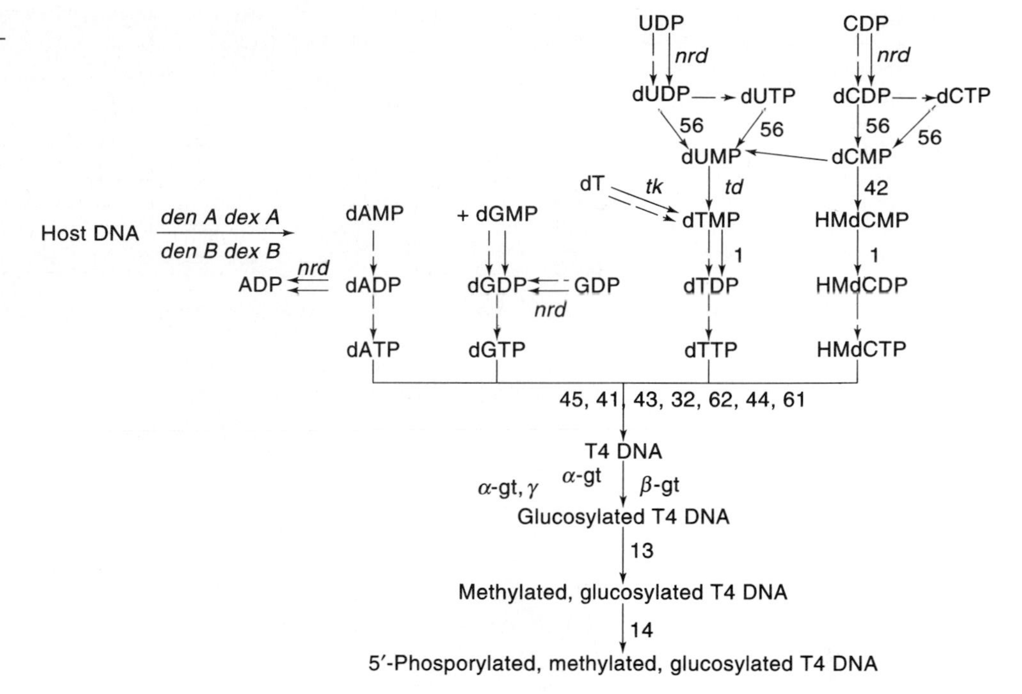

FIGURE 8.25

Metabolic pathways required for T-even phage DNA synthesis. Dashed arrows are activities present in *E. coli.* Those supplied by the phage genome (solid arrows) are identified by gene names or numbers. (Modified from C. K. Mathews. (1977) In *Comprehensive virology,* ed. H. Fraenkel-Conrat and R. R. Wagner. Vol. 7. Plenum Press, New York)

Not only are the genes of such functional groups close to one another, but the cooperating enzymes also tend to become associated with one another in particulate complexes. It is noteworthy that despite their great number of genes, these phages, like other viruses, make extensive use of their host's cellular machinery. Thus, the early phase of T-even phage development, is dependent upon *E. coli* RNA polymerase to synthesize phage mRNAs. The *E. coli* translation apparatus is needed throughout. Soon after infection, the phage does shut down transcription of the host cell DNA, thereby making sure that all the mRNAs translated in the cell are from then on its own.

Finally, as with all more complex viruses, genes are transcribed in a definite and advantageous order or program. Thus, the replication cycle of these phages, only 20 to 30 minutes long, is usually described as immediate-early, delayed-early, quasi-late, and true-late. Among the very early functions is the production of a phage-specific DNA polymerase. Enzymes are also made early that degrade the host cell DNA and bring the resultant deoxynucleotides to the triphosphate level. The formation of HMC and its triphosphate is concomitant with the deamination of any excess deoxycytidylic acid. The enzyme biochemistry of T-even phages is summarized in Fig. 8.25.

Virion Assembly The assembly of T4 phage particles has been intensely studied, and these studies have been greatly facilitated by the availability of many mutants that are defective in different components of the assembly process of the phage. Table 8.3 lists the genes that play a part in T4 virion assembly. Extracts from cells infected with these mutants may contain an accumulation of an assembly. Intermediate as illustrated in Figure 8.26. By mixing extracts of *E. coli* infected with different mutant phages and observing the

TABLE 8.3

T4 genes involved in virion structure and assembly

Class	*Gene*	*Function*
A	34–38, 57, 63	Tail fiber and tail fiber joining to tail plate
B	5–12, 25–29, 48, 51, 53, 54	Tail plate and tail plate assembly
C	3, 18, 19	Tail core and sheath
D	2, 4, 13–15, 50, 64, 65, *wac*	Collar, head–tail joining, head completion
E	16, 17, 20–24, 31, 40, 49, *hoc, soc*	Head and head filling

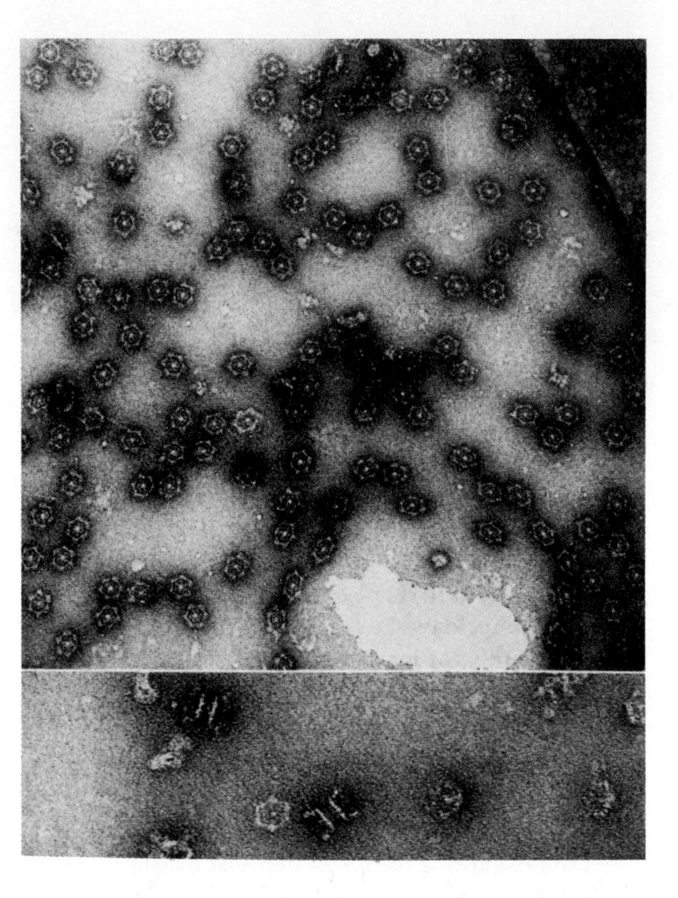

FIGURE 8.26

Electron micrographs of base plates from T4 mutant-infected cells. The upper panel shows precursor base plates from cells infected with a mutant defective in gene 19, which codes for the major tail-tube subunit. The side-to-side base-plate diameter is about 40 nm. The lower panel shows dimers of base-plate precursors joined by their inner faces and lying on their sides to reveal the spikes that are prominent on the mature phage. These base plates apparently dimerize because of lack of gene product 48, needed for initiation of tail-tube polymerization. [From W. B. Wood and J. King. (1979) In *Comprehensive virology,* ed. H. Fraenkel-Conrat and R. R. Wagner. Vol. 13. Plenum Press, New York]

process of *in vitro* phage assembly and its arrest at various stages by electron microscopy and other techniques, a remarkably clear picture of this complex process has been obtained for the T-even and certain other phages (Fig. 8.27).

Phage assembly is a much more complex process than the assembly of simple RNA viruses like TMV (see Fig. 2.14). While many phage genes code for structural proteins, others specify various enzymatic activities and **scaffolding proteins** required for particular assembly. These various components do *not* associate with one another at random. Instead, assembly follows a morphogenetic pattern in which each step provides the substrate for the next. In order, the intermediate stages of T-even phage assembly are:

1. Assembly of the base plate, leading to
2. Assembly of the tail tube and sheath
3. A separately initiated assembly of the head
4. An attachment of tail to head, and finally
5. An attachment of the separately assembled tail fibers to that particle.

As shown in Fig. 8.27 base-plate assembly is probably the most complex assembly process in virology. As many as 20 proteins are involved and certain enzymatic activities, namely, **dihydrofolate reductase** and **thymidylate synthetase**, remain associated with the final products. Assembly proceeds by a process of helical aggregation similar to that which produces rod-shaped viruses, but what limits the phage tail assembly to a specific length remains unknown. In the rod-shaped or fibrous viruses it is the length of the internal nucleic acid molecule that determines the length of the virus particle, but in the case of the phage tails the length-determining component is unknown. The prohead is formed largely from the ultimate capsid protein gp 23 and the assembly core protein gp 22. The main internal protein IPIII also enters the prohead during the early stage of assembly. Gp 23 is then

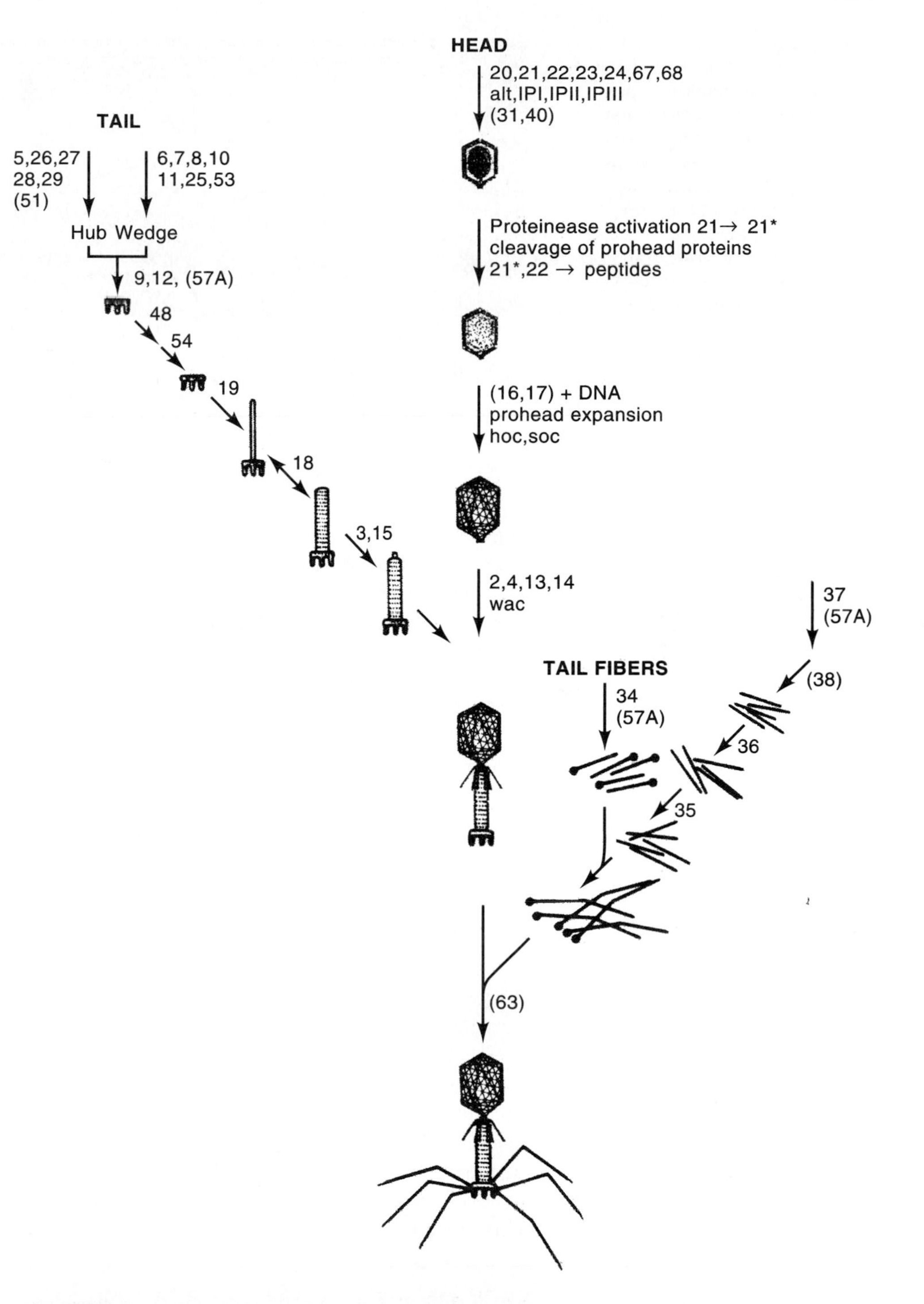

FIGURE 8.27

Pathway of gene product interaction in T4 tail assembly. The two major structural intermediates in base-plate assembly are shown schematically as a wedge-shaped arm and a circular plug. Sedimentation coefficients are shown beneath each intermediate. Gene products (gp) listed inside each intermediate represent those proteins found in the structure. Proteins shown in boldface type are those that contribute the major portion of the mass of the completed structure. The two columns at the lower right indicate the gene products present in the completed base plate and in the completed tail, respectively. Most of the proteins in each of the two pathways are specified by clusters of neighboring genes. [From W. B. Wood and J. King. (1979) In *Comprehensive virology*. ed. H. Fraenkel-Conrat and R. R. Wagner. Vol. 13. Plenum Press, New York]

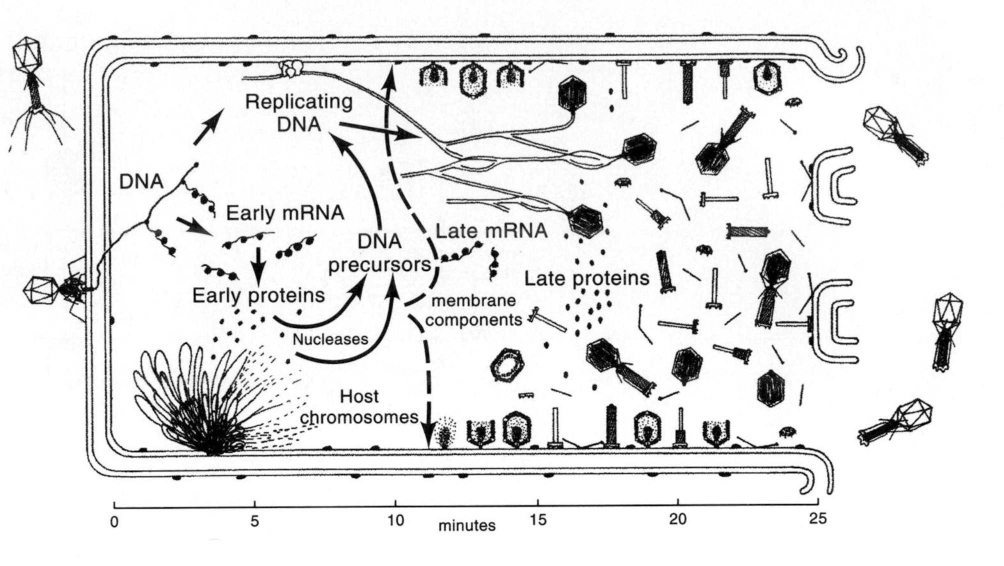

FIGURE 8.28

Schematic presentation of the infection cycle of T4. Note that host cell DNA breakdown provides some DNA precursors; that the replicating DNA is much longer than virion DNA; that glucose residues are added after polymerization of DNA; that several phage-coded proteins become associated with the cytoplasmic membrane of the host; and that maturation of the head occurs at a membrane site. (From C. K. Mathews. (1977) In *Comprehensive virology,* ed. H. Fraenkel-Conrat and R. R. Wagner. Vol. 7. Plenum Press, New York)

cleaved to gp 23*, and gp 22 is degraded completely. At least four additional proteins play a role in the maturation of the phage head.

The attachment of the tail to the head (step 4) appears to be a spontaneous event, requiring no enzyme action. The attachment of the fibers (step 5), however, requires the presumably enzymatic action of gp 63. Since this occurs only after the head-to-tail attachment, it appears that the whiskers that are part of the head must play a role in positioning the tail fiber properly.

The incorporation of the DNA into the phage head is the least well understood step in the T-even phage assembly. Considerable conformational change is required for any nucleic acid molecule to overcome the ionic repulsion inherent in packaging it into the limited space inside the virion. In the case of the DNA phages, most of the nucleic acid becomes an orderly folded coil in the course of this process and the prohead enlarges changes and from a spherical to a more icosahedral shape. In most phages, certain proteins leave the head or become proteolytically altered or degraded as the DNA enters. Proteolytic processing of the ultimate capsid proteins also frequently accompanies this maturation step. What forces make the DNA enter the head of the T-even phages as well as what forces make it later leave the head upon transfer to the host cell are unclear.

Lysis The T-even viruses carry a gene that codes for **lysozyme,** a lytic enzyme. The function of this enzyme is to attack and weaken the peptidoglycan layer in the bacterial cell wall. Lysozyme action is not essential for phage development, since mutants lacking the lysozyme function are also viable. Several other genes are involved in the timing of this event and consequent lysis of the cell. The replicative cycle of the T-even phages is schematically presented in Figure 8.28.

SECTION 8.7.4 FURTHER READING

Original Papers

Ellis, E. L., and M. Delbrück. (1939) The growth of bacteriophage. *J. Gen. Physiol.* 22:365.

Streisinger, G., et al. (1964) Chromosome structure in phage T4. I. Circularity of the linkage map. *Proc. Natl. Acad. Sci. USA* 51:775.

Edgar, R. S., and W. B. Wood. (1966) Morphogenesis of bacteriophage T4 in extracts of mutant-infected cells. *Proc. Natl. Acad. Sci. USA* 55:498.

Simon, L. D., and T. F. Anderson. (1967) The infection of *Escherichia coli* by T2 and T4 bacteriophages as seen in the electron microscope. I. Attachment and penetration. *Virology* 32:279.

Edgar, R. S., and I. Lielausis. (1968) Some steps in the assembly of bacteriophage T4. *J. Mol. Biol.* 32:263.

FARID, S. A. A., and L. M. KOZLOFF. (1968) Number of polypeptide components in bacteriophage T2L contractile sheaths. *J. Virol.* 2:308.

INOUYE, M., and A. TSUGITA. (1968) Amino acid sequence of T2 phage lysozyme. *J. Mol. Biol.* 37:213.

TSUGITA, A., and M. INOUYE. (1968) Complete primary structure of phage lysozyme from *Escherichia coli* T4. *J. Mol. Biol.* 37:201.

KIKUCHI, Y., and J. KING. (1975) Genetic control of bacteriophage T4 baseplate morphogenesis. I. Sequential assembly of the major precursor, *in vivo* and *in vitro. J. Mol. Biol.* 99:645.

ISHII, T., and M. YANAGIDA. (1977) The two dispensable structural proteins (soc and hoc) of the T4 phage capsid: Their purification and their binding with the defective heads *in vitro. J. Mol. Biol.* 109:487.

KING, J., et al. (1978) Control of the synthesis of phage P22 scaffolding protein is coupled to capsid assembly. *Cell* 15:551.

ARISAKA, F., J. TSCHOPP, R. VAN DRIEL, and J. ENGLE. (1979) Reassembly of the bacteriophage T4 tail from the core-baseplate and the monomeric sheath protein P18: A cooperative association process. *J. Mol. Biol.* 132:369.

DREXLER, K., et al. (1986) Morphogenesis of the long tail fibers of bacteriophage T2 involved in proteolytic processing of the polypeptide (gp 37) constituting the distal part of the fiber. *J. Mol. Biol.* 191:267.

LIN, T.-C., et al. (1987) Cloning and expression of T4 DNA polymerase. *Proc. Nat. Acad. Sci. USA* 84:7000.

HINTERMANN, E., and A. KUHN. (1992) Bacteriophage T4 gene 21 encodes two proteins essential to phage maturation. *Virology* 189:474.

HONG, Y. R., and L. W. BLACK. (1993) Protein folding studies *in vivo* with a bacteriophage T4 expression-packaging-processing vector that delivers encapsidated fusion proteins into bacteria. *Virology* 194:481.

MAKHOV, A. M., et al. (1993) The short tail-fiber of bacteriophage T4: Molecular structure and a mechanism for its conformational transition. *Virology* 194:117.

Books and Reviews

EISERLING, F. A. (1979) Bacteriophage structure. In *Comprehensive virology.* ed. H. Fraenkel-Conrat and R. R. Wagner. Vol. 13, chap. 8, 543–80. Plenum Press, New York.

MATHEWS, C. K. (1979) Production of large virulent bacteriophages. In *Comprehensive virology.* ed. H. Fraenkel-Conrat and R. R. Wagner. Vol. 7, 179–294. Plenum Press, New York.

WOOD, W. B., and J. KING. (1979) Genetic control of complex bacteriophage assembly. In *Comprehensive virology.* ed. H. Fraenkel-Conrat and R. R. Wagner. Vol. 13, chap. 9, 581–633. Plenum Press, New York.

MATHEWS, C. K., et al. (1983) *Bacteriophage T4.* American Society for Microbiology, Washington, D.C.

KARAN, J. D. (Ed.) (1994) *Molecular biology of bacteriophage* T4. American Society for Microbiology, Washington, D.C.

chapter

9

Subviral Pathogens and Other Viruslike Infectious Agents

Viruses were initially regarded in essentially negative terms; they were *not* visible in the microsope, *not* retained by bacterial filters, and *not* cultivatable on known microbiological media (see Sec. 1.2.2). To appreciate the differences between viruses and the several types of subviral pathogens to be discussed in this chapter, it may be helpful to realize that these latter agents have also been identified and studied on the basis of their differences from conventional viruses.

In particular, recall that virus replication is directed by a genomic nucleic acid and that introduction of this genome into a suitable host cell leads to the production of new virus particles. Among the subviral pathogens, only viroids and prions replicate independently; the prion particle, however, appears to lack a genomic nucleic acid. Satellite viruses and satellite RNAs contain conventional nucleic acid genomes, but their replication, like that of defective interfering (DI) RNAs (see Sec. 4.1.4) is dependent on the presence of a **helper virus.** Unlike DI RNAs, however, satellite viruses and RNAs do not exhibit substantial sequence homology with their helper viruses. In certain cases, expression of their genetic information also requires the synthesis of satellite-specific protein(s). The properties of defective interfering particles, viral satellites, and viroids are compared in Table 9.1.

TABLE 9.1
Comparison of defective interfering particles, viral satellites, satellite RNAs, and viroids

Property	*DI Particle*	*Satellite Virus*	*satRNA*	*Viroid*
Replicates only in presence of helper	Yes	Yes	Yes	No
Encapsidated in specific coat	No	Yes	No	No
Encapsidated in helper coat	Yes[a]	No	Yes	No
Suppresses replication of helper	Yes	Yes	Yes	—
Sequence homology with helper	Yes	No[b]	No[c]	—
RNA stability *in vivo* and *in vitro*	Low	Low	High	High

[a]Variable, depending on the location of the deletion. Polio DI RNA is replicated but not encapsidated (see Sec. 2.2).

[b]The 3′-termini of STMV and its helper virus exhibit substantial sequence and structural similarities.

[c]Certain chimeric satRNAs may exhibit extensive 3′ sequence homologies with the genomes of their helper viruses.

TABLE 9.2
Plant satellite viruses

Virus	Genome Size (nucleotides)	Protein Products (amino acids)
Satellite tobacco necrosis virus	1239	195
Satellite tobacco mosaic virus	1059	159, 58
Satellite panicum mosaic virus	826	157
Satellite maize white line mosaic virus	1168	218

9.1 SATELLITE VIRUSES

Satellite viruses were first observed in plants. Their replication is dependent on the presence of a helper virus that provides the replicase, but the satellite virus is *not* required for helper virus replication. Furthermore, there is little or no sequence similarity between the genomes of satellite viruses (or most satellite RNAs) and those of their helper viruses. As shown in Table 9.2, all known plant satellite viruses encode their own capsid proteins. This interdependence is reminiscent of that between hepatitis delta virus and HBV, adeno-associated virus and adenovirus (see Sec. 8.1) as well as the P2–P4 system (see Sec. 9. 5.).

The helper virus for satellite tobacco necrosis virus (STNV), is tobacco necrosis virus (TNV), a typical small (30 nm) icosahedral plant virus. TNV normally infects plant roots, where it replicates independently of other viruses. Prior to 1962, certain TNV isolates were known to contain substantial amounts of a smaller, virus-like particle. In that year Kassanis showed that replication of this 18 nm particle was dependent on the presence of TNV. The STNV genome contains only 1239 nt and appears to act as a **monocistronic mRNA** for the synthesis of its 22 kDa coat protein.

The genome of a second satellite virus, satellite tobacco mosaic virus (STMV; Fig. 9.1) is somewhat smaller and contains two ORFs. *In vitro* translation of STMV RNA in either a rabbit reticulocyte or wheat germ cell-free system yields two major products—the 17.5 kDa STMV coat protein plus a serologically unrelated 6.8 kDa polypeptide derived from the 5′ ORF. It is not known whether or not this smaller polypeptide is synthesized *in vivo*, and its role in STNV replication (if any) remains to be determined.

STNV is an **obligatory parasite** of its helper, dependent for its replication on the presence of TNV in the cells it enters. The specificity of this dependence is illustrated by (1) the inability of plant viruses other than TNV to act as helper and (2) variation in the ability of certain TNV strains to support the replication of different strains of STNV. Although the exact nature of this interdependence is unknown, it is assumed that the RNA replicase of TNV is also able (and necessary) to replicate STNV. Replication of TNV is

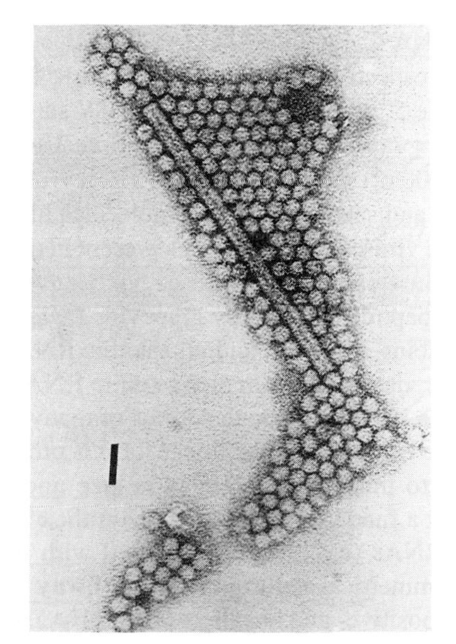

FIGURE 9.1

Electron micrograph of a negatively stained preparation of satellite tobacco mosaic virus (STMV). Note the difference in size and morphology between the small (17 nm) icosahedral particles of STMV and the long (18 × 300 nm) helical rods of its helper, tomato mild green mosaic virus (TMGMV). (From Valverde and Dodds. (1987) *J. Gen. Virol.* 68:965)

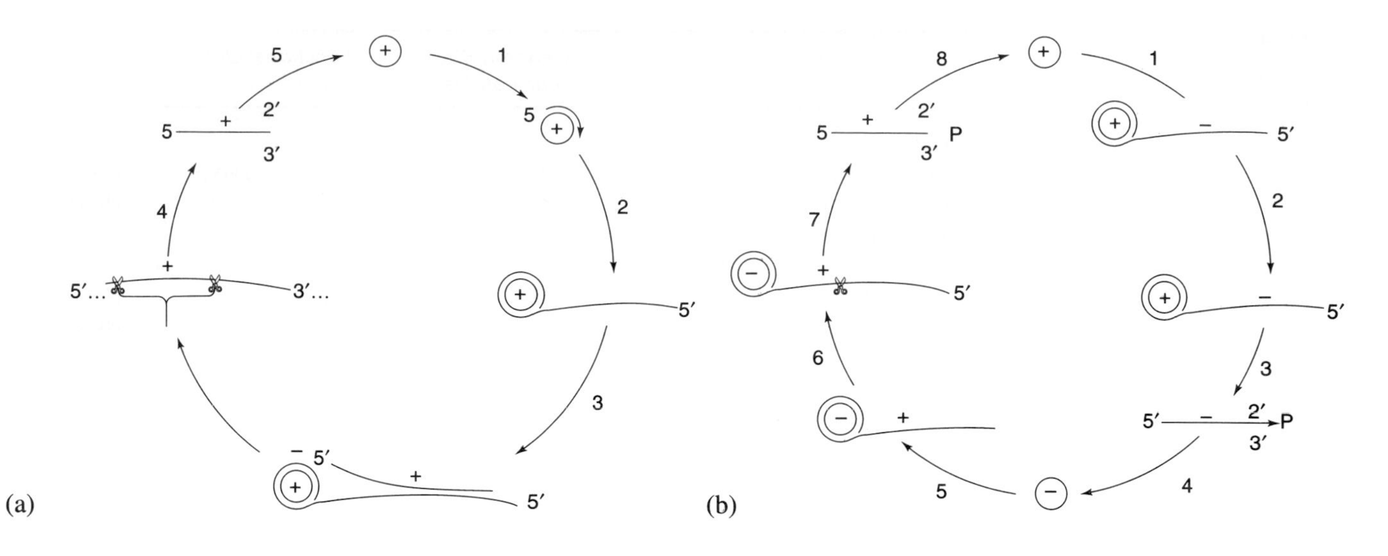

FIGURE 9.4

Asymmetrical and symmetrical models for rolling-circle replication. Beginning at the top of each diagram, an infecting positive-sense RNA (+) becomes a template for negative-sense RNA (–) synthesis. In the asymmetrical cycle (a), the resulting multimeric negative-sense RNA is copied into a multimeric positive-sense precursor (step 3) that is then cleaved (step 4) and circularized (step 5). In the symmetrical cycle (b), the multimeric negative-sense RNA is first cleaved to unit length (step 3) and circularized (step 4). The resulting circular negative-sense RNA then serves as a rolling-circle template for synthesis of the multimeric positive-sense precursor. (From Branch and Robertson. (1984) *Science* 223:450)

to release monomeric linear molecules. Circularization of this linear, negative-sense RNA provides the template for synthesis of the positive-sense progeny. Self-cleavage of both positive- and negative-sense strands may proceed via formation of a common "hammerhead" structure. Cleavage of negative-sense sTobRSV multimers, however, involves a completely different structure (see Fig. 9.5). A second group of satellite RNAs contains both large and small RNAs, but none of these molecules are able to undergo spontaneous self-cleavage. Instead, satellite RNAs such as CARNA 5 appear to use a replicative pathway that is virtually identical with that used by their helper viruses.

The molecular mechanisms responsible for these self-cleavage reactions, as well as the overall structural organization of satellite RNAs, are currently under intensive investigation. The results of these studies lie beyond the scope of our discussion, but it may soon be possible to assign biological (as well as biochemical) functions to certain structural features of satellite RNAs. We again consider the ability of certain RNA molecules to spontaneously **self-cleave** in Chapter 13, when we discuss the possible use of "**ribozymes**" to engineer virus disease resistance.

The inability of satellite RNAs to replicate in the absence of their helper viruses and their often pronounced preference for certain helper virus strains point to the involvement of one or more virus-encoded proteins in satellite RNA replication. Also the linear, double-stranded form of CARNA 5 that accumulates in infected tissue *in vivo* contains an unpaired guanosine residue at the 3′ terminus of the negative-sense strand—just like the corresponding RFs for the three CMV genomic RNAs. The nature of this dependence remained a matter for speculation, however, until purified preparations of CMV RNA replicase were shown to catalyze the replication of its satellite RNA *in vitro;* that is, the synthesis of full-length positive- as well as negative-sense RNAs starting from a CARNA 5 positive-sense RNA template.

Biological Activities of Satellite RNAs The presence of a satellite RNA often has a dramatic effect on the symptoms observed in the infected host. For example, Figure 9.6 illustrates the lethal necrosis induced by certain CARNA 5 sequence variants in

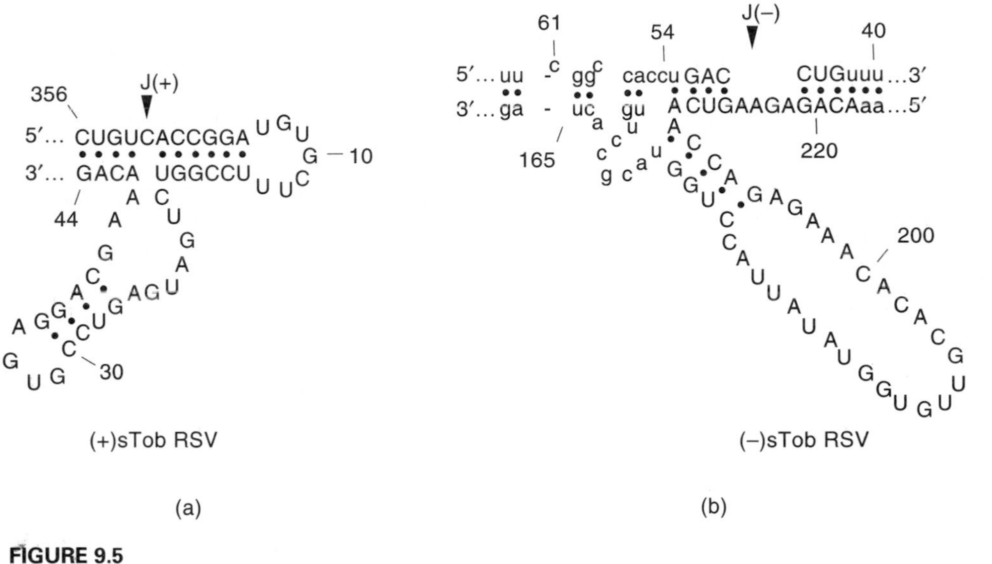

FIGURE 9.5

Secondary structures for the self-cleaving sequences of sTobRSV positive- and negative-sense RNAs. The satellite RNA associated with the budblight strain of TobRSV contains 359 nucleotides. (a) Cleavage of the positive-sense RNA occurs between C359 and A1 within a typical "hammerhead" cleavage site, which contains approximately 50 contiguous nucleotides. (b) Cleavage of the negative-sense RNA occurs between positions 48 and 49 in a "paper clip" structure formed from nucleotides derived from two separate portions of the satellite genome. Lowercase letters designate nucleotides that are not essential for the self-cleavage reaction. (From G. Bruening. (1990) *Seminars in virology.* 1:127)

tomato. Just as striking, however, is the **attenuation** of symptoms observed when the host plant is tabasco pepper. Many factors other than the choice of assay host—sequence changes in the satellite RNA, use of a different helper virus strain, and changes in environmental conditions, to name but a few, can produce a similar "Dr. Jekyll and Mr. Hyde" effect on symptom expression. The presence of a satellite RNA often depresses replication of its helper virus, but the large nepovirus satellite RNAs usually have little or no effect on either helper virus accumulation or symptom expression.

FIGURE 9.6

Modulation of cucumber mosaic virus disease symptoms by its satellite RNA. The upper row of plants was infected by virus alone, the lower row by virus plus CARNA 5. In tabasco pepper (left), CARNA 5 attenuates disease symptoms; in tomato (right), CARNA 5 induces a lethal necrosis. (Courtesy of J. M. Kaper)

FIGURE 9.7

Schematic representation of two CMV satellite RNA sequences: Y-CARNA 5, which is incapable of inducing lethal necrosis, and D-CARNA 5, which is lethal to tomato within 3 weeks (see Fig. 9.6). The two satellites have large regions of identical sequence (solid lines) that include the 3′ "necrogenic determinant" (solid block). The 5′ portion of this sequence homology however is interrupted by the replacement of an approximately 70 nucleotide (nt) stretch of D-CARNA 5 by about 100 nt of unrelated sequence in Y-CARNA 5 (shaded block). Other sequence differences that may influence necrogenicity differences between D- and Y-CARNA 5 are shown in the 3′ half of Y-CARNA 5. The arrow points to a position where site-directed mutagenesis had a major effect on necrogenic expression. (Courtesy of J. M. Kaper)

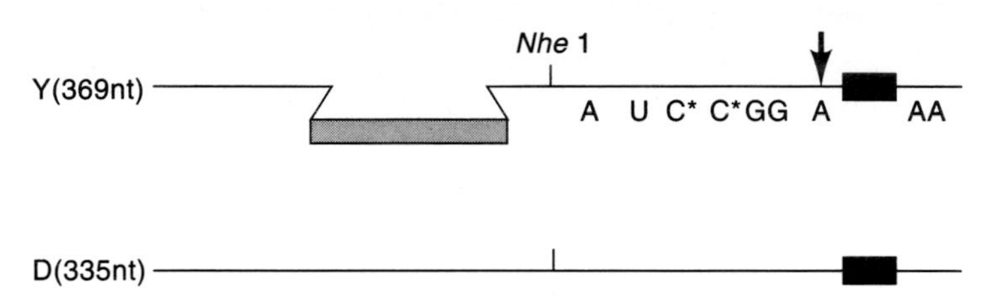

How do satellite RNAs exert their often dramatic biological effects? By comparing the nucleotide sequences of CARNA 5 variants that are able to induce a lethal necrotic response with those of variants that lack this ability, a "necrogenic determinant" within the 3′ portion of the molecule has been identified. The role of individual nucleotides within this determinant has been further investigated by **site-directed mutagenesis.** As one might expect, however, even this comparatively simple "all or none" response has proved much more complicated than initially anticipated.

As shown in Figure 9.7, characterization of additional CARNA 5 sequence variants has revealed the presence of another necrosis "determinant" in the 5′ half of the molecule. Participation of this determinant in the necrotic response has been confirmed by constructing the appropriate satellite chimeras and comparing their biological properties. However, the absence of a reliable model for the structure of CARNA 5 has frustrated further progress toward understanding how the interactions between CARNA 5, its helper virus, and the host plant modulate disease development.

9.3 VIROIDS

Viroids are the smallest known agents of infectious disease—small (246–375 nucleotides), highly structured, single-stranded RNA molecules lacking both a protein capsid and detectable messenger RNA activity. Whereas viruses have been termed "obligate parasites of the cell's translational system" and supply some or most of the genetic information required for their replication, viroids can be described as "obligate parasites of the cell's transcriptional machinery."

The first viroid disease to be studied by plant pathologists was *potato spindle tuber disease.* In 1923 its infectious nature and ability to spread in the field led Schulz and Folsom to group potato spindle tuber disease with several other "degeneration diseases" of potatoes. Most of these maladies—long attributed to senility, "reversion," or loss of vigor caused by prolonged asexual reproduction—are now known to be caused by infection with conventional RNA viruses. And, nearly 50 years were to elapse between the first published descriptions of potato spindle tuber disease and Diener's 1971 demonstration of the fundamental differences between the structure and properties of its causative agent, **potato spindle tuber viroid** (PSTVd), and those of conventional plant viruses:

1. Viroids exist *in vivo* as nonencapsidated, low-molecular-weight RNAs;
2. Infected tissues do not contain virus-like particles;
3. Only a single species of low-molecular-weight RNA is required for infectivity;
4. Viroids do not code for any proteins;
5. Despite their small size, viroids are replicated autonomously in susceptible cells (i.e., no helper virus is required).

Although these criteria alone provide ample justification to classify viroids as a taxon separate from viruses, subsequent studies of their structural and functional properties have shown this disparity to be even greater than initially imagined. The recent decision to add a "d" to all viroid acronyms (i.e., PSTVd rather than PSTV for potato spindle tuber viroid) was intended to draw attention to these differences.

9.3.1 Genome Structure and Classification

Efforts to understand how viroids replicate and cause disease without the assistance of any viroid-encoded polypeptides prompted detailed analysis of their structure. Viroids possess rather unusual properties for single-stranded RNAs (e.g., a pronounced resistance to digestion by ribonuclease and a highly cooperative thermal denaturation profile), which led to the early realization that they might have an unusual higher-order structure. Their small size also made viroids tempting objects for detailed structural investigation. As a result, only the structure of tRNA is better understood.

To date, nearly 20 different viroid species plus a number of sequence variants have been completely sequenced (see Table 9.4). All known viroids are single-stranded circular RNAs that contain 246 to 375 unmodified nucleotides and appear to lack any unusual 2′, 5′-phosphodiester bonds or 2′-phosphate moieties. Electron microscopy, optical melting, and other physicochemical studies—as well as theoretical calculations of their lowest free energy secondary structures—indicate that most viroids assume a highly base-paired rod-like conformation *in vitro*. Figures 9.8 and 9.9 provide two different views of the so-called native structure of PSTVd.

TABLE 9.4
Viroid species of known nucleotide (nt) sequence

Grouping[a]	*Viroid*	*Abbreviation*	*Length (nt)*
Potato Spindle Tuber Viroids ("Pospiviroids")			
	Chrysanthemum stunt	CSVd	354, 356
	Citrus exocortis	CEVd	369–375
	Citrus viroid IV	CVd-IV	284
	Coconut cadang-cadang	CCCVd	246, 247
	Coconut tinangaja	CTiVd	254
	Columnea latent	CLVd	370
	Cucumber pale fruit[b]	CPFVd	303
	Hop latent	HLVd	256
	Hop stunt	HSVd	297–303
	Indian bunchy top[c]	CEVd-t	372
	Potato spindle tuber	PSTVd	356–360
	Tomato apical stunt	TASVd	360
	Tomato planta macho	TPMVd	360
Apple Scar Skin Viroids ("Apscaviroids")			
	Apple scar skin	ASSVd	330
	Australian grapevine	AGVd	369
	Citrus bent leaf	CBLVd	318
	Grapevine yellow speckle	GYSVd	367
	Grapevine yellow speckle-2	GYSVd-2	363
	Pear blister canker	PBCVd	315
***Coleus blumei* Viroid**			
	Coleus blumei	CbVd	248
Self-cleaving Viroids			
	Avocado sunblotch	ASBVd	246–250
	Peach latent mosaic	PLMVd	337
	Carnation stunt associated	CarSAVd	275

[a]Modified from those of Koltunow and Rezaian (1989) *Intervirology* 30:194. Groupings in parentheses are based on a phylogenetic analysis of viroids, viroidlike satellite RNAs, and the viroidlike domain of hepatitis δ virus RNA. (Elena et al. (1991) *Proc. Natl. Acad. Sci. USA* 88:5361)

[b]Actually a HSVd variant

[c]Actually a CEVd variant

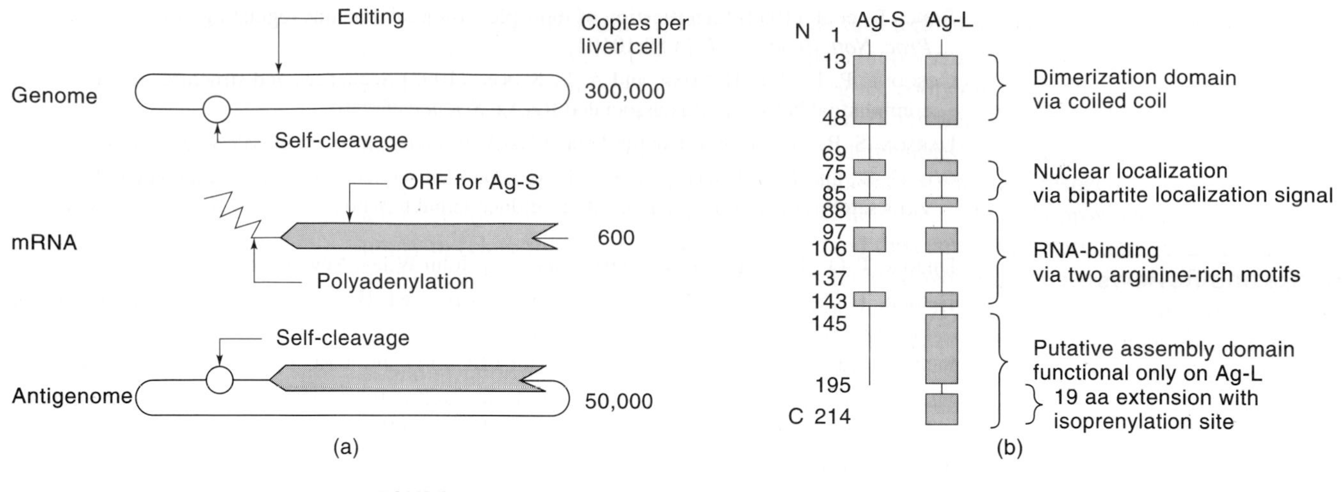

FIGURE 9.10

Structural and functional features of the HDV genome. (a) Features of the three RNAs of HDV. The genomic RNA and its complement, the antigenome, are shown in a circular rod-shaped conformation. The genome contains an indicated site of RNA editing, which as a result of RNA-dependent RNA synthesis, alters not only the antigenomic RNA but also the polyadenylated mRNA species. As a consequence, the termination codon is mutated in the indicated open reading frame (ORF) of the mRNA for δAg-S, so as to allow the synthesis of δAg-L. Both the genomic and antigenomic RNAs contain a single site, as indicated, at which self-cleavage and self-ligation occur. The relative abundance of the three RNAs in the liver of an infected animal, are indicated at the right side. (b) Features of the two forms of delta antigen. δAg-S and δAg-L are represented as containing a series of autonomously functioning domains. With respect to the packaging property of δAg-L, the 19 aa unique to this protein may activate an adjacent proline/glycine rich domain that is responsible for hepadnavirus sAg interaction by providing a site of isoprenyl addition. (Modified from Lazinski, D. W. and J. M. Taylor (1994) Advances in Virus Research, Academic Press, San Diego.)

WANG, K.-S., et al. (1986) Structure, sequence and expression of the hepatitis delta viral genome. *Nature* 323:508.

PONZETTO, A., et al. (1987) Titration of the infectivity of hepatitis D virus in chimpanzees. *J. Infect. Dis.* 155:172.

WEINER, A. J., et al. (1988) A single antigenomic open reading frame of the hepatitis delta virus encodes the epitope(s) of both hepatitis delta antigen polypeptides p24 and p7. *J. Virol.* 62:594.

KUO, M. Y.-P., et al. (1988) Characterization of self-cleaving RNA sequences on the genome and antigenome of human hepatitis delta virus. *J. Virol.* 62:4439.

KUO, M. Y.-P., M. CHAO, and J. M. TAYLOR. (1989) Initiation of replication of the human hepatitis delta virus genome from cloned cDNA: Role of delta antigen. *J. Virol.* 63:1945.

WU, H. N., and M. M. C. LAI. (1989) Reversible cleavage and ligation of hepatitis delta virus RNA. *Science* 243:652.

CASEY, J. L., et al. (1992) Structural requirements for RNA editing in hepatitis delta virus: Evidence for a uridine-to-cytidine editing mechanism. *Proc. Natl. Acad. Sci. USA* 89:7149.

RYU, W.-S., et al. (1993) Ribonucleoprotein complexes of hepatitis delta virus. *J. Virol.* 67:3281.

Reviews

BAROUDY, B. M., et al. (1988) Molecular biology of the hepatitis delta virus. In *Viral hepatitis and liver disease.* ed. A. J. Zuckerman. 395 Alan R. Liss, New York.

PURCELL, R. H., and J. L. GERIN. (1990) Hepatitis delta virus. In *Virology,* 2d ed., ed. B. N. Fields and D. M. Knipe Vol. 2, 2275–2287. Raven Press, New York.

TAYLOR, J. M. (1992) The structure and replication of hepatitis delta virus. *Ann. Rev. Microbiol.* 46:253.

LAZINSKI, D. W., and J. M. TAYLOR. (1994) Recent developments in hepatitis delta virus research. In *Adv. Virus Res.* 43. 187–231. (in Press).

FIGURE 9.11

Electron micrograph of a mixture of P2 and P4 bacteriophages. (Courtesy of R. C. Williams)

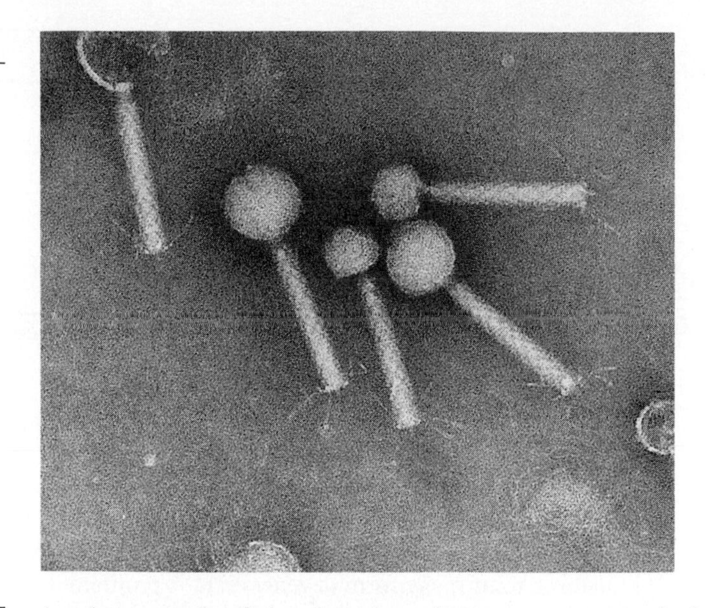

9.5 COLIPHAGES P2 AND P4

Another example of virus interdependency occurs among the bacteriophages. For this reason, the P2 and P4 phages are discussed in this chapter. Bacteriophage P2 is a temperate phage, a member of the **Myoviridae,** with a contractile tail and tail fibers. Its DNA genome (23×10^6 Da) has cohesive termini 19 nucleotides long. Its mode of productive replication differs from that of other lysogenic phages, such as lambda, in that it lacks the bidirectional (or θ) phase of DNA replication. During the σ stage of replication, its DNA tail stops after one round—similar to the situation with ϑX174. P2 is also of interest for another reason: It acts as an essential helper in the replication of a *satellite* or *parasitic phage* known as P4. As shown in Figure 9.11, P4 is similar in appearance to P2 but has a smaller head (45 vs. 60 nm in diameter). Both phages have 135 nm contractile tails.

The P4 virion contains the same proteins as that of P2 but relatively fewer of the main head proteins, and certain P2 proteins are required for P4 maturation. The DNA genome of P4 is only one-third as long as that of P2, but has the same cohesive ends. It can infect and lysogenize *E. coli,* but lacking essential structural genes, P4 is replicated only in the presence of P2. Surprisingly, the mode of P4 DNA replication differs from that of P2 DNA in that it proceeds bidirectionally from one origin. It also appears to utilize functions other than the usual *E. coli* enzymes and gene products required for DNA replication.

SECTION 9.5 FURTHER READING

FLENSBURG, J., and R. CALENDAR. (1987) Bacteriophage P4 DNA replication. Nucleotide sequence of the P4 primase gene and the *cis* replication region. *J. Mol. Biol.* 195:439.

BERTANI, L. E., and E. W. SIX. (1988) The P2-like phages and their parasite, P4. In *The bacteriophages.* ed. R. Calender. Vol. 2, 73. Plenum Press, New York.

9.6 PRIONS

A group of infectious agents, tentatively termed **prions** (*pro*teinaceous *in*fectious agents), exhibit properties that distinguish them from both viruses and viroids. The only macromolecule found to be associated with infectivity in sheep suffering from scrapie, and in humans with Creutzfeldt-Jakob disease, was a protein of about 27 kDa present in amyloid fibrils in the brain. Extensive studies on this agent have so far failed to identify any nucleic acid associated with infectious materials, yet biochemical studies to date cannot conclusively rule out the requirement for nucleic acid for infectivity. Whether or not prions contain RNA or DNA, their known biological and physical properties can best be described and appreciated in comparison with those of other small infectious agents (e.g., RNA species) described in the preceding sections.

TABLE 9.5
Diseases thought to be caused by prions[a]

Disease	*Natural host*
Scrapie	Sheep and goats
Transmissible mink encephalopathy (TME)	Mink
Chronic wasting disease	Mule deer and elk
Kuru	Humans (Foré tribe)
Creutzfeldt-Jakob disease (CJD)	Humans
Gerstmann-Straussler syndrome (GSS)	Humans
Fatal familial insomnia	Humans

[a]Alternative terminologies include *subacute transmissible spongiform encephalopathies* and *unconventional slow virus diseases.*

There are now seven diseases of animals or humans that are thought to be caused by prions (see Table 9.5). In several cases, little physical data about the infectious agent exist, and the classification is essentially based on similarities in epidemiologic and pathological findings. Both genetic and sporadic forms of the diseases have been categorized by some as **spongiform encephalopathies,** because they are associated with a loss of neurons resulting from spongiform degeneration. A proliferation of astrocytes can also be observed, and myeloidlike plaques often appear. Most notable is the absence of any detectable immune response in affected hosts, and the pathological effects are confined to the nervous system. Incubation periods can be months to decades prior to clinical illness, and a progressive clinical course of weeks to years invariably leads to death. The diseases all show specific cytopathic effects on selected types of brain cells. The agent can also be found without pathologic effects in lymph tissues and salivary glands.

Scrapie, which was described in sheep and goats over 50 years ago, is still found in most parts of the world, despite attempts to eradicate the agent by destroying infected flocks. The disease is characterized by incoordination and unsteady gait, which becomes progressively worse until death. The infectious agent, in the form of extracts from infected brains, has been passed experimentally to mice, hamsters, ferrets, mink, and monkeys, but apparently is not infectious for chimpanzees or rabbits. The disease induced by the scrapie agent in mink is analogous to that in sheep and is indistinguishable from a transmissible mink encephalopathy (TME) found naturally in mink. Likewise, chronic wasting disease (CWD) of mule deer and elk is similar both in pathology and known physical properties to the scrapie agent. A recently-described condition in cattle, the so-called "mad cow disease," is believed caused by a prion transmitted in contaminated feed supplements derived from infected sheep and goats. And concern has been raised that this agent might cause disease in humans eating contaminated meat.

Kuru, the first slow infectious disease of humans to be identified, has been found in only a small region of the Eastern Highland Province of Papua New Guinea, chiefly in a tribe of people called the Foré. The incidence of disease in the Foré has dropped radically since they abandoned their practice of ritualistic cannibalism and the associated contact with infected tissue. It appears that the infectious agent of the disease was transmitted from generation to generation by consumption of the brains of deceased elders or rubbing the body with diseased tissue, customs that were thought to instill wisdom. The neurological symptoms of kuru in humans are similar to those in sheep with scrapie, but since kuru can be transmitted only to apes, research on the agent has been limited.

Creutzfeldt-Jakob disease (CJD), occurs rarely throughout the world, and typically presents as a dementia in either men or women at least 60 years old. Some cases are clustered in certain families, and the fact that some of these families also have an apparently higher incidence of Alzheimer's disease has led to the supposition that the two diseases may be related. Only CJD is readily transmissible to apes (and occasionally, rodents) however. CJD also appears to be transmissible to humans by contaminated products. A case was recently reported to of a young man who died of CJD that is believed to have been transmitted in preparations of **human growth hormone** administered during his childhood. So convinced were the officials at the U.S. National Institutes of Health that the infection came from the hormone, which was prepared from extracts of human pituitaries, that they banned further use of this material for clinical applications. This action served to accelerate government approval of human growth hormone produced by genetic engineering in *E. coli,*. This material had already undergone unusually protracted clinical trials.

Other reports of transmission of CJD have been linked to corneal transplants taken from individuals dying from this disease. Young recipients later developed neurological symptoms of CJD, and this agent, like scrapie, has been found to be extremely stable. Brains obtained from individuals who died from CJD and stored in formalin for over 10 years have been shown to contain the infectious agent when inoculated as extracts into rodents. Another human disease, which progresses more rapidly than CJD and is also transmissible to apes, is the **Gerstmann-Straussler syndrome (GSS).**

As yet there are no biochemical or cell culture assay systems for either scrapie agent or any other of these agents. Transmission of the disease to mice or hamsters allows end-point titration of biological activity to be performed, albeit slowly. With the finding that the interval from inoculation to onset of disease is inversely proportional to the dose of scrapie agent injected intracerebrally into hamsters, it has become possible to assay samples with the use of four animals in 60 to 70 days—provided that the titers of scrapie agent in the samples are sufficiently high. This assay has been used to help purify the infectious principle and investigate its structure on the molecular level.

Variants of the scrapie agent have been isolated from standard stocks by passage in mice and hamsters at limiting dilution. Some of these exhibit different neuropathological effects, including markedly different incubation times and tissue tropism. These "strains" seem to be stable and do not revert or interconvert with detectable frequency. The existence of such stable variants is most easily explained by the operation of a genetic mechanism based on the presence of a nucleic acid genome in the infectious agent. However as noted above, no such genome has been found and no viruslike particles have been observed in infectious samples.

The scrapie agent (like viruses) passes through filters of 22 to 100 nm. It can be inactivated by strong alkaline solutions (1 *M* NaOH) but is highly resistant to nonionic detergents. Its resistance to nucleases and relative sensitivity to proteases led to its consideration as an infectious protein [prion].

In sucrose gradients, infectivity comigrates with rod-shaped particles that resemble so-called **amyloid fibrils** found by electron microscopy in infected brains. Polyacrylamide gel electrophoresis of the proteins in the infectious fraction from sucrose gradients has revealed a single protein termed the *prion protein* (PrP) with a relative molecular weight of 27-30 kDa. Further analyses have revealed that the scrapie PrP or PrP^{sc} is a **sialoglycoprotein.**

A great advance in the field came with the derivation of a monoclonal antibody that recognizes PrP^{sc}. In addition to the prion-associated protein that is readily detected in infected tissue, a cross-reacting normal cellular analogue (PrP^{c}) has also been identified using this antibody. The latter protein has homologues in many animal species including mammalian and most likely avian.

Purification and partial sequence analysis of PrP^{sc} (Fig. 9.12) allowed the design of synthetic oligonucleotide DNA probes complementary to all possible codons that can encode the PrP amino acid sequence. These probes in turn, have led to the cloning of mRNAs encoding PrP and detection of such mRNAs in both normal and scrapie-infected hamster tissues. Because DNAs of both normal and infected cells contain a single copy of a gene for this mRNA, it appears that PrP is a component of normal brain tissue that remains closely associated with scrapie agent infectivity during extensive purification.

The amino acid sequence of the normal and scrapie PrP is the same and PrP^{c} expressed in uninfected cells is the precursor for PrP^{sc}. The ability of the PrP^{sc} to form aggregates of fibers in the brain distinguishes it from the normal protein. Moreover, the normal protein is slightly smaller and more easily digested by proteinase K than is PrP^{sc}.The **conformational** differences between the two proteins distinguish them and are determined

10.1 PROPERTIES OF TRANSFORMED CELLS IN CULTURE

10.1.1 Fibroblasts

Until recent years, about the only type of normal animal cell that could be cultured in the laboratory was the **fibroblast,** a relatively unspecialized component of all connective tissues in the body. Whole embryos as well as certain organs from adult animals, such as kidneys, are minced and digested with proteases to provide cell suspensions for cultures. Even though many different cell types obviously exist in such tissues, the resulting cultures generally produce only thin, flat, spindle-shaped fibroblasts, whereas cells of other types fail to survive or replicate, regardless of the tissue of origin. This selection of cell type results from the medium and culture conditions used, since they were initially developed to grow fibroblasts. Currently, new techniques with different culture media are leading to the successful cultivation of epithelial cells with various differentiation properties as well as the growth of white blood cells, hematopoietic stem cells, and others.

The most obvious sign of neoplastic transformation induced in fibroblasts by tumor viruses is morphological: The cells become rounded and more refractile under light (Fig. 10.1). Other changes in cell growth behavior are apparent, however. One of the most important for studying tumor viruses is the loss of what is termed **density-dependent inhibition** of cell division. Normal fibroblasts grow in culture attached to a surface, either glass or specially treated plastic; and they grow only until they form a confluent sheet with the thickness of a single cell, or a *monolayer.* In contrast, transformed fibroblasts do not stop dividing when a monolayer is achieved. Instead, they continue dividing, with the result that the cells form multilayered cultures with much higher densities of cells per unit surface area than are found in monolayers of normal cells. The density-dependent inhibition of cell division exhibited by normal fibroblasts is often referred to as *contact inhibition;* but strictly speaking, however, this term was devised to indicate that normal fibroblasts, which have a tendency toward ameboid movement in culture, cease to move when contact with another cell is established. Since it is still not clear that intercellular contact as opposed to close proximity is the actual cause of inhibition of cell division in dense monolayers of normal cells, the term **topoinhibition** (from the Greek word *topo* meaning "place") has been coined to provide a more neutral term than contact inhibition.

The combination of the morphological changes and the loss of topoinhibition is used to advantage in quantitative assays for transformation by tumor viruses. When only a few particles of transforming virus are added to a culture of many more fibroblasts, a limited

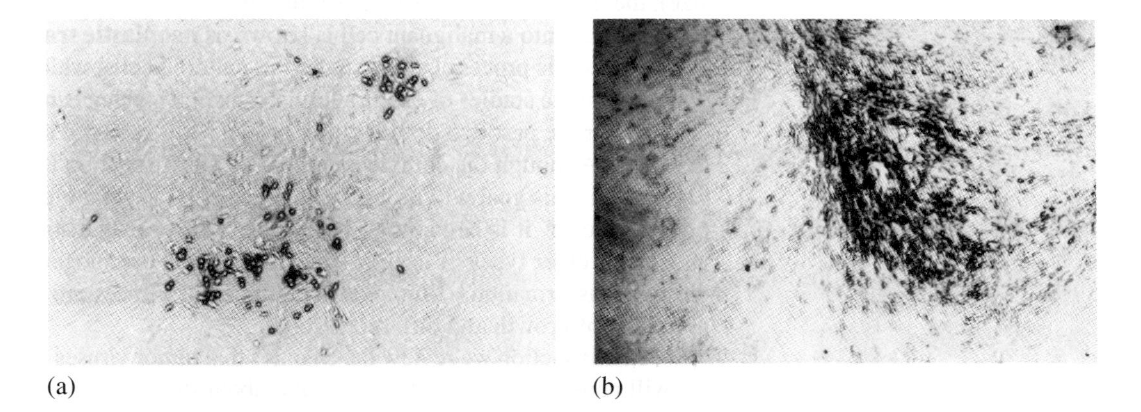

FIGURE 10.1

Mouse sarcoma virus transformation of monolayer cells. (a) Several foci of morphologically transformed normal rat kidney (NRK) cells can be detected by their piling up on top of the normal monolayer cells (× 25). (b) Mouse embryo fibroblast cells transformed by the retrovirus. Note the irregular crossing over and piling up of malignant cells on top of the normal contact-inhibited cells (× 40). The transformed cells are obviously more retractable than the normal fibroblasts.

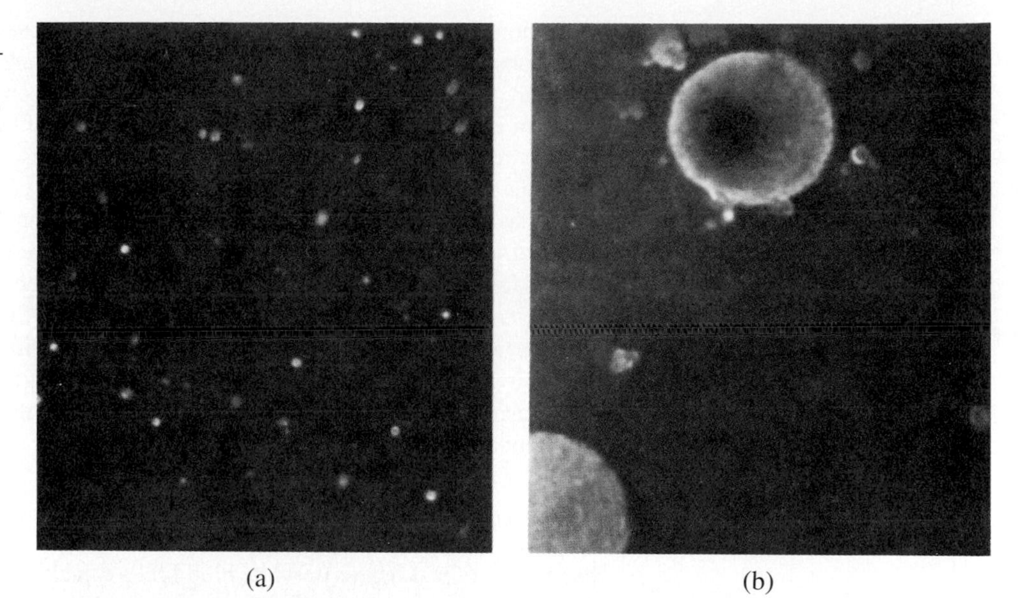

FIGURE 10.2

Photomicrographs of rat cells in "soft agar." (a) Normal rat embryo fibroblasts were seeded in growth medium containing 0.3 percent agar and incubated for 3 weeks at 37°C. Only individual cells are evident; none has formed colonies. (b) Rat embryo cells transformed by murine leukemia virus plus a chemical carcinogen (see Sec. 10.4) were seeded and incubated as in (a). Parts of two large colonies and several smaller ones are visible, in addition to some individual cells that did not divide and therefore appear not to be transformed. Overall magnification is about 640× in both panels.

number of infected cells will form small areas of transformed cells that resemble miniature tumors on top of the monolayer of normal cells (Fig. 10.1). Each little tumorlike mass is called a **focus** of transformed cells, and the virus assay is therefore known as a **focus formation assay.** The other principal cell culture assay for tumor viruses that transform fibroblasts depends on the fact that only transformed cells are able to grow in suspension, unattached to a culture surface. To aid quantitation of the number of transformed cells in an infected culture, cells are suspended in a support medium such as a low concentration of agar gel, or "soft" agar, to keep individual colonies separated (Fig. 10.2).

A long list of other changes induced in fibroblasts by various tumor viruses can be compiled from the volumes of literature on the subject. Some of these are recognized only in terms of growth or social behavior in culture, although many biochemical differences be-

TABLE 10.1

Altered properties of cultured fibroblasts transformed by tumor viruses

Morphology and Behavior
Become more rounded and refractile
Attach to surfaces less firmly
Disorganized pattern of cells ("crisscross")
Lose contact inhibition of movement (topoinhibition)
Grow in multiple layers
Grow in suspension (lose anchorage dependence)
Grow to higher densities per culture
Require less serum
Microtubules and microfilaments of the "cytoskeleton" disaggregate
Can form tumors in appropriate animals
Surface Properties
Glycosylation of proteins (increased hyaluronic and decreased sialic acid)
Lectin agglutination increased
Large external transformation sensitive (LETS) protein (250 kDa) disappears
Lipid ganglioside content decreases
Surface protein mobility increased
Fetal antigens expressed
Virus-specific T antigens appear
Sugar transport increases
Other Biochemical Changes
Proteases released
Transcription patterns change
Peptide growth and transformation factors produced and released
Phosphorylation patterns of proteins change
Increased use of glycolysis vs. aerobic processes for energy

tween normal and transformed fibroblasts have also been noted. Some of the more commonly observed alterations in transformed fibroblasts are presented in Table 10.1. For the most part, the importance of each of these changes for converting a normal cell into one that can form a tumor is not known. Nor is it clear which of the myriad of changes is a primary or direct result of interaction with tumor virus functions, as opposed to some secondary effect of the initial change(s). In fact, for many years one of the biggest problems in cancer research has been that too many differences between normal and transformed cells have been observed, making it impossible to determine which are most important. One of the major reasons for studying tumor viruses is the hope that, because of the relative simplicity and specificity of action of these viruses in cells compared with chemicals or other agents inducing cancer, the molecular basis of transformation will be more easily discovered.

Viral agents also offer the opportunity to apply genetic analyses to the transforming mechanism, and this approach may prove essential in the identification of those most crucial events that initiate the cancer process.

10.1.2 Lymphocytes and Macrophages

Some types of leukocytes or white blood cells can also be transformed in culture by tumor viruses. These cells, primarily lymphocytes and macrophages, are important members of the immune system (see Sec. 12.3). Most white blood cells do not attach to the surface of culture vessels, and they do not divide readily in culture. With the present availability of **cytokines** (e.g., IL-1, IL-2) many normal white cells can now be maintained for months in cell culture. After infection with appropriate tumor viruses, however, some lymphocytes and macrophages replicate autonomously in culture without the need for growth factors. They may even form colonies in soft agar. Therefore, transformation of these cells can also be studied *in vitro.* Methods for demonstrating virus-induced neoplastic transformation of most of the hundreds of other cell types in animal tissues have yet to be developed. Nevertheless, cells of some tissues are not transformed by any known virus in culture or in the intact animal.

SECTION 10.1 FURTHER READING

CLARKSON B., and R. BASERGA, eds. (1974) *Control of proliferation in animal cells.* Cold Spring Harbor Press, Cold Spring Harbor, New York.

SMITH, J. R., and L. HAYFLICK. (1974) Variation in the life span of clones derived from human diploid strains. *J. Cell Biol.* 62:48.

BARIGOZZI, C., ed. (1978) *Origin and natural history of cell lines.* Alan Liss, New York.

PRESCOTT, D. M., and A. S. FLEXER. (1982) *Cancer: The misguided cell.* Charles Scribner's Sons, New York.

HARRIS, C. C. (1987) Human tissues and cells in carcinogenesis research. *Cancer Res.* 47:1.

10.2 DNA TUMOR VIRUSES

Most families of animal DNA viruses include members that are able to induce benign or malignant tumors in some host, at least under limited experimental conditions. These include the papovaviruses, adenoviruses, some herpesviruses, and a few poxviruses. The simplest of these are the **papovaviruses,** best typified by polyomavirus, frequently found in wild mice, and SV40, the simian vacuolating agent (see Sec. 7.2). Human papovaviruses similar to polyoma and SV40 can also induce transformation in animal cells (e.g., BK and JC). In contrast, attempts to demonstrate replication or transformation in cultured cells by the papilloma-inducing members of the papovavirus family have so far been difficult. Recent success, however, with the bovine papillomavirus and some human papilloma viruses (HPV) has been reported using cultured keratinocytes. These viruses can induce skin tumors and probably some carcinomas in other tissues, and benign growths such as human warts (see Secs. 7.2 and 12.7). Also the larger adenoviruses (Sec. 8.1) can be repli-

TABLE 10.2
Characteristics of the oncogenic DNA and RNA tumor viruses

DNA	*RNA*
Productive infection with cell lysis.	Productive infection leads to transformation. No cell lysis.
Abortive infection leads to cell transformation with little or no virus production.	Abortive infection leads to transformation with no virus production.
Transformation is inefficient (10^6–10^7 virus particles/transformation event).	Transformation fairly efficient (10^2–10^4 virus particles/transformation event).
Some induce cellular DNA synthesis.	Induce cellular DNA synthesis.

cated in cultured fibroblasts, and most adenovirus strains that can induce tumors will also transform fibroblasts from some animal species. Several herpesviruses appear to transform lymphocytes in the host (Chapter 12), but most of these viruses do not replicate in these cells. No neoplastic transformation of cultured cells has been demonstrated for the Yaba monkey virus, one of the few known tumor-inducing poxviruses (Sec. 8.3).

Table 10.2 compares the different characteristics of DNA and RNA tumor viruses when they interact with cells. As we have stated in previous chapters, the normal replicative cycle of DNA viruses leads to cell lysis and death; thus, transformation occurs only when **abortive infection** takes place. In contrast, the RNA tumor viruses (i.e., retroviruses) can complete an infectious cycle with progeny virus production in the course of cell transformation. In this regard, any approach that decreases replication of the DNA viruses can enhance their transformation capability. Such treatments have included temperature changes during virus infection, UV irradiation, and chemical exposure. This fact became very apparent with measures introduced to control herpes simplex virus infections of humans. Initially, infections of the genital and oral areas were treated with neutral red dye and light. These conditions were found to block virus replication in cells. Subsequent studies, however, indicated that these procedures did not kill the virus but actually enhanced in cell culture the **transforming potential** of several herpesviruses. Thus, this clinical treatment for herpes is now only rarely used for severe cases. Fortunately for those treated, no cancers have appeared.

10.2.1 Hepadnaviruses

Hepadnaviruses do not carry an oncogene, nor do they transform cells in culture. They also do not appear to integrate into cellular genome domains near known proto-oncogenes (see Sec. 10.3.6). However, the suggestion that the ground squirrel liver tumors are associated with a possible **insertion promotion process** (see below) involving the *c-myc* proto-oncogene merits further attention. Moreover, many human hepatocellular carcinomas contain integrated HBV, and the viral X region appears to have pleiotropic effects that could be involved in the oncogenic process. In some or many cases the relationship of this family of viruses to induction of hepatic cancer could involve initial liver cell necrosis caused by HBV infection. The subsequent hepatic cellular proliferation results in genetic mutations that eventually lead to carcinoma. A similar concept is now being considered to explain the association of liver cancers and hepatitis C infection (see Sec. 14.3). Such a promoter role for EBV in B cell lymphoma has been proposed, although in that situation, EBV appears to induce directly the replication (not destruction) of the B lymphocytes that later become transformed (see below).

10.2.2 Papilloma, Polyoma and Related Viruses

As mentioned earlier, the cytopathic effects of many viruses are strikingly dependent on the species of origin of cultured host cells. This fact is an especially important consideration for observing transformation of fibroblasts by essentially all DNA tumor viruses, since they are generally capable of a full lytic cycle of reproduction in cells of many species. To study the transformation mechanism of a DNA tumor virus in cell culture, a species of cell or other conditions must be found that allow only partial expression of the viral genome. Viral gene expression must be limited by the cells to early functions to facilitate establishment of a persistent, nonproductive infection of the transformed cell. Otherwise the cell replicating the virus undergoes lysis (see Table 10.2). As noted above, animal DNA viruses do not seem to produce repressors like those made by **temperate bacteriophage** (see Sec. 8.7). Their replication is limited if certain cellular factors are missing. Polyoma and SV40 viruses, for example, replicate efficiently in cells of the natural host species (mouse and monkey) but undergo abortive infection in cells of other rodents or humans (see Sec. 7.2). This nonpermissive state, however, can be overcome if the infected cells are co-cultivated with permissive cells (see Sec. 11.8). This observation, and those using mouse human **somatic hybrid cells,** indicate that certain factors required for virus replication are lacking in nonpermissive cells.

Papillomavirus A great deal of new information has been obtained on the role of papillomaviruses in human cancer. Using recombinant DNA technology, researchers have been able to detect the presence of subtypes of the human wart virus in carcinomas of the cervix, anus, genital area, and other tissues of the body.

Cancer development is a rare consequence of HPV infection. Usually, the latent period is 20 to 50 years. In some malignancies, such as cervical cancer, environmental factors including cigarette smoking and contraceptive use can promote this transformation. The correlation of HPV type 16 with development of cervical cancer is now used as a predictor of high-grade cervical cancer through cytological examination of cervical smears. Similar studies can also predict cancer development in the anal canal. Currently, over 70 different human papillomaviruses have been identified (see Sec. 7.2), and certain subtypes (e.g., HPV 6 and 16) have been highly associated with human cancers. Malignant transformations of rodent cells have been demonstrated with HPV-16, 18, 31, and 33.

The model for the human wart virus is the **Shope papilloma virus** in rabbits. This virus is found in benign warts on the skin of the rabbit and is also integrated into the genome of malignant carcinomas developing within the benign tumors. As noted in Section 7.2, the lack of production of virions appears to be characteristic of this virus in the transformed cell. As noted above (Table 10.2), its oncogenic potential is expressed when replication of the virus does not occur.

In studies of HPV, cultured human skin has been infected with HPV, type 11, which was extracted from genital warts. Subsequent placement of these cells beneath the renal capsule of athymic mice, led to morphological transformation resulting in the development of benign tumors identical to those that occurred spontaneously in the patients. The tumors contained the intranuclear group-specific antigen of papillomaviruses and HPV DNA. This approach could prove useful in defining the mechanism of papillomavirus transformation and replication. Moreover, as noted above, Keratinocytes have been affected by certain HPV strains.

In determining the mechanism of transformation by the papillomaviruses, an early gene product (E5) has been found most likely responsible for the initial proliferation of the epithelial cells that leads to **papilloma** development. A 44–amino acid protein is coded by this gene, and appears sufficient to transform murine cells in culture. This is thus the smallest "transforming" protein identified. It has a region that localizes to the membrane fraction of transformed cells and one that induces cellular DNA synthesis in resting cells. One mechanism for this activity could also be binding to cellular proteins involved in proliferation (Fig. 10.3). In this regard, proteins encoded by the E6 and E7 regions of the papillomaviruses are also associated with cell transformation. E6 binds to the p53 protein and increases its proteolytic degradation. E7 attaches to the retinoblastoma gene product RB (Fig. 10.3a). The N-terminal region of E7 resembles domains in the adenovirus E1A proteins (see below) that are also linked to malignancy. These virus-host cell protein interactions, as noted above, most probably result in an increase in cell growth and eventual cell transformation. Recent evidence has supported the role of the suppressor genes in preventing cervical cancer. In a study of papillomavirus-negative cervical cancers, a mutation in the p53 suppressor gene has been noted.

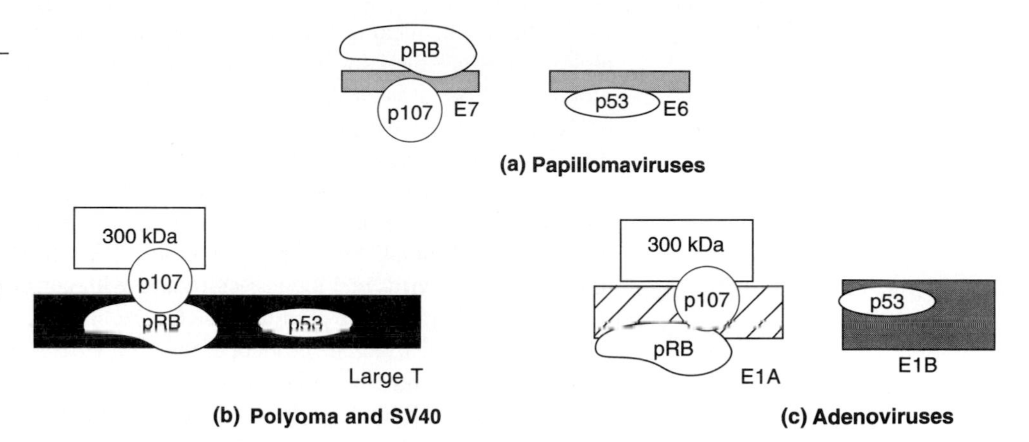

FIGURE 10.3

Interaction of viral proteins with tumor suppressor gene products and other cellular proteins. (a) Papilloma virus E6 and E7 proteins bind to the retinoblastoma (RB) and p53 proteins as well as the p107. (b) SV40 large T protein binds the RB and p53 gene products as well as the p107 and 300 kDa cellular proteins. (c) Adenovirus E1A and E1B proteins bind to the RB and p53 proteins, as well as other cellular proteins. These interreactions are considered potentially important in the transformation of cells by these viruses. (Modified from a figure provided by P. Howley.)

A similar process for cell transformation by polyoma viruses and SV40 has also been considered. In the case of SV40, the large tumor-specific (T) antigen of this virus (see below) binds to both p53 and the RB gene products (Fig. 10.3b). Thus, it alone has the activities of the E6 and E7 proteins of the papillomaviruses.

Polyoma and SV40 Observations on the malignant properties of polyoma and SV40 form a model for biological events that have been observed as well with the easily replicated human papovaviruses JC and BK. Polyoma and SV40 are nononcogenic in their natural hosts, but the former produces tumors in hamsters, rats, and rabbits, and the latter in hamsters. However, DNA sequences similar to SV40 have recently been found in ependymomas and choroid plexus tumors in children. A papovalike virus might therefore have an etiological role in these neoplasms. Because of the small size of papovaviruses, attempts at identifying the specific transforming region of their genome have been feasible. The production of T antigens appears to be essential for establishment of transformation, but the specific mechanism for this event, as is true for other transforming viruses, has not yet been fully elucidated. In this regard, with polyoma as with other DNA viruses, certain early viral functions are always expressed in transformed cells (see Sec. 7.2). The early region of polyoma, which consists of about 3000 base pairs, or half of the entire genome, produces three related polypeptides. These were originally distinguished as related tumor-specific or **T antigens** in transformed cells and have been designated "small t," "middle T," and "large T." The three polypeptides share a common N-terminal amino acid sequence and are otherwise partially homologous, indicating that they are produced from overlapping genes via multiple splicings of RNA transcripts (see Sec. 7.2). The early genes of SV40 are organized in similar fashion, and in fact, considerable sequence homology exists between the early polypeptides of these two viruses. However, SV40 apparently produces only two distinct polypeptides, small t and large T. Since part of the large T of SV40 contains amino acid sequences similar to those of middle T of polyomavirus, evidently SV40 large T performs the functions of both the middle and large T proteins from polyomavirus.

The biological effects and biochemical functions of the early genes of polyoma and SV40 viruses have been investigated by a variety of techniques, usually involving mutant viruses that have lost some early functions. Polyomavirus mutants defective in the large T gene are unable to initiate viral DNA synthesis in cells permissive for productive infection; they also fail to initiate late RNA synthesis, which occurs normally after DNA replication, as with many other DNA viruses. In addition, polyoma large T, which is found in the nucleus of transformed cells, is required for initiation of neoplastic transformation, but it is not always needed to maintain that transformed state. Cells can be transformed at a permissive temperature by a mutant with temperature-sensitive large T product. Raising the temperature to a nonpermissive level (at which the large T product fails to stimulate viral DNA synthesis) causes some properties of the transformed cells to be lost, whereas others are not, depending on the specific host cell.

The biochemical activity of the polyomavirus large T product that accounts for the biological effects has not been completely established. Similar to the analogous large T product of SV40, it is a *DNA-binding protein,* and it is found primarily in the nucleus. The SV40 protein attaches specifically to three sites contained within 120 nucleotides around the origin of replication of the SV40 DNA molecule. The same region overlaps the starting points for both early and late mRNA transcription (Sec. 7.2). Thus, by influencing early and late gene expression, the DNA-binding activity appears to be required for both viral replication and establishing transformation. When nonpermissive cells, for example, are infected with the virus and then co-cultivated with permissive cells, virus progeny production occurs only if large T protein is made.

The small t early gene product of polyomavirus is probably concentrated primarily in the cytoplasm of transformed cells. Nevertheless, deletion mutants of SV40 lacking most of the small t gene have provided evidence that small t product may be involved in the stimulation of host DNA synthesis. These mutants are able to transform only cells that are actively replicating and dividing at the time of infection, whereas wild-type virus can transform fibroblasts that are "resting," that is, not dividing due to either topoinhibition or removal of serum from the culture medium. Once growing cells are transformed by mutants lacking the small t product, however, they appear to behave no differently than cells transformed by the wild-type virus. Moreover, the SV40 deletion mutants without small t are able to cause tumors when injected into hamsters, although tumor formation seems to be somewhat slower. So far no biochemical activity has been associated with the small t product of either polyoma or SV40 viruses.

The biological effects of polyoma middle T product have been difficult to distinguish due to a lack of suitable mutants with only altered middle T product. This lack probably results from the overlapping of the middle and large T genes. However, polyoma middle T product has been localized in the plasma membrane of transformed cells. A *protein kinase activity* has been found associated with the middle T product but appears to reside not in the middle T product itself but in another polypeptide that binds to it. Nevertheless, the kinase activity fits well with current ideas of gene regulatory proteins, as explained in the discussion of the action of interferon on cells (see Sec. 12.4), and the role of some oncogenes (e.g., *src)* (see Sec. 10.3). Polyoma middle T product is itself phosphorylated by the associated kinase activity, once again raising the possibility of self-regulation through autophosphorylation.

A similar kinase activity is also associated with purified SV40 large T product, which contains amino acid sequences analogous to both the middle T and the large T products of polyomavirus. The SV40 large T phosphoprotein also appears to have an ATPase activity, possibly involved in the phosphorylation action of the associated kinase activity. It has recently been linked to a helicase that unwinds viral DNA and binds to both RB and p53.

Finally, the presence of virus-specific antigens at various sites, particularly on the surface of transformed and tumor cells, was known long before the molecular biology of the papovaviruses came under study. It now appears very probable that these various antigens (termed TSTA for **tumor-specific transformation antigen**), T, S, and U, are identical with, or derived by processing from, the viral gene products large, middle, and small T or t. A list of the properties of the SV40 large T antigen is given in Table 10.3. Similar features are associated with the T antigens of polyomavirus and will most likely be found linked to related proteins found with other papovaviruses.

In summary, current thinking about regulation of the cell cycle in eukaryotes envisions several steps that resting cells must take to begin DNA synthesis. The various early gene functions of polyoma and SV40 viruses may be aimed at the different processes that eventually can determine if a cell replicates the virus or is transformed by the virus. The interaction of papovavirus proteins with cellular suppressor gene products can influence the final outcome. The question of whether or not integration of the DNA of polyoma and other similar papovaviruses is absolutely required for cell transformation has not been answered. Chromosomal DNA from transformed cells does contain integrated viral DNA, but the location(s) of the viral genome can vary considerably in chromosomes of different clones of

TABLE 10.3
Properties of the SV40 large T antigen

1. Product of the A gene
2. A phosphoprotein of 100 kDa
3. Found primarily in the cell nucleus
4. Stimulates host cell DNA synthesis
5. Binds to SV40 DNA at origin of replication
6. Essential for initiation of viral DNA synthesis
7. Essential for establishment and maintenance of transformation
8. Has ATPase activity associated with a helicase (i.e., unwinds DNA)
9. Plays a role in virus replication; its expression is necessary for activation of virus production in nonpermissive cells
10. Facilitates replication of human adenovirus in monkey cells
11. Induces ribosomal RNA synthesis
12. Binds to p53 and RB suppressor gene products
13. Binds DNA polymerase α
14. Binds cellular heat shock protein, hsp70

transformed cells. Moreover, the related papillomaviruses do not appear to integrate but remain in plasmid forms in the cell nucleus. Thus, it seems unlikely that integration of the viral genome at any one particular chromosomal site is needed to initiate or maintain the transformed state. Nevertheless, a sufficient amount of viral DNA is most probably required, since low copy numbers of extrachromosomal viral DNA have been associated with a *nontransformed* state. Moreover, it appears that integration is not necessary for expression of the early genes of polyoma and its close relatives. All these functions are also present at high levels during lytic infections, when most of the viral DNA does not integrate. Thus the early viral products associated with cell transformation could be made without viral integration. Therefore, integration is generally needed primarily to prevent loss of the viral DNA as the transformed cells multiply; otherwise, transformation would not be maintained.

10.2.3 Adenoviruses

In 1962 Trentin and associates announced their observation that the common adenoviruses, when inoculated into hamster, were oncogenic. This startling news stopped the efforts toward developing a live vaccine against adenoviruses. It was soon learned that human adenoviruses could be grouped according to their malignant potential in animals. Highly oncogenic types (Group A) are 12, 18, and 31; intermediate (Group B) are 3, 7, 11, 14, 16, and 21; and nononcogenic (Group C) are 1, 2, 5, and 6. Others from the 37 known types make up Groups D and E.

The adenoviruses have the common feature of transforming cultured fibroblasts in a manner similar to that of the papillomaviruses discussed above; only nonpermissive cells become transformed. One exception is that these transformed cells do not release infectious virus after co-cultivation with permissive cells. This observation supports results of later experiments, indicating that the complete adenovirus genome is generally not found in the transformed cell. Since adenovirus genomes are almost an order of magnitude larger than SV40, progress on defining the **transformation mechanism** has been slower but recent studies are encouraging. At least a portion of the adenovirus genome must be present in the transformed cell, but no special integration site in the host genome appears to be required. According to the conventional linear maps of adenovirus genomes, only the left end, coding for early gene products of E1A and E1B, seems to be necessary [see Fig. 8.6(a)] for all viral transformation functions. One of these E1A proteins induces cellular DNA synthesis in quiescent cells. More direct evidence for this localization of transforming genes comes from the observation with adenovirus DNA cleaved with appropriate bacterial restriction nucleases. A purified DNA fragment comprising 11 to 12 percent of the entire DNA molecule, starting from the left end, is able to transform fibroblasts alone almost as efficiently as do whole DNA molecules.

The tumor or T antigens in adenovirus-transformed cells were first detected using fluorescein-labeled antibodies from tumor-bearing animals. These polypeptides are expressed early in the infectious cycle, as has been observed with T antigens of other DNA viruses. They are found in both the nucleus and cytoplasm of transformed cells. Moreover, as expected from the studies cited above, they are encoded in the left 11 to 12 percent of the genome as judged from *in vitro* translation of viral mRNAs obtained during the early cycle of virus infection. The transforming genes of some adenoviruses have been sequenced, and the major early region (E1A, E1B) mRNAs coding for transforming proteins have been identified. The T antigens that are encoded by the viral E1A mRNAs of 13 S and 12 S have molecular weights of 32 kD and 26 kD, respectively, and possess identical amino and carboxy termini as deduced from the sequence of the cDNA copies of the E1A mRNAs. Thus, the two E1A T antigens differ only in internal sequences. Three major E1B T antigens include a protein of 53 kDa, translated from the 22 S mRNA, a protein of 19 kDa from the 22 S and 13 S mRNAs, and a protein of 20 kDa from another mRNA. The E1B 20 kDa and 53 kDa T antigens share some amino acid sequences and thus must be translated in the same reading frame [see Fig. 8.4(a)]. At present the terminology of T antigens is reserved just for polyoma and SV40 viruses. The formerly designated T antigens for adenoviruses are now identified as E1A or E1B proteins.

The transforming region of an adenovirus is more complex than that of papillomaviruses and SV40. As noted above, two transcription units encode at least five distinct antigens (32, 26, 19, 20, and 53 kDa). Using adenovirus transformation-defective mutants, researchers have determined that the two E1A antigens and the E1B 19 kDa antigen play a role in the transformation process. An activity that can cause primary cells to permanently proliferate is solely encoded by the E1A gene. In conjunction with E1B, complete transformation of the normal cell takes place. Like E6 from HPV and the SV40 T antigen, E1B has been shown to bind the cellular tumor suppressor protein p53 (Fig. 10.3c). Moreover, the E1A gene product, the SV40 large T antigen, and the human papillomavirus E7 proteins described above, share a short amino acid sequence that is required for transformation. Because they all bind to the RB gene product (Fig. 10.3), this common mechanism for transformation has been considered.

Recent studies have shown that the adenovirus E1A protein can dissociate complexes of the RB product from the cellular E2F transcription factor. This activity is observed also with the papillomavirus E7 protein and the SV40 large T antigen. The interruption of this E2F-RB interaction could be an important step in human cell carcinogenesis, stimulating resting cells to a proliferative state and eventual transformation. The activity may be induced normally by the virus to provide a replicating cell for virus production. As noted above, if lytic infection does not occur, oncogenic transformation may result.

Differences occurring naturally in the E1A gene of adenoviruses of various serotypes can determine whether or not transformed cells are oncogenic in immunocompetent rodents. Another E1A effect that potentially influences tumorgenicity is the suppression of expres-

TABLE 10.4
Properties of DNA viruses associated with cell transformation

	Binds p53	Binds RB	Binds pp2A	Binds $pp60^{c\text{-}src}$	Transformation of Primary Cells	Transformation of Established Cell Lines
Human papillomavirus E6	+	−	−	−	[+][a] (E6 and E7)	+
E7	−	+	−	−		+
Polyoma, large T antigen	−	−	−	−	[+][a] (large, middle and small T)	−
middle T antigen	−	−	+	+		+
small T antigen	−	−	+	−		−
SV40, large T antigen	+	+	−	−	+	+
small T antigen	−	−	+	−	−	−
Human adenovirus E1A	−	+	−	−	[+][a] (E1A and E1B)	+
E1B	+	−	−	−		+

[a] Transforms if other transformation-associated proteins are expressed. Table provided by E. Sawai.

sion of major histocompatibility complex proteins on the cell surface. Thus, the cells could be protected from recognition by immunologically active cytotoxic T cells. Studies using mutants of adenovirus 12 with deletions in the E1A or E1B genes have shown that the E1A mutant can remain fully competent for replication and growth in cells and can stimulate the expression of all early adenovirus genes. However, these E1A-deleted viruses cannot transform cells in culture. These types of genetic approaches should eventually lead to identification of the particular genes responsible for transformation by adenoviruses.

From the above discussion it is evident that a common theme in cell transformation involves an inhibition of tumor suppressor genes as shown by studies of the DNA tumor viruses. The various binding properties of their transforming genes that appear associated with cell transformation are summarized in Table 10.4. The ability of each protein alone to transform primary cells into established cell lines can distinguish them, but many can transform cells if they are coexpressed with another transformation-associated protein.

10.2.4 Adenovirus-SV40 Hybrids

In the initial development of vaccines for polio and adenovirus, both agents were grown in monkey kidney cells. Subsequently, SV40 was found to be a contaminant of the primate cells and, in fact, was needed for the efficient replication of the adenovirus. Further studies demonstrated that in some cases a genetic recombination had occurred in the cultured cells between the SV40 and adenovirus genomes. The resultant viruses generally had deletions in adenovirus information, and thus they required an accompanying intact adenovirus for completion of the replication cycle. A similar requirement for a replication-competent virus has been described for growth of the parvovirus, adeno-associated virus, and for hepatitis D virus (see Sec. 9.4). Similarly, replication-defective sarcoma viruses of the retrovirus family require a helper virus (see Sec. 6.1). However, some replication-competent recombinant adeno-SV40 viruses have also been isolated. Cells infected with these hybrid agents are transformed and carry the T antigen of SV40. The recombinant viruses have the envelope and morphology of the adenovirus and its sensitivity to heat; they contain no SV40 coat antigen. These viruses are highly oncogenic and have been useful for mapping regions of the SV40 and adenovirus genomes. They have also provided important information on the process of recombination between two distinct viral DNAs in infected cells, a process that has not been found to occur in nature.

10.2.5 Herpesviruses

Herpesviruses have been implicated in a variety of human cancers and are well established as the cause of animal tumors, including Marek's disease of birds and lymphoma of monkeys. Table 10.5 summarizes some of the major herpesviruses linked to cancer and other diseases in animals and humans. A herpesvirus has even been associated with tumor formation in frogs and fish. It is noteworthy that the Lucké virus was one of the first herpesviruses found linked to a malignancy, when it was detected in renal adenocarcinomas of frogs. This virus has adapted well to its host and is malignant only at 4°C and not at higher temperatures. At room temperature, for instance, the **Lucké herpesvirus** replicates without transforming the cells. The **Marek's disease virus,** associated with neurolymphomatosis in chickens, is noteworthy because it was responsible for millions of dollars lost in poultry farms throughout the world. About 15 years ago, a turkey herpesvirus was found with serologic properties similar to the Marek's virus, but it did not cause malignancy. This turkey virus has been used as a live vaccine in chicken farms throughout the world. The virus replicates in the chicken, causes no disease, but prevents the malignant changes associated with the Marek's disease virus.

Another interesting group of herpesviruses are those found in primates and humans. *Herpes saimiri* and *herpes ateles* cause lymphomas only in New World monkeys. These viruses appear to be the counterpart to the Epstein-Barr virus (EBV) that is found in humans

TABLE 10.5
Herpesviruses and cancer

Virus	*Host*	*Clinical Syndrome*	*Tumor Induction*	*Tissue Culture Cell Transformation*
Lucké	Frog	Renal adenocarcinoma	Frog	—
Oncorhynchus mosou	Salmon	Carcinoma	—	—
Marek	Chicken	Neurolymphomatosis	Chicken	Chicken
Herpesvirus sylvilagus	Cotton tail rabbit	Lymphoma	Rabbit	—
Herpesvirus saimiri	Squirrel monkey (New World)	Lymphoma and acute lymphocytic leukemia	Monkeys, rabbits	Monkey leukocytes
Herpesvirus ateles	Spider monkey (New World)	Lymphoma	Marmoset, monkey	Monkey leukocytes
Guinea pig herpeslike virus	Guinea pig	Leukemia	—	Hamster, guinea pig leukocytes
Herpes-like virus	Cow	Lymphosarcoma	—	—
Herpes-like virus	Sheep	Pulmonary adenomatosis	—	—
Equine herpesvirus	Horse	—	?Hamster lymphoma	Hamster
Epstein-Barr	Human	Burkitt's lymphoma Nasopharyngeal carcinoma	Owl monkey, marmoset	Human leukocytes
Herpesvirus type 1	Human	—	—	Hamster, mouse
Herpesvirus type 2	Human	—	Hamster	Hamster, mouse, human
Cytomegalovirus	Human	—?Kaposi's sarcoma	—	Hamster, human

SOURCE: Adapted from J. A. Levy. (1977) *Cancer Res.* 37:2957

and Old World monkeys. The primate viruses infect mainly T lymphocytes, whereas EBV infects and can cause transformation of B lymphocytes. EBV, first found associated with Burkitt's lymphoma in Africa, was later linked to infectious mononucleosis and nasopharyngeal carcinomas. However, how this herpesvirus is involved in these tumors, is still not clear; virus infection occurs many years prior to development of the cancer.

EBV appears to give a spectrum of diseases, depending on the immune response of the host. Thus, in individuals with strong cellular immunity against the virus, either no symptoms or infectious mononucleosis may occur. In others the virus can over time lead to malignancies such as B cell lymphomas and nasopharyngeal cancer (Fig. 10.4). The latter cancer is particularly prominent in Chinese people living in southern China. A correlation of this cancer with the use of herbal medicines taken from the Crutus tree has been suggested. The product of this plant is croton oil, which is a known **tumor promoter.**

The **human herpesvirus type 2** was initially linked to cervical carcinomas because of a higher prevalence of antibodies to the virus in women with this cancer than in healthy females. However, recently, this cancer has been more directly associated with papillomaviruses (e.g., HPV 16; see Sec. 7.2, 10.2) Finally, **cytomegalovirus** has been isolated from patients in Africa with a cancer of the endothelial cells lining blood vessels, called Kaposi's sarcoma. This disease has become well-known because of its prominence in individuals suffering from acquired immune deficiency syndrome (AIDS). The possible association of cytomegalovirus (CMV) with this cancer is still being considered, but no direct proof of its playing a role in the malignancy has been found. Only a few Kaposi's sarcoma tissues contain evidence of CMV genetic material or proteins.

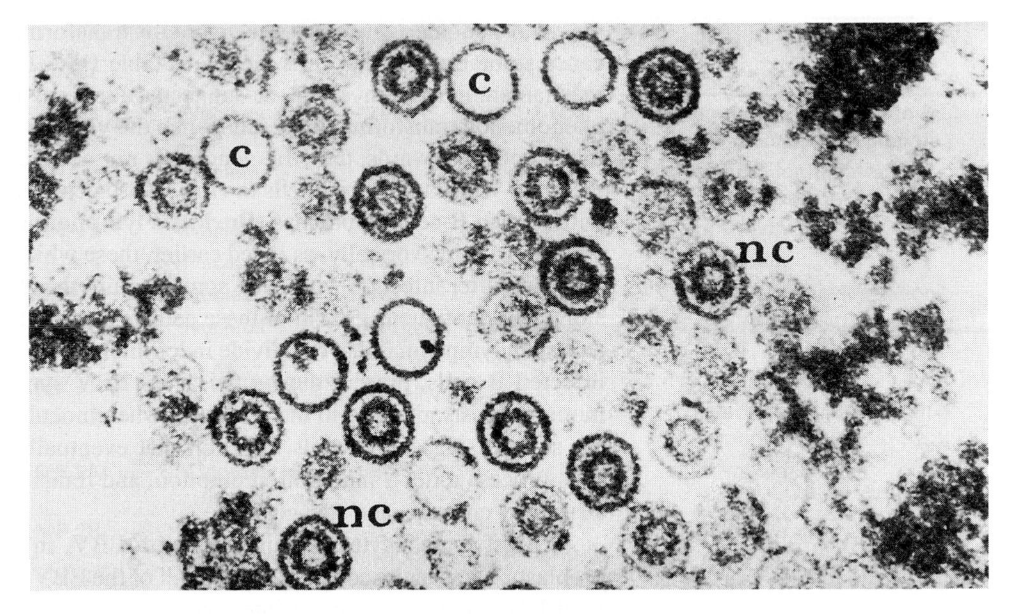

FIGURE 10.4

Factory of Epstein-Barr virus production in B cells cultured from a patient with nasopharyngeal carcinoma. Note the nucleocapsids (nc) containing viral DNA as well as virions that lack the nucleocapsid (c). *X* 80,000.

We have discussed the fact that transformation by DNA viruses requires either a non-permissive cell or a defective virus introduced into a permissive cell (Table 10.2). In the case of herpesviruses, the former pathway of transformation has provided the means of showing their oncogenic potential. The process of neoplastic transformation by herpesviruses has been studied in cell culture mainly using human herpesviruses. Two strains of **herpes simplex viruses** (HSV), types 1 and 2, which are antigenically distinct, have been found under certain conditions to transform rodent fibroblasts. Since these cells are permissive for late HSV functions and therefore lysis, transformation was accomplished when the late HSV functions were inactivated in the virus, usually by UV irradiation of the virions, or by introducing temperature-sensitive mutations into the viral genome. Many hamster fibroblast cell lines, for example, have been morphologically transformed in culture by UV-inactivated HSV, and most of these cells are tumorigenic in hamsters. Efforts to understand the transformation process induced by HSV have now been directed at revealing the viral genes and products present in these transformed, tumorigenic rodent cells.

Early studies showed that the entire viral genome need not be present in the transformed cell. However, because the HSV genomes are so large (about 10^8 Da), it proved difficult to pinpoint the exact fraction of the genome left in transformed cells, using nucleic acid hybridization with intact viral DNA as probe for viral DNA in the cells. Soon after transformation by UV-inactivated HSV, viral DNA could be readily detected by nucleic acid hybridization, but much less HSV DNA was present later in cells from tumors produced by injection of the cells into hamsters. Hybridization studies have utilized probes consisting of specific fragments of HSV DNA generated by the action of bacterial restriction endonucleases on the whole viral DNA molecule. Thus, it became clear that most HSV-2 transformed hamster fibroblasts eventually retain only a few copies of specific viral DNA sequences from just two relatively small, distinct regions of the linear HSV-2 genome. Genes for late viral functions are not found.

Transformation of cells can be induced by fragments of HSV-1 and HSV-2. These are located in the midsection of the genome in a region that encodes an HSV glycoprotein (a DNA-binding protein), and an origin of viral DNA replication. Only one of the regions of the genome is found naturally in tumor cells transformed by UV-inactivated virus. The transforming regions of HSV-1 and HSV-2 are not homologous and do not code for any of the early viral genes. This observation suggests that herpesviruses transform cells by mechanisms different from those of the other DNA viruses. None of the protein products of the transforming genes of HSV have been found; thus we have no idea yet what enzymatic activities might be involved in transformation. Finally, whether

tumors. The gene is related to the *fos/jun* family of oncogenes (Table 10.7). Thus, in this one unusual case, an oncogene in a herpesvirus might be responsible for the lymphomas that develop in these animals. A short latency period between infection and malignancy favors this mechanism for transformation. Its relevance to human herpesvirus merits further study.

In summary, the events involved in transformation by the herpesviruses have not yet been fully elucidated. These viruses appear to transform cells in culture by a variety of mechanisms, including the production of proteins with mutagenic properties, and proteins with the potential for stimulating rearrangements or amplifications of DNA sequences of the normal cell. As noted above, their integration next to certain cellular genes may also elicit a transformation process.

10.2.6 Poxviruses

Poxviruses have also been associated with tumors, generally benign fibromas in rabbits, (the **Shope fibroma virus**) and in monkeys (Yaba virus). Nevertheless, in recent years, a malignant rabbit fibroma virus giving rise to lethal tumors was isolated and found to be a recombinant of the Shope fibroma virus and the myxoma virus of rabbits. A replacement of a small region of the myxoma virus genome with related sequences from the Shope virus produced the malignant properties of the virus. The relationship of genes within these viruses that can encode proteins with some homology to epidermal growth factor may be related to their tumor-inducing properties (see Sec. 10.3).

SECTIONS 10.1 AND 10.2 FURTHER READING

Original Papers

LUCKÉ, B. (1934) A neoplastic disease of the kidney of the frog, *Rana pipiens*. *Am. J. Cancer* 20:352.

ROUS, P., and J. W. BEARD. (1935) The progression to carcinoma of virus-induced rabbit papilloma (Shope). *J. Exp. Med.* 62:523.

BURKITT, D. (1958) A sarcoma involving the jaws in African children. *Br. J. Surg.* 46:218.

TRENTIN, J. J., et al. (1962) The quest for human cancer viruses. *Science* 137:835.

EPSTEIN, M. A., B. G. ACHONG, and Y. M. BARR. (1964) Virus particles in cultured lymphoblasts from Burkitt's lymphoma. *Lancet* i:702.

HUEBNER, R. J., et al. (1964) Induction by adenovirus type 7 of tumors in hamsters having the antigenic characteristics of SV40 virus. *Proc. Natl. Acad. Sci. USA* 52:1333.

ZUR HAUSEN H., et al. (1970) EBV DNA in biopsies of Burkitt tumor and anaplastic carcinomas of the nasopharynx. *Nature* 228:1056.

DUFF, R., and F. RAPP. (1971) Oncogenic transformation of hamster cells after exposure to herpes simplex virus type 2. *Nature* 233:48.

GIRALDO, G., E. BETH, and F. HAGUENAU. (1972) Herpes-type virus particles in tissue culture of Kaposi's sarcoma from different geographic regions. *J. Natl. Cancer Inst.* 49:1509.

ORTH, G., et al. (1978) Viral sequences related to a human skin papillomavirus in genital warts. *Nature* 275:334.

LASSAM, N., et al. (1979) Tumor antigens of human ad5 in transformed cells and in cells infected with transformation-defective host-range mutants. *Cell* 18:781.

GALLOWAY, D., et al. (1980) Analysis of viral DNA sequences in hamster cells transformed by herpes simplex virus type 2. *Proc. Natl. Acad. Sci. USA* 77:880.

JARIWALLA, R. J., et al. (1980) Tumorigenic transformation induced by a specific fragment of DNA from herpes simplex virus type 2. *Proc. Natl. Acad. Sci. USA* 77:2279.

LOWY, D. R., et al. (1980) *In vitro* tumorigenic transformation by a defined subgenomic fragment of bovine papillomavirus DNA. *Nature* 287:72.

GALLOWAY, D. A., and J. K. MCDOUGALL. (1983) The oncogenic potential of herpes simplex viruses: Evidence for a "hit-and-run" mechanism. *Nature* 302:21.

MONTELL, C., et al. (1984) Complete transformation by adenovirus 2 requires both E1A proteins. *Cell* 35:951.

NELSON, J. A., et al. (1984) Structure of the transforming region of human cytomegalovirus AD169. *J. Virol.* 49:109.

YANG, Y.-C., et al. (1985) Bovine papillomavirus contains multiple transforming genes. *Proc. Natl. Acad. Sci. USA* 82:1030.

KREIDER, J. W., et al. (1986) *In vivo* transformation of human skin with human papillomavirus type II from condylomata acuminata. *J. Virol.* 59:369.

LEIBOWITZ, D., et al. (1987) An Epstein-Barr virus transforming protein associates with vimentin in lymphocytes. *Mol. Cell. Biol.* 7:2299.

MORAN, E. (1988) A region of SV40 large T antigen can substitute for a transforming domain of the adenovirus E1A products. *Nature* 334:168.

SMITH, K. T., and M. S. CAMPO. (1988) "Hit and run" transformation of mouse C127 cells by bovine papillomavirus type 4: The viral DNA is required for the initiation but not for maintenance of the transformed phenotype. *Virology* 164:39.

WHYTE, P., et al. (1988) Association between an oncogene and an anti-oncogene: The adenovirus E1a proteins bind to the retinoblastoma gene product. *Nature* 334:124.

WATANABE, S., et al. (1989) Human papillomavirus type 16 transformation of primary human embryonic fibroblasts requires expression of open reading frames E6 and E7. *J. Virol.* 63:965.

BERGSAGAL, D. J., et al. (1992) DNA sequences similar to those of simian virus 40 in ependymomas and choroid plexus tumors of childhood. *New Engl. J. Med.* 326:988.

CHELLAPPAN, S., et al. (1992) Adenovirus E1A, simian virus 40 tumor antigen, and human papillomavirus E7 protein share the capacity to disrupt the interaction between transcription factor E2F and the retinoblastoma gene product. *Proc. Natl. Acad. Sci. USA* 89:4549.

CUZICK, J., et al. (1992) Human papillomavirus type 16 DNA in cervical smears as predictor of high-grade cervical cancer. *Lancet* 339:959.

SIRCAR, S., et al. (1992) Adenovirus transformation revertant resistant to retransformation by E1 but not by SV40-T and HPV16-E7 oncogenes. *Virology* 191:187.

VON HOYNINGEN-HUENE, V., M. KURTH, and W. DEPPERT. (1992) Selection against large T-antigen expression in cells transformed by lymphotropic papova virus. *Virology* 190:155.

FAHA, B., E. HARLOW, and E. LEES. (1993) The adenovirus E1A-associated kinase consists of cyclin E-p33^{cdk2} and cyclin A-p33^{cdk2}. *J. Virol.* 67:2456.

HOPPE-SEYLER, F., and K. BUTZ. (1993) Repression of endogenous p53 transactivation function in HeLa cervical carcinoma cells by human papillomavirus type 16 E6, human mdm-2, and mutant p53. *J. Virol.* 67:3111.

MOORTHY, R. K., and D. A. THORLEY-LAWSON. (1993) Biochemical, genetic, and functional analyses of the phosphorylation sites on the Epstein-Barr virus–encoded oncogenic latent membrane protein LMP-1. *J. Virol.* 67:2637.

Books and Reviews

EPSTEIN, M. A., and B. G. ACHONG. (1979) *The Epstein-Barr virus.* Springer-Verlag, Berlin.

TOOZE, J., ed. (1980) DNA tumor viruses, 2d ed. Cold Spring Harbor Press, Cold Spring Harbor, New York.

RAPP, F., and F. J. KENKINS. (1981) Genital cancer and viruses. *Gyn. Onc.* 12:S25.

HAYWARD, G. S., and G. R. REYES. (1983) Biochemical aspects of transformation by herpes simplex viruses. In *Advances in viral oncology.* ed. G. Klein. Vol. 3, 271–306. Raven Press, New York.

HOWLEY, P. M. (1983) The molecular biology of papilloma virus transformation. *Am. J. Pathol.* 113:414.

KIEFF, E., et al. (1983) Epstein-Barr virus transformation and replication. In *Advances in viral oncology,* Vol. 3, ed. G. Klein. 133–182 Raven Press, New York.

KLEIN, G. (1983) Specific chromosomal translocations and the genesis of B-cell-derived tumors in mice and men. *Cell* 32:311.

PETTERSSON, U., and G. AKUSJÄRVI. (1983) Molecular biology of adenovirus transformation. In *Advances in viral oncology,* Vol. 3, ed. G. Klein. 83–131 Raven Press, New York.

CROCE, C. M., and P. C. NOWELL. (1985) Molecular basis of human B-cell neoplasia. *Blood* 65:1.

ORTH, G., and M. FAVRE. (1985) Human papilloma viruses: biochemical and biologic properties. *Clin. Dermatol.* 3:27.

GREEN, M. (1986) Transformation and oncogenesis: DNA viruses. In *Fundamental virology.* ed. N. B. Fields and D. M. Knipe. 183–234 Raven Press, New York.

GREEN, M. (1989) When the products of oncogenes and anti-oncogenes meet. *Cell* 56:1.

ZUR HAUSEN, H. (1989) Papillomaviruses as carcinomaviruses. In *Advances in viral oncology.* ed. G. Klein, Vol. 8, 1. Raven Press, New York.

LEVINE, A. J. (1992) The p53 tumor-suppressor gene. *New Engl. J. Med.* 326:1350.

10.3 RNA TUMOR VIRUSES

If we were to devote space in this book to the various topics in proportion to the amount of work in the research literature, the size of the book might be doubled by the RNA tumor viruses. This wealth of information is a consequence of more than a decade of special research programs sponsored primarily by the U.S. government in hopes of finding the cause of human cancers. As we will see in Chapter 12, these efforts led to the discovery of human **oncogenic retroviruses** and opened the field of oncogene research (see below). Moreover, they provided the important background for the discovery of the AIDS virus (see Sec. 14.1). Researchers studying a variety of animal RNA tumor viruses have established an important unifying concept regarding neoplastic transformation with these viruses. They have discovered that cancers can result from alteration in the expression and/or structure of normal host cell genes. RNA tumor viruses have provided the means to isolate such cellular genes (called **proto-oncogenes** or *c-onc)* and to study their regulation (see Table 10.7). In this section we focus on some developments in this field that both directly demonstrate the oncogenicity of these RNA viruses and support the concept that some cellular genes can play a role in transformation.

10.3.1 Types of RNA Viruses That Transform Cell Cultures

All the RNA-containing viruses that induce tumors in animals are members of the family Retroviridae (Sec. 6.1) They are named for their ability to form a DNA copy of their RNA through their reverse transcriptase enzyme. Representative viruses from several different animal species are able to transform cultured fibroblasts, but the pioneer work in the field has been done with retroviruses of avian or murine (mouse or rat) origins. In the appropriate host animal, most RNA viruses that transform fibroblasts in culture induce only solid tumors in connective tissues, composed of transformed fibroblasts (i.e., **sarcomas**); hence, these viruses are called **sarcoma viruses.** Other RNA tumor viruses induce some form of cancer of blood cells and are therefore known as **leukemia viruses.** Recently, cell culture systems for studying transformation by some of these leukemia viruses have been established.

Based on present knowledge, a few general statements can be made about transformation by the RNA tumor viruses. First, in contrast with DNA viruses, these viruses can all establish chronic infections in which virus production occurs and is not lethal to the cell (see Table 10.2). Therefore, it is not necessary to find replication-defective viruses or host cells that are nonpermissive for viral replication to demonstrate transformation, as it is for DNA tumor viruses. Also, the RNA tumor viruses transform very efficiently. Infection of a cell with a single infectious virion may induce transformation. We can generalize, too, about the role in transformation of integration of proviral DNA into the host cell genome. It is widely accepted that integration is required for transcription of viral genes for all retroviruses. However, members of the lentivirus subfamily possibly replicate without integration (see Sec. 14.1).

Recent studies using bacterial restriction endonucleases (see Sec. 7.5.1) for mapping of proviral sequences in host cell DNAs have permitted detailed questions on integration to be answered. In this approach, DNAs from infected cells are digested with a given restriction enzyme (such as *Eco* RI or *Hind* III), and the resulting specific fragments are partially separated by electrophoresis in agarose gels. Separated DNA fragments are transferred to filter paper by soaking or "blotting" the DNA out of the gel onto the paper (see Appendix 1, Fig. A.3). Then radioactive viral RNA (or DNA made with viral RNA and reverse transcriptase) is allowed to form double-stranded hybrids with those cellular DNA fragments containing viral sequences. It is impossible to place all the various distinct cell DNA fragments into a linear map like those obtained for isolated viral DNAs, because typically 1 million or more different fragments are obtained by digesting a mammalian cell DNA. However, the patterns of radioactive RNA or DNA hybridized to cell DNA fragments on the filter, as revealed by autoradiography, can be used to compare integration locations of viruses in different cells.

The general conclusion from such studies is that more than one site for integration exists for a given virus in most cells. Although several of these sites may allow transcription of the integrated provirus, integration at other sites may result in reduced transcription, which could prevent transformation by the virus. Little is known about the integration mechanism, but restriction mapping studies have shown that there need not be any sequence homology between the transforming virus and a site of integration that allows transformation. The location of **virus integration,** is random but may be important for induction of the transformation process (see below). At present, most attention has been focused on expression of viral genes involved in transformation. To discuss this topic we will consider specific types of viruses in more detail.

10.3.2 Avian Sarcoma Viruses (ASV)

The first sarcoma virus to be discovered was found in a chicken tumor by Peyton Rous in 1911. Over half a century elapsed before viruses with similar biological activities were discovered in mammals. Even today much of what we know about this type of tumor-inducing agent derives from initial work on the Rous sarcoma virus and other related virus strains, mostly found in birds raised for commercial purposes. Most isolates of *a*vian *s*arcoma *v*iruses (ASV) can be replicated only by chicken or other avian cells, but some strains will cause tumors in rodents and can transform cultured rodent fibroblasts. However, the cells are nonpermissive to ASV replication, so that production of virions by the transformed rodent cells is low or undetectable.

Much is now known about the transforming mechanism of ASV, although the story is not yet complete. The relative simplicity of the small viral genome has greatly facilitated both molecular and classic genetic analyses of the ASV transforming functions. Analyses of **temperature-sensitive viral mutants** have proved that some viral gene product must function continuously, not only to initiate transformation but also to maintain the transformed state in infected cells. Spontaneously arising deletion mutants, which are able to replicate normally but cannot transform, have also been used to identify a region of the viral genome required for transformation (see Sec. 6.1). These viral transformation-associated sequences, initially designated *src* (for sarcoma), constitute about 15 percent of the entire 9 kb ASV genome and appear to include the locations of all transformation-defective, temperature-sensitive mutations identified so far. Several independent isolates of ASV all contain closely related, if not identical, *src* **sequences,** now termed oncogenes *(onc).*

Genetic analyses of ASV mutants that have lost their transforming property suggested that only one gene in the *src* region was linked to transformation and a single *src* gene protein of about 60 kDa has been isolated. This *src* gene product was first identified by Raymond Erikson and his colleagues by specific precipitation from tumor cell extracts using antibodies taken from tumor-bearing rabbits. Subsequently, investigators were able to produce the *src* gene protein by *in vitro* translation of ASV RNA. The identity of this product was confirmed by showing that the transformation-defective ASV mutants produce no protein or altered forms of the viral gene product.

The **src** protein was found to have a **protein kinase** activity; in fact, this gene product was actually the first tumor virus protein kinase to be discovered. It was subsequently shown to transfer phosphate groups to the amino acid tyrosine. This phosphokinase activity is unusual, because generally serine and threonine are phosphorylated. The kinase activity appears essential for transformation, since mutant viruses in which the *src* gene product is temperature-sensitive cannot transform cells in culture. The association of phosphoproteins with transformation is noteworthy, since they are known to play a role in several viruses' replication cycles.

DNA complementary to *src* sequences in ASV RNA can be made using viral reverse transcriptase. When this DNA was allowed to hybridize with DNAs from many different species of normal animals, it was discovered that all higher animal genomes, including human, contain DNA sequences remarkably similar to those of the ASV *src* gene. These cellular sequences are referred to as *sarc* or c-*src* sequences. Antisera that precipitate the viral *src* (v-*src*) product also react with a similar protein found in uninfected chicken cells, and this cellular product also possesses protein kinase activity. It thus appeared that the transforming gene of ASV is derived in some way from the host cell genome by recombination with a nontransforming virus. Therefore, ASV may be viewed as a result of specialized transducing viruses, by analogy with bacteriophage counterparts (see Sec. 12.10).

10.3.3 Viral Oncogenes

Based on these initial observations with the ASV, researchers have identified several different oncogenes carried by a variety of transforming retroviruses (Table 10.7). These oncogenes have been linked to specific proteins of varying sizes (21 kDa to 140 kDa) that may have protein kinase activity, resemble growth factor receptors, or be growth factors themselves. The findings indicate that cell transformation can be affected by several different proteins or combinations of proteins. Thus, we are beginning to see a joining together of aspects of normal and malignant growth via cellular proto-oncogenes. Clearly, the degree of regulation of their expression can play an important role in determining whether a tumor cell emerges.

Another example is the *ras* gene. It was first identified in the Harvey and Kirsten rat sarcoma viruses (i.e., H-*ras,* K-*ras*) that can transform a variety of monolayer cells. The linking of the cellular counterpart for *ras* with malignancy was noted following transformation of established mouse NIH-3T3 cells by the introduction of DNA from extracts of human cancers. Presently, the human *ras* gene family consists of N-*ras* (on chromosome 1), H-*ras* (on chromosome 11), and K-*ras* (on chromosome 12). They code for very similar proteins of about 21 kDa that are localized on the inner side of the plasma membrane. The *ras* gene proteins have chemical properties similar to G proteins, which are involved in signal transduction processes within the cell. The membrane receptors for activation of *ras* are not yet known. In some cases, it appears that the cellular *ras* gene can become transforming by single-point mutations in the coding region. In that case, the biochemical properties of the *ras* gene protein are modified and affect normal intracellular control mechanisms.

As with **ras** and **src,** the most direct effect in transforming cells would be the production of a gene product (e.g., kinase, G-binding protein) that can interrupt the normal function of the cell and cause autonomous growth and development of the malignancy. As noted above, other types of oncogenes can code for receptors for growth factors and thus make them more sensitive to the growth-inducing properties of these cytokines. An example of this process is the **human T-cell leukemia virus** that induces IL-2 production and has enhanced IL-2 receptor expression on the surface of infected cells (see Sec. 14.2). Finally, growth factors themselves, such as platelet-derived growth factor (PDGF) (the *sis* oncogene), can have a direct effect on the cells in an autocrine fashion, permitting their autonomous growth.

10.3.4 Murine Sarcoma Viruses (MSV)

Several different MSV have been recovered from mice and rats experimentally infected with retroviruses isolated from mice with spontaneous leukemia, or with well established-murine leukemia viruses (MuLV). Typically, a rodent injected with MuLV will develop leukemia of some sort, but occasionally sarcomas arise in connective tissues at the site of MuLV injection. These rare tumors have been found to contain new viruses that induce sarcomas efficiently and also transform rodent fibroblasts in culture. As in the case of ASV, the MSV all appear to be recombinants of an existing retrovirus (in this case an MuLV) with host cellular genes that are probably responsible for transformation of fibroblasts. One major difference between some ASV and all known MSV, however, is that some ASV strains encode all viral functions needed for replication, whereas MSV repre-

TABLE 10.7
Oncogenes associated with retroviruses

Oncogene	*Virus*[a]	*Transforming Protein*	*Location*	*Function*[b]	*c-onc Gene Product*[c]	*Location*	*Function*[b]
Class 1: Growth Factors							
sis	SSV	$gp28^{env\text{-}sis}$	Secreted and cell associated	PDGF-like growth factor	PDGF β chain	Secreted	Growth factor
Class 2: Receptor and Nonreceptor Tyrosine Protein Kinases							
src	RSV	$pp60^{v\text{-}src}$	Plasma and cytoplasmic membranes	Tyrosine PK	$pp60^{c\text{-}src}$	Plasma and cytoplasmic membranes	Tyrosine PK
yes	Y73	$P90^{gag\text{-}yes}$	Plasma and cytoplasmic membranes	Tyrosine PK	$p62^{c\text{-}yes}$	Plasma and cytoplasmic membranes	Tyrosine PK
fgr	GR-FeSV	$P70^{gag\text{-}fgr}$	Plasma and cytoplasmic membranes	Tyrosine PK	$p57^{c\text{-}fgr}$	Plasma and cytoplasmic membranes	Tyrosine PK
fps	FSV	$P140^{gag\text{-}fps}$	Cytoplasm	Tyrosine PK	$p98^{c\text{-}fps}$	Cytosol	Tyrosine PK
fes	ST-FeSV	$P85^{gag\text{-}fes}$	Cytoplasm	Tyrosine PK	$p92^{c\text{-}fes}$	Cytosol	Tyrosine PK
abl	Abelson MuLV	$P160^{gag\text{-}abl}$	Plasma membrane and cytoskeleton	Tyrosine PK	$p150^{c\text{-}abl}$	Cytosol and nucleus	DNA-binding tyrosine PK
ros	UR2	$P68^{gag\text{-}ros}$	Membrane associated	Tyrosine PK	?	Plasma membrane?	Receptor tyrosine PK?
*erb*B	AEV	$gp68/74^{erbB}$	Plasma and cytoplasmic membranes	Truncated EGF receptor tyrosine PK	EGF receptor ($gp180^{c\text{-}erbB}$)	Plasma membrane	EGF receptor tyrosine PK
fms	SM-FeSV	$gP180^{gag\text{-}fms}$	Plasma and cytoplasmic membranes	Mutant CSF-1 receptor tyrosine PK	CSF-1 receptor ($gp150^{c\text{-}fms}$)	Plasma membrane	CSF-1 receptor tyrosine PK
kit	HZ4-FeSV	$P80^{gag\text{-}kit}$	Cytoplasm	Truncated SCF receptor tyrosine PK	SCF receptor ($gp160^{c\text{-}kit}$)	Plasma membrane	SCF receptor tyrosine PK
sea	S13	$gP155^{env\text{-}sea}$	Plasma membrane	Tyrosine PK	?	Plasma membrane?	Receptor tyrosine PK?
ryk	RPL30	$gP69^{env\text{-}ryk}$	Plasma membrane	Tyrosine PK	?	Plasma membrane?	Receptor tyrosine PK?
Class 3: Membrane-associated G Proteins							
H-*ras*	Ha-MuSV	$p21^{H\text{-}ras}$	Plasma membrane	Mutant GTP-binding protein (low GTPase)	$p21^{c\text{-}H\text{-}ras}$	Plasma membrane	GTP-binding signaling protein (high GTPase)
K-*ras*	Ki-MuSV	$p21^{K\text{-}ras}$	Plasma membrane	Mutant GTP-binding protein (low GTPase)	$p21^{c\text{-}K\text{-}ras}$	Plasma membrane	GTP-binding signaling protein (high GTPase)
Class 4: Protein-serine Kinases							
raf	3611-MuSV	$P90^{gag\text{-}raf}$	Cytoplasm	Serine PK	$p74^{c\text{-}raf}$	Cytoplasm	Signaling serine PK
mil	MH2	$P100^{gag\text{-}mil}$	Cytoplasm	Serine PK	?	Cytoplasm?	Signaling serine PK?
mos	Mo-MuSV	$p37^{v\text{-}mos}$	Cytoplasm	Serine PK	$p37^{c\text{-}mos}$	Cytoplasm	Cytostatic factor
akt	AKT8	$P105^{gag\text{-}akt}$	Cytoplasm/membranes?	Serine PK	?	Cytosol	Serine PK

Oncogene	Virus[a]	Transforming Protein	Location	Function[b]	c-onc Gene Product[c]	Location	Function[b]
Class 5: Cytoplasmic Regulators							
crk	CT10	P47$^{gag\text{-}crk}$	Cytoplasm	SH2/SH3 domain protein	p35$^{c\text{-}crk}$	Cytoplasm?	SH2/SH3 domain signaling protein
Class 6: Nuclear Transcription Factors							
myc	MC29	P110$^{gag\text{-}myc}$	Nucleus	Sequence-specific transactivator (bHLH) (heterodimer with *max)*	p62/64$^{c\text{-}myc}$	Nucleus	Sequence-specific transactivator (bHLH) (heterodimer with *max)*
myb	AMV	P48$^{gag\text{-}myb\text{-}env}$	Nucleus	Mutant sequence-specific transactivator	p75$^{c\text{-}myb}$	Nucleus	Sequence-specific transactivator
fos	FBJ-MuSV	p55$^{v\text{-}fos}$	Nucleus	Sequence-specific transactivator (bZIP) as a heterodimer with c-*jun*	p55$^{c\text{-}fos}$	Nucleus	Sequence-specific transactivator (bZIP) as heterodimer with c-*jun*
jun	ASV17	p65$^{gag\text{-}jun}$	Nucleus	Mutant sequence-specific transactivator (bZIP)	p44$^{c\text{-}jun}$	Nucleus	Sequence-specific transactivator (bZIP) (AP-1)
erbA	AEV	P75$^{gag\text{-}erbA}$	Nucleus	Dominant negative mutant transcription factor	Thyroxine (T_3) receptor (p46$^{c\text{-}erbA}$)	Nucleus	Sequence-specific T_3-regulated transactivator
rel	REV-T	p59$^{v\text{-}rel}$	Nucleus/ cytoplasm	Mutant sequence-specific transactivator (NF-κB family) (dominant negative?)	p70$^{c\text{-}rel}$	Cytoplasm	Sequence-specific transactivator (NF-κB family)
ets	E26	P135$^{gag\text{-}myb\text{-}ets}$	Nucleus	Mutant transcription factor with twin DNA-binding domains	p56$^{c\text{-}ets}$	Nucleus	Sequence-specific transactivator *(ets* family)
ski	SKV	P125$^{gag\text{-}ski}$	Nucleus	Mutant sequence-specific DNA-binding protein	?	Nucleus?	Sequence-specific DNA-binding protein

[a]Where there are several retroviral isolates that contain the same oncogene, only a typical example is listed. RSV, Rous sarcoma virus (chicken); Y73, Yamaguchi 73 virus (chicken); GR-FeSV, Gardner-Rasheed feline sarcoma virus; FSV, Fujinami sarcoma virus (chicken); ST-FeSV, Snyder-Theilin feline sarcoma virus; Abelson-MuLV, Abelson mouse leukemia virus; UR2 (chicken); AEV, avian erythroblastosis virus; SM-FeSV, Susan McDonough feline sarcoma virus; HZ4-FeSV, Hardy-Zuckerman-4 feline sarcoma virus; S13, erythroblastosis virus (chicken); MuSV, mouse sarcoma virus; MH2, Mill Hill-2 avian leukemia virus; Mo-MuSV, Moloney mouse sarcoma virus; SSV, simian sarcoma virus (woolly monkey); Ha-MuSV, Harvey mouse sarcoma virus; Ki-MuSV, Kirsten mouse sarcoma virus; FBJ-MuSV, Finkel-Biskis-Jinkins mouse sarcoma virus; MC29, avian myelocytoma virus; AMV, avian myeloblastosis virus; REV-T, reticuloendotheliosis virus (turkey); SKV, Sloan Kettering virus (chicken); ASV17, avian sarcoma virus 17 (Junana); E26 virus (chicken); RPL30 = Regional Poultry Laboratory sarcoma virus (chicken); AKT8 = AKR T cell lymphoma virus (mouse); CT10 = Chicken tumor number 10 (Claude's agent).

[b]PK, protein kinase; EGF, epidermal growth factor; PDGF, platelet-derived growth factor; CSF, colony stimulating factor; SCF = stem cell factor or steel factor; bZIP = basic region/leucine zipper; bHLH = basic region/helix-loop-helix; AP-1 = family of transcription factors composed of *jun* family homodimers and *jun* family/*fos* family heterodimers; SH2/SH3 = sarc homology regions; NF–κB = nuclear factor kappa β.

[c]The protein products of the cellular gene or c-*onc* gene corresponding to be viral oncogene are given according to convention, with the number representing the size of the protein in kDa and the superscript giving the genetic locus involved in the makeup of the protein in order from N to C terminus. The *fes* and *fps* oncogenes represent homologous cat and avian genes. Likewise, the *raf* and *mil* genes are homologous mouse and avian genes.

SOURCE: Prepared by Tony Hunter, Salk Institute, San Diego, Calif.

sent **recombinant forms** of MuLV in which some MuLV replication genes are replaced with cellular genes (Fig. 6.9). Thus, MSV genomes are dependent on the presence of a MuLV genome in the same cell to help with replication of the MSV genome; they provide reverse transcriptase and other virion proteins (see Sec. 6.1).

There is no apparent nucleotide sequence homology shared among the oncogene sequences of ASV and those of several MSV from rats or mice. Nevertheless, several MSV examined so far produce phosphoproteins, some of which are enzymes with protein kinase activities similar in function to that of the ASV *src* gene product (see Fig. 6.9). For example, the nucleotide sequence of the Moloney MSV oncogene (called *mos*) has been shown to encode a polypeptide with a high degree of amino acid sequence homology (45 percent in one region) to the avian RSV **src** gene product, even though the *nucleotide sequence* homology is much less. Thus, it appears that the Moloney sarcoma virus *mos* gene is derived from a mouse gene analogous to the avian *src* gene. It is noteworthy, however, that the *mos* product has a serine and not tyrosine kinase activity. These findings are compatible with the hypothesis that several different but possibly interdependent cellular kinases may play important parts in the overall growth regulatory systems of cells.

In one particular case it was shown that combination of the moloney MSV promoter for transcription with the cellular counterpart of the MSV transforming gene *mos* (i. e. the cellular proto-oncogene cloned by genetic engineering) resulted in a DNA molecule that was able to transform cells just as efficiently as the MSV proviral DNA. Therefore, in this case, expression of the cellular oncogene at an abnormally high level may have been sufficient to cause transformation, whereas the corresponding protein was probably present in uninfected cells only in low amounts. However, it is possible that other factors were involved when the cellular gene was placed under the regulation of the viral promoter.

In the transformed cell, the kinase of this same MSV isolate, the Moloney strain, seems to associate with actin and myosinlike microtubular proteins in the cytoplasm. These cellular proteins make up a scaffolding system inside the cytoplasm, called the **cytoskeleton,** which is responsible for maintaining the flat shape of a fibroblast attached to a surface (see Fig. 10.5).

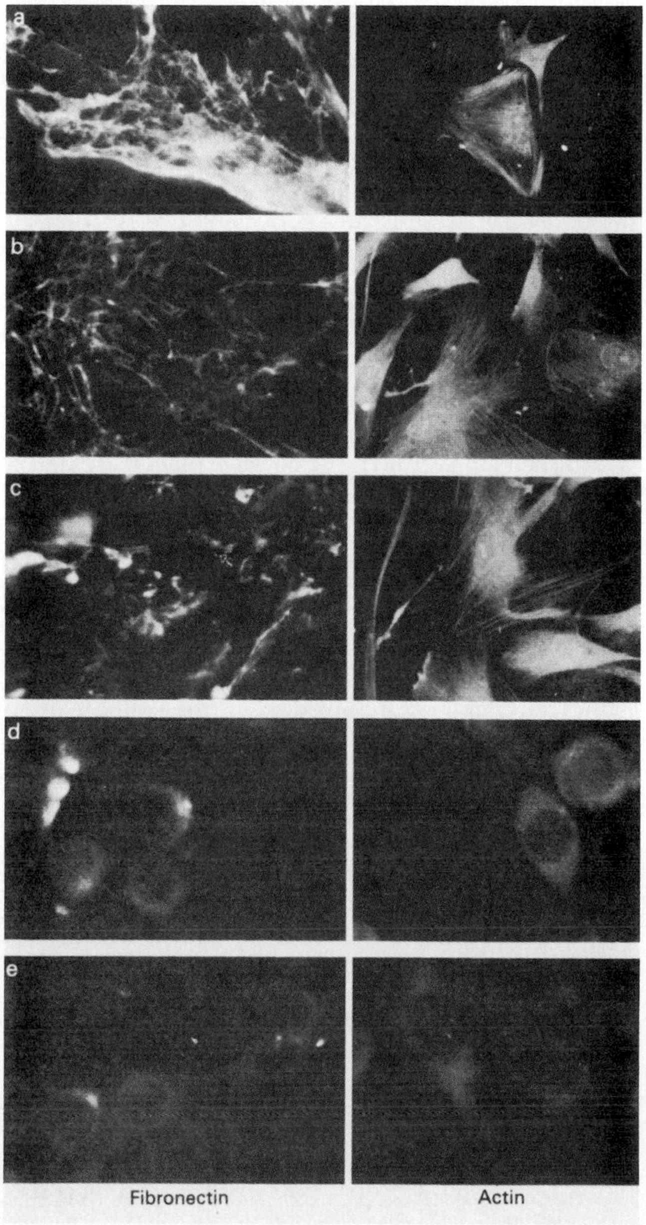

FIGURE 10.5

Effects of neoplastic transformation on the organization of the cytoskeleton. Antibodies to two protein components of the cytoskeleton, fibronectin and actin, were coupled to a fluorescent dye to reveal their locations in the cell. (a) Normal fibroblasts from the embryo of a field vole, a small rodent. Note the highly ordered filamentous network of both actin and fibronectin, forming the cytoskeleton that maintains the flat morphology of the cells. (b) and (c) Clones of cells that were transformed by Rous sarcoma virus (RSV) but then lost the transformed morphology during cultivation. Accordingly, the cytoskeletons appear essentially normal. (d) and (e) Clones of cells transformed by RSV. Note the total disruption of the cytoskeleton network, leading to rounding of the cells. Actin and fibronectin are visible in soluble form throughout the cytoplasm. (Courtesy of A. F. Lau)

It is known that neoplastic transformation involves disruption of polymeric actin fibers and microtubules comprising the cytoskeleton. This disruption causes the cell to assume a nearly spherical shape, the characteristic gross morphology of a transformed fibroblast in culture. Under certain circumstances phosphorylation of actin and microtubular proteins prevents their association to form the cytoskeleton, for instance, during mitosis. It would thus be exciting if the MSV kinase was shown to possess the ability to phosphorylate specifically actin and/or microtubular proteins. However, tests of this MSV kinase have failed to demonstrate such an activity.

One other aspect of transformation by MSV seems likely to provide an important avenue for understanding the molecular basis of the process. Transformation by certain MSV strains results in release from the cells of several small polypeptides with rather powerful biological activities. When nanogram (1 billionth of a gram) amounts of these polypeptides are included in the culture medium of normal fibroblasts, the cells soon become morphologically indistinguishable from cells transformed by MSV. Moreover, the treated cells pile up in multilayers like transformed cells, and they even form colonies in soft agar. Removal of these **tumor growth factors** (TGF) from the culture medium reverses all these effects. At least one of these polypeptides appears to bind to cell surface receptors for a known hormone, **epidermal growth factor** (EGF). In the case of EGF it is recognized that binding to the receptor stimulates a membrane-associated kinase that also phosphorylates tyrosine. Further details of the mechanism of action of EGF, TGF, and other growth factors are certain to be forthcoming in the near future. Their interaction provides the arena for understanding normal and malignant growth.

10.3.5 Leukemia Viruses

All known retroviruses associated with leukemias can be divided into two groups on the basis of biological activity in the animal and the ability to be replicated alone. In both chickens and rodents, those leukemia viruses that rapidly produce acute cases of blood cancer (**acute leukemia viruses**) have been found to be defective for replication. For progeny production, they require a helper retrovirus in much the same manner as do MSV (see Sec. 6.1). It also seems that these defective viruses have origins similar to those of MSV, because they are all recombinant forms of independently replicating retroviruses and some cellularlike genes. Several different cellular-like genes, all unrelated to known *src* genes, are found in different acute leukemia viruses from each host species (e.g., chicken; see Table 10.7). It is remarkable that most of these supposed leukemia viruses can also induce other types of tumors, including sarcomas and solid tumors in epithelial tissues, depending on the host strain. Many also transform fibroblasts in culture. Thus, not only can different modified cellular genes transform the same target cell, but a single transforming gene may be able to transform widely different types of cells from various tissues.

One of the acute murine leukemia viruses, the Abelson strain, has been shown to encode a **tyrosine phosphokinase,** the cellular target of which is not yet known. This virus

TABLE 10.8
Proto-oncogenes and human tumors

Proto-oncogene	*Human Malignancy*	*Genetic Change*
c-*abl*	Chronic myelogenous leukemia	Translocation
c-*myc*	Burkitt's lymphoma	Translocation
	Small cell carcinoma of the lung	Amplification
	Breast carcinoma	Amplification
c-*ras*	Bladder carcinoma	Point mutation
c-*erb*B	Squamous cell carcinoma	Amplification
	Glioblastoma	Amplification

Examples of viral oncogenes whose cellular proto-oncogenes have been linked to human cancer. The genetic change in the c-*onc* region noted in the tumors is listed. Adapted from Varmus (1984) *Nature* 314:581 and Bishop (1987). *Science* 235:305.

was isolated from a mouse treated with corticosteroids and inoculated with Moloney mouse leukemia virus. It induces null and B-cell lymphomas in mice and can transform fibroblasts in culture. Its oncogene (*abl*) appears to be involved in B-cell proliferation and recently has been linked to leukemias in humans (Table 10.8).

Replication-defective retroviruses have been shown to transform cultured fibroblasts, but only a few can transform white blood cells *in vitro.* One of these is the **avian erythroblastosis virus** (AEV). Work with this acute leukemia virus and its temperature-sensitive, transformation-defective mutant suggest that a viral gene product is required to initiate and maintain the growth of transformed cells in soft agar. Erythrocytes in chickens, unlike mammalian counterparts, contain a nucleus and therefore are able to grow and divide under appropriate conditions. Transformation with AEV not only stimulates red blood cell multiplication in suspension cultures but also inhibits production of hemoglobin in the infected cells. When cells transformed with the temperature-sensitive AEV mutant are shifted from the permissive low temperature to the nonpermissive high temperature, the virus function is inactivated and the cells resume production of the hemoglobin protein. It has been suggested, therefore, that AEV transforms the cell, at least in part, by shutting off cellular genes for specialized red blood cell functions. These genes are normally expressed as the erythrocyte matures via cell division through a specific chain of precursor cell types. In other words, an AEV gene product seems to inhibit proper differentiation of **erythroblasts,** the cell type directly converted to erythrocytes. Erythroblasts transformed by AEV in the intact chicken therefore continue to proliferate excessively rather than differentiate into functional erythrocytes that make hemoglobin and do not normally divide; hence, the AEV-induced disease is called *erythroblastosis.*

The findings with the AEV mutant present one of the most unambiguous cases of an RNA tumor virus directly controlling expression of cellular genes that are not directly involved in cellular growth and reproduction. This important observation emphasizes the **balance between cell differentiation and malignancy.** Cancer in essence reflects an arrest in the normal developmental process. This concept is also well illustrated in monolayer cell cultures that undergo differentiation. For instance, chicken cells that differentiate into muscle can be transformed readily by a temperature-sensitive mutant of ASV when they are grown at the permissive temperature. When a nonpermissive temperature is used, the cells undergo normal differentiation because of the lack of production of the *src* gene product. It is noteworthy that once differentiation has occurred, cells returned to the permissive temperature do not transform but remain as muscle cells or die. Moreover, experiments have shown that blood cells transformed *in vivo* by mouse chronic leukemia viruses can be induced to differentiate by a variety of substances, including dimethylsulfoxide and phorbol esters. In some experimental systems these "reverted" cells can then be used as donor cells for lethally irradiated animals. All these examples indicate the interplay between the enhanced cell multiplication characteristic of cancer and the slowing of growth and acquisition of function that are part of the differentiation process.

Most of the replication-defective leukemia viruses have been recognized relatively recently. Classically, leukemia viruses are able to replicate without help from another virus. However these viruses do not generally induce leukemias with the speed and severity that are characteristic of the replication-defective leukemia viruses. For this reason they are called **chronic leukemia viruses,** and their ability to directly induce neoplasms and transformation has been difficult to demonstrate. For many years all retroviruses with type C virion morphology were called "leukemia" viruses as long as they did not transform fibroblasts (see Fig. 10.6). However, the discovery of the acute leukemia viruses, with their oncogenes, led to the separation between the acute and chronic viruses. Moreover, other observations have now indicated that some of these replicating retroviruses are nonpathogenic and could instead play a role in normal developmental processes (see Sec. 6.4). The long incubation period for leukemia development associated with the chronic leukemia viruses suggests that

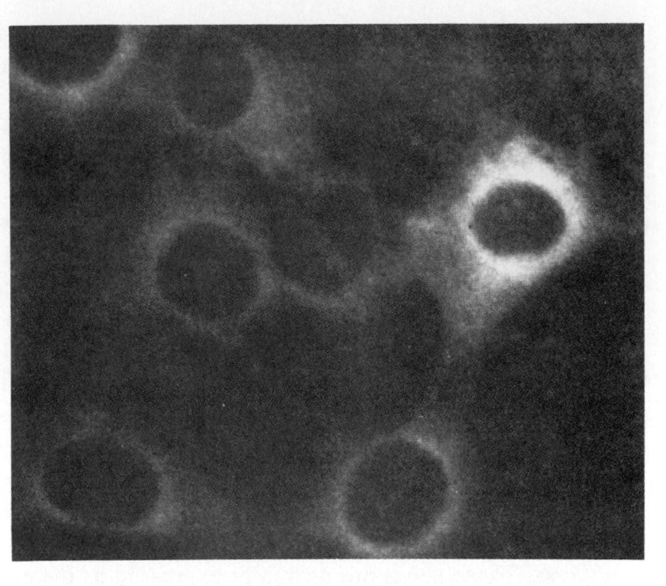

FIGURE 10.6

AKR mouse embryo fibroblasts infected with the AKR strain of murine leukemia virus. The infected cells are not transformed morphologically, but the presence of virus can be demonstrated by fluorescent antibodies directed against viral surface antigens, as shown by the halos of light covering the cytoplasm and surrounding nuclei of several cells. (Courtesy of John Rice)

they do not carry transforming genes (Fig. 6.9), as does the fact that they do not transform target cells in culture. One example is the Friend leukemia virus that induces splenomegaly and erythroid leukemia in mice (Fig. 10.7); the process takes several months to develop. As noted below, the viral oncogenic potential may be expressed by a different mechanism than an oncogene.

10.3.6 Transformation by Chronic Leukemia Viruses

We have discussed the mechanism by which acute transforming viruses, through *onc* genes, cause malignancies, either by enhanced expression of a **transduced cellularlike gene** or by expression of an altered protein coded by that gene. The chronic leukemia retroviruses appear to cause cancer after a long incubation period by different methods. Most of them do not have a definitive oncogene but may induce transformation by one of the following processes.

1. **Insertion: promotion** The viruses may *insert a viral promoter* (e.g., LTR) near normal cellular genes and enhance the production of a normal cell product leading to transformation. An example of this event is the B-cell lymphoma in chickens, in which the integration of the avian leukosis virus near the cellular proto-oncogene of myc (the c-*myc* gene) is associated with enhanced expression of the c-*myc* RNA.

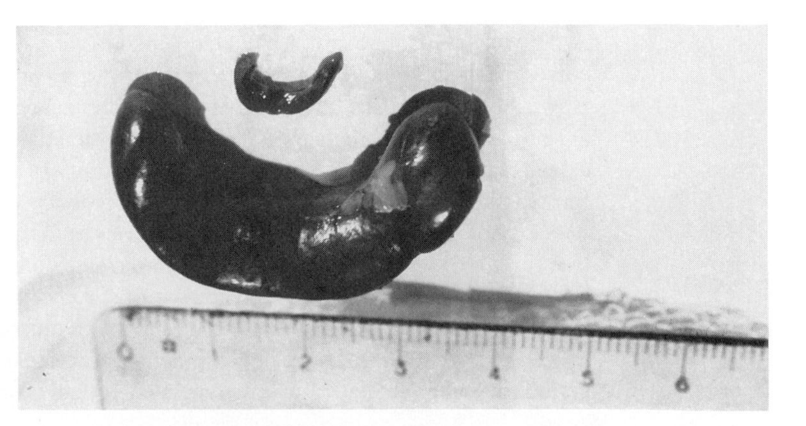

FIGURE 10.7

Massively enlarged spleen removed from a mouse infected with the Friend murine leukemia virus. A normal mouse spleen is presented for comparison. (Courtesy of S. Levy)

2. **Transactivation** The "slow transforming" oncogenic viruses may code for trans-acting proteins that diffuse through the cell and induce the viral LTR to enhance promotion of gene transcription. This product may also influence cellular genes and eventually lead to a transformed state. Transactivation may be responsible for the induction of malignancy by the human T cell leukemia virus (HTLV-I). (Sec. 14.2)
3. **Interruption of cellular control** Another possible mechanism for transformation by the leukemia viruses is *"hit and run."* An example of this process is the action of the Abelson mouse leukemia virus that induces B-cell lymphomas but then no longer remains in the tumor. Apparently, the infection disrupts normal cellular control mechanisms and permits the proliferation and transformation of the B cells. The virus is then not required for maintenance of the transformed state.
4. **Cellular stimulation** The viruses may act as *mitogens* for the T or B cell that they infect. The constant presence of viral protein(s) that interact with specific cellular receptors induces *an increased proliferation* of the cells, with subsequent transformation.

In this regard, as discussed in Chapter 6, a replication-defective retrovirus has recently been linked to the polyclonal expansion of B lymphocytes in the immune deficiency in mice termed MAIDS. The defective leukemia-inducing virus can only replicate in a cell with its accompanying helper virus. Once integrated into the cell (e.g., like MSV) this virus can eventually cause transformation of the nonvirus-producing cell or other cells in the host. The proposed mechanism of action for MAIDS is induction of cytokine production (IL-6) by CD4+ T lymphocytes responding to stimulation by the viral core protein expressed on the surface of infected cells. The IL-6 causes an increased proliferation of B cells that eventually leads to transformation, perhaps similar to the effect of EBV on B cells. A replication-defective FeLV virus causing feline immune deficiency has also been identified. The mechanism for disease induction is not well-defined but is associated with the viral envelope and the ability of the virus to cause cytopathic effects in cultured T cells.

10.3.7 Virus-related Host Genes in Chemical Carcinogenesis

For many years it has been known that certain chemicals can induce cancer in animals and humans. Most of these tumors originate in epithelial tissues rather than in connective or blood cell types. Such tumors are called **carcinomas,** and agents inducing epithelial tumors are referred to as **carcinogens.** Over the past decade it has been established that most known potent chemical carcinogens are powerful mutagens, in a variety of organisms. Therefore, these agents presumably transform cells by inducing mutations in cellular genes. Because so many different transforming genes of RNA tumor viruses seem to be acquired by recombination with the host genome, several attempts have been made lately to implicate these cellular genes in neoplastic transformation processes induced by chemicals. Thus far, only one virus-related gene system has produced evidence suggesting involvement in chemical carcinogenesis.

A number of specialized cell lines, mostly derived from rodent embryo fibroblasts cultured for many generations, can be transformed into tumor cells by chemical carcinogens. However, mouse or rat fibroblasts freshly explanted from the embryo are resistant to transformation by chemicals, unless they are infected with an independently replicating murine leukemia virus (MuLV) prior to carcinogen exposure (see Fig. 10.8). However, when cells from older animals are used, or the cells are passed for a long period of time in culture, transformation by chemicals can be achieved without virus. Finally, cells from aging rats or after long-term passage will often undergo spontaneous transformation. Thus, in this system the retrovirus itself does not appear to transform the fibroblasts but increases their susceptibility to the transforming event associated with chemical carcinogens or with the aging process.

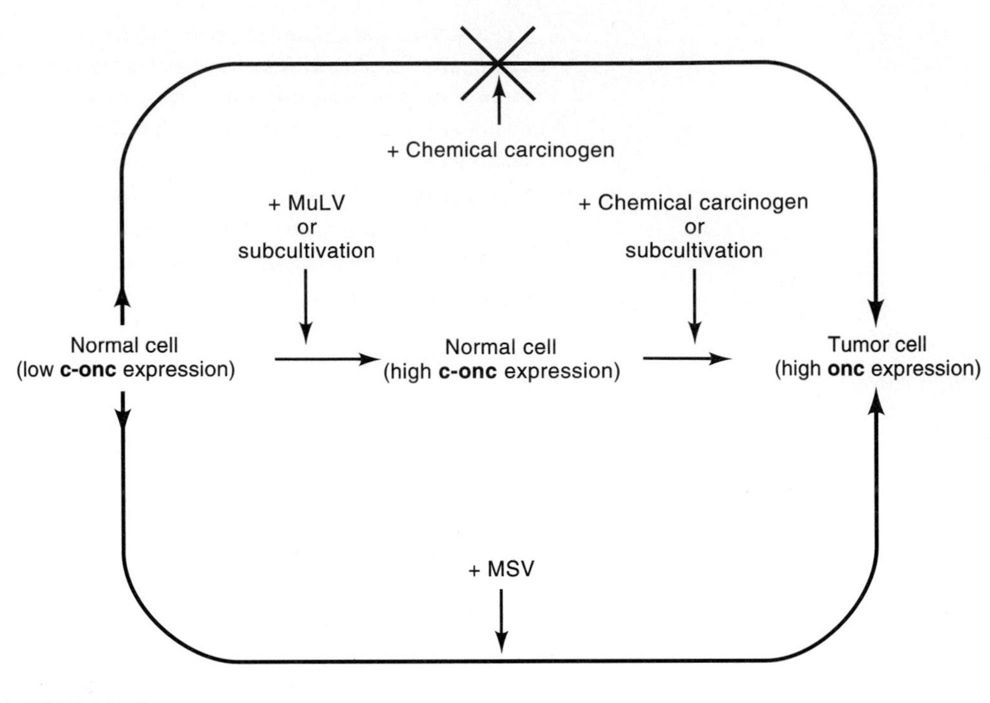

FIGURE 10.8

Neoplastic transformation of cultured rat fibroblasts by chemical carcinogens and/or viruses. Normal cells cannot be transformed into tumorigenic cells by chemical carcinogens alone (top pathway). After infection with murine leukemia virus (MuLV) or extensive subcultivation, cells become sensitive to transformation by chemical carcinogens; extended subcultivation alone also eventually produces tumorigenic cells in culture (middle pathway). Murine sarcoma virus (MSV) can directly transform normal cells into tumorigenic cells (bottom pathway). Relative levels of expression of cellular genes (c-*onc*) that resemble the MSV transformation genes (v-*onc*) are indicated at each stage of transformation.

The extent of cellular c-*onc* expression, as illustrated in Fig. 10.8, reflects changes occurring after virus infection or subcultivation, but clearly, activation of cellular proto-oncogenes does not alone result in transformation of the cells. Although they contain as much sarcoma virus–related RNA as cells transformed by MSV, only after application of a chemical carcinogen or long-term passage do tumor cells result. This observation indicates that the oncogenes of MSV must be different from the related rat cell genes, and perhaps the chemical mutagen transforms the cell by inducing a change in the cellular *onc*-related gene. This mutation would then convert the cellular gene into the viral *onc*-like gene.

This possibility is illustrated in related studies in which rat tumor cells induced by chemicals were co-cultivated with rat embryo cells producing a nontransforming rat type C retrovirus. Whereas the tumor cells alone did not release any transforming agent, the coculture produced a new rat sarcoma virus (RaSV) that transforms fibroblasts readily. This recovery, through the helper activity of the **nontransforming virus,** was the first example of an *in vitro* "rescue" of a transforming retrovirus. It clearly suggested that the rat type C virus transduced the gene responsible for transformation from the original chemically induced tumor cells. Subsequent work demonstrated that the transforming gene in this rat sarcoma virus (i.e., *ras*) was similar to that found in the Harvey and Kirsten mouse sarcoma virus (Table 10.7). These viruses, interestingly, had been passed previously through rats. This *ras* oncogene has now been linked with several human and mouse cellular genes found associated with cancer (Tables 10.7 and 10.8). The findings support the suggestions that genes from tumor cells can be transduced by retroviruses and that oncogenes exist. Nevertheless, as noted in the studies of chemical transformation and in molecular studies, the viral oncogenes (e.g., Ha-*ras*) differ sufficiently from the cellular proto-oncogenes (c-*ras*) to indicate that more than one modification in a normal cellular gene occurred in the production of the transforming gene. The events that have given rise to the identification of about 20 different potential cellular oncogenes appear to be rare in nature and thus are unlikely to be a major cause for the development of naturally occurring tumors. Further work in the field of oncogenes should clarify the possibility that cellular genes resembling viral transforming genes are responsible for tumors in animals and humans. One important aspect of oncogene study that has emerged is their association with growth factors and normal differentiation processes.

In conclusion, animal RNA tumor viruses are serving as tools for identification and purification of many cellular genes that could cause cancer when their levels of expression or perhaps the structure of their products is altered by environmental agents. The fact that the same *src* gene has been found in two different sarcoma viruses that were isolated from hosts as evolutionarily distant as the chicken and the cat (Table 10.8) suggests that many of the possible viral oncogenes have been identified. Thus, the total number of cellular genes with oncogenic potential yet to be discovered may be sufficiently limited to permit complete elucidation of growth regulatory circuits in some types of cells in the next few years.

SECTION 10.3 FURTHER READING

Original Papers

ROUS, P. (1911) Transmission of a malignant new growth by means of a cell-free filtrate. *J. Amer. Med. Assoc.* 56:198.

GROSS, L. (1951) "Spontaneous" leukemia developing in C3H mice following inoculation, in infancy, with AK-leukemic extracts, or AK-embryos. *Proc. Soc. Exp. Biol. Med.* 76:27.

HARVEY, J. J. (1964) An unidentified virus which causes the rapid production of tumors in mice. *Nature* 204:1104.

HARTLEY, J. W., and W. P. ROWE. (1966) Production of altered cell foci in tissue culture by defective Moloney sarcoma virus particles. *Proc. Natl. Acad. Sci. USA* 55:780.

HUEBNER, R. J., et al. (1966) Rescue of the defective genome of Moloney sarcoma virus from a non-infectious hamster tumor and the production of pseudotype sarcoma viruses with various murine leukemia viruses. *Proc. Natl. Acad. Sci. USA* 56:1164.

HUEBNER, R. J., and G. J. TODARO. (1969) Oncogenes of RNA tumor viruses as determinants of cancer. *Proc. Natl. Acad. Sci. USA* 64:1087.

SVOBODA, J., and R. DOURMASHKIN. (1969) Rescue of Rous sarcoma virus from virogenic mammalian cells associated with chicken cells and treated with Sendai virus. *J. Gen. Virol.* 4:523.

BALTIMORE, D. (1970) Viral RNA-dependent DNA polymerase. *Nature* 226:1209.

TEMIN, H. M., and S. MIZUTANI. (1970) RNA-dependent DNA polymerase in virions of Rous sarcoma virus. *Nature* 226:1211.

WHITMIRE, C. E., and R. A. SALERNO. (1972) RNA tumor virus gs antigen and tumor induction by various doses of 3-methylcholanthrene in various strains of mice treated as weanlings. *Cancer Res.* 32:1129.

HOLTZER, H., et al. (1975) Effect of oncogenic virus on muscle differentiation. *Proc. Natl. Acad. Sci. USA* 72:4051.

RASHEED, S., M. B. GARDNER, and R. J. HUEBNER. (1976) *In vitro* isolation of stable rat sarcoma viruses. *Proc. Natl. Acad. Sci. USA* 75:2972.

STEHELIN, D., et al. (1976) DNA related to the transforming gene(s) of avian sarcoma viruses is present in normal avian DNA. *Nature* 260:170.

BRUGGE, J. S., and R. L. ERIKSON. (1977) Identification of a transformation-specific antigen induced by an avian sarcoma virus. *Nature* 269:346.

DELARCO, J., and G. TODARO. (1978) Growth factors from murine sarcoma virus–transformed cells. *Proc. Natl. Acad. Sci. USA* 75:4001.

GRAF, T., N. ADE, and H. BEUG. (1978) Temperature-sensitive mutant of avian erythroblastosis virus suggests a block of differentiation as mechanism of leukemogenesis. *Nature* 275:496.

SPECTOR, D., H. VARMUS, and M. BISHOP. (1978) Nucleotide sequences related to the transforming gene of avian sarcoma virus are present in DNA of uninfected vertebrates. *Proc. Natl. Acad. Sci. USA* 75:4102.

WITTE, O. N., et al. (1978) Identification of an Abelson murine leukemia virus–coded protein present in transformed fibroblast and lymphoid cells. *Proc. Natl. Acad. Sci. USA* 75:2488.

POIESZ, B. J., et al. (1980) Detection and isolation of type C retrovirus particles from fresh and cultured lymphocytes of a patient with cutaneous T-cell lymphoma. *Proc. Natl. Acad. Sci. USA* 77:7415.

BLAIR, D. G., et al. (1981) Activation of the transforming potential of a normal cell sequence: A molecular model for oncogenesis. *Science* 212:941.

VAN BEUEREN, C., et al. (1981) Nucleotide sequence and formation of the transforming gene of a mouse sarcoma virus. *Nature* 289:258.

WATERFIELD, M. D., et al. (1983) Platelet-derived growth factor is structurally related to the putative transforming proteins p28^{sis} of simian sarcoma virus. *Nature* 304:35.

WEINBERGER, C., et al. (1986) The c-*erb*-A gene encodes a thyroid hormone receptor. *Nature* 324:641.

DEZELEE, P., et al. (1992) Small deletion in v-*src* SH3 domain of a transformation defective mutant of Rous sarcoma virus restores wild-type transforming properties. *Virology* 189:556.

ISFORT, R., et al. (1992) Retrovirus insertion into herpesvirus *in vitro* and *in vivo. Proc. Natl. Acad. Sci. USA* 89:991.

LEPRINCE, D., et al. (1993) A new mechanism of oncogenic activation: E26 retrovirus v-*ets* oncogene has inverted the C-terminal end of the transcription factor c-*ets*-λ. *Virology* 194:855.

ZHANG, J., and H. M. TEMIN. (1993) 3' junctions of oncogene-virus sequences and the mechanisms for formation of highly oncogenic retroviruses. *J. Virol.* 67:1747.

TIKHONENKO, A. T., and M. L. LINIAL. (1993) Transforming variants of the avian *myc*-containing retrovirus FH3 arise prior to phenotypic selection. *J. Virol.* 67:3635.

Books and Reviews

HUNTER, T. (1981) Oncogenes and proto-oncogenes: How do they differ? *J. Natl. Cancer Inst.* 73:773.

WEINBERG, R. A. (1982) Fewer and fewer oncogenes. *Cell* 30:3.

HUNTER, T. (1984) The proteins of oncogenes. *Sci. Am.* 251:70.

VARMUS, H. (1984) Reverse transcriptase rides again. *Nature* 314:581.

BALTIMORE, D. (1985) Retroviruses and retrotransposons: The role of reverse transcription in shaping the eukaryotic genome. *Cell* 40:481.

BISHOP, J. M. (1985) Viral oncogenes. *Cell* 42:23.

WEINBERG, R. A. (1985) The action of oncogenes in the cytoplasm and nucleus. *Science* 230:770.

LEVY, J. A. (1986) The multifaceted retrovirus. *Cancer Res.* 46:5457.

BISHOP, J. M. (1987) The molecular genetics of cancer. *Science* 235:305.

DUESBERG, P. H. (1987) Cancer genes: Rare recombinants instead of activated oncogenes. *Proc. Natl. Acad. Sci. USA* 84:2117.

SACHS, L. (1987) Cell differentiation and by-passing of genetic defects in the suppression of malignancy. *Cancer Res.* 1981.

CATLEY, L. C., et al. (1991) Oncogenes and signal transduction. *Cell* 64:281.

MORSE, H. C., III, et al. (1992) Retrovirus-induced immunodeficiency in the mouse: MAIDS as a model for AIDS. *AIDS* 6:607.

chapter

11

Consequences of Virus Infection to the Cell

11.1 VIRUS-CELL INTERACTIONS

Many practical problems in medicine, veterinary science, and agriculture can be traced to viruses that infect eukaryotic organisms. In search of solutions to these problems, scientists have increasingly attempted to apply to eukaryotes the methods of cellular and molecular biology that have been so successful with prokaryotes. The usual strategy for investigation of a virus newly isolated from a higher organism is to study the most accessible questions first, such as the structure of the virion and the mechanism for replication of the virus in cell cultures. In this way it is hoped to find basic principles that can be extrapolated to explain the biological effects of the virus on the intact host and to develop approaches to therapy.

Several factors combine to limit the rate of progress in research on eukaryotic cells and their viruses. First, there are theoretical considerations. The volume of a typical animal cell is on the order of 1000-fold greater than that of an *E. coli* cell, and the animal cell genome contains about 10,000 times as much DNA as a bacterial chromosome. These two factors, **size** and **complexity,** have made the study of viral or cellular processes in eukaryotes more difficult experimentally than studies with prokaryotes. Often, viral products represent only a tiny fraction of all constituents of the infected eukaryotic cell, and the existence of a distinct nucleus and other complicated organelles make reconstruction of intracellular processes in a test tube much more difficult than for bacteria.

Practical aspects of research with eukaryotic cells also are more restrictive than with bacteria. Animal cell cultures require complex media supplemented with animal sera and antibiotics, which can raise the cost of each experiment with an animal virus (e.g., a plaque assay) by a factor of 10 or more compared to the equivalent experiment with bacterial viruses. In addition, the generation times of animal cells are at least an order of magnitude greater than that for bacteria, and the times required for replication of viruses in the two types of systems also show similar disparities. Thus, there are theoretical and practical limitations to the accumulation of knowledge on viruses in eukaryotes, as compared with prokaryotes. A great deal of progress has nevertheless been made.

There is yet another level of influence on the rate of progress in this type of research, one that may be characterized as socioeconomic or political rather than scientific. As noted above, much of the motivation for research on viruses of eukaryotes stems from the desire to control diseases of prominent **medical** or **agricultural** significance. So far the most successful means for such control has been either by eradication of the means of transmission (such as insect vectors), elimination of the virus, or by immunization of the natural host (see Sec. 13.2). None of these methods generally requires detailed knowledge of the molecular mechanisms of virus replication, let alone pathogenesis. Moreover, once a given virus has been eliminated as a medical or economic problem, government agencies that support most virological research are understandably unenthusiastic about funding further studies on that virus. For example, limited research efforts are currently being expended on the molecular mechanisms of the effects of poliovirus infection on the host cells, although this virus still stands as the **prototype** for studying cytocidal virus–animal cell interactions. Unfortunately, it is becoming increasingly evident that the simple control strategies of the past will not always work with virus diseases, and we will have to gather much more basic information before suitable approaches can be developed for preventing or treating newly discovered virus diseases. One of science's greatest challenges is to find an effective antiviral drug that mirrors the achievements made in control of bacterial and even fungal diseases with antibiotics. (See Chapter 13)

In this chapter we will discuss the various effects of viruses on eukaryotic cells (primarily animal cells) and on prokaryotic cells. These include **cytopathic effects** (CPE), or signs of infection in the diseased cell that can be recognized as morphological or gross structural changes, as well as effects on cellular processes required for macromolecular synthesis. For example, a cytopathic virus might cause cell fusion, arrest of cell division or induce vacuole formation in cells. We will also review mechanisms of interference with virus replication at the cellular level.

Any virus infection that results in death of the host cell is called a **cytocidal infection.** If this infection results in cell disruption or lysis, the infection or the virus is also **cytolytic.** When infection is studied in a population of cells, either in culture or in intact organisms, the process can be relatively brief and self-limiting (i.e., **acute**). Sometimes only a fraction of the cells can be infected at any time, and the rest can continue to prolif-

TABLE 11.1
Characteristics of virus infections in cell cultures

Type of Infection		*State of Viral Genome*	*Fate of Cells*	*Selected Examples*
Cytocidal (cytolytic)	Acute	Completely active in all cells	All killed	Many RNA and DNA viruses in permissive hosts
			Without lysis	Polio in mitotic cells; enveloped viruses (especially some paramyxoviruses, rhabdoviruses)
			With lysis	Nonenveloped viruses (e.g., picornaviruses); polyoma in mouse cells; poxviruses
	Chronic	Completely active in some cells, may be latent in the rest	Only cells actively producing virus are killed; the rest proliferate	Rhabdo-, toga-, and bunyaviruses (especially in insect cells)
Persistent	Productive	Completely active in all cells; may be integrated into host DNA	All survive and proliferate	Some paramyxoviruses; retroviruses; rhabdo-, toga-, bunyaviruses in insect cells
			Also may be neoplastically transformed	Retroviruses
	Nonproductive	Latent in all cells (early functions active but little or no viral nucleic acid and proteins made)	All survive and proliferate	Herpesviruses; retroviruses; papovaviruses in nonpermissive hosts (e.g., polyoma in hamster)
			Also may be neoplastically transformed	Herpesviruses (especially Epstein-Barr virus) in human tumors; retroviruses

erate. This situation results in a prolonged infection that neither kills all the cells nor results in elimination of the virus. Such infections are called **chronic** and chronically infected cultures or, especially, organisms are known as virus **carriers** (see Sections 11.1 and 12.6).

Some viruses are not cytocidal, at least in some cells. They may therefore, establish a **persistent** infection of each individual cell. Persistently infected cells may produce virus particles continuously, or they may be *nonproductive.* These various responses are summarized, with examples in Table 11.1.

Persistent nonproductive infections are somewhat analogous to lysogeny in bacteria, and in some cases the viral genome is integrated into the chromosome of the eukaryotic host cell. Thus far, however, no viral regulatory mechanism equivalent to the phage immunity system based on a repressor, has been identified in eukaryotic viruses. No universally accepted single term exists for these nonproductive persistent infections, although the word **latent** has often been used to denote this specific situation. The terms *persistent, chronic,* and *latent* have also been used more or less interchangeably to describe various cellular responses to viral infections (see Table 11.1). **Cellular latency,** however, should be reserved for infections in which no virus particles are produced. The same terms are used as well for whole organisms, in which the virus is not rapidly eliminated but remains somewhere in the organism in an unknown state of replication. In this case, **clinical latency** is generally the preferred designation. These various responses to virus infection are discussed in more detail in Sec. 12.4.

Inoculation of cultures of susceptible animal cells with many (although not all) types of animal viruses, results in a pattern of cell destruction and death that is often characteristic for a given **virus-host combination.** These changes can sometimes be detected by the naked eye, or more often by using a microscope, with or without stains to emphasize various cellular constituents. The range of species and tissues, the cells of which are susceptible to a given virus isolate, and the time course of infection (measured in hours or days) as well as the histological details, are all useful in identifying the type of virus in the inoculum. Certain groups of animal viruses, however, produce essentially no morphological cellular changes that are detectable by light microscopy. In these cases, the presence of the virus can be observed only by indirect means (e.g., hemadsorption, antibody reactions) or by using the electron microscope to visualize virions and viral components in infected cells. These examples illustrate the point that massive cellular damage is generally not an integral part of the virus replication process.

For most animal viruses that do cause visible CPE, the mechanism underlying this effect is not clear, since very little of the cellular mass is converted into virions—especially in comparison with some bacteriophage, insect, and plant virus systems. In these latter systems, 10 percent or more of the dry weight of the infected cell may be viral matter. The small amounts of precursors required for synthesis of viral macromolecules should be readily obtained from preexisting pools in animal cells. This reserve eliminates any need for the virus to degrade host structures for additional precursor supplies. Gross morphological damage by some cytocidal viruses can, under certain conditions, be prevented without inhibition of the replication process. For example, poliovirus replication in cultures of human epithelial tumor (HeLa) cells usually causes a distinctive pattern of nuclear and cytoplasmic damage concluding with cell lysis and death. HeLa cells infected during mitosis, however, show none of these typical signs of poliovirus infection and release new virions without lysis. Undoubtedly, we can not understand the role of cellular damage in the replicative cycle of cytocidal animal viruses—at least until we know the molecular basis of such damage.

11.2 VIRUS CELLULAR RECEPTORS AND CELL MEMBRANE CHANGES

11.2.1 Mode of Infection

The cell surface plays an important role in the early stages of infection of eukaryotes, especially animal cells. This fact is due to the requirement of animal viruses to interact specifically with some surface component that serves as a virus **receptor.** This requirement can be dramatized by demonstrations that certain cells resistant to virions of a given virus

TABLE 11.2
Comparisons of virus-specific receptors on cells of prokaryotes and eukaryotes

Organism	*Nature of Receptor*	*Approximate Number/Cell*	*Virus Example*
Animals	Any plasma membrane glycoprotein with sialic acid	$>10^5$	Ortho- and paramyxoviruses
	Plasma membrane glycoproteins[b]	$>10^5$	Most enveloped viruses with glycoprotein spikes
	Plasma membrane (glyco)proteins[c]	10^4	Most nonenveloped viruses (e.g., picornaviruses) and retroviruses
Plants	None known	—	
Bacteria	Cell wall polysaccharide	$>10^5$	ε phages in *Salmonella*
	Cell wall lipoproteins	$>10^5$	T-even phages
	F-pilus	1 (male only)[a]	Small RNA phages; small filamentous DNA phages

[a]More than one virion may attach to a single pilus: RNA phages attach to sides of pilus, DNA phages, to the end.
[b]Most receptors for enveloped viruses probably are more specific in terms of glycosylation patterns than in the case of myxoviruses.
[c]No proven virus-specific receptors have yet been isolated, but characterization of some shared receptors, such as the CD4 molecule for the AIDS virus, has been accomplished.

can nevertheless be infected by naked viral RNA or RNA encapsidated in the coat protein of another virus (see Sec. 11.7). By way of comparison (see Table 11.2), we have already seen that many **bacteriophages** (e.g., T-even; Sec. 8.7.4) initially attach to specific components in the bacterial wall, external to the cell membrane.

Plant viruses, in contrast, seem to lack specific cellular receptors, since they enter cells only either through wounds that penetrate both the wall and membrane below, or via cell-to-cell transfer through special intercellular connections. Even in cultures of plant cell **protoplasts,** after enzymatic removal of the impenetrable wall, no evidence for virus-specific receptors in the cell membrane has been obtained so far.

The early interaction of an **animal virus** particle with the cellular receptor is mediated by the same sort of weak forces (hydrogen bonds, ionic attractions, van der Waals' forces) that promote self-assembly of virions or cellular organelles from subunits. As in those analogous assembly processes, the interacting molecules must have complementary shapes to facilitate a close "lock and key" fit. This interplay allows the weak forces to come into play, since their effective ranges are short compared with the overall dimensions of molecules as large as proteins. The initial interaction between virion and cell may involve only one viral protein molecule and one receptor molecule, usually a protein. Therefore, the virion is, at first, not tightly bound to the cell and may detach. Irreversible binding probably occurs when multiple sites on one virion bind to multiple receptors.

Studies on this transition from reversible binding to irreversible binding have revealed a general difference between receptors for many enveloped viruses as opposed to *naked virions.* Irreversible binding occurs in the cold for the enveloped viruses but not for those lacking envelopes. The explanation seems to be that receptors for enveloped viruses are glycoproteins that are often extremely abundant on cell membrane surfaces, whereas receptors for nonenveloped viruses may be less abundant membrane components, also probably glycoproteins. For example, myxoviruses may bind to any glycoprotein with sialic acid residues and thus to any of millions of molecules per cell, whereas there are probably only about 10,000 poliovirus receptors per host cell. To achieve binding of one polio virion to multiple receptors probably requires receptors to migrate in the fluid plasma membrane matrix into the vicinity of the initially bound virion. The membrane lipids are, however, less fluid in the cold, and thus movement of embedded proteins through the matrix is prevented.

It is difficult to imagine that cells contain receptors that are specifically designed for infectious viruses. Most likely, the viruses have evolved to use these natural cell surface proteins for entry. Several studies have supported this possibility (Table 11.3): For

TABLE 11.3
Cellular receptors for viruses

Receptor	Virus
Complement (CR2, glycoprotein)	EBV
Acetylcholine	Rabies
Epidermal growth factor	Vaccinia
CD4 molecule	HIV
Ia antigen (mouse)	Lactic dehydrogenase virus
ICAM-1	Rhinovirus
IgG-like molecule	Poliovirus
VLA-2 (integrin)	Echovirus type1
β-adrenergic receptor	Reovirus
Sialic acid	Influenza
Permease	Murine leukemia virus
Fibroblast growth factor	?Herpes simplex

example, the receptor for the Epstein-Barr virus in B lymphocytes appears to be associated with the complement receptor; the mouse Ia antigens are receptors for lactic dehydrogenase virus; the rabies virus may use the acetylcholine receptor; vaccinia may enter via the epidermal growth factor receptor; and the AIDS virus appears to bind to the CD4 molecule, the marker for helper T lymphocytes. For most viruses, the natural function of the cellular receptor is not known.

What happens after irreversible binding of the virion to allow the animal viral genome to enter the host cell? This question is not answered in molecular detail for any virus, and it seems obvious that there must be different answers for specific types of viruses. Various possibilities are summarized schematically in Fig. 11.1. For simple nonenveloped virions, electron microscopy studies suggest that the nucleocapsid does not enter intact into the cytoplasm. Biochemical studies suggest that the host cell membrane somehow causes a rearrangement of capsid proteins, perhaps through limited **proteolysis,** which frees the nucleic acid and allows it to pass through the membrane by an undetermined mechanism. Similar processes may be operating for infections by all nonenveloped virions that contain positive-sense RNA or have DNA, both of which are infectious in pure nucleic acid form. Whether the naked DNA or RNA passes through the plasma membrane by the same route used when it is delivered by a virion remains to be determined. The nucleic acid of most DNA viruses must also pass through the nuclear membrane to the site of DNA replication.

As mentioned in connection with paramyxoviruses (Sec. 4.2), most enveloped virions probably gain access to the cell through fusion of the envelope with the plasma membrane. This process may be the reverse of the "budding" process that releases newly made virions. The result of such fusion is to deposit the nucleocapsid more or less intact into the cytoplasm of the host cells. With most viruses, this process is pH-independent. In some cases, the entire virion enters by **endocytosis.** For negative-sense RNA viruses and the retroviruses, this aspect of entry is crucial, since virion-associated enzymes, either a transcriptase or reverse transcriptase activity, must accompany their nucleic acid templates to initiate infection. In all these viruses, the enzymatic activity appears to be associated with the complete nucleocapsid core rather than being a soluble protein, so that the core must get inside the cell essentially intact.

For some of the larger, structurally more complex virions, such as those of poxviruses, herpesviruses, or reoviruses, electron micrographs suggest that virions may be actively ingested by the host cell by the same process used to ingest other particles, including inert carbon granules, namely **phagocytosis.** They may also enter by receptor-mediated endocytosis, as described above. Virions are often observed inside membranous vesicles that form by invagination of the plasma membrane, engulfing the particle. Within the vesicles, cellular digestive enzymes released from lysosomes could digest away outer layers of complex nucleocapsids to release the nucleic acid.

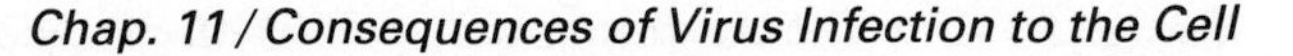

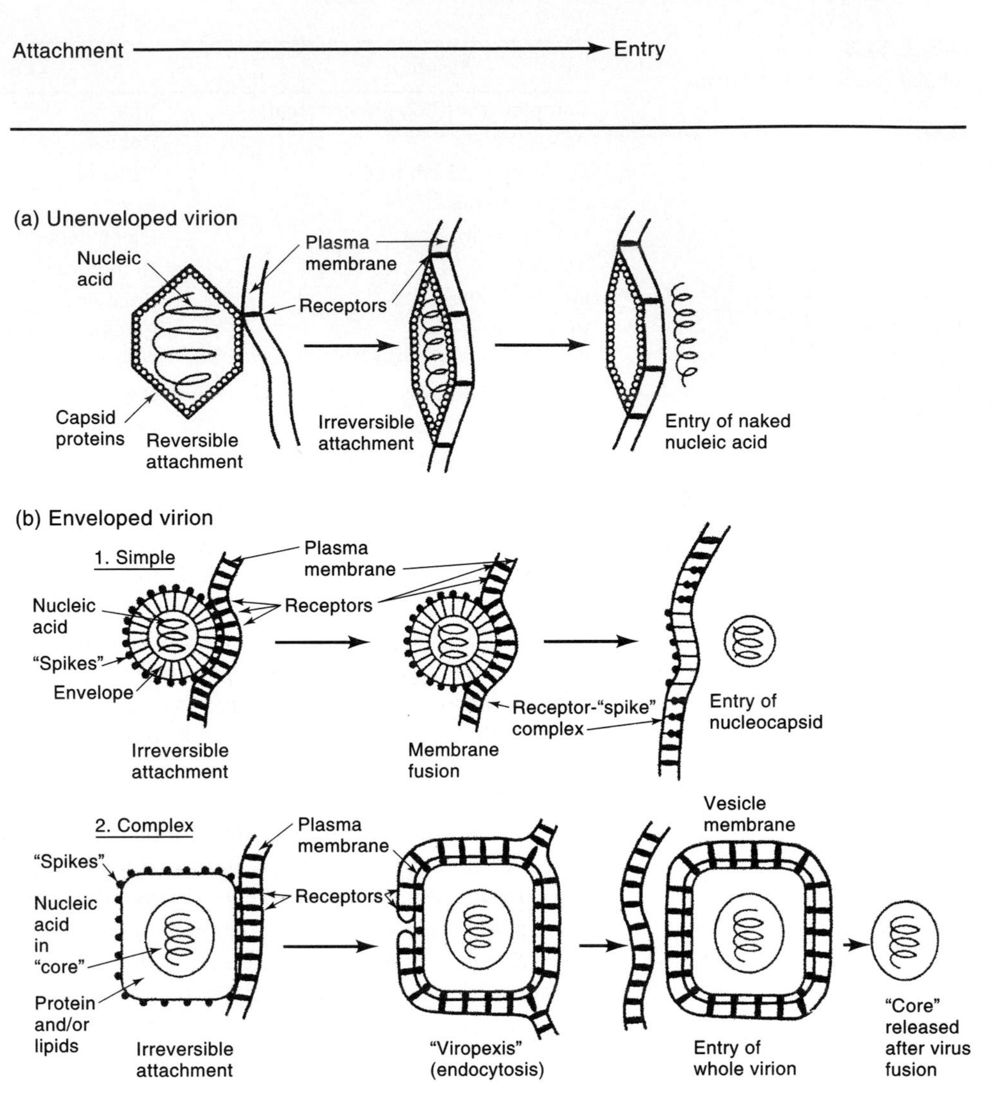

FIGURE 11.1

Schematic comparison of early events of infection by various types of animal viruses. (a) Unenveloped virions attach permanently only after migration of receptors to the virion. Irreversible attachment partially rearranges the virion, allowing nucleic acid to pass through the plasma membrane. (b) Enveloped virions attach irreversibly to multiple receptors immediately. For simple virions (case 1) fusion of the plasma membrane with the envelope deposits the nucleocapsid (with transcriptase, if any) in the cytoplasm. For complex virions (case 2), it is believed that the entire virion enters by phagocytosis; subsequently the viral core enters the cytoplasm after viral fusion with the vesicle membrane. See the text for further details and examples of each type of virion.

Limited proteolysis in these vesicles probably activates the reovirus transcriptase associated with the core, but how the viral nucleic acid escapes from the vesicles without degradation from known lysosomal nucleases remains a mystery. In some viral systems, fusion with the vesicle membrane brings the core into the cytoplasm. This process is generally at a low pH. However, it is not absolutely certain that the vesicle-associated virions actually are initiating infection; they simply could be degraded while other virions entering by another means could start the replication process.

The fact that virus-specific receptors play such a key role in determining whether or not a virion can enter an animal cell strongly suggests that ordinary phagocytosis alone cannot initiate infection for most viruses. Nor does giving a special name (**viropexis**) to this putative phagocytic virion entry process establish that it truly exists. Most likely, more than one method is used in virus infection. Further work on the complicated reactions in virion–cell membrane complexes will be needed to understand the molecular changes occurring at the cell surface during viral entry. Perhaps these are initiated by the allosteric structural changes favoring the binding of viral proteins to host cell receptors.

One other feature of virus-cell interaction that is observed in a variety of virus families is the importance of cell-associated proteases. In paramyxoviruses, the F protein must be activated by a trypsinlike molecule. For lentiviruses, the envelope gp120 may require cleavage by a cellular protease for fusion and entry to take place. Finally, with astroviruses, pancreatic trypsin in the intestinal mucosa activates virus infectivity.

11.2.2 Cell Surface Changes and Related Effects

Many virus infections of animal cells result in changes in the cell surface late during the replication cycle. This effect is especially true for enveloped viruses that mature by budding from the cell membrane. As we have seen in earlier chapters, these viruses insert viral proteins into the cell membrane, thereby displacing cellular membrane proteins, to form an envelope composed primarily of viral proteins and cellular lipids. However, the viral proteins in the membrane of the infected cell often change the biology of the host cell in various ways. They may also elicit an immune response by the infected host (see Secs. 12.3.1 and 12.4.5). Sometimes the changes on the cell surface are subtle and can be detected only with sensitive immunological tests. Other viral proteins produce more dramatic effects, such as hemadsorption, hemagglutination, aggregation, or cell fusion.

As noted previously, the virions of many viruses have an ability to adsorb to the surface of erythrocytes as well as to receptors on their true target cells. This adsorption can cause **hemagglutination** or clumping of the red blood cells that are coated with virions. The viral protein responsible for this activity (and most likely for attachment to host cell receptors) is called a **hemagglutinin,** and obviously it must be located on the virion surface. In orthomyxoviruses, for example, the hemagglutinin comprises one of the proteins in the "spikes" protruding from the envelope (see Sec. 4.3). This fact is probably also true for other virions with "spikes."

During infection of the usual host cells, insertion of the hemagglutinin into the infected host cell membrane prior to budding allows added erythrocytes to adhere to the host cell surface. This effect is known as **hemadsorption,** and it can sometimes be used as a tool to detect infections in cell cultures that otherwise result in minimal CPE (see Fig. 11.2). Individual viruses have varying abilities to bind erythrocytes from different animal species. Therefore, hemadsorption or hemagglutination reactions can be used in clinical virology to detect specific viruses. For example, reovirus type 3 can be distinguished from types 1 and 2 by its ability to agglutinate human O erythrocytes. Moreover, hemagglutinin variants of this virus may show altered cell tropism. Cells infected with paramyxo-, toga-, rhabdo-, corona-, and poxviruses all exhibit hemadsorption under certain conditions.

Virions of many more viruses, including some without envelopes, cause hemagglutination but are unable to cause hemadsorption to infected cells. For the nonenveloped virus of this type, such as picorna-, polyoma-, adeno-, parvo-, and some reoviruses, the lack of hemadsorption is due to the fact that viral proteins are not inserted into the cell membranes, since the virion has no envelope. Certain enveloped viruses that hemagglutinate may not cause hemadsorption under the physiological conditions needed for struc-

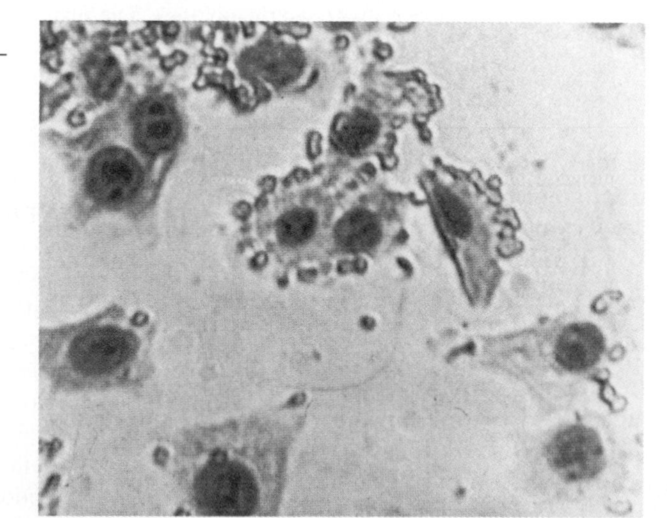

FIGURE 11.2

Human HEP-2 cells persistently infected with measles virus, showing hemadsorption of smaller Rhesus monkey red blood cells to the surface of the infected fibroblasts. (Courtesy of John Rice)

TABLE 11.4
Properties of some commonly used lectins

Lectin	Source	Binding Specificity	Application
Concanavalin A	Jackbean *(Canavalia ensiformis)*	α-D-Mannopyranosyl or α-D-glucopyranosyl residues	Mitogen for lymphocytes; detection of changes on cell surfaces
Phytohemagglutinin	Kidneybean *(Phaseolus vulgaris)*	*N*-Acetyl-D-galactosamine	Mitogen for T lymphocytes
Wheat germ agglutinin	Wheat germ *(Triticum vulgare)*	*N*-Acetyl-D-glucosamine	Agglutinates transformed cells but not normal counterparts, unless protease-treated
Soybean lectin	Soybean *(Glycine max)*	*N*-Acetyl-D-galactosamine α-D-galactosamine	Agglutinates transformed cells; binds to normal lymphocytes; mitogenic for neuraminidase-treated lymphocytes
Pokeweed	*(Phytolacca americana)*	?	Mitogen for T and B lymphocytes

tural integrity of the infected cell. Alternatively, they may bud through nuclear or intracytoplasmic (e.g., Golgi) membranes, so the cell surface membrane is not modified.

Viral hemagglutinins or other glycoproteins inserted into the cell surface membrane can also be bound by proteins other than those embedded in the surface of the target cell or erythrocytes. **Lectins** are a class of mainly plant glycoproteins that bind to specific carbohydrate moieties of glycoproteins from various sources, including erythrocytes and virions (see Table 11.4). Each lectin recognizes a particular combination of sugars with specific chemical linkages. Since all animal cells have surface membrane glycoproteins, they all react with some lectins—even without viral infection. Viral envelope proteins, however, may intensify lectin binding or introduce binding sites for lectins that are unable to bind to the uninfected cells. In some cases, especially in cells transformed into tumor cells by viruses, cellular membrane glycoproteins may be chemically or physically modified so that binding of certain lectins is altered, usually increased. Because many lectins have more than one binding site on each protein molecule, they are able to form intercellular linkages resulting in **lectin-induced agglutination.** Animal cells infected with many viruses (e.g., pox-, adeno-, orthomyxo-, and paramyxoviruses) tend to adhere to each other or **aggregate** in the absence of lectins, presumably due to interactions between viral proteins and cell surface proteins, including the virus-specific receptors.

Cell fusion as a result of virus infection may be considered an extreme form of intercellular binding of surface membranes. The resulting fused cells, which contain several nuclei within one common cytoplasm, are called **syncytia** (Fig. 11.3). This phenomenon is generally limited to certain strains of paramyxoviruses (including measles and respiratory syncytial viruses), but some herpesviruses, a few vaccinia virus isolates, a coronavirus (avian infectious bronchitis), and some retroviruses can also induce syncytia, either in cultured cells or in cells in the intact animal. The molecular basis of this biological effect has been studied in most detail with **paramyxoviruses** such as the avian Newcastle disease virus (NDV) and the murine Sendai virus. Two types of fusion are recognized operationally, although their biochemical basis may well be the same. Late during infection NDV causes *fusion from within* the cell. Addition of large amounts of virions to uninfected cells (i.e., infection at high multiplicities of infection, m.o.i.), even after inactivation of viral infectivity, results in rapid cell fusion known as *fusion from without.*

The molecular mechanism of cell fusion by paramyxoviruses has not yet been fully established, but the critical viral envelope protein, called the fusion factor, has been identified (see Sec. 4.2). As noted, this fusion factor is synthesized as a single glycosylated polypeptide, F, which must be cleaved to generate infectious and fusion-inducing particles. At least some of the cellular receptors for attachment of NDV are known to be glycopro-

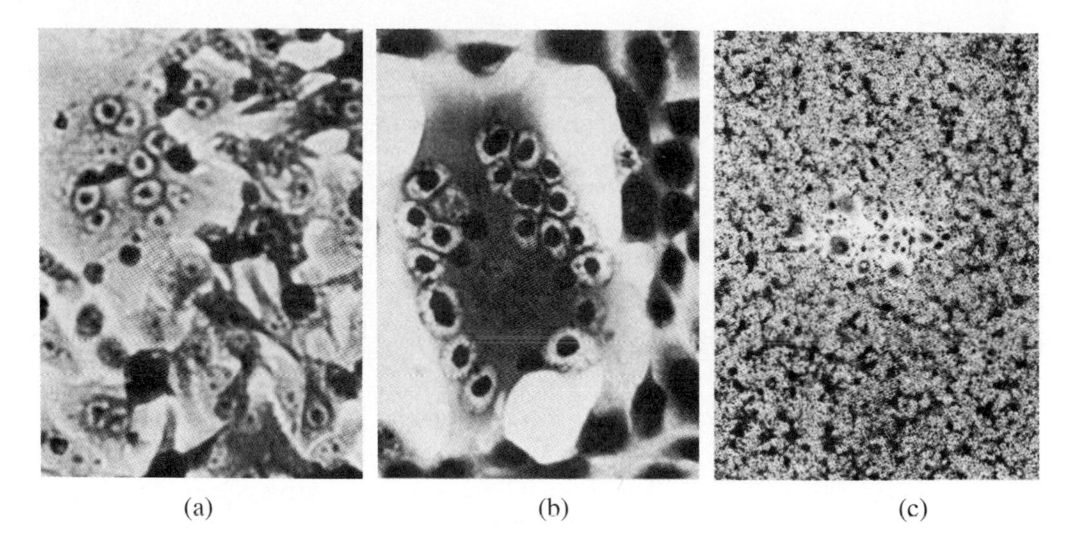

(a) (b) (c)

FIGURE 11.3

Cell fusion and nuclear inclusions induced by an RNA and a DNA virus. (a) Human HEP-2 cells infected with Edmonston measles virus. A large syncytial cell with several nuclei is centrally located. The nuclei in the syncytium have large darkly stained inclusions that are absent from nuclei of several apparently uninfected cells in the upper left corner. The smaller dark bodies in uninfected nuclei are the nucleoli. (Courtesy of John Rice) (b) Rabbit kidney cells infected with equine herpesvirus type 2, strain LK. A large syncytial cell is centrally located. Also note the large dark nuclear inclusion bodies in the nuclei of the syncytium and some peripheral cells. Other peripheral nuclei (lower left) appear normal. (Courtesy of James Blakeslee, Jr.) (c) A plaque in the XC rat sarcoma cells induced by a murine leukemia virus (retrovirus). The virus was transferred to the rat cells by an infected mouse cell. It fused with the XC cells to produce large cells containing up to 100 nuclei of both rat and mouse origin. This plaque assay has

teins containing *N*-acetylneuraminic or **sialic acid residues.** Evidently, these glycoproteins are not the only receptors, however, because enzymatic stripping of sialic acid from the cell surface does not inhibit attachment and infection by at least some strains of NDV. However, stripping of sialic acid does prevent fusion. Also, mutant cells have been isolated that are resistant either to Sendai-induced fusion or to binding and resultant killing by certain lectins. These mutant cells will adsorb and replicate Sendai virus or NDV normally, and the active fusion factor is produced during the infection, but no fusion is observed in either type of mutant cell. One of these mutant cells is defective in a transferase for *N*-acetylglucosamine, but the membrane substrate for glycosylation by this enzyme has not yet been identified. Presumably, these substrate glycoproteins, other glycoproteins, or, perhaps, glycolipids interact with a viral fusion factor soon after attachment of the virion. This event causes a configurational change in the envelope membrane that can be observed with fluorescent antibodies to viral envelope proteins. The fusion factor then enters the cell membrane matrix and diffuses away from the virus attachment site, thus propagating the configurational change throughout the cell surface membrane. This change in both the envelope and the cell membrane then causes fusion of these two membranes, as well as fusion of any two adjacent cell membranes exposed to the fusion factor.

These observations indicate that some viruses such as NDV have two points of interaction with the cell surface: an attachment site and a receptor for the fusion factor. Infection is probably more efficient when both points of virus binding are active, but clearly the fusion protein is not required for virus infection. It aids this process, and antibodies against the fusion factor can limit the efficiency of virus infection. This phenomenon has been observed with paramyxoviruses and has been implicated in studies of the HIV lentivirus (Chap. 14). Antiviral neutralizing antibodies and antibodies that block cell fusion by the virus can be distinguished. This fact suggests that two proteins can play a role in virus infection.

Thus, besides mutations in the cells and the use of certain lectins, antiviral antibodies, and antifusion factors can be employed to influence—inhibit or enhance—the fusion process. With a better understanding of membrane fusion will come knowledge about virus-induced cell lysis and insights into the more general problems of membrane biosynthesis, structure, and function.

11.2.3 Cell Fusion as a Multi-Purpose Tool

Although the mechanism of virus-induced fusion is not thoroughly understood, the process has been used extensively to study other questions in virology and cell biology. This technique is, for instance, very helpful for introducing certain materials (including virus particles) into cells that do not ordinarily take them up. Some virions could be trapped between fusing cells and be included in the resulting syncytium. This process provides a way of circumventing the usual need for a virus-specific receptor, if one wishes to study the replication of a virus by cells that it normally could not enter. Cell fusion has also been used to detect certain infectious viruses in tissue culture by a plaque assay [Fig. 11.3(c)].

Another application of cell fusion is based on the fact that in some syncytial cells two or more of the nuclei can also join together into one nucleus, and the resulting **hybrid** cell may be viable. This **somatic cell hybridization** method has been used to examine dominance of one set of chromosomes over another in a single nucleus. For example, in cells of some species the small DNA virus polyoma can cause only a lytic infection, but in others it is not replicated as a free genome but is able to integrate into the host cell DNA (see Sec. 7.2). Unlike many temperate bacteriophages, release of polyoma virions cannot be directly induced after integration; but fusion of cells carrying polyoma with cells that are permissive for lytic infection results in production of virus from the hybrids. This process is called **co-cultivation.** The dominance of the permissive cell chromosomes in allowing virus replication in hybrids suggests that the cells that do not allow lytic infection cannot provide some function essential for virus replication rather than that the virus produces a soluble repressor in the nonpermissive cells (see below).

Similar experiments have been conducted to map genes or integrated viruses onto particular chromosomes. This approach relies on the finding that interspecies somatic cell hybrids tend to be unstable in chromosome content, with chromosomes of one or the other parent cell being gradually and preferentially lost as the hybrid cells multiply. In this procedure, cells of one species carrying an integrated viral genome are fused with those of another species that is also nonpermissive for lytic infection. Resulting hybrid cells are examined for karyotypes and viral antigen expression at various times and multiple replication cycles after fusion. This procedure allows determination of which specific chromosome(s) of the originally infected cells must remain to produce viral antigens in hybrid cells and, thus, which chromosome(s) carry the integrated viral genome. Many nonviral genes have also been mapped in a similar fashion.

In the past, Sendai virus has been used in most of these cell hybridization studies, but there have been problems associated with the frequent presence of avian retroviruses contaminating Sendai stocks grown in chicken embryos. Even though the Sendai virus stocks are inactivated with UV light or chemicals, the presence of other viral genetic information can interfere with the studies. The need for continuous preparation of uniformly inactivated Sendai virus is a further disadvantage. Presently, most cell fusion experiments employ pure chemical fusion factors, such as polyethylene glycol or lysolecithin, which work efficiently, are easier to control, and cannot introduce any extraneous viral information into the fused cells.

SECTIONS 11.1 AND 11.2 FURTHER READING

Original Papers

Rosen, L. (1960) A hemagglutination-inhibition technique for typing adenoviruses. *Am. J. Hyg.* 71:120.

Cords, C. E., and J. J. Holland. (1964) Alteration of the species and tissue specificity of poliovirus by enclosure of its RNA within the capsid of coxsackie B1 virus. *Virology* 24:492.

Rowe, W. P., W. E. Pugh, and J. W. Hartley. (1970) Plaque assay techniques for murine leukemia viruses. *Virology* 42:1136.

Duff, R., and F. Rapp. (1973) Oncogenic transformation of hamster embryo cells after exposure to inactivated herpes simplex virus type I. *J. Virol.* 12:209.

ROBINSON, H. L. (1976) Intracellular restriction on the growth of induced subgroup E avian type C viruses in chicken cells. *J. Virol.* 18:856.

POLOS, P. G., and W. R. GALLAGHER. (1979) Insensitivity of a ricin-resistant mutant of Chinese hamster ovary cells to fusion induced by Newcastle disease virus. *J. Virol.* 30:69.

LENTZ, T. L., et al. (1982) Is the acetylcholine receptor a rabies virus receptor? *Science* 215:182.

BURNESS, A. T. H., and I. U. PARDOE. (1983) A sialoglycopeptide from human erythrocytes with receptor-like properties for encephalomyocarditis and influenza viruses. *J. Gen. Virol.* 64:1137.

PASTAN, I., and M. C. WILLINGHAM. (1983) Receptor-mediated endocytosis: Coated pits, receptosomes, and the Golgi. *Trends Biochem. Sci.* 8:250.

ROGERS, G. N., et al. (1983) Single amino acid substitutions in influenza haemagglutinin change receptor binding specificity. *Nature* 304:76.

SPRIGGS, D. R., et al. (1983) Hemagglutinin variants of reovirus type 3 have altered central nervous system tropism. *Science* 220:505.

ABRAHAM, G., and R. J. COLONNO. (1984) Many rhinovirus serotypes share the same cellular receptor. *J. Virol.* 51:340.

FINGEROTH, J. D., et al. (1984) Epstein-Barr virus receptor of human B lymphocytes and the C3d receptor CR2. *Proc. Natl. Acad. Sci. USA* 81:4510.

INADA, T., and C. A. MIMS. (1984) Mouse la antigens are receptors for lactate dehydrogenase virus. *Nature* 309:59.

CO, M. S., et al. (1985) Isolation and biochemical characterization of the mammalian reovirus type 3 cell-surface receptor. *Proc. Natl. Acad. Sci. USA* 82:1494.

EPSTEIN, D. A., et al. (1985) Epidermal growth factor receptor occupancy inhibits vaccinia virus infection. *Nature* 318:663.

CHOI, A. H. C., and P. W. K. LEE. (1988) Does the β-adrenergic receptor function as a reovirus receptor? *Virology* 163:191.

BERGELSON, J. M., et al. (1992) Identification of the integrin VLA-2 as a receptor for echovirus 1. *Science* 255:1718.

KEWALRAMANI, V. N., A. T. PANGANIBAN, and M. EMERMAN. (1992) Spleen necrosis virus, an avian immunosuppressive retrovirus, shares a receptor with the Type D simian retroviruses. *J. Virol.* 66:3026.

PETIT, M.-A., et al. (1992) PreS1-specific binding proteins as potential receptors for hepatitis B virus in human hepatocytes. *Virology* 187:211.

SCHMID, A., et al. (1992) Subacute sclerosing panencephalitis is typically characterized by alterations in the fusion protein cytoplasmic domain of the persisting measles virus. *Virology* 188:910.

ANDERSON, R., and F. WONG. (1993) Membrane and phospholipid binding by murine coronaviral nucleocapsid N protein. *Virology* 194:224.

GUIRAKHOO, F., et al. (1993) Selection and partial characterization of dengue 2 virus mutants that induce fusion at elevated pH. *Virology* 194:219.

NI, Y., and R. F. RAMIG. (1993) Characterization of avian reovirus-induced cell fusion: The role of viral structural proteins. *Virology* 194:705.

TANABAYASHI, K., et al. (1993) Identification of an amino acid that defines the fusogenicity of mumps virus. *J. Virol.* 67:2928.

Books and Reviews

LONBERG-HOLM, K., R. L. CROWELL, and L. PHILIPSON. (1976) Unrelated animal viruses share receptors. *Nature* 259:679.

CROWELL, R. L., and B. J. LANDAU. (1979) Receptors as determinants of cellular tropism in picornavirus infections. In *Receptors and human disease,* ed. A. G. Bearn and P. W. Choppin. p. 1. Josiah Macy Fdn. Press, New York.

BUKRINSKAYA, A. G. (1982) Penetration of viral genetic material into host cell. *Adv. Virus Res.* 27:141.

DIMMOCK, N. J. (1982) Initial stages in infection with animal viruses. *J. Gen. Virol.* 59:1.

LENARD, J., and D. MILLER. (1982) Uncoating of enveloped viruses. *Cell* 28:5.

TARDIEU, M., R. L. EPSTEIN, and H. L. WEINER. (1982) Interaction of viruses with cell surface receptors. *Int. Rev. Cytol.* 80:27.

MARSH, M. (1984) The entry of enveloped viruses into cells by endocytosis. *Biochem. J.* 218:1.

WHITE, J. M., and D. R. LITTMAN. (1989) Viral receptors of the immunoglobulin superfamily. *Cell* 56:725.

WHITE, J. M. (1990) Viral and cellular membrane fusion proteins. *Ann. Rev. Physiol.* 52:675.

11.3 VIRAL HOST RANGES

After introduction of the virus into the host and the virus interaction at the cell membrane, the next steps in infection encompass the actual mode of **entry** of virus particles into a susceptible cell, their replication, and their passage into neighboring cells. In this regard, the importance of the term *susceptible cell,* should be stressed again. Certain viruses are very selective in that thay are replicated only in one host species, whereas others can infect almost all mammals as well as insects. Examples of viruses that are highly specific for humans are most flu viruses, poliovirus, and measles virus. As you know from the earlier discussions of these viruses, they have nothing in common except this host specificity. Nonselective viruses include the rabies and other rhabdoviruses, several encephalitis viruses, and poxviruses. The factors that determine the range of **host specificity** are largely unknown, but the chemical affinities of cell surface and virus surface proteins probably play a major role. For example purified polio viral RNA is able to infect nonhuman cells that are resistant to the intact virus. Thus, susceptibility to infection by a virus is largely a genetic property of a given animal species.

One classification scheme for host range virus variants has been established for retroviruses (see Table 11.5), based on the ability of these viruses to infect cells of their own species **(ecotropic)** or those of heterologous species **(xenotropic).** Because the xenotropic viruses cannot infect cells of their own species, they provide biological evidence for the germline inheritance of infectious retroviruses. In some animal species they only can be detected by using approaches that activate endogenous viruses. Both ecotropic and xenotropic viruses can be found in certain animals (e.g., mice, cats). Moreover, retroviruses with both xenotropic and ecotropic **cellular host ranges** have been identified in mice. Such viruses have a distinct envelope glycoprotein and are classified as a third subgroup, the **amphotropic** viruses. Some ecotropic viruses preferentially infect thymus-derived cells; they are called **thymotropic.** Finally, any virus formed by recombination between ecotropic and xenotropic viral genes would have the amphotropic host range but the envelope properties of both ecotropic and ecotropic viruses. These viruses have been termed **dualtropic** or **polytropic** retroviruses. All but the xenotropic viruses have been linked to leukemia. The xenotropic viruses cannot reinfect host cells that have been considered agents that could play a role in normal developmental processes.

As explained in Sec. 11.1, there are specific virion surface proteins that interact with the receptors on the cell surface and it appears that sometimes only a small change in the viral protein may allow it to expand its host range. This finding is especially evident

TABLE 11.5
Cellular host range classification of retroviruses

Name	*Derivation*	*Host Range*	*Example*
Ecotropic	Greek, *oikos,* "home"; *tropos,* "turning"	Productively infects cells of the host species	Gross murine leukemia virus, feline leukemia virus, woolly monkey leukemia virus
Thymotropic		Selective replication in thymocytes	Radiation leukemia virus (mouse)
Xenotropic	Greek, *xenos,* "foreign"	Productively infects only cells from an animal species different from the host	Mouse xenotropic virus, cat RD114 virus
Amphotropic	Greek, *amphos,* "both"	Productively infects cells of both the host species and from heterologous species	Mouse amphotropic virus
Dualtropic (polytropic)	Recombination between ecotropic and nonecotropic genetic information	Same as amphotropic	Mink cell focus-forming virus (MCF) (mouse)

SOURCE: J. A. Levy (1986). *Cancer Res.* 46:5457.

with flu viruses, for which the hemagglutinin protein initiates the attachment process. A number of outbreaks of human influenza have been traced to people working with influenza-virus-infected swine. Although many flu viruses are species-specific, some swine strains appear to have acquired the ability to infect humans. Analyses of the hemagglutinins of both swine and human isolates suggest that the change in host range may be the result of point mutations affecting only one or a few bases in the hemagglutinin gene. Similar alterations in the host range of avian influenza strains are also suspected to have started other human infections (see Sec. 4.3).

11.4 MORPHOLOGICAL ALTERATIONS IN CELLS

11.4.1 Intracellular Inclusions

Besides cell fusion, other changes can be induced in cells after virus infection. As might be expected, some localized changes in intracellular architecture arise at sites of synthesis or assembly of viral components. These discrete foci of viral activity frequently are called **virus factories.** They are often visible by light microscopy as distinct **inclusion bodies,** especially when the cells are stained specifically for nucleic acids or with fluorescent antibodies for viral proteins. Depending on the virus, there may be one or many inclusions of various sizes. The shape and location within the cell are also characteristic of the causative virus.

Because of their large genomes, the poxviruses provide a good example. They are able to replicate in the cytoplasm, independently of the host cell DNA synthetic machinery in the nucleus (see Sec. 8.3). The infected cells therefore contain large stainable cytoplasmic patches where viral replication occurs (see Fig. 11.4). These areas have been termed **Guarnieri's bodies** after the Italian physician who first discovered their diagnostic significance in relation to infections with the smallpox agent. Similarly, infection with adenoviruses, that replicate in the cell nucleus, results in formation of characteristic intranuclear inclusions that consist of crystalline arrays of virions or viral proteins. Rabies virus, and to a certain extent other rhabdoviruses, produce intracytoplasmic inclusion bodies that are thought to represent accumulations of nucleocapsids; for some reason these fail to reach the cell membrane where the process of envelopment and secretion of completed virions takes place (see Fig. 11.5). The rabies inclusion, named **Negri bodies** after another Italian physician, are still considered final proof of rabies infection in human or animal brain tissue in many cases, although virus-specific fluorescent antibodies are rapidly displacing all such earlier histological diagnostic tools.

Some of the most obvious viral inclusions are found in infections of insect cells. Often virions are embedded in polyhedral crystals held together by a virus-coded matrix protein. Thus, the diseases induced by these viruses are called **polyhedroses,** and either nuclear or cytoplasmic polyhedroses can be found, depending on the virus and insect host (Fig. 11.6). Two important diseases of silkworm larvae are a nuclear polyhedrosis caused by a double-stranded DNA virus and a cytoplasmic polyhedrosis caused by a double-stranded RNA virus of the Reoviridae family (see Secs. 5.1 and 8.5). Other insect virus infections are characterized by principally cytoplasmic granuloses, accumulations of grains or capsules of viral protein with one or several DNA-containing virus particles. Some insect viruses produce no inclusions at all, however.

Inclusion bodies of animal viruses do not always represent only masses of pure viral materials. Reovirus-infected cells contain unique crescent-shaped cytoplasmic bodies that nearly surround the nucleus. These structures contain the disrupted mitotic spindle apparatus of the cell, aggregated with dsRNA and reovirus virions. The function of this association in viral replication, if any, is unclear. Similarly, cytoplasmic inclusions in arenavirus-infected cells contain clumps of ribosomes that are also regularly enclosed within virions for unknown reasons. Still other inclusions could have little or no association with virion components. For example, herpesviruses are replicated in the nucleus, but most of the changes there during the active replication phase are too subtle to be seen without the electron microscope. The virions are completed by addition of envelopes during budding through the nuclear membrane. This event results in major morphological alteration of the

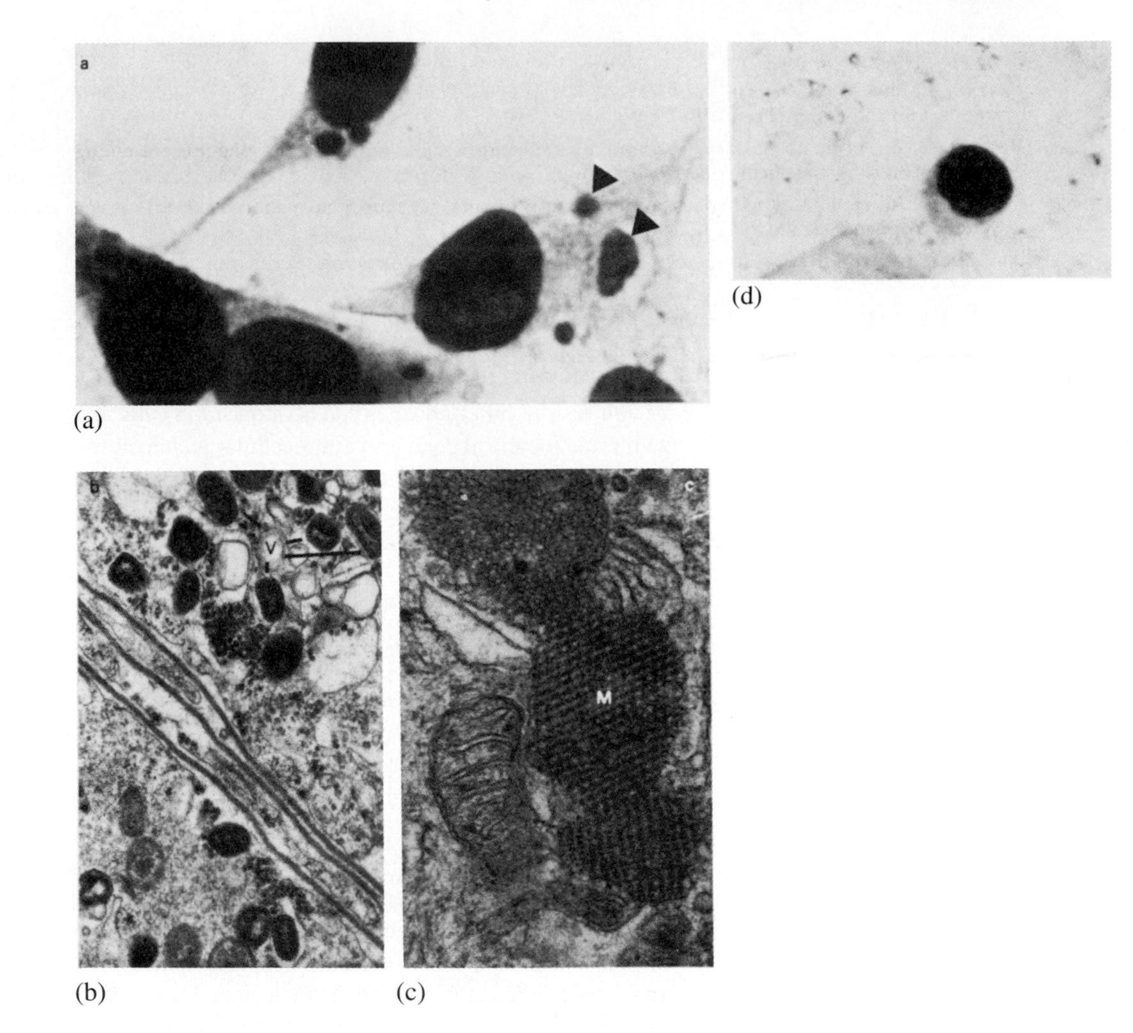

FIGURE 11.4

Cytopathic effects of viruses. Pox viruses replicate in "factories" in the cytoplasm, called Guarnieri's bodies. (a) Light microscopy of mouse L cells infected with vaccinia virus (× 1200). Arrows indicate large inclusions. Electron microscopy reveals that some poxvirus inclusions contain virions, whereas others contain only a matrix of proteinaceous materials. In monkey cells infected with the Yaba monkey tumor poxvirus, some inclusions (b) contain virions with typical poxvirus morphology. A few are indicated by (V) (× 39,000). Other inclusions, e.g., (c), contain no visible virus particles but only an orderly array of matrix (M) of presumed virion components (× 69,700). (Courtesy of Adalbert Koestner) (d) Typical intranuclear inclusion observed in large epithelial cell infected by cytomegalovirus (× 40).

nuclear membrane. After the virions have left the nucleus, however, the cellular chromatin clumps into typical intranuclear inclusion bodies that can be easily viewed by light microscopy of stained cells. The cytomegalovirus inclusion body is a good example and is diagnostic of the infection (see Fig. 11.4d). Poliovirus also causes major morphological changes in infected cells, but only after its replication in the cytoplasm is essentially over. These changes include **nuclear pyknosis,** shrinkage of the nucleus and condensation of the chromatin (see Sec. 11.4.3).

Some viruses induce seemingly irrelevant proliferations of cellular membranes. Poliovirus stimulates synthesis of intracytoplasmic membranes, whereas alphaviruses of the Togaviridae family, which also replicate in the cytoplasm, stimulate perinuclear membrane proliferation. Certain viruses in which both DNA replication and virion assembly occur in the nucleus nevertheless cause pronounced **vacuoles** or bubbly-looking vesicles in the cytoplasm (Fig. 11.7). SV40 derives its original name, simian vacuolating agent no. 40,

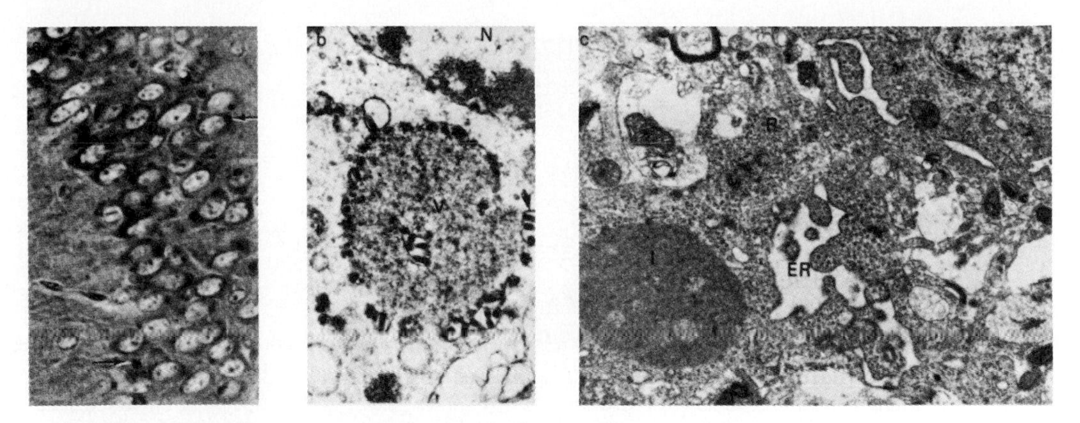

FIGURE 11.5

Cytopathic effects of rabies virus. (a) Infection of cells with rabies virus induces typical cytoplasmic inclusion bodies called Negri bodies, seen as dark bodies (arrows) by light microscopy in brain tissue of an infected mouse (× 800). Some inclusions contain whole virions, whereas others contain only virus-related materials. (b) Neuron in the brain of an infected Rhesus monkey contains a cytoplasmic inclusion (V) with matrix material surrounded by mature viral particles (× 79,000). Some particles are intimately attached to ribosome-linked membranes (arrow). N = Nucleus. (c) Neuron from the brain of an infected mouse containing an inclusion body (I) consisting of matrix without mature virions (× 17,000). Although the endoplasmic reticulum (ER) is dilated compared with that of normal cells, the cytoplasm is rich in ribosomes (R), indicating the cell had probably not yet been killed by the virus. (Courtesy of Adalbert Koestner)

from this characteristic effect. Moreover, the spumavirus genus of retroviruses received its name from this biological effect in cell culture (see Sec. 6.1).

11.4.2 Ultimate Results of Virus Infection

Essentially all phages except the filamentous ones such as fd, are released by lysis, a process that necessarily also entails death of the cell (see Chap. 8). Many animal viruses are also *cytocidal* and *cytolytic*, but other viruses of eukaryotes (especially of these animals) are not cytolytic. Their mature virions have a membranous envelope consisting of viral proteins embedded in **cellular lipids** of the plasma membrane (see Sec. 11.2). Virion release occurs by a budding process that does not disrupt the cell and therefore need not be lethal. In fact, some of the enveloped animal viruses cause little or no cytopathic effects.

The retroviruses are a good example of this phenomenon. The progeny of most of these viruses are produced by budding from the cell membrane without causing cell death. Nevertheless, the AIDS virus generally kills CD4+ lymphocytes during replication in these cells. Several togaviruses can replicate in cultures of both invertebrate (arthropod) cells and vertebrate (avian or mammalian) cells. While these viruses produce acute and highly cytocidal infections in the vertebrate cells, infections of invertebrate cells can be nearly symptomless, even though about the same high yield of virus is produced initially in both types of cells. After the initial burst of virus production in invertebrate cells, **chronic infections** are established. These result in a low level of virus production, probably by only a few cells at any given time. The rest of the cells may contain the virus in a *latent* form and divide normally. Part of this inhibiting process is mediated by **interferon** production (see Sec. 12.4). Thus, the overall culture can proliferate more or less indefinitely. Many paramyxoviruses can establish *persistent* steady-state infections in which essentially all cells continuously produce large amounts of virus without cell death or CPE. In some cases, such as with rubella virus, replication of chronically infected cells may be slowed though not prevented. These various types of essentially noncytocidal, noncytolytic infections will be considered again in relation to possible virus disease mechanisms in higher organisms in Sec. 12.6.

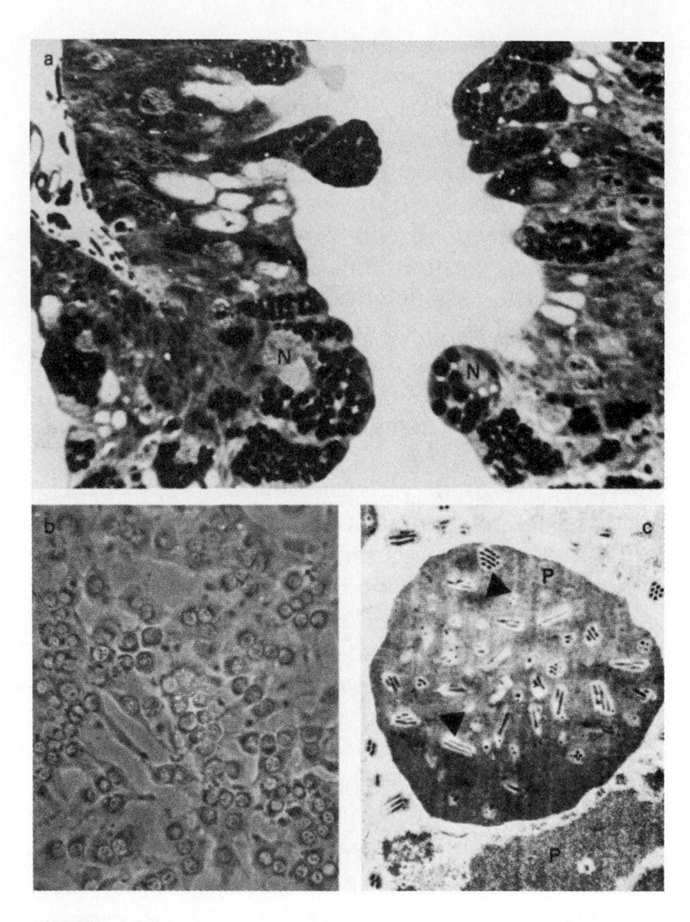

FIGURE 11.6

Cytoplasmic versus nuclear inclusions induced by insect viruses. (a) Midintestinal epithelium of a Salt Marsh caterpillar larva infected with cytoplasmic polyhedrosis virus (× 700). Note the dark crystals in the cytoplasm of several cells. N = nucleus. (Courtesy of Gordon Stairs) (b) Nuclear polyhedrosis virus in cultured cells from ovarian tissue of a moth *(Tricoplasia ni).* The clustered cells in the center exhibit highly refractile nuclear inclusion bodies or polyhedra bodies (× 260). (c), These nuclear "polyhedral bodies" (P) contain clusters of rodlike virions (arrows) embedded in a matrix protein (× 17,000). As infection progresses, bundles of virions may be released by degradation of the inclusion matrix, as shown by the disrupted inclusion in the lower right-hand corner. [(b) and (c) courtesy of Walter F. Hink, Jr.]

For animal viruses that are cytolytic, it is not necessary to postulate a viral gene product with direct disruptive action on the cell membrane. As described below, many of these viruses have functions that kill the cell in other ways and all cells killed by any sort of nonviral metabolic poisons (including chemicals) exhibit essentially the same lethal processes. For cells growing attached to a surface, these signs include rounding and detachment from the culture vessel, disruption of intracellular morphology, and **autolysis.** This autolysis is due to the presence of a variety of degradative enzymes that normally are contained in **lysosomes,** special cytoplasmic vesicles within the cells. At least 50 different lysosomal enzymes have been identified, including nucleases, proteases, and other enzymes capable of breaking down lipids or carbohydrates. Lysosomal enzymes seem to play a role in degradation of materials taken into the cell by **phagocytosis,** since lysosomes often appear to fuse with the vesicles formed around the ingested particles. It is clear that late during many cytolytic virus infections the membranes of lysosomes become destabilized, allowing enzymes to leak out into the cytoplasm. Thus, it has been proposed that some virus-induced lysis is due to lysosomal enzymes of the cell rather than directly to a viral gene product. Proof of this idea remains elusive, and a specific viral gene product responsible for destabilization of the lysosomal membranes has not been identified.

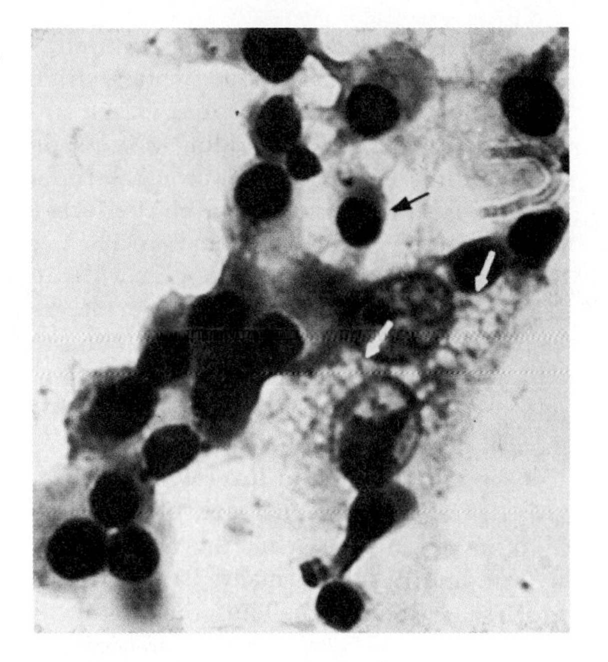

FIGURE 11.7

Cytopathic effects of SV40 virus in monkey (VERO) cells. White arrows show "bubbly" or vacuolated cytoplasm for which the virus was originally named "simian vacuolating agent no. 40." The black arrow designates one of several pyknotic or condensed nuclei (× 1140). (Courtesy of James Blakeslee, Jr.)

However, although certain viral mutants that do not lyse the cell also fail to disrupt the lysosomes, destabilization of lysosomes and cell lysis might still be independent events triggered by a single viral function. Finally, recent cell death has been reported to result from **apoptons** (programmed cell death) induced by some viruses (e. g. HIV). This process involves endonuclease-directed degradation of cellular chromosomes and is a normal event in lymphocyte development.

11.4.3 Picornavirus CPE

Originally, picornaviruses (and especially poliovirus) were favored for studies on CPE of animal viruses due to the medical significance of poliomyelitis. Despite the development of effective vaccines (Sec. 13.2), poliovirus still serves as the prototype cytolytic virus in most of the few current studies on the molecular basis of CPE. As noted earlier, the principal forms of morphological damage usually produced by this virus in cultured cells, occur rather late in infection, after most viral replication; these include nuclear pyknosis, cytoplasmic membrane proliferation that creates intracytoplasmic vesicles called **cisternae,** lysosomal disruption, and finally, lysis (Fig. 11.8).

What viral product(s) can trigger CPE? From numerous studies with metabolic inhibitors of either cellular or viral protein or RNA synthesis, it is evident that neither cellular RNA nor cellular protein synthesis is required for development of CPE, whereas inhibition of early viral proteins or of viral RNA synthesis prevents all aspects of CPE. Also, mutants in viral RNA polymerase, an early viral protein, do not induce CPE. These mutants cannot produce the **replicative intermediate form** of viral RNA. Since purified double-stranded RNA is itself highly cytotoxic, these mutant studies might be taken to suggest that putative double-stranded regions of a replicative intermediate of RNA are the viral trigger for CPE. This hypothesis neglects the fact that most of the viral proteins, as well as the positive- and negative-sence RNAs, are made without any accumulation of double-stranded RNA (see Sec. 2.2). One of these viral products would therefore appear to be the probable direct cause of CPE. Yet purified virion protein or virions inactivated with UV light are not cytotoxic when added to cell cultures.

Whatever the viral trigger may be, there is much circumstantial evidence to indicate that lysosomes are the key cellular target for initiation of CPE. Even the cisternal membrane proliferation might be accounted for by release of certain lysosomal **phospholipases.** Paradoxically, these enzymes degrade lipids rather than synthesize them; but certain products of lipid degradation are known to stimulate membrane synthesis.

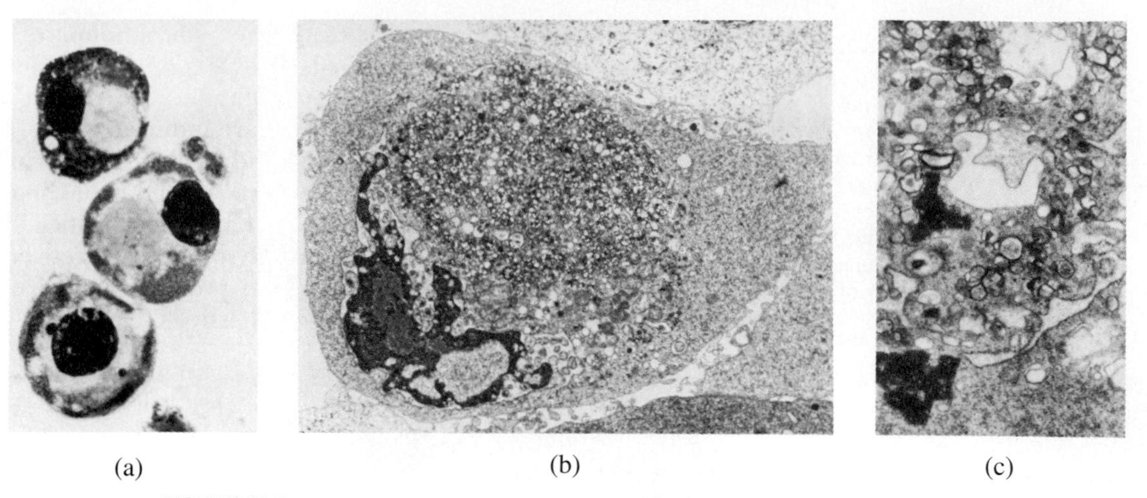

FIGURE 11.8

Cytopathic effects (CPE) of poliovirus in cultured human (Hep-2) cells. (a) Light microscopy reveals nuclear pyknosis (condensation). Cells were stained with Giemsa stain late (at 9 hours) after infection, when lysis was imminent. The upper two cells show dark, displaced nuclei at the left and right cell respectively, whereas the lower cell appears to be normal with a centrally located nucleus (× 1000). (b) Details of the lightly stained masses that displace the nuclei [in part (a)] can be seen by electron microscopy. The flattened nucleus (lower left part of cell) is compressed by a mass of intracytoplasmic membranous vesicles *(cisternae).* Some larger vesicles are seen surrounding the nucleus, within the dilated perinuclear space (× 7500). (c) Occasional crystals of poliovirus particles can be seen by electron microscopy, amidst the vesicles or in adjacent cytoplasm (lower left) of infected cells (× 24,000). Disruption of lysosomes, which also occurs late in the infection cycle, and is not visible with these techniques. It can be seen with stains that detect lysosomal enzymes. (Courtesy of Kurt Bienz)

The selective proliferation of cytoplasmic membranes rather than the perinuclear membrane might be a consequence of virus replication that occurs in complexes bound to the cytoplasmic membranes only. In fact, when phospholipase is artificially delivered into the cytoplasm of normal cells using experimental lipid vesicles called **liposomes,** which tend to fuse with membranes, morphological alterations with a striking resemblance to poliovirus CPE are induced.

It is obvious that any explanation for poliovirus-induced CPE still remains speculative. For example, in enucleate cells (cells from which the nuclei have been artificially removed) poliovirus produces CPE without disruption of lysosomes. Moreover, as we mentioned earlier, mitotic HeLa cells are found to produce poliovirus without CPE, whereas the same cells in interphase show massive CPE.

SECTIONS 11.3 AND 11.4 FURTHER READING

Original Papers and Reviews

BOLANDE, R. P. (1961) Significance and nature of inclusion-bearing cells in the urine of patients with measles. *New Engl. J. Med.* 265:919.

BIENZ, K., D. EGGER, and D. WOLFF. (1973) Virus replication, cytopathology and lysosomal enzyme response of mitotic and interphase Hep-2 cells infected with poliovirus. *J. Virol.* 11:565.

LEVY, J. A. (1973) Xenotropic viruses: Murine leukemia viruses associated with NIH Swiss, NZB, and other mouse strains. *Science* 182:1151.

CHEVILLE, N. F. (1975) *Cytopathology in viral diseases.* S. Karger, Basel, pp. 1–22; and 100–147.

HINK, W. F., E. S. STRAUSS, and W. A. RAMOSKA. (1977) Propagation of *Autographa californica* nuclear polyhedrosis virus in cell culture: Methods for infecting cells. *J. Invertebr. Pathol.* 30:185.

LEVY, J. A. (1978) Xenotropic type C viruses. *Curr. Top. Microbiol. Immunol.* 79:111.

BOSSART, W., and K. BIENZ. (1979) Virus replication, cytopathology and lysosomal enzyme response in enucleated Hep-2 cells infected with poliovirus. *Virology* 92:331.

LENK, R., and S. PENMAN. (1979) The cytoskeletal framework and poliovirus metabolism. *Cell* 16:289.

HOWE, C., J. E. COWARD, and T. W. FENGER. (1980) Viral invasion: Morphological and biophysical aspects. *Comp. Virol.* 16:1.

GARRY, R. F., E. T. ULUG, and H. R. BOSE, Jr. (1982) Membrane-mediated alterations of intracellular Na^+ and K^+ in lytic-virus-infected and retrovirus-transformed cells. *Biosci. Rep.* 2:617.

GALLAHER, W. R. (1987) Detection of a fusion peptide sequence in the transmembrane protein of human immunodeficiency virus. *Cell* 50:327.

OLDSTONE, M. B. A. (1989) Viruses can cause disease in the absence of morphological evidence of cell injury: Implication for uncovering new diseases in the future. *J. Infect. Dis.* 159:384.

GROUX, H., et al. (1992) Activiation-induced death by apoptosis in CD4 + T cells from human immunodeficiency virus-infected asymptomatic individuals. *J. Exp. Med.* 125:331.

PEREZ, L., and L. CARRASCO. (1992) Lack of direct correlation between p220 cleavage and the shut-off of host translation after poliovirus infection. *Virology* 189:178.

CHEN, P. H., D. A. ORNELLES, and T. SHENK. (1993) The adenovirus L3 23-kilodalton proteinase cleaves the amino-terminal head domain from cytokeratin 18 and disrupts the cytokeratin network of IIeLa cells. *J. Virol.* 67:3507.

RICE, S. A., V. LAM, and D. M. KNIPE. (1993) The acidic amino-terminal region of herpes simplex virus type 1 alpha protein ICP27 is required for an essential lytic function. *J. Virol.* 67:1778.

11.5 VIRUS EFFECT ON HOST CELL METABOLISM

Besides causing gross changes in cell structure and behavior, cytocidal viruses in particular, usually have the ability to interfere with synthesis of host macromolecules. This interference may be caused directly by some virus function, or it may result indirectly from the effects of the virus on some other host cell function. For example, poliovirus causes direct and rapid inhibition of host cell protein synthesis, but host cell DNA synthesis declines slowly over several hours, suggesting some kind of indirect inhibition mechanism. As discussed above, many viruses have little or no effect on the host cell, as demonstrated by a lack of CPE and little inhibition of host cell multiplication. In contrast, some viruses actually stimulate host cell synthetic functions. In this section we consider a wide range of effects on different host cell functions by a variety of viruses. (Table 11.6)

11.5.1 Inhibition of Host Cell Protein Synthesis

The extent and timing of the inhibition of host cell protein synthesis vary for different viruses, most probably due to different mechanisms of inhibition. A number of viruses can rapidly and completely shut off production of host cell proteins. These include most picorna- and togaviruses, and some paramyxo- and herpesviruses. With other viruses, such as pox- and adenoviruses, inhibition is more gradual and appears to be due to progressive displacement

TABLE 11.6
Summary of effects of animal viruses on host cell macromolecules during cytocidal infections

Function	Effect	Virus Examples
Protein synthesis	Direct inhibition (viral product)	Most picorna- and togaviruses; some paramyxo- and herpesviruses
	Indirect inhibition (mRNA competition)	Poxviruses, adenoviruses
RNA synthesis	mRNA inhibition—rapid	Togaviruses
	mRNA inhibition—gradual	Picornaviruses and most cytocidal viruses except adeno-and herpes viruses
	rRNA processing—rapid	Picornaviruses
	rRNA processing—gradual	Adeno- and herpesviruses
DNA synthesis	Inhibition—gradual (indirect from lack of proteins)	Picorna-, paramyxo-, adeno-, and herpesviruses
	Stimulation	Polyoma, SV40, retroviruses
Intermediary metabolism	Stimulation of dehydrogenases	Togaviruses

of host cell mRNA from polysomes by increasing amounts of viral mRNA relatively late in infection. Of those viruses that rapidly inhibit host cell protein synthesis, poliovirus has been studied most intensively regarding the mechanism. In this case it is clear that synthesis of all host cell proteins is equally depressed. Host cell mRNA is not degraded by poliovirus; rather, initiation of translation is the critical step that is blocked by virus infection.

A number of studies using cell-free extracts have been directed at identifying the specific poliovirus function that prevents host cell protein synthesis. Double-stranded viral RNA inhibits initiation of translation of host cell mRNA *in vitro,* but viral mRNA is equally affected under these conditions. In any case, inhibition *in vivo* occurs after the earliest viral protein synthesis but before much if any double-stranded viral RNA is made.

The viral polypeptide(s) responsible for inhibition of host cell protein synthesis have not been fully identified, but the target host cell macromolecule for this viral function has been found in the case of poliovirus (see Sec. 2.2). Poliovirus infection brings about an inactivation of a large cellular protein complex *(cap binding protein)* that transfers the 5′ terminal cap structure onto host cell mRNA. This cap structure is required for proper ribosome–host cell mRNA interaction but not for poliovirus mRNA translation. Thus, preferential translation of viral mRNA results from inactivation of this protein complex.

Cells infected by influenza virus also utilize 5′-cap structures and adjacent oligonucleotides on host cell mRNA as primers for virus-specific mRNA synthesis and RNA replication. These are transferred to influenza RNA by a unique virus-specific enzyme. Thus, it appears that synthesis of host cell protein is inhibited in part because their capped mRNAs are consumed by viral replication (see Sec. 4.3).

For other viruses causing rapid inhibition of host cell protein synthesis, it is known only that some viral proteins are made early during infection. Effects on the **cap binding factor** cannot be invoked for inhibition by most other viruses, however, since they have mRNAs 5′-methylated caps like those of the host cell. It has been noted that virion components of certain viruses, for example the fibers of adenovirus virions, are cytotoxic to cells in purified form. Nevertheless, the presence of fibers on entering virions is not likely to account for inhibition of host cell protein synthesis during infection, because the inhibition does not begin immediately after entrance of the virion. New fibers produced inside the cell might become inhibitory a few hours after infection is initiated.

11.5.2 Inhibition of Host Cell RNA Synthesis

Cytocidal viruses may interrupt host cell RNA synthesis as well as protein synthesis, but the effect is usually less rapid and dramatic. In the case of poliovirus, for example, there is a striking effect on the manufacture of ribosomes that results from rapid interference with processing of the 45 S precursor for the 28 S and 18 S ribosomal RNAs. The mechanism and advantage (if any) of this interference by poliovirus is not understood. Mengovirus, another picornavirus, can shut off host cell RNA synthesis more rapidly than poliovirus, but the speed of inhibition is dependent on the host cell type and does not appear to influence virus replication. Many other viruses, including adeno- and herpesviruses, also interrupt synthesis and processing of cellular RNA, but usually only after several hours of infection.

Cytocidal viruses often also depress synthesis of host cell mRNA. For poliovirus all cellular DNA-dependent RNA synthesis is halted soon after inhibition of host cell protein synthesis. However, no modification of host cell **RNA polymerases** has yet been detected. Thus, the problem may be related to subtle changes in the chromatin as noted earlier in the discussion of CPE. Certain togaviruses cause rapid inhibition of all host cell RNA synthesis, and viral mutants that fail to make viral mRNA also fail to inhibit host cell RNA synthesis. But no further definition of the viral function responsible for host cell RNA inhibition has been possible so far. Prevention of host cell mRNA synthesis is not mandatory by any means, because adeno- and herpesviruses, the DNAs of which are replicated in the nucleus, do not limit host cell RNA production.

Selective inhibitory effects on cellular tRNA synthesis or processing have not been reported. In particular, no virus-specific tRNAs, comparable to those produced by bacteriophage T4 (See Sec. 8.7), are known to exist for any virus of eukaryotes.

11.5.3 Inhibition of Host Cell DNA Synthesis

The rate of host cell DNA synthesis generally undergoes a gradual decline in cells infected with cytocidal animal viruses (e.g. picorna-, paramyxo-, adeno-, and herpesviruses). This inhibition may well be an indirect effect of depressed host cell protein and/or RNA synthesis earlier during the infection. Synthesis of both these macromolecular constituents is required for normal DNA replication. Protein synthesis is needed to provide enzymes and other factors for initiation of cellular DNA synthesis as well as for optimal chain elongation rate. Moreover, newly made RNA is required as primer for DNA polymerase to begin to synthesize DNA strands (Sec. 7.15). Thus, there is little reason to suspect that any animal virus has specific functions aimed directly at inhibition of host cell DNA replication. Nevertheless in some cases, such as reoviruses, DNA synthesis appears to decline somewhat more rapidly than host cell protein synthesis. Certainly there is no eukaryotic virus analogue for the T-even bacteriophages that totally solubilize host cell DNA to reutilize the constituents for viral DNA synthesis (Sec. 8.7). Pox- and herpesviruses do produce nucleases, but these enzymes cause only moderate fragmentation of host chromatin, and appear to provide no real benefit to viral replication. Cell division may also be prevented by cytocidal viruses. Again, this effect most probably reflects the inhibition of protein synthesis. Infected cells sometimes divide, but only once and then only during the first few hours of infection.

11.5.4 Stimulation of Cellular Functions

Not all effects of viruses on macromolecular synthesis in eukaryotes are inhibitory. We have already considered stimulation of membrane proliferation during infections with poliovirus and other cytocidal viruses. Certain viruses also induce specific cellular enzymes that in some cases may be needed for replication of the virus. One of the clearest examples of specific enzyme induction is seen with certain small DNA viruses, such as SV40. Large DNA viruses, which encode functions for viral DNA replication, may have no effect on host cell DNA synthesis or even inhibit it, as noted above. Small DNA viruses often rely entirely on host cell enzymes for replication of viral DNA, since their genomes are too small to accommodate all the necessary functions. Thus **parvoviruses,** which contain somewhat less genetic information than SV40, can be replicated only in cells that are in S phase at the time of infection.

SV40, however, is able to induce the resting host cell to initiate a new round of replication, beginning with production of enzymes for DNA synthesis, such as DNA polymerases and thymidine kinase that are necessary for DNA synthesis. After host cell DNA replication and concomitant production of the histones needed to form new chromatin, the infected cells may finally go through mitosis if they are not susceptible to killing by the virus. Infection with SV40 of cells that are nonpermissive for lytic infection may result in their transformation into tumor cells. The mechanism of this transformation, which involves viral integration, and the role of induction of host DNA replication in that process are discussed in Section 10.1.

Retroviruses also generally require cellular DNA synthesis to complete their replication cycle. Using their virion-associated enzyme, reverse transcriptase, these viruses make a DNA copy of their genomic RNA and this C DNA is integrated into the chromosome of dividing cells to establish the infection. Without cellular DNA synthesis, the infection is abortive. Nevertheless, some retroviruses (especially those of the lentivirus subfamily) have been found to complete their replicative cycle without cell division. This phenomenon is particularly evident with visna virus of sheep and HIV in humans where the viruses replicate in nondividing macrophages.

Few examples of the induction of specific cellular genes by eukaryotic viruses have been documented. Certain togaviruses stimulate production of glucose-6-phosphate dehydrogenases and other components of intermediary metabolism, but the mechanism and function of this induction are not known. We noted in Sec. 10.3 that some DNA tumor viruses, besides SV40, are now known to induce specific cellular genes, probably as part of the mechanism of transformation. Obviously much remains to be learned about the normal gene regulatory processes as well as about the ways in which viruses disturb them. One particular gene system that can be triggered by a wide variety of viruses in animal cells is the **interferon** system (see Sec. 12.4).

11.6 VIRAL INTERFERENCE

In addition to inhibiting host cell metabolism, many animal viruses also inhibit the replication of other viruses (see also Secs. 8.7 and 11.11). In principle, one virus could interfere with another by competition for cellular resources or by specific antiviral mechanisms active at any level of the virus replication cycle. At present only a few types of interference phenomena have been observed.

Viruses can generally prevent infection by members of the same subgroup via competition for **cell surface receptors.** However, examples in which diverse viruses interfere with one another have also been reported. For instance, rubella can block infection of monkey kidney cells by Coxsackie virus A9 or Echo virus 11 because they share the same receptor. Rubella interferes with Sindbis virus in human cells. This interference assay led to the recognition of the rubella virus in 1962. Within the Retroviridae it is this interference pattern that helps distinguish specific virus substrains. Since these viruses establish chronic productive infections, cell receptors on the surface of infected cells are saturated by newly released virions and a related strain of virus cannot superinfect. Interference patterns have been useful in classifying the mouse and avian type C retroviruses. In the case of HIV, interference takes place when the infecting virus causes a reduction in the expression of the CD4 cellular receptor molecule on the cell surface.

Inside the cell, viruses can interfere with each other in several ways. For cytocidal viruses, the inhibition of other viruses most often seems to result from the same mechanisms that inhibit host macromolecular synthetic processes. RNA viruses (be they positive-sense, negative-sense, or double-stranded) and some DNA viruses (especially herpesviruses) are subject to a form of self- or homologous interference due to spontaneous generation of **defective interfering** (DI) **particles,** which are naturally occurring deletion mutants. These were discussed in some detail in Sec. 2.2 (picornaviruses), Sec. 4.1 (rhabdoviruses), and Sec. 4.3 (orthomyxoviruses). Here we simply remind the reader that they arise at high frequencies, when cells are infected at high multiplicities of infection. Large quantities of DI particles in a stock of a cytocidal virus can block virus replication and reduce the severity of lethal effects of infection. As we shall see in Section 12.6, DI particles are also thought to play an important role in the establishment of some chronic infections in animals and humans, In some cases, this effect is partially mediated by the induction of interferon (see Sec. 12.4).

Finally, in some animal systems, infection with one virus may actually favor superinfection and growth by another virus. This phenomenon of reverse interference has been shown, for instance, with poliovirus and the Coxsackie or Echo viruses. It has been cited for the interaction of adenoviruses and papovaviruses (Sec. 10.2.4) and has been observed with Shope fibroma virus (a poxvirus) and VSV. Reverse interference can also be demonstrated with unrelated retroviruses. The mechanisms for these effects are not known, but they probably involve activation of transcription processes.

SECTIONS 11.5 AND 11.6 FURTHER READING

Original Papers

von Magnus, P. (1954) Incomplete forms of influenza virus. *Adv. Virus Res.* 2:59.

Rubin, H. (1960) A virus in chick embryos which induces resistance *in vitro* to infection with Rous sarcoma virus. *Proc. Natl. Acad. Sci. USA* 46:1105.

Parkman, P. D., et al. (1962) Recovery of rubella virus from army recruits. *Proc. Soc. Exp. Biol. Med.* 111:225.

Hackett, A. J. (1964) A possible morphological basis for the autointerference phenomenon in vesicular stomatitis virus. *Virology* 24:51.

Steck, F. T., and H. Rubin. (1966) The mechanism of interference between an avian leukosis virus and Rous sarcoma virus. I. Establishment of interference. *Virology* 29:628.

Sarma, P. S., et al. (1967) A viral interference test for mouse leukemia viruses. *Virology* 33:180.

AUBERTIN, A., et al. (1970) The inhibition of vaccinia virus DNA synthesis in KB cells infected with frog virus 3. *J. Gen. Virol.* 8:105.

HUNT, J. M., and P. I. MARCUS. (1974) Mechanism of Sindbis virus–induced intrinsic interference with vesicular stomatitis virus replication. *J. Virol.* 14:99.

DUBOVI, E. J., and J. S. YOUGNER. (1976) Inhibition of pseudorabies virus replication by vesicular stomatitis viruses. *J. Virol.* 18:526.

CHEN, C., and N. A. CROUCH. (1978) Shope fibroma virus–induced facilitation of vesicular stomatitis virus adsorption and replication in nonpermissive cells. *Virology* 85:43.

DE GROOT, R. J., R. G. VAN DER MOST, and W. J. M. SPAAN. (1992) The fitness of defective interfering murine coronavirus DI-a and its derivatives is decreased by nonsense and frameshift mutations. *J. Virol.* 66:5898.

LE GUERN, M., and J. A. LEVY. (1992) Human immunodeficiency virus (HIV) type 1 can superinfect HIV-1-infected cells: Pseudotype virions produced with expanded cellular host range. *Proc. Natl. Acad. Sci. USA* 89:363.

FURUYA, T., T. B. MACNAUGHTON, N. LA MONICA, and M. M. C. LAI. (1993) Natural evolution of coronavirus defective-interfering RNA involves RNA recombination. *Virology* 194:408.

MOSCONA, A., and R. W. PELUSO. (1993) Persistent infection with human parainfluenza virus 3 in VC-1 cells: Analysis of the role of defective interfering particles. *Virology* 194:399.

11.7 MODIFICATION OF VIRUSES

Besides producing changes in the cell itself or in its sensitivity to entry by other viruses, infection can also modify the virus itself. Some examples have been cited in previous chapters and are summarized here.

11.7.1 Phenotypic Mixing

When a cell is infected by more than one virus, replication of both agents can lead to an exchange of their envelope coats. Some of the resulting progeny virus can then appear to be, as defined by neutralization and ability to infect cells, the virus of the other type. By this mechanism of *phenotypic mixing* (Figure 11.9), viruses that were once unable to infect certain cells can now expand their host range. After gaining entrance into a cell via the outer envelope, the inner core of the phenotypically mixed particle can release its own genetic material and replicate that virus in the cell. This phenotypic mixing or **pseudo-type formation** can occur among viruses of the same genus or among viruses of different genera. Thus, rhabdo-, herpes-, retro-, and paramyxoviruses can all undergo phenotypic mixing among themselves and have the potential of enhancing their transmission to different animals and even humans. With retroviruses this fact has been dramatically shown by infecting mouse or avian cells with mouse and avian retroviruses. The mouse xenotropic virus (Chapter 6) and the avian leukemia virus can both infect duck cells. A cell infected with these viruses gives rise to progeny that now can transfer the avian virus into a wide variety of mammalian cells, and the mouse retrovirus into several different avian cells that once were resistant. This phe-

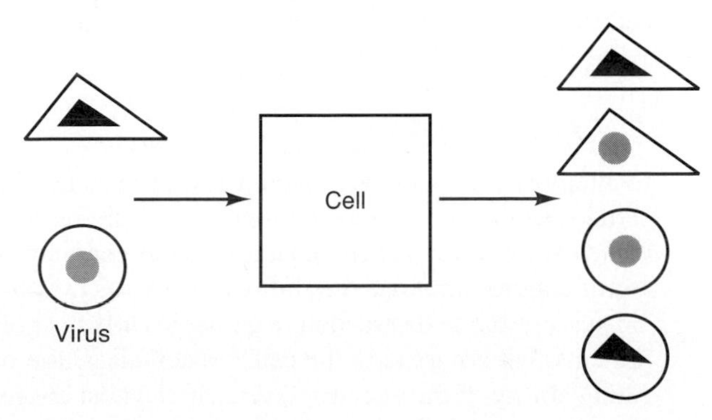

FIGURE 11.9

When two viruses of different types enter the same cell, the progeny produced will consist of the initial viruses and also viruses that have exchanged their outside coat. These new viruses then have the host range of the virus from which they derived their envelope. (From J. A. Levy. (1977) In *Autoimmunity.* ed. N. Talal. Courtesy Academic Press, New York.)

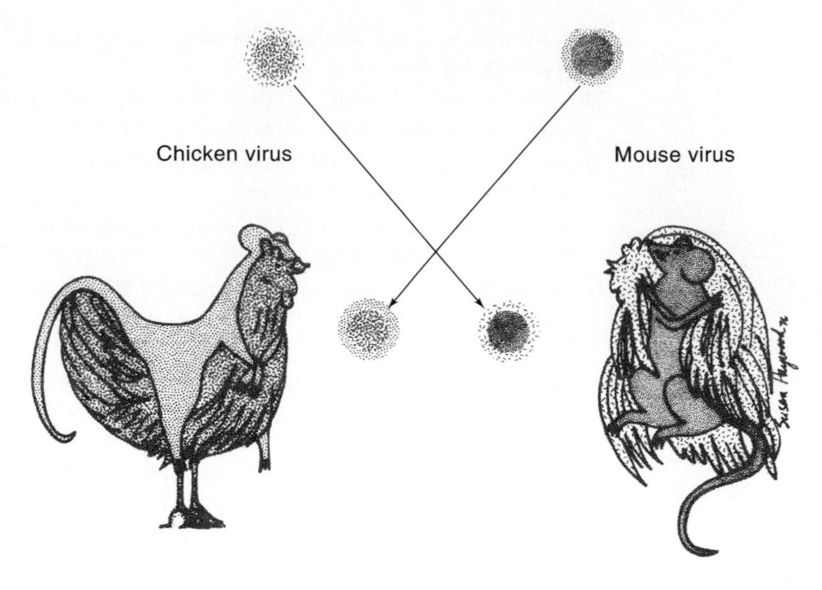

FIGURE 11.10

Phenotypic mixing of C type retro viruses. When a cell is infected by a chicken and mouse type C virus, exchange of the virus envelopes can occur. The avian virus genome carries the mouse virus coat, and the mouse virus bears the coat of the chicken virus. (J. A. Levy. (1978) *Curr. Top. Microbiol. Immunol.* 79:111. Courtesy of Springer-Verlag Co. Illustration by Susan Haywood.)

nomenon can take place only when the receptor for penetration of the cell is the sole block for replication of the virus. Figure 11.10 demonstrates in cartoon fashion the exchange that can take place among diverse viruses through phenotypic mixing.

11.7.2 Recombination/Reassortment

Another mechanism by which viruses can be modified when replicating within the same cell is that of **recombination.** Under this condition, the genomes of the viruses interact and exchange actual genetic information, so that the emerging virus genome represents a genetic mixture of the two virus types. Recombination is almost exclusively limited to members of the same genus or subfamily. Recently however, the process was reported to involve a retro- and herpesvirus. Viral genetic changes can occur readily with influenza viruses and among retroviruses where entire RNA segments can either **reassort** or recombine to form new virus particles. Reassortment takes place by the exchange of individual segments in the viral core; it does not involve a genetic change in the viral genome. The resulting virion has different biological and perhaps host range properties than the initial parental virus.

11.7.3 Transduction

One important potential effect of virus infection of the cell is the **incorporation of cellular genes** into the virion genome. In this process the virus, after integrating into the cell DNA, takes part of the cellular sequences into its own genome. When the virus progeny are then produced, they carry these cellular genes into new recipient cells (see Fig. 11.11). This process of **transduction** was first discovered in bacteriophages. It has recently received attention because of the evidence suggesting that oncogenic viruses have incorporated normal cellular genes into their genome. With some modifications (by procedures as yet unknown), these sequences have been converted into **oncogenes** (see Secs. 10.3.7 and 11.7.3).

11.7.4 Attenuation

In attempts at developing infectious-virus vaccines, scientists recognized that passage of certain pathogenic viruses through cell culture reduces their virulence. This **attenuation** results from small genetic changes in the genome of the virus (see Fig. 11.11). These modifications eliminate the ability of the virus in some cases to replicate within cells that are susceptible to destruction (e.g., nerve cells for poliovirus) or reduce its production of proteins that are toxic to the cell. Nevertheless, the outside envelope proteins and replicating ability of the attenuated virus in the host are maintained. Thus, antibodies or cell-

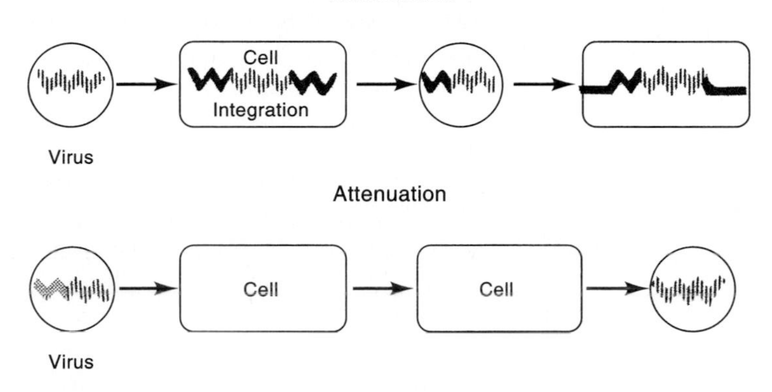

FIGURE 11.11

Changes in a virus after infection of a cell. **Transduction.** When a virus enters a cell and integrates into the host cell genomes, the progeny may take with it (transduce) cellular genetic information. This new virus can then transfer this cellular information to other recipient cells. **Attenuation.** When a virus is passed for several generations through cells, its virulent properties may be lost through genetic changes in the virus. Although incapable of causing disease, the resultant agent may still induce a strong immune response in the host.

mediated immunity by the infected host against the initial wild-type isolate can be elicited. Examples of attenuated viruses include the Sabin polio, the Edmondston measles, and the Cendehill-51 or RA 27/3 rubella vaccine strains (see Sec. 12.3).

SECTION 11.7 FURTHER READING

Original Papers

CHOPPIN, P. W., and R. W. COMPANS. (1970) Phenotypic mixing of envelope proteins of the parainfluenza virus XV5 and vesicular stomatitis virus. *Virology* 5:609.

HUANG, A. S., et al. (1974) Pseudotype formation between enveloped RNA and DNA viruses. *Nature* 252:743.

LEVY, J. A. (1977) Murine xenotropic type C viruses. III. Phenotypic mixing with avian leukosis and sarcoma viruses. *Virology* 77:811.

WEISS, R. A., and A. L. WONG. (1977) Phenotypic mixing between avian and mammalian RNA tumor viruses: I. Envelope pseudotypes of Rous sarcoma virus. *Virology* 76:826.

FIELDS, B. N., and D. R. SPRIGGS. (1982) Attenuated reovirus type 3 strains generated by selection of haemagglutinin antigenic variants. *Nature* 297:68.

CANIVET, M., et al. (1990) Replication of HIV-1 in a wide variety of animal cells following phenotypic mixing with murine retroviruses. *Virology* 178:543.

HU, W.-S., and H. M. TEMIN. (1990) Retroviral recombination and reverse transcription. *Science* 250:1227.

BERGMANN, M., A. GARCIA-SASTRE, and P. PALESE. (1992) Transfection-mediated recombination of influenza A virus. *J. Virol.* 66:7576.

JENSEN, M. J., and D. M. MOORE. (1993) Phenotypic and functional characterization of mouse attenuated and virulent variants of foot-and-mouth disease virus type O_1 campos. *Virology* 193:604.

KOST R., et al. (1993) Retrovirus insertion into herpesvirus: characterization of a Marek's disease virus harboring a solo LTR. *Virology* 192.

Reviews

VOGT, P. K. (1967) Phenotypic mixing in the avian tumor virus group. *Virology* 32:708.

LEVY, J. A. (1978) Xenotropic viruses. *Curr. Top. Microbiol. Immunol.* 79:111.

BOETTIGER, D. (1979) Animal virus pseudotypes. *Prog. Med. Virol.* 25:37.

LEVY, J. A. (1986) The multifaceted retrovirus. *Cancer Res.* 46:5457.

OLDSTONE, M. B. A. (1987) Molecular mimicry and autoimmune disease. *Cell* 50:819.

11.8 INTRACELLULAR CONTROL OF VIRUS INFECTION

In Chapter 7, we discussed the influence of the host cell on virus replication. One cell may be **permissive** for the growth of one virus (e.g., monkey cells for SV40), but **nonpermissive** for another (e.g., monkey cells for polyoma virus). In the latter case an abortive infection occurs. As previously stated, transformation by DNA viruses can occur only when replication is blocked. These observations indicate that approaches limiting virus replication in the cell must be conducted with caution because of the potential for enhancing transformation.

The ability of polyomalike viruses to replicate to high titers in cells of their host species was discussed in Sec. 7.2. Thus, SV40 and polyoma virus can be grown to substantial levels in monkey and mouse cells. Respectively, BK and JC viruses can be replicated in human cells, but the latter has a selective tropism for human glial cells. These viruses can infect other mammalian cells, but because they cannot complete their replication, they have the potential to transform these cells. Studies with all the papovaviruses illustrate the fact that only abortive infection can lead to transformation with DNA viruses, since replication generally leads to cell death. Animals inoculated with these human viruses will develop tumors in which no replicating virus can be detected. However, some viral proteins (e. g. **T proteins**) can be detected in these tumors (see Sec. 10.2). Generally, fetuses or newborn animals are more susceptible to such transformation processes. In most cases the entire papova viral genome is present in the transformed cell, since co-cultivation of these cells with permissive cells can lead to recovery of infectious virus (see Chapter 7). Rescue of the entire virion has not been achieved for cells transformed by larger DNA viruses (see Chapter 8). Only portions of these large DNAs appear to be integrated into the host cellular DNA.

What are the biochemical events that differentiate the productive from the transforming activity of almost all oncogenic DNA viruses? Briefly stated, they are as follows: Virus is adsorbed and taken up by nonpermissive cells as by permissive ones, with movement of the viral DNA to the nucleus. Transcription of the early genes then takes place and is soon followed by a great stimulation of all the host cell's synthetic activities required for and leading to cellular DNA replication and mitosis. After about 12 hours (but only in permissive cells) the late phase starts, characterized by viral DNA replication, late gene transcription, and virion production. This phase requires the presence of large T in SV40 and probably of middle T of about 55 kDa in polyoma-infected cells. Actually, host cell DNA synthesis is considerably increased in permissive cells early upon infection under the influence of T protein. At the beginning of the late phase, DNA synthesis ceases and the cell lyses.

In nonpermissive cells, the early genes are also expressed, and cell DNA synthesis and mitosis are triggered, but no replication of viral DNA can be detected, nor any late gene transcription leading to synthesis of the viral coat proteins. As noted above, *viral DNA replication and transformation are mutually exclusive,* since viral replication leads to cell death (Secs. 11.1 and 12.2). Rather than replicating the viral DNA, the nonpermissive cells incorporate it into their genome. The genes for the T proteins are transcribed in transformed cells, and these proteins can usually (but not always) be detected in such cells. Viral coat proteins are never detectable (see Sec. 10.1).

11.9 LATENCY

Certain DNA viruses, such as parvo–, papova–, and herpes viruses, can establish a **latent state,** generally through integration into the host cell DNA. In some cases, such as with the herpes- or adenoviruses, transformation without replication may occur, which indicates that certain portions of the viral genome (e.g., **oncogenic proteins**) are being transcribed. In others, such as parvoviruses, no evidence for the expression of the virus may be found, and this latent state can be activated only by superinfection with a virus that replicates in the host cells (see Sec. 7.1). Both types of viruses are then produced. Similarly, latency both oncogenic and nononcogenic types, of retroviruses is characterized by a state in which there is little or no expression of viral proteins or RNA and no infectious virus is produced (see Sec. 11.1). One mechanism for this latency appears to be the methylation. This process of regions of the viral genome is most likely mediated by cellular enzymes. With the human AIDS virus induction of a latent state in some cases may be related to expression of a viral gene (e.g., *nef*) that can down-regulate virus production. Studies with chemicals that can activate genes have shown that viral latency can be broken through incorporation of base analogues (e.g., iododeoxyuridine) or through demethylation procedures (e.g., with 5-azacytidine). Initially observed with mouse leukemia viruses present in latent form in fibroblasts, this activation of retrovirus production has been demonstrated with a variety of viruses, including the AIDS retrovirus. It is not yet clear what factors can activate an integrated virus from its latent state in nature. For further discussion of these topics see Sec. 12.6.

SECTIONS 11.8 AND 11.9 FURTHER READING

Original Papers

CASSELLS, A. C., and D. C. BURKE. (1973) Changes in the constitutive enzymes of chick cells following infection with Semliki Forest virus. *J. Gen. Virol.* 18:135.

GABELMAN, N., et al. (1974) Alterations in macromolecular synthesis and cellular growth in mouse embryo fibroblasts infected with Friend leukemia virus. *Int. J. Cancer* 13:343.

MACALLISTER, P. E., and R. R. WAGNER. (1976) Differential inhibition of host protein synthesis in L cells infected with RNA temperature-sensitive mutants of vesicular stomatitis virus. *J. Virol.* 18:550.

POSTEL, E. H., and A. J. LEVINE. (1976) The requirement of simian virus 40 gene A product for the stimulation of cellular thymidine kinase activity after viral infection. *Virology* 73:206.

PLOTCH, S. J., et al. (1979) Transfer of 5′-terminal cap of globin mRNA to influenza viral complementary RNA during transcription *in vitro. Proc. Natl. Acad. Sci. USA* 76:1618.

TRACHSEL, H., et al. (1980) Purification of a factor that restores translation of vesicular stomatitis virus mRNA in extracts of poliovirus-infected HeLa cells. *Proc. Natl. Acad. Sci. USA* 77:770.

SHARPE, A. H., and B. N. FIELDS. (1981) Reovirus inhibition of cellular DNA synthesis: Role of the S1 gene. *J. Virol.* 38:389.

KWONG, A. D., and FRENKEL, N. (1987) Herpes simplex virus–infected cells contain a function(s) that destabilizes both host and viral mRNAs. *Cell Biol.* 84:1926.

BRATANICH, A. C., N. D. HANSON, and C. J. JONES. (1992) The latency-related gene of bovine herpesvirus 1 inhibits the activity of immediate-early transcription unit 1. *Virology* 191:988.

FRASER, N. W., T. M. BLOCK, and J. G. SPIVACK. (1992) The latency-associated transcripts of herpes simplex virus: RNA in search of function. *Virology* 191:1.

CHOWDHURY, M. I. H., et al. (1993) cAMP stimulates human immunodeficiency virus (HIV-1) from latently infected cells of monocytemacrophage lineage: Synergism with TNF-α. *Virology* 194:345.

Books and Reviews

TAMM, I. (1975) Cell injury with viruses. *Am. J. Pathol. 81:163.*

MINOR, P. D., O. KEW, and G. C. SCHILD. (1982) Poliomyelitis—epidemiology, molecular biology and immunology. *Nature* 299:109.

RUECKERT, R. R. (1986) Picornaviruses and their replication. In *Fundamental virology,* ed. B. N. Fields and D. M. Knipe. 357–390. Raven Press, New York.

FOLKS, T. M., and D. P. BEDNARIK. (1992) Mechanisms of HIV-1 latency. *AIDS* 6:3.

11.10 PLANT VIRUS INFECTIONS

Plant virus infections become economically important only when their presence causes substantial disruption of the normal patterns of plant growth and development. The macroscopic disease symptoms commonly associated with virus replication that are discussed in Section 12.9, often reflect virus-induced histological changes—necrosis, hypoplasia, and hyperplasia—within the infected plant. In this section, after briefly describing the cytologic effects associated with plant virus infections, we will consider certain virus-induced modifications of the host cell that facilitate both short- (i.e., cell-to-cell) and long-distance virus movement. We end with a discussion of factors that may control the host range of plant viruses.

11.10.1 Cytopathological Effects

Although the cytopathological effects associated with plant virus infection are often related to the site(s) of virus replication or accumulation, virtually every plant organelle may be affected—especially late in infection. Whereas many viruses have no detectable effect on nuclear structure, accumulation of pea enation mosaic virus particles in the nucleus is accompanied by disintegration of the nucleolus and formation of vesicles in the perinuclear space. As mentioned in Section 4.8.1, the cores of some plant rhabdoviruses also accumulate in the perinuclear space. Geminiviruses replicate within the nucleus, leading to a marked hypertrophy of the nucleolus as well as the formation of fibrillar rings and an accumulation of virus particles within the nucleus. Potyviruses replicate in the cytoplasm, but two polypeptides released by proteolytic cleavage of their polyprotein precursors (i.e., the 49 and 58 kDa proteins) accumulate within **nuclear inclusion bodies.** The 49 kDa protein is a serine protease, whereas the 58 kDa protein exhibits certain sequence similarities to other viral RNA polymerases.

Although plant viral infections are frequently accompanied by nonspecific degenerative changes in chloroplast structure, the small peripheral vesicles and other changes that occur in and near the chloroplasts in plants infected by turnip yellow mosaic virus appear to be intimately related to viral replication (see Figure 11.12). These vesicles have been shown to contain both TYMV RNA-dependent RNA polymerase and viral RI, and a comprehensive model for TYMV assembly involving these vesicles has been proposed. Curiously, large quantities of viral coat protein and empty TYMV particles also accumulate in the nucleus.

Most plant viruses (particularly RNA viruses) replicate in the cytoplasm, and a wide variety of virus-induced regions (or **viroplasms)** have been seen in the cytoplasm of virus-infected cells. Granular inclusions containing the 126/183 kDa components of the TMV RNA-dependent RNA polymerase can be detected in TMV-infected tobacco leaves. Later in infection, TMV particles accumulate in large crystalline inclusions in the cytoplasm and vacuole (see Fig. 11.13). Caulimovirus DNA synthesis and virion assembly take place in cytoplasmic inclusion bodies, and many icosahedral viruses form crystalline arrays in the cytoplasm of infected cells.

Potyviruses induce the formation of characteristic cylindrical inclusions in the cytoplasm of infected cells. As illustrated in Figure 11.14, these inclusions are complex structures containing a central tubule from which radiate curved "arms" to give an overall pinwheel appearance. These inclusions are formed from a 70 kDa virus-encoded polypeptide that contains the highly conserved nucleotide triphosphates binding motif present in most viral RNA-dependent RNA polymerases. The central tubule is located directly over the **plasmodesmata,** cytoplasmic connections through the plant cell wall (Fig. 11.15). The cylindrical inclusion may continue from cell to cell. Virus particles are closely associated with these inclusion bodies, particularly early in infection.

11.10.2 Cell-to-Cell and Long-Distance Movement

If a virus is able to replicate in the initially infected cells but cannot move to neighboring cells, its replication may escape detection. Such situations have been termed **subliminal infections.** A limited ability to move from cell to cell is often associated with the formation of **local lesions** (small areas of infected cells that often become necrotic) or the failure of the virus to move out of the inoculated leaf. A **systemic infection** occurs when a virus is able to move freely in the infected plant, both from cell to cell and long distances through the vascular tissues.

In a series of classic experiments on plant virus movement and distribution, Samuel dissected TMV-infected tomato plants into appropriate pieces at various times after inoculation and then inoculated suitable indicator plants with extracts made from these tissue pieces to detect the presence of infectious virus. As shown in Fig. 11.16, the

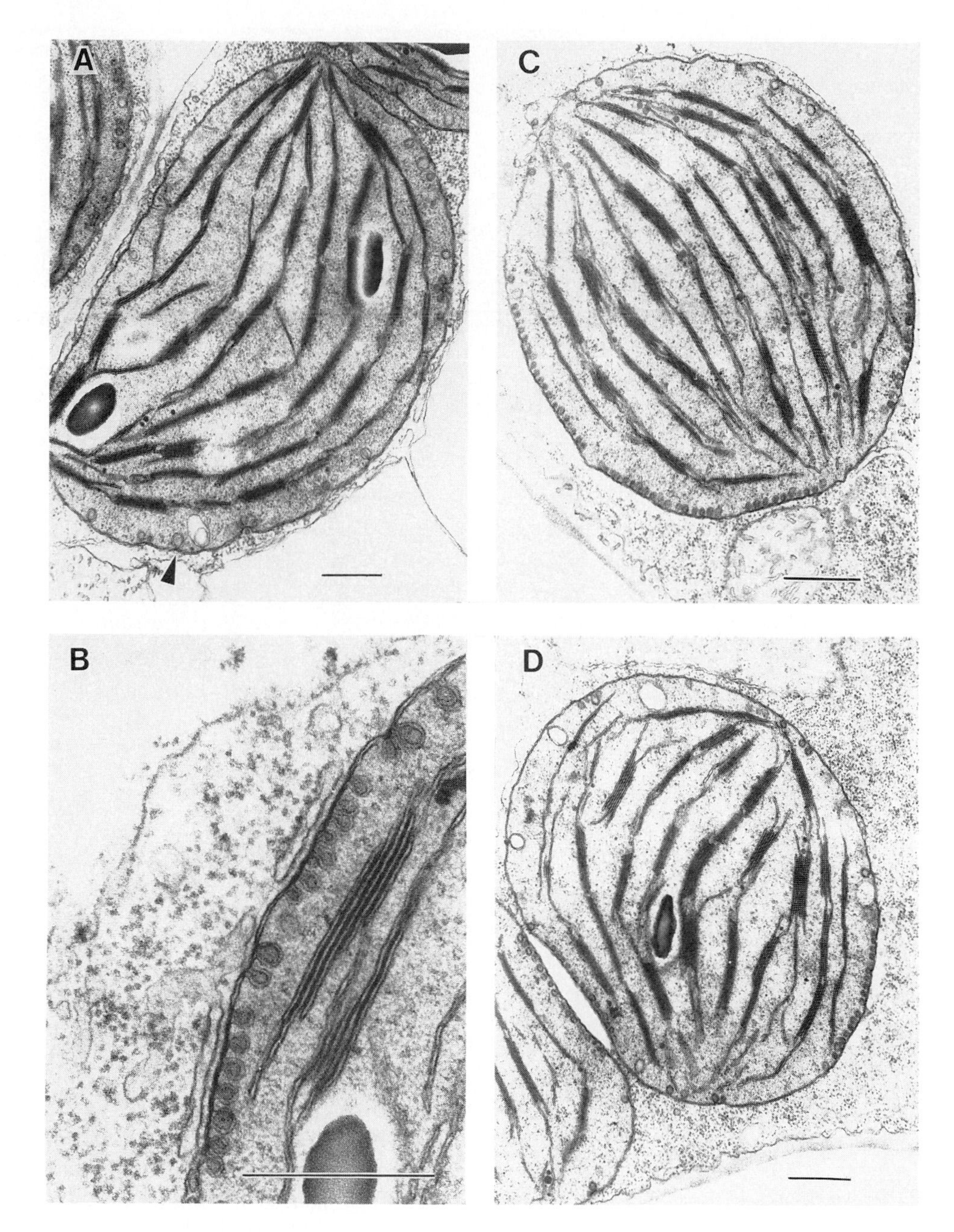

FIGURE 11.12.

The sequence of cytological changes induced by TYMV infection in Chinese cabbage cells. (a) Scattered small peripheral vesicles; chloroplasts otherwise normal; (b) Chloroplasts now swollen. Scattered vesicles are still present, but many clusters of vesicles are also present. Endoplasmic reticulum is present in the cytoplasm over the clustered vesicles. (c) Electron-lucent areas appear in the cytoplasm over the clustered vesicles. This material reacts with antibodies directed against TYMV coat protein, and freeze-fracture electron microscopy suggests the presence of coat protein hexamers and pentamers. (d) The chloroplasts have become clumped with the electron-lucent areas in contact. At later stages, the electron-lucent material is replaced by virus particles. (From Hatta and Matthews. (1974) *Virology* 59:383.)

typical pattern of plant virus movement follows the **source-to-sink** movement of photosynthate in the phloem: First, out of the inoculated leaf and down to the roots; next, up to young leaves and apical meristem; and, finally, to some or all of the older leaves. To move systemically, the virus must be able to (1) move out of the initially infected

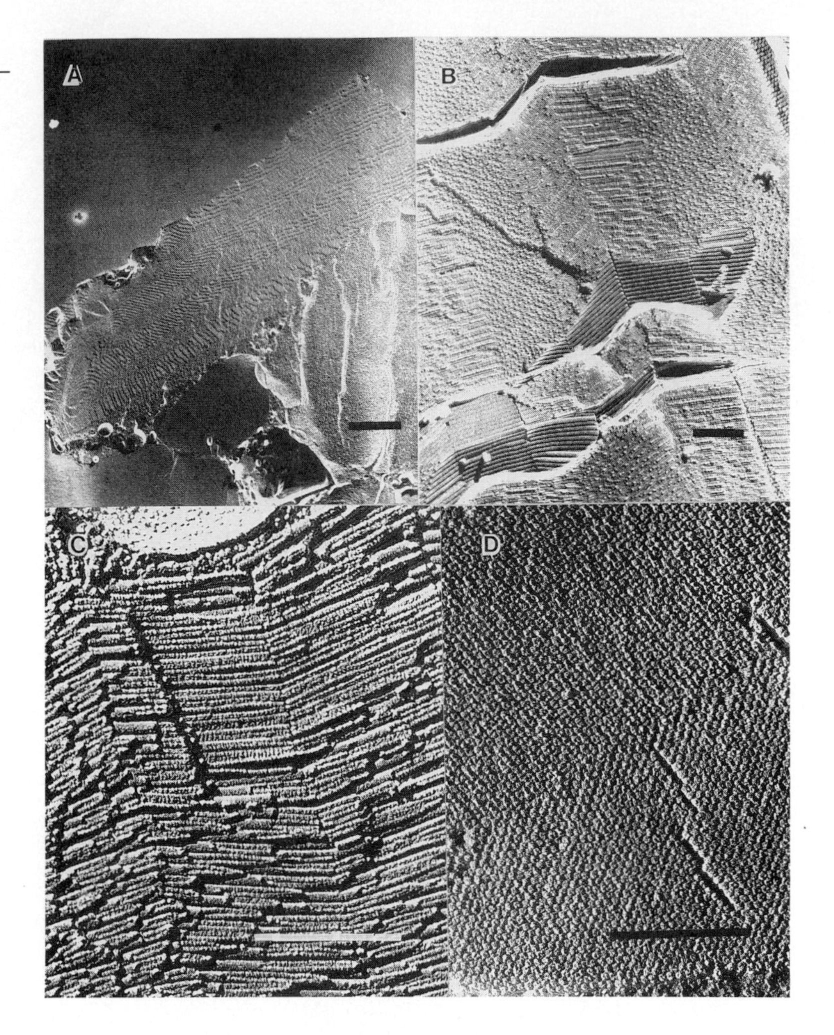

FIGURE 11.13.

Crystalline array of TMV particles in a freeze-etched mesophyll cell at successively higher magnifications. Note the herringbone arrangement of the individual particles visible in (a)–(c). The interior of individual virions is visible in (c). (d) An "end-on" view of the crystal showing the hexagonal arrangement of TMV particles within the crystal. Bar = 1 μm. (Courtesy of R. L. Steere and E. Erbe)

cells, (2) move from parenchyma cells into the vascular system (usually the **phloem**), and (3) move out of the vascular system and return to the parenchyma. Recent molecular studies involving both TMV and several other viruses have provided a reasonably detailed description of virus movement.

Cell-to-cell movement of TMV is dependent on the synthesis of a virus-encoded 30 kDa "movement protein." Initial evidence for the active participation of plant viruses in their own movement came from genetic studies with temperature-sensitive mutants of TMV. Peptide mapping studies showed that the only detectable difference between wild-type TMV and one such mutant strain (LsI) was located in the 30 kDa protein. Nucleotide sequencing studies conducted by other investigators subsequently identified this change as a serine → proline substitution. Furthermore, **site-directed mutagenesis** of full-length TMV cDNAs has confirmed the ability of mutations within the 30 kDa protein to confer a temperature-sensitive phenotype upon TMV. Such mutants replicate normally in protoplasts but are unable to produce either local lesions or systemic infections in intact plants. Transgenic tobacco plants that constitutively express wild-type 30 kDa protein are also able to complement the LsI mutation in *trans*. In addition to supporting systemic replication of the mutant TMV at nonpermissive temperatures, plants that constitutively express the 30 kDa protein also allow more rapid virus movement at lower temperatures.

How does the TMV 30 kDa protein mediate virus movement? Virion formation is *not* required, because TMV mutants unable to synthesize coat protein move with the same efficiency as wild-type virus. Association of the TMV 30 kDa protein with the cellular plasmodesmata is accompanied by as much as a 10-fold increase in their molecular exclusion limit. Even such modified plasmodesmata are much too small to permit the passage of con-

FIGURE 11.14

Potyvirus pinwheel inclusions. (a) and (b) Two models for wheat streak mosaic virus inclusions; (c) A twisted hyperboloid model for the cytoplasmic inclusions of tobacco etch virus. (From Mernaugh et al. (1980) *Virology* 106:273)

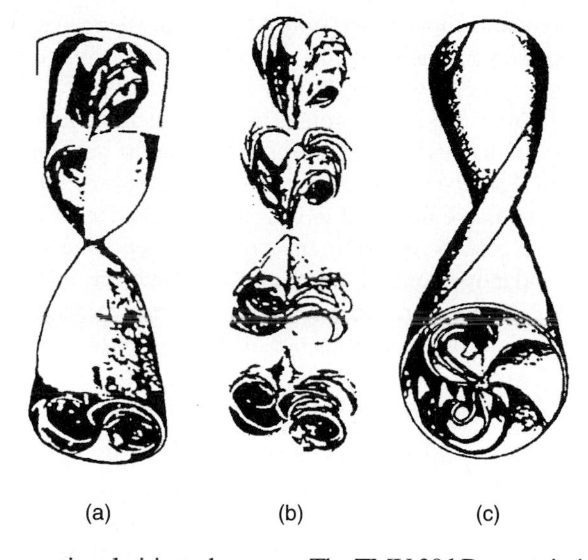

ventional virions, however. The TMV 30 kDa protein is also able to bind cooperatively to single-stranded nucleic acids. As shown in Fig. 11.15, it is possible that one domain in this protein increases the functional size of the plasmodesmata and another coats the viral RNA into an unfolded form able to move through the still-narrow channel. Putative movement proteins have also been identified within several other viral genomes, and the product of CaMV ORF1 (see Fig. 6.15) has also been shown to have single-stranded nucleic acid binding activity.

At present, less is known about long-distance or **systemic movement** of virus in infected plants. As mentioned above, most viruses appear to move in the phloem—at rates measured in centimeter per hour rather than the μm per hour characteristic of cell-to-cell movement. The movement of many viruses (including TMV) through the phloem appears to require a complete virion. Although virus mutants unable to synthesize functional coat

FIGURE 11.15

Structure and function of plant plasmodesmata. Above, longitudinal (a) and transverse (b) views of a simple plasmodesmata. Below, hypothetical translocation pathway for plant virus nucleic acids through plasmodesmata. Movement protein molecules are shown as shaded circles; wavy line, single-stranded nucleic acid; CW, cell wall; R, ribosome. Step 1, formation of movement protein–single-stranded nucleic acid transport complex; step 2, targeting of the transport complex to the plasmodesmata; step 3, movement protein–induced increase in plasmodesmatal permeability; step 4, translocation across the plasmodesmatal channel. (Courtesy of V. Citovsky)

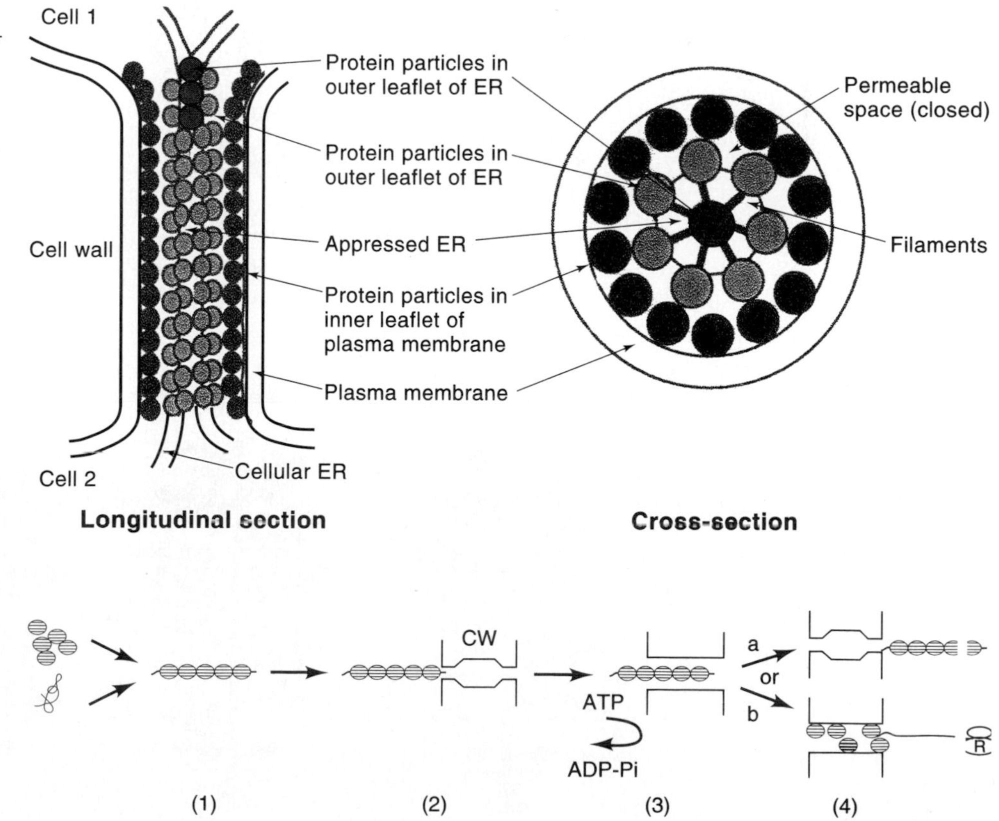

protein sometimes move from leaf to leaf, their long-distance movement is usually sporadic and inefficient. Virion formation is *not* required for the systemic movement of some geminiviruses, however.

The role of coat protein in long-distance virus movement is almost certainly more complex than its function in encapsidating the genomic RNA and protecting it from degradation. For example, a tobamovirus isolated from orchids is able to replicate in tobacco, but it cannot move from leaf to leaf. Exact replacement of the coat protein gene of TMV with the coat protein gene from this orchid virus produced a hybrid virus that was able to replicate and move from cell to cell like TMV but was unable to move systemically. Entry and/or exit of virions from the vascular system may depend on a precise interaction between the virions and some component of the phloem cells.

11.10.3 Factors Controlling Host Range

Why does a particular virus infect one plant species but not another? This question has long been a matter of speculation among plant virologists, but an increasing appreciation of the complex molecular interactions between viruses and their hosts has recently begun to provide some definite answers. Viruses might be prevented from causing systemic disease at any of four stages: (1) during the initial events that end with viral uncoating, (2) during replication in the initially infected cells, (3) during movement from those initially infected cells, or (4) by an active host response to these early stages of virus replication. Barriers to systemic replication arising at stages 1 through 3 can be considered essentially passive phenomena, reflecting a basic "incompatibility" between virus and host.

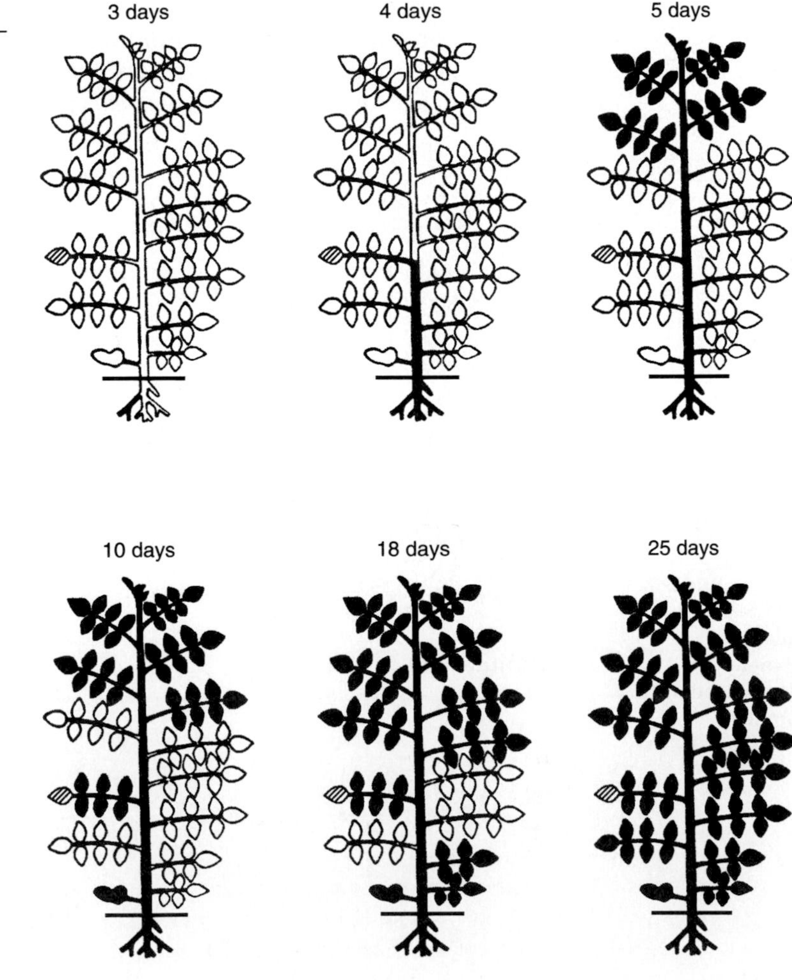

FIGURE 11.16

TMV spread in a young tomato plant. The inoculated leaf is shaded, and systemically infected tissues are shown in black. (From G. Samuel. (1934) *Ann. Appl. Biol.* 21:90)

Surrounded by thick cellulose cell walls, plant cells appear to **lack the specific protein receptors** present on the surface of animal or bacterial cells. Thus, whereas the envelope or surface proteins of plant rhabdo- and reoviruses almost certainly play a critical role in recognizing susceptible cells in their insect vectors, the coat proteins of other plant viruses appear to play little if any positive role in recognizing potential host cells. Plant viruses usually enter their hosts through wounds on the plant surface, and there seems to be no specificity in the uncoating process. For example, TMV particles are uncoated and express their RNA in *Xenopus* oocytes.

The viral genes most likely to function as host range determinants are those involved in replication and cell-to-cell movement. In Chapter 2 we saw that **viral RNA replicases** contain both host- and virus-encoded subunits. Any plant species in which a virus is unable to form a functional replicase in the initially infected cells will be *immune* to infection. However, the ability of many viruses to replicate in protoplasts derived from species in which there are no visible signs of infection following mechanical inoculation of intact leaves suggests that the relative ability to move from cell to cell is a more common host range determinant. For example, 65 of 1031 cowpea lines tested were operationally immune to cowpea mosaic virus; that is, visible disease did not develop and infectious virus could not be recovered from inoculated leaves. When protoplasts were prepared from 55 of these "immune" lines, however, all but one of them were susceptible to cowpea mosaic virus. The complex molecular interactions between the host cell and virus movement proteins provide ample opportunity for "mismatches" to arise. Several active plant responses to virus infection that limit the extent of virus spread will be described in Section 12.9.

SECTION 11.10 FURTHER READING

Original Papers

SAMUEL, G. (1934) The movement of tobacco mosaic virus within the plant. *Ann. Appl. Biol.* 21:90.

BEIER, H., et al. (1977) Survey of susceptibility to cowpea mosaic virus among protoplasts and intact plants from *Vigna sinensis* lines. *Phytopathology* 67:917.

WOLF, A., et al. (1989) Movement protein of tobacco mosaic virus modifies plasmodesmatal size exclusion limit. *Science* 246:377.

CITOVSKY, V., et al. (1990) The P30 movement protein of tobacco mosaic virus is a single-stranded nucleic acid binding protein. *Cell* 60:637.

ROMERO, J., et al. (1993) Characterization of defective interfering RNA components that increase symptom severity of broad bean mottle virus infections. *Virology* 194:576.

Books and Reviews

ATABEKOV, J. G., and M. E. TALIANSKY. (1990) Expression of a plant virus-coded transport function by different viral genomes. *Adv. Virus Res.* 38:201.

CITOVSKY, V., and P. ZAMBRYSKI. (1991) How do plant virus nucleic acids move through intercellular connections? *Bioessays* 13:373.

MATTHEWS, R. E. F. (1991) *Plant virology,* 3d ed. Chs. 7, 8, and 10. Academic Press, New York.

DAWSON, W. O. (1992) Tobamovirus-plant interactions. *Virology* 186:359.

11.11 BACTERIOPHAGE INFECTIONS

Earlier chapters have introduced a variety of bacterial viruses, or bacteriophages. Variations in virion morphologies and nucleic acids, genetic information contents, and modes of replication were considered. As a consequence of these variations, the different viruses also show wide variations in biological effects on the host cell. Infections with some phages cause minimal effects, producing an infected cell with biological properties barely distinguishable from those of the uninfected cell. However, for many phage infections the biological consequences for the cell are disastrous: The bacterial cell is destroyed in the process of producing virus. Owing to the killing and/or disruption of the cell, such infections are called **cytocidal or lytic** as they are with animal viruses. In this activity of this objective, various phages employ different functions to inhibit bacterial cell processes. Host defense mechanisms against the virus are systematically incapacitated, and cellular resources of energy and materials are diverted from replication of the cell to production of viral components.

Other phages have evolved mechanisms that control their replicative and destructive functions so that they can maintain a relatively stable, long-term association with the host cell. The result can either be a symbiotic relationship, or a state of **lysogeny** (that involves the integration of the phage genome into that of the host cell).

11.11.1 Productive Phage Infections

Inhibition of Host Cell Functions Some of the simplest phages with the minimum amount of genetic information needed for replication are replicated without drastic interruption of the host cell physiology. For example, phage fd, a filamentous single-stranded DNA virus (**Inoviridae**), establishes a productive infection without killing or lysing the cell (see Sec. 7.6). The infected cells continue to replicate themselves as well as the virus, although the drain on cellular resources does slow bacterial growth. The virions are released by a nonlethal process of secretion (see Sec. 7.6). This essentially nondestructive mode of replication and release is rare among bacterial viruses, but many plant and certain animal viruses utilize analogous strategies in dealing with their hosts. Conversely, most bacteriophages eventually kill the host cell during release of progeny virus particles, if not before. This is true for the small RNA phages (**Leviviridae**; Sec. 2.7) as well as generally for the isometric single-stranded DNA phages (**Microviridae**; Sec. 7.5) and for all the double-stranded DNA phages (Sec. 8.7). Therefore, most phages inhibit at least some specific host cell functions during infection. Major examples of such inhibitions, discussed below are summarized in Table 11.7.

Double-stranded DNA phages are generally complex, often carrying genes for nonessential functions, including specific products aimed at neutralizing defense mechanisms of the host. T4, one of the largest phages, has several of these weapons, some active or offensive, others passive or defensive. For example, the product of phage gene *2* (i.e., gp 2) enters the cell with the DNA as it is injected by the syringelike virion (see Sec. 8.7). This protein protects the ends of the phage DNA strands from a bacterial defense against the virus, a specific exonuclease (exo V) that shortly after injection, in the absence of gp 2, will degrade the phage DNA, starting at a free end. Another phage function, which is expressed early upon infection, directly inhibits exo V enzymatic activity so that it cannot attack newly synthesized phage DNA.

As pointed out in the discussion on replication of these viruses (see Sec. 8.7.4), T4 and other T-even phages also inhibit transcription of host cell genes into mRNA soon after infection starts. This process is accomplished by means of a phage-coded enzyme that adds one adenosine diphosphate linked to ribose (ADP-ribose) to a component of the host cell RNA polymerase enzyme. This modification not only interrupts host cell gene ex-

TABLE 11.7

Major examples of inhibition of specific host cell functions by bacteriophages

Virus	*Host Function*	*Inhibition Mechanism*
T4	Transcription of host genes	ADP-ribosylation of RNA polymerase modifies promoter recognition
T7	Transcription of host genes	Phosphorylation of RNA polymerase inactivates enzyme
T-even	Host protein synthesis	Indirectly by inhibition of transcription
T-even	DNA synthesis	Nuclease degrades template lacking HMC
T-even	DNA synthesis	Inhibitors of synthesis of cytosine nucleotides
PBS2	Uracil-DNA glycosylase (removes uracil from DNA)	Inhibitors of DNA glycosylase
T-4	Exo V	Inhibitor synthesized early to protect new DNA
T-even	Anti-T-even DNA restriction nuclease	Glucose on HMC-DNA
T4	Anti-T4 DNA restriction nuclease (soluble)	Inhibitor
T4	Anti-T-even DNA restriction nuclease (membrane bound)	Attachment of virion to cell inhibits nuclease
T3	Restriction nuclease dependent on S-adenosylmethionine (SAM)	SAMase

pression but also helps to divert the host cell enzyme into production of phage mRNA, possibly by allowing the polymerase to recognize promoters of only certain phage genes. Somewhat less complex double-stranded DNA phages, such as T7, also inhibit host cell transcription by modification of the cellular RNA polymerase. In this case however the modification seems to be only phosphorylation, which eliminates its enzymatic activity. An entirely new phage-coded RNA polymerase is produced by the phage (see Sec. 8.7.1).

Host cell protein synthesis is also halted early during infection with T-even phages, which is probably due to Ty reg A inhibition of translation. In bacterial cells mRNA is normally unstable: Degradation by cellular nucleases generally allows host cell mRNA a half-life of only 1 or 2 minutes. Therefore, host cell protein synthesis stops in only a few minutes when no new mRNA is made to replace that which is degraded. Although some workers have suggested more specific mechanisms for inhibition of host cell protein synthesis by phages, no convincing evidence for specific translation inhibition by T-even or other phages has been presented.

T-even phages also make nucleases that rapidly destroy host cell double-stranded chromosonal DNA. This action means that there is soon no template left to make new host cell DNA or RNA, but the main purpose of this attack appears to be a requisitioning of all possible precursors for phage DNA. In addition to the nucleases, degradation of bacterial DNA involves other phage functions that are likewise dispensable for phage replication. These functions include the unfolding of the bacterial chromosome from its normal condensed nucleoprotein structure prior to DNA degradation. It should be noted that most simpler phages are not as aggressive as T-even phages, in that they do not usually have the ability to consume preexisting cellular DNA.

As mentioned earlier, T-even phages contain hydroxymethylcytosine (HMC) in place of cytosine in their DNA, and several phage functions are required for synthesis of this precursor. However, to conserve resources, certain phage gene products also inhibit synthesis of cytosine nucleotides. Moreover, the nucleases that destroy host cell DNA act only on DNA containing cytosine and not HMC. Other phages with unusual DNA bases must also produce special gene products to facilitate synthesis of their DNAs. For example, the DNA of phage PBS2, which infects *Bacillus subtilis,* contains uracil instead of thymine. This host normally contains an enzyme, uracil-DNA glycosylase, that removes uracil from DNA; thus, obviously, PBS2 must carry a weapon to inhibit this specific host cell enzyme if uracil is to be incorporated into new phage DNA. However, the same host cell enzyme will destroy parental phage DNA shortly after its injection unless this enzyme is neutralized by an early phage function. It also appears that one of the few genes of phage T5 that is carried in the small DNA portion injected first in this unique two-step injection process (see Sec. 8.7) codes for an inhibitor of host uracil-DNA glycosylase. However, the significance of this activity for T5 replication is unclear, since there is no uracil in the DNA of this phage.

Host Cell Restriction Systems Among the more wide-spread antiviral defense mechanisms of bacteria are sequence-specific DNA endonucleases or restriction enzymes (see Sec. 7.5). These enzymes usually distinguish bacterial DNA from "foreign" DNA of viruses (or from any source) by virtue of methyl groups placed on key nucleotides in the recognition sequence for a **restriction endonuclease.** This modification of host cell DNA is accomplished by bacterial *methylases,* which often share with the companion nuclease, polypeptide subunits that convey the sequence specificity. The intermediate-size DNA phage, T3, has a strategy for neutralizing some host cell restriction enzyme systems that depend on methylation of the host cell DNA for protection. However, this strategy is somewhat indirect. Obviously, there must be some chemical source of methyl groups used by the sequence-specific modifying enzyme, methylase, which is paired with each restriction nuclease. In many bacteria this methyl group donor is the amino acid methionine, in the form of a high-energy intermediate, **S-adenosylmethionine (SAM).** The high-energy bond between the sulfur and an adenosine nucleoside provides the driving force for transfer of the methyl group from the sulfur to the DNA. Because of the close relationship between the modifying and restricting components of the system, which often includes shared subunits, some restriction nucleases require SAM as a cofactor to be active However, SAM does not serve as a methyl donor as it does when used by the modifying enzyme. T3 has the unusual ability to produce an enzyme, SAMase, which hydrolyzes SAM. This strange T3 gene product also has other effects on the host, to be discussed below.

In addition to the methylation-based systems, there are also many other types of bacterial DNA restriction systems. For example, it was pointed out previously that the HMC in T-even DNA has glucose residues attacked by phage enzymes. These glucose residues render T-even DNA resistant to degradation by another type of bacterial restriction nuclease. For some reason, this enzyme seems to attack only T-even DNA and not cell DNA or other phage DNA, even though the latter DNAs lack glucose residues. The glucose is added to phage DNA only after HMC is incorporated, so that newly synthesized phage DNA would be highly vulnerable to this type of restriction system. Therefore, T4 has developed one more offensive gene product to directly inhibit this anti-T-even restriction nuclease, which affects only T4 DNA, prior to phage DNA replication.

E. coli also produces yet another **restriction nuclease,** which seems to combat not only T4 DNA lacking glucose, but also that of any other T-even phage. However, this second anti-T-even nuclease is located in the cell membrane rather than in the cytoplasm, so that it cannot affect replicating forms of T-even DNA. Although incoming T4 DNA is usually glucosylated and therefore protected from this second nuclease, T4 has discovered a way to inhibit the membrane-associated nuclease merely by attachment of the virion to the outside of the cell. Thus, T4 has at least four specific mechanisms to combat host cell functions that appear, in turn, to be designed primarily, to ward off virus infections.

Many DNA phages do not seem to have the ability to inhibit host cell restriction systems and their DNA can therefore be degraded shortly after infection. But if by chance the DNA is modified by the host cell system before it is restricted, the DNA can replicate, and all progeny will also be modified. Then the new mutated phage will not be restricted upon infection of another cell with the same restriction system.

Release of Phage Particles After production of all phage components and their assembly into new virus particles inside the cell, there must be some mechanism for release of virus progeny. Usually, this activity involves complete disruption of the cell or lysis. However, lysis is not mandatory for release of all phages; as noted above, productive infection with the filamentous phage fd does not kill the host (see Sec. 7.5). As the single-stranded phage DNA passes through the membrane (Fig. 11.17), the coat protein displaces another phage gene product *(gp V)* associated with the DNA inside the cell. The presence of phage coat protein in the cell surface alters the immunological properties of the cell, creating new antigens that can be recognized by antibodies generated in animals injected with phage-infected bacteria. Although this mode of release is rare for phages, many animal viruses leave the cell by an analogous process known as **budding** (See Chapter 6).

It might seem that only thin virions such as the filaments of fd could exit the cell without major disruption of the membrane, but the somewhat more cumbersome icosahedral capsid of the single-stranded DNA phage φX can also leave a bacterial cell without complete lysis. However, this process can occur only at temperatures somewhat below the optimum for cell growth. It is not clear whether this event requires two different mechanisms for release of φX, depending on temperature. Perhaps the enzymes involved in lysis at the higher temperature are less active at the lower temperature and thus only limited cell disruption occurs (gene *E* in the case of φX).

In the usual lytic infections there are at least two morphologically and chemically distinct obstacles to virus release that must be overcome: the cell membrane and an outer cell wall. Presumably, passage of phage fd DNA is facilitated by special properties of membrane sites loaded with coat proteins. Thus this transit does not harm the membrane and the narrow fd virion probably fits through existing openings in the cell wall. But large virions that are completely formed inside the cell cannot freely pass the lipid-rich membrane, which normally retains even soluble proteins in the cytoplasm. Moreover large virus particles are also prevented from escaping the cell by the rigid outer wall that gives bacterial cells their distinctive

FIGURE 11.17

Schematic diagram of virion of ssDNA phage fd leaving an infected cell without lysis. The A protein passes through the cell membrane first at a site where coat protein molecules are embedded. The intracellular ssDNA circle is coated with dimers of gp5 protein that are displaced by coat protein as the DNA passes through the intact membrane.

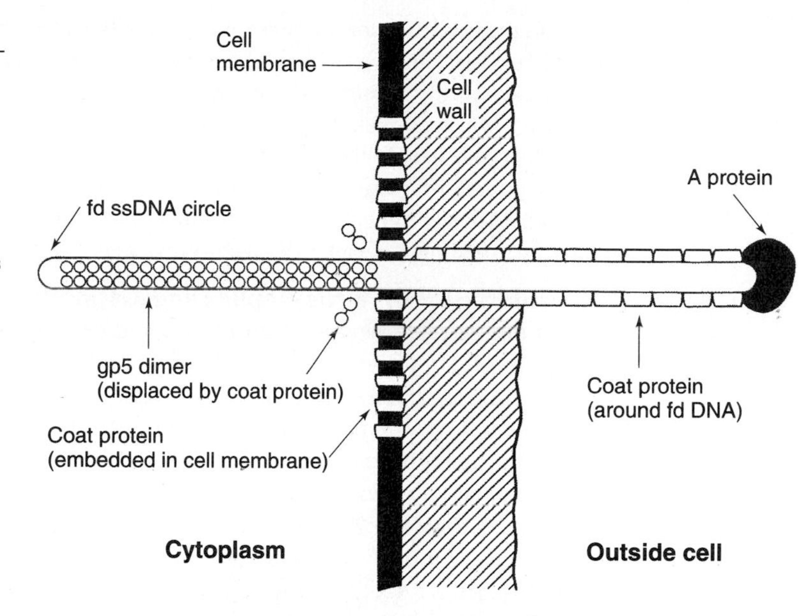

shapes (see Fig. 11.18). In some gram-positive bacteria, (e.g., *Bacillus subtilis*), this wall consists mainly of a thick cross-linked matrix of peptides and polysaccharides **(peptidoglycan);** in gram-negative bacteria (e.g., *E. coli*), there is less peptidoglycan, and an additional outer wall layer composed of **lipopolysaccharides** (LPS) is present. Moreover, large phages must encode different enzymes to disrupt the membrane and the cell wall.

Some viruses can actually lyse the cell without replicating in it. For example, T4 virions contain an enzymatic activity that partially degrades the outer LPS layer of the *E. coli* wall. Apparently, some of the low-molecular-weight digestion products trigger a change in the phage tail that results in release of the phage DNA into the cell. Passage of one or a few DNA molecules through the cell membrane evidently does not destroy its integrity because the cell cytoplasm does not leak out. But at high multiplicity of infection when many virions (i.e., several thousand) attach to one cell and inject their DNA at once, the partial hydrolysis of the LPS layer weakens the wall, and the multiple punctures from injected DNA disrupt the membrane. The result, called **lysis from without,** is a nearly instantaneous explosion of the infected cell. Obviously, the injected phage DNA is not replicated in this situation.

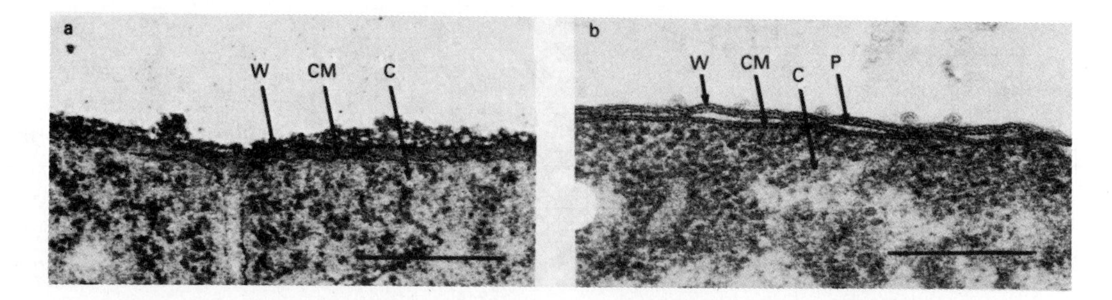

FIGURE 11.18

Electron micrographs of sections of the surface layers of (a) a gram-positive bacterium *(Bacillus megaterium)* and (b) a gram-negative bacterium (*E. coli*), illustrating the differences in structure. (W, wall; CM, cell membrane; C, cytoplasm; P, peptidoglycan). In the gram-positive bacterium, the wall consists of a single, thick, continuous layer. In the gram-negative bacterium, the wall is multilayered and only loosely attached to the cell membrane. In *E. coli* and some other gram-negative bacteria, there is an intermediate layer of peptidoglycan visible as a single layer between the wall and bilayered membrane. Scale bars correspond to 250 mμ. (Courtesy of Mr. John Hanson, Department of Microbiology)

In many of the more complex phages, control genes have evolved to regulate the activities of enzymes involved in lysis. A product of the T4 **rapid lysis gene,** rII, becomes embedded in the cell membrane and thereby inhibits the function of *gp t;* thus, the phage-coded lysozyme, which is made long before lysis, cannot immediately pass through the membrane to reach its substrate in the wall. T4 has another gene for controlling lysis, gene *sp.* When gp sp is missing, T4-infected cells will lyse at the usual time—*even if the phage does not produce lysozyme.* Apparently, gp sp somehow prevents bacterial enzymes in the dying cell from prematurely rupturing the wall, perhaps by preventing their passage through the membrane.

Inhibition of Virus-Replicating Capability Many phages exhibit another type of biological effect on the host that could be considered a form of inhibition: inhibition of the ability to support replication of other viruses. Phages that are greatly different in replication mechanisms may inhibit or modify different cellular functions. Consequently, after infection by one phage, the cell might no longer be able to support replication of a different phage. This phenomenon is called **mutual exclusion.** Similar phenomena in animal and plant virus systems are referred to more generally as forms of **viral interference** (see Sec. 11.6). Mutual exclusion or interference does not necessarily involve inhibition of host cell functions. Thus, one virus might produce products that directly inhibit or degrade products of another virus. Some viruses can even be self-inhibitory under certain circumstances. For example, a second virion of T4 is unable to infect a cell after a first T4 DNA molecule enters and begins transcription. The second phage DNA molecule is degraded after injection by enzymes made from the phage that got there first. This phenomenon is called **superinfection exclusion.**

11.11.2 Nonproductive Infections: Lysogeny

Once inside the bacterial cell, the phage can assume a state of latency or silent infection. **Lysogeny**, as it is called, involves an extensive regulation of phage functions that allows the viral genome to remain for an indefinite time inside the infected cell without killing it and without producing virions. Moreover, the replication of the phage genome becomes coordinated with replication of the host genome. Thus, when the infected cell divides, at least one copy of the viral genome is retained in each daughter cell. As we shall see, there is more than one way for a phage to achieve coordination between replication of its genome and that of the host.

Phage DNA may physically integrate into host cell DNA molecules, or it may remain physically independent of host cell DNA. Its replication can be linked to cell DNA replication solely through the actions of gene products. Only phages that possess the necessary strict controls over their replicative cycle can establish lysogeny. As mentioned before, these are often called **temperate** phages as opposed to the **virulent** phages, that can only carry out productive infections. The advantages of this alternative existence are clear. Virus is able to multiply and spread in the environment while remaining entirely within the sanctuary of the host cell. Nevertheless, certain lethal insults to the cell can induce the usual process for productive infection so that new virions can escape from the doomed host. Although the temperate phage genomes in **lysogens** usually do not affect cellular replication functions, they may cause other cellular changes that can have profound biological effects not only on the lysogens but also on higher organisms serving as hosts to these bacteria.

Immunity Bacteria carrying the genome of a particular temperate phage can be identified by testing for a key property of all lysogens, immunity to superinfection with another particle of the phage of the same type. This **immunity** is due to the diffusible nature of certain elements in the phage control processes. The process differs from **superinfection exclusion** by T4, since the DNA injected by a second phage particle is not degraded immediately. Instead it is prevented from expressing all replicative functions. The control element responsible for immunity is a protein called a **repressor,** which prevents phage gene expression by binding to certain sites in the DNA. This binding inhibits initiation of transcription of almost all phage genes except, under most conditions, that of the repressor gene itself.

Mechanism of Lysogeny The central region of the linear λ DNA in the virion contains all the regulatory elements involved in establishment and maintenance of lysogeny (see Fig. 8.18). The coding sequence for the chief repressor protein, as well as its binding sites, are in this same region, which is, therefore, known as the **immunity region.** Normal plaques of λ on a lawn of susceptible bacteria appear turbid in the center due to overgrowth by immune lysogens. In contrast, λ mutants, which lack repressor, cannot establish lysogeny and thus produce plaques that, like those of a virulent phage, are entirely clear. Hence the symbol cI, for clear plaque mutant class I, was chosen for the repressor gene. In this regard, other classes of clear plaque mutants have defects in genes *c*II or *c*III.

As mentioned in the description of the productive replication cycle of λ (Sec. 8.7.2), there are two main promoters where gene transcription can be initiated by the host RNA polymerase alone. These are located on opposite sides of the cI gene. Next to these promoters, P_{left} and P_{right}, lie the binding sequences for *c*I repressor (see Fig. 11.19). These sequences are called **operators,** by analogy to control sequences with a similar role in the inducible bacterial gene system (or operon) coding for several products involved in lactose metabolism. Each operator, O_{left} and O_{right}, is actually composed of **three similar but distinct sites** (e.g., O_L1, O_L2, O_L3) to which the repressor can bind, primarily in the form of a protein dimer. Thus, up to six monomeric molecules of repressor can bind to each operator. The strongest repressor binding occurs with sequences O_L1, and O_R1. These regions are covered by RNA polymerase when it binds to its promoters next to either the cro gene (O_R) or the N gene (O_L). The polymerase also covers a few sequences at the very ends of those two genes to be transcribed initially, as illustrated for O_R in Fig. 11.20. Thus, the repressor prevents expression of early λ genes by directly interfering with binding of RNA polymerase at the O_R 1 and O_L 1 sequences.

When a λ DNA molecule first enters a cell not already carrying the viral genome, no repressor is present and transcription begins immediately at P_R and P_L. Transcription to the left starting at P_L produces mRNA for gp N. This protein is a **positive control protein** that must be present to allow RNA polymerase to continue transcription from P_L and P_R past the ends of the *N* or *cro* genes, respectively, and into the next adjacent early genes. The gene next to *N* is another regulatory gene, *c*III, which is transcribed as soon as the first gp N is made. The first gene transcribed to the right beyond *cro,* once gp N is present, is called *c*II. As noted above, gp cII and gp cIII are needed besides gp cI to establish lysogeny efficiently. Somehow these two products act in concert in a positive regulatory role. This action allows RNA polymerase to initiate transcription somewhere to the right of the DNA region encoding the cro mRNA. This transcription continues leftward, back through the *cro* gene, in the direction opposite from *cro* transcription, and eventually results in production of mRNA for *c*I repressor.

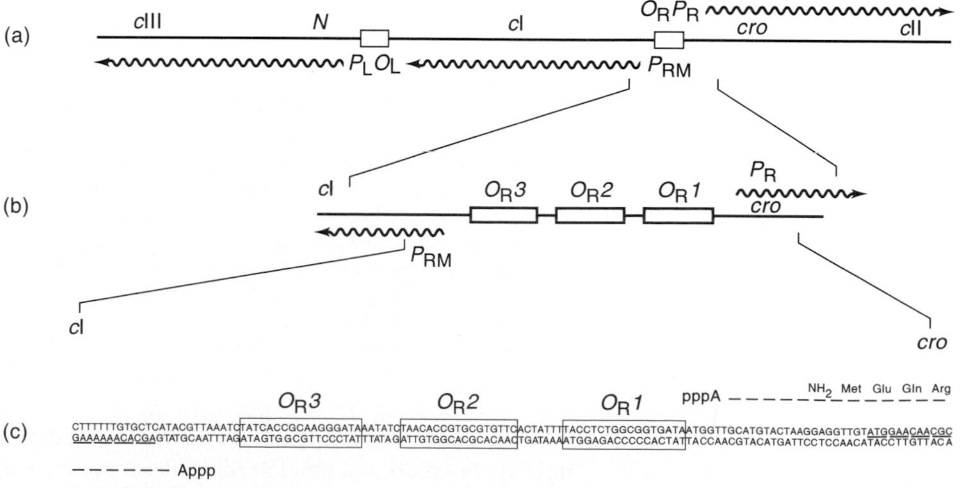

FIGURE 11.19

λ genes and regulatory elements. (a) A portion of the λ genome. The arrows indicate the directions and start points of transcription of various genes. O_LP_L and O_RP_R are the left and right operator and promoter regions. P_{RM} is the cI promoter active in a lysogen. (b) Expanded diagram of the λO_R region. O_R1, O_R2 and O_R3 are repressor and *cro* binding sites, each 17 bp long. The start points of transcription from P_R and P_{RM}, which are located outside the operators, are indicated. As described in the text, polymerase bound to either promoter overlaps repressor binding sites in the operator. (c) Nucleotide sequence of the λO_R region. The start points of the cI and *cro* mRNAs are shown, as are the amino terminal portions of the corresponding protein sequences. (From Ptashne et al. (1980) *Cell* 19:1)

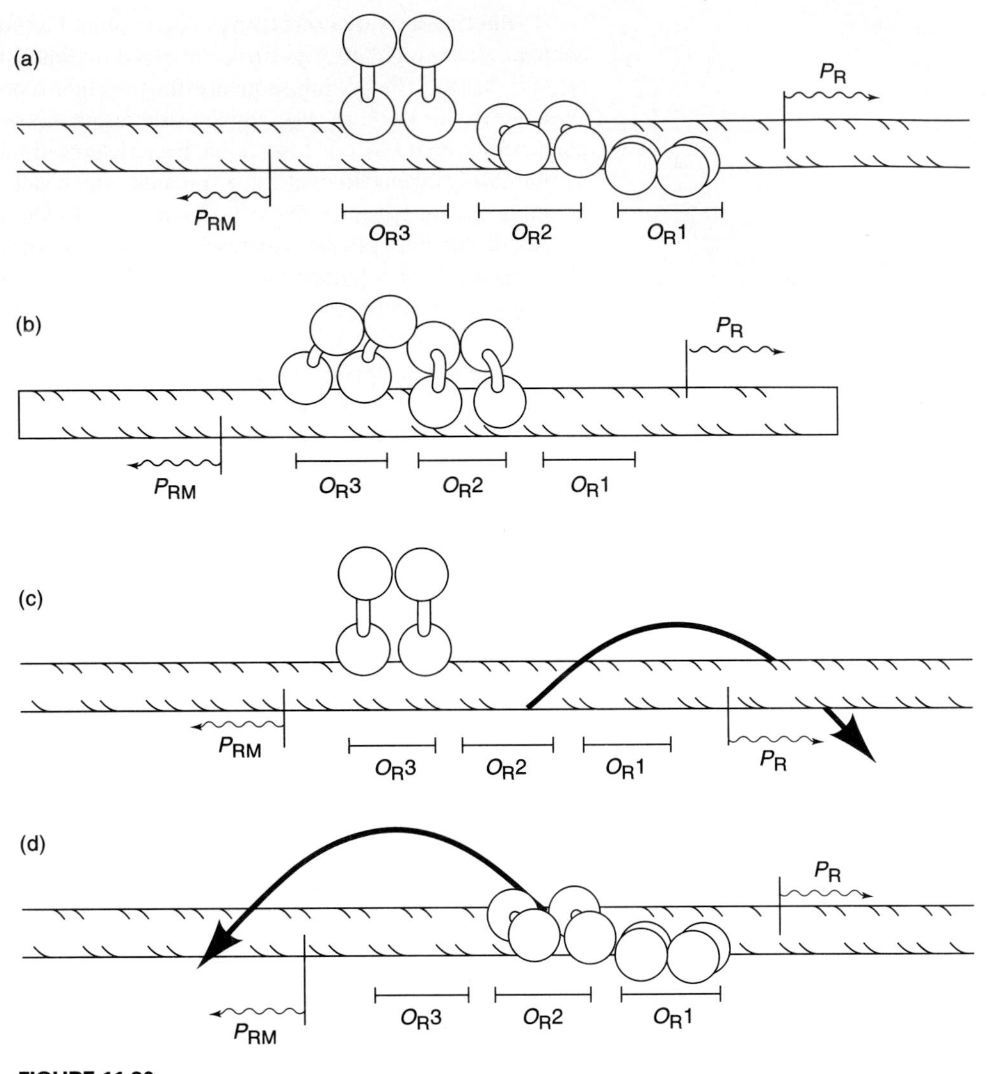

FIGURE 11.20

Arrangements of repressor and RNA polymerase molecules bound in the vicinity of O_R. Repressor dimers are shown bound to one or more of the sites in O_R, and RNA polymerase molecules are shown bound to P_R and P_{RM}. The start points of transcription of P_R and P_{RM}, located on different strands on opposite sides of the helix, are shown. (a) Repressor dimers at all three sites in O_R, with those bound at O_R1 and O_R2 interacting cooperatively. (b) O_R1 is mutant so that it cannot bind repressor, and the alternate pairwise interaction between repressor dimers at O_R2 and O_R3 is shown. (c) Repressor at O_R3 and polymerase at P_R. Note that repressor at O_R3 only should have no effect on polymerase at P_R. (d) Repressor at O_R1 and O_R2 and polymerase at P_{RM}. Note that an amino terminal domain of a repressor bound at O_R2 could plausibly touch a polymerase at P_{RM}. The components of the figure are drawn roughly to scale. An RNA polymerase molecule is drawn as a shape that implies its direction of transcription. A repressor monomer is drawn to indicate its two domains and the connector that joins them. Repressor dimers are stabilized by contacts between the carboxy terminal domains, and DNA is contacted by amino terminal domains. In (a), (b), and (d) repressor dimers are shown interacting through additional contacts, presumed to be between carboxy terminal domains. The DNA helix is drawn in B form with 10.4 bp per turn. The positioning of bound polymerase molecules was deduced from a comparison of the P_R and P_{RM} sequences with those of promoters on which chemical probe experiments have been performed. (From Ptashne et al. (1980) *Cell* 19:1)

As soon as gp cI begins to accumulate, it binds to O_R and O_L and stops production of mRNA for gp cII and gp cIII. Since these products are unstable, they soon are no longer present to initiate cI transcription. But this does not stop production of the repressor, because binding of repressor to the O_R2 site also has another effect: It allows RNA polymerase to initiate transcription at a previously **silent promoter** for the cI gene. This second cI promoter, called P_{RM} (*p*romoter for *r*epressor *m*aintenance), is located beside the O_R3 repressor binding site (see Fig. 11.19). Because the repressor has a greater affinity for the O_R1 and O_R2 sites than for the O_R3 site, the first repressor made after infection has a tendency to bind only to the O_R1 and O_R2 sites, this process stimulates production of more repressor from the newly activated promoter.

What inhibits λ from establishing lysogeny every time it infects a cell? Infection is prevented by the product of the *cro* gene. Gene *cro* (for *c*ontrol of *r*epressor *o*peron) produces a **DNA-binding protein** that acts much like a repressor. It turns out, in fact, that gp cro binds to λ DNA at the O_L and O_R sites to which repressor binds; however, within O_R, gp cro binds first and most tightly to O_R3. By this action it interferes with RNA polymerase attempting to initiate transcription of the cI gene at the promoter near O_R3. When large amounts of gp cro are present, as in the absence of cI repressor, all three sites in O_R as well as O_L become occupied with gp cro. This event inhibits transcription from the two main promoters, including transcription of the repressor gene and even the cro gene itself. In cells committed to the lytic cycle, this device limits production not only of repressor but also of the other early gene products from the immunity region. All of these proteins are needed only in small amounts, if at all, for virus production.

As mentioned above, *cro* is one of the first two genes to be transcribed, together with N, which is needed to transcribe the cII and cIII genes. Thus, gp cro is made before cII and cIII products are present to start cI transcription. How λ manages to establish immunity and form a lysogen is an important question. Current thinking puts cII at the center of the decision. The first molecule of gp cro bound to O_R3 may fall off and allow a molecule of RNA polymerase to proceed into the cI gene. Moreover, until the concentration of gp cro becomes much higher than that of repressor, the O_R1 site will still be open to bind repressor. This becomes the first step toward shutdown of all phage productive functions, including *cro*. It appears that there is a real competition between the *cro* and cI systems to gain control of the λ genome early during infection.

The various regulatory genes of λ and their functions are summarized in Table 11.8. It is not important at this point to remember all the names and detailed organization of the controlling elements in phage λ, but this intricate network does exemplify certain principles

TABLE 11.8
Summary of bacteriophage λ control genes and their functions

Gene	*Product*	*Function(s)*[a]
cII and cIII	Positive regulatory proteins for cI gene	1. Required together for initiation of transcription of cI starting to the right of *cro* and proceeding leftward early after infection
cI	Repressor of main phage operons	1. Binding at O_R1 inhibits transcription of *cro* 2. Binding at O_L1 inhibits transcription of N 3. Binding at O_R2 allows transcription of cI starting left of O_R 4. Binding at O_R3 inhibits transcription of cI starting left of O_R
cro	Controller of λ repressor	1. Binding at O_R3 inhibits transcription of cI left of O_R 2. Binding at O_L and O_R inhibits transcription of *cro* and other early genes
N	Positive regulatory protein	1. Required for transcription of cII and other genes left of N and cIII and other genes right of *cro*
cII and cIII	Positive regulatory proteins for cI gene	1. Required together for initiation of transcription of cI starting to the right of *cro* and proceeding leftward early after infection

[a]Binding sites of regulatory proteins to DNA are listed in order of decreasing affinity.

and be pertinent to other biological systems. For example there is an overall similarity to systems of cellular differentiation in higher organisms. Moreover, different cells in the embryos of higher organisms all arise from one fertilized egg, but at some early developmental point each becomes committed to one stable pattern of gene expression or another, and thus a variety of cell types arise. Although we do not yet know the mechanisms involved in these processes, it would not be surprising if **autogenous gene regulation** and interacting regulatory circuits turn out to be as important here as they are for the much simpler phage λ.

Integration When the repressor function has been overcome and the commitment to lysogeny has been made, the phage DNA cannot be replicated in the usual fashion, for this involves several relatively late phage genes (see Sec. 8.7). However, integration of λ into the host cell DNA allows it to be replicated and distributed to daughter cells just like any other part of the host chromosome. As noted earlier, the linear λ DNA in the virion circularizes upon entering the cell, owing to complementary sequences in the single-stranded "sticky" ends. For attachment to host cell DNA, the λ circle is broken at a specific site near the middle of the linear form of the molecule (see Fig. 11.21). The much bigger but also circular host chromosome is similarly cleaved in one specific site, and then the two molecules are joined into one large circle.

There is very little nucleotide sequence homology between the attachment sites in the phage and host chromosomes (only 15 bases). The usual genetic exchanges between two cells or two viruses involve much more similarity in the sequences of the DNA molecules

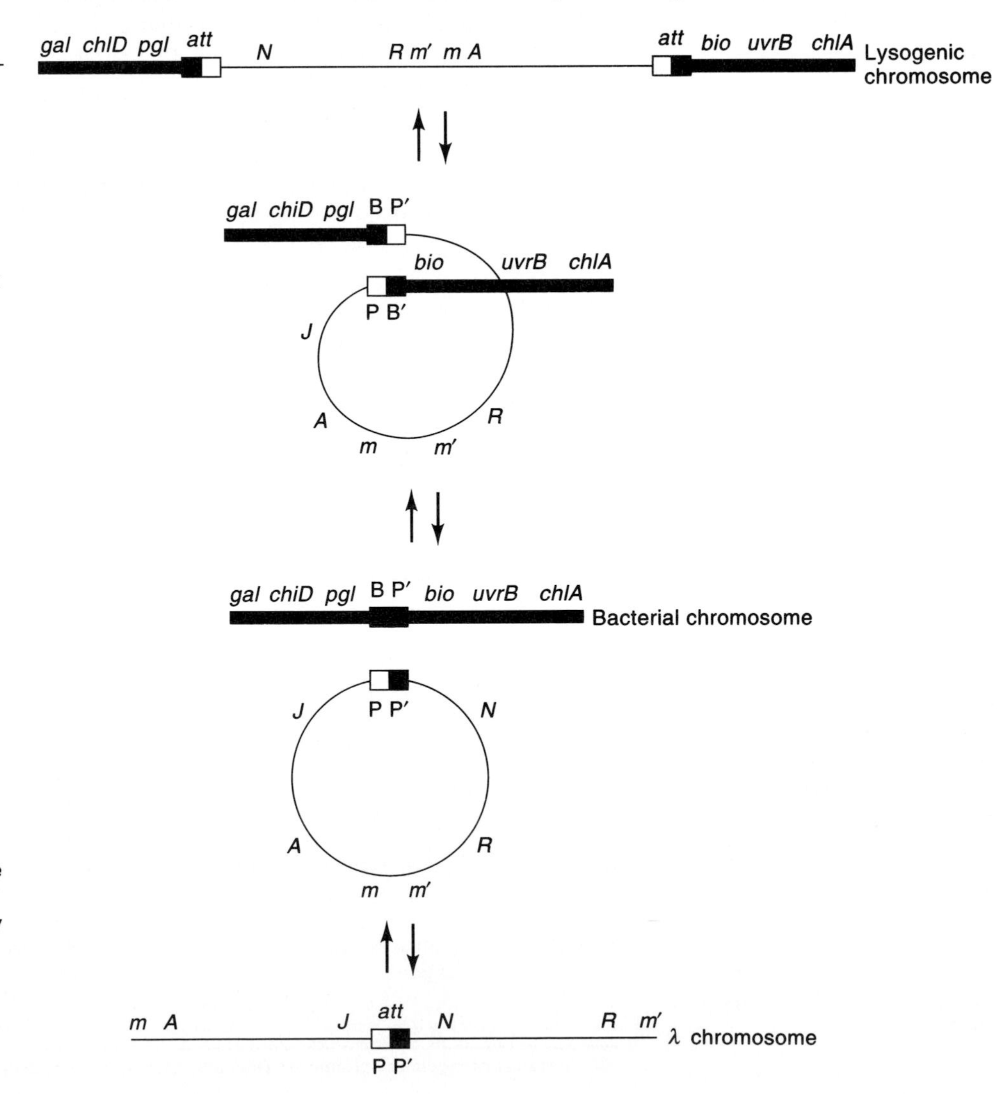

FIGURE 11.21

Insertion and excision of λ. The bottom line represents λ DNA as it occurs in the virion. After infection and circularization, λ DNA is broken at a specific site (P.P′) and rejoined to bacterial DNA at a specific site (B.B′), thus splicing λ DNA into the continuity of the host chromosome. In excision, the final step is reversed, regenerating the bacterial chromosome and the circular λ DNA molecule, which may then replicate and generate progeny virions. Bacterial genetic symbols are gal, a cluster of three genes whose products function in the metabolism of the hexose sugar galactose; *chlD* and *chlA,* genes whose products are components of the nitrate reductase system, absence of which renders the cell resistant to chlorate under anaerobiosis; *pgl,* structural gene for phosphoglucololactonase; *bio,* a cluster of five genes coding for enzymes that function in the biosynthesis of the vitamin biotin; and *uvrB,* a gene whose product functions in repair of DNA damage such as that caused by ultraviolet light. (From "Genetic structure" by Allan Campbell. 1971 In *The bacteriophage lambda,* ed. A. D. Hershey. " Cold Spring Harbor Laboratory)

undergoing **breakage and reunion.** Therefore, integration of λ DNA at the specific host attachment site would be extremely inefficient were it not for the fact that the λ genome carries a gene, *int,* that encodes the **integrase** that catalyzes the unusual site-specific recombination reaction between phage and cell DNA. This process requires both the *int* gene and *c*I with separable promoters and is controlled by *c*II. Once integrase has done its job, the integrated λ genome, which is now referred to as the **prophage,** need only produce *c*I repressor to remain a passenger in the host cell indefinitely. This process of λ integration resembles that of certain mammalian DNA and RNA viruses (e.g., herpes and retroviruses).

In recent years, some very small genetic elements (600 to 1500 bp), called **insertion sequences (IS),** have been found in bacterial genomes. These IS elements are able to relocate in the host genome, either alone or as part of larger elements called "transposable elements" or **transposons.** The IS elements appear to encode enzymes that promote the random-site integration of IS and associated transposons. (see Sec. 6.4) Similarly, the terminal repeated sequences of retrovirus proviral DNA closely resemble those of bacterial IS elements. Transposons have been found in many eukaryotic genomes and are actually considered to be possible evolutionary precursors to the retroviruses. As discussed in Section 6.4, retroviruses integrate using biochemical mechanisms similar to transposons (see Sec. 6.1).

Induction of a Lysogen Part of the survival advantage for the temperate phage lies in the ability of the prophages of many, although not all, lysogenic phages to resume production of virions when the host is no longer able to survive. Again we return to λ to consider this *induction* process, for here it is best known and that process has a counterpart in the induction of animal viruses from a latent state (Sec. 11.9).

Occasionally, the regulatory circuit seems to fail spontaneously, so there are always a few cells in any lysogenic culture that will produce phage and lyse. However, agents that cause extensive damage to host cell DNA, such as UV light or the strand cross-linking chemical **mitomycin C,** have the ability to induce the prophage in nearly every cell of an exposed culture. Among the several cellular genes induced under these conditions are those that code for nucleases and other enzymes that attempt to remove the faulty DNA components and repair any breaks in the host chromosome. The overall effects of this system of emergency genes has earned them the nickname **SOS repair system.** It appears that derepression of this group of host cell genes is under the control of a single regulatory gene, *Rec*A, which produces a protease. This protease can attack and destroy the repressors, *Lex*A of the host SOS genes, which is probably the cause of initiation of their transcription. Evidently the *c*I repressor has evolved into a form that is sensitive to this specific host cell protease, and in this fashion the prophage is activated from the traumatized cell and begins to produce new virus particles.

Another early gene, *xis,* located next to *int,* produces the *"excisionase"* that together with *int* is needed to disengage the prophage from the host chromosome. This process again involves reciprocal recombination, which results in circularization of the λ DNA.

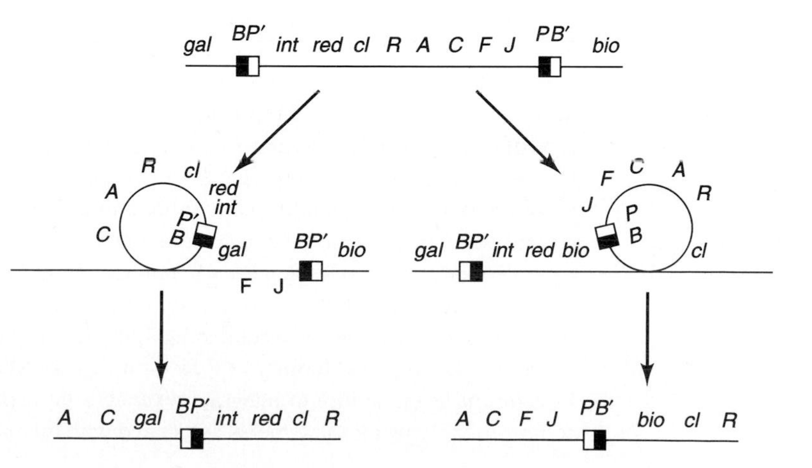

FIGURE 11.22

Genesis of λgal (left) and λbio (right) transducing phages, by rare abnormal excision from a λ lysogen. Recombination is assumed to occur at the bottom of the circular loops. Symbols are described in the legend to Figure 9.11.21 (From Max E. Gottesman and Robert A. Weisberg. 1971. "Prophage insertion and excision. In *The bacteriophage lambda,* ed. A. D. Hershey." Cold Spring Harbor Laboratory)

Although the excision reaction is ordinarily quite precise, occasionally the process goes awry and excision occurs somewhere else besides the attachment sites. In addition to phage DNA, the resulting circular molecules may contain cellular DNA from regions adjoining the integrated prophage (see Fig. 11.22). In the usual λ host these regions code for genes involved in galactose metabolism *(gal),* on one side of the attachment site, and biotin synthesis *(bio)* on the other. Since the "sticky" ends of the phage virion DNA are joined together in the middle of the integrated prophage, many of the incorrectly excised circles still contain these ends and can be packaged into phage virions, provided that the total length of their DNA is not too much greater than the original linear phage molecule. Because of this limitation, most phage particles carrying the *gal* genes have lost some phage sequences and are therefore defective in some vital functions. However, some such particles can be replicated and the resulting proviruses can even integrate, with the aid of a normal prophage. In contrast, the *bio* genes are closer to the attachment site; therefore, fully competent derivatives of λ carrying *bio* genes can be isolated.

When either of these recombinant forms of λ DNA integrates into a new chromosome, the attached host cell genes may be expressed independently of the rest of the prophage. If the infected cell already has its original copy of the *gal* or *bio* genes, the lysogen will then have two copies of these host genes. In other words, the lysogen will be partially diploid in that region of the genetic map. The existence of such special phage genomes has been exploited extensively in analyses of gene structure and function. In particular, this type of system can be most useful for comparing defects (mutations) in the cellular and phage copies of the host cell genetic sequences. If the two copies contain defects in two different genes, both affecting the same overall function (e.g., galactose utilization), each copy will provide the defective or missing gene product the other cannot, and the partially diploid cell will regain the overall capability that neither copy could provide. In that case the two copies of the host cell DNA sequences are said to **complement** each other; this type of test is commonly used in various microbial systems to determine the number of distinct genes controlling a particular cellular trait. The process of transferring cellular genes from one cell to another via a phage has been termed **transduction.** Moreover, because for prophages that integrate at only one site the genes that can be transduced are usually limited to those adjoining the attachment site in the host chromosome, this type of gene transfer is called **specialized transduction.**

Phages in a lysogenic state have sometimes been discovered in the same manner as have virulent phages, by testing cultures for plaque-forming virus, which is nearly always present at low levels in cultures of lysogens. However, after the discovery that several prophages, including that of λ, could be induced efficiently by UV light or mitomycin C, many more temperate phages were discovered by induction of a variety of bacterial cultures (Fig. 11.23).

Lysogeny without Integration Whereas immunity is a universal part of the process whereby phages become part of the host, integration of the prophage DNA is not. The DNA of phage P1 is found in lysogenic *E. coli* as a circular prophage replicating free from the host chromosome. However, its replication is closely coordinated with cell division, so that each cell contains one and only one copy of the P1 prophage per host chromosome. This coordinated replication of circular DNA is quite similar to that of another type of self-replicating circular DNA molecule found in many bacteria. Such molecules are known as **plasmids,** and they can express some genes involved in their own replication. Papillomavirus DNA, and possibly other animal virus DNAs, also become either integrated into cellular DNA or exist as mammalian cell plasmids.

Bacterial plasmids often carry a number of other noteworthy genes, including some that confer on the host cell, the resistance to various antibiotics. Some plasmids are transmissible from one cell to another via a cellular mating process known as **conjugation.** The best known of these is the so-called fertility or F-factor in *E. coli*. Moreover, some plasmids, including the F-factor, have the ability to integrate reversibly into specific sites in the main host chromosome, probably using IS elements associated with the plasmid and/or cell chromosome.

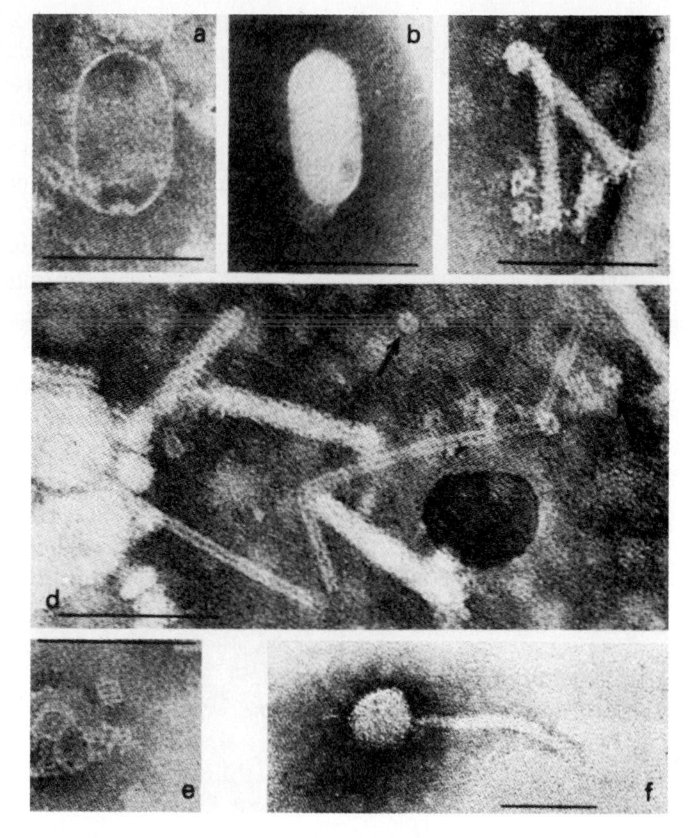

FIGURE 11.23

Electron micrographs of phagelike particles. (a), (b), (c), (d), and (e) Products of mitomycin C induction of *Aerobacter cloacae* × 333,000. (f) Particle from mitomycin C induction of *Alacalegenes faecalis* × 200,000. Scale bars correspond to 100 mμ. (Courtesy of David E. Bradley)

SECTION 11.11 FURTHER READINGS

Original Papers

KRUEGER, D. H., et al. (1975) Biological functions of the bacteriophage T3 SAMase gene. *J. Virol.* 16:453.

THURM, P., and A. J. GARRY. (1975) Isolation and characterization of prophage mutants of defective *Bacillus subtilis* bacteriophage PBSX. *J. Virol.* 16:184.

DHRMALINGAN, K., and E. B. GOLDBERG. (1976) Mechanism localization and control of restriction cleavage of T4 and lambda chromosomes in vivo. *Nature* 260:406.

ROBERTS, J. W., C. W. ROBERTS, and N. L. CRAIG. (1978) Escherichia coli recA gene product inactivates phage lambda repressor. *Proc. Natl. Acad. Sci. USA* 75:4714.

BIDWELL, K., and A. LANDY. (1979) Structural features of lambda site-specific recombination at a secondary att site in gal T. *Cell* 16:397.

GOFF, C. G., and J. STEZER. (1980) ADP-ribosylation of Escherichia coli RNA polymerase is nonessential for bacteriophage T4 development. *J. Virol.* 33:547.

WARNER, H. R., et al. (1980) Early events after infection of Escherichia coli by bacteriophage T5. III. Inhibition of uracil-DNA glycosylase activity. *J. Virol.* 33:535.

GROCNEN, M. A. M., and P. VAN DE PUTTE. (1986) Analysis of the ends of bacteriophage Mu using site-directed mutagenesis. *J. Mol. Biol.* 189:597.

ROMERO, A., R. LOPEZ, and P. GARCIA. (1992) The insertion site of the temperate phage HB-746 is located near the phage remnant in the pneumococcal host chromosome. *J. Virol.* 66:2860.

REISINGER, G. R., et al. (1993) Lambda kil-mediated lysis requires the phage context. *Virology* 193:1033.

CHAUDHURI, B., et al. (1993) Isolation, characterization, and mapping of temperature-sensitive mutations in the genes essential for lysogenic and lytic growth of the mycobacteriophage L1. *Virology* 194:166.

DELECLUSE, H. J., et al. (1993) Episomal and integrated copies of Epstein-Barr virus coexist in Burkitt's lymphoma cell lines. *J. Virol* 67:1292.

BRADLEY, D. E. (1967) Ultrastructure of bacteriophages and bacteriocins. *Bacteriol. Rev.* 31:230.

GARRO, A. J., and MARMUR. (1970) Defective bacteriophage. *J. Cell. Physiol.* 76:253.

Books and Reviews

BARKESDALE, L., and S. B. ARDEN. (1974) Persisting bacteriophage infections, lysogeny, and phage conversions. *Ann. Rev. Microbiol.* 28:265.

HOWE, M., and E. G. BADE. (1975) Molecular biology of bacteriophage Mu. *Science* 190:624.

SINGER, R. A. (1976) Lysogeny and toxinogenicity in Corynebacterium diphtheriae. In Mechanisms in bacterial toxinology, ed. A. W. Bernheimer. 1–30. John Wiley, New York.

LEWIN, B. (1977) Phage lambda: Development. In Gene expression. Vol. 3. 412–533. John Wiley, New York.

LEWIN, B. (1977) Phage T4. In Gene expression, ed. B. Lewin. Vol. 3. 536. John Wiley, New York.

CALOS, M. P., and J. H. MILLER (1980) Transposable elements. *Cell* 20:579.

COHEN, S. N., and J. S., SHAPIRO. (1980) Transposable genetic elements. *Sci. Am.* 242 (no. 2):40.

PTASHNE, M., et al. (1980) How the lambda repressor and *cro* work. *Cell* 19:1.

MATHEWS, C. K., et al. (1983) Bacteriophage T4. American Society for Microbiology, Washington, D.C.

ECHOLS, H. (1986) Bacteriophage λ: Temporal switches and the choice of lysis or lysogeny. *Trends Gen.* 2:26.

MIZUNCHI, K., and R. CRAIGIE. (1986) Mechanism of bacteriophage Mu transposition. *Am. Rev. Gen.* 20:385.

chapter

12

Consequences of Virus Infections in Animals and Other Organisms

The subject of this chapter could be illustrated in the form of a large tree. Its roots lie in Chapters 2 through 9, which describe viruses, and its stem in Chapters 10 and 11 which deal with the effect of viruses on cells. Its branches then represent the many vast fields of human and veterinary medicine, plant pathology, immunology, chemotherapy, epidemiology, and ecology. To keep this book within reasonable dimensions, we can obviously select only a few examples of the effects of virus infection on higher animals and plants and microorganisms to illustrate the many possible consequences of such infections in humans, animals, plants, and societies. A number of specific diseases associated with the various groups of viruses have been mentioned in each of the previous chapters, and these are summarized in Table 12.1.

12.1 EPIDEMIOLOGY OF VIRUS INFECTIONS

Epidemiology encompasses the study of the transfer and persistence of viruses in populations of organisms and may also be viewed as the ecology of viruses. Some of the key elements in the understanding of virus ecology are knowledge about routes of infection, and the species and geographic distributions of host populations. Host defense mechanisms and the ability of many viruses to persist in the individual host organism for long periods of time can also affect the infection pattern in the host populations. These will be considered in more detail in subsequent sections.

12.1.1 The Infection Process in the Animal Kingdom

The first stage of virus infection is the introduction of the virus into the organism. Viruses generally cannot pass through the skin, the surface of which is frequently covered by hair, feathers, or scales, and always by a thick and impermeable layer of dead keratinized cells. Viruses can get to the surface of living cells at skin wounds, which may be almost too small to be noticed. The small wounds created by biting insects such as mosquitoes, ticks, fleas, and lice represent a very frequent means of infection. The role of insect bites is illustrated by the story of the role of the **yellow fever virus** in blocking the heroic French efforts to build the Panama Canal, which makes fascinating reading. So does the later American enterprise there, that led to the elucidation of the mode of transmission, control of the virus, and incidentally to the completion of the canal. The role of major traumas is illustrated by the transmission of rabies through animal bites. Rabid dogs and cats used to be largely responsible for spreading rabies, but through extensive vaccination and leash laws this danger has been greatly diminished in most countries. At present, foxes have become overabundant in Europe and represent a dangerous reservoir in the spread of rabies, but in America, skunks and bats are now the most frequent transmittors of this disease.

TABLE 12.1
Some diseases caused by animal viruses

Family (Viridae)	*Virus Subfamily or Genus*	*Diseases*	
		Humans, Monkeys[a]	*Other Animals*
RNA Viruses			
Picorna	Entero	Enteritis, occasionally CNS (polio)	Enteritis
	Cardio	—	Encephalomyocarditis
	Rhino	Common cold (many serotypes)	Respiratory
	Calci	Enteritis	
	Aphtho	—	Foot-and-mouth disease
Corona	—	Respiratory and enteric	Many different diseases in different animals
Toga	Alpha	Rare encephalitis	Equine, etc., encephalitis
	Flavi	Yellow fever, encephalitis	Equine, etc., encephalitis
	Rubi	Skin rash, German measles (rubella)	—
	Pesti	Occasionally congenital diseases	Mucosal disease (cattle)
Retro	Type C	T-cell leukemia (HTLV-I); sarcoma (monkey)	Avian, murine, and other animal leukemias and sarcomas Paralysis (mice)
	Type B	—	Murine mammary tumors
	Type D	Immune deficiency (monkey)	
	Lenti	AIDS, encephalopathy Immune deficiency (monkey)	Immune deficiency (cats), maedi-visna (sheep); encephalopathy, arthritis (goats)
Rhabdo	Vesiculo	—	Stomatitis (cattle, swine)
	Lyssa	Rabies	Rabies (monkeys)
Filo	—	Hemorrhagic fever (Marburg, Ebola)	—
Arena	—	Hemorrhagic fever (Lassa)	Lymphocytic choriomeningitis (mice)
Bunya	Bunyamwera	Encephalitis (Calif. enceph.)	Many diseases
Paramyxo	Paramyxo	Childhood respiratory, croup (parainfluenza) Salivary gland (mumps)	Newcastle disease (birds)
	Morbilli	Skin rash (measles)	Rinderpest (cattle), distemper (dogs)
	Pneumo	Childhood lower respiratory Pneumonia (respiratory syncytial)	
Orthomyxo	Type A	Influenza (flu)	Respiratory
	Type B	Influenza	—
Reo	Orthoreo	—	—
	Orbi	Diarrhea	Diarrhea (blue-tongue of sheep)
	Rota	Children's diarrhea	Enteric
	Cytoplasmic polyhedrosis	—	Lethal insect infection

Family (Viridae)	Virus Subfamily or Genus	Host: Humans, Monkeys[a]	Host: Other Animals
DNA Viruses			
Parvo	—	Aplastic anemia (humans) Fifth disease (B19)	Enteritis (dogs, cats) Encephalopathy (rats) Prenatal infections
Hepadna	—	Hepatitis; liver cancer	Same (woodchucks, squirrels, ducks)
Papova	Polyoma	Malignant tumors under certain specific conditions; encephalopathy	Same
	Papilloma	Warts, carcinomas	Warts, at times malignant (Shope papilloma)
Adeno	(Many serotypes)	Acute respiratory diseases, conjunctivitis	Same (occasionally oncogenic)
Herpes	Alphaherpes	Skin rash: chickenpox, varicella Cold sores, shingles (herpes simplex 1) Venereal, congenital (herpes simplex 2)	— Bovine mammilitis, etc.
	Betaherpes	Congenital malformations (cytomegalo) Exanthem subitum (herpes virus 6)	Respiratory and congenital diseases
	Gammaherpes	Infectious mononucleosis, etc. (Epstein-Barr), possibly cancers	Marek's disease (chickens)
Baculo	(Nuclear polyhedrosis, granulosis)	—	Lethal insect infections
Pox	(Many genera)	Smallpox; Yaba (monkey)	Pox, myxomatosis (rabbits)

[a]Name of specific virus is in parentheses in several instances.

By far the more frequent modes of transmission of viruses among urban, particularly affluent, populations is not by bites, but by intake of air and food. The **epithelial cells** lining the **respiratory tract**—nose, pharynx, bronchi, and lungs—constitute the major portal for many viruses. The size of the inhaled particles determines the depth of penetration into the system and therefore may influence the nature of the resulting disease, if any. For example, nasal drops of liquid containing certain Coxsackie viruses may cause a nasal cold, but inhalation of a contaminated mist will cause a more severe chest cold. Release of virus from the infected host is also an important part of transmission, and the symptoms of respiratory infections—coughs and sneezes—help to generate highly infectious aerosols. Viruses that frequently initiate infections in the respiratory tract include adeno-, herpes-, rhino-, myxo-, paramyxo-, and coronaviruses. Other viruses that cause generalized diseases without respiratory symptoms may also enter via the respiratory route. These include the viruses that cause African swine fever, varicella, smallpox, mumps, measles, rubella, and the arenavirus responsible for lymphocytic choriomeningitis (LCM) in mice. It must be noted, however, that most rhinoviruses (the cause of the common cold) are passed more often by contact with hands that have touched respiratory secretions than by inhalation of infectious aerosols.

Another group of viruses initiates infections through the **epithelium** of the **alimentary tract**—the mouth, esophagus, and intestinal mucosa. In the case of HIV, direct infection of the rectal mucosa may be the initial portal in some cases. Again, some of these cause only local symptoms or no disease, such as certain adeno-, picorna-, toga-, corona-, reo-, and parvoviruses. Others produce generalized diseases, including many enteroviruses of the picornavirus family, such as polio and the virus responsible for hepatitis A. Transmission of many of these viruses that enter via the mouth is facilitated by stability of the virions in feces and sewage, since contamination of water supplies allows the cycle of infection to be repeated.

Other viruses may be transmitted by **oral secretions,** including most herpesviruses (e.g., Epstein-Barr, the causative agent of infectious mononucleosis, often transmitted by kissing) and, of course, rabies. Recent studies of a newly described human disease of the tongue associated with AIDS, oral hairy leukoplakia, have shown the presence of replicating EBV in epithelial cells. This is the first example of EBV maturation in any cell but B lymphocytes.

Few viral infections start in the **eye,** although disease symptoms may appear there due to prior spread of the virus through the bloodstream. Localized infections of the eye by adenoviruses and sometimes by Newcastle disease virus can lead to inflammatory conditions called "pink eye." These are generally limited to the eye but at times can also involve the respiratory tract. Moreover, herpes simplex virus can give severe eye infections requiring treatment. Furthermore, documented cases of transmission through corneal transplants of Jakob-Creutzfeldt disease, caused by an as yet undescribed infectious particle considered a prion (see Sec. 9.6), as well as rabies have been described.

Several viruses are passed **venereally,** for example, herpes simplex type 2 and papillomaviruses. The latter have been linked to cervical cancer and are responsible for venereal warts. Furthermore, the human retroviruses, HTLV-1 and the AIDS virus, HIV, can be passed through sexual contact, most likely by virus-infected cells. Most commonly these viruses, particularly HIV, pass via genital-anal contact through abrasions in, or infection of, the rectal epithelium. The infected cells can gain entrance through the cervical os into the uterus and infect the female. Or the virus can also pass via lesions in the vaginal canal caused by concomitant venereal infections. Moreover, the AIDS virus appears to be able to infect the male through breaks in the mucosal lining of the urethra or through infection of white cells present in the urethra. Finally, viruses that do not enter the body through the **urinary tract** can be excreted in urine. For example, cytomegalovirus and polyomalike viruses can be isolated from the urine. The latter have been recovered particularly from the urine of cancer patients or immunosuppressed individuals. Many RNA viruses, especially toga-, paramyxo-, arena-, and some retroviruses are also excreted in urine. Finally, some viruses, such as HTLV-I and HIV, can be passed to newborns through **breast milk.**

12.1.2 Vertical Versus Horizontal Transmission

All the infection routes mentioned so far involve transfer of virus from an infected individual to another and are called **horizontal transmission** routes. They are distinguished from special cases of transfer only from parents to offspring, at times over several generations, or **vertical transmission.** One method of vertical transmission is direct passage of virions from the blood of the infected mother through the placenta. This route has been shown to occur for cytomegaloviruses, HIV, and the rubella and LCM viruses. LCM viruses and possibly others can also be transmitted in the cytoplasm of the ovum. Mouse mammary tumor viruses can be transmitted in the milk, since they are eventually replicated in the mammary epithelium when infection is initiated by ingestion. However, these type B retroviruses, as well as many retroviruses of type C morphology, are also vertically transmitted in most if not all mice in a more direct manner, through "inheritance." The viral genome becomes integrated into the host's germ cell DNA. Finally, some viruses, such as the herpes simplex virus, are passed congenitally, since they can be found in the vaginal canal.

The DNAs of all cells in many animal species contain base sequences for complete genomes of retroviruses that are, therefore, known as inherited or **endogenous viruses** (Chap. 6). Generally, there are multiple related, but often not identical, genome copies present. Virus expression, either partial or complete, is controlled by the cell and may be expressed in certain tissues during normal development of the individual. Cells of a given species are usually resistant to horizontal infection with their own endogenous retroviruses (i.e., **xenotropism**), but cell cultures or even tissues *in vivo* from some species may continuously express all or some of the endogenous virus genes. The resistance may be related to expression of low amounts of the endogenous retroviral envelope on the cell surface receptors (see Sec. 11.6). In the case of mouse xenotropic viruses (see Sec. 11.3), it appears that the receptor for the virus is absent on mouse cells but present on a wide variety of cells from other animal species.

Sometimes the partial expression of a virus is detected by complementation of a defect in an exogenously added virus, such as Rous sarcoma virus by a chicken leukosis virus (see Sec. 6.1). The latter is an endogenous virus expressed in embryonic fibroblasts of certain chickcn strains that "hclps" rcplication-dcfcctivc strains of RSV by providing an cnvelope protein needed for infection.

Studies of the human placenta suggest that endogenous retroviruses may also be found in humans. Virus-related reverse transcriptase has been detected in extracts of human placentas, particularly in the syncytial trophoblast layer that joins the maternal and fetal portions of the placenta. Furthermore, purified material banding at 1.16 g/ml (typical of retroviruses) contains particles resembling retroviruses. Nevertheless, cultivation of replicating virus in culture has not been achieved. Thus, these results only suggest an actual endogenous human retrovirus. Other work with human tissues has shown sequences in human genes with some homology to several animal retrovirus genomes (see Chapter 6).

When endogenous viruses of closely related animal species are compared by nucleic acid hybridization, often the viral DNA sequences are also closely related. Evolutionarily distant species may share endogenous virus sequences, but usually the homology is only partial. In other words, the evolution of many endogenous viruses appears to parallel that of the host species. In an attempt to apply this type of evolutionary analysis in reverse, George Todaro and associates examined the relatedness of endogenous virus genes in primate species. From this approach it appears that there are basically two distinct groups of primates. Among the apes, which are considered the highest primates, chimpanzees and gorillas appear to be closely related, whereas gibbons, orangutans, and humans form a separate group with a common set of endogenous virus sequences. Since the gibbon and orangutan are thought to have originated in Asia, these results have been taken as evidence for an Asian origin of humans. The evolutionary beginnings of the other apes are known to be in Africa. Africa is generally considered the most likely region for most of human evolution, based on the morphological and anatomical resemblance of humans to the African apes, and because most of the recent paleontological discoveries have occurred in Africa. Thus, whether the study of viruses can adequately contribute to our knowledge of anthropology awaits further analysis, particularly if a human endogenous virus can be identified.

Occasionally it appears that the endogenous virus of one species has been transmitted horizontally to another species. This process is evident from the total absence of endogenous virus sequences in certain species that are considered to be close evolutionary ancestors of the species presently carrying the endogenous virus, whereas an unrelated species has essentially the same virus in its genome. For instance, it has been deduced that an endogenous virus now found in cats and baboons was transmitted to an ancestor of the domestic cat, probably relatively recently on the evolutionary scale—about 3 to 10 million years ago, in Africa or the Mediterranean basin, where the appropriate primate and feline ancestors originated. Recently, the process of introducing a new endogenous virus has been achieved in the laboratory by microinjection of a mouse leukemia virus into mouse embryos in *utero*. In contrast to natural endogenous viruses or even the same virus transmitted horizontally, the artifactual endogenous virus was continuously expressed in all tissues, and this created new disease symptoms, including a change of hair color. A similar insertion of a viral gene into chickens has also recently been achieved. It will permit analysis of what effect these inherited agents can have on development.

12.1.3 Zoonosis

Any multiple-host infection that is naturally transmitted from animals to man is known as **zoonosis.** In the case of yellow fever mentioned earlier, it was long thought that the virus was transmitted only among human beings by mosquitoes. This may be true in urban areas, but it is now clear that in the jungles of Africa the virus is also transmitted in monkey populations. The infected monkeys show no signs of disease and most become immune within one week of infection. Whether these animals during acute infection could serve as virus **carriers** that reintroduce yellow fever into human populations via mosquitoes merits consideration. Persistent infections with other viruses exist (see Sec. 12.6). In contrast, in the South and Central American jungles, the principal primate species suffer high mortality from yellow fever, as do human populations. To survive in the Americas, therefore, the yellow fever virus must constantly infect new susceptible populations of primate hosts, since it is not generally transmitted among mosquitoes. Recently, however, yellow fever virus has been found in mosquito eggs and in male mosquitoes that do not suck blood; thus the virus might be maintained in this vector. Incidentally, the ability of the African primates to survive yellow fever virus infection has been taken to mean that this host-virus system has been undergoing mutual evolution for many generations (see below). Thus, the yellow fever virus may have originated in Africa and may have been introduced into the Western world only recently, perhaps during the slave-trading years.

In considering transmission of viruses from animal to animal by insects, there is an important distinction to be made between passive or **mechanical transmission** as opposed to **propagative transmission.** In the former, virions are transferred directly without replication in the insect, whereas propagative transmission, as the name implies, involves virus replication in the insect vector. In general, there are a number of other distinctions between the two types of insect transmission, all of which tend to make the propagative mode more effective for transmission in the long range, both in time and space (see Table 12.2). In propagative transmission, it is not enough for the virus to be replicated just anywhere in the insect or the host. The insect becomes initially infected in the gut from ingesting blood of the host. However, unless the virus is able to spread to the insect's salivary gland, it will not be transmitted during feeding. This aspect of virus biology, that viruses can be adapted to deal with such widely different metabolic environments as warm-blooded human and cold-blooded mosquito, represents a remarkable illustration of nature's ingenuity. It must also be noted that such viruses do not cause disease or death in the transmitting insect. Obviously, this is again an important feature in ensuring safe transit of a virus from one mammalian host animal to another. The modes by which persistence of a virus for long periods in a host may be achieved are discussed later.

TABLE 12.2
Comparison of propagative and mechanical virus transmission by insects

Feature	*Propagative*	*Mechanical*
Virus replicated in vector	Yes	No
Virus source in animal	Blood	Skin lesions or blood
Vector specificity	Relatively species-specific	None other than biting preferences
Incubation period required in insect before transmission	Yes	No
Virus examples	Toga-, bunya-, orbi-, and some rhabdoviruses	Bovine leukemia virus (retrovirus), papillomaviruses, poxviruses, Hepatitis B (rare)

SECTION 12.1 FURTHER READING

Original Papers

FENNER, F., and G. M. WOODROOFE. (1965) Changes in the virulence and antigenic structure of strains of myxoma virus recovered from Australian wild rabbits between 1950 and 1964. *Austral. J. Exp. Biol. Med. Sci.* 43:359.

GROSS, L., and Y. DREYFUSS. (1967) How is the mouse leukemia virus transmitted from host to host under natural life conditions? In Carcinogenesis: A broad critique. *20th Annual Symposium on Fundamental Cancer Research,* Williams and Wilkins, Baltimore.

BENTVELZEN, P., et al. (1968) Genetic transmission of mammary tumor inciting viruses in mice: Possible implications for murine leukemia. *Bibl. Haematol.* 31:101.

HUEBNER, R. J., et al. (1970) Group-specific (gs) antigen expression of the C-type RNA tumor virus genome during embryogenesis: Implications for ontogenesis and oncogenesis. *Proc. Natl. Acad. Sci. USA* 67:366.

KALTER, S. S., et al. (1973) Brief communication: C-type particles in normal human placentas. *J. Natl. Cancer Inst.* 50:1081.

BENVENISTE, R. E., and G. J. TODARO. (1976) Evolution of type C viral genes: Evidence for an Asian origin of man. *Nature* 261:101.

KENDAL, A. P., G. R. NOBLE, and W. R. DOWDLE. (1977) Swine influenza viruses isolated from man and pig contain two coexisting subpopulations with antigenically distinguishable hemagglutinins. *Virology* 82:111.

NELSON, J., J. LEONG, and J. A. LEVY. (1978) Normal human placentas contain virus-like directed RNA-DNA polymerase activity. *Proc. Natl. Acad. Sci. USA* 75:6263.

BEASLEY, R. P., et al. (1981) Hepatocellular carcinoma and hepatitis B virus: A prospective study of 22,707 men in Taiwan. *Lancet* 2:1129.

MARTIN, M. A., et al. (1981) Identification and cloning of endogenous retroviral sequences present in human DNA. *Proc. Natl. Acad. Sci. USA* 78:4892.

PLUMMER, F. A., et al. (1991) Cofactors in male-female sexual transmission of human immunodeficiency virus type 1. *J. Infect. Dis.* 163:233.

COFFIN, J. M. (1992) Structure and classification of retroviruses. In *Retroviridae.* ed. J. A. Levy, Vol. 1, 19. Plenum Press, New York.

DUNN, D. T., et al. (1992) Risk of human immunodeficiency virus type 1 transmission through breastfeeding. *Lancet* 340:585.

OU, C. Y., et al. (1992) Molecular epidemiology of HIV transmission in a dental practice. *Science* 256:1165.

Books and Reviews

ROWE, W. P. (1973) Genetic factors in the natural history of murine leukemia virus infection: G. H. A. Clowes Memorial Lecture. *Cancer Res.* 33:3061.

EVANS, A. S. (1982) *Viral infections of humans.* Plenum, New York.

LONDON, W. T. (1983) Hepatitis B virus and primary hepatocellular carcinoma. In *Advances in viral oncology.* ed. G. Klein. Vol. 3, 325–341. Raven Press, New York.

BAER, G. M. (1984) Control of rabies infection in animals and humans. In *Control of virus diseases.* ed. E. Kurstak and R. G. Marusyk. Marcel Dekker, New York.

LEVY, J. A. (1986) The multifaceted retrovirus. *Cancer Res.* 46:5457.

LEVY, J. A. (1993) The transmission of HIV and factors influencing progression of disease. *Am. J. Med.,* 95:86.

12.2 PATHOGENICITY AND VIRULENCE OF VIRUSES

In Table 12.1 we list some of the diseases caused by different families of animal viruses. With a few exceptions, the consequences of virus infections are quite similar to those of many bacterial infections. Many viruses remain localized near the site of primary infection, causing the well-known symptoms of colds, coughs, sore throats, indigestion, diarrhea, general malaise, and so on. This early stage of virus infection can also be almost symptomless and is often overlooked. It may not lead to further pathological manifestations, although it usually can be demonstrated by testing for virus-specific antibodies (see the next section).

With other viruses, or under other circumstances, the viruses may, after such primary infection, enter the lymphatic and blood circulatory systems, a stage that is termed **viremia** and leads to generalized infection. The time lag between these two stages is termed the **incubation period** and can be variably long and of more or less specific duration for different viruses (see Fig. 12.1). Skin rashes are often an early symptom of generalized infection, although the cause of this phenomenon is not well understood. Immunological response probably plays a role in some of these rashes (e.g., immune complexes) that are characteristic of many common children's diseases. Most of the affected cells, both at the primary sites of infection and at the secondary lesions, die. They usually release toxic substances causing fever (**pyrogens**) regardless of the infecting agent (viral, bacterial). This response is the first defense of the organisms, since the replication of many viruses is depressed at higher temperatures, perhaps via the accompanying production of interferon (see Sec. 12.4).

Certain viruses have definite affinities for specific organs and become localized there after viremic spreading. Examples are the salivary glands for mumps virus, the brain for the encephalitis viruses, the sensory nerves and at times the central nervous system for certain herpesviruses, and, rarely, the motor nerve ganglial cells for polio. Actually, initial infection with polio that occurs in the alimentary tract appears to cause at all ages a disease that is most frequently nonapparent or very minor. This initial infection always leads to viremia, antibody formation, immunity, and generally to complete recovery. Only 1 to 2 percent of the polio virus infections lead to more or less widespread involvement of the motor nerves of the central nervous system and paralysis, both in children and adults. The reason why polio invades the motor nerves in only a few cases is not known, but the situation suggests another instance of **multifactorial pathogenesis.** Development of disease may involve selective passage along nerve channels.

A similar situation is noted with many other viral infections. For instance, in rabies, one of the most virulent virus diseases, 30 percent of individuals bitten on the hand by a rabid animal will develop neurological findings, and 70 percent if bitten on the head. Nevertheless, there are always some individuals who do not develop disease and yet show antibodies to the virus. Lassa fever, which is caused by a rodent virus, recently gave rise to fatal **hemorrhagic fever** in infected individuals coming to Africa. This arenavirus, however, does not cause disease in many people indigenous to Africa. They develop antibodies to the virus, without symptoms. Clearly, a difference in the host response to the agent is involved. The situation illustrates the selection of a more resistant host in the environment in which the virus has been present for a long time (see Sec. 12.6).

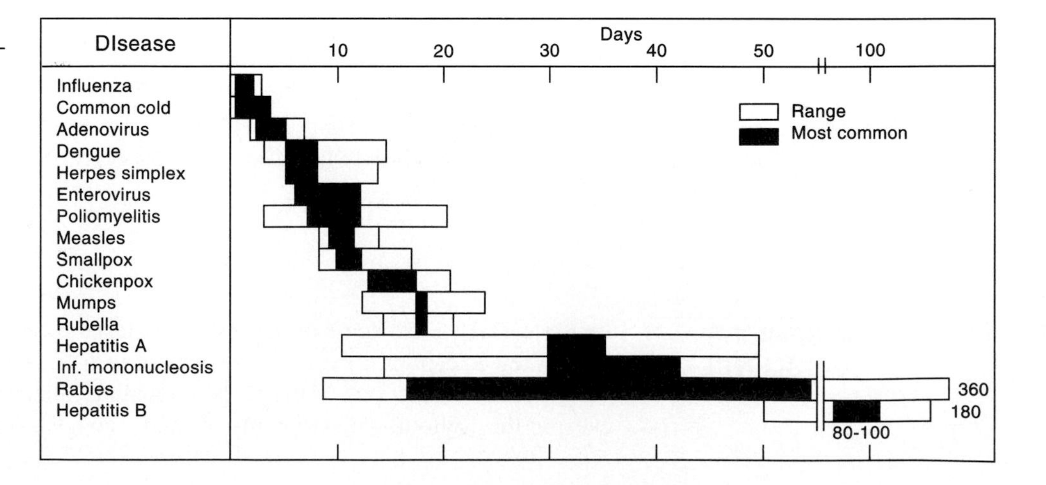

FIGURE 12.1

Incubation periods of virus diseases. (From A. S. Evans, ed. (1978) *Viral infections of humans.* Plenum Press, New York)

In general, the molecular or even biological bases for organ **tropisms** (affinity) of viruses have not been determined. There may be tissue-specific receptors on the cell surface in some cases, or tissue-specific differences in cell metabolic machinery. An example of the latter types is provided by the **feline panleukopenia virus.** This parvovirus has a tropism for highly mitotic tissues, and therefore its effects are chiefly on bone marrow cells, lymphoid tissue, and epithelium of the intestine. It is clear that this pattern of tropism is a direct result of the requirement of parvoviruses for host cell DNA replication enzymes that are active only in cells actively replicating their own DNA (Sec. 7.1).

Besides fever, antibody formation, cellular immune response (see Sec. 12.3), and interferon action (discussed in Sec. 12.4) are the most important defense mechanisms that limit virus production and may lead to recovery of the patient. Nevertheless, these responses do not necessarily mean complete eradication of the virus from the host. Moreover, some viruses may induce immune reactions that are detrimental to the host (see Sec. 12.3.1). It must also be noted that many viruses that cause no definite symptoms and for which no pathogenicity is known are detectable in man and animals. Particularly surprising is the situation for western equine encephalitis virus, which causes serious diseases in horses and humans in the western United States but not in the eastern states, even though it is present there as well.

The term **virulence** is often used in evaluating the pathogenicity of a virus. The virulence of a virus is, however, always relative. A virus may cause no symptoms in one host or a vector, and kill another species. The physiological state of the host also plays an important role. In persons treated with immunosuppressive drugs, a technique used for organ transplants, cryptic virus infections (such as those often caused by herpes-, as well as hepatitis B viruses) may become highly pathogenic. The virulence of a virus is often due to a single surface protein that may cause cessation of synthesis of host cell macromolecules. Thus, different strains or serotypes of a virus may lose their virulence through a mutation affecting that protein. Virulence also correlates with temperature resistance of a virus strain. Not only in animals but also in plants, resistance favors their withstanding the feverish state of the host.

12.2.1 Complex Etiologies

We cannot leave the topic of viral epidemiology without discussing some of the main kinds of data that are used in analyses of often complex disease patterns associated with viruses. These complexities are nowhere more apparent than for viruses associated with human cancers, such as the now-classic example of Epstein-Barr virus, and the more recently recognized cases of the papilloma-herpes simplex type 2, and hepatitis B viruses. Infections by these viruses rarely, if ever, result in cancers. Yet it is evident that these viruses frequently are associated with particular cancers. In the case of hepatitis B virus (HBV), it has been shown that in those areas of the world where the incidence of primary liver cancer is particularly high, such as parts of Africa and Asia, so is the prevalence of HBV, especially with chronic infections (see Sec. 10.2.1). Moreover, HBV antigens are detected in the blood of many patients with hepatoma—up to 90 percent in some populations. Further, liver cancer patients are also very likely to show signs of extensive liver damage or cirrhosis, usually associated with long-standing HBV infections (or chronic alcoholism in the West). Therefore, an hypothesis has been advanced that infection with HBV leads first to chronic liver disease, resulting eventually in cirrhosis, which in turn can progress to hepatoma. The nature of the virus itself, however, suggests that an oncogenic protein (e.g. x) may be made after virus integration. Nevertheless, not all people infected with HBV develop liver tumors; in fact, only a small percentage do. Therefore, the hypothesis includes a role for some other factors to work with HBV as a **co-carcinogen** in some patients who, by chance, happen to fall victim to both HBV and the other unknown factor(s). Such factors could be dietary (e.g., aflatoxin or alcohol) or from some other source.

Because of the long incubation period, the association of Epstein-Barr virus with **Burkitt's lymphoma** and nasopharyngeal carcinomas (NPC) has suggested that other factors contribute to the development of these malignancies. For instance, children developing Burkitt's lymphoma have been found to have been infected by EBV up to 10 years previously. In Africa, it is believed that malaria acts as a cofactor in the development of Burkitt's lymphoma by compromising the function of macrophages. In China, exposure to phorbol ester present in the leaves of the *Crutus* tree may be a cofactor in NPC. That immunological response is an important variable in determining the outcome of EBV infection is supported by observation on an X-linked immunodeficiency syndrome (Duncan's kindred). Males born with this genetic defect are unable to control EBV infection; their cellular immune response is specifically unable to respond to EBV. As a result, they develop a variety of disorders, including autoimmune disease, aplastic anemia, and B-cell lymphomas.

In such complex disease systems, it is extremely difficult to rule out by statistical methods of association only, that those people who are predisposed to a particular type of tumor, perhaps for the same reason, also may be predisposed to infection by the associated virus. Even the recent finding of HBV DNA in liver tumors (see Sec. 6.2.3) does not rule out this possibility. The fact that herpes simplex virus type 2 can cause neoplastic transformation of animal cells (Sec. 10.2.5) does not prove that the virus can cause human tumors. Ultimately, the epidemiologist may only be able to make a strongly suggestive case for causal involvement when a rare disease syndrome is associated with a widespread virus in a multifactorial etiology in which no single agent alone is *the* cause.

Another example is the human T-cell leukemia virus, type 1 (HTLV-1), which has been linked to T-cell leukemias in individuals living in the southern islands of Japan, the southeastern United States, and the Caribbean. Most people infected with this virus do not develop the disease or do so 30 to 40 years after infection. Thus, HTLV-1 may initiate the oncogenic process but other steps are needed to maintain it. In recent studies cytokines produced by the infected cells appear to act as promoters for the autonomous growth of eventually malignant cells (see Sec. 14.2). Finally, there are many types of T-cell leukemias that are not linked to this virus. Therefore, as with other viruses associated with cancer, the infectious agent may be a promoter of the oncogenic process that can occur even in its absence. Thus, elimination of the virus could reduce the incidence of this cancer but not prevent its development by other means.

Prevention of virus infection through vaccination (see Sec. 13.2) may be the only final proof that the virus is part of the cause in such situations. This approach has been proposed for the Epstein-Barr virus and Burkitt's lymphoma in Africa and may be attempted during the next decade.

Finally, Reye's syndrome, a disease in children, first described in 1963, has been recognized to be associated with certain virus infections. It is characterized by encephalopathy and fatty liver, and its onset has been noted to coincide with an epidemic of influenza B virus. Subsequently, the syndrome was found linked to infections in children by adenoviruses and herpesviruses. The illnesses included chickenpox, respiratory diseases, and gastrointestinal disorders. **Reye's syndrome** is not associated with impaired cellular immune responses but clearly represents a disease with complex etiology. One common finding in this condition was the use of aspirin (salicylic acid) during the virus infections. Administration of this common drug for symptoms of virus infection in children is now not recommended, since a large number of Reye's syndrome cases have been attributed to salicylate use.

12.3 IMMUNOLOGICAL RESPONSES TO VIRUS INFECTION

Vertebrates have developed a powerful defense mechanism against many foreign agents, be they infectious or noninfectious, living like bacteria, parasitic like viruses, or nonliving like toxic proteins. This mechanism is the immune system. Their lymphocytes produce specific antibodies that bind to and usually inactivate the foreign agents, termed **antigens** in this context. Other blood cells destroy tumor and virus-infected cells by direct cell-to-cell contact.

The most effective antiviral antibodies are **neutralizing,** which means that they inactivate the virus and thus arrest the disease process. If active virus is thereby totally removed, the patient is cured. However, all progress carries risks, and all new defenses call for new and better offensive weapons. Thus, certain viruses may elicit no antibody response, probably by jettisoning their protein coats; other viruses remain infectious in the antibody-complex stage; and others actually become powerful pathogens only through causing excessive antibody formation. Deposition of virus-antibody complexes clog the small arteries in the kidney and elsewhere, and lead to what has been termed immune complex disease, often part of an **autoimmune disease.** Still other viruses, such as the AIDS virus (see Sec. 14.1) have evolved more active offenses against the immune system. With this virus, and others (e.g. dengue) antibodies may be produced that bind to the virus but do not neutralize it. Instead the antibody-virus complex can infect cells via the cellular **Fc receptor** that interacts with the antibody (see Sec. 14.1). Moreover any virus that kills lymphocytes, such as HIV the feline panleukopenia virus and many others, reduces the ability of the immune system to respond either with antibodies or activated T cells. The lymphocytes need not be killed to inhibit immune responses. Measles and LCM viruses release factors during the infection that suppress normal immune activities. Also, Epstein-Barr virus, which persists in lymphocytes and may even transform them into cancerous cells, often interferes with the immune responses of the host. A virus need not even be replicated in cells of the immune system to inhibit them, as startlingly revealed in initial attempts to immunize cats with killed virions or purified surface proteins of feline leukemia virus: The cats became much more susceptible to subsequent infection with the live virus because a viral surface protein inhibited the immune functions of certain lymphocytes. This unanticipated result revealed new problems in harnessing the immune system to eradicate certain types of viruses by vaccination.

Typically the neutralizing immune response occurs in the bloodstream. There large antibody molecules (IgM, 900 kDa) first appear, and later the smaller but more abundant antibodies (IgG, 150 kDa). These antibodies are specific for a given viral surface protein and bind to the virus during viremia, thus preventing it from entering new cells (**humoral immunity**). Such complexes are dissociable without damage to the virus or antibody. However, they are normally taken up by leukocytes and particularly macrophages and are digested. The **complement system,** a series of interacting and partly enzymatic serum proteins that also interact with antigen-antibody complexes and facilitate their degradation, often plays an important role in these processes. The destruction by complement of enveloped viruses in the antibody complex can be seen on electron micrographs. Complement alone may also inactivate certain viruses directly.

Another important arm of the immune system is **cell-mediated immunity** (CMI). CMI involves appropriate cell: cell recognition, particularly by macrophages, NK cells, and lymphocytes. This interaction can lead to production of suppressor factors (e.g., lymphokines) that help an antiviral response of the cell or to cytotoxic activity of immune cells in which the infected cell is destroyed. Some of these cellular responses are mediated via antibody production. Thus, the interplay of both humoral and cellular immunity is important in countering virus infection. In general, neutralizing antibodies are particularly helpful in preventing the initial infection by the virus (e.g., by vaccination), but cell-mediated immunity is most important after infection when spread and eventual elimination of the virus are affected by the host's ability to kill virus-infected cells.

Finally, as noted below, interferon production by infected cells or immune cells can play an important role in arresting virus infections. The rise in temperature observed with acute virus illness may in fact be of benefit because it is associated with the production of increased levels of interferon.

12.3.1 Autoimmune Phenomenon

Viruses replicating in the cell can express their viral proteins on the cell surface. This event may then lead to an antibody response against the infected cells. The result can be either destruction of these virus-infected cells (Fig. 12.2) or production of autoantibody(ies) to

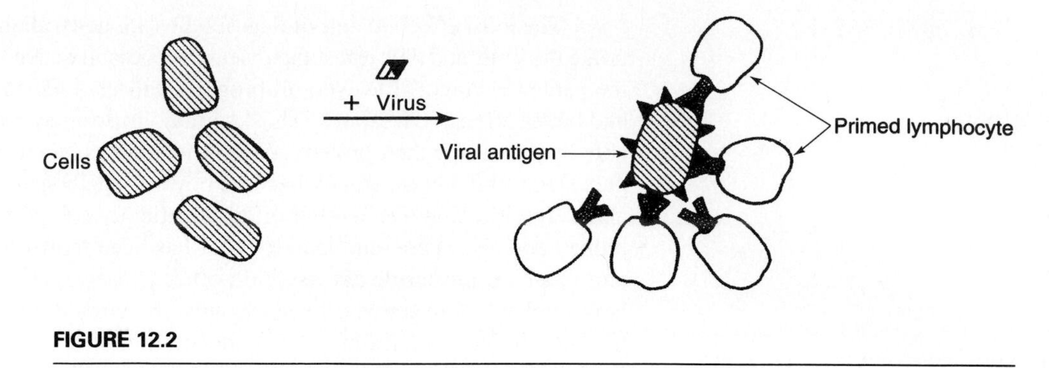

FIGURE 12.2

Infected cells will produce viral antigens on the surface against which an immune response can be directed. This *heterogenization* of the cell membrane can lead to destruction of the infected cells—an effective means of arresting virus infection. (Adapted from J. A. Levy (1977). In *Autoimmunity.* ed. N. Tala. Courtesy of Academic Press, New York)

normal host cell proteins. By the latter phenomenon, the viral protein on the cell surface acts as a "carrier" for a normal cell protein (see Fig. 12.3) that is normally not immunogenic in the individual. For example, Lindenmann and colleagues noted that mouse tumor cells were not rejected by the animal until they were infected first with the mouse influenza virus. Both the infected cell and the virus budding from the tumor cell became immunogenic. Animals immunized against either of these materials rejected not only the infected tumor cell but also the uninfected tumor cell. Thus, virus infection uncovered or enhanced the immunogenicity of the malignant cell. This kind of phenomenon associated with viruses can be responsible for induction of antibodies to normal self-antigens. It is one of the proposed mechanisms for production of **autoantibodies** in individuals developing autoimmune diseases such as rheumatoid arthritis, systemic lupus erythematosus, or even multiple sclerosis. In these cases, perhaps the virus inducing the disease has already been eliminated by the host, so that the causative agent cannot be recognized. This phenomenon has been termed "hit and run" and has been considered in the induction of certain cancers (see Sec. 10.3.6). The infectious agent is gone, but the autoantibodies that have been induced by it lead to the secondary immunological diseases observed.

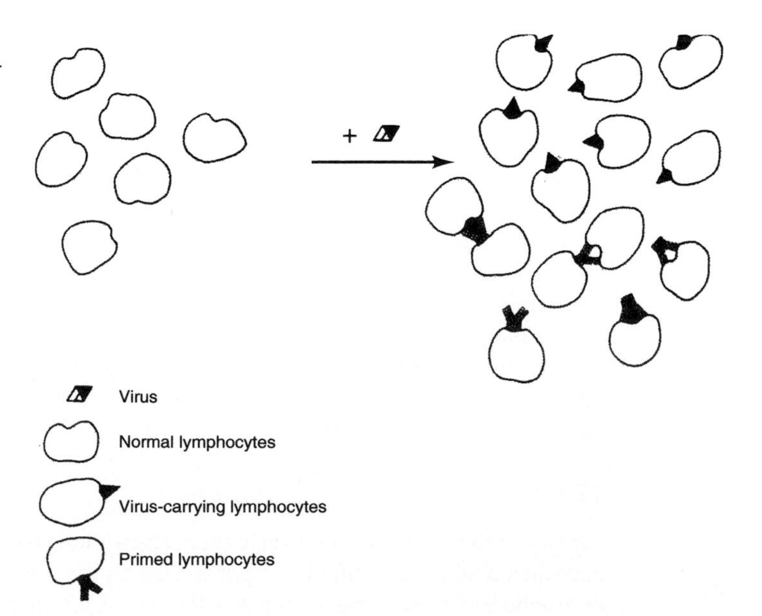

FIGURE 12.3

The virus, on infecting a cell, may produce its antigens on the surface of the cell and induce an immune response by activated (primed) lymphocytes. This response may lead to a reaction of the lymphocytes against not only the viral antigen but also normal antigens that have been revealed as a result of the viral infection. In this sense, the virus acts as a "carrier" for a small protein on a normal cell and makes it immunogenic. (Adapted from J. A. Levy (1977) In *Autoimmunity.* Ed. N. Talal. Courtesy of Academic Press, New York)

12.4 INTERFERON

A major cellular response to virus infection is production of the antiviral protein interferon. It can act locally or conceivably widely in the host through its presence in the circulation. Interferon was discovered by Isaacs and Lindenmann in the late 1950s during studies on viral infection in which an indirect interference process was found. Interference between viruses in bacteriophage systems is discussed in Section 11.11 and differs from the interference mechanism detected with animal cells. Isaacs and Lindenmann observed that fluids from influenza virus–infected chick cell cultures contained proteins that could render cells resistant to infection to many viruses. It was subsequently shown that the virus-induced proteins were not viral in origin. Rather, they were cellular products secreted in response to infection with a wide variety of RNA and DNA viruses. This natural antiviral substance, which its finders named **interferon,** has been the object of intensive research over the last three decades. Although interferons were first recognized for their potent antiviral properties, it has now been shown that interferons regulate cellular processes including metabolism, growth, and immune responses. Recent advances in recombinant DNA technology have provided the opportunity for the cloning and production of large quantities of purified recombinant products for therapeutic trials. Clinically, trials of interferon in virus infections and cancer have reached the stage of licensing for specific uses in human diseases.

12.4.1 The Nature of Interferons

Interferons are usually glycoproteins that are produced by many, if not all, vertebrate cells. They are regarded as prototypes of **cytokines,** which are signalling peptides responsible for cell-to-cell communication. Cytokines play crucial roles in many biological processes including microbial infections, inflammation, immunity, and hematopoiesis. Cytokines are produced by many cells, including macrophage/monocytes, fibroblasts, endothelial cells, and tumor cells. The complex interaction of cytokines forms a cascade of cellular processes that result in programmed growth, and death, development, and differentiation of cells and tissues. Being prototypes of cytokines, interferons possess antitumor and immunomodulatory activities in addition to their antiviral effects. Due to structural differences in individual interferons from different animals, biological activities of interferons are in general species-specific. Human interferons can, however, induce antiviral activity in some animal cells.

On the basis of their biochemical properties and protein structures, interferons can be classified into three different types, namely, alpha, beta, and gamma (Table 12.3). They have molecular weights ranging from 17 to 25 kDa. In response to inducers, they are produced *de novo* and secreted by cells into the extracellular fluid. Each type of interferon is produced primarily by specific host cells and differs in the agents that induce it. The α and β interferons, previously known as type I interferons, are similar in protein structure. However, interferon γ (type II) is quite distinct from the other two. It has been shown that interferon α and β bind to the same receptor, whereas interferon γ binds to its own receptor.

Interferon α comprises at least 15 subtypes. All of them are often induced as a group by foreign cells, viruses and viral envelopes, bacteria, and tumor cells. They have the common characteristic of being produced primarily by B lymphocytes, null lymphocytes, and macrophages. Mitogens for B cells may also stimulate interferon α production. The interferon

TABLE 12.3
Characteristics of the human interferons (IFN)

Type	*Number of Subtypes*	*Major Cell Source*	*Inducer*	*Mol Wt*	*Acid Stability (pH 2)*
α	At least 15	B lymphocytes; Macrophages/monocytes	Viruses dsRNA	17–20 kDa	Yes[a]
β	1	Fibroblasts; macrophages; Epithelial cells	Viruses dsRNA	17 kDa	Yes
γ	1	T lymphocytes	Antigens Mitogens	21–25 kDa	No

Note: NK cells can also produce IFN-α and -γ.
[a]Some acid-labile forms have been demonstrated.

α subtypes are stable to low pH. However, an unusual **acid-labile** interferon α has recently been identified in individuals with systemic lupus erythematosus, rheumatoid arthritis, pemphigus, and AIDS. The pathophysiology of these diseases may be due to the enhanced effects of this interferon on the immune system, and the presence of interferon may reflect the existence of an infectious agent. Unlike other interferons, the α interferons are not glycosylated.

Viral and other foreign nucleic acids induce the synthesis of human interferon β in fibroblasts, as well as in macrophages and epithelial cells. It also is resistant to low pH. The previously known interferon β 2 has been reclassified as interleukin-6, since its putative antiviral activity actually is due to the **induction** of interferon β.

The γ interferon is made primarily by T lymphocytes along with other lymphokines in response to foreign antigens and T-cell mitogens. Under some conditions, NK cells may also produce interferon γ. It differs from the other human interferons in its primary cell of origin, its primary amino acid sequence, the gene's chromosome location (human chromosome 12), and its lack of stability at low pH. In general, it is not induced by viruses or double-stranded RNA. Interferon γ appears to be mainly an immunomodulator instead of an antiviral protein and it potentiates the action of the other interferon subtypes. Compared with the other interferons, interferon gamma is more effective in inhibiting intracellular microorganisms other than viruses, for example, some rickettsia, bacteria (e.g. *Listeria*), and protozoa.

The gene encoding the **receptor** for human interferon α and β has been localized to chromosome 21q, whereas that for the human interferon γ receptor has been mapped to a different chromosome (6q). This fact further illustrates the differences between the α, β, and γ interferons in addition to their dissimilar cellular origins and biological properties.

12.4.2 The Mechanism of Induction of Interferons

That a variety of both RNA and DNA viruses can induce interferons, even after inactivation of virus infectivity by UV light and other agents, led Isaacs to speculate initially that any foreign nucleic acid entering a cell would activate the antiviral system. Later studies with pure nucleic acids of various types showed that double-stranded RNA is the efficient nucleic acid inducer of interferon. A possible solution to the apparent paradox that both DNA and RNA viruses can elicit the interferon response was revealed by the discovery that at least one family of DNA viruses, the poxviruses, produces partially double-stranded RNA as a minor component of the products of transcription carried out by the virion-associated transcriptase. The function of this double-stranded RNA is still not clear, but all normal cells of both animals and higher plants also appear to produce small amounts of double-stranded RNA (about 0.01 percent of the total cell RNA). These viral and cellular double-stranded RNAs, and even synthetic RNA duplexes of homopolymers (e.g., poly(rA):poly(rU)), are potent inducers of interferon when added in pure form to cultured cells. Thus, neither the sequence nor source of double-stranded RNA is important for induction.

The extreme potency of double-stranded RNA for this biological activity was dramatically illustrated by the demonstration of interferon induction by a single virion containing presumably double-stranded RNA in which the two strands were reported to be covalently cross-linked, so that RNA replication was impossible. The intracellular target of the double-stranded RNA has not been identified, but it is clear that interferon expression requires RNA as well as protein synthesis. The double-stranded RNA by some mechanism turns on the interferon genes.

The induction of expression of different subtypes of interferon appears to be controlled at multiple levels. Various types of interferons are produced by different cell types in response to different inducers. When cells are induced to express interferons, an increase of transcription of the appropriate interferon gene(s) occurs rapidly by a process that does not require protein synthesis. Following a brief increase, the transcription of interferon genes stops and the mRNA synthesized may be selectively degraded with cessation of interferon production. Using molecular biology techniques, it has been found that a complex system of regulatory elements exists in the promoter of each interferon gene. The interferon gene activation or inhibition of expression appears to be tightly regulated by the specific binding of newly synthesized or modified proteins (nuclear transcription factors) to the regulatory regions of interferon genes.

12.4.3 The Antiviral Activity of Interferon

The antiviral state permits virus infection of cells but prevents virus assembly and replication. With retroviruses, for instance, interferon generally blocks the formation of budding virus particles at the cell surface, although certain viral antigens can be produced and released into the culture medium. This broad range of inhibition of different viruses suggests that more than one mechanism of antiviral action is set in motion in response to this cytokine.

Interferon itself has no direct activity on virus particles, but cells treated with interferon develop an antiviral state. Many studies suggest that interferon primes the cell to respond directly to the presence of a virus invader. Pretreatment of cells with very low doses of interferon prepares them to establish an antiviral state. To elicit its antiviral and biological effects, interferon binds to its cognate receptor and enters the cells by a process called *internalization.* However, internalization of the **interferon-receptor complex** is not a requirement for interferon to exert its biological activities. Following the binding to its specific receptor on the cell surface, interferon induces transmembrane signalling including the activation of a tyrosine kinase and possibly phospholipase A2. These events result in forma-

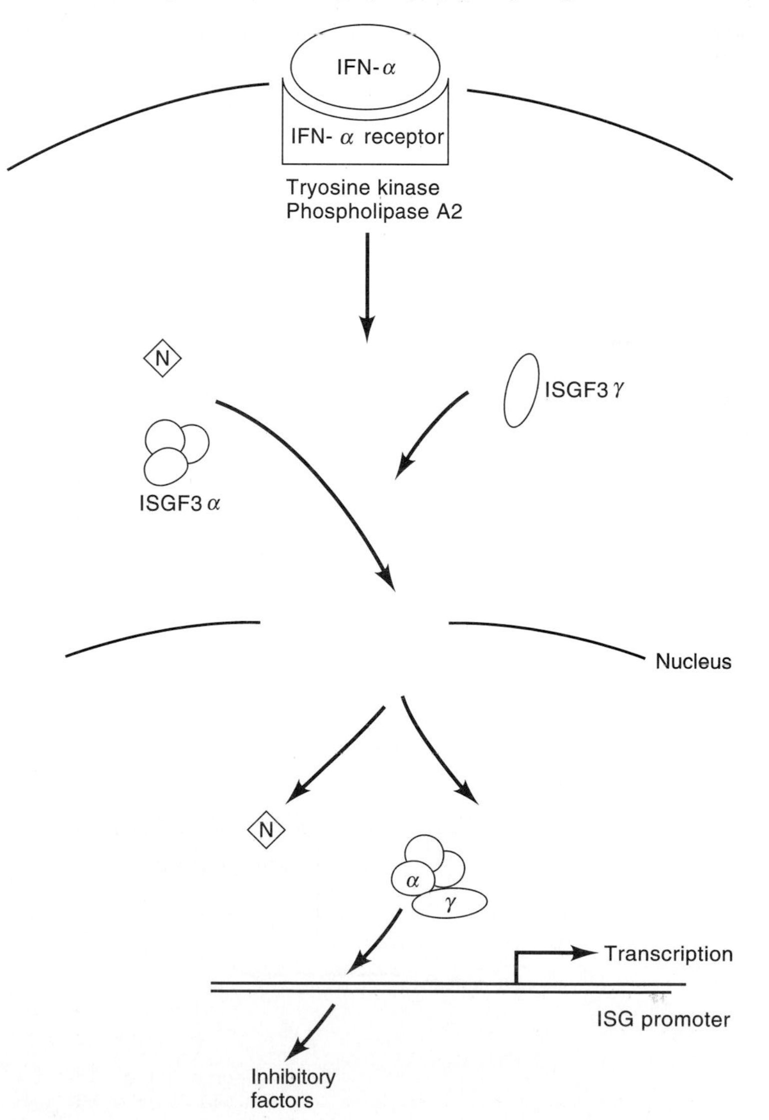

FIGURE 12.4

Interferon-α induction of interferon stimulated gene factors (ISGF). Upon binding of interferon α to its cognate receptor, transmembrane signalling involving the tyrosine kinase and/or phospholipase A2 activity results in association of a nuclear shuttle protein N with the ISGF3α complex. This combined complex consequently interacts with ISGF3γ to form a high affinity transcription-activation complex that is translocated to the nucleus. Inside the nucleus, the transcription factor complex binds to specific regions of the ISG promoter, namely the interferon stimulated responsive element (ISRE). This binding of ISGF3 complex to the ISG promoters results in initiation of transcription. The nuclear shuttle protein is not yet identified, although nucleolin is a possible candidate. Following translation, the ISG products mediate antiviral activity by inhibiting the replication of DNA and RNA viruses of most virus families to greater or lesser extents. (Modified by A. Lau from figure by B. Williams)

tion and translocation of transcription factors specific for interferon-stimulated genes (ISG) into the nucleus. A number of these ISG factors have been identified, including ISGF-1, ISGF-2, and ISGF-3. The steps involved in interferon action are summarized in Figure 12.4.

The action of the interferons against viral infections is thought to be mediated by at least three metabolic pathways that degrade RNA and inhibit protein synthesis. As, summarized in Figure 12.5, the binding of interferon to its cellular receptor leads to the induction of an unusual intracellular polymerase, which links together AMP residues of ATP by means of 2′-to-5′ phosphodiester linkages. This 2′–5′–linked oligoadenylate (2–5A) synthetase must be activated by binding to double-stranded RNA to produce oligonucleotides with the structure $pppA(2'p5'A_n)$, where *n* ranges from 1 to 10. The trinucleotide and larger species, in turn, serve to reversibly activate another enzyme, a ribonuclease. This 2–5A dependent ribonuclease cleaves viral or cellular mRNAs with about equal efficiencies in cell-free extracts, whether the RNA is free or associated with polyribosomes. However, in intact cells exposed to interferon, accumulation of viral mRNA is preferentially inhibited.

Besides 2–5A synthetase, a second interferon-induced enzyme is a 68 kDa protein kinase. Protein kinases phosphorylate proteins using ATP as a source of phosphate groups and energy. Protein kinases are extremely important components of gene regulation and cell growth control in eukaryotes. This particular p68 kinase, like the 2–5A–dependent ribonuclease mentioned above, is active only in the presence of double-stranded RNA. This activation also requires ATP and involves autophosphorylation of the kinase protein by itself. This self-phosphorylation is a common autoregulatory feature of other known protein kinases. Once activated by phosphorylation consequent to interferon action, the p68 kinase is able to catalyze the phosphorylation of eukaryotic initiation factor 2α (eIF2α), thus rendering eIF2α inactive. Since eIF2α is required for the initiation of all protein synthesis in eukaryotes, this action results in inhibition of translation of viral as well as host cellular proteins.

Therefore, it appears that the 2–5A–dependent ribonuclease and the p68 kinase systems act in concert to inhibit viral replication. Initially, the 2–5A–dependent ribonuclease is induced to degrade viral double-stranded RNA and mRNA, whereas the p68 kinase sys-

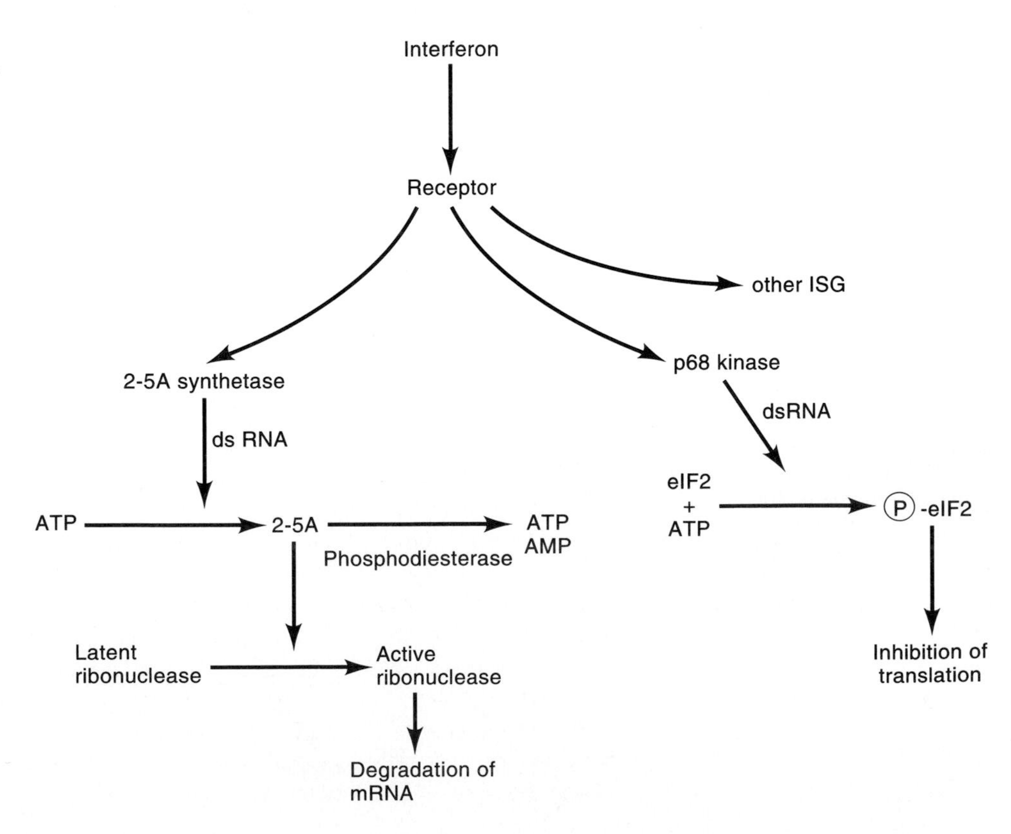

FIGURE 12.5

Activation of antiviral pathways in cells exposed to interferon. The binding of interferon to its cellular receptor leads to the induction of 2'-5'-A synthetase that triggers the activation of a ribonuclease leading to degradation of viral mRNA. The interaction of this cytokine with a receptor also induces other interferon stimulated genes (ISG) and kinase activities that lead to inhibition of translation of viral mRNA. (Courtesy of A. Lau)

tem is activated later to block the synthesis of virion structural components. With activation of these interferon-regulated enzymes, infected cells cannot synthesize their own proteins, in addition to being damaged by lethal viral products made earlier during infection. Finally, other polypeptides have recently been found to be induced by γ interferon. These γ interferon–regulated gene products may be involved in additional pathways affecting the metabolism of cells.

12.4.4 Virus Evasion of Interferon Action

Although interferons have a broad spectrum of antiviral activity against many viruses, not all viruses are equally susceptible to interferon. DNA-containing viruses are in general less sensitive to interferons than are RNA viruses. This observation may reflect in part the limited, if any, double-stranded RNA is synthesized during DNA virus infection. In addition, some DNA viruses have evolved specific strategies for evading the action of interferons. For instance, herpes simplex virus induces the synthesis of 2–5A analogues that compete with the natural 2–5A for binding with the 2–5A–dependent ribonuclease, thereby rendering the nuclease inactive against viral RNA. Similarly, both SV-40 and vaccinia viruses are known to subvert the 2–5A system by making nonfunctional 2–5A analogues. In contrast, RNA viruses inhibit the 2–5A system and the ribonuclease activity by a different mechanism. This process is not well understood but appears to be independent of the 2–5A analogue competitors.

12.4.5 Immune Regulatory Effects

All interferons appear to be capable of regulating the immune response, depending on their dose, time of administration, and the presence of other immune modulatory or hormonal factors. They can enhance the function of cytotoxic T lymphocytes and increase production of cytotoxic factors made by natural killer cells. Interferons can alter the expression of certain cell-surface antigens and receptors including MHC antigens and the IL-2 receptor. Interferons may play a role in inducing production of certain growth factors such as IL-2 and other cellular products (e.g., cytokines) by lymphocytes and macrophages. These changes all affect cell recognition and immune cell functions. At times the interferons compete with one another in modulating immune responses. For instance, interferon γ enhances the function and differentiation of macrophages, whereas interferon α inhibits macrophage maturation. The effect of interferon γ on macrophages led to this lymphokine's alternative name, macrophage activating factor (MAF). The relative influence of interferons on the immune response may therefore reflect the particular cell active at the moment for the initiation or continuation of the immune reaction (e.g., macrophages versus lymphocytes). Finally, the interferons (particularly α and β), can affect antibody formation, either through activation of suppressor T cells that reduce IgG and IgM production or through direct inhibiting effects on B cells.

12.4.6 Antiproliferative Effects

Given the pleiotropic effects of interferons on cellular processes, it is perhaps not surprising that cellular replication can be inhibited by interferon. What is remarkable, however, is that this inhibition often appears to be selective for tumor cells. As mentioned above, interferons are often produced by leukocytes in response to tumors, probably through leukocyte recognition of foreign antigens on the tumor cell lines. The antitumor effects of interferons may be exerted through several mechanisms including direct antiproliferative effects on tumor cells. Other mechanisms are indirect and are mediated through the activation of cytotoxic effectors of the host immune system, including natural killer cells, activated macrophages, and sensitized T lymphocytes. In addition, interferon γ enhances the production of cytokines, including tumor necrosis factor, by macrophages. The induction of cytokines results in a cascade of cellular processes that may further perpetuate the antitumor activity of the host immune system.

The antiproliferative activity of interferon may involve the same pathway as the antiviral effect, particularly inhibition of host cell protein synthesis. For instance, interferon suppresses the expression of enzymes, such as ornithine decarboxylase, that are important for cell growth. In addition, interferons may reduce DNA synthesis by processes still unknown. The antitumor effect of interferon was discovered in animal models some years ago, due initially to studies on the ability of interferon to inhibit oncogenic viruses. Soon it was realized that nonviral tumors were also inhibited by interferon.

12.4.7 Clinical Application of Interferons

Clinical trials of interferon have been aided by recent advances in genetic engineering techniques that allow biotechnology companies to produce mass quantities of purified recombinant interferon for therapeutic uses. Prior to the advent of biotechnology, a single culture of about 1 million cells was needed to produce only 1 pg of interferon following painstaking purification processes. Billions of cells were required to obtain a microgram of interferon for clinical studies. All three types of interferon can now be produced in *E. coli* in large amounts.

By 1992, interferons received FDA approval for five clinical indications in the United States. Interferon α was approved for **hairy cell leukemia,** condyloma acuminata (human papillomavirus), and hepatitis C virus infection as well as for Kaposi's sarcoma (in patients with HIV infection). Interferon γ was recently approved for prophylactic use in patients with chronic granulomatous disease to prevent recurrence of bacterial infections. This signified the first approval of any interferon as an immunomodulatory agent. In addition, interferons have demonstrable efficacy in laryngeal papillomatosis; hepatitis B virus infection; early stages of HIV infection; chronic myelogenous leukemia; multiple myeloma; carcinoid, basal cell carcinoma; and cutaneous squamous cell carcinoma.

12.5 CELLULAR OR VIRAL PRODUCTS AFFECTING THE IMMUNE SYSTEM

12.5.1 Cytokines

During infection, some viruses induce the immune system to produce cellular products called **cytokines** that can modulate virus production. As discussed above, interferon suppresses virus production; in contrast, tumor necrosis factor (TNF) can sometimes increase virus production. Alternatively, direct infection of specific cells such as macrophages or lymphocytes can cause enhanced endogenous production of some of these cytokines. As an example, HIV infection of macrophages can produce IL-1 and TNF. Moreover, infection of T cells and macrophages can produce IL-6, which enhances B cell proliferation. The IL-6 production may in fact be responsible for the proliferation and eventual transformation of B cells into lymphomas. In other situations, the release of factors such as **tumor growth factor** β (TGF-β) can increase the transforming abilities of viruses (e.g., retroviruses). Finally, the cellular products, although made in defense against infectious organisms can themselves be the cause of many of the symptoms of virus infections, including malaise, muscle aches, headaches, swollen lymph glands, or fatigue (see also Chapter 14 and Table 14.8). In large quantities, they can be toxic to the cells.

12.5.2 Viral Products Affecting the Immune System

The ability of viruses to code for proteins resembling products of the immune system has recently been linked to their evasion of the host response. The first example of this phenomenon was the production by the Shope fibroma virus of a protein, T2, that resembles the TNF cellular receptor. This viral gene product is secreted into the fluid and binds to the cytokine, TNF, that is part of the immune response against virus infection. TNF is thereby inactivated. This recognition of the production of so-called **decoys** by viruses provides a new insight into how viruses can evade immune cells. In other examples, the human adenoviruses encode for three proteins that can protect virus-infected cells from TNF. Other viruses such as the herpes- and poxviruses can bind to MHC Class I molecules and prevent their expression on the cell surface. Thus, the infected cell can be protected from recognition by the cellular immune system. EBV codes for an analogue of IL-10 that can affect the antiviral response. Finally, certain retroviruses have transmembrane envelope proteins (e.g., murine p15E) that suppress the immune system most likely through inhibition of signal transduction. All these recent observations suggest that viruses have mechanisms for countering the host's immune defense. They offer new directions for potential antiviral therapies.

SECTIONS 12.4 AND 12.5 FURTHER READING

Original Papers

ISAACS, A., and J. LINDENMANN. (1957) Virus interference I. The interferon. *Proc. Royal Soc. London B* 147:258.

LINNEMANN, C. C., JR., et al. (1974) Association of Reye's syndrome with viral infection. *Lancet* ii:179.

TOWNSEND, A.R.M., et al. (1986) The epitopes of influenza nucleoprotein recognized by cytotoxic T lymphocytes can be defined with short synthetic peptides. *Cell* 44:959.

WHITTON, J.L., et al. (1988) Molecular definition of a major cytotoxic T-lymphocyte epitope in the glycoprotein of lymphocytic choriomeningitis virus. *J. Virol.* 62:687.

MOLINA, J. M., et al. (1990) Production of cytokines by peripheral blood monocytes/macrophages infected with human immunodeficiency virus type 1 (HIV-1). *J. Infect. Dis.* 161:888.

PIRCHER, H. P., et al. (1990) Viral escape by selection of cytotoxic T cell-resistant virus variants *in vivo*. *Nature* 346:629.

BARROW, P., TISHON, A., and OLDSTONE, M. B. A. (1991) Infection of lymphocytes by a virus that aborts cytotoxic T lymphocyte activity and establishes persistent infection. *J. Exp. Med.* 174:203.

ODERMATT, O. O. et al. (1991) Virus-triggered acquired immunodeficiency by cytotoxic T-cell dependent destruction of antigen-presenting cells and lymph follicle structure. *Proc. Natl. Acad. Sci. USA* 88:8252.

WILLIAMS, B. R. G. (1991) Transcriptional regulation of interferon stimulated genes. *Eur. J. Biochem.* 200:1.

OPGENORTH, A, et al. (1992) Deletion of the growth factor gene related to EGF and TGFα reduces virulence of malignant rabbit fibroma virus. *Virology* 186:175.

GILBERT, M. J., et al. (1993) Selective interference with Class I major histocompatibility complex presentation of the major immediate-early protein following infection with human cytomegalovirus. *J. Virol.* 67:3461.

KHAN, M. A., et al. (1993) Inhibition of growth, transformation, and expression of human papillomavirus type 16 E7 in human keratinocytes by alpha interferons. *J. Virol.* 67:3396.

KYBURZ, D., et al. (1993) Virus-specific cytotoxic T cell-mediated lysis of lymphocytes *in vitro* and *in vivo*. *J. Immunol.* 150:5051.

ROUTES, J. M. (1993) Adenovirus E1A inhibits IFN-induced resistance to cytolysis by natural killer cells. *J. Immunol.* 150:4315.

Books and Reviews

NOTKINS, A., MERGENHAGEN, S., and HOWARD, R. (1970) Effect of virus infections on the function of the immune system. *Ann. Rev. Microbiol.* 24:525.

DOHERTY, P.C., and ZINKERNAGEL, R.M. (1974) T cell-mediated immunopathology in viral infections. *Transplant. Rev.* 19:89.

LINDENMANN, J. (1974) Viruses as immunological adjuvants in cancer. *Biochem. Biophys. Acta* 49:355.

DENT, P. B. (1975) Immunodepression by oncogenic viruses: Mechanisms and relevance to oncogenesis. In *The immune system and infectious diseases.* S. Karger, Basel 95.

ROUSE, B. T., and HOROHOV, D. W. (1986) Immunosuppression in viral infections. *Rev. Infect. Dis.* 8:850.

WELSH, R. M. (1986) Regulation of virus infections by natural killer cells. A review. *Nat. Immun. Cell Growth Regul.* 5:169.

KAWADE, Y., and S. KOBAYASHI, eds. (1988) *The biology of the interferon system.* Kodansha Scientific Ltd., Tokyo, Japan.

TANIGUCHI, T. (1988) Regulation of cytokine expression. *Ann. Rev. Immunol.* 6:439.

TATSUYAMA, T., et al. (1991) Cytokines and HIV infection: Is AIDS a tumor necrosis factor disease? *AIDS* 5:1405.

BARON, S., ed. (1992) *Interferon: Basic science and medical application.* Univ. Texas Medical Branch Press, Galveston, Texas.

BORDEN, E. C. (1992) Interferons: Expanding therapeutic roles. *New Engl. J. Med.* 326:1491.

GOODING, L. R. (1992) Virus proteins that counteract host immune defenses. *Cell* 71:5.

MERRILL, J. E. (1992) Cytokines and retroviruses. *Clin. Immunol. Immunopathol.* 64:23.

LEVY, J. A. (1993) Pathogenesis of human immunodeficiency virus infection. *Microbiol. Rev.* 57:183.

12.6 PERSISTENT AND LOW-LEVEL VIRUS INFECTIONS

Sudden epidemics causing widespread deaths are the most dramatic expressions of virus biology, but they are not at all the most typical. It appears probable that under normal circumstances most viruses have a symbiotic relationship with their normal host, and may cause no evident harm or disease. Obviously, death of the host also means the "death" of the virus, since viruses are dependent on the functioning of living cells. Most often, virus infection leads to a mild or undetectable disease that elicits the formation of antibodies. In many instances, low levels of virus replication and virus neutralization persist throughout the normal life span of the animal. This situation can produce lifelong immunity of the animal to that virus. Children's diseases such as measles and mumps are well-known examples of this occurrence. Infection with EBV and other herpesviruses represent the same phenomenon.

Human civilization has introduced over the years hygienic precautions that protect most children against viral and bacterial infections. This approach, however, makes them susceptible to infection at a later stage when for not clearly understood reasons several of these diseases are not as well handled by the adults. Thus, rubella virus, the agent causing German measles in children, is responsible for serious malformations of the fetus upon infection of the mother early during pregnancy. And mumps virus can affect the testes of adults and cause sterility.

The fact is that most adult and healthy animals carry many viruses and virus-specific antibodies. This situation is called **endemic** or **enzootic** infection *(demos* is for the human as contrasted with *zoos,* for the animal world). When such viruses are occasionally transmitted to a species or a population that is not normally a host for this virus, or that has been protected against or freed from that virus, a devastating **epidemic** (or **epizootic**) may be initiated. This epidemic can lead to the detriment of both the virus and the host population. Thus, smallpox and measles may have contributed to the demise of the Roman empire, and surely to those of the Aztecs and Incas. However, there are always some survivors of viral infections and when the population density is low as a consequence of the mass-killing epidemic, the rate of disease spread is also slowed down; evolution then has time to take a lead. Mutants of the virus that are less deadly are favored, as are mutants of the host that are more resistant, and gradually an endemic situation can again become established.

That this is not a hypothetical situation is illustrated by several well-documented examples. Thus, when the European rabbit, introduced into Australia, had increased and multiplied to a point where rabbits caused considerable damage to agriculture, a deadly and very specific rabbit myxomavirus was intentionally introduced. It was fantastically effective. However, within a few years its effect had worn off and the number of rabbits increased. Now not only is the Australian myxomavirus less virulent, but also the Australian rabbit population is more resistant to infection by the original myxomavirus strain.

12.6.1 Persistent Virus Infections

When a virus can infect and persist within the host without causing disease, it is considered to be in a state of **balanced pathogenicity.** In this condition, neither the host nor the virus is in serious danger. Thus, it is through balanced pathogenicity that new viruses can spread from animal to animal and from animals to humans. Recent examples include Lassa fever, which was transmitted from a rodent population in Nigeria to humans, and Marburg and

Ebola fevers, which moved from monkeys to humans. In many cases the existence of carrier virus states, such as with **hepatitis B virus,** represents balanced pathogenicity in humans. Another example is the animal spumaviruses. The infected hosts do not become ill, but their tissues and blood can contain virus and transmit infections to others.

Symbiotic relationships of virus and host may be called **persistence** and can actually have several different mechanisms and expressions. Besides the situation illustrated by measles and polio virus infection where lifelong immunity is best explained by a lifelong persistence of a few virus particles in the organism, there is also the situation of the common cold, which we all catch again and again. The symptoms of the common cold can be caused by over 100 different strains (serotypes) of the **rhinovirus,** as well as by many unrelated viruses. Thus, one might surmise that each cold is the consequence of an infection by a new virus against which we have not yet maintained an effective immune response. However, there are many instances in which we catch a cold in the absence of any potential donor of the virus. It seems much more likely that the cold viruses persist and occasionally, because of weather and other environmental circumstances, become replicated in the host, leading to a brief period of the well-known disease. The high mutability of RNA viruses may play a contributory role in this, as it also certainly does in the case of recurrent attacks of the **influenza virus** (see Fig. 4.10).

There are other consequences of persistent virus infection that are more or less rare. *Varicella zoster* or shingles, the unpleasant and painful skin rash localized along the area of a particular sensory nerve path on one side or the other of the face or body, is an occasional and then recurring consequence years after the acute infection with **varicella virus,** the herpes virus that also causes chickenpox. Recurring fever blisters and related skin symptoms also are expressions of the persistence of herpesvirus, and healthy persons can also continuously excrete virus while having detectable circulating antiherpes antibodies.

Epstein-Barr virus and cytomegalovirus are two other herpesviruses that appear to be endemic almost worldwide without generally causing symptoms. EBV persists mostly in the lymphoid tissue, releasing some virus and continuously evoking antibody formation. At times (usually of immune deficiency) the virus becomes active and can cause a variety of diseases. As noted previously, EBV, under such conditions in the host, can give rise to the malignant tumor first described in African children, Burkitt's lymphoma. The virus is also associated with nasopharyngeal carcinomas and some other rare cancers (see Sec. 10.2.5).

The **cytomegaloviruses** persist in fibroblasts of many species and are usually replicated very slowly. In humans they probably persist in some lymphocytes and granular leukocytes. The viruses tend to become reactivated upon any change in the immune system and/or the presence of any foreign antigen in the host. This activation can cause severe birth defects of various kinds, and death, in newborn children, even though the mothers show no symptoms of disease. Recently the use of attenuated vaccines during pregnancy has shown some promise in warding off this viral condition. The range of effects that herpesviruses can cause in humans and animals—include acute diseases, chronic or recurring diseases, malignancy, birth defects, and death.

Measles virus is another virus that produces wide-ranging symptoms, although fortunately it is only very rarely chronic and deadly. The slowly developing neural disease **subacute sclerosing panencephalitis** appears many years after measles infections and has been definitely shown by immunological tests to be due to that virus, possibly as a temperature-sensitive mutant. The role, if any, of measles virus in the etiology of multiple sclerosis, a slow degenerative human disease, is not yet clear.

Human **hepatitis B** virus, possibly present in half of the world's population, can cause severe disease, but it also persists, with viremia and antibodies detectable in seemingly healthy individuals for years. In most, the virus remains present in the liver, and can seed the blood, which is the reason blood transfusions are checked for evidence of HBV infection. How hepatitis B virus can also directly cause or contribute to the appearance of liver cancer is under active study (see Sec. 12.2.1). Other persistent infections result from human papillomaviruses, including the **BK** and **JC** papovaviruses.

Persistence is not clearly differentiated from certain other modes of virus-host symbiosis such as **latency, chronic virus disease,** and **slow virus disease** (see Sec. 11.1). The term *persistence* is often restricted to the situation where virus continues to be replicated and released, although at a low level and without causing disease in the host, but with detectable antibody formation. Chronic infection is the same situation, but the term is generally used with cases involving symptoms of the infection. **Latency,** in contrast, is a state where the virus is for long periods "silent." In the case of cells, little or no viral proteins, or infectious viruses are produced. In the host, low to absent antibodies, and no symptoms are detected (Sec. 11.1) The situation of cellular latency has recently been documented with the human AIDS virus; latently infected lymphocytes can be found in the lymph node and blood of infected individuals. It is a common event with HTLV infection (see Chapter 14). However, as with most definitions and classifications, the actual situations rarely correspond exactly with the defined terms. For example, many years of HIV or HTLV infection may precede symptoms of the virus: **clinical latency**. Nevertheless, within the infected individual, limited virus production, is taking place.

12.6.2 Mechanisms of Virus Persistence

What are the causes and mechanisms that account for virus persistence or latency? A surprisingly large number of different biological situations have been implicated, with more or less evidence in each specific case. Those now favored, in approximate order of their importance, are:

1. The **integration of the viral genome,** DNA directly and RNA after reverse transcription, *into the host chromosone.* This mechanism surely occurs in many instances and best accounts for persistence and latency. Latency may evolve through molecular events that inhibit viral genome transcription.
2. Viruses that are **nonantigenic,** such as uncoated infectious nucleic acids, might persist because they avoid the most powerful antiviral cellular response. Related to these are viruses that are poor antigens because of the high lipid content of their envelope, and viruses that encounter "immunologic tolerance," a particularly frequent occurrence with congenital infection. Also, certain viruses, principally those carried in lymphatic cells (e.g., some retroviruses), inhibit antibody formation or may induce *nonneutralizing* blocking antibodies (e.g., respiratory syncytial virus).
3. Many viruses tend to mutate frequently in the host, as shown by antigenic drift, and avoid immune recognition. **Temperature-sensitive (ts) mutants** may be replicated at too slow a rate to cause symptoms or to trigger an immune response. They can become activated by a back mutation or a further mutation that can enable them to become again productive at the host's body temperature. Alternatively the viruses can replicate following a change in that temperature. Evidence that several persistent viruses are *ts* mutants is strong (e.g., measles).
4. Viruses that have an affinity for neural tissue can become persistent by becoming **sequestered in the central nervous system.** Because of the absence of the typical lymphoid system and presence of a unique blood circulatory system, this largely nondividing tissue is not readily infected by viruses. It also does not favor virus replication. But when infected, neural and brain tissue retains viruses for long time periods (e.g., herpesvirus). The slow diseases discussed below illustrate this situation.
5. Just as the relative levels of virus and (neutralizing) antibody can determine whether infection is aborted or maintained (item 2 above), so also can the balance between virus infectivity and **interferon-stimulating activity** become a determining factor.
6. The appearance of **DI (defective interfering) particles** seems to accompany most, if not all, animal virus infections. These particles have been shown to render infection mild and persistent in host as well as in cell cultures.

The term **slow infections** was coined for a poorly defined group of seemingly viral diseases that have gradually been found to be caused by several quite different viruses or, to be more cautious, infectious agents. They nevertheless share the properties of causing disease very slowly, affecting largely the central nervous and/or the lymph system, and usually, in the course of several or many years, leading to death. Many induce no immunological response (see Chapters 9 and 14).

Clearly identified as retroviruses are the **maedi** or **progressive pneumonia** and the **visna** viruses of sheep (see Chap. 6). Since these viruses carry reverse transcriptase, their mode of slow progression can involve integration into the host's genome (item 1 above). These viruses, in contrast with most other retroviruses, do not cause malignant tumors or leukemias. They belong to the genus of retroviruses called **lentivirinae** (Latin *lenti,* "slow"; see Sec. 6.1) because of their long incubation period. They are associated with a wide variety of pathological conditions including pneumonias, arthritis, encephalitis, and AIDS (the human lentivirus) (see Table 12.2). These retroviruses have been found in a large number of animal species including monkeys and, recently, cats.

Two rare human diseases, **kuru** and **Creutzfeldt-Jakob disease,** as well as a brain disease of mink, and scrapie (the long-known disease that kills sheep and goats), have in common the fact that no virus particles or known microorganisms have been found to produce them, although the cause is clearly infectious. It has repeatedly been suggested that small nucleic acids may be responsible for these diseases, possibly similar in nature to the viroids that cause diseases of plants (see Sec. 9.3). However, there is also evidence against this hypothesis, particularly for the scrapie agent, which is resistant to intense irradiation and nucleases. The existence of an infectious protein particle (prion) has been suggested (see Sec. 9.6).

That the similarly slowly developing neurological disease, **subacute sclerosing panencephalitis** (SSPE) has been identified as a rare consequence of chronic persistent infection by **measles virus** has been mentioned. Another rare disease of this nature, progressive multifocal leukoencephalopathy (PML), has been attributed to human **polyomalike** viruses, such as BK or JC. For several other slowly developing neurological diseases of humans, such as multiple sclerosis, causative viruses, such as a coronavirus or measles virus, have been considered but not demonstrated.

In a Darwinian sense, viruses are also influenced by a struggle for survival. Persistence is the appropriate result of this phenomenon. As noted before, the most pathogenic viruses would soon kill their host and thus eliminate their prolonged existence. The ecological pressure on them is therefore to adapt by replicating to a lesser extent, and reducing their **pathogenicity.** The result is latent or chronic infections, attenuated viral forms, and agents that are more controllable by a host's immune system. Likewise, the host will in time modify its reaction to the virus; it may mount a stronger or weaker immune response. Soon, as noted above, a state of balanced pathogenicity is established—both the virus and the host coexist without great compromise.

The primary evolutionary pressure on a virus, with living organisms, is to produce as much progeny as possible. However, in nature such an event is counteracted to produce a balance of growth. For example, the high-replication-rate retroviruses can lead to cancers and cell death. Thus, their high production rate threatens their own existence. Elimination of their host creates one balance to the uncontrolled growth of a virus. Genetic mutations in the agent that reduce its replicative and pathogenic properties would be another. Only with moderation and equalization among living organisms can nature provide for the existence of all living forms—whether they are parasites such as viruses, or self-sufficient living creatures such as humans.

SECTION 12.6 FURTHER READING

Original Papers

BARINGER, J. R., and P. SWOVELAND. (1974) Persistent herpes simplex virus infection in rabbit trigeminal ganglia. *Lab. Invest.* 30:230.

DUTKO, F. J., and M. B. A. OLDSTONE. (1981) Cytomegalovirus causes a latent infection in undifferentiated cells and is activated by induction of cell differentiation. *J. Exp. Med.* 154:1636.

CHINCHAR, V. G., et al. (1993) Productive infection of continuous lines of Channel catfish leukocytes by Channel catfish virus. *Virology* 193:989.

LLOYD, R. E., and BOVEE, M. (1993) Persistent infection of human erythroblastoid cells by poliovirus. *Virology* 194:200.

MOSCONA, A., and PELUSO, R. W. (1993) Persistent infection with human parainfluenza virus 3 in CV-1 cells: Analysis of the role of defective interfering particles. *Virology* 194:399.

Reviews

YOUNGER, J. S., and O. T. PREVELE. (1980) Viral persistence: Evolution of viral populations. *Compr. Virol.* 16:73.

LEVY, J. A. (1986) Multifaceted retroviruses. *Cancer Res.* 46:5457.

OLDSTONE, M. B. A. (1991) Molecular anatomy of viral persistence. *J. Virol.* 65:6381.

12.7 FORMATION OF TUMORS

As indicated in Sections 7.2 and 11.5, some viruses cause increased synthesis of cellular components, which leads to cell division and cell proliferation. The result of these events in the animals can be tumor formation. In many tumors, be they caused by viruses or other agents, the character of the cells is not much altered, and thus these growths generally do not endanger the organism. We call these tumors **benign.** The biochemical mechanisms of their formation are not well understood. The warts caused by several papillomaviruses in humans, rabbits, and so on, and several tumors induced by poxviruses in chickens, rabbits, monkeys, and humans, are examples of that type of cell response to virus infection **(fibromas, myxomas).** They also appear to reflect the ability of the virus to be expressed in the cells (see Sec. 8.3).

Cellular proliferation can also become **malignant:** The character of the tumor cells is altered with particular reference to the cell membranes and the cytoskeleton in general. Also, chromosome changes can occur, giving a stable new phenotype. Thus, malignant cells grow into and across adjacent tissue in an **invasive** manner, perhaps due to release of proteases that disrupt intercellular connections in surrounding tissues and allow tumor cells to enter that tissue. Cells growing into blood vessels become detached and are carried elsewhere in the body, causing new tumor foci, which are called **metastases.** The malignant tumors or cancers are called **carcinomas** if they originate from epithelial tissue, and **sarcomas** if they originate from connective tissue. Uncontrolled proliferations of the cells of the lymphatic system are called **leukemias** (circulating cells) or **lymphomas** (solid tumors), and these, like the carcinomas and sarcomas, usually cause the death of the organism.

Several viruses cause such malignant tumors in their natural host (Secs. 6.1, 10.2, and 10.3). The most intensively studied of these are members of the Retroviridae family that can be found in many diverse species. The oncogenic types are frequently found in chickens, mice, cats, and monkeys, in which they induce mostly leukemias, lymphomas, and sarcomas. The domestic cat seems to be a natural mammalian host few cancers that are infectious (i.e., transmitted horizontally by the highly infectious feline leukemia and sarcoma viruses); the viruses are spread by saliva. Recently, an antileukemia vaccine has been prepared for cats, not by vaccination with the virus (see above), but by using a tumor-specific surface antigen (Sec. 13.2). The applicability of this approach to controlling human cancers remains to be tested.

The role of viruses in human cancer is becoming more apparent as techniques of detecting these viruses improve, both in cell culture and through molecular analysis. Herpesviruses, papillomaviruses, and hepatitis B virus can now be linked to several cancers in humans as they have been in animals (see Table 12.1). Moreover, the discovery of the human T cell leukemia virus type 1 and its association with T cell leukemia suggest that retroviruses may be associated with other human malignancies. Viruses such as HIV may compromise the host so that cancers can develop. Moreover, as discussed earlier, the relationship of oncogenic portions of retroviruses to cellular genes may provide insights to further our general understanding of the development of tumors.

Finally, we cannot forget that tumor formation in mammals is a much more complex process than **neoplastic transformation.** It has been estimated that thousands of cells in any one body become transformed during a lifetime, but few, if any, progress to become malignant tumors. Immune responses to tumor antigens and, perhaps, interferon may destroy the nascent tumors. The malignant cell may be sloughed off during the normal replacement of epithelial lining cells, or the cancers may simply develop so slowly as not to be apparent until postmortem examination. As noted before, malignancy develops by multiple steps and has multifactorial etiologies.

12.8 INTERACTIONS AMONG PLANT VIRUSES

In nature, virus-infected plants often contain more than one virus. As discussed in Chapter 9, such **mixed infections** may contain a replication-competent "helper virus" plus one or more viral satellites—either satellite RNA(s) or satellite viruses—unable to replicate in the absence of their helpers. In many cases, however, each of the co-infecting viruses may be replication-competent. These mixed infections often involve specific pairs of viruses and produce distinctive diseases (e.g., carrot mottley dwarf, groundnut rosette, and lettuce speckles) whose symptoms are quite different from those produced by the individual viruses.

In each of these three examples, one virus is an aphid-transmissible luteovirus, and the second is a mechanically transmissible virus that becomes aphid-transmissible only from doubly infected plants. This change in the specificity of vector transmission results from **heterologous encapsidation,** a phenomenon that can also facilitate the long-distance (systemic) movement of certain viruses. In addition to their obvious importance for understanding virus disease epidemiology, such interactions may also shed light on a more fundamental question: the origin of multicomponent RNA plant viruses.

Aphid transmission of luteoviruses, including barley yellow dwarf virus (BYDV), is highly vector-specific. For example, the RPV strain of BYDV is efficiently transmitted by *Rhopalosiphum padi* but not by *Sitobion avenae.* Just the opposite is true for the MAV strain. *R. padi* rarely transmits MAV from oats infected with MAV alone, but readily transmits both RPV and MAV from doubly infected plants. Whereas injection of anti-MAV antiserum into *S. avenae* before the insects were allowed to feed on infected plants blocked transmission of MAV, *R. padi* injected with the same antisera were able to transmit both RPV and MAV. These results are best explained by assuming that the MAV genome has become encapsidated in a coat of RPV protein and is thereby protected from inactivation by the anti-MAV antiserum (see Fig. 12.6). The failure to detect encapsidation of the RPV genome within MAV capsids illustrates the specificity inherent in such heterologous encapsidation reactions.

Heterologous encapsidation can also involve unrelated viruses. For example, whereas barley stripe mosaic virus (BSMV) systemically infects barley and replicates to relatively high titers, tobacco mosaic virus (TMV) replicates poorly in barley and only in the inoculated leaves. When barley plants are inoculated with a mixture of BSMV and TMV, however, both viruses replicate to high titers and move systemically. Although phenotypically BSMV, up to 8.5 percent of the virus particles in doubly infected plants contained TMV RNA. As discussed in Chapter 11, the viral capsid protein plays a key role in the long-distance transport of many viruses. In the case of the interaction between BSMV and TMV, however, it is not known whether heterologous encapsidation *per se* or some other consequence of the coinfection allows TMV to move systemically.

Yet another example of heterologous encapsidation involves ST9-associated RNA and the coat protein of beet western yellows virus (BWYV), another luteovirus. Virions of the ST9 isolate of BWYV, originally isolated from broccoli plants showing unusual redden-

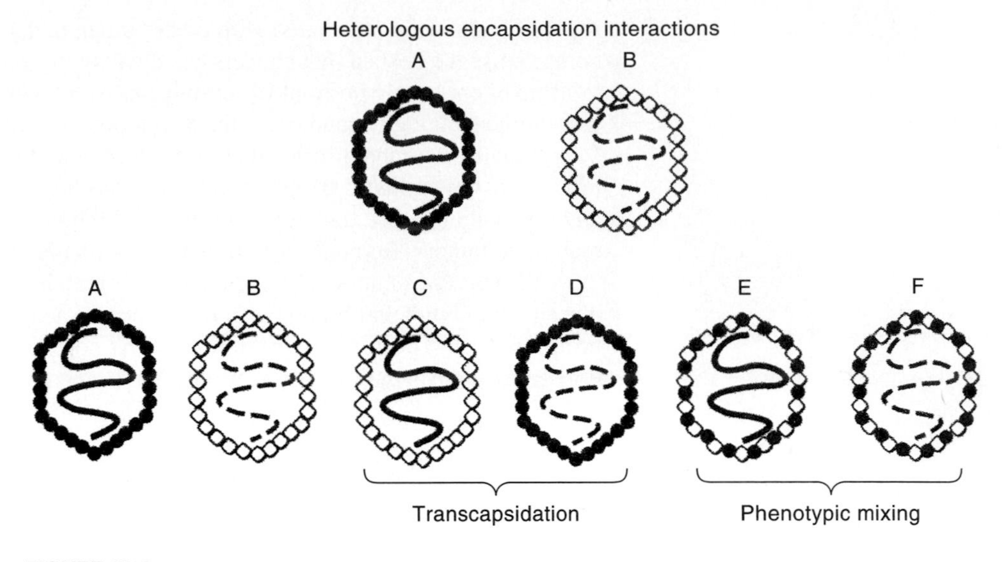

FIGURE 12.6

Possible results of heterologous encapsidation interactions when two viruses co-infect the same host cell. Outer solid circles or empty squares represent capsid protein subunits, and the inner solid or dashed lines represent the parental genomes. Upper virions (labeled A and B) are the co-infecting parental viruses. Possible progeny are shown in the lower row (A–F): A and B are parental types; C and D are virions where the genome of one parental virus has been **transencapsidated** by protein subunits derived from the other; and E and F are **phenotypically mixed** particles whose capsids contain protein subunits from both parental viruses. (Courtesy of Bryce Falk)

ing symptoms, contain two RNAs—the approximately 5600 nt BWYV genomic RNA and a smaller (2843 nt) ST9-associated RNA. Cloning and sequence analysis of the ST9-associated RNA revealed the presence of three large and one small ORFs, and the deduced amino acid sequences of the three large ORFs exhibit similarities to the RNA-dependent RNA polymerases of carmolike viruses (but *not* that of BWYV). Because RNAs transcribed from cloned ST9 cDNAs are able to replicate independently when electroporated into protoplasts, ST9-associated RNA cannot be considered either a satellite virus or a satellite RNA (see Sec. 9.1 and 9.2). ST9-associated RNA is also able to replicate independently in mechanically inoculated *Capsella bursa-pastoris* plants, but systemic movement requires encapsidation by BWYV coat protein. Several naturally occurring BWYV isolates are able to support systemic replication of ST9-associated RNA *in planta,* and the appearance of the severe symptom phenotype requires the presence of both BWYV and the ST9-associated RNA.

Our final example of viral interaction(s), this time on an evolutionary time scale, involves two members of the furovirus (i.e., *fu*ngus-borne *ro*d-shaped virus; see Table 2.2) group. The prototype member of this group, soil-borne wheat mosaic virus (SBWMV), contains a bipartite genome, whereas field isolates of beet necrotic yellow vein virus (BNYVV) contain 4 or 5 distinct RNA species. Sequencing studies have shown that (1) the RNA 1 of each virus encodes a ≥200 kDa replicase-related polypeptide and (2) the capsid protein is located near the 5′ terminus of RNA 2. As shown in Figure 12.7, these studies have also revealed several major differences in the organization of the SBWMV and BNYVV genomes.

The most striking differences involve the structure of the 3′ terminus and proteins believed responsible for cell-to-cell movement. Whereas BNYVV RNAs are polyadenylated, the 3′ untranslated regions of both SBWMV RNAs can be folded into a $tRNA^{val}$-like structure similar to those of tymo- and tobamoviruses. The 37 kDa protein encoded by SBWMV RNA 1 exhibits certain sequence similarities with the cell-to-cell movement proteins of tobamo- and tobraviruses and is probably expressed via a 3′-coterminal subgenomic mRNA *in vivo*. The BNYVV cell-to-cell movement protein is a potexvirus triple-gene block-type protein that is encoded by RNA 2. Based on sequence similarities and

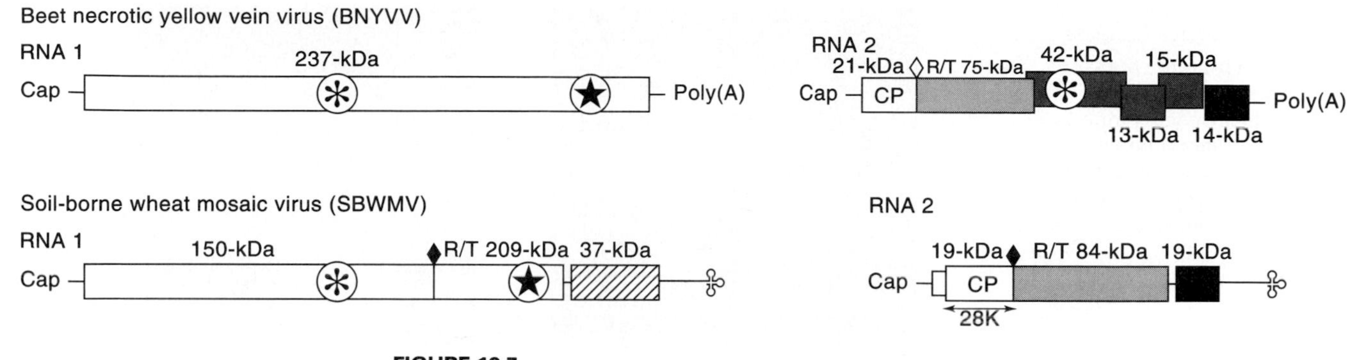

FIGURE 12.7

Comparison of the genome organization of soil-borne wheat mosaic and beet necrotic yellow vein virus RNAs 1 and 2. [pattern] putative RNA replicase; [pattern] TMV-type cell-to-cell movement protein; [pattern] triple gene block movement proteins; [pattern] cysteine-rich protein; CP, capsid protein; [pattern] capsid-read-through region possibly involved in fungus transmission; cloverleaf, a 3′ terminal tRNA-like structure; and filled and open diamonds, suppressible termination codons. Locations of the NTP-binding helicase and RNA polymerase motifs within the putative RNA replicases are indicated by asterisks and stars, respectively. (From Shiraku and Wilson. (1993) *Virology* 195:16)

genome organization, SBWMV is a member of the "Sindbis-like" superfamily of plant and animal viruses (see Postscript) and is more closely related to the hordei-, tobra-, and tobamoviruses than to BNYVV. Although members of the same virus grouping, SBWMV and BNYVV seem to have evolved independently via RNA recombination between different "ancestral" viruses.

12.9 PLANT RESPONSES TO VIRUS INFECTION

The range of possible molecular interactions between plant viruses and their hosts appears to be only slightly less complex than those described for animal viruses. Certain interactions essential for replication and systemic spread were considered in Section 11.10, but our discussion of the effects of virus replication on the host was limited to effects on individual cells or cellular organelles. Here we turn our attention to the entire host plant. This section begins with a brief description of some of the more common visible symptoms associated with virus infection. Next, we discuss two physiological responses by the host to virus infection—the synthesis of **pathogenesis-related (or PR) proteins** and a phenomenon known as **systemic acquired resistance.** Finally, we consider the genetics of host resistance to virus infection and the nature of the virus-coded signals that induce disease.

12.9.1 Disease Symptoms

For many years, plant virus research depended heavily on the ability of viruses to cause disease. Most virus names describe either an important disease symptom or the host in which the virus was first recognized. Even today, the ability of a virus to induce a characteristic disease syndrome continues to be very useful—especially during the initial stages of purification and characterization of a virus for which other specific assays are not available.

It is important to realize, however, that virus infection is not always accompanied by the appearance of visible disease. Naturally occurring **mild strains** have been described for many viruses (e.g., the Holmes "masked" strain of TMV), and these may replicate just as vigorously as more severe strains of the same virus. Plants infected with a mild virus strain may show much reduced or even no visible symptoms. Certain **tolerant hosts,** although able to support normal levels of virus replication, show little or no visible signs of infection. Finally, a large number of apparently healthy plants have also been shown to harbor **cryptic viruses** (see Sec. 5.4.2).

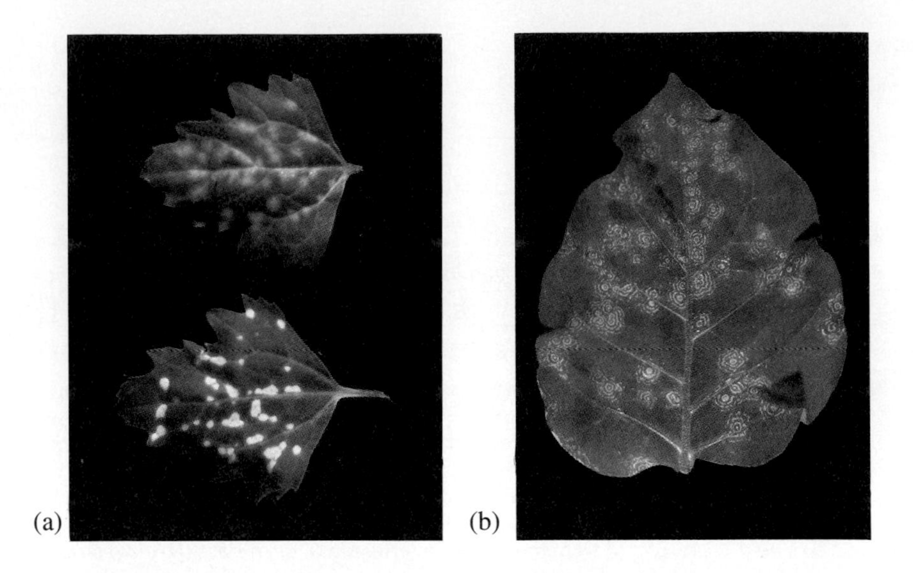

FIGURE 12.8

Localized symptoms of plant infection. (a) Necrotic (left) and chlorotic local lesions (right) caused by different strains of cucumber mosaic virus in *Chenopodium quinoa* (Courtesy of Y-H Hsu) (b) Ring spot lesions caused by tomato ringspot virus in tobacco. (Courtesy of E. V. Podleckis)

Although of little or no economic significance, the **local lesions** that frequently develop near the site of virus entry on inoculated leaves are very important for biological assays. As shown in Fig. 12.8, the virus-infected cells may either (1) lose their chlorophyll and/or other pigments, thereby producing a chlorotic lesion, or (2) die, producing a necrotic lesion. In 1929 F. O. Holmes showed that the number of necrotic local lesions appearing on the leaves of *Nicotiana glutinosa* plants after mechanical inoculation with TMV could be used to estimate the relative virus titer. Even today, local lesion assays remain the most convenient method for determining the proportion of viable particles in a virus preparation.

As replication continues and the virus spreads to other parts of the host, a variety of **systemic symptoms** may appear in the uninoculated tissue. Commonly observed systemic symptoms include stunting, disruption of normal pigmentation patterns in the foliage, and various types of necrotic responses. Several examples are shown in Fig. 12.9 (see also Fig. 1.5). Necrotic responses may lead to the death of certain tissues, whole organs, or even the entire plant—if the apical meristem is affected. Virus-infected plants also show a wide range of developmental abnormalities, including the appearance of **enations** (outgrowths from the upper or lower surfaces of leaves associated with veinal tissue) and **tumors** (see Fig. 5.11). More than one type of systemic symptom is often observed, and disease development may involve the sequential appearance of several different types of symptoms.

Virus infection often interferes with normal chloroplast development, leading to patterns of light and dark green areas in the leaves that are known as **mosaics** when they occur in dicots, or **streaks** and **stripes** when monocots are affected. Flowers, fruits, and seed coats may also show similar abnormal pigmentation patterns (see Fig. 1.5).

FIGURE 12.9

Systemic symptoms associated with plant virus and viroid infections. Mosaic disease: (a) Mosaic in leaf from a rose infected with rose mosaic virus; (b) Variegation or "color break" in a flower from a Darwin tulip infected with tulip color break virus (see also Fig. 1.5). Growth abnormalities affecting the foliage: (c) Uneven growth of leaf lamina in TMV-infected tomato; (d) Enations (outgrowth of vascular tissue) on a virus-infected grape leaf. Reduction in size: (e) Healthy and tomato ringspot virus (TomRSV)-infected grapevines; (f) fruit clusters from healthy and TomRSV-infected grapevines. Symptoms affecting fruit: (g) "Dapple apple" in apples harvested from trees infected with apple scar skin viroid; (h) "Stony pit" in pears infected with an agent believed to be a virus. In panels e–h, healthy controls are on the left. (Courtesy of E. V. Podleckis)

(a)

(b)

(c)

(d)

(e)

(f)

(g)

(h)

12.9.2 Physiological Responses to Virus Infection

The microscopic and macroscopic symptoms of virus disease must originate from biochemical disturbances induced by the virus, but we are almost totally ignorant about the molecular mechanisms involved. Virus-infected plants often show (1) a decreased rate of photosynthesis associated with lower amounts of chlorophyll, chloroplast ribosomes, and ribulose bisphosphate carboxylase; (2) an increased rate of respiration; (3) increased activity of certain enzymes, including polyphenoloxidase; and (4) altered levels or ratios of various plant hormones. Although important for understanding disease development, these are almost certainly *secondary effects* of virus infection and may obscure the initial interaction(s) between viruses and their hosts.

The appearance of necrotic local lesions is the most visible manifestation of the **hypersensitive response,** a complex series of host responses aimed at localizing the invading pathogen. Shortly after such lesions begin to appear, synthesis of a series of PR proteins begins, and the host becomes resistant to further virus infection (i.e., so-called **systemic acquired resistance**). Possible connections between these two phenomena are not yet clear.

Two different genes control the ability of tobacco plants to form local lesions in response to inoculation with TMV. In the presence of either the **N gene** from *Nicotinia glutinosa* or the **N' gene** from *N. sylvestris,* the mosaic symptoms that normally appear on both inoculated and systemically infected leaves are replaced by necrotic local lesions on the inoculated leaves alone. As defined by Ross and co-workers, systemic acquired resistance refers to the reduction in lesion size and/or number that is observed when Samsun NN tobacco previously inoculated with TMV on its lower leaves is subsequently challenged by a second inoculation with virus. Resistance is detectable within 2 to 3 days post inoculation, reaches a maximum after 7 days, and persists for approximately 20 days. Interesting features of this resistance include (1) the ability of TMV-induced resistance to protect against challenge with viruses other than TMV and (2) the ability of exogenous application of salicylate to induce a comparable level of resistance. In fact, salicylic acid may be the natural "second messenger" by which the plant signals the presence of any pathogen able to induce a necrotic response.

TABLE 12.4

PR proteins induced in Samsun tobacco (NN genotype) by TMV infection (From Bol et al. (1990) Ann. Rev. Phytopathol. 28:113)

	Acidic PR Proteins		*Basic PR Proteins*		
Group	*Name*[a]	*Mol Wt (kDa)*	*Name*	*Mol Wt (kDa)*	*Function*
1	1a	15.8	16 kDa	16	Unknown
	1b	15.5			
	1c	15.6			
2a	2	39.7	Gluc.b	33	β-1,3-Glucanase
	N	40.0			
	O	40.6			
	Q′	36.0			
2b	O′	25.0			β-1,3-Glucanase
3	P	27.5	Ch.32	32	Chitinase
	Q	28.5	Ch.34	34	
4	s1	14.5			Unknown
	r1	14.5			
	s2	13.0			
	r2	13.0			
5a	R	24.0	Osmotin	24	Unknown
	S	24.0			(TL-proteins[b])
5b			45 kd	45	Unknown

[a]Nomenclature according to Fritig et al. (1989) In *NATO ASI Ser. H: Cell Biol.* Vol. 36, 161. Springer-Verlag, Vienna.

[b]Thaumatin-like proteins

As summarized in Table 12.4, TMV infection of Samsun tobacco homozygous for the N gene also induces the synthesis of two series of PR proteins. Acidic PR proteins are protease-resistant and located in the intercellular spaces, whereas their basic homologues accumulate in the vacuoles. Several of these proteins have been shown to possess enzymatic activities that could participate in a direct attack on an invading fungal and bacterial pathogens that are also capable of inducing the hypersensitive response. Other PR proteins may help localize the pathogen at the site of infection via lignification of the cell wall or merely reflect host cell adaptation to disease-imposed stress (e.g., superoxide dismutase). Although the evidence accumulated to date is certainly suggestive, the exact role of the PR proteins in disease resistance remains to be determined. Virus resistance precedes the detectable synthesis of PR proteins, and transgenic plants that constitutively express levels of single PR proteins comparable to those present in TMV-infected plants are just as susceptible to infection by TMV or alfalfa mosaic virus as are the untransformed controls.

12.9.3 Genetic Determinants of Virus Disease

Where applicable, breeding for disease resistance is undoubtedly the most effective strategy to control plant virus disease. As summarized in Table 12.5, most examples of host resistance to virus infection involve only a single (usually dominant) locus. Dominant alleles tend to be associated with resistance mechanisms involving local lesion formation and virus localization (i.e., the hypersensitive response discussed above). In contrast, gene-dosage-dependent resistance is strongly associated with mechanisms providing only partial virus localization. To date, no resistance genes have been cloned, but some information about their interactions with viral gene products is available.

Dominant genes specifying a potent inhibitor of an early step in virus replication may confer immunity on the host. A possible example is the gene that controls the constitutive synthesis of a protease inhibitor in "Arlington" cowpeas. As shown by Bruening and coworkers, protoplasts prepared from "Arlington" cowpeas are virtually immune to infection by cowpea mosaic virus, and the protease inhibitor specifically inhibits the processing of the cowpea mosaic virus polyprotein. Even in this case, however, resistance is not complete and may involve more than inhibition of the viral protease.

From a molecular viewpoint, the best-characterized resistance genes are the *Tm-1*, *Tm-2*, and *Tm-2*^{2} genes in tomato that control resistance to tomato mosaic virus (i.e., TMV-L, a tobamovirus). Although resistance specified by *Tm-2* and *Tm-2*^{2} is not expressed at the protoplast level and appears to involve inhibition of cell-to-cell movement, protoplasts prepared from tomatoes containing the *Tm-1* gene are resistant to infection by wild-type

TABLE 12.5
Genetics of host resistance to plant virus infection

Genetic Basis	*Host-Virus Combinations*	*Examples*[a]	*Resistance Phenotype*
Monogenic	69		
Single dominant gene	38	*N*/TMV (*N. glutinosa*)	Hypersensitive localization
		N'/TMV (*N. sylvestris*)	" "
		*/CPMV (cowpea)	Inhibition of viral protease
Incompletely dominant	13	*Tm-1*/TMV (tomato)	Inhibition of viral replicase?
Apparently recessive	18	*zym*/ZYMV (cucumber)	Resistance to virus multiplication
Possibly oligogenic	18	Two genes/CMV (squash)	Resistance to virus spread from inoculated leaf
		Polygenic/RMV (rye)	Resistance to infection
Total	87		

[a]Form of each entry: **Resistance gene designation**/virus affected (host plant). Abbreviations used: TMV, tobacco mosaic virus; CPMV, cowpea mosaic virus (comovirus); ZYMV, zucchini yellow mosaic virus (potyvirus); CMV, cucumber mosaic virus (cucumovirus); and RMV, ryegrass mosaic virus. *, unnamed gene.

From Fraser, R. S. S. (1990) *Ann. Rev. Phytopathol.* 28:179.

TMV-L. Sequence analysis of several resistance-breaking strains of TMV-L has shown that (1) amino acid changes in the 30 kDa movement protein are able to overcome the resistance conferred by the *Tm-2* gene, and (2) similar changes in the 126 and/or 183 kDa replicase proteins allow the virus to overcome resistance specified by *Tm-1*. The latter result suggests that the **ability of host proteins to interact with viral replicase proteins** is an important determinant of virus host range.

How do virus-encoded macromolecules regulate symptom expression? Although information at the nucleotide sequence level has only recently become available, it is already clear that several different viral genes may be involved. Once again, TMV provides a convenient place to begin. Most naturally occurring strains of TMV induce a systemic mosaic in the developing leaves of tobacco or tomato. Several groups have selected and characterized spontaneous mutants that produce greatly reduced or even no disease symptoms, and their results show that a single amino acid substitution in the 126/183 kDa replicase proteins may cause dramatic changes in symptom expression. Whether it is the replicase itself or changes in the rate of virus replication that interfere with chloroplast development remains to be determined.

We have already discussed the importance of the hypersensitive response in limiting virus replication and spread. Studies of resistance controlled by the *N′* gene of *Nicotiana sylvestris* have shown that the viral coat protein also plays a critical role in virus-host interaction. The *N′* gene controls a hypersensitive response directed against many tobamoviruses, but because the product of this gene fails to recognize certain tobacco strains of TMV, these viruses are able to cause systemic disease in *N. sylvestris.* Characterization of spontaneous mutants able to induce the hypersensitive response has shown that this phenotype maps to the coat protein gene. Analysis of approximately 30 amino acid substitutions in the coat protein of TMV-U1 showed that almost any alteration within a region where adjacent coat protein molecules interact is able to induce the hypersensitive response (HR) and the appearance of local lesions (see Fig. 12.10). Expression of the altered coat protein in transgenic plants is sufficient to induce a similar phenotype, demonstrating that neither viral replication nor other viral proteins are required.

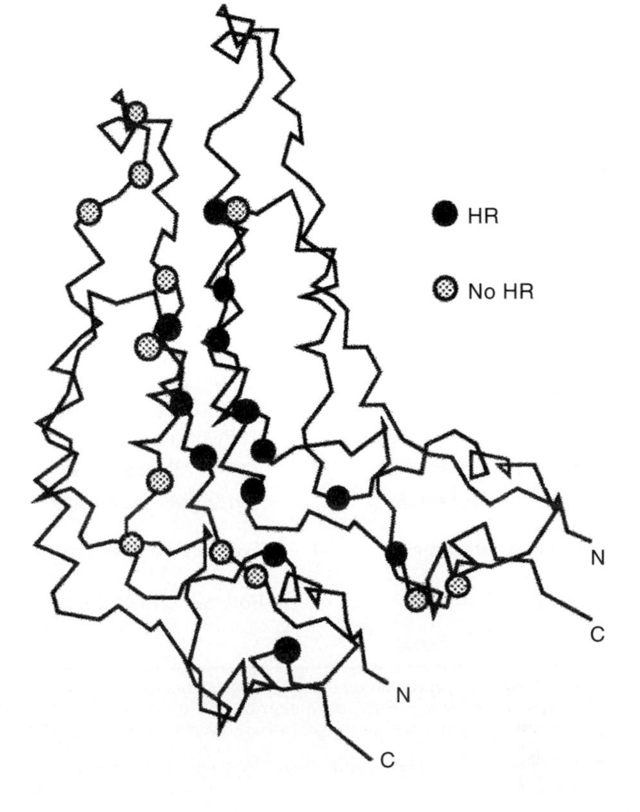

FIGURE 12.10

Locations of HR-eliciting and non-HR-eliciting amino acid substitutions within the alpha carbon outline of two TMV coat protein subunits. (Courtesy of J. N. Culver, W. O. Dawson, and G. J. Stubbs)

Viral genes other than those encoding the coat protein have also been shown to play important roles in the disease process. For example, TMV is able to induce a hypersensitive response in tobacco cultivars containing the *N* gene from *N. glutinosa.* Although the viral protein responsible for inducing the *N*-gene-related hypersensitive response has not been identified, the coat protein has been eliminated as a possible elicitor. In the case of cauliflower mosaic virus (CaMV), expression of ORF VI in transgenic tobacco plants may cause the appearance of viruslike symptoms—but only when the transformed species is *not* a system host for CaMV. CaMV ORF VI encodes the most abundant protein component of the viroplasms present in the cytoplasm of virus-infected cells (see Section 6.3.1), and symptom expression appears to be related to the level of gene *VI* expression. For systemically susceptible species, there was no visible difference between transformed plants and the untransformed controls. Construction of hybrid CaMV genomes containing DNA fragments derived from mild and severe virus isolates have identified a number of genetic determinants, in addition to gene *VI,* that affect disease development and it is likely that most or all of the plant virus genome is somehow involved in the disease process.

SECTIONS 12.8 AND 12.9 FURTHER READING

Original Papers

HOLMES, F. O. (1929) Local lesions in tobacco mosaic. *Bot. Gaz.* (Chicago) 87:39.

VAN LOON, L. C., and A. VAN KAMMEN. (1970) Polyacrylamide disc electrophoresis of the soluble leaf proteins form *Nicotiana tabacum* var. "Samsun" and "Samsun NN." II. Changes in protein constitution after infection with tobacco mosaic virus. *Virology* 40:199.

BEIER, H. et. al. (1977) Survey of susceptibility to cowpea mosaic virus among protoplasts and intact plants from *Vigna sinensis* lines. *Phytopathology* 67:917.

GIANINAZZI, S., C. MARTIN, and J.-C. VALLEE. (1979) Hypersensibilité aux virus, temperature et proteines solubles chez le *Nicotiana* Xanthi n.c. Apparition de nouvelles macromolecules lors de la repression de la synthese virale. *C.R. Hebd. Seances Acad. Sci., Ser. D* 270:2383.

BAUGHMAN, G. A., J. D. JACOBS, and S. H. HOWELL. (1988) Cauliflower mosaic virus gene VI produces a symptomatic phenotype in transgenic tobacco plants. *Proc. Natl. Acad. Sci. USA* 85:733.

MESHI, T., et al. (1988) Two concomitant base substitutions in the putative replicase genes of tobacco mosaic virus confer the ability to overcome the effects of a tomato resistance gene, Tm-1. *EMBO J.* 7:1575.

LINTHORST, H. J. M., et al. (1989) Constitutive expression of pathogenesis-related proteins PR-1, GRP, and PR-S in tobacco has no effect on virus infection. *Plant Cell* 1:285.

MESHI, T., et al. (1989) Mutations in the tobacco mosaic virus 30-kd protein gene overcome TM-2 resistance in tomato. *Plant Cell* 1:515.

STRATFORD, R., and S. N. COVEY. (1989) Segregation of cauliflower mosaic virus symptom genetic determinants. *Virology* 172:451.

NELSON, R. S., et al. (1993) Impeded phloem-dependent accumulation of the masked strain of tobacco mosaic virus. *Mol. Plant-Microbe Interact.* 6:45.

Books and Reviews

BOL, J. F., H. J. M. LINTHORST, and B. J. C. CORNELISSEN. (1990) Plant pathogenesis-related proteins induced by virus infection. *Ann. Rev. Phytopathol.* 28:113.

FRASER, R. S. S. (1990) The genetics of resistance to plant viruses. *Ann. Rev. Phytopathol.* 28:179.

CULVER, J. N., A. G. C. LINDBECK, and W. O. DAWSON. (1991) Virus-host interactions: Induction of chlorotic and necrotic responses in plants by tobamoviruses. *Ann. Rev. Phytopathol.* 29:193.

MATTHEWS, R. E. F. (1991) *Plant virology,* 3d ed. Chaps. 11 and 12. Academic Press, New York.

DAWSON, W. O. (1992) Tobamovirus-plant interactions. *Virology* 186:359.

12.10 BACTERIOPHAGES AND THEIR HOST

There are a number of reasons for interest in the consequences of bacterial virus infections aside from basic curiosity about the diversity of biological phenomena. Some bacteriophage systems have served as models to guide studies on animal viruses that cause tumors. Other phages are important for medical reasons, either directly through involvement in pathogenic mechanisms or indirectly, because of selective associations with pathogenic hosts that can be exploited for identification purposes. Phages were once used to identify subtypes of bacteria judged by their sensitivity to lysis by the phage. This **phage typing** permitted the recognition of transfer or dominance of certain bacterial strains in a community or hospital. This latter technique has now been almost completely replaced by testing for antibiotic resistance and selected genetic markers by a variety of molecular procedures (see Appendix 1). As noted below, they can also induce toxin production by bacteria. Bacteriophages are a major concern in certain industries, whenever the physiological processes of bacteria are harnessed to process foods or other materials. Constant vigilance and evasive action are required to keep economic losses due to phages under control.

Pathogenic Conversions Remarkable and medically important changes in the phenotype of a bacterial host are brought about by prophages in *Corynebacterium diphtheriae* and *Clostridium botulinum.* These pathogens produce two of the most potent protein poisons or **toxins** known and are responsible for the severe pharyngitic disease *diphtheria* and the food poisoning known as *botulism,* respectively. In each case examinations of nontoxinogenic variants of the bacteria have led to the discovery that these toxic proteins are actually products of prophages residing in the toxinogenic strains. Phages producing diphtheria toxin, members of the β group have been examined most closely.

In this system toxin production is remarkably free of **regulatory influences** of the prophage or cell. Repressed lysogens produce toxin as efficiently as lytically infected cells. Moreover, when nutritional limitations inhibit cell growth and reduce the overall rate of protein synthesis by as much as 10-fold, the lysogen still produces toxin at the normal rate. Only the level of iron in the culture medium has been identified as a modulating influence on expression of the *tox* gene.

This independence of *tox* gene expression may be related to its location in the β prophage genome (see Fig. 12.11). The overall genomic configuration is much like that of integrated λ prophage, with an immunity region specifying a repressor located in the middle. The *tox* gene is located at one end of the inserted prophage, in an area that is not directly repressed in integrated prophage λ but that is not active due to repression of the positive regulatory gene *N.* Evidently, the *tox* gene has its own promoter, which is recognized by the bacterial RNA polymerase without help from phage functions.

This independent transcription and the location of the *tox* gene at the end of the prophage DNA suggest that the β phages carrying the *tox* gene are actually some sort of specialized transducing phage. No active toxin-producing gene has been detected in the bacterial chromosome. Thus, the *tox* gene might have undergone considerable change through evolution since it entered the genome of the β phage. Alternatively, the *tox* gene might have arisen by duplication of some structural phage gene and further evolution of one of the copies. This possibility is suggested by the discovery that one of the β phage virion proteins is closely related to the toxin protein. Whatever the original source of the *tox* gene, it has somehow spread to a variety of unrelated phages despite the fact that it provides no direct benefits to the phages. They are all completely viable without the *tox* gene. Nevertheless, to have survived and spread through such a large segment of the *Corynebacterium* population, the *tox* gene must confer some advantage on the host cell, at least under certain environmental conditions. This notion finds support in the observation that following the widespread immunization of human populations with inactive (mutant) forms of the toxin protein to prevent diphtheria, detectable *C. diphtheriae* has been eliminated in the normal bacterial flora of the human throat and nasopharynx.

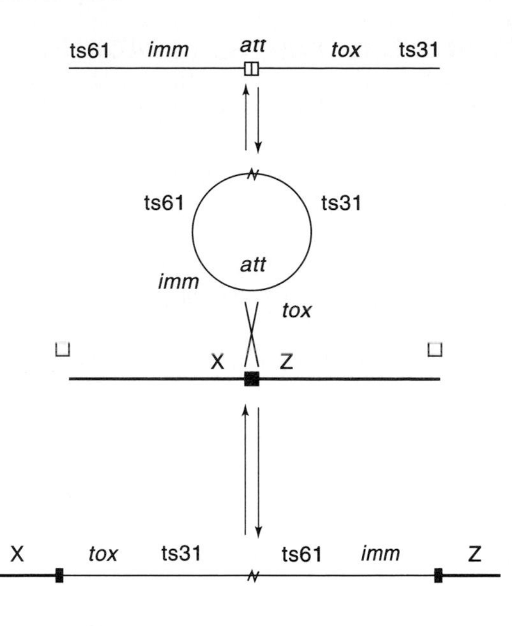

FIGURE 12.11

Hypothetical mechanism for the integration of β prophage into the chromosome of C. *diphtheriae.* The general mechanism of phage DNA (thin line) circularization and integration into the bacterial chromosome (thick line), and the permuted prophage gene order thereby obtained, is supported by genetic evidence. Phage genetic markers: ts31 and ts61, temperature-sensitive mutants defective in replication; *imm,* immunity region; *tox,* toxin gene. X and Z are host genes. (From Singer (1976), in *Mechanisms in bacterial toxinology*)

Phage-induced **conversion** may be a fairly common aspect of bacterial pathogenesis. Certain streptococci (Group A, β-hemolytic) elaborate a toxin that is **erythrogenic** (i.e., able to induce reddening in afflicted tissues), and this toxin is responsible for the disease condition called *scarlet fever.* Since two related temperate phage strains have been found that can convert nonlysogenic Group A streptococci to toxinogenic strains, this toxin is probably also a phage gene product. Certain strains of staphylococci produce hemolytic toxins, but in this situation conversion by lysogenic infection with certain phage results in shut-off of toxin production. In this case, another cellular phenotypic change occurs simultaneously: induction of a staphylokinase that can activate plasminogen and thereby trigger the anticlotting system in the blood of an infected person. It is not clear exactly how this phage regulates either of these functions.

Since several toxins classified as bacterial in origin have now been shown to be phage gene products, it will not be surprising if others are added to this list in the near future. Lysogeny with particular phages has been associated with production of the *cholera* toxin, for example, although a specific phage gene for that protein has yet to be identified. So far we have considered only pathogenic mechanisms involving soluble protein toxins released from the bacterial cell, or **exotoxins.** Many gram-negative bacteria however produce their characteristic disease symptoms with toxic configurations of LPS on the cell surface. Considering the known modifications of LPS by prophages in *Salmonella* and other bacteria, certainly some examples of this type of cell-associated poison, known as an **endotoxin,** are bound to be specified or modulated by phage enzymes that affect the LPS layer. Other noteworthy effects of phages on bacterial characteristics have been noted in the literature, such as effects on pigment production or colonial morphology, but the molecular basis and importance of these changes for the host or phage have not been determined. Finally, as mentioned above, certain phage can carry genes that code for enzymes such as penicillivase or resistance factors that make bacteria **resistant** to **antibiotics.**

SECTION 12.10 FURTHER READING

Original Papers

OGG, J. E., M. D. SHRESTHA, and L. POUDAYL. (1978) Phage-induced changes in *Vibrio cholerae:* Serotype and biotype conversions. *Infect. Immun.* 19:231.

CHAUDHURI, B., et al. (1993) Isolation, characterization, and mapping of temperature-sensitive mutations in the genes essential for lysogenic and lytic growth of the mycobacteriophage L1. *Virology* 194:166.

SNAPPER, S. B., et al. (1988) Lysogeny and transformation in mycobacteria: Stable expression of foreign genes. *Proc. Natl. Acad. Sci. USA* 85:6987.

Books and Reviews

GARRO, A. J., and J. MARMUR. (1970) Defective bacteriophage. *J. Cell. Physiol.* 76:253.

BARKESDALE, L., and S. B. ARDEN. (1974) Persisting bacteriophage infections, lysogeny, and phage conversions. *Ann. Rev. Microbiol.* 28:265.

GEIDUSCHECK, E. P. (1991) Regulation of expression of the late genes of bacteriophage T4. *Ann. Rev. Genet.* 25:437.

LEVY, S. B. (1992) *The Antibiotic Paradox. How miracle drugs are destroying the miracles.* Plenum, New York.

chapter

13

Viruses as Tools in Medicine and Biotechnology

We have thus far discussed the general aspects of virology and the various major virus families in vertebrate and invertebrate species, plants, and microorganisms. The effects of these virus infections at the cellular level and on the total organism have also been considered. We now turn to the ways in which viruses can be used as tools in medicine and biotechnology.

13.1 DIAGNOSTIC USE OF IMMUNE RESPONSE

Besides the role of the immune system in defending the host against viral diseases, antibody production itself can be extremely valuable in the diagnosis of such illnesses and, in some cases, their prognosis. A variety of tests are available in the laboratory that can detect the IgM antibodies characteristic of acute infection with a virus as well as the IgG antibodies occurring over time in the infected host. One procedure that is easily automated is the **enzyme-linked immunosorbent assay** (ELISA), which involves the reaction of an individual serum with fixed total viral antigens (see Fig. A.6, Appendix 1). The binding of the serum antibodies is detected by the use of an enzyme-labeled antibody to the serum antibodies; a positive reaction is measured by a colorimetric method.

Another conventional procedure for detection of antiviral antibodies is indirect **immunofluorescence** (IFA), in which infected cells are fixed to slides. Serum is reacted with the cells and the attachment of antibodies is detected by a second fluorescein-labeled antibody. The presence and pattern of the host antibody reaction with the virus-infected cells is then read with a fluorescent microscope (see Fig. 13.1).

The IFA procedure has permitted the identification of antibodies to early antigens (EA) in EBV infection and can denote potential progression of Burkitt's lymphoma when antibodies to the early restricted antigen (EA-R) are found, or nasopharyngeal carcinoma when antibodies to the early diffuse antigen (EA-D) are present. Past infection by EBV is noted by the presence of EBNA antibodies (Fig. 13.1) that develop 6 months after the acute infection (see Sec. 8.2.3). Reactivation of CMV, EBV, or other viruses can sometimes be detected by the recurrence of IgM antibodies.

Finally, the recent development of the **immunoblot** (Western blot) technique permits the detection of antibodies to selected viral proteins (see Appendix 1). Purified virus is disrupted by detergent, and the proteins are placed on a gel to migrate according to size. These proteins are subsequently transferred to nitrocellulose filters, and serum is then incubated with the filters so that antibodies react with the separated viral proteins. The distribution of antibodies to these viral antigens is then read using a second labeled antibody that binds to

FIGURE 13.1

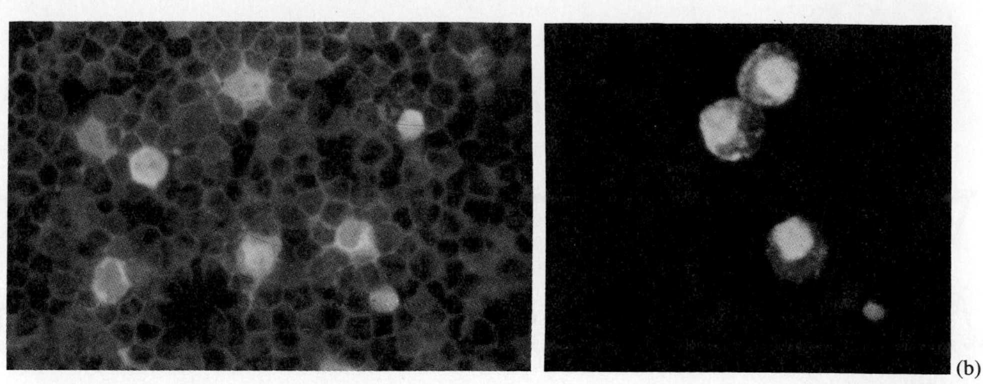

Immunofluorescent staining of Burkitt's lymphoma cells using antibody to Epstein-Barr virus (EBV). (a) Note the distribution of viral capsid antigen in histiocytes scattered through the B cells. (b) Immunofluorescent staining for EBV nuclear antigen (EBNA). Note the intense staining of the nucleus of these EBV-containing cells.

the host antibody. This procedure can also be helpful in prognosis. Thus, the fluctuation in antibodies to selected viral protein, such as the core p25 protein of HIV, may be predictive of disease. When antibodies to this HIV protein decrease, as observed by immunoblot procedures, individuals have been found to progress to a more severe HIV infection.

13.2 VIRAL VACCINES

A major use of the immune reaction is for the prevention and cure of virus diseases by **vaccination** or **immunization.** Active immunization represents the stimulation of the immune response by introducing an antigen (vaccination). In the case of viruses, this procedure can be used either with an "inactivated" virus or an attenuated or mild form of a "live" virus.

Inactivated vaccines are viruses that have lost their infectivity through damage to their nucleic acid by various agents such as formaldehyde or UV irradiation. These inactivating agents and treatments must be carefully selected and controlled so that they will not affect the antigenic specificity of the protein surface of the respective virus.

TABLE 13.1

Viral vaccines recommended for use in humans

Disease	*Vaccine Strain*	*Cell Substrate*	*Attenuation*	*Inactivation*	*Route*
Smallpox	Vaccinia	Calf or sheep skin	+	–	Intradermal
Yellow fever	17D	Chick embryo	+	–	Subcutaneous
Poliomyelitis[a]	Sabin 1, 2, 3	WI-38	+	–	Oral
Measles	Schwarz	Chick fibroblasts	+	–	Subcutaneous
Rubella	RA 27/3	WI-38	+	–	Subcutaneous
	Cendehill	Rabbit kidney	+	–	Subcutaneous
Mumps	Jeryl Lynn	Chick fibroblasts	+	–	Subcutaneous
Chicken pox	Varicella/zoster	Human diploid fibroblasts	+	–	Subcutaneous
Rabies	Pitman-Moore	WI-38	–	β-Propiolactone	Intramuscular
Influenza	A/HK/68 (H3N2)	Chick embryo	–	Formaldehyde and deoxycholate	Subcutaneous
Adenovirus[b]	4, 7	WI-38	–	Live, in enteric-coated capsules	Oral
Serum hepatitis[c]	Hepatitis B	Human plasma	–	–	Subcutaneous
		Yeast	–	–	Subcutaneous
Hepatitis[b]	Hepatitis A	Human diploid lung	–	Formaldehyde	Intramuscular

SOURCE: Modified from A. S. Evans, ed. (1982) *Viral infections of humans.* Plenum Press, New York. A great variety of different viral strains and cellular substrates are used in different countries; the selection listed is not comprehensive.

[a]Killed virus also used

[b]Live attenuated vaccine is under investigation

[c]Viral envelope protein obtained from patient's plasma or from yeast through a recombinant DNA process.

"Live" virus vaccines use infectious virus strains that through passage through atypical host animals or cultured cells have been allowed to mutate and have become "mild" or lack virulence. These **attenuated** strains (Sec. 11.7) are inoculated into the species for vaccination. Jenner's use of cowpox virus to provoke antibodies that would upon any future infection with the human smallpox virus cross-react to some extent with that virus exemplifies such a live virus vaccination. These experiments represented the beginning of both virology and immunology as sciences. In fact, the term vaccination was derived from the name for cowpox, vaccinia (Latin: *vacca,* cow).

Both modes of vaccination have surely saved millions of lives and turned serious and often epidemic diseases into minor discomforts. Tables 13.1 and 13.2 list commonly used human and veterinary vaccines of both the inactivated and attenuated type.

TABLE 13.2
Some veterinary viral vaccines

Disease	*Animal Viral Genus*	*Nature of Vaccine*
	Dog	
Canine hepatitis	*Adenovirus*	Live attenuated
Canine distemper	*Paramyxovirus*	Live attenuated
Rabies	*Rhabdovirus*	Live attenuated
	Cat	
Panleukopenia	*Parvovirus*	Formalin-inactivated, in adjuvant
Leukemia	*Retrovirus*	Viral protein
	Cattle	
Infectious bovine rhinotracheitis	*Herpesvirus*	Live attenuated
Foot-and-mouth disease	*Picornavirus*	Formalin-inactivated
Bovine virus diarrhea	*Flavivirus*	Live attenuated
Rinderpest	*Paramyxovirus*	Live attenuated
Vesicular stomatitis	*Rhabdovirus*	Live attenuated
	Sheep	
Sheep pox	*Poxvirus*	Live virulent (special site) or live attenuated
Scrabby mouth	*Poxvirus*	Live virulent (special site)
Louping ill	*Flavivirus*	Formalin-inactivated
Rift Valley fever	*Bunyamweravirus*	Live attenuated
Blue-tongue	*Orbivirus*	Live attenuated
	Horse	
Equine rhinopneumonitis	*Herpesvirus*	Live attenuated
Venezuelan equine encephalitis	*Alphavirus*	Live attenuated
Western equine encephalitis	*Alphavirus*	Live attenuated
Infectious arteritis	*Alphavirus*	Live attenuated
Equine influenza	*Orthomyxovirus*	Live attenuated
African horse sickness	*Orbivirus*	Live attenuated
	Pig	
African swine fever	*Poxvirus*	Live attenuated
Teschen disease	*Enterovirus*	Live attenuated
Swine fever (hog cholera)	*Flavivirus*	Live attenuated, with antiserum
Transmissible gastroenteritis	*Coronavirus*	Formalin-inactivated, in adjuvant
	Chicken	
Fowlpox	*Poxvirus*	Live attenuated
Marek's disease	*Herpesvirus*	Live attenuated
Infectious laryngotracheitis	*Herpesvirus*	Live attenuated
Fowl plague	*Orthomyxovirus*	Formalin-inactivated
Newcastle disease	*Paramyxovirus*	Live attenuated
Infectious bronchitis	*Coronavirus*	Live attenuated

SOURCE: A. S. Evans, ed. (1982) *Viral infections of humans.* Plenum Press, New York.

Other vaccines expected in the near future are directed against hepatitis C, hepatitis E, respiratory syncytial virus, parainfluenza virus, and rotavirus.

13.2.1 Problems with Vaccines

It must be noted, however, that each type of vaccine can present problems and dangers. Live vaccines may be too mild to elicit an immunological response, or they may retain sufficient infectivity to cause the disease. They may also contain other contaminating unattenuated viruses.

Inactivated vaccines have the danger that a minuscule proportion of virus particles may "survive" the chemical treatment and produce the disease that the vaccine is supposed to prevent. This problem caused the death of some children during the early days of mass vaccinations with the formaldehyde-treated polio virus, the Salk vaccine. Moreover, the monkey polyoma-type virus SV40 was at first present in the Salk polio virus vaccine, which was made by growing polio in monkey kidney cells. The monkey virus was resistant to the formalin procedures used to inactivate polio virus. Fortunately, SV40 is not oncogenic in humans, since millions of children were given this vaccine before this contamination was discovered and abolished.

Table 13.3 summarizes some of the problems encountered with the use of vaccines. Many of these situations may vanish when the new types of vaccines become available. They are produced with virion-surface proteins made from the respective cloned genes. Nevertheless, new problems may still arise when these vaccines come into widespread use.

One well-known complication of a vaccine involved respiratory syncytial virus. The formaldehyde-inactivated vaccine did not prevent spread of this virus, because the fusion protein of the virus was inactivated during the process. Thus, the host was not protected against cell-to-cell transmission. In fact, in several cases, a more severe respiratory disease resulted because of the immune response against the virus in cells in the lungs. This problem emphasizes the need to understand the basic events involved in virus infection in order to develop the most appropriate vaccines. It certainly is an important feature to recognize with vaccines in which virus-cell fusion (particularly by enveloped viruses) is involved.

TABLE 13.3
Some problems encountered during the development and use of viral vaccines

Vaccine	*Problem*
Live virus vaccines	
Poliovirus (Sabin)	Back-mutation to virulence (type 3)
	Interference by endemic enteroviruses
	Contaminating viruses (e.g., SV40)
Smallpox	Inadequate attenuation leading to complications
	Overattenuation leading to lack of protection
Measles	Inadequate attenuation leading to fever and rash
Rubella	Inadequate attenuation leading to arthritis in adult females
Yellow fever	Contaminating hepatitis B virus in "stabilizing" medium
Inactivated virus vaccines	
Poliovirus (Salk)	Residual live virulent virus
	Contaminating virus (SV40) resisting formaldehyde
Influenza	Induction of fever
Rabies	Allergic encephalomyelitis
Measles	Hypersensitivity reaction on subsequent natural
Respiratory syncytial	infection or live vaccine booster

SOURCE: Modified from A. S. Evans, ed. (1982) *Viral infections of humans.* Plenum Press, New York.

13.2.2 New Methods for Developing Viral Vaccines

In recent years modern molecular biological technologies have been considered for the development of viral vaccines. Four different approaches have been investigated: (1) production of viral proteins from cloned genes expressed in microbial cells; (2) insertion of cloned genes from viruses into the genomes of other micro-organisms such as salmonella, or vaccinia, baculoviruses, poliovirus, and adenoviruses; (3) chemical synthesis of peptides representing small segments of the virus proteins against which neutralizing antibodies react, and (4) development of **antiidiotype antibodies.**

Since antibodies that neutralize the infectivity of viruses generally recognize the surface proteins, emphasis has been placed on viral envelope proteins for use in vaccines. Genes for the external proteins are cloned through bacterial systems and then introduced through vectors, generally bacteriophage, into the genome of *E. coli.* Yeast has also been used successfully. These organisms can then produce large quantities of the selected viral protein. This approach gave rise to a vaccine against some of the picornavirus serotypes causing **foot-and-mouth disease,** as well as to the first human recombinant vaccine approved (in 1986)—the cloned hepatitis B virus envelope gene made in yeast. This latter vaccine is effective because the protein is expressed by yeast in the same form as it appears on the virion. The advantages of these cloned gene vaccines over those using live or "killed" viruses are the complete assurance of absence of any infectious virus and the lower cost of the process. For instance, recombinant strains in *E. coli* can produce proteins from cloned genes up to 10 percent of cellular dry weight. This amount corresponds to a yield of over a gram of pure protein per liter of culture.

A vaccine against the **fish rhabdovirus** infectious hepatitis necrosis virus (IHN) has also recently been developed using a genetically engineered protein produced in bacteria. This protein is administered to the fish in the water through the killed bacterial vector that penetrates the fish through the gills. It can produce immunity in young fishlings for their lifetime. Another means of vaccination in fish includes the spraying of killed virus or viral antigens at high pressure onto their skin, where the proteins can penetrate through the linings of the gills.

Nevertheless, there have been problems in producing viral proteins by this cloned gene methodology. In some cases, the resulting product is not in a form capable of inducing strong neutralizing antibodies. Through expression in bacterial systems some posttranslational modifications such as formation of disulfide bridges between cystene residues, specific cleavage of the peptide chain by proteases, and addition of phosphates, lipids, or sugars may fail to occur. Part of the problem with **glycosylation** has been countered through the use of yeast, but even this microorganism may not produce the exact envelope protein appearing on the virion. For this reason, Chinese hamster ovary and other mammalian monolayer cells have been used to receive the cloned gene, and procedures have been developed to enhance secretion of the viral protein. Finally, the use of purified envelope proteins may not be able to induce a sufficiently strong cell-mediated response against virus-infected cells.

The second approach, the use of **recombinant viral vectors,** involves integration of the viral protein into specific sites in the genome of a live virus, such as vaccinia, or recently poliovirus and varicella-zoster. During infection with the recombinant vector, surface proteins produced from the cloned genes become inserted into the cell membrane as if it were infected with the virus itself. In this manner the new viral protein can be recognized by the host immune system and stimulate antibodies that neutralize that virus itself. A vaccinia vector carrying the gene for the envelope protein of vesicular stomatitis virus (VSV) has already been shown to protect cattle from disease. Attention has focused on the use of vaccinia recombinants with the AIDS virus in which the antibody response to **HIV envelope protein** was elicited in primates as well as humans inoculated with this virus. In fact, in recent studies, vaccinia viruses carrying the HIV envelope or the core proteins have been cotransfected into cultured cells. Particles are then produced by the cells that consist of HIV cores carrying the viral envelope gp120 but devoid of the HIV genome. These particles may be of use as immunogens in vaccines. Finally, an oral hepatitis B vaccine may soon be available in which the envelope gene of the virus has been introduced into adenovirus.

The use of a vaccinia vaccine offers these advantages:

1. Production costs are relatively low, and the procedures are simple, since the virus can be produced efficiently in chicken embryos available worldwide.
2. The virus can be lyophilized and remains relatively stable at room temperature.
3. The vaccine can be administered quickly and inexpensively to large populations by a scratch of the skin and even faster by jet gun.
4. Generally, a single dose is sufficient to produce long-term immunity and eliminate the need for booster injections.
5. A minor scar is left by the vaccination, providing proof of the procedure, so that reliable medical record keeping is not required.
6. The vaccination procedure requires very little material and is superficial, so that the likelihood of dangerous side effects is reduced.

Nevertheless, the use of any vaccinia-based immunization procedure is not completely safe. Vaccination against smallpox has produced lethal side effects in a small percentage of recipients, especially in those **immunocompromised.** Some individuals with HIV infections have, after a standard smallpox vaccination, shown severe side effects, particularly involving the skin. Finally, it is uncertain whether the introduction of these new viral agents into a population will not bring hazards that were not easily detected in small-scale trials. Thus, we would expect to see the vaccinia vectors applied first to commercially important animal diseases before being used in humans.

The third approach is a synthetic peptide vaccine. It is sometimes possible to raise antibodies against an intact virus particle using small peptides (e.g., 15 to 20 amino acids) synthesized in the laboratory from selected sequences of the viral surface protein. This procedure offers the ultimate in safety, specificity, and economy, since the chemical synthesis of the vaccine product provides a better opportunity for quality control than other production methods. Not only is it impossible for the vaccine to contain any infectious agents, but also the chance of contamination with toxic materials found in cells is eliminated. This third approach, however, poses the critical issue of whether or not small peptides can induce **long-lasting immunity** by vaccinations that are acceptable for human or even animal applications. Generally, peptides of less than 20 amino acids are poorly immunogenic by themselves and thus must be given with either a larger protein or a powerful adjuvant. Freund's adjuvant, a mixture of bacterial extracts, is commonly used to enhance immune responses to antigens in experimental animals. This material, however, is toxic and is not acceptable in human and several animal recipients. Thus, many companies are devising new adjuvants that can be used in humans. Some consist of lipid bases, such as muramyl dipeptide.

The fourth technique being developed for viral vaccines completely avoids the use of any viral material. By this procedure, neutralizing antibodies against the virus are used as the immunogen in an animal host. Antibodies induced against this antibody (**antiidiotype antibodies**) should resemble the epitope (e.g., binding site) on the virus that is susceptible to the neutralizing antibody. This antiidiotype, therefore, when used to immunize another host, should elicit neutralizing antibodies against itself and by analogy against the infectious virus itself. This approach has been applied successfully in an experimental hepatitis B vaccine and is one approach being considered for development of an HIV vaccine.

13.2.3 Post-infection Immunization

A recent antiviral procedure that has received attention is immunizing individuals already infected with a virus with the viral proteins in attempts to increase the immune response. This approach has been tried in HIV infection with the envelope protein. No sign of toxicity has occurred, but the effect on the clinical condition is not yet known. Postinfection immunization has been used with rabies. In this case, there is a window period of about 1 week after the initial infection when vaccination can help prevent or reduce the pathogenic course. Vaccination with a live attenuated varicella virus is also being examined as an approach at preventing outbreaks of herpes zoster in previously infected elderly individuals. Enhanced cellular immune responses have been observed. These two examples, however, involve a known effective vaccine. It is still not clear whether such immunization procedures would be effective in curing or modulating infection by other viruses.

13.2.4 Ideal Properties of a Vaccine

Vaccination programs must consider features of the virus that can affect the success of a campaign (see Table 13.4). For HIV, for example, the infected cell is a major source of transmission, and it can remain latent and express very few viral antigens. Thus, it may not be recognized even if an appropriate immune response has been elicited against the virus. Moreover, lentiviruses have the capacity of changing their envelope protein frequently and thereby "escaping" an immune response. Viruses such as HIV that can spread through genital contact also pose the challenge of finding approaches for inducing an immune response at the local sites for infection, such as the anal canal and the vagina. Finally, some viruses may share regions of their envelope with that of normal proteins. This **molecular mimicry,** noted in Section 12.3, can induce **autoantibodies** following the vaccination. This response can lead to more detrimental effects in the immunized individuals.

In general, the major immune response that prevents infection by a free infectious virus is neutralizing antibodies. If the virus is able to establish infection (one infectious cycle), then cellular immune responses become the important mechanism for its control. Virus-infected cells must be recognized and either destroyed or prevented from releasing virus that can spread through the body. Some virus infections, such as dengue and HIV, induce antibodies that attach to the virus but do not inactive it. These antibodies often enhance infection by bringing the virus-antibody complex into cells through the Fc portion of the antibody. This antibody region interacts with Fc receptors on the surface of macrophages and other cells. In some studies the antibody-antigen complex binds complement and can enter cells via the complement receptor. In dengue virus, immunization or infection by one virus subtype may produce antibodies that neutralize that subtype but enhance infection and spread by other subtypes that are not inactivated after binding to the antibody. A similar possibility must be considered in the development of HIV vaccines. Therefore, this **antibody-dependent enhancement** (ADE) represents another potential feature of vaccine programs that should be recognized and prevented. As an example, the ideal properties expected of an anti-HIV vaccine are listed in Table 13.5.

TABLE 13.4
Features of HIV infection that affect vaccine development

- The infected cell is a major source of virus transmission.
- Virus infection involves integration of the viral genome into the chromosome of the infected cell. This cell becomes a reservoir for persistent virus production.
- The infected cell can transfer the virus by cell-to-cell contact.
- The infected cells can remain "latent" and express very few viral proteins.
- Several independent serotypes and subtypes of HIV can be identified.
- HIV infection occurs at specific sites in the host (e.g., the rectum, the vagina).
- Portions of HIV proteins resemble normal cellular proteins. Could induce autoimmune response.

TABLE 13.5
Ideal properties of an anti-HIV vaccine

- Elicits neutralizing antibodies that react with all HIV strains and subtypes.
- Induces immune cytotoxic responses against virus-infected cells.
- Induces immune responses that recognize latently infected cells.
- Does not induce antibodies that enhance HIV infection.
- Induces local immune response at all sites of HIV entry in the host.
- Is safe, showing no toxic effects.
- The effect of the vaccine is long-lasting.

13.3 OTHER ANTIVIRAL APPROACHES

There are two principal approaches to the prevention and abolishment of virus diseases: vaccination that makes everybody immune, or eradication of the virus or at times the insect vector that transmits it. We have mentioned some of the problems with vaccination. These include the fact that many diseases, such as the common cold, are caused by a great variety of viruses and their strains and that many viruses, such as influenza, change frequently through antigenic drift. These situations make vaccination programs against certain virus diseases very difficult and of dubious success. One would therefore think that all efforts should be directed toward eradication of pathogenic viruses. The technique to achieve this objective is presently only by means of "starving the viruses to death" through a worldwide obligatory vaccination program. This approach is reported to have been successfully achieved for the smallpox virus. In contrast, the building of the Panama Canal was possible only after strict measures were taken to get rid of *Aëdes aegypti,* the mosquito that transmits yellow fever. However, complete eradication of such a vector has never been, and probably can never be, achieved. Nevertheless, through natural attrition, certain common viruses seem to have occasionally disappeared in isolated locations, such as on rarely visited islands.

Measles virus, for instance, must find a new host every two weeks, since from acute infection to its control by an individual takes 14 days. Thus, unless measles virus is found in a community with sufficient susceptible individuals (e.g., high birth rates), it will gradually be lost. Other viruses, such as the herpes- and papovaviruses, that can remain latent in the body can be maintained in a population for many years. For this reason, certain island communities show no sign of immune response to measles but have continual presence of antibodies to the DNA viruses.

The question is; Does the disappearance of virus X at site Y signal the end of the war or only that a battle has been won? And this question cannot be unambiguously answered. Such a population freed from virus X is also free from any antibodies to virus X (and this may soon be true for smallpox virus). Any small mistake, any aboriginal coming to town from the inaccessible jungle, any monkey with monkey pox that has undergone a mutation may infect a *Homo sapiens,* a human, for example, in the city. Then not only is this individual and all friends and relations extremely susceptible to that virus, but the doctor who has never seen this disease may not diagnose it. Such a serious epidemic could travel around the world by jet. This is not science fiction, but has taken place to a limited extent. The onset of AIDS appears to be a recent example.

We come to the conclusion that eradication of pathogenic viruses, be it through vector control or through prophylactic vaccination of almost all 5 billion people on earth, even if it were possible, might not offer the best solution.

13.3.1 Antiviral Drugs

What other means exist for the prevention and cure of viral diseases? The answer is not very optimistic. Although a detailed discussion of virus therapy is beyond the scope of this book, we will briefly mention some of the measures that can be taken against some specific viruses. As evident from earlier discussions, most virus diseases are fortunately self-limiting through the host's immune system (active immunity). Newborn animals that have not yet developed the capability of producing antibodies are usually protected by congenital maternal antibodies, and injection of antiviral antibodies is sometimes used to prevent or treat virus diseases in persons with known recent exposure such as after a bite from a known or suspected rabid animal (passive immunity). Such immunization represents a means of assisting the natural process. So could interferon (see Sec. 12.4) be helpful as it becomes available for widespread use. Its use in nasal spray form for rhinoviruses has already shown promise in clinical trials.

Bacterial diseases have become largely curable through the use of **antibiotics** and **chemotherapy.** Are these methods applicable to viruses? The antibiotics that are used against bacterial infections attack critical metabolic processes, and since viruses only utilize the host cell's metabolism, this type of drug is generally not applicable. But certain chemicals seem to inhibit preferentially viral, as contrasted with host, replication processes, and these have found limited use in controlling some virus diseases (see Table 13.6). Since certain DNA viruses cause unusually high rates of DNA synthesis in the infected cell, DNA synthesis–inhibiting chemicals may have antiviral activity. These products are usually analogues of deoxyribonucleosides, such as 5-iododeoxyuridine, and bases attached to arabinose instead of deoxyribose, such as ara-A (adenine arabinoside). The same types of agents, for the same reason, are also anticancer drugs (Table 13.6).

Acyclovir, [9-(2-hydroxyethoxy) methyl guanine], a guanosine analogue, appears to be specifically effective against alphaherpesviruses. This is because the inhibitory activity of this compound toward the herpes-coded thymidine kinase and DNA polymerase is much stronger than toward the corresponding host cell enzymes. Because of its low toxicity, acyclovir has replaced ara-A in use. Compounds that stimulate interferon production, such as double-stranded polynucleotides (polyI/polyC) are also used therapeutically. **Azidothymidine** (AZT) has helped in controlling the spread of the AIDS virus, HIV, by interfering with the reverse transcriptase enzyme. However, the side effects of the drug can limit its usefulness. Moreover, one of the difficulties in using antiviral drugs is the development of resistance. Acyclovir resistance results in mutations in either the viral thymidine kinase or DNA polymerase genes. AZT-resistant viruses have mutations in the reverse transcriptase gene.

Gancyclovir [9-(1,3-dihydroxy-2-propoxy) methyl guanine], another guanosine analogue, resembles acyclovir and is active against cytomegalovirus as well as HSV. It acts against the viral DNA polymerase and resistance to gancyclovir occurs through mutations in the DNA polymerase gene. Another guanosine analogue, **ribavirin,** acts against several viruses including influenza virus, respiratory syncytial virus, parainfluenza virus, Lassa fever virus, and in some studies, HIV. It has recently shown promise with HCV infection. Ribavirin apparently inhibits several steps in virus replication, particularly viral mRNA translation.

Amantidine has a cagelike structure and is primarily a symmetrical amine. It prevents influenza A virus uncoating and release of the viral RNA into the cytoplasm. Rimantadine is a related compound with a similar activity. Resistance to these drugs develops from mutations in the matrix protein.

Finally, **foscarnet** (phosphonoformate) is a pyrophosphate analogue that acts against herpesviruses, hepatitis B virus, and HIV. It is a noncompetitive inhibitor of viral DNA polymerases and of the reverse transcriptase of HIV. It has been used to treat infections caused by acyclovir- or gancyclovir-resistant viruses and has been particularly helpful in CMV retinitis. Resistance develops from mutations in the viral DNA polymerase gene.

TABLE 13.6
Antiviral therapies

Agent	*Clinical Use*
Interferon	Hepatitis B, Hepatitis C, rhinovirus infections
Amantadine	Prophylaxis and treatment of influenza A virus infection
Acyclovir	HSV and VZV infections
Gancyclovir (DHPG)	CMV infection
Ribavirin	Respiratory syncytial virus infection
Foscarnet	Treatment of infection by acyclovir-resistant HSV and VZV, and gancyclovir-resistant CMV

It appears that natural immunization remains the most useful tool in combatting virus disease in humans. In the case of acute epizootics (epidemics of animals) of domesticated animals, killing of the infected herd is usually resorted to as the most effective means of arresting the spread of the disease. We previously discussed this procedure for the control of African swine fever virus. This approach is obviously difficult with wild animals, and these continue to represent dangerous reservoirs of viruses that can and do become transmitted to domesticated animals and humans. Vector control can play an important role here.

An important weapon against the spread of all types of viruses across the seas used to be the public health agencies' quarantine regulations. However, with the present rapid movement of millions of people, this method of control is becoming rather ineffectual, at least as far as diseases carried by humans are concerned. International monitoring of new outbreaks of disease should be further emphasized.

SECTIONS 13.1 TO 13.3 FURTHER READING

Original papers

SABIN, A. B. (1957) Properties and behavior of orally administered attenuated poliovirus vaccine. *J.A.M.A.* 164:1216.

NATHANSON, N., and A. D. LANGMUIR. (1963) The Culter Incident, Poliomyelitis following formaldehyde-inactivated poliovirus vaccination in the United States during the spring of 1955. I. Background. *Am J. Hygiene* 78:16.

HORVATH, B. L., and F. FORNOSI. (1964) Excretion of SV40 virus after oral administration of contaminated polio vaccine. *Acta Microbiol. Acad. Sci. Hung.* 11:271.

HEDING, L. D., et al. (1976) Inactivation of tumor cell–associated feline leukemia virus: A model for producing an infectious virus tumor cell vaccine. *Cancer Res.* 36:1647.

OLSEN, R. G., et al. (1977) Abrogation of resistance to feline oncornavirus disease by immunization with killed leukemia virus. *Cancer Res.* 37:2082.

MEYER, H. M., Jr., et al. (1978) Control of measles and rubella through use of attenuated vaccines. *Am. J. Clin. Pathol.* 70:128.

SISSONS, J. G. P., and M. B. A. OLDSTONE. (1980) Killing of virus-infected cells: The role of antiviral antibody and complement in limiting virus infection. *J. Infect. Dis.* 142:442.

EDMAN, J. C., et al. (1981) Synthesis of hepatitis B surface and core antigens in *E. coli. Nature* 291:503.

KEW, O. M., et al. (1981) Multiple genetic changes can occur in the oral poliovaccines upon replication in humans. *J. Gen. Virol.* 56:337.

ENNIS, F. A., et al. (1982) Antibody and cytotoxic T lymphocyte responses of humans to live and inactivated influenza vaccines. *J. Gen. Virol.* 58:273.

DIETZSCHOLD, B., et al. (1983) Characterization of an antigenic determinant of the glycoprotein that correlates with pathogenicity of rabies virus. *Proc. Natl. Acad. Sci. USA* 80:70.

MIYANOHARA, A., et al. (1983) Expression of hepatitis B surface antigen gene in yeast. *Proc. Natl. Acad. Sci. USA* 80:1.

FERGUSON, M., et al. (1985) Induction by synthetic peptides of broadly reactive, type-specific neutralizing antibody to poliovirus type 3. *Virology* 143:505.

LOWE, R. S. (1987) Varicella-zoster virus as a live vector for the expression of foreign genes. *Proc. Natl. Acad. Sci. USA* 84:3896.

WANDELER, A., et al. (1988) Oral immunization of wild-type virus against rabies: Concept and first field experiments. *Rev. Inf. Dis.* 10:5649.

DE CLERCQ, E. (1989) New acquisitions in the development of anti-HIV agents. *Antiviral Res.* 12:1.

MANNING, D. S., and J. C. LEONG. (1990) Expression in *Escherichia coli* of the large genomic segment of infectious pancreatic necrosis virus. *Virology* 179:16.

OBERG, L. A., J. WIRKKULA, D. MOURICH, and J. C. LEONG. (1991) Bacterially expressed nucleoprotein of infectious hematopoietic necrosis virus augments protective immunity induced by the glycoprotein vaccine in fish. *J. Virol.* 65:4486.

YARCHOAN, R., et al. (1991) Anti-retroviral therapy of human immunodeficiency virus infection: Current strategies and challenges for the future. *Blood* 78:859.

ISSEL, C. J., et al. (1992) Efficacy of inactivated whole-virus and subunit vaccines in preventing infection and disease caused by equine infectious anemia virus. *J. Virol.* 66:3398.

LOPEZ, J. A., *et al.* (1992) Recombinant vaccine for canine parvovirus in dogs. *J. Virol.* 66:2748.

WERZBERGER, A., et al. (1992) A controlled trial of a formalin-inactivated hepatitis A vaccine in healthy children. *New Engl. J. Med.* 327:453.

JOHNSTON, M. I., and HOTH, D. F. (1993) Present status and future prospects for HIV therapies. *Science* 260:1286.

PRIEL, E., *et al.* (1993) Inhibition of retrovirus-induced disease in mice by camptothecin. *J. Virol.* 67:3624.

SALK, J., *et al.* (1993) A strategy for prophylactic vaccination against HIV. *Science* 260:1270.

Books and Reviews

FENNER, F. (1974) Prevention and treatment of viral diseases. In *The biology of animal viruses,* ed. F. Fenner, B. R. McAuslan, C. A. Mims, J. Sambrook, and D. O. White. 543–586 Academic Press, New York.

ARSON, R. (1980) Chemically defined antiviral vaccines. *Ann. Rev. Microbiol.* 34:593.

NATHANSON, N. (1982) Eradication of poliomyelitis in the United States. *Rev. Infect. Dis.* 4:940.

PROVOST, P. J., et al. (1982) Progress toward a live, attenuated human hepatitis A vaccine (41387). *Proc. Soc. Exp. Biol. Med.* 170:8.

KEATING, M. R. (1992) Antiviral agents. *May Clin. Proc.* 67:160.

LEONG, J. C., and J. L. FRYER. (1993) Viral vaccines for aquaculture. In *Annual review of fish diseases.* Vol. 3. 225. Pergamon Press, New York.

13.4 ANIMAL VIRUSES AND BIOTECHNOLOGY

As we noted in the discussion of vaccines, a variety of animal viruses have been harnessed for use in selected expression of either cellular or other viral genes. Vaccinia, polio-, retro-, and adenoviruses, as well as the *Ty* elements of yeast, have all been manipulated by molecular techniques so that cellular or viral genes can be inserted into their genomes and expressed in the infected cells. These approaches may eventually lead to development of viral vaccines free from the dangers associated with the presence of a complete (and therefore potentially infectious) viral genome—a very important consideration in the development of an HIV vaccine.

One common approach for inserting cellular genes into cells has been the use of packaging mutants of an amphotropic or polytropic murine retrovirus (see Sec. 6.1). Such viruses have the envelope proteins needed to infect a wide variety of different hosts (particularly human) and thereby to bring the foreign gene into the host cell. After integration into the cellular chromosome, the desired protein is expressed via its associated promoter, but the virus vector cannot reproduce itself. Thus, unable to repackage itself as an infectious virion, such murine retrovirus mutants provide a useful means to bring about **somatic mutation.** This process has recently been used to introduce the adenosine deaminase (ADA) enzyme into the bone marrow of infants suffering from immunological disorders resulting from ADA deficiency. It can also be utilized to transfer the multidrug resistance gene (*P*-glycoprotein) into the bone marrow of cancer patients scheduled for chemotherapy. Other possible targets include the normal genes needed to compensate for a variety of genetically inherited diseases. One potential drawback of this strategy is the possibility that virus integration may disrupt normal gene expression and lead to malignant transformation. This possibility is currently under careful scrutiny.

To illustrate some of the many additional applications of animal virus vectors in biotechnology, let us briefly consider the **baculovirus gene expression system.** As described in Section 8.5, the large, double-stranded, covalently closed circular DNA genomes of insect baculoviruses vary in size from 88 to 200 kb. Virus gene expression is divided into four phases (immediate-early, delayed-early, late, and very late), and in general, expression levels rise with each succeeding phase. Transcription of the very late genes occurs while the viral occlusion bodies are being assembled within the nucleus of the infected cells, and very late genes such as those encoding **polyhedrin** or **p10 protein** can be deleted without affecting virion production. Because the polyhedrin and p10 gene promoters are extremely efficient and the resulting proteins may represent as much as 50 percent of the total cell protein late in infection, most efforts to develop baculovirus expression vectors have focused on these two genes.

The choice of an appropriate gene expression system—prokaryotic, yeast, mammalian, or baculovirus—depends on both the nature of the recombinant protein to be produced and its final use. Advantages of the baculovirus system include the following:

1. **Dispensable virus gene products** Synthesis of either polyhedrin or p10 protein is *not* required for production of infectious recombinant virus.
2. **Strong gene promoters** The strength of polyhedrin and p10 gene promoters ensures the synthesis of large amounts of the foreign protein.
3. **Timing of gene expression** Expression of the foreign protein very late in infection (i.e., after the maturation of budded, infectious virus) allows synthesis of even cytotoxic proteins without interfering with viral replication.
4. **Ability to accommodate large amounts of foreign DNA** The viral nucleocapsid simply elongates to accommodate the presence of a larger genome.
5. **Posttranslational processing** Replication in eukaryotic cells ensures faithful processing of foreign gene products. Differences in glycosylation patterns have been reported, however.
6. **Safety** Because baculoviruses replicate only in invertebrates (i.e., insects and a few crustaceans), there is no known risk to the user.
7. **Easy scale-up** Large volumes of insect cell cultures can be produced in fermenter systems.

Disadvantages of the baculovirus expression system include the discontinuous nature of foreign gene expression (i.e., virus infection results in death of the host cell) and sometimes important differences in protein glycosylation. On balance, however, the advantages of a baculovirus system far outweigh its disadvantages. The most widely used virus-cell line combination is AcMNPV *(Autographica californica* multiple nuclear polyhedrosis virus) originally isolated from the alfalfa looper, and the Sf9 insect cell line derived from the Fall army worm *(Spodoptera frugiperda).*

The large size of the baculovirus genome prevents its direct manipulation in a manner similar to that used with bacterial or yeast vectors. Instead, foreign DNA sequences are cloned in a transfer vector containing appropriate baculovirus promoter and transcription termination signals and then transferred to the virus genome by **homologous recombination** *in vivo.* Recombination between a circular baculovirus genomic DNA and the transfer vector is a comparatively rare event following cotransfection, but its frequency can be considerably enhanced by linearizing the baculovirus DNA. In the absence of transfer vector DNA, the linear baculovirus DNA is noninfectious—which favors the isolation of the desired virus recombinant.

Studies with recombinant baculoviruses have shown that the insect cell is able to carry out many of the processing events—glycosylation, phosphorylation, fatty acid acylation, amidation, proteolytic processing, and cellular targeting/secretion—required for the formation of biologically active, heterologous proteins. The latter is especially important for large-scale production of certain foreign proteins. Yields have ranged from less than 1 mg to approximately 500 mg/L of cell culture, with yields of membrane-associated glycoproteins tending toward the lower end of this scale.

One of the most useful features of the baculovirus expression system is its ability to support **tertiary and quaternary structure formation** by the recombinant protein(s). By expressing two or more proteins simultaneously, it is possible to study a variety of protein-protein interactions in such a system. For example, expression of the entire coding region of poliovirus type 3 is accompanied by the assembly of noninfectious, intact virions containing VP0, VP1, and VP3. Expression of the capsid coding region alone, however, leads to synthesis of only the uncleaved precursor, thereby demonstrating the involvement of a virus-encoded protease in the processing pathway (see Sec. 2.2.2).

The ability to specifically engineer recombinant baculoviruses has also led to an increased interest in the use of baculoviruses as **insecticides.** Advantages of such a control strategy include the **specificity** of most baculoviruses for one (or a very few) insect species, their **virulence** for that host, and their proven **safety** (lack of harmful effects on nonhosts, including humans). Although it has sometimes been possible to achieve an economically significant degree of insect control by simply extracting viruses from naturally infected insects and spraying the viruses back onto the crops, such an approach requires a high-titered virus and has not been widely used.

Because baculoviruses do not kill their hosts as quickly as chemical sprays do, their use may eventually prove to be most suitable for **integrated pest management** schemes in which the long-term population control achievable through the use of baculoviruses is supplemented by the use of either microbial (e.g., *Bacillus thurengensis*) or chemical insecticides as required. Modification of baculoviruses to express insecticidal proteins such as the *B. thurengensis* delta endotoxin is also being examined as a possible means of improving their efficacy.

SECTION 13.4 FURTHER READING

Original Papers

URAKAWA, R., et al. (1989) Synthesis of immunogenic, but non-infectious, poliovirus particles in insect cells by a baculovirus expression vector. *J. Gen. Virol.* 70:1453.

KITTS, P. A., M. D. AYRES, and R. D. POSSEE. (1990) Linearization of baculovirus DNA enhances the recovery of recombinant virus expression vectors. *Nucl. Acids Res.* 18:5667.

LASALLE, G. L., et al. (1993) An adenovirus vector for gene transfer into neurons and glia in the brain. *Science* 259:988.

Books and Reviews

PODGWAITE, J. D. (1985) Strategies for field use of baculoviruses. In *Viral insecticides for biological control,* ed. K. Maramorosch and K. E. Sherman. 775–797. Academic Press, New York.

BLAESE, R. M., and CULVER, K. W. (1992) Gene therapy for primary immunodeficiency disease. *Immunodeficiency Rev.* 3:329.

KING, L. A., and R. D. POSSEE. (1992) *The baculovirus expression system: A laboratory guide.* Chapman ′ Hall, New York.

13.5 PLANT VIRUSES AND BIOTECHNOLOGY

Plant viruses are generally considered to be second only to fungi in terms of the disease losses they cause. Virus-related losses vary greatly—from region to region, from crop to crop, and, particularly, from year to year. Accurate figures for virus-related losses are not available, but estimates of annual worldwide losses due to all plant disease (i.e., fungal, viral, and bacterial) are as high as $60 billion (U.S.). After a brief discussion of the circumstances under which virus diseases may become economically significant, we consider several strategies that have been used to control virus diseases in crop species. Our discussion ends with a more detailed consideration of the newest of these disease control strategies, the creation of **disease resistance via genetic engineering.**

As is true for many animal viruses, a number of serious plant virus disease problems are the direct or indirect result of human activities. Such activities include: (1) introduction of viruses or virus vectors into new areas, (2) introduction of susceptible hosts into areas where the virus is endemic, (3) large-scale replacement of traditional genetically heterogeneous crop varieties with genetically uniform "improved" varieties, (4) repeated replanting with the same crop, and (5) modified cultural practices. Because no truly effective (i.e., curative) chemotherapy protocols are currently available, conventional control strategies attempt to prevent either **initial infection** or **subsequent spread** of the virus within the susceptible population. Unlike the strategies employed against human diseases, those used to control plant viruses emphasize the health of the population over that of the individual. The disease status of an individual plant rarely has any particular significance.

Losses due to plant virus infection may become important in 1) **perennial crops,** where there is a considerable investment in time and land (e.g., fruit trees); 2) **annual crops** grown from seed, where the disease incidence may vary widely (e.g., vegetables or grains); and 3) **vegetatively propagated crops,** where infection by mild strains of the virus is widespread, and yields may be only slightly reduced. As discussed in Section 12.9, the effects of virus infection on plant growth and yield can be quite complex, but certain general trends are often observed. Infections occurring late in the growing season usually have a smaller effect on yield than those occurring earlier, and virus spread is often slower and less efficient among older plants. Furthermore, infection with two or more viruses often causes more than additive (i.e., **synergistic**) effects. A classic example of such a synergistic interaction involves potato viruses X, A, and Y. Depending on the virus strain, PVX alone may replicate without apparent effect on plant vigor. In the presence of potato virus A or Y, however, the titer of PVX increases, and severe symptoms (crinkling, rugosity, and necrosis) appear in the foliage. Crop quality as well as quantity may be affected.

Potential **disease-control strategies** include (1) conventional breeding for disease resistance, (2) control of virus vectors (e.g., insects, but also nematodes and fungi), (3) production of virus-free plants, and (4) *de novo* creation of disease resistance through genetic engineering or gene transfer techniques. When successful, these strategies block initial infection of the susceptible host, minimize subsequent virus spread from plant to plant, or prevent the development of systemic disease. Detailed consideration of the first three strategies lies beyond the scope of this text, but we do note that despite the current intense interest in engineering resistance to virus disease, conventional plant breeding remains the preferred approach for developing virus-resistant plants. Exposure of substantial populations of resistant plants to natural virus infection often leads to the appearance of more aggressive "resistance-breaking" virus strains, but resistance to bean common mosaic *Potyvirus* has not broken down after more than 45 years.

In a few cases, most notably with tobacco mosaic virus infections of tomatoes grown in greenhouses, or citrus tristeza virus infection of citrus, useful levels of disease control have been achieved by "vaccinating" the plants with carefully selected mild strains of a particular virus. Multiplication of the mild strain occurs with little or no damage to the host and, most importantly, acts to subsequently protect against subsequent infection by more severe virus strains. First described by H. H. McKinney in 1929, this **cross-protection** phenomenon can also be observed when plants are successively inoculated with mild and severe isolates of a satellite RNA or viroid. In fact, vaccination of tomato and pepper seedlings with a mild strain of cucumber mosaic virus plus a non-necrogenic strain of CMV satellite RNA is widely used in China to protect against catastrophic crop losses due to subsequent infection by CMV strains carrying necrotic satellite RNAs (see Sec. 9.2).

Despite its seemingly attractive features, this approach to disease control has not been widely adopted. Possible disadvantages of cross-protection strategies include (1) mutation of the dominant mild strain to a more severe strain in the field, (2) the occurrence of serious disease following the introduction of an unrelated virus (e.g., the synergism observed between potato viruses X plus Y or A discussed above), and (3) the possibility that the infected crop may act as a virus reservoir from which other, more sensitive crops may become infected.

13.5.1 Genetically Engineered Virus Resistance

Since the early to mid 1980s, use of the Ti plasmid of *Agrobacterium tumefaciens* to transfer foreign genes into plant cells has become a standard technique in many plant molecular biology laboratories. As a result, it is now possible to introduce almost any foreign gene into a variety of plants and study its expression. Among the many genes studied have been a number of plant virus genes, including the coat protein gene of TMV. Expression of these viral genes often leads to varying degrees of disease resistance. Before discussing several examples of such **genetically engineered resistance,** it is useful to briefly consider the background against which the early experiments were carried out.

FIGURE 13.2

Coat protein–mediated protection against virus infection. *N. benthamiana* plants were photographed 5–6 weeks after challenge inoculation with turnip mosaic virus. Left, transgenic plants expressing the coat protein of bean yellow mosaic virus (another potyvirus); middle, nontransformed plants; and right, uninoculated controls (nontransformed plants that were not challenged with turnip mosaic virus). [Courtesy of J. Hammond]

Even today, the molecular mechanism(s) responsible for **cross protection** is not completely understood. In the early 1980s it was known that the coat protein could not be the *only* viral gene product involved in cross protection, and there was considerable interest in determining more precisely what its role might be. Together with the ability to use *Agrobacterium* to transform a variety of host plants, the cloning of the TMV genome in 1982 and determination of its complete nucleotide sequence defined the coat protein gene necessary to carry out such experiments. Additional impetus for such experiments was provided by the possibility of deriving disease resistance genes from the pathogen's own genome; i.e., that "key gene products from the parasite, if present in a dysfunctional form, in excess, or at the wrong developmental stage, should disrupt the function of the parasite while having minimal effect on the host." This statement made by Sanford and Johnston (1985) has subsequently proven to be prophetic.

In 1986, transformed tobacco plants in which the TMV coat protein gene was under the transcriptional control of the cauliflower mosaic virus 35 S promoter were reported to be resistant to infection by TMV. This resistance could be partially overcome by inoculation with isolated TMV RNA rather than intact virions. And it was later shown that disease resistance requires synthesis of functional coat protein. Thus, TMV cDNAs that generate untranslatable mRNAs do not confer virus resistance. Similar results have now been reported for a number of plant viruses. Figures 13.2 and 13.3 provide an indication of the resistance levels that have been achieved.

In general, plants that express a viral coat protein are resistant to infection by both that virus and related viruses. Resistance is usually related to the extent of coat protein gene expression, and the site of inhibition seems to be an early event in virus replication—perhaps an interference with virus uncoating. Virus entry *per se* seems unlikely to be affected, because plant cells appear to lack receptors similar to those responsible for the recognition and entry of animal viruses. The levels of coat protein–mediated resistance now available are so promising that commercial introduction of resistant selections now awaits only regulatory approval and decisions about proprietary protection for engineered cultivars.

A variety of other strategies for engineering virus resistance—including expression of antisense RNAs, noncoat viral genes, and viral satellite RNAs—have also been examined, some with quite encouraging results. For example, transformation of tobacco with the gene encoding the putative 54 kDa replicase-related protein from the U1 strain of TMV (see Sec. 2.6.3 and Fig. 2.14) yielded plants that were virtually immune to infection by either TMV-U1 virions or isolated RNAs. Unlike coat protein–mediated resistance, howev-

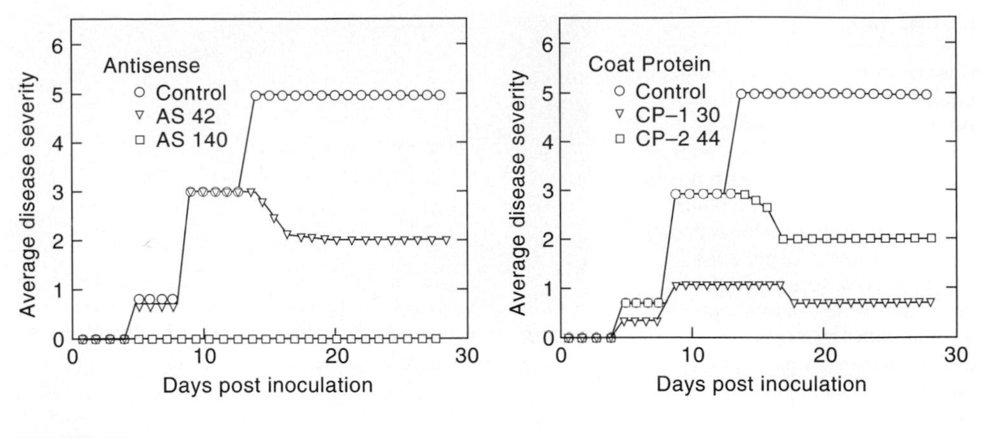

FIGURE 13.3

Severity of bean yellow mosaic virus (BYMV) infection in *Nicotiana benthamiana* plants expressing either BYMV antisense RNA or coat protein and challenged with 100 μg/ml BYMV. Antisense panel: o—o, nontransformed controls; ∇—∇, plants from line AS 42 were initially susceptible to infection but later recovered; and □—□, AS 140 plants did not support BYMV replication and appeared to be immune. Coat protein panel: o—o, nontransformed controls; ∇—∇, only three of 8 plants from line CP-1 30 were initially susceptible to virus infection, and these later recovered; and □—□, all plants from line CP-2 44 were initially susceptible but later recovered. After recovery from infection, the asymptomatic younger foliage of transgenic plants did not contain detectable levels of BYMV. [Courtesy of J. Hammond]

er, this resistance was *highly strain specific.* Furthermore, the role of viral protein is unclear. Although the resistant plants contain mRNA for the 54 kDa protein, efforts to detect the protein itself have not been successful.

With only a few exceptions, only low levels of resistance are generally observed in plants expressing various virus-complementary **antisense RNAs.** Such negative results may be due to the choice of heavily transcribed 5′ and 3′ untranslated regions and internal promoter binding sites as target sequences for these antisense RNAs. In the case of tobacco etch virus (a *Potyvirus*), untranslatable transcripts of the coat protein gene sequence have been shown to interfere with virus replication in both intact plants and isolated protoplasts. As RNA-RNA and RNA-protein interactions during virus replication become better understood, antisense resistance strategies (directed at either positive- or negative-sense viral RNAs) may become more widely applicable. Although not discussed in detail in this book, similar approaches have also been used to block the replication of animal viruses (e.g., HIV).

A variety of other antiviral strategies are also possible, including the construction of plants that express either **ribozymes** designed to cleave infecting viral RNAs or **virus-specific antibodies** able to block virus uncoating. One particularly interesting strategy targets the insect vectors that are often responsible for plant virus transmission in the field rather than the virus itself. In some cases, aphid transmission of certain plant viruses has been shown to depend on the synthesis of a virus-encoded **helper component** (for potyviruses) or **aphid acquisition factor** (for cauliflower mosaic virus). These proteins appear to be bifunctional, interacting with sites on both the virus surface and sites in the aphid stylet or anterior gut. Constitutive expression of modified proteins that are able to interact with only one site/domain could interfere with the transmission process.

13.5.2 Plant Viruses as Gene Expression Vectors

Several attempts have been made to use plant viruses as RNA-based vectors for **transient expression** of foreign genes in plant cells. Such vectors could have a number of uses—as a means of producing certain proteins or chemicals on a commercial scale or as tools for studying the molecular biology of the individual viruses themselves. Advantages of RNA viruses as vectors include (1) genomes that can be extensively modified without loss of infectivity, (2) an ability to infect and multiply in most host cell types, and (3) the high virus titers and levels of virus proteins produced. As much as 70 percent of the protein synthesis in TMV-infected leaves is viral coat protein.

In early experiments demonstrating the feasibility of this approach, the coat protein genes of brome mosaic and tobacco mosaic viruses were replaced by bacterial chloramphenicol acetyltransferase. Such free-RNA viruses are unable to systemically invade the host plant, and have only limited potential as expression vectors. In the case of TMV, this problem can be solved by *adding* the foreign gene to the viral genome rather than *substituting* it for one of the normal viral genes. If care is taken to avoid duplication of subgenomic mRNA promoter sequences, such constructs are stable. Thus the foreign gene is not removed by homologous RNA recombination. Nevertheless, the expression levels achieved thus far are much lower than those of TMV coat protein. Higher levels of gene expression may be obtained from genes present at the 5′ end of a replicating RNA, from a dependent (i.e., satellitelike) replicon associated with a helper virus, or in transgenic plants constitutively expressing one or more viral genes.

13.5.3 Use of Viruses to Control Plant Disease

The virtual disappearance of the **American chestnut tree** as the result of a fungal blight provides a classic example of a plant disease epidemic caused by the introduction of an exotic organism. Between 1905, when chestnut blight was first reported on trees growing within the New York Zoological Gardens, and the mid-1950s, this disease spread throughout the entire natural range of the American chestnut (from Maine to Alabama and westward to the Mississippi River) and destroyed several billion mature trees. Somewhat later, a similar epidemic of chestnut blight broke out near Genoa, Italy, and began to spread throughout Europe. As discussed below, the possibility of controlling chestnut blight via a viruslike agent that debilitates the fungus may eventually restore the American chestnut to its former prominence as a major source of timber, tannin, and food for both human and animal consumption.

The causative agent of chestnut blight is *Cryphonectria (Endothia) parasitica,* an ascomycete fungue apparently introduced into North America on nursery stock of oriental chestnut species that are naturally resistant to the disease. After entering through a wound in the bark, *C. parasitica* invades the vascular tissue, thereby limiting both nutrient movement through the phloem as well as the tree's ability to produce new growth. Although this "girdling" of the stem causes distal portions of the tree to wilt and die, the root system may survive for many years and continue to generate new shoots. Because these new shoots remain susceptible to chestnut blight, they rarely reach maturity, however. Where it was once the dominant forest species the American chestnut now survives as an understory shrub.

The first evidence that biological control of this disease might be possible came from the discovery of several chestnut trees growing near Genoa that had survived initial infection by the fungus. Rather than the deeply indented cankers normally found on dying trees, these trees showed only superficial cankers that appeared to be healing. In the early 1960s, French mycologists showed that these unusual symptoms were due to infection by altered forms of *C. parasitica* rather than to any inherent disease resistance on the part of the host. Later, a consistent correlation was noted between reduced fungal virulence (i.e., **hypovirulence**) and the presence of viruslike double-stranded RNAs. The hypovirulence phenotype can be transmitted to virulent strains via anastomosis (i.e., hyphal fusion followed by exchange of cytoplasmic material). In the case of the European (but *not* the American) chestnut, naturally occurring hypovirulent strains of *C. parasitica* have been successfully used to control chestnut blight. Within 40 years after its initial appearance, natural dissemination of these hypovirulent strains had eliminated chestnut blight as a threat to chestnut cultivation in Italy.

The vegetative incompatibility system that controls the ability of different strains to undergo anastomosis is the most important factor restricting the spread of hypovirulent strains of *C. parasitica* in North America. Compatibility is controlled by five to seven nuclear genes, and the number of different vegetative compatibility types is significantly higher in North American than in European populations of *C. parasitica*. Fortunately, results from recent molecular studies of the viruslike double-stranded RNAs associated with hypovirulence have suggested a possible way around this problem.

Hypovirulent strains of *C. parasitica* contain several species of dsRNA that are localized within pleomorphic vesicles, but only the largest species (L-dsRNA) remains constant in both size (12–13 kbp) and concentration upon repeated subculturing of the fungus. Cloning and sequence analysis of L-dsRNA isolated from two different hypovirulent strains has shown that this molecule contains two closely spaced ORFs, and phylogenetic analysis suggests that L-dsRNA and the plant potyviruses share a common ancestry. Consistent with such a relationship, **autoproteolysis** has been shown to play a prominent role in regulating the expression of both L-dsRNA ORFs. Transformation of virulent strains of *C. parasitica* with a full-length cDNA copy of L-dsRNA is followed by the appearance of a cytoplasmically replicating form of L-dsRNA and conversion to hypovirulence. Most importantly, the presence of an integrated cDNA copy of L-dsRNA bypasses the limitations imposed by vegetative incompatibility upon dissemination among wild population of *C. parasitica*—that is, L-dsRNA can now spread via sexual crossing as well as anastomosis.

The viruslike elements present in the hypovirulent strains of *C. parasitica* have been shown to alter a number of fungal processes, including host-pathogen interaction. Intensive efforts are now underway to engineer more effective hypovirulent strains of *C. parasitica*. It is conceivable that their release could lead to the restoration of the American chestnut to its former position of prominence in the forests of the eastern United States. The feasibility of such a strategy, however, is contingent on a thorough assessment of such issues as possible changes in host range of the engineered strains, their ecological fitness, the meiotic stability of the integrated cDNA, and the mitotic stability of the resurrected L-dsRNA. Virulence-modulating viruses and unencapsidated dsRNAs have been reported for a large number of other plant pathogenic fungi (see Sec. 5.5), and virus-mediated biocontrol strategies similar to that being developed for chestnut blight may also prove effective against these pathogens.

SECTION 13.5 FURTHER READING

Original Papers

French, R., M. Janda, and P. Ahlquist. (1986) Bacterial gene inserted in an engineered RNA virus: Efficient expression in monocotyledonous plant cells. *Science* 231:1294.

Powell-Abel, P., et al. (1986) Delay of disease development in transgenic plants that express the tobacco mosaic virus coat protein gene. *Science* 232:738.

Takamatsu, N., et al. (1987) Expression of bacterial chloramphenicol acetyltransferase in tobacco plants mediated by TMV RNA. *EMBO J.* 6:307.

Cuozzo, M., et al. (1988) Viral protection in transgenic tobacco plants expressing the cucumber mosaic virus coat protein or its antisense RNA. *Bio/Technology* 6:549.

Dawson, W. O., P. Bubrick, and G. L. Grantham. (1988) Modifications of the tobacco mosaic virus coat protein gene affecting replication, movement, and symptomatology. *Phytopathology* 78:783.

Ward, A., P. Etessami, and J. Stanley. (1988) Expression of a bacterial gene in plants mediated by infectious geminivirus DNA. *EMBO J.* 7:1583.

Powell, P., et al. (1989) Protection against tobacco mosaic virus in transgenic plants that express tobacco mosaic virus antisense RNA. *Proc. Natl. Acad. Sci. USA* 86:6949.

Golemboski, D. B., G. P. Lomonosoff, and M. Zaitlin. (1990) Plants transformed with a tobacco mosaic virus non-structural gene sequence are resistant to the virus. *Proc. Natl. Acad. Sci. USA* 87:6311.

DOLJA, V. V., H. J. MCBRIDE, and J. C. CARRINGTON. (1992) Tagging of plant potyvirus replication and movement by insertion of β-glucuronidase into the viral polyprotein. *Proc. Natl. Acad. Sci. USA* 89:10208.

LINDBO, J. A., and W. G. DOUGHERTY. (1992) Untranslatable transcripts of the tobacco etch virus coat protein gene sequence can interfere with tobacco etch virus replication in transgenic plants and protoplasts. *Virology* 189:725.

Books and Reviews

SANFORD, J. C., and S. A. JOHNSTON. (1985) The concept of parasite-derived resistance: Deriving resistance genes from the parasite's own genome. *J. Theor. Biol.* 113:395.

FULTON, R. W. (1986) Practices and precautions in the use of cross protection for plant virus disease control. *Ann. Rev. Phytopathol.* 24:67.

BEACHY, R. N., S. LOESCH-FRIES, and N. E. TUMER. (1990) Coat protein–mediated resistance against virus infection. *Ann. Rev. Phytopathol.* 28:451.

HULL, R. (1990) Non-conventional resistance to viruses in plants: Concepts and risks. In *Proceedings, 19th Stadler conference,* ed. G. P. Gustafson. 289–303. Plenum Press, New York.

MATTHEWS, R. E. F. (1991) *Plant virology,* 3d ed. Chap. 16. Academic Press, New York.

NUSS, D. L. (1992) Biological control of chestnut blight: An example of virus-mediated attenuation of fungal pathogenesis. *Microbiol. Rev.* 56:561.

KUMAGAI, M. H., et al. (1993) Rapid, high-level expression of biologically active α-trichosanthin in transfected plants by an RNA viral vector. *Proc. Natl. Acad. Sci. USA* 90:427.

13.6 BACTERIOPHAGES IN BIOTECHNOLOGY

We discussed in Chapter 12 how bacterial viruses that do not integrate into the host chromosome and remain as plasmids in the cytoplasm can be used to express selected foreign proteins. By this procedure, gene products can be amplified for a variety of purposes, not only to obtain viral vaccines but also to produce many of the enzymes used in biotechnology research. The process involves the selective use of endonucleases that can cut certain sequences of the plasmid, thereby allowing for insertion of a desired specific gene sequence. Transfection of this plasmid into bacteria leads to its autonomous replication and enhanced expression of the foreign gene product.

This procedure has recently been used to express antibody gene products on the surface of phage and also in bacteria. These methods may lead to a more rapid generation of monoclonal antibodies as well as a variety of other proteins. One problem that often arises with this approach is the inability to have the engineered protein produced and secreted by the bacterium. Thus, in most cases, the bacteria must be extracted to obtain the desired protein product(s). New vectors that may permit such expression are currently under investigation.

SECTION 13.6 FURTHER READING

Original Papers

PARMLEY, S. F., and G. P. SMITH. (1988) Antibody-selectable filamentous fd phage vectors: Affinity purification of target genes. *Gene* 73:305.

BURTON, D. R., et al. (1991) A large array of human monoclonal antibodies to type 1 human immunodeficiency virus from combinatorial libraries of asymptomatic seropositive patients. *Proc. Natl. Acad. Sci. USA* 88:10134.

Reviews

SCOTT, J. K. (1992) Discovering peptide ligands using epitope libraries. *Trends in Biochem. Sci.* 17:241.

chapter

14

Virologic Challenges of the 1990s: The Role of Viruses or Possible Viruslike Agents

Having discussed the various virus families that have been recognized in nature, we review in this final chapter five human diseases that are of major importance today. Two of them are associated with a human retrovirus (HIV, HTLV), and the others either with a not fully classified virus (hepatitis C) or with agents that have not yet been identified. Nevertheless, in the latter two cases the clinical course and the nature of the host response, as well as the pathology observed in the host tissues, suggest a virus or viruslike agent as the cause. The five clinical entities described are noteworthy challenges facing physicians and scientists in the next decade.

14.1 ACQUIRED IMMUNE DEFICIENCY SYNDROME (AIDS)

Acquired immune deficiency syndrome (AIDS) is a disease currently challenging many investigators and physicians throughout the world. First recognized as a distinct entity in 1981, cases were noted as early as 1978 in New York and Port-au-Prince, Haiti. Studies subsequently conducted have indicated that AIDS first occurred in individuals returning in 1976 from Central Africa, where the virus is thought to have emerged, perhaps as early as the 1950s. The origin of the AIDS virus is still unknown, although some researchers believe it was passed from monkeys to human some time in the recent past. This conclusion is based on a cross reactivity of proteins among the simian and certain human immunodeficiency viruses (see Fig. A.4, Appendix 1). It is also possible that the progenitor for animal lentiviruses (see Chapter 6) evolved with the species over thousands of years and gave rise to the large number of animals that are known to have their own lentiviruses, which all differ in genomic sequences: sheep, cattle, goats, horses, cats, monkeys, and perhaps dogs.

14.1.1 Epidemiology

The number of AIDS cases in the United States accumulated during the period 1981–1993, is over 200,000, and many more, perhaps 1 million, have occurred in Africa. By the year 2000, epidemiologists predict that 30 million to 100 million people will be infected world wide. In the United States about 1 million people are currently believed to be infected by the AIDS retrovirus, called the human immunodeficiency virus (HIV) (see also Chapter 6). Based on gender distribution of cases, it is estimated that 1 in 100 males and 1 in 800 females in this country are infected.

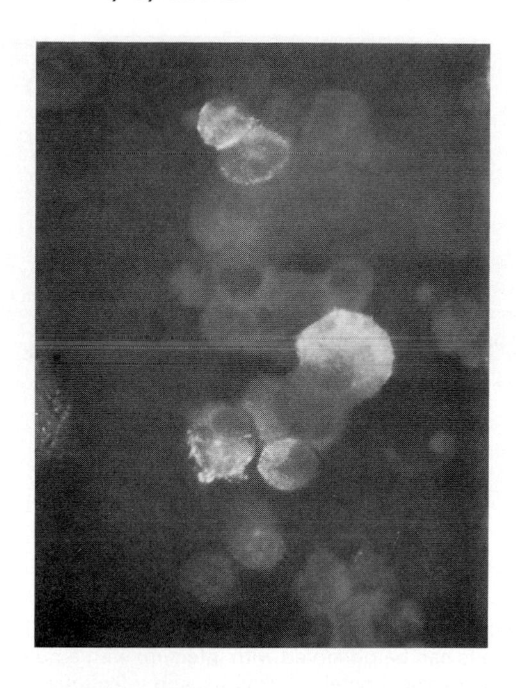

FIGURE 14.1

An HIV-infected T cell demonstrated by immunofluorescent staining (white) in contact with an uninfected cell. Virus can be transferred and neutralizing antibodies would have no effect. × 40. (Courtesy of Evelyne Lannette)

HIV is spread by blood and sexual contact as well as by transmission from mother to child, either before or during birth. There is no evidence for casual transmission. The most effective means of virus transfer is through virus-infected cells, which can pass HIV by cell-to-cell contact from lymphocytes to epithelial cells or by cell-to-cell fusion (Fig. 14.1). In some infected individuals, many HIV-containing white cells can be found in genital fluids (see Appendix 1, Fig. A.11) and can transfer the virus to mucosal lining cells in the anal or vaginal canals. It is this means of transfer that makes the control of HIV, and ultimately a vaccine, so difficult (see Chapter 13). HIV is the one disease, aside from malignancy, in which immune responses against the affected cell itself must be effective.

Two subtypes of HIV are now recognized: HIV-1 and HIV-2. HIV-2 was first found associated with AIDS in West Africa and is spreading through Africa and parts of Europe. Isolated infections have been found in the United States but are uncommon. Nevertheless, because the envelope proteins of HIV-1 and HIV-2 can differ, blood banks must now test for both subtypes, because the early ELISA assay for HIV-1 (Appendix 1, Fig. A.6) could miss some HIV-2 infections. Like HIV-1, HIV-2 can give rise to the same spectrum of diseases caused by immune destruction, but some researchers believe the infection course is more protracted and that this virus type is not transmitted so readily as HIV-1.

14.1.2 Clinical Findings

Clinical signs of acute (primary) HIV infection can be noted in about 40 percent of individuals. The findings include a red, slightly raised rash, lymph node swelling, low- or high-grade fevers, pains in the eyes, throat, and muscles, and general malaise. Some infected individuals develop gastrointestinal and neurological symptoms. In addition, in the second week, individuals with primary HIV infection can show many atypical lymphocytes (abnormal CD8+ cells) (Fig. 14.2) that often accompany other virus infections, such as EBV, CMV, and infectious hepatitis. The rash can distinguish AIDS from these other diseases. During this early period, the CD4/CD8 cell ratio (normally 1.5 to 2) decreases, possibly reflecting destruction of CD4+ lymphocytes by HIV as well as an increase in the number of CD8+ cells, a characteristic of many acute virus infections.

TABLE 14.1
Characteristics common to lentiviruses

Clinical

1. Association with a disease with a long incubation period
2. Association with immune suppression
3. Involvement of hematopoietic system
4. Involvement of the central nervous system
5. Association with arthritis and autoimmune disorders

Biological

1. Host species-specific
2. Exogenous and nononcogenic
3. Cytopathic effect in certain infected cells, e.g., syncytia (multinucleated cells)
4. Infection of macrophages—usually noncytopathic
5. Accumulation of unintegrated circular and linear forms of viral cDNA in infected cells
6. Latent or persistent infection in some infected cells
7. Morphology of virus particle by electron microscopy: cone-shaped nucleoid

Molecular

1. Large genome (≥ 9 kb).
2. Truncated *gag* gene; several processed *Gag* proteins
3. Highly glycosylated envelope gene
4. Polymorphic, particularly in the envelope region
5. Novel central open reading frame in the viral genome separating the *pol* and *env* regions

FIGURE 14.4

Structure of the human immunodeficiency virus. The outside envelope contains two glycoproteins: gp120, the surface (SU) protein of 120,000 Da, and gp41, the transmembrane (TM) protein. Just below the lipid bilayer is the matrix (MA) protein of 17,000 Da (p17). The nucleoprotein is made up of a capsid (CA) protein of 24–25,000 Da (p24, p25). Inside the core are various nucleocapsid proteins (p9, p7), as well as the polymerase enzyme containing reverse transcriptase (RT, p63), the protease (PR, p15), and the integrase (IN), p11. (Reprinted from Levy, *Microbiol. Rev.* 57:183, 1993)

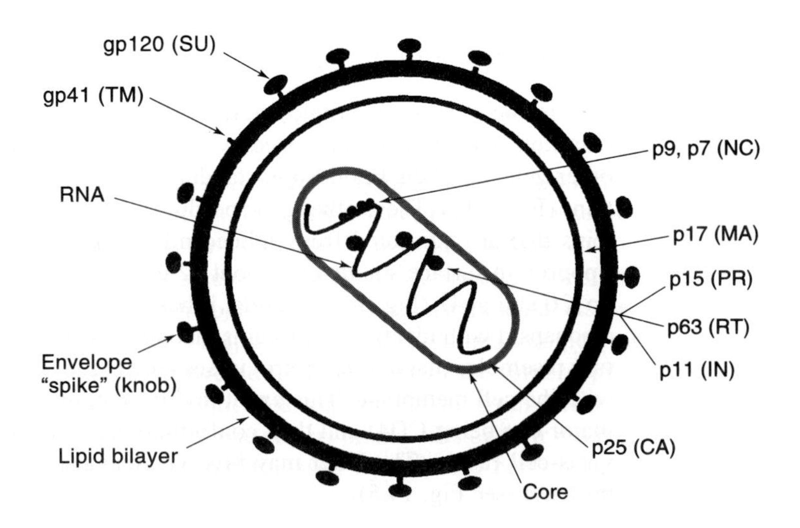

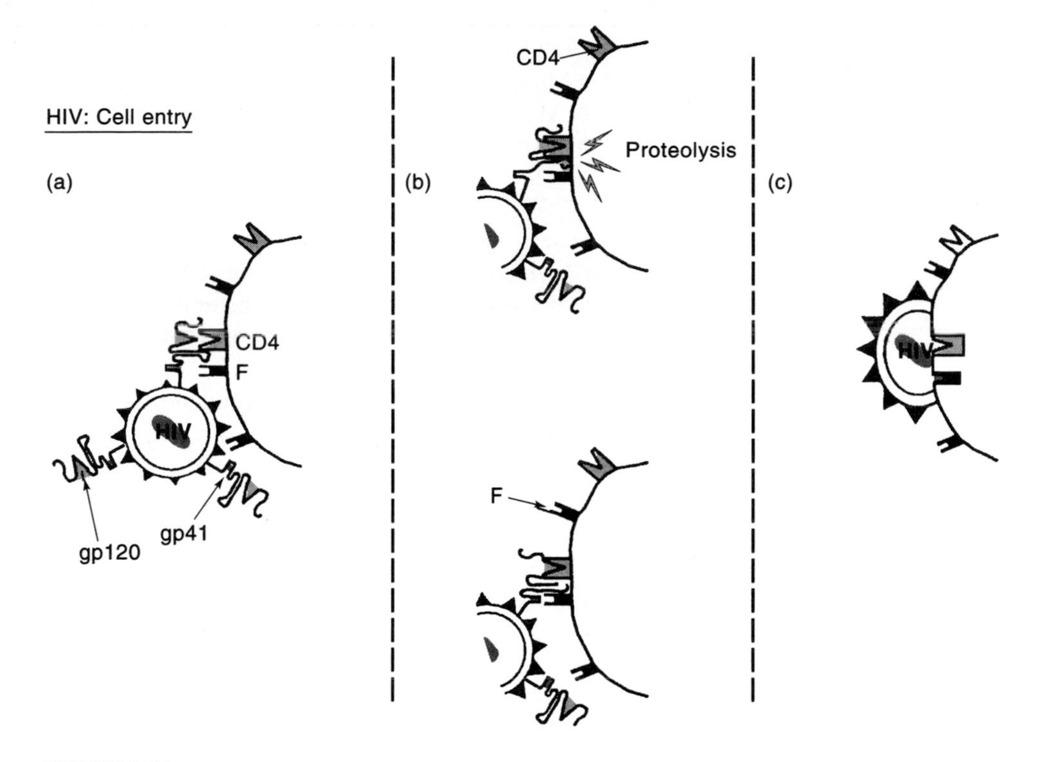

FIGURE 14.5

Proposed steps involved in HIV entry into cells. (a) HIV interaction with the cell surface. The viral surface envelope protein, gp120, interacts with a domain on the cell surface receptor, CD4. (b) The upper diagram shows that this virus: cell interaction causes a conformational change in gp120 (and perhaps CD4) resulting in potentially a proteolytic cleavage of gp120, most probably in the V3 loop. The lower diagram shows that this process results in a subsequent interaction of the external portion of gp41 (a fusion domain) with the proposed fusion receptor (F) on the cell surface. (c) HIV fuses with the cell. Subsequently entry of the nucleocapsid and virus infection takes place. (Reprinted from Levy, Microb. Rev. 57: 183, 1993)

HIV strains exist as a large number of variants with different biological and serologic features. These include the cells they infect and how well they replicate in them. Nearly every HIV strain isolated from an individual will have characteristics that are somewhat different from those of other viral isolates. This heterogeneity defines its challenge to antiviral therapies and vaccines (see Table 14.2). The various properties have been linked to the ability of the virus to destroy the immune system and to cause neurological disease as well as bowel disorders. Moreover, as the individual advances to disease, the virus often takes on more features associated with virulence in the host as defined by laboratory studies (see Table 14.3). These include the ability to infect many different cells, to grow rapidly and to high levels in these cells, and to readily kill CD4+ cells. These observations suggest that after initial infection the virus goes through various sequence changes secondary to the mutations caused by the error-prone nature of its

TABLE 14.2
Features of HIV heterogeneity

1. Cellular tropism
2. Replication kinetics
3. Level of virus production
4. Cytopathicity
5. Induction of latency
6. Sensitivity to neutralizing/enhancing antibodies
7. Genetic structure

TABLE 14.3
Characteristic of HIV strains associated with virulence in the host

1. Enhanced cellular host range
2. Rapid kinetics of replication
3. High titers of virus production
4. Disruption/alteration of cell membrane permeability
5. Efficient cell killing

reverse transcriptase (see Chapter 6). This enzyme is reported to cause up to 10 base changes during each replicative cycle. Many mutations are lethal, but eventually some strains are selected that "escape" neutralizing antibodies or cellular immune responses. They replicate rapidly to high titer and give rise to the immune abnormalities leading to the clinical conditions observed. The replication cycle resembles that of other retroviruses (Fig. 6.6). The steps involved are summarized in Figure 14.6.

During replication, the greatest changes in the virus appear to occur in the envelope and regulatory genes, *tat, rev* and *nef.* The Tat protein interacts with the viral LTR to up-regulate virus replication (see Chapter 6). The Rev protein prevents splicing of the viral mRNA so that large mRNA species can be transported into the cytoplasm to make the structural proteins needed for virion progeny production. The Nef protein appears to be pleiotropic. With some viruses, it suppresses virus replication, whereas with others, it can enhance replication in certain cells. The basic observation is that an interaction of these regulatory factors determines whether a virus can replicate to high levels or establish a latent infection in cells depending on the cell type infected. These findings with the regulatory proteins of AIDS virus can be relevant to normal cellular events, since an interaction of these viral genes with intracellular factors is involved. Similar activities may exist for cellular proteins.

FIGURE 14.6

Replication cycle of HIV. HIV attaches to a cell surface receptor (step 1) and after conformation, fuses with the cell membrane (step 2) and the nucleocapsid enters the cytoplasm of the cell. The RNA within the nucleocapsid is reverse transcribed, using the RNA-dependent DNA polymerase, into a DNA copy (cDNA) (dotted line) (step 3) and, using the same enzyme, a double-stranded DNA is formed (step 4). This non-covalently bound circular DNA enters the nucleus where it integrates into the cellular chromosome (step 5). From this proviral cDNA, both the viral RNA and messenger RNA (mRNA) are produced (steps 6 and 7). The mRNA codes for viral proteins that come to the cell surface, where the RNA is encapsulated in the core proteins (step 8). This core buds from the cell surface capturing the envelope glycoproteins (step 9). Infectious virus is then produced. (Reprinted from Levy *Microb. Rev. 57: 183,*1993)

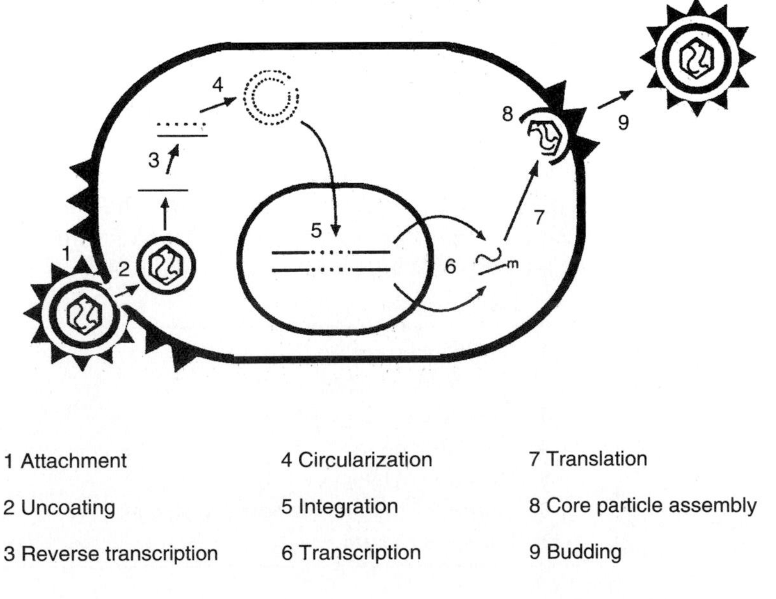

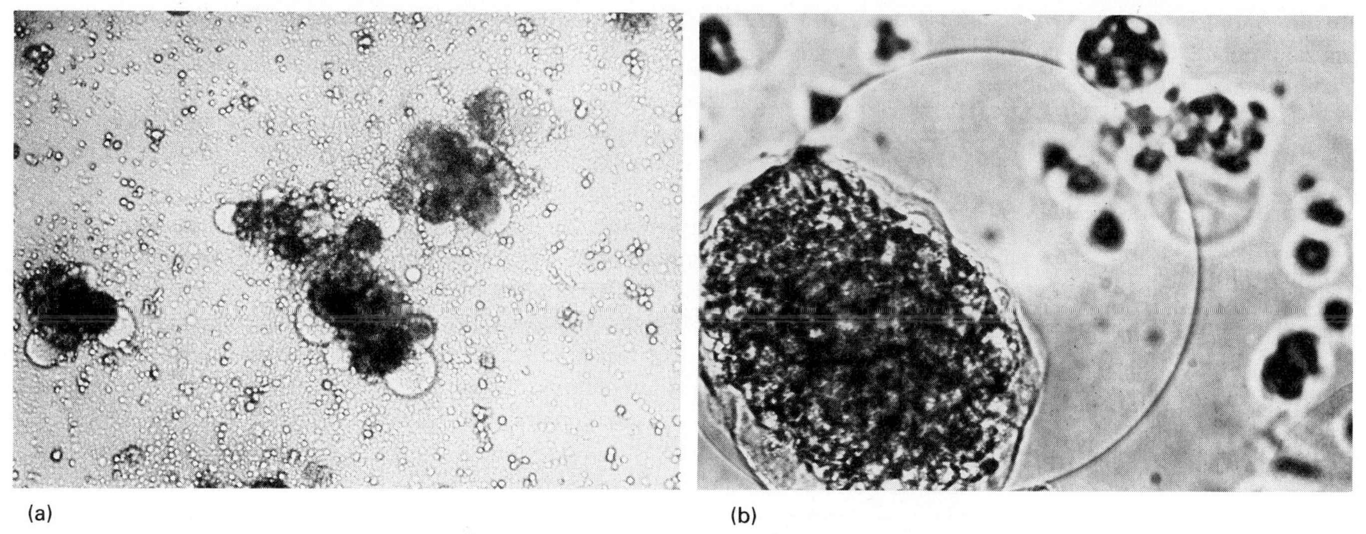

(a) (b)

FIGURE 14.7

Cytopathology induced in peripheral blood mononuclear cells by HIV. Note the (multinuclear cell formation syncytia) and balloon cell degeneration that most likely result from membrane changes induced by the virus. An influx of sodium, potassium, and calcium into the cell along with water causes the balloon cell degeneration and eventual cell death. Moreover, membrane changes induced by the viral envelope protein lead to the syncytia that are characteristic of this infection. (a) × 40, (b) × 250.

HIV preferentially replicates in CD4+ T-helper lymphocytes, but it can infect a wide variety of other hematopoietic cells, particularly macrophages, in the body. Upon acute infection of CD4+ cells in cell culture, the virus causes marked cytopathic effects consisting of balloon cell degeneration and syncytia (multinuclear cell) formation (Fig. 14.7). This cytopathic effect can be quantitated via the induction of plaques in a lawn of CD4+ T cells (Fig. 14.8). It also infects cells in the brain, gastrointestinal tract, lung, heart, and kidney. Moreover, as with other lentiviruses and retroviruses, HIV can establish a latent infection (see Sec. 11.9) that could be responsible for the long incubation period noted from initial infection to onset of disease.

FIGURE 14.8

Plaques formed in a human T lymphocyte (MT-4) cell by cytopathic strain of HIV-1. (A) Cytopathology induced by HIV-1_{SF33} in peripheral blood mononuclear cells, × 65. (B) HIV-1_{SF33}-induced foci of syncytia and cell death detected by staining as plaques in the MT-4 cell monolayer, × 40. (Courtesy of C. Cheng-Mayer. Reprinted from Levy *Microb. Rev.* 57: 183 1993)

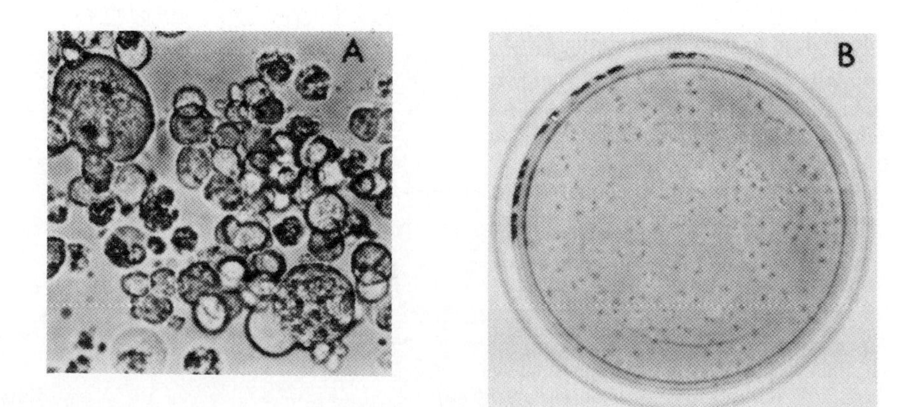

14.1.4 Immune Response Against HIV

Infected individuals show immune reactions against HIV as evidenced by both the production of anti-HIV antibodies and cell-mediated immune responses. The antibodies are usually detected within 1 to 4 weeks after primary infection, by ELISA, immunoblot, and immunofluorescent assays (see Chapter 13 and Appendix 1). Rarely, an infected person will take a year to seroconvert. In most individuals, these immunological reactions are not protective, since neutralizing antibodies for the virus within the host are low. The observation could reflect an "escape" from the immune response. In some cases, the antibodies bind to the virus and enable it to infect other cells via the Fc portion of the antibody. This antibody-dependent enhancement has been shown to involve both complement and Fc receptors on the surface of cells (Fig. 14.9). The enhancing antibodies are usually found in the blood of individuals as they progress to disease. They are a response to the virus that must be considered in any vaccination program against HIV.

The cellular immune response appears to be the most promising means for controlling HIV infection, but because the virus directly infects cells of the immune system, their ability to contain HIV can be limited. Moreover, with any agent that affects the immune system (as most viruses do) several abnormal immune responses can occur, such as production of autoantibodies (some against CD4+ cells) or emergence of cytotoxic cells that could attack CD4+ cells. These factors, in addition to direct virus infection, could be involved in the loss of CD4+ cells that is characteristic of AIDS (Table 14.4). Autoantibodies appear to be responsible for the decrease in red cells, neutrophils, and platelets observed in some infected individuals, because plasmapheresis (removal of antibodies) often reduces these clinical conditions in patients.

Despite the effect of HIV on the immune system, a great body of evidence indicates that CD8+ T lymphocytes can play a role in suppressing virus replication and preventing disease in some virus-infected individuals. In this sense, the response resembles the control by these lymphocytes of EBV infection in patients with acute infectious mononucleosis. The CD8+ cell response is demonstrated through direct killing of the virus-infected cells or through production of a cellular factor (cytokine) that suppresses HIV replication (Fig. 14.10).

Long-term survivors have four conditions that appear to mirror these observations in the laboratory: They are infected with a virus that is not highly cytopathic, they do not show high levels of virus in their blood, they do not have enhancing antibodies, and they have strong CD8+ cell antiviral activity. The latter characteristic probably is responsible for the low virus levels and the lack of emergence of virulent strains.

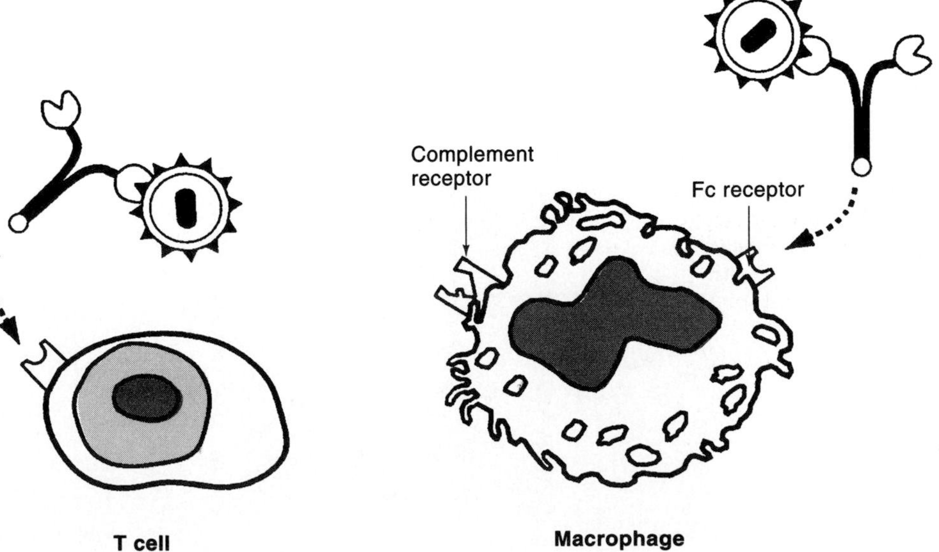

FIGURE 14.9

Antibody (shown in the upper right corner) attaches to a virus but does not inactivate it. Instead the virus/antibody complex is brought into macrophages or lymphocytes via the Fc receptor, or if this complex binds to complement, it can enter by the complement receptor. This process enhances the ability of the virus to spread in these cells. (Reprinted from Levy *Microb. Rev.* 57: 183, 1993)

TABLE 14.4
Factors involved in HIV-induced immune deficiency

1. Direct cytopathic effects of HIV and its proteins on CD4+ cells: cell destruction, effect on stem cells, effect on cytokine production, effect on electrical potential of cells, enhanced fragility of CD4+ cells
2. Effect of HIV on signal transduction and cell function; induction of apoptosis
3. Cell destruction via circulating envelope gp120 attachment to normal CD4+ cells: ADCC, CTL
4. Immunosuppressive effects of immune complexes and viral proteins (e.g., gp120, gp41, Tat)
5. Anti-CD4+ cell cytotoxic activity (CD8+ and CD4+ cells)
6. CD8+ cell suppressor factors
7. Anti-CD4+ cell autoantibodies
8. Cytokine destruction of CD4+ cells

ADCC, antibody-dependent cellular cytotoxicity
CTL, cytotoxic T lymphocytes

14.1.5 Approaches to Treatment

Much research has been directed at finding methods for arresting the virus after infection, but as noted in Chapter 13, there are only a few true antiviral drugs, and they have specific activities. Treatment of the opportunistic infections and cancers has helped AIDS patients, and promising new antiviral drugs are being developed (see Sec. 13.3). Vaccines will require novel approaches (see Sec. 13.2), particularly since HIV can mutate and be transmitted effectively by infected cells. Furthermore, some HIV proteins (e.g., some portions of the envelope), can suppress the immune system so that the use of recombinant protein vaccines might be difficult. Drugs against the viral reverse transcriptase, such as AZT, have helped delay symptoms in individuals infected by HIV and in some cases have prolonged life. These types of therapies, however, do not provide a cure, most probably because they can block the virus only in the early stages of the infection cycle. What is needed is an antiviral approach that kills the virus-infected cell or blocks its ability to replicate virus. In this case, newer approaches, including the use of certain cytokines, could be of value. Most importantly, as noted above, because HIV is transferred effectively by virus-infected cells, the immune system must be stimulated to recognize infected cells soon after infection, before they can pass the virus readily to cells within the host.

At this time, it is uncertain how soon HIV can be controlled. Education is the major objective for prevention. In some countries in Africa, up to 20 percent of the population over the age of 15 are already infected. The message for all people is to prevent infection through elimination of contaminated bloods, sterilization of needles, heat treatment of clotting factor products used for hemophiliacs (already required in most countries), and safe sexual practices that avoid contact with genital fluids.

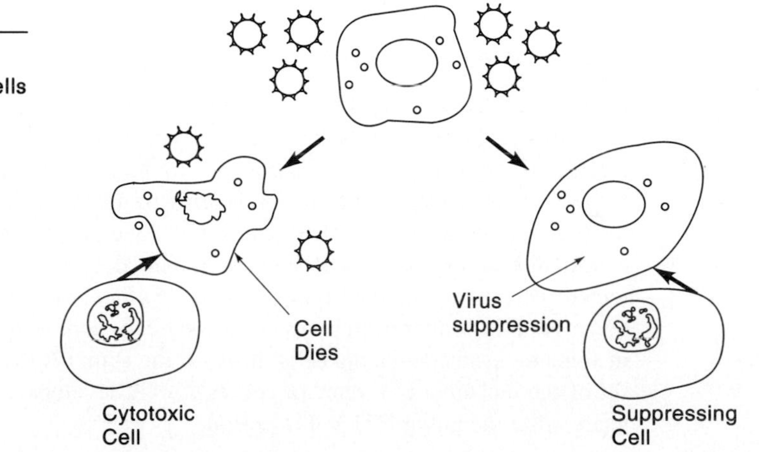

FIGURE 14.10

CD8+ cells have been described that interact with HIV-infected cells (top) to kill the cells (cytotoxic response) or to inhibit HIV replication in the cells (suppressing response).

14.1.6 AIDS Virus in Other Animals

The recent focus on this human lentivirus and immune deficiency has led to a great deal of research on the primate lentivirus, SIV, and the feline agent, FIV. The latter two lentiviruses, in contrast with others (such as visna-maedi, caprine, and equine infectious anemia virus) that primarily infect macrophages (see Chapter 6), are similar to HIV in their preferential infection and destruction of CD4+ lymphocytes. SIV has had a major detrimental effect on primate colonies in the United States, and FIV has been recognized in a number of household pets that developed infections without the presence of the feline leukemia virus. Obviously, any successful approach to prevention of HIV, SIV or FIV infection will apply to all three lentiviruses.

14.2 ADULT T-CELL LEUKEMIA

Another disease associated with a human retrovirus is adult T-cell leukemia. In the early 1980s, the human T-cell leukemia virus (HTLV), or adult T-cell leukemia virus (ATLV) as it was known in Japan, was linked to this human cancer. The virus was initially detected in the lymphomas of individuals living in the southeastern islands of Japan and subsequently in people from the Caribbean and southeastern United States.

14.2.1 Epidemiology

The development of antibody tests and improved procedures for virus isolation have indicated that infection by HTLV-I and the other subsequently isolated subtype, HTLV-II, can be found in many populations throughout the world, particularly in Africa. HTLV-II shares many genomic and serologic characteristics with HTLV-I but can be distinguished by sequence diversity from this other subtype. Of particular concern in the United States is the rise in HTLV-II infection in intravenous (IV) drug users. This virus has not been found associated with any known disease and differs from HTLV-I in its infection primarily of CD8+ cells. HTLV-I preferentially infects CD4+ cells. Nevertheless, the similarity of HTLV-II to the other subtype suggests that its relation to some malignant or neurological state may eventually be found. Interestingly, a related virus, the simian T-cell leukemia virus (STLV), is also associated with this type of malignancy in primate populations.

HTLV-I is transmitted generally from mother to child, by blood transfusion, and sexually from male to female. There is no clear evidence of sexual transmission from female to male. Thus, the limited routes for transfer appear to curtail the overall spread of this infection. Similar routes of transmission for HTLV-II are expected but not yet shown. Moreover, as noted above, the recent increase of transmission of this virus among IV drug users is noteworthy. In the newborn, HTLV is transferred primarily from the mother after delivery, most likely through breast milk containing virus-infected cells. Nevertheless, transmission *in utero* is still a possibility. Recently, a concern about transmission of the virus by blood transfusion has been raised. It is now mandatory in Japan and certain other countries to test blood for the HTLV viruses in addition to HIV.

14.2.2 Clinical Findings

Adult T-cell leukemia (ATL) was first identified as a disease entity in 1977. It was recognized by the cluster of cases in southern Japan, particularly the islands of Kyushu, Shikoku, and Okinawa. ATL consists of the following features: (1) it begins generally in adulthood; (2) the disease course can be acute, leading to death, but sometimes lasts a long time; (3) the transformed white cells (leukemia cells) have properties of T cells (mostly CD4+), and certain distinct morphological features (indented, loculated nuclei) (Fig. 14.11); (4) the infected patient has enlarged lymph nodes; (5) there are frequently destructive changes in bones resulting either from invasion by the tumor cells or toxicity of their cell products; (6) in some cases, the leukemia cells infiltrate the skin; (7) the median survival time after onset of the leukemia is 1 year. In general, ATL develops a long time (often more than 30 years) after the initial HTLV-I infection.

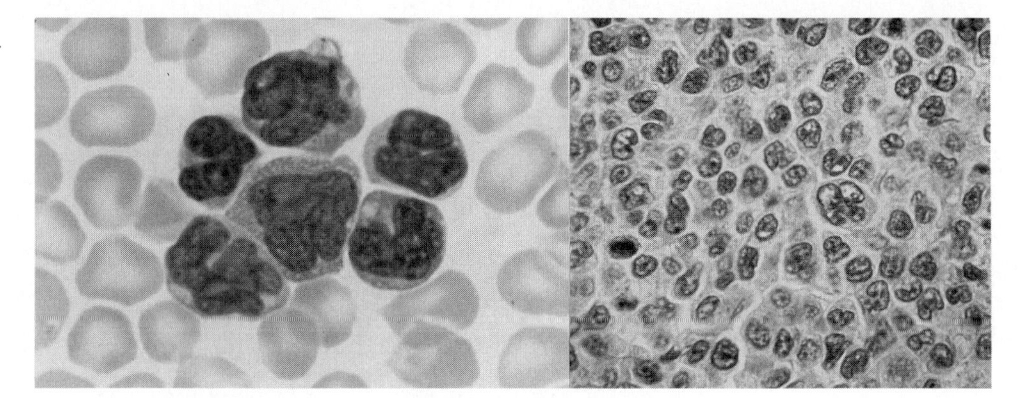

FIGURE 14.11

ATL cells in peripheral blood (right) and in lymph node (left). The cell smear was stained with May-Giemse (right) and with hematoxylin-cosin (left). (Courtesy of Dr. M. Hanaoka. Reprinted from Sugamura and Hinuma, *The Retroviridae,* Vol. 2, ed. J. Levy, 1993 Plenum Press, New York)

HTLV-I has also been found associated with cases of tropical spastic paraparesis (TSP). Most of the patients with these neurological problems have come from tropical islands in the Caribbean, but TSP can also be found in Central and South America, Africa, and India. A related form of chronic spastic disease of muscles and nerves has also been reported in Japan and is known as HTLV-I-associated myopathy, or HAM. The disease occurs more frequently in females than in males. HAM and TSP are the same entities. The onset of TSP is gradual, but occasionally acute cases have been described. In one individual, TSP developed 3 years after receipt of an HTLV-I-contaminated blood transfusion. In most individuals showing TSP, bilateral signs are present. The progression is slow and may take several years to reach its most severe condition. Symptoms include muscle weakness, pain, numbness, and bowel and bladder dysfunction. Mental abilities are usually normal, as is the function of the cranial nerves. In contrast with AIDS, both ATL and TSP patients have usually normal CD4/CD8 ratios, or the CD4+ lymphocyte number can be elevated. The latter situation, at high levels, could reflect the transformation of CD4+ cells by the virus.

14.2.3 The Virus

The HTLV viruses are complex retroviruses, as is HIV, but they belong to a different retrovirus genus (see Chapter 6 and Appendix 2): HTLV previously was placed in the *oncovirinae* subfamily. Its overall structure resembles other retroviruses (Figs. 6.4, 6.10 and 14.4) as does its replicative cycle (Fig. 6.6). By electron microscopy, however, HTLV can be distinguished from the AIDS virus, HIV, as well as from the other known human retrovirus, a spumavirus, that has not been linked to any disease process (see Sec. 6.1) (Fig. 14.12). The other member of the HTLV group is the bovine leukemia virus (BLV) that has a genomic structure and malignant process similar to that of HTLV. These agents are exogenous, and their genomes contain a unique region (pX) that produces proteins essential for virus replication.

The HTLV genome codes for the same products as other retroviruses: the core protein (gag), a protease, polymerase, and envelope proteins, in addition to the pX region products. The mRNA of the pX is spliced twice to give rise to three pX proteins: $p40^{tax}$, $p27^{rex}$, and p21. The $p40^{tax}$ transactivates the transcription of viral message directed by the LTR (see Figs. 6.6 and 14.13), which is essential for replication of the virus. The Tax protein appears to act via cellular factors to bind to enhancer regions on the viral LTR. It is a major determinant of cell transformation because it also transactivates several cellular genes, some of which are involved in cell proliferation. The Rex protein is another regulatory gene product. It resembles the Rev protein of HIV in its role in the eventual replication of the virus. It controls the differential transport from the nucleus of mRNA and its stabilization. Although the similarities of the *tax* and *rex* genes to *tat* and *rev* of HIV have been demonstrated in function, their nucleotide sequences are not homologous, nor are the amino acids coded for by these sequences. The function of p21 is not known.

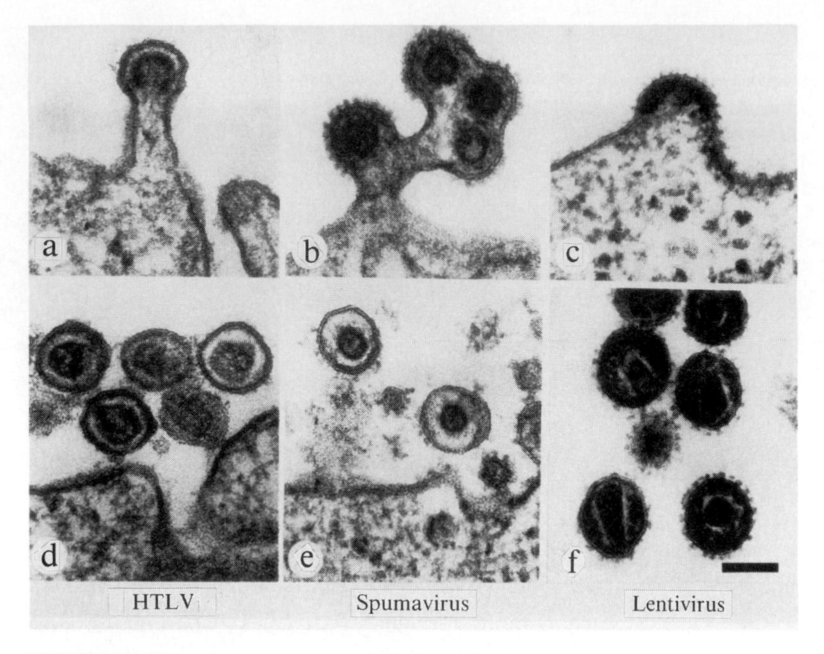

FIGURE 14.12

Electron microscope pictures of three different genera of human retroviruses: HTLV-I, spumavirus or foamy virus, and lentivirus, the subgroup of the AIDS virus. The top panels show budding particles, and the bottom panels show a cross section of the virion structure. Note the prominent spikes on the spumavirus and to some extent on the lentiviruses, as well as the cone-shaped cores of the lentivirus versus the ovoid nucleoids in the HTLV-I and the spumavirus (Courtesy of R. Munn.). Bar: 100nm

FIGURE 14.13

Genomic structure of HTLV with major proteins that are produced. Suggested functions of various proteins are listed. The LTR has effects on cellular genes suggested in Figure 14.14. The action of p40^{tax} on the LTR (U_3RU_5) leads to transactivation of virus expression. The p27^{rex} gene acts to block splicing so that the long mRNA species from the *gag, pol* and *env* regions can be transferred from the nucleus to the cytoplasm where they express the corresponding proteins. (Courtesy of Sugamura and Hinuma, *The Retroviridae,* Vol. 2, ed. J. Levy, 1993 Plenum Press, New York)

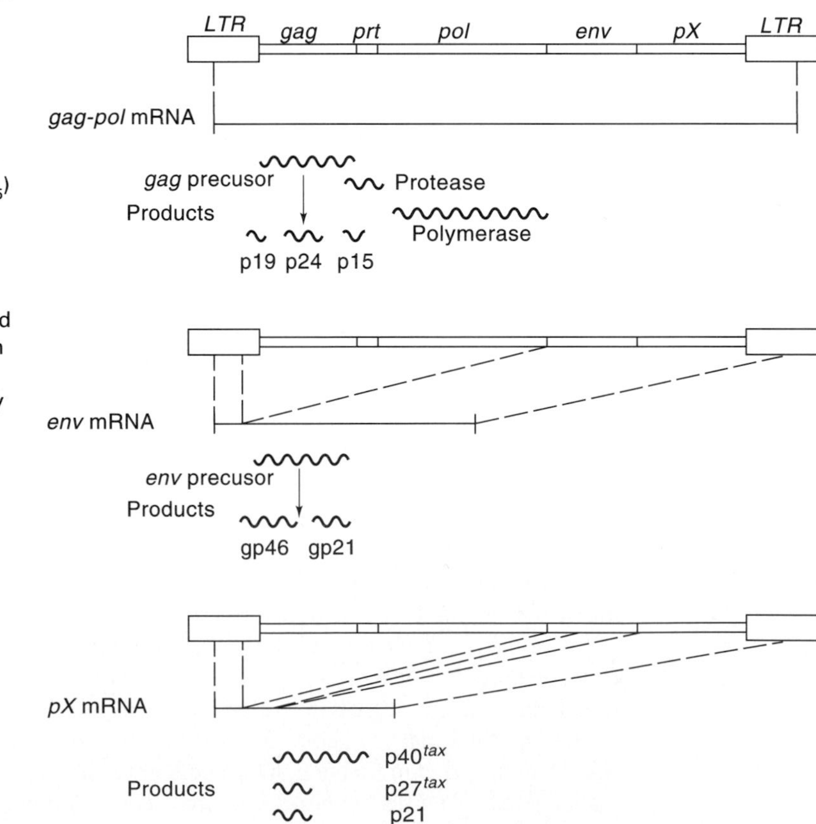

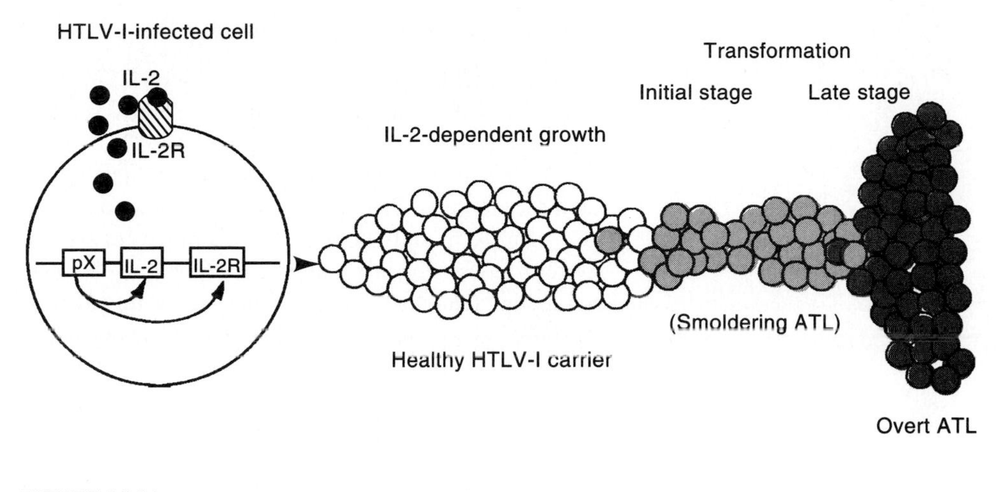

FIGURE 14.14

Hypothetical model of HTLV-I cell induced transformation and acute T cell leukemia (ATL) development. In HTLV-I infected T cells, $p40^{tax}$ is produced by the pX region and transactivates expression of the IL-2 and IL-2R genes. Subsequently, HTLV-I infected T cells preferentially proliferate in response to autocrine and paracrine IL-2 production. During IL-2 dependent proliferation, most cells are eliminated by the host's immune response, but some cells acquire genetic changes that result in IL-2 independent growth *in vitro,* or evolve into a pre-leukemic state (smoldering ATL) *in vivo.* Finally, a single clone of cells acquires further genetic (e.g., chromosomal) changes to complete the malignant transformation resulting in overt ATL. (Courtesy of Sugamura and Hinuma, *The Retroviridae,* Vol. 2, ed. J. Levy, 1993 Plenum Press, New York)

14.2.4 The Virus and Malignancy

The genomes of HTLV-I and II do not contain any known oncogenes, so the promoter-insertion model for tumor development has been considered (see Sec. 10.3.5). However, since the HTLV proviral genome appears **integrated randomly** in the DNA of HTLV-I-immortalized cell lines, this mechanism does not seem involved. Most recently, the induction of receptors on CD4+ cells for IL-2 has been considered an important step in leukemia development (Fig. 14.14). These cells become very responsive to the growth-inducing properties of IL-2. It appears that the pX gene products induce the IL-2 receptor and are essential for this leukemogenic process.

Host immune activity may be important in controlling expression of leukemia in HTLV-I-infected individuals. It is generally concluded that only 1 percent of infected people develop leukemia after a long latency period. Most studies suggest that the cellular immune response is the major mechanism for reducing the expression of HTLV-associated leukemia.

14.2.5 The Virus and Neurological Disease

In contrast with ATL, as mentioned above, TSP may develop more rapidly in individuals infected with the virus. Lymphocytes may play a role in this disorder by penetrating the central nervous system and causing demyelinization of the spinal cord and brain. Nerve cells may be directly infected by the virus, but inflammation and autoimmune responses seem most likely to be involved, particularly since high levels of antibodies to HTLV are associated with TSP. This latter result is believed to reflect the hyperimmune response of the individual against the virus as well as possibly a high level production of cytokines that can be toxic to nerves or increase the autoimmune responses of other cells.

14.2.6 HTLV and Other Diseases

Although HTLV has been associated with TSP, there is no clear evidence of an etiologic link of HTLV with other neurologic diseases, such as multiple sclerosis. Nevertheless, recent evidence suggests that integration of HTLV-like viruses in certain hematopoietic cells might be responsible for malignancies that have not elicited an immune response in the host against the virus. Perhaps the agent is not recognized in its latent or silent state. (see Secs. 11.9 and 12.6) Only through the techniques of polymerase chain reaction and other procedures for detecting integrated viral genomes (see Appendix 1) as the association of this virus with other malignancies been suggested. If transformation was caused by the detected virus, it could have occurred by interruption of the normal function of the cells as a result of virus integration in the cellular genome. Further work is required to evaluate this possibility.

14.2.7 Treatment

There is no specific treatment for ATL or TSP. Chemotherapy is used for the leukemia, but long-term survival is rare. TSP patients have shown improvement with steroids, and certain antiviral drugs are being considered.

14.3 HEPATITIS C

Non-A, non-B hepatitis is a major disease often associated with blood transfusion. The majority of cases are now known to be caused by hepatitis C virus (HCV). After the discovery of hepatitis B virus (Chapter 6) the identification of the non-HBV, non-HAV agent involved in other forms of hepatitis was an important challenge. Hepatitis C virus (HCV) can be distinguished from the other known hepatitis viruses as summarized in Table 14.5. This agent was identified for the first time in virology by molecular techniques and not by the usual isolation in cell culture or by inoculation of animal species. HCV identification came by pelleting from the blood of chimpanzees the presumed virus particles that were present in large quantities. The agent could not be grown in culture, but through extraction of nucleic acids and by random priming with PCR procedures, a cDNA library of the virus was created in an expression vector.

TABLE 14.5
Characteristics of hepatitis viruses

Virus	*A*	*B*	*C*	*D*	*E*
Genome	RNA	DNA	RNA	RNA	RNA
(Family)	(Picorna)	(Hepadna)	(Flavi/Toga)	(Viroidlike)	(Calici)
Transmission route	Fecal-oral	Parenteral; sexual; vertical	Parenteral; vertical ?; sexual	Like HBV	Fecal-oral[a]
Disease course	Acute[b]	Acute and chronic	Mild acute, mostly chronic	Acute or chronic[c]	Acute
Latency period for disease	2–6 weeks	3–5 months	2–9 weeks	Like HBV	1 month
Diagnostic tests	Antibody	Antibody, Antigen	Antibody, Antigen	Antibody	None[d]
Time to sero-conversion	3–4 weeks	3–4 weeks	3–6 months[e]	Like HBV	?
Virus secretion: site, time after infection	Stool, 2–3 weeks	Blood, 3–4 weeks	Blood, 3–4 weeks	Blood, 3–4 weeks	Stool, 1 month

[a]India and Asia
[b]Lasting usually 1 to 3 weeks
[c]Associated with HBV; often more fulminant course
[d]Genome cloned, tests expected
[e]Can be longer

FIGURE 14.15

Virionlike and nucleocapsidlike particles recovered from the blood of HCV-infected patients. Note the viruslike particle (center of the photograph) with a diameter of 55 nm observed in a potassium bromide density gradient centrifugation fraction with the density of 1.115 g/m. (Courtesy of Mishiro Takahashi et al. *Virology* 1992, 191:431–434)

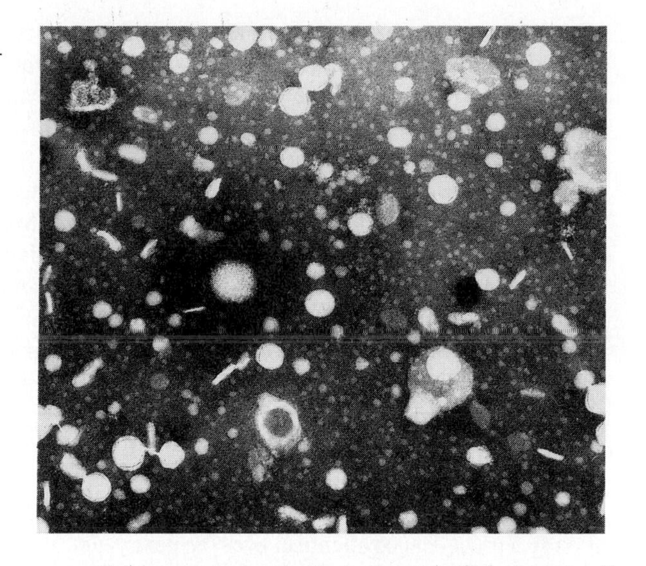

100 nm

Subsequent transfection of susceptible target cells with the "viral" cDNA led to production of HCV-specific proteins identified by antibodies in the sera of infected chimpanzees and humans. Using a cDNA probe to a portion of HCV, researchers then obtained the entire viral genome from the liver of an infected chimpanzee and defined it (Fig. 14.15).

14.3.1 Epidemiology

The successful expression of HCV proteins has led to an ELISA and other assays to detect antibodies to this hepatitis agent. These tests have revolutionized blood bank surveillance by helping to detect individuals carrying this agent and preventing its transmission through blood transfusions. Two envelope glycoproteins, E1 (gp33) and E2 (gp72) (a nonstructural protein), have been identified. These proteins resemble the envelope products of flaviviruses. Since these latter proteins protect individuals from flavivirus infection, a similar result with the HCV proteins might be expected.

One important problem with testing for hepatitis C virus is the long latency period (up to 1 year) that can occur before production of antibodies to the agent (Table 14.5). Thus, testing blood donors for liver enzyme abnormalities also appears necessary at this time, although newer assays for detecting the viral genome in the blood are under evaluation. Transmission of HCV is generally by transfusion of blood or blood products and, as expected, in intravenous drug users. Sexual transmission appears to be uncommon. Whether transfer from mother to child takes place is still not certain but has been reported.

14.3.2 Clinical Findings

HCV is a major cause of transfusion-associated non-A, non-B hepatitis throughout the world. Individuals infected with the agent can show intermittent bouts of hepatitis characterized by abnormally high liver enzyme levels in the blood. Other infected patients have persistent liver abnormalities and a course of chronic liver disease. The latter condition occurs most often.

Hepatitis C, because of its induction of chronic hepatitis, has been associated with liver cancer, as has hepatitis B virus. Like other hepatitis viruses, a chronic inflammation and cell destruction concomitant with renewed replication of liver cells leading to chromosome changes are the most likely causes for the malignancy.

14.3.3 The Virus

HCV is 30 to 60 nm in diameter with a lipid-containing envelope and an RNA genome (Fig. 14.15). At present, the virus is recognized as sharing some properties with both toga- and flaviviruses, but it may eventually be placed in its own virus family.

One noteworthy feature of HCV is its high variability; at least five distinct genotypes exist. As with HIV, the envelope of HCV has several variable regions that can change readily, permitting "escape" from neutralizing antibodies and other immune responses. Thus, development of a vaccine will be as daunting as that of an HIV vaccine and not as straightforward as with other viruses, such as hepatitis B. Most data suggest weak immune responses to HCV infections. In this regard, HCV-infected chimpanzees were found susceptible to reinfection by the same or a different strain. Moreover, because hepatitis C virus cannot yet be grown readily in cell culture, its replicative cycle is difficult to study. Recent reports suggest that an *in vitro* culture system using liver cells or certain monkey cells lines will be developed. Therapeutic approaches directed at interrupting the ability of HCV to reproduce in the host now depend on the use of primate models that are costly and not amenable to controlling for all the variables involved in virus production.

14.3.4 Conclusions

Because of its frequent association with severe liver disease and cancer, HCV will be receiving a great deal of attention over the next years. Attempts to grow it consistently in culture will be increased. Interferon appears to offer some hope for antiviral therapy. To students of virology the discovery of HCV could initiate other attempts using molecular techniques to define nonculturable agents potentially involved in other human diseases, such as chronic fatigue syndrome and multiple sclerosis. However, this approach will depend on availability of sufficient quantities of the presumed etiologic agent in blood or certain tissues in the body.

14.4 CHRONIC FATIGUE SYNDROME

Unlike the previous diseases we have discussed, chronic fatigue syndrome has not been linked to a definitive infectious agent, but the characteristics of the disease suggest that a virus could be involved. In 1984 an outbreak of a flulike illness occurred in Lake Tahoe, California, and spread to over 300 individuals. The disease was characterized by fluctuating episodes of sore throat, malaise, muscle aches, headache, and low-grade fever with extreme fatigue lasting longer than 6 months, and sometimes for several years. Neurological findings were common, including memory loss, sensory or motor abnormalities, and disorders of cognitive function. The episodic debilitating nature of chronic fatigue syndrome (CFS) has baffled researchers and physicians as to its cause. The description of this disease in the Lake Tahoe area led to the recognition of other cases throughout the United States and many parts of the world, including Europe, Australia, and the Far East.

14.4.1 Epidemiology

In historical perspective, CFS is similar to diseases that have been reported in the literature over the last 200 years. In the 1700s it was known as Fabricula, in the 1800s as **neurasthenia,** and in the early 1900s as **da Costa's syndrome,** when involvement of the heart was suspected. Moreover, throughout its history a variety of agents have been linked to this clinical entity, including brucella, influenza virus, and candida.

CFS is commonly found in the age group 30 to 45 (average age 35). Although initially described in women of middle class socio-economic level, it is now recognized in males and females from all phases of society and all ethnic and racial groups. The higher prevalence of CFS in women than in men (3:1) has supported the theory of an autoimmunelike illness; autoimmunity is more common in females. In addition, the recurrences of chronic fatigue and other symptoms appear to correlate with exposure of the individual to antigens in the air, odors, or other products that restimulate a hyperresponding immune system (see below). CFS as a disease is not contagious, since it is unusual to find the disorder in members of the same household. Nevertheless, the causative agent could be worldwide with only a few individuals showing symptoms of the infection. This latter characteristic is common to many viruses (e.g., polio) (see Chapter 12).

TABLE 14.6
Research case criteria for the chronic fatigue syndrome

A case of chronic fatigue syndrome must fulfill major criteria 1 and 2 and the following minor criteria: 6 or more of the 11 symptom criteria and 2 or more of the 3 physical criteria or 8 or more of the 11 symptom criteria.

Major Criteria

1. New onset of persistent or relapsing, debilitating fatigue in a person with no previous history of similar symptoms: fatigue that does not resolve with bed rest and is severe enough to produce or impair average daily activity for at least 6 months
2. Other clinical conditions that may produce similar symptoms must be excluded by thorough evaluation based on history, physical examination, and laboratory findings

Minor Criteria

Symptom criteria (those that began at or after onset of fatigue; they must have persisted or recurred over at least 6 months)

1. Mild fever (oral temperature of 37.5°–38.6°C) or chills
2. Sore throat
3. Painful lymph nodes in the anterior or posterior cervical or axillary distribution
4. Unexplained generalized muscle weakness
5. Muscle discomfort or myalgia
6. Prolonged (24 hours or longer) generalized fatigue after levels of exercise that would have been easily tolerated in the patient's premorbid state
7. Generalized headaches different from ones that patient may have had in the premorbid state
8. Migratory arthralgia without joint swelling or redness
9. Neuropsychological complaints (e.g., photophobia, transient visual scotomata, forgetfulness, excessive irritability, confusion, difficulty in thinking, inability to concentrate, depression)
10. Sleep disturbance (hypersomnia or insomnia)
11. Description of the main symptom complex as initially developing over a few hours to a few days

Physical Criteria

1. Low-grade fever
2. Nonexudative pharyngitis
3. Palpable or tender anterior or posterior cervical or axillary lymph nodes

SOURCE: Data from Holmes et al., *Ann. Int. Med.*

14.4.2 Clinical Findings

CFS has the following prominent features: (1) extreme fatigue that occurs after only mild exercise; (2) muscle aches and pains, particularly during the acute onset (as noted above); and (3) neurological disorders including loss of sensory and motor activity in some patients and reduction in memory. Some patients may have frank dementia. This disease course usually fluctuates with periods lasting from one month to several months, and recurrences appear intermittently, but often following exposure to antigens or a common virus infection. Diagnosis is made from criteria established by the Centers for Disease Control (Table 14.6) and most importantly includes the ruling out of any other disorder that could give rise to the overwhelming fatigue that is a major characteristic of this disease. Although depression is often found in individuals with CFS, it is usually not one of the presenting symptoms, but occurs after the prolonged disease course.

Initially, the symptoms of CFS were considered secondary to Epstein-Barr virus (EBV) infection, which has an associated postviral fatigue syndrome in some infected individuals. This latter condition can occur with other viruses (e.g., Coxsackie). Antibodies to EBV were found in high titer, and thus chronic EBV syndrome (CEBV) was a name initially given to the disorder. Subsequently, several individuals were found without this increase in EBV antibodies, and the condition was renamed for its major symptom, chronic fatigue. Later, the term

TABLE 14.7

CD8+ cell surface markers in CFS patients and healthy controls

	Group A (n = 67)	*Group B (n = 59)*	*Group C (n = 21)*	*Total CFS (n = 147)*	*Healthy controls (n = 80)*
% CD8+ cells expressing					
CD11b	16 ± 10[a]	23 ± 12	25 ± 9	19 ± 16	25 ± 10
CD38	53 ± 15[a]	42 ± 8	38 ± 16	47 ± 20	35 ± 12
HLA-DR	32 ± 10[a]	21 ± 10	20 ± 15	22 ± 18	14 ± 6
CD25	04 ± 1	02 ± 1	03 ± 1	03 ± 1	04 ± 2
CD57	26 ± 10	27 ± 10	29 ± 12	28 ± 17	28 ± 12
Leu8	48 ± 12	45 ± 11	46 ± 10	47 ± 15	50 ± 12

Data are obtained by flow cytometry using specific monoclonal antibodies against cell surface proteins.
[a]Statistical analysis (Mann-Whitney U test) showed significant difference ($P < .01$) when group A patients were compared with group C patients and healthy controls.
Source: Landay, A. L., et al. (1991) *Lancet.*

chronic fatigue immune dysfunction syndrome (CFIDS) was used for the disorder by some groups because of the abnormalities in the immune system described below. In Japan the disease is known as chronic NK cell deficiency syndrome because of that immune disorder.

14.4.3 Immune Abnormalities and Pathogenesis of CFS

Several laboratories have demonstrated that the immune function in individuals with CFS can be abnormal. Flow cytometry demonstrates that the number of CD4+ and CD8+ lymphocytes can be normal, but the number of activated (stimulated) CD8+ cells can be increased dramatically. Moreover, the number of CD8+ suppressor cells is often decreased (Table 14.7). In addition, as mentioned above, the ability of NK cells to kill target cells is markedly reduced. In some studies, responses of T and B cells are also diminished. These observations have suggested that CFS most probably represents an immune reaction against an infection, but chronic activation occurs as an abnormal consequence. Perhaps the inability of NK cells to dampen the overactive response is the cause.

It is proposed that in the pathogenesis of CFS an infectious agent, probably a virus, enters the host and creates an activated immune response characteristically observed in acute virus infection. As shown in Figure 14.16, during an acute virus infection the number of NK cells and CD8+ cells generally increase, leading to a decrease in the helper-suppressor lymphocyte ratio. You will recall that a similar reduction occurs in AIDS but

FIGURE 14.16

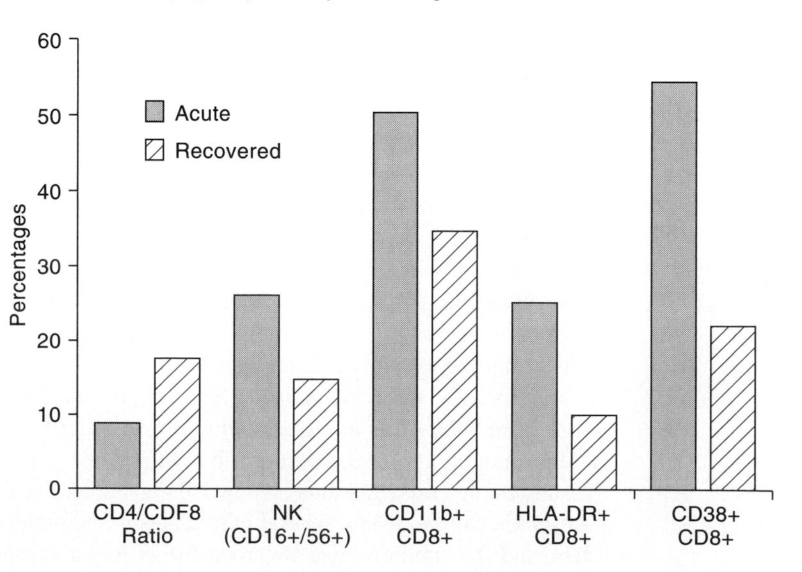

Lymphocyte analyses at day 2 of an acute viruslike illness in a representative individual and at the recovery stage 10 days later. The percentage of peripheral white blood cells expressing the antigens listed is presented. For CD4/CD8 ratio, 10 and 20 = 1.0 and 2.0. The results were obtained using monoclonal antibodies to the cell subsets noted and were measured by flow cytometry. (From Levy, J. et al. *Viral Infections* 1992 eds. M. Sande and R. Root, Churchill Livingstone, New York)

TABLE 14.8
Symptoms during cytokine therapy that can occur during virus infections

Fever	Anorexia	Tachycardia
Myalgias	Fatigue	Hypotension
Nausea	Lethargy	Confusion

is related to a destruction of CD4+ cells and not a rise in CD8+ cells. One week to 10 days after a common virus infection, the immune system returns to normal with a CD4/CD8 ratio approaching 2.0, and a return in NK cell number (Fig. 14.16). Moreover, during acute virus infection, NK cell function is generally elevated. In CFS, in contrast to an acute viral illness the hyperactive response by CD8+ cells does not decrease and NK cell activity is reduced. With this disease, most people are probably infected with the causative agent and it is eventually eliminated in a short period of time; the immune system then returns to normal. However, in the individuals showing CFS, the agent either is not eliminated, or if it is, the immune system still continues to show a stimulated response against the infection. In this latter case, it resembles the "hit-and-run" concept discussed in relation to cancer induction (Sec. 10.3.6). The CFS agent may enter a host, create the immunological disorder, and be eliminated. However, the immune dysfunctions continue. The symptoms, including fatigue, could then be explained by the production of **cytokines** by the activated cells (e.g., TNFα, IL-1) that are known to cause these types of clinical problems in individuals with other infections, or during therapy for cancer (Table 14.8). An obvious control for the disease would be suppression or elimination of the agent, or the return of the immune system to balance. The inability to do so may be reflected in the reduced NK cell function or CD8+ suppressor cell numbers. In some ways, this activated state resembles an autoimmune-like illness in which hyperactive autoreactive cells and antibodies to normal proteins are made. In this regard, autoantibodies to the thyroid and other cellular proteins can be detected in CSF.

The existence of CFS as a clinical entity is often challenged by those psychiatrists and physicians who believe that the condition reflects depression and not a true physical illness. But the immunological abnormalities, particularly the activated immune system, cannot be explained by depression alone. Conceivably, a stressful condition affects the extent of the immune response of a host, so that infection by a virus or another infectious agent could have a detrimental effect. It could cause an overresponsive state and induce the chronic immune abnormalities observed. As noted above, in most individuals an active immune response to infection by this putative agent would be appropriate, but then the immune system would return to normal.

14.4.4 Search for the Cause

The search for a causative viral agent has focused on enteroviruses, herpesviruses, and retroviruses as well as on nonviral agents such as mycoplasma and even prions. The latter agent has been considered because of the neurological symptoms observed. Thus far, no virus has been directly linked to this disorder, although as described above, it resembles conditions that can follow infection by several viruses, particularly herpesviruses and enteroviruses (e.g., CMV, EBV, Coxsackie B, polio).

14.4.5 Conclusions

The present objectives of researchers are to identify the possible agent(s) responsible for CFS and to develop approaches to treatment. The therapeutic directions would be aimed at the agent or at the pathogenic pathway, such as the overproduction of cytokines, loss of NK cell activity, or the hyperreactive cellular immune response. A variety of therapies have been tried by physicians, but none has proved effective for most individuals. Moreover, diagnostic tests for CFS would be extremely helpful and await the recognition of the causative agent or a common immunological abnormality that could be evaluated routinely in the laboratory.

14.5 ALZHEIMER'S DISEASE

An illness that baffles many individuals who have family members suffering from the disorder is Alzheimer's disease (AD). Many investigators believe it too may be caused, at least in some cases, by a virus. AD has become the fear of people approaching the age of 60 and noting a reduction in memory, in most cases a normal consequence of aging. Approximately 5 percent of individuals over the age of 65, and 20 percent over 80 suffer from this condition, but it can occasionally occur in people as young as 40. With people living longer today, AD could reach epidemic proportions.

14.5.1 Clinical and Pathologic Findings

AD usually begins with forgetfulness and loss of interest in usual pursuits by affected individuals; some become depressed for no obvious reason. A history of impaired learning, reduction in recent or remote memory, and other symptoms of dementia can be observed. No fever is present. The disease can progress slowly over 5 to 10 years but then results in a patient who is bedridden and in a state likened to that of a child of a few months to 2 years of age. Death comes from recurrent infectious diseases. There is no known treatment.

The diagnosis of AD is made by exclusion, although recently the finding in the cerebrospinal fluid of reduced levels of the precursor protein for the amyloid protein of AD (see below) could be helpful. In this disease the cortical neurons and their axons degenerate; the white matter of the brain shrinks, and the cerebral hemispheres enlarge. These findings are most often present in the temporal, frontal, and parietal lobes. Neuronal loss from neurofibrillary degeneration can be observed microscopically. This pathology is characteristic of AD and other nontransmissible brain amyloidoses. Concentrated amyloid B peptide (ABP) is deposited extracellularly with polymerization to insoluble fibrils and neurofibrillary tangles. These are toxic to blood vessels and neuronal cells. These types of amyloid deposits accumulating in the brain and blood vessels can occur normally with age; they are not usually seen in prion-induced diseases (e.g. scrapie, kuru; see Sec. 9.6). The number of plaques correlates with extent of dementia. The reason for increased amyloid formation in AD is not known. In some cases it has been most recently linked to production of an abnormal type of amyloid that accumulates in the cell perhaps because it is not sensitive to normal degradation processes (e.g., cleavage by an amyloid precursor protein secretase). The importance of that observation to the etiology of other AD cases requires further study.

14.5.2 Search for a Cause

A direct relationship of AD to the prion-induced amyloidoses cannot be made. Transmission of AD to animals (primates) has not been achieved, and, differences can be appreciated between the amyloid deposits in AD made up of amyloid B protein and those of a prion, in which PrP is found (Fig. 14.17; see Sec. 9.6). Nevertheless, in both cases, production of an abnormal protein (i.e., ABP, PrP) appears to be an important part of the disease process. Inability to cleave that protein could be involved in both pathogenic pathways.

Neuronal cell degeneration in the prion-induced amyloidoses is different from that observed on AD. The pathology appears to be induced by the agent that initiates the structural change of the normal PrP precursor protein to an insoluble protease-resistant amyloid subunit (see Sec. 9.6). Furthermore, the amyloid plaques in these prion-associated conditions are immunologically distinct from those in AD (Fig. 14.18). Finally, PrP is located on a different chromosome than is the AD amyloid protein (which in some patients is chromosome 21). Despite these differences, many researchers believe that the uncovering of the cause of prion-induced neurological symptoms will provide answers to the eventual treatment of other human dementias such as Alzheimer's. Moreover, it is possible that this disease, as well, will have a viral or viruslike infectious agent etiology.

FIGURE 14.17

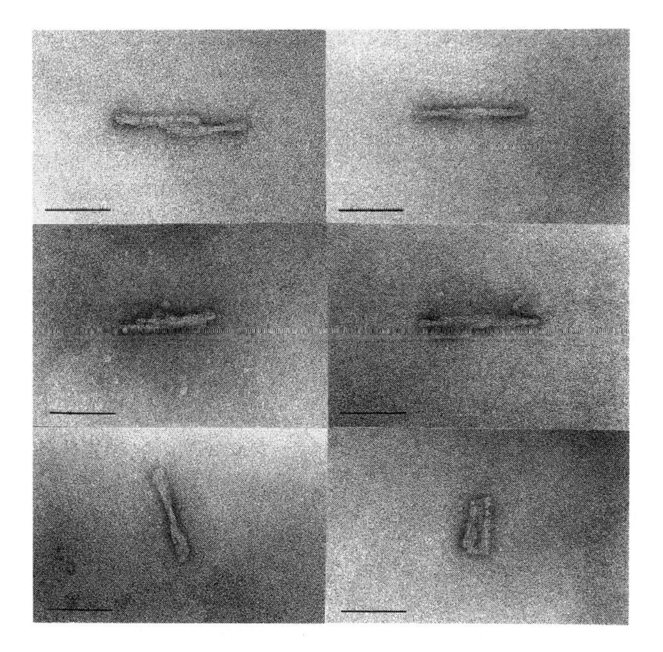

Electron micrographs of small aggregates containing a few prion rods. Preparations were negatively stained with uranyl formale. Bars are 100nm. (Courtesy of Prusiner et al. *Cell* Vol. 35, 1985 Cell Press, The New England Journal of Medicine)

FIGURE 14.18

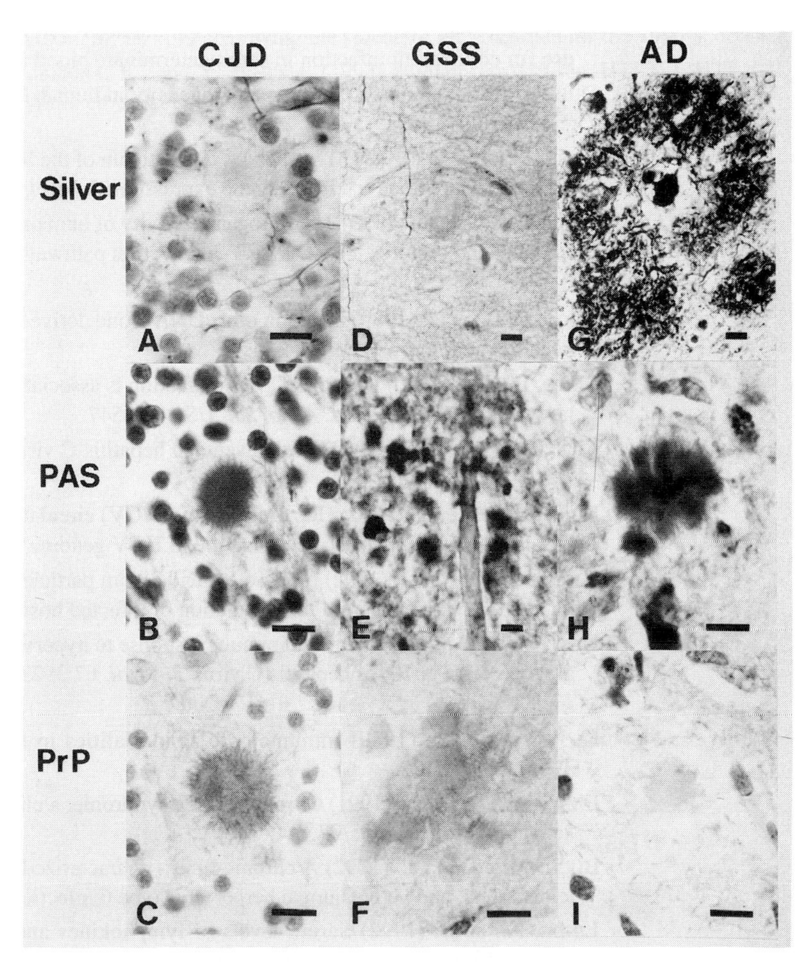

Characteristics of amyloid plaques in Creutzfeldt-Jacob *(CJD)*, Gerstmann-Straussler syndrome, *(GSS)*, and Alzheimer's disease *(AD)*. The Bielschowsky silver stain for neurites shows that dystrophic neurites are not characteristic of amyloid plaques in CJD *(A)* or GSS *(D)* while they are the hallmark of the senile plaques of AD *(G)*. The Periodic Acid Schiff *(PAS)* histochemical reaction stains all amyloid plaques intensely *(B, E,* and *G)*. The plaques of GSS *(E)* are more dense than the plaques of CJD *(B)* and similar in density to the amyloid of senile plaques *(H)*. The plaques of GSS are typically perivascular *(E)*. Peroxidase immunohistochemistry with an antiserum specific for the scrapie prion protein *(PrP)* causes strong specific staining of plaques in CJD *(C)* and less intense, but specific, staining of plaques in GSS *(F)*. Amyloid in AD *(I)* fails to react with the PrP antiserum *(C, F,* and *I* were counter stained with hematoxylin). Bar: 25 um. (Courtesy of S. Prusiner and S. DeArmand, *Laboratory Investigation* 56:349 1987 The U. S. and Canadian Academy of Pathology)

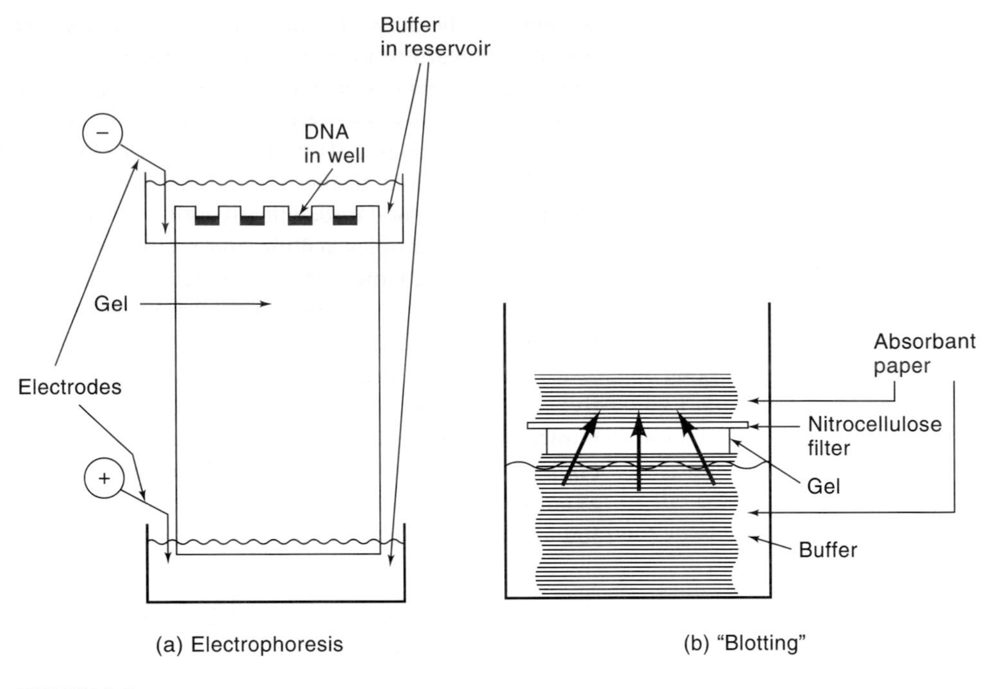

FIGURE A.2

Schematic representation of analysis of DNA fragments by electrophoresis and "blot" hybridization (Southern). (a) DNA digested with restriction enzymes is loaded into a well formed in a slab of agarose gel that has each end in contact with buffer in a reservoir with an electrode. A voltage source is applied to the system, with the positive pole at the end of the gel farthest from the DNA, which is negatively charged. After current flows through the gel for an appropriate time, the gel is stained with a fluorescent dye, which allows visualization of the DNA fragments, as in Figure A.3. The gel is then immersed in alkaline solution to separate the strands of the double-stranded DNA fragments. After neutralization the DNA is "blotted" out of the gel onto a nitrocellulose filter paper, with buffer flowing by capillary action up through the gel, as indicated by the arrows in the side view (b). After several hours of "blotting," the filter is first baked to affix the single-stranded DNA permanently, and then it is incubated with a radioactive DNA or RNA probe to detect those DNA fragments with the same sequences as the probe (e.g., viral sequences). After unreacted probe is removed, a sheet of X-ray film is exposed to the filter to reveal the DNA fragments with which the radioactive probe formed double-stranded hybrids (for example, see Figure A.3).

Methods utilizing gel electrophoresis rely on the same principle as does chromatography on ion exchange resins or celluloses carrying charged groups. Other column chromatographic methods differentiate primarily on the basis of particle size by using more or less inert gels made up of beads of different porosity. Thus, the principle of Sephadex fractionation has been extended to particles of virus dimensions by using beads of agar or preferably agarose instead of dextran gels. In all methods with gels, however, ion exchange effects must also be considered.

Gel electrophoresis has been very helpful in separating viral DNA (Southern blot) (Figs. A.2 and A.3), RNA (Northern blot), and proteins (Western blot) (Fig. A.4). For all these procedures, extraction of nucleic acids or proteins is conducted by standard techniques, and the materials are added at optimal concentrations to individual wells on an **agarose** (DNA or RNA) or **polyacrylamide** (proteins) gel. The gels are electrophoresed, separating the nucleic acids or proteins according to their molecular size, and the molecules are transferred to a nitrocellulose membrane. To detect virus-specific nucleic acids, a radioactive probe to viral nucleic acids is hybridized to the nucleic acids separated on the nitrocellulose membrane. Viral nucleic acids are visualized as bands on X-ray film. Similarly, antibodies to viral proteins can be used to detect the presence of viral proteins on the nitrocellulose filter (Fig. A.4). This procedure, called

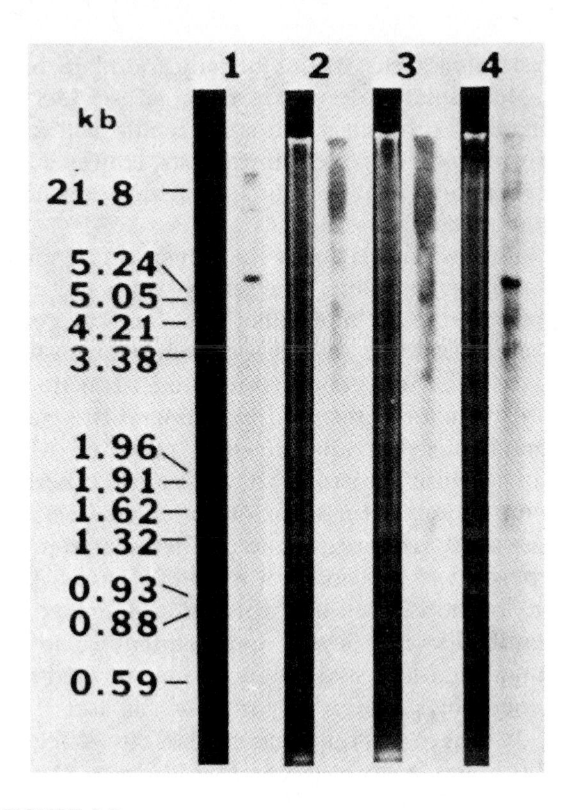

FIGURE A.3

Southern blot. Hybridization analyses of integrated retroviral DNA. Various DNA samples were digested with the restriction endonuclease *Eco* RI, and resulting fragments were separated by electrophoresis through an agarose gel. The DNAs were then "blotted" out of the gel onto a nitrocellulose filter, which was then incubated with a radioactive DNA probe. After exposure of X-ray film, DNA fragments on the filter that formed hybrids with the radioactive DNA are visualized as dark bands. In this case, the probe was a small piece of DNA isolated from a cloned DNA copy of the Harvey strain of murine sarcoma virus (Ha-MSV). The left column of each numbered lane represents the photograph of the gel stained with fluorescent dye; the right column shows the exposed X-ray film. DNA migrated from the top to the bottom in the figure. The numbers to the left of the figure indicate positions of various sizes (in kilobases) of marker DNA fragments from a lane not shown here. Lane 1 contains about 100 pg of DNA from a recombinant λ phage that carries the Ha-MSV DNA. This amount of DNA is too small to be detected by the fluorescent dye, but several bands of the incompletely digested phage DNA as well as the predominant band of about 10 pg of pure Ha-MSV DNA (at about 5.5 kb) are detected on the exposed X-ray film. Lane 2 represents a digest of 10 μg of mouse DNA from uninfected cells. The dye shows a continuum of different DNA fragments (about a million), from those so large that they remain in the well to those at the bottom edge of the gel. The two most intense bands (at about 1.3 and 2.0 kb) represent cellular DNA sequences that are present in many copies per genome. The hybridization analysis shows some evidence of homology of the probe with large cell DNA fragments, but no distinct bands are visible. Lane 3 represents 10 μg of DNA from mouse cells infected with another MSV, the Kirsten strain (Ki-MSV), which also contains the probe sequences. Again, no distinct bands of hybridization are seen, indicating that each cell has Ki-MSV DNA integrated into a different fragment of its DNA. Lane 4 shows the DNA (10 μg) of a single clone of cells infected with Ki-MSV. Since the DNAs of all the cells in this culture are identical, a single fragment of about 5.3 kb, which contains about 10 pg of Ki-MSV DNA integrated into the cellular DNA, is clearly visible in the hybridization analysis.

FIGURE A.4

Western blot. HIV-1 (lane 1) and simian immunodeficiency virus (SIV)-infected cells (lane 2) were extracted and the protein extracts placed in the wells of an acrylamide gel as described in the immunoblot procedure. The migration of virus-specific proteins was determined using antibodies specific to HIV-1 or HIV-2. Antibodies to HIV-2 cross-react with SIV. Protein sizes are given on the left side of the figure. The studies indicated that antibodies to HIV-1 detect HIV-1-specific proteins present in lane 1. Antibodies to HIV-2 detect proteins that are specific for SIV (lane 2). Also noted in the studies were the cross reactivity of the viral core p25 antigen in the two primate lentiviruses. (Courtesy of A. S. Evans et al.)

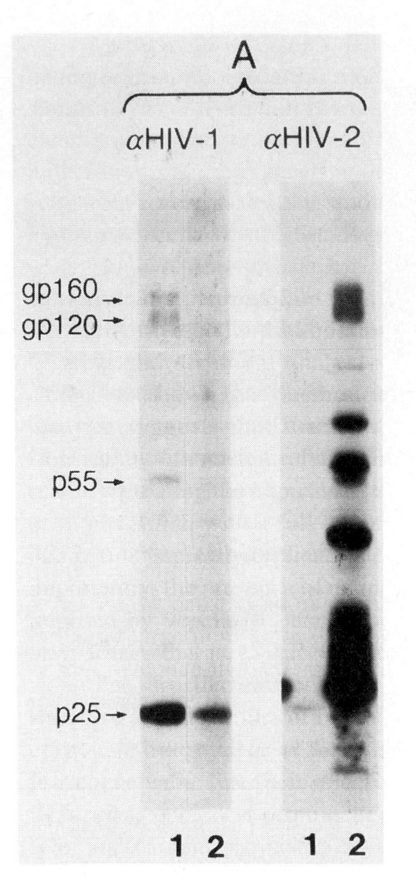

Western or **immunoblot analysis,** can demonstrate the presence and relative amounts of viral proteins. The latter reaction can also be shown by a radioactive Staphylococcal protein A procedure. Immunohistochemistry and *in situ* hybridization detects viral proteins and nucleic acids directly in the cell itself. Thus, these procedures can reveal the relative abundance and distribution of the molecules in the cell. Finally, a NorthWestern procedure was recently developed in which proteins that bind to specific nucleic acids can be detected in a similar manner by hybridizing a radiolabeled protein probe to nucleic acids separated on a nitrocellulose membrane.

D. Column Purification

Viruses can also be purified on appropriate columns in which antibodies to the virus are bound to Sephadex beads or, if the envelope to the virus reacts with specific lectins, to these lectins placed on the beads. It is important to note that virus purification usually inactivates many particles. Thus, the majority of purified virus particles are not infectious but can be used to study nucleic acids and antigenic properties of the virus.

E. Assessing the Purity of Virion Preparations

It is valuable to establish the degree of virion homogeneity or heterogeneity as carefully as possible before using a virus for physicochemical, biochemical, or biological experiments. Many of the preparative methods used in the purification of a virus also may be used to assess the homogeneity of a virion preparation. If a virus forms a single sharp band in a sucrose gradient, this finding indicates approximate homogeneity in terms of sedimentation but not necessarily of other properties. And, as stated above, many related or unrelated substances can have the same sedimentation rate. Thus, if the virus is purified only by repeated cycles of **sucrose gradient centrifugation,** faster- or slower-sedimenting material being removed each

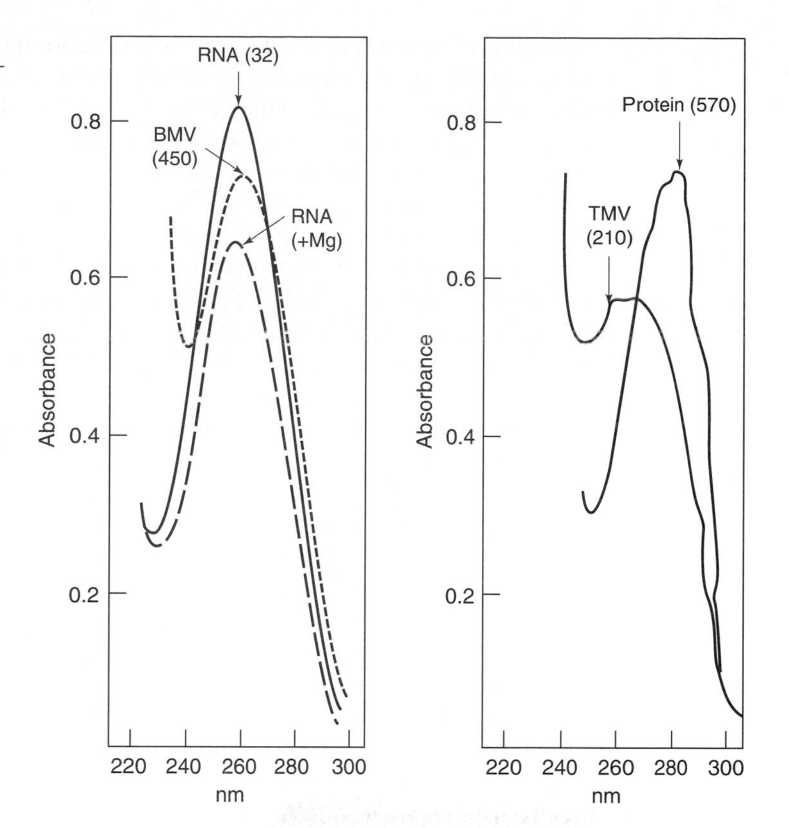

FIGURE A.5

Ultraviolet absorption spectra of viruses and viral components. On the left is an isometric virus containing 21 percent RNA (BMV) and viral RNA (TMV RNA) in hyperchromed state in water, and the same sample on addition of $MgCl_2$ to $10^{-3}M$. On the right is TMV (whole virus, 5 percent RNA) and TMV protein. The numbers refer to the concentrations (in μg/ml) of the respective materials used. Note that the maximal absorbance per milligram is about 1:20:20 for TMV protein: TMV: TMV RNA. BMV, Bromemosaic virus; TMV, tobacco mosaic virus.

time, the uniformity of sedimentation has little discriminatory importance. In other words, only by the use of a variety of methods that rely on different principles in the course of purification and/or testing for homogeneity can substantial evidence be presented to show that a homogeneous preparation of particles has been obtained. Examples of such distinct criteria are sedimentation behavior, buoyant density, electrophoretic mobility, electron microscopic appearance, serology, infectivity, stability (all previously discussed), and UV absorbancy.

Since **recording spectrophotometers** are available in most laboratories, the plotting of the UV absorption spectrum of a virus to determine purity has become technically simple. Only 0.05 to 0.2 mg of virus is needed, and the material is not used up by this test. The details of the shape of the absorption curve reveal much information about the approximate composition of the virus in terms of protein and nucleic acid. Both the 260/280-nm absorbancy ratio and the max/min ratio represent measures of the proportion of nucleic acid to protein in the sample, and a constancy in the details of the spectrum on subjection of the virus preparations to further purification and fractionation attempts is an additional indication of the purity of a virus preparation (Fig. A.5).

F. *Characterization of Virions by Other Procedures*

Serological Methods Most viruses are effective immunogens when injected into rabbits or other suitable animals. The high specificity and sensitivity of detection of virions and virion components with antibodies has been applied for virus identification, virion quantitation, and for mapping structural features of virions. Depending on the procedure used, antibodies that react principally with the intact virion or with the dissociated coat proteins of the virion can be obtained. In the case of complex virions that contain envelopes, such as those of influenza and measles viruses, the antibodies react primarily with the glycoproteins of the envelope. Only by prior degradation of the envelope, which can be achieved by treatment with diethylether, are the antigenic properties of the capsid proteins revealed. In the case of simple viruses the antibodies elicited by the intact virion are different from those induced by the isolated structural subunits.

The principal means of detecting **viral protein-antibody reactions** are: (1) precipitation and agglutination, including the formation of lines of precipitation appearing in agar plates between the wells containing antibodies and viral proteins. After the two components diffuse or electrophorese toward one another; (2) neutralization (i.e., inactivation) of viral infectivity by the specific antibody; (3) complement fixation, that is, removal of the normal serum component, complement, as a consequence of the interaction of viral protein and antibody; (4) immunoblot procedures (described above), and (5) detection of the presence of viruses or viral components microscopically or electron microscopically through their reaction with a specific antibody. The antibody is made visible by coupling it with fluorescein or is made electron-opaque by coupling it with ferritin, an iron-rich protein showing a typical pattern on electron micrographs. Excellent reviews of the use of serological methods in virus research are available.

Great refinements in the use of serology in the identification and quantitation of viruses have been due to the use of monoclonal antibodies, *e*nzyme-*l*inked *i*mmunosorbent *a*ssays (ELISA) and *r*adio*i*mmune *a*ssay methods (RIA). ELISA has been an innovative approach for rapid screening of serum for antibodies to different viruses. Purified viruses or their viral proteins are placed in the bottom of 96-well plates (Fig. A.6), and serum is added to these plates. After appropriate incubation and washing, enzyme-linked antibodies against human antibodies are added, and the binding of the serum antibodies to viral proteins is detected by a color reaction involving the enzyme (e.g., peroxidase). This approach can lead to false-positive results if the virus is not totally pure. For example, retroviruses often have cellular proteins contaminating the preparation and anticellular antibodies would be detected. For this latter reason, purified viral proteins produced by genetic and molecular techniques can substantially reduce the false-positive results. Moreover, because these viral proteins can be concentrated, the test can be more sensitive and give fewer false-negative results.

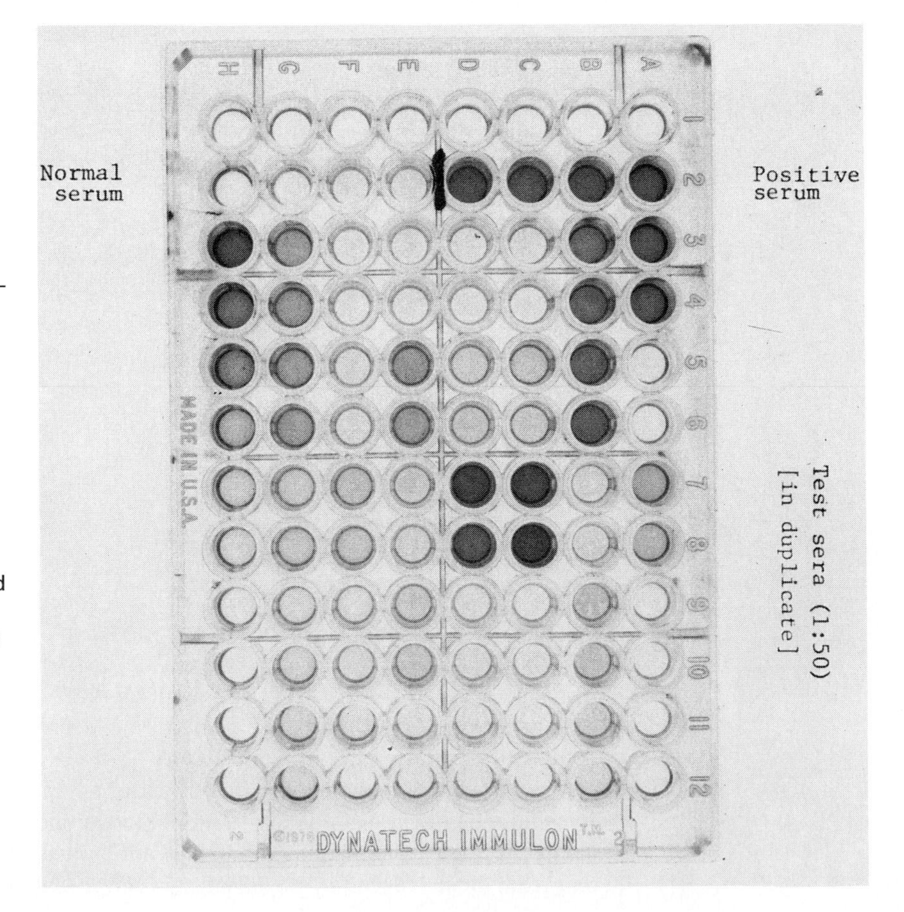

FIGURE A.6

ELISA test for HIV antigens. Purified HIV-1 has been placed in the well of a 96-well plate. After appropriate drying, dilutions of sera with potential antibodies to HIV-1 are placed in the wells. Following incubation and washing, immunoperoxidase-labeled antihuman antibodies are added to the wells. After a 1-hour incubation, this material is washed out and plates are exposed to the enzymic reaction that gives a dark color to those wells containing antibodies to the human antibodies. The results indicate a reaction of the serum immunoglobulins with viral proteins. (Courtesy of J. Homsy)

The RIA method involves labeling viral proteins through *in vitro* cultures with ^{35}S-methionine or ^{14}C-leucine and interacting these proteins after extraction with monoclonal antibodies attached to a solid phase, usually a plastic microtiter dish. The binding of the labeled protein with the antibody is detected with a scintillation counter. Dilutions of the protein-containing material can provide an estimation of the amount of virus-specific protein present. In **competition RIA,** unlabeled protein or a competitor peptide of the protein is used in the reaction, and the decrease in the amount of radioactivity bound to the antibody is quantitated. Another procedure, **radioimmune precipitation assay** (RIPA), utilizes a similar labeling technique, but antibodies are used to precipitate the extracted viral proteins that are subsequently placed on an SDS-polyacrylamide gel. Those proteins that are precipitated are separated by size, and the relative amounts of the protein in the cellular extract can be quantitated by densitometry or by direct protein analysis. Alternatively, immunoprecipitation can be performed with a monoclonal antibody or polyclonal serum and the proteins obtained analyzed by the Western blot procedure.

The main value of the ELISA, RIA, and RIPA techniques is that they provide great increases in sensitivity and ease of performing quantitative immunoassays. Moreover the development of monoclonal antibodies has markedly improved the specificity of immunological reagents and allowed major advances in understanding the structure of viruses and other macromolecular structures.

Monoclonal Antibodies One important limitation in the use of antibodies for comparison of antigenic properties among different viruses has been the fact that ordinary antisera are polyclonal. The sera contain multiple antibody species with different antigen binding sites. Each antibody species can bind to only a single localized site (e.g., as few as four contiguous amino acids in a protein sequence), which is known as an **epitope** on the antigen. Because ordinary antisera contain mixtures of antibodies recognizing different epitopes, when cross-reactions with different antigens occur, it is often difficult to determine whether all or only some of the different antibodies are responsible for a reaction. Thus, a cross-reaction may indicate only the partial identity of two proteins, that is, they may share certain epitopes recognized by the polyclonal serum but not others. Immunologists have believed for some time that each different antibody species in a polyclonal antiserum, which recognizes only a small portion of the whole virus, is produced by a different B cell within the immune system. Each single cell can then give rise to daughter cells by cell division, and the resulting groups of cells are called *clones.*

One of the most important technical developments in the immunological field was the discovery of methods to use cultured cells to produce antibodies to a single epitope or **monoclonal antibodies.** Two procedures are currently being used. In the first, the B cells are fused with mouse, or sometimes human, myeloma cells using fusigenic procedures (see Chapters 4, 11). The resulting **hybridomas** (composed of a B cell and a myeloma cell) are stimulated through the genetic nature of the malignant cell to produce large amounts of the antibody normally made by the fused B cell. The second method involves the transformation of the antibody-producing B cell with Epstein-Barr virus (see Chapter 8) so that it can replicate indefinitely in culture. Large numbers of cells derived from one infected cell (a clone) can then be used to produce a single species of antibody with unique specificity. This fact makes these monoclonal antibodies powerful new reagents for comparing and elucidating protein structures, for diagnosing pathogens, and, in some cases, for therapy.

Another recent development in immunology combines the high specificity of monoclonal antibodies with the ability to control the antibody specificity. When a complex macromolecular structure, such as a virus particle, is presented to the vertebrate immune system, the system appears to recognize and respond to it by producing antibodies against only a few selected epitopes. When an isolated portion of the structure, for example a purified surface protein, is presented to the same immune system, antibodies to epitopes different from those recognized on the intact virus may be produced. Similarly, a fragment of the viral protein may induce still a different group of antibodies, only some of which may be the same as some of those induced by the intact protein.

Taking this reductionist approach even further, immunologists have found that antibodies can be raised against small peptides (5 to 20 amino acids) derived from amino acid sequences of larger proteins. In a surprising number of cases, these antipeptide antibodies have been found to react with the complete native protein or virus particle containing the same amino acid sequence. When polyclonal antisera against the intact virus are examined for reactivity with the pure peptides that induce antiviral antibodies, there is generally no evidence of the antibody reaction with the peptide. Evidently, the immune system preferentially recognizes more complex three-dimensional or discontinuous epitopes, consisting of amino acids from several different portions of a protein or even from different proteins. The conformational structure with its folding and aggregation of the polypeptide chains bring these epitopes into proximity within the intact virion.

Since the antipeptide antiserum can be directed to a highly limited region (often a single epitope) of a selected protein, antibodies produced in this way have been said to have **predetermined specificity.** This approach does not work with every randomly selected peptide from a virus. In fact, it is most successful when something about the structure of the virion is known (e.g., what protein regions are exposed to antibodies by being on the virion surface). Nevertheless, the usefulness of this approach has been amply illustrated in many virus systems. It has even been successfully utilized to make antibodies that react with proteins that have never been directly detected by other means (e.g., putative gene products defined by gene sequencing only).

Electron Microscopy The electron microscope is one of the virologist's most powerful tools. The image is the result of the scattering of the electron beam by the electrons in the specimen and is thus favored by the presence of heavy atoms in the sample. For this reason, most routine electron microscopy is of samples that have been "negatively stained" with salts containing a heavy metal or "shadowed" with a heavy metal. The necessary drying of the virion-containing aqueous solution often makes the presence of nonvolatile salts quite undesirable. By freeze-drying the sample, the flattening of soft or hollow objects that occurs on drying is considerably diminished.

The classical method of obtaining an image of the gross shape of a virion is by shadow casting, which consists in monodirectionally depositing vaporized heavy metals on the specimen, thereby also creating shadows behind it, this process reveals much of the three-dimensional shape of a particle. One of the most instructive illustrations of this technique is provided by the image of the large insect virus *Tipula* iridescent virus (IIV), where shadowing from two angles reveals its icosahedral shape (see Fig. 1.7).

The staining procedures generally used differ from shadowing in that they do not increase the dimensions of the objects and show considerably more detail. So-called **negative staining** rapidly yields quite clear and informative images. It consists of adding the heavy-atom stain (sodium phosphotungstate or **uranyl acetate**) to the neutral aqueous virus solution and allowing the mixture to dry on the specimen film. Under these conditions the stain does not combine with the viral components but stains the background, including all holes and crevices, so that the particles are revealed in great detail. Our understanding of the structure of isometric viruses was increased greatly by this technique and others, such as the **freeze-fracture** and replica methods. Thin sectioning is a particularly useful technique for revealing virons within cells as well as their entry into and release by the cell. To observe individual viral nucleic acid molecules, various **spreading techniques** have been developed as well. Electron microscopy serves also as a quantitative tool, in that it allows the counting of typical virus particles (Fig. A.7). Quantification is achieved by adding a definite number of indicator particles to the solution. For example, from the ratio of TMV rods to polystyrene latex spheres in a number of microdrops the absolute number of TMV rods per milliliter can be calculated if the number of indicator spheres per milliliter is known.

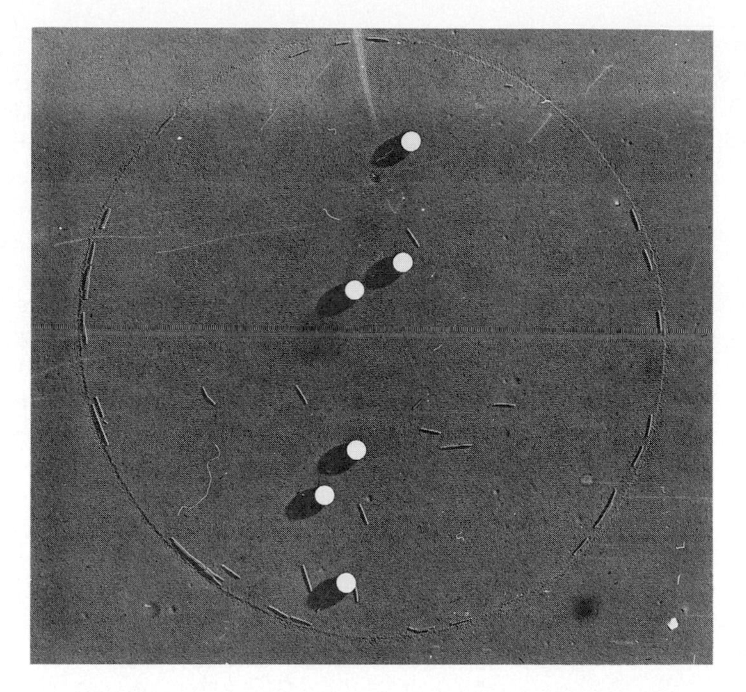

FIGURE A.7

Electron micrograph of a microdrop of a solution containing TMV particles and a known concentration of polystyrene latex particles. The ratio of the two types of particles permits calculation of the concentration of the virus particles. Fragments of rods, shorter than the standard 300 nm, can be seen. (Courtesy of R. C. Williams)

A new approach to electron microscopy of virions, **cryoelectron microscopy,** exploits thin, frozen sheets of concentrated virion solutions, low electron beam density, and no stain for the sample. Intensive computer analysis of the images allows a three-dimensional pattern of electron density to be reconstructed for the virion sample. A resolution of 2 to 3 nm has been obtained for the best samples studied by this technique.

X-ray diffraction analysis represents a powerful tool for elucidating the structure of substances that can be obtained in a state of orderly molecular alignment. This is the case for the true crystals formed by many isometric viruses, as illustrated for poliovirus in Figure A.8. The X-ray analysis of TMV gave the first indication that the structure of a virus is composed of many regularly arranged subunits. The location of the nucleic acid within virus particles and many details of capsomer topography and other structured features also were established by X-ray diffraction. More recently, **neutron-scattering** techniques have provided additional insight into virion structure.

(a) 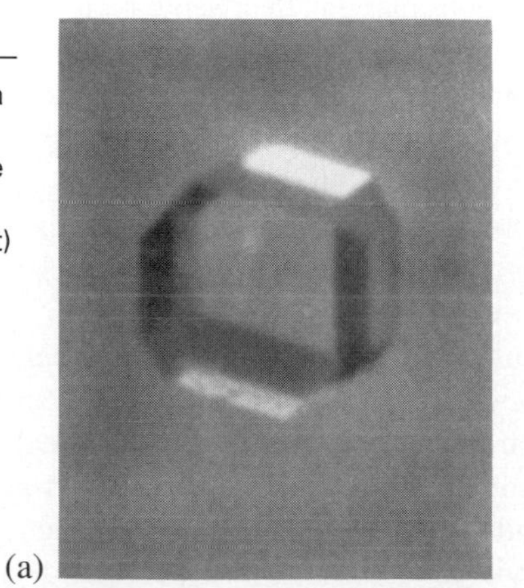

(b) 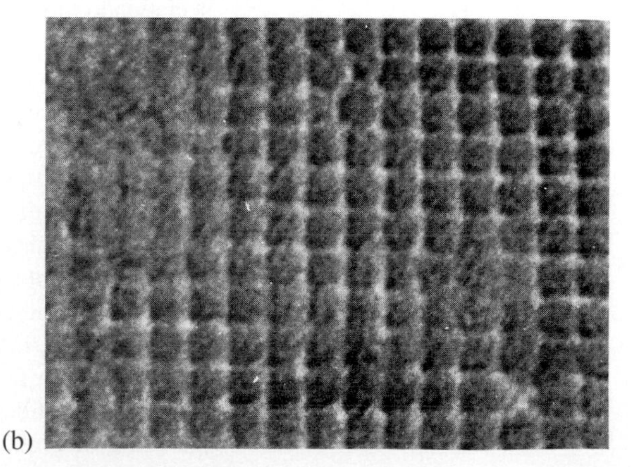

FIGURE A.8

A light-microscope picture of (a) a polio virus crystal and (b) an electron microscopic image of the orderly arrangement of polio virions. (Courtesy of C. E. Schwerdt)

II. METHODS FOR RECOVERY OF VIRAL PROTEINS AND NUCLEIC ACIDS

A. *Identification*

All but the simplest viruses contain more than one species of protein, and many carry multiple nucleic acid species. Standard biochemical methods are then required to isolate single species of these macromolecules as well as the lipids obtained from the envelopes. Among these are **gel exclusion chromatography** using Sephadex, Biogel, or Ultragel, ion exchange chromatography, electrophoresis on paper or polyacrylamide gel, sedimentation, salt precipitation, and other methods. Most of these procedures are the same as those mentioned for virus purification.

Heat, acid, alkali, and many kinds of denaturants dissociate virions. From the resultant mixtures proteins and nucleic acids can often be separated on the basis of differences in their solubility, at various concentrations of salts or organic solvents. However, such preparations may well be irreversibly denatured or partially degraded. If the purpose is the isolation of intact and preferably native viral components, carefully controlled conditions must be used and tailored for each specific virus component.

With viruses, as with many other biological materials, the amounts that are available for study are often very small. Radioactive labeling is then the method of choice. Viruses can be labeled by allowing the host cells to incorporate a radioactive metabolite containing ^{3}H, ^{14}C, ^{35}S, or ^{32}P. These may be amino acids, nucleosides or bases, carbon dioxide, phosphate, sulfate, and so on. The isolated viral components can also be made radioactive by chemical modification. Iodination with $^{125}I_2$ or $^{131}I_2$ is frequently used for proteins but can also be employed with nucleic acids, although for the latter, enzymatic addition of [^{32}P]phosphate to chain ends is usually preferable.

Of particular advantage for the characterization of viral components is the use of different radioactive metabolites during their biosynthesis. Thus, proteins found to be labeled with ^{32}P are phosphoproteins; those labeled when [^{14}C] fucose or [^{3}H] glucosamine is given to the cells are glycoproteins, and so on.

B. *Analysis of Virion Components*

The procedures for characterizing viral proteins and nucleic acids are no different from those used with other biologically active macromolecules. As usual, the methodologies employed to isolate them and separate them from one another already contributes much to that characterization. Properties such as molecular weight, isoelectric point, and the presence of carbohydrates and phosphate groups can become evident during their isolation.

Further characterization after separation of individual components usually concerns the chemical, physicochemical, and biological (i.e., functional) properties of these macromolecules or molecular complexes. Among the techniques useful for this purpose are electron microscopy, X-ray diffraction, and determination of number of subunits (if any). Moreover, dissociation to monomeric molecules and their separation, if not identical, by sedimentation and polyacrylamide gel electrophoresis (PAGE) without or with prior denaturation can be used. In addition, amino acid composition and sequences, nucleotide composition and sequences, and location and structures of associated polysaccharides, lipids, and phosphates can be helpful. Nucleic acids must also be analyzed for associated small proteins, and possibly vice versa.

The functional property of nucleic acids to be tested first is their infectivity. The *in vitro* messenger and/or template activities of many nucleic acids, as well as possible tRNA activities, can also be assessed.

At present the **nucleotide sequence** of viral nucleic acids can often be rapidly determined, using either virion RNA or DNA or recombinant DNA clones of the viral genetic material. DNA may be sequenced directly by **chemical degradation methods** (Maxam and Gilbert), or DNA or RNA may be sequenced by **enzymatic synthesis methods** using dideoxynucleotides and DNA polymerases (Sanger method). In the best cases, partial sequencing of proteins via chemical methods (Edman degradation) is used to confirm the amino acid sequences of proteins predicted by the open reading frames (ORFs) of the nucleic acid sequences translated by computer.

Many viral proteins have a structural role and can be tested for their ability to aggregate in a specific manner and re-form virus particles *in vitro*, with or without the viral or any nucleic acid. Certain proteins of the more complex viruses may be able to interact to form specific organelles, such as the spikes of enveloped viruses or the tail or base plate of certain bacteriophages. Other proteins isolated from viruses may have enzymatic activities that can be assayed, such as the neuraminidase of many animal viruses, the reverse transcriptase of retroviruses, and so on. Protein interactions and enzymatic activities are often interrelated phenomena. Thus, the enzymatic activity can be a property of a complex of several identical or different peptide chains, some of which may also play a structural role.

C. Isolation of Intact Nucleic Acids

Obtaining intact nucleic acids is now usually achieved by extraction of an aqueous, neutral, or slightly alkaline virion solution with phenol at 0° to 37°C, with or without pretreatment with the detergent sodium dodecylsulfate (SDS, 0.1 to 1 percent) at 37°C. This strong anionic detergent alone suffices for the degradation of TMV. Frequently, the degradation of viruses is facilitated by treatment with a protease, often in the presence of SDS. For this purpose, the bacterial protease K is now preferred. Many nucleic acids are known to carry a covalently linked protein molecule at their 5′ end, and if this protein is to be retained, proteinase must obviously be avoided. The nucleic acid can be precipitated with two to three volumes of ethanol if the coat protein has been removed by phenol extraction. Stepwise precipitation with salt solution can also be employed to separate protein and nucleic acid. Moreover, polyvinyl sulfate, heparin, and other charged materials are often used during the purification of viral RNA to inhibit the action of ribonuclease. Alkali, particularly, must be avoided with RNA, and acid with DNA. The specific conformations of single-stranded viral nucleic acids, such as hairpin structures and other folding, are lost if very low salt concentrations and/or metal chelating agents, such as EDTA, are used during their isolation. This process leads to UV hyperchromicity but does not cause any loss of biological activity, since such denaturation is apparently readily reversible (see Fig. A.5).

D. Isolation of Viral Proteins

The isolation of disaggregated **viral proteins,** as do all disturbances of protein/protein interactions, involves some conformational changes. Such partial denaturation is also generally (or at least frequently) reversible, and functional proteins can be recovered. The most useful agents for nucleic acid isolation, phenol and chloroform/bultanol, cannot be used in the purification of proteins. Many proteins can be renatured after exposure to gentle alkaline or acid conditions or high concentrations of urea. The glycoproteins of viral envelopes, such as the neuraminidase and hemagglutinin of influenza virus, may be obtained in undissociated form by removing the associated lipid by treatment with ether or other organic solvents. Carefully controlled treatment with detergents, including SDS, can also be used to isolate envelope proteins from this type of virus without loss of its biological activity. The active envelope proteins of the paramyxoviruses or retroviruses are preferentially prepared by treatment with a gentle nonionic detergent such as Triton X-100 or by guanidine hydrochloride at a concentration of 1 to 2 M. Digestion of envelope proteins by proteolytic enzymes can at times facilitate isolation of the internal proteins.

III. DETECTION AND QUANTITATION OF VIRUS ACTIVITY

The effects of virus infection on host organisms, tissues, or cells can range from nondetectability to severe disease symptoms and death. Plants and animals may become generally affected, which is termed **systemic** infection, or they may show only a localized response at the site of application. The **local lesion** response to plant virus application on a leaf (Fig. A.9) corresponds to that produced on the chorioallantoic membrane of the chick embryo within the egg by many animal viruses. It is also analogous to the **plaques** produced by viruses on monolayers of animal cells in culture (Figs. 11.3 and 14.8) and by bac-

FIGURE A.9

Local lesions produced on Turkish tobacco, cultivar *Xanthi nc*. Each leaf has common TMV on one half-leaf (the large lesions), and a strain that produces small lesions (HR) on the other half-leaf; lesions were observed 6 days after inoculation.

teriophages on bacterial lawns (Fig. A.10). In each case the size and appearance of the lesions or plaques, each representing a focus of infection (reflected by cell death or morphological changes), can give important information about the virus responsible, and the number of such lesions reflects the number of infectious particles applied to the test object. Thus, the plaques on bacterial and animal cell cultures and the local lesion response of plants are useful for both diagnosis and quantitation.

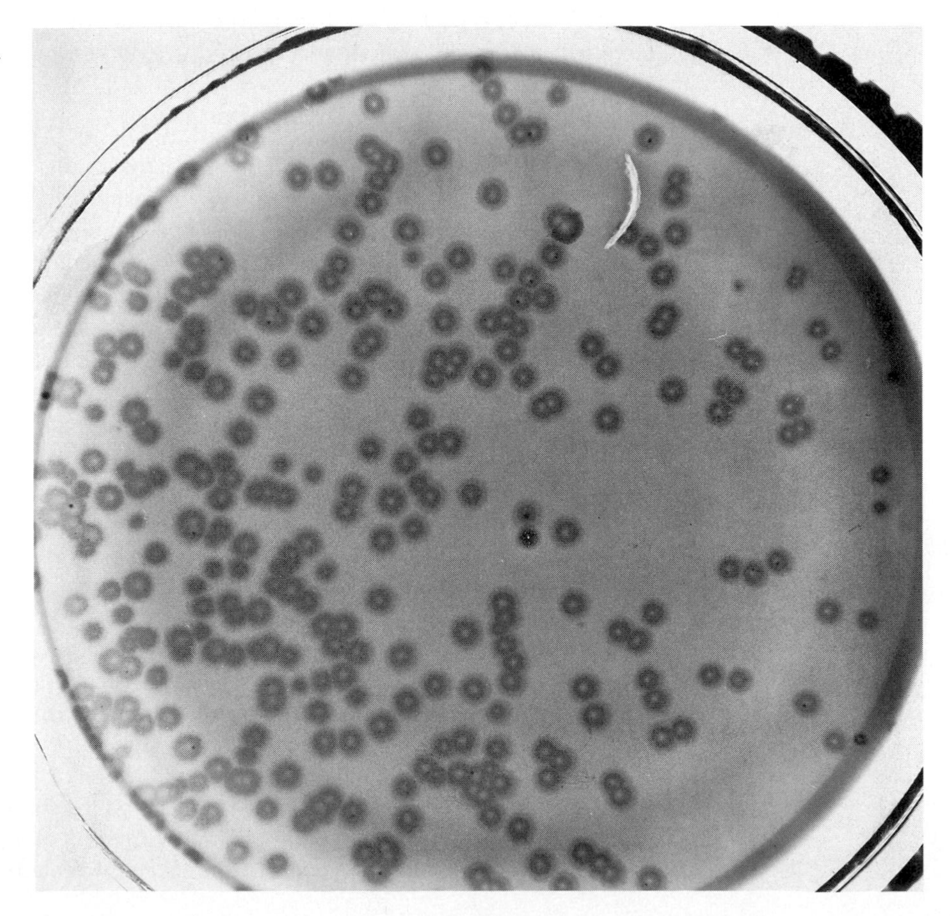

FIGURE A.10

Petri dish covered with a lawn of *E. coli* (Br +) and inoculated with T2(r) phage. A few phage mutants (r) can be detected (one near top). (Courtesy of G. Stent)

The difficulties inherent in the detection, identification, and propagation of viruses are often grossly underestimated. Although it is easy to achieve these objectives for well-known viruses, the characterization of a new virus, available in only trace amounts in a tissue sample or biological fluid, is a difficult and often unsuccessful task. To visualize a virus directly by electron microscopy requires at least 1 million particles. Thus, sufficient replication and survival of the agent must occur. Many viruses, however, replicate to low titers and are very labile; they may decompose during attempts to isolate or transmit them. Some viruses only slightly perturb their infected host and produce either no pathological changes or barely detectable changes. Others may be replicated in only one type of tissue (see Chapter 11). In these instances, the amounts of inoculum are likely to be very limited, and the inoculum may well be exhausted before a host for detection by propagation has been found.

A. Quantitation Assays

An important prerequisite for most biological research is the availability of one or more quantitative tests. Great advances in our knowledge of both plant and animal viruses often were preceded by the development of reliable means of quantitating viral infectivity. In a typical **plant local lesion assay** system, tobacco mosaic virus (TMV) is applied to fully-expanded leaves with an abrasive powder (needed to break the cell wall) such as carborundum, and the inoculum in 0.1M phosphate buffer of pH 7.0. Inoculum concentrations typically will be varied over a 1000-fold or greater range, and the numbers of local lesions observed will be proportional to the inoculum concentration over an approximately 50-fold range (Fig. A.9).

Tobacco (*Nicotiana tabacum,* var. *Xanthi nc*) shows lesions corresponding to virus concentration. Thus, 0.1 ml of a 0.01 μg/ml (10^{-8} g/ml) solution of TMV applied to a leaf may, in a given test, produce 50 necrotic lesions (dark brown spots); one-tenth that concentration of virus (0.001 μg/ml), 5 lesions; and a fivefold higher concentration (0.05 μg/ml), 250 lesions. Naturally, such numbers are obtained only as the average of many infected leaves. To minimize the variability of the response, unknown samples or solutions are usually compared with a known concentration of the pure virus, if it is available, and both solutions are distributed over equivalent leaf areas. Often, half of a leaf is used for the unknown, and the opposite half for the known (standard) sample. The relative infectivity of the unknown can in this manner be determined on about six plants (30 half-leaves for unknown and standard), and with an accuracy of about ± 15 percent. Figure A.9 shows examples of local lesions caused by a TMV mutant giving smaller lesions on that strain of tobacco (see also Sec. 12.9).

Quantitation of virus assays is complicated by the fact that, for practical reasons, relatively few lesions can be scored per assay. This problem introduces uncertainties into the calculated concentration of **infective units** (IU), due to sampling variations. Such variations can be estimated using statistical methods. In particular, under the usual assumptions about sampling variations, the statistical reliability of the IU concentration increases in proportion to the number of samples assayed and the total number of lesions counted. The level of certainty of a determined number of IUs is therefore usually expressed as the average level detected in several samples, plus or minus a percentage (e.g., ± 15 percent in the TMV assay above). This means roughly that, given the observed lesion counts, the actual IU level is expected to be within ± 15 percent of the calculated level, for example, 99 of the 100 times that the assay is performed. More extensive and precise **statistical considerations** of virus assays are beyond the scope of this book, but they are, nonetheless, essential for the proper practice of experimental virology as well as other microbiological disciplines, such as bacteriology.

Bacteriophage-containing solutions are tested by mixing them with an excess of susceptible bacteria and nutrient agar, plating the mixtures on petri dishes, and incubating them (see Fig. A.10). Most animal viruses can similarly be tested on tissue culture plates containing a monolayer of susceptible cells. In these assays each infectious particle induces an area or focus of cell killing or transformation (see Figs. 11.3 and 14.8). Direct proportionality between virus particle count and number of foci of infection is generally accepted for the bacterial and animal viruses. Nevertheless, biological variability, chance, and other factors affect all these methods. They are not comparable to the titration methods of the chemist and preferably should be termed virus assays rather than virus titrations. This suggestion is particularly advisable because in most systems it is not the absolute number of virus particles that is determined but a number lacking direct physical significance, namely, the number of particles that happen to be able to cause an infection [infectious units (IU) in general or, more specifically, for example, plaque-forming units (PFU)] under the conditions of the test (See Chapter 1).

In the case of the T-even bacteriophages, the ratio of particles to PFUs has been shown under the most favorable conditions to be between 1 and 2, meaning that almost every virus particle starts an infection. For animal and plant viruses this ratio can range from 4 to 10,000 or more. This fact could be and has often been interpreted as signifying that in some instances most virus particles are not infectious. However, the more probable explanation is that all typical (i.e., nondefective) virus particles are potentially infectious but that many or most of them happen not to be able to initiate an infection under the test conditions used. The previously mentioned fact that many bacteriophages possess an **infection-favoring organ** but that animal and plant viruses infect only through certain cell receptor sites or wounds suggests chance as an important factor in explaining the varying and generally lesser efficiency of infection by plant and animal viruses.

It must be stressed that quantitative evaluation of the infectivity in terms of virus particles is possible regardless of the ratio of physical particles to lesions, infective centers, or PFUs, provided that a solution containing purified virus at a known concentration of particles is available and its infectivity can be compared with that of the unknown infectious material. Furthermore, it should be noted that the proportionality between concentration and lesion number, in other words, the linear **dose-response curve,** that can be observed with most viruses over a certain range of concentrations is a strong indication that infection can result when several or many particles are in these instances required for infectivity. However, this fact does not preclude the entry and replication of several infectious particles in one host cell. Actually, in the case of certain specific plant viruses, and also a few animal virus infections, two or three virus particles are required for infection and replication (see Chapter 1, Sec. 6.1.5 and Figs. 2.8 and 6.11). For these viruses the dose-response curves are nonlinear and complex. In such cases, the number of particles required in a single cell to initiate infection can, in principle at least, be determined from the shape of the dose-response curve using a statistical tool known as the Poisson distribution.

B. Poisson Distribution

One of the most important factors in the outcome of infections is the ratio of input virus to target cells, that is, the **multiplicity of infection (m.o.i.).** When this ratio is high, viral host cells may be killed immediately without replicating the virus, and the natural defenses of whole organisms may also be overwhelmed more easily than in cases of infections beginning with a low m.o.i. Alternatively, defective particles may be present that interfere with normal virus replication (see Sec. 4.1.4)).

The **average m.o.i.** in a culture of many cells is simply the number of infectious units (IU) divided by the number of cells in the culture. Infection at a high m.o.i. (e.g., greater than 10 IU/cell) generally results in essentially all cells receiving at least one infectious virion. However, owing to the particulate nature of viruses (a cell cannot be infected by half a virion), different cells will receive different numbers of particles. Once again, statistical methods can be used to advantage in virology, this time to predict the outcome of infection at any given m.o.i. In particular, the **Poisson equation** describes the sampling distribution for the number of cells receiving 0, 1, 2, ...virions as a function of m.o.i.:

$$P_r = \frac{s^r e^{-s}}{r!}$$

where r is the number of particles in a given sample (i.e., in a given cell for this application), s is the average number of particles per sample (i.e., the m.o.i. here), e is the base of the natural logarithm system, and $r!$ ("r factorial") is the product of $r \times (r-1) \times (r-2) \times \cdots \times 3 \times 2 \times 1$. P_r is then the probability that a given sample will contain r particles (or in this case, the fraction of cells receiving r infectious virions). As a specific example, it can be calculated that with an average m.o.i. of 1 IU per cell, only about one-third the cells receive one infectious particle, one-third receive two or more, and the remainder of the cells are not initially infected at all.

Historically, the Poisson analysis played an important role in the establishment of the concept that bacteria have genes, that is, that they have heritable traits that can change only rarely through chance mutation. Prior to the famous experiments that were conducted in the 1940s by S. E. Luria and M. Delbrück, it was generally thought that each individual bacterial cell could "adapt" to any new environmental condition, such as introduction of an antibiotic. Luria and Delbrück used statistical analyses of the spontaneous occurrence of drug-resistant variants in the absence of the drug to demonstrate that known genetic principles were also operating in bacteria. The Poisson analysis permitted these authors to predict a genetic mutation rate of 10^{-8} that can be applied to all genes. The impact of these experiments was at first quite small, because most biologists were not familiar with statistical concepts. The great advancement of virology and other biological sciences today, however, is due in large measure to the introduction of mathematics by physicists such as Delbrück and others who were intent on making biology into a more quantitative and precise science.

C. *Other Procedures for Detection and Quantitation*

When plaque or local-lesion assays are not available for a given plant or animal virus, several less quantitative means of estimating the concentrations of infectious virus must be considered. One of these techniques determines by **serial dilution** the minimal amount or dose required to elicit the typical systemic disease symptoms. In other instances, the time necessary to produce a certain response, as compared with a known virus titer, can be used as a measure of virus concentration. Combinations of these two methods of estimating viral infectivity are also used at times.

To obtain statistically meaningful quantitation by this approach, the dose required to infect 50 percent of a group of test organisms is determined; this is called the **infectious dose$_{50}$** or, in the special case of death of the host as the assay end point, the lethal dose$_{50}$ (LD_{50}). In cell culture studies, it is the tissue culture infective dose ($TCID_{50}$). Once again, the reliability of the $TCID_{50}$ estimate from any assay depends on the number of replicate samples with infections detected in the assay.

The virus $TCID_{50}$ is determined by diluting out a virus preparation in replicate plates that contain susceptible target cells (e.g., peripheral CD4+ lymphocytes for HIV). 2-, 4-, or 10-fold dilutions are made in a sufficient number of wells so that the number of positives can be used to determine the point when 50% of the wells have a probability of being infected. The cell culture supernatant or the cells themselves are tested for the presence of virus, and the $TCID_{50}$ dose is determined by the procedure of Reed and Muench.*

The preceding procedures measure infectious viruses, because they record the ability to infect and replicate in the cell. Other methods for quantitation include the ability of viruses to bind to red cells **(hemagglutination).** In this example, only the viral envelope is needed, not an infectious virion. This approach has enabled researchers to estimate the number of infectious particles per total virions present in a virus preparation. In some cases, more than 1000 to 10,000 noninfectious particles accompany the infectious parti-

*Reed, L. J., and H. Muench, (1938) *Am. J. Hyg.* 27:493.

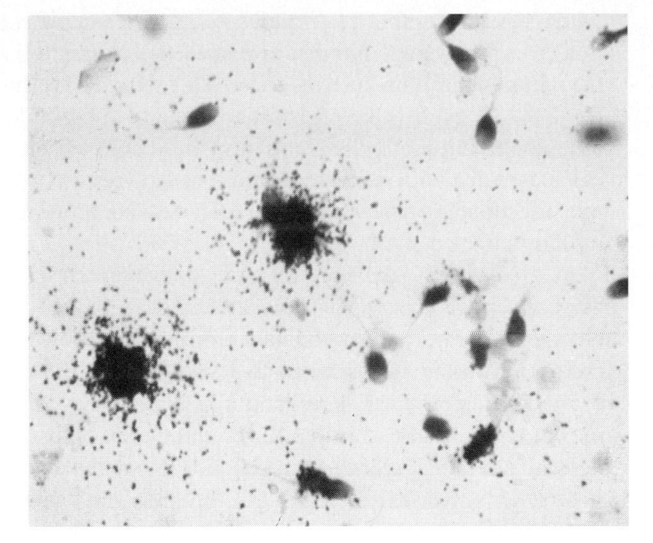

FIGURE A.11

In situ hybridization of HIV-1 in seminal fluid. For these studies, a radioactive cDNA probe to HIV-1 RNA is used to interact with cells from seminal fluid fixed on a slide. The extent of HIV-1 RNA production is monitored by the grains of radioactivity, indicating the interaction of the radioactive probe with the virus RNA. Two cells among the spermatozoa are seen producing HIV-1. (From Levy, J. (1988) JAMA 259:3037)

cles. These defective viruses can interfere with infection and are observed in a variety of virus families (see Chapters 2 and 4). They can be reduced by inoculating cells with a high dilution of the virus-containing preparation (i.e. low m.o.i.).

In situ **hybridization** is another mechanism for detection and quantification of virus in tissue or within infected cells (Fig. A.11). By this procedure, a radioactive probe to the viral genetic material is added to a specimen fixed with paraformaldehyde or in frozen section or to cells acetone-fixed on a slide. The interaction of these probes with viral nucleic acids within the cells can be detected, either by radiographic film or by immunoperoxidase staining. Quantitation is then possible by counting the positive cells. It is important to recognize that these techniques detect and quantitate virions and not infectious particles. Infectious virus can be quantitated only by a biological assay using standard cell culture procedures or animal experimentation.

Most recently, **polymerase chain reaction (PCR)** techniques have been developed for detecting small amounts of a virus in a tissue or in cells. There involve the amplification of the viral genetic material using small primers (approximately 20 base pairs) that bridge a larger portion of the genetic sequence of the virus. Using a process in which heat-resistant DNA polymerase deoxynucleotide triphosphates and primers are added to the DNA extracted from cells or tissue, the primers hybridize to their target sequence and allow synthesis of a complete segment copy of the genome. Subsequent heating separates these primer pair products, and the cycle can be repeated. After 30 cycles, up to a million copies of the initial genetic material can be obtained. The sequence can then be detected by the Southern blot procedure using specific labeled probes to this particular genetic region of the virus. This amplification technique enables the detection of minute numbers of copies of viruses (or parts of them) or other organisms in a specimen. It can be used to detect the number of virions in a body fluid. In the case of RNA viruses, recent modifications take advantage of the ability of reverse transciptase to convert RNA to DNA and then to proceed with the usual PCR process. Most recently, PCR has been developed to amplify genetic material inside one cell for detection by *in situ* hybridization. Thus, the infectious agent can be identified in a particular cell recognized by immunohistology. The PCR process promises to be a great help in microbiology for detection of known infectious organisms or related ones suspected to be responsible for clinical disease. It is most helpful for the quantification of viruses in the host in different tissues and for estimating the number of copies present within a cell. Because of its value, the developer of PCR, Mullis, won the Nobel Prize in 1993.

appendix

2

Vertebrate and Insect Virus Families

Adenoviridae: A very uniform family of icosahedral nonenveloped virions about 80 nm in diameter composed of 252 capsomers; buoyant density in CsCl, 1.34 g/cm^3. The 12 vertex capsomers (pentons) carry strain-characteristic glycoprotein fibers (10–30 nm long) with knobs at the ends; the rest is made up of hexons. The genome is linear double-stranded DNA of 20 to 25 $\times 10^6$ Da in the mammalian and 28 to 30 $\times 10^6$ Da in the avian viruses. At least 10 proteins of 5 to 120 $\times 10^3$ Da make up the virion. (*Example:* human adenovirus, types 1-30) (*adenus,* "gland")

Arenaviridae: Enveloped pleomorphic, though predominantly round, virions 100 to 200 nm in diameter (325 to 580 S; density in sucrose, 1.2 g/cm^3.) consisting of a core containing ribosomelike particles and a lipid bilayer envelope with surface projections. Two viral ambisense RNAs of about 1.1 and 2.7 $\times 10^6$ Da and smaller ribosomal RNAs are present. The nucleocapsid protein is about 63 $\times 10^3$ Da; two glycoproteins are somewhat smaller. (*Examples:* lymphocytic choriomengitis virus, Lassa fever virus) (*arena,* "sand.")

Baculoviridae: Insect viruses. The virions are usually rod-shaped nucleocapsids, with lipid bilayer envelopes, frequently in bundles occluded in "crystalline" protein bodies. The nucleocapsids are about 50 $\times$ 300 nm and have a density in CsCl of 1.47 g/cm^3, compared with 1.21 g/cm^3 for the enveloped virion. The genomes are circular, double-stranded DNAs of 60 to 110 $\times 10^6$ Da, and there are 10 to 25 virion proteins, including the single virus-coded matrix protein, termed *polyhedrin Autographica californica.* (*Example:* multiple nuclear polyhedrosis virus) (*baculus,* "stick")

Birnaviridae: Icosahedral particles 60 nm in diameter (at times resembling Reoviridae) (435 S; density in CsCl, 1.32 g/cm^3) contain two linear double-stranded RNAs of 2.5 and 2.3 $\times 10^6$ Da and four proteins of 29 to 105 $\times 10^3$ Da. Viruses mostly of fish and molluscs; some in insects (Example: infectious pancreatic necrosis virus of fish) (*bi*, two)

Bunyaviridae: Large family of viruses. Oval to spherical virions about 95 nm in diameter (400 S, density in CsCl, 1.2 g/cm^3) containing three major envelope glycoproteins, one minor large nucleocapsid protein forming long helical nucleocapsids (2.0 to 2.5 nm in diameter) with three negative-sense RNAs (3 to 5, 1 to 2, and 0.3 to 0.8 $\times 10^6$ Da). Some members have been shown to use an ambisense expression strategy. Reassortment between the RNAs of different bunyaviruses or strains that carry the information for the proteins of equivalent size has been observed. The viruses have lipid-rich envelopes and a wide host range. [*Examples:* Bunyamwera virus. Haanton virus (Bunyamwera is African site of first virus identification)]

Caliciviridae: This virus family somewhat resembles enlarged Picornaviridae. The isometric particles with cup-shaped depressions have a diameter of 38 nm (183 S; density in CsCl, 1.37 g/cm^3) and contain 18 percent positive-sense RNA of 2.8 $\times 10^6$ Da with poly(A) at the 3′ end and a 10 to 15 $\times 10^3$ Da protein bound to the 5′ end. The virions are composed of 180 molecules of a single protein of about 67 $\times 10^3$. Their particle weight is thus 15 $\times 10^6$ Da, compared with 8 $\times 10^6$ Da for the picornaviruses. [*Example:* vesicular exanthema of swine virus, (possibly Norwalk agent, Hepatitis E)] (*calix,* "cup", "goblet").

Coronaviridae: Spherical and pleomorphic particles, 60 to 220 nm in diameter (buoyant density in sucrose, 1.18 g/cm^3) with characteristic club-shaped, widely spaced projections about 20 nm long. They give the particle the so-called coronalike appearance. The helical nucleocapsid consists of a positive-sense RNA of about 6×10^6 Da carrying poly(A) and a phosphoprotein of about 55×10^3 Da at the 3′ and 5′ ends, respectively. Three or four proteins and glycocoproteins as well as lipids make up the envelope, and a large glycoprotein, the peplomers. (*Example:* avian infectious bronchitis virus) (*corona,* "crown").

Filoviridae: Heterogeneous long (up to 14 um) filaments, 80 nm in diameter with surface projections 10 nm apart (7 nm long). One negative-sense RNA of 4.5×10^6 Da. Proteins (L, G, N, four capsid proteins) and lipids; resemble Rhabdoviridae (*filo,* "thread," or "filament") in their replication. Infection can be very deadly in humans. (*Examples:* Marburg, Ebola viruses)

Flaviviridae: Spherical, 40 to 60 nm in diameter with lipoprotein envelope consisting of two or three virus-coded proteins and host lipids. Positive-sense single-stranded genomic RNA is capped at 3′ end, and particles contain core protein of 15×10^3 Da. Mammals, birds, and arthropods are hosts. In the latter, the viruses are generally not pathogenic. (*Example:* yellow fever, tick-borne encephatitis virus) (*flavi,* "yellow")

Hepadnaviridae Spherical particles are 45 nm in diameter. A thin detergent-sensitive envelope covers an icosahedral nucleocapsid. A partially double-stranded circular DNA of 1.6×10^6 Da (one negative-sense of strand 3200 bases; the other, a short positive-sense strand of 1700 to 2800 bases). A 22×10^3 Da protein makes up the core, and there are three additional minor proteins, including a reverse transcriptase-RNAse H. The viruses are highly host-specific. The human hepatitis B virus occurs in the blood as the Dane particle of 42 nm diameter and as a core of 22 nm diameter. (*Examples:* Human hepatitis B virus, duck hepatitis virus, ground squirrel hepatitis virus) (*hepadna,* "liver DNA virus")

Herpesviridae: More than 80 herpesviruses have been isolated from many different hosts. The virion, 120 to 200 nm in diameter (density in CsCl, about 1.25 g/cm^3), consists of a nucleocapsid core, an icosahedral shell (10^5 nm in diameter) of 162 capsomers, the tegument, and the envelope with surface projections. The molecular weight of the linear double-stranded DNA ranges from 80 to 150 $\times 10^6$ Da for different genera, and those of the more than 20 proteins from 12 to 200×10^3 Da. Several proteins are phosphorylated. The envelope contains lipid and glycoproteins. (*Examples:* varicella virus, Epstein-Barr virus, cytomegalovirus, Marek's disease virus) (*herpes,* "creep").

Iridoviridae: A rather heterogeneous group of large icosahedral particles 130 to 300 nm in diameter (1300 to 4500 S; density, 1.16 to 1.35 g/cm^3). They contain one or two molecules of double-stranded linear DNA of 100 to 250×10^6 Da. The viruses also contain, besides the DNA, 15 to 20 proteins (about 75 percent) and lipid (about 6 percent). (*Example: Tipula* iridescent virus) (*iridos,* "iridescent").

Nodaviridae: Insect virus. Isometric particles 29 nm in diameter (135 S; density, 1.34 g/cm^3; stable in detergents and at pH 3) consisting of two positive-sense RNAs, lacking poly(A), of about 0.46 and 1.15×10^6 Da, both required for infectivity, and an intracellular RNA of 0.15×10^6 Da, derived from the large RNA. The 104×10^3 Da gene product is probably a component of the RNA polymerase; the 46×10^3 Da is the coat protein precursor. (*Example:* Nodamuravirus) (named for site of isolation; Japanese village called Nodamura)

Orthomyxoviridae: A monogenic family of enveloped negative-sense RNA viruses, containing types A, B, and C influenza virus. The particles are pleomorphic, about 100 nm in diameter (about 750 S; density in sucrose, 1.19 g/cm^3); very large filamentous particles also occur. The lipid-rich envelope carries many of two (or three) types of glycoprotein spikes, the hemagglutinin and the neuraminidase of about 80 and 60×10^3 Da. The RNA of types A and B is in eight segments of 0.2 to 1×10^6 Da (total about 5×10^6 Da). RNAs are single-stranded and carry at least 10 genes; the shortest RNA contains two overlapping genes. They are covered by the nucleocapsid protein as helical filaments of varying lengths. The main protein makes up the matrix that underlies the envelope. Three large proteins are associated with RNA polymerase and other enzymatic functions. (*Example:* influenza virus) (*myxa,* "mucus," referring to the glycolipoprotein envelope)

Papovaviridae: Two genera make up this DNA virus family: the larger the *papilloma* and the smaller the *polyoma* viruses. Both are nonenveloped and icosahedral, 55 and 45 nm in diameter (300 S and 240 S), respectively; density in CsCl, 1.32 g/cm^3. They consist of one molecule of circular double-stranded DNA of 3 to 5×10^6 Da, and seven and five proteins, respectively. Papovaviruses have been found in many mammals, including humans. They give rise to warts and some cancers (*Example:* papilloma human virus) (Name derived from viruses in what were formerly regarded as three genera: *Pa*pilloma, *po*lyoma, and *va*cuolating agent, now SV 40.

Paramyxoviridae: Pleomorphic enveloped RNA viruses about 150 to 300 nm in diameter (about 1000 S; density in sucrose, 1.19 g/cm^3). The lipid-rich envelope carries 8-nm-long spikes consisting of two glycoproteins. The RNA is a single negative-sense molecule of about 6×10^6 Da that is found within a flexuous helix. There are seven proteins: the largest is the RNA polymerase; the most abundant is the matrix protein supporting the envelope. (*Examples:* Newcastle disease virus, mumps virus, measles virus) (name: see Orthomyxoviridae).

Parvoviridae: This family consists of small (18–26 nm diameter) icosahedral particles (about 115 S; density in CsCl, 1.4 g/cm^3) that contain single-stranded DNA of 1.5 to 2.0×10^6 Da. The viruses are stable over the pH range 3 to 9, and up to 60°C. The three proteins of the virion may result from the processing of a single translation product. There are also two nonstructural proteins. (*Example:* Kilham rat virus, human B19 virus, feline and canine parvovirus) (*parvus,* "small")

Picornaviridae: Small spherical nonenveloped particles about 27 nm in diameter (140 to 165 S; density in CsCl, 1.33 to 1.44 g/cm^3) consisting of one molecule of positive-sense RNA of 2.5×10^6 Da with a 5′ small terminal VpG protein and 3′-terminal poly(A). Have equal numbers (60) of four capsid proteins (three of 24 to 41×10^3, and one of about 10×10^3 Da). Most of the picornaviruses are very host-specific. (*Examples:* polio virus, foot-and-mouth disease virus, hepatitis A, rhino (cold) virus) (*picorna,* "small RNA")

Polydnaviridae: Particles of 85 × 330 nm, covered by two envelopes, the inner one being virus-coded virions contain. Several supercoiled DNAs and many proteins of 10 to 200×10^3 Da make up the particle. Marked specificity for insect hosts (*Examples:* ichnovirus, bracovirus) (*polydna,* "many DNAs")

Poxviridae: A family of large viruses with complex structure infecting most vertebrates and many invertebrates. The ovoid enveloped particles of about 400 × 200 nm contain a single molecule of circularized double-stranded DNA of 90 to 200×10^6 Da. At least 40 structural proteins, glycoproteins, and phosphoproteins, forming specific internal organelles (e.g. lateral bodies), and lipid that is not near the surface. The viruses are generally very host-specific; they are transmitted by aerosol, by contact, and by insects. (*Examples:* smallpox virus, vaccinia virus) (*poc* or *pocc,* "pustle" or "ulcer" (Old English))

Reoviridae: This family is characterized by large icosahedral particles, at times covered by an outer protein shell (60 to 80 nm in diameter; density, about 1.38 g/cm^3), containing 10 to 12 molecules of linear double-stranded RNAs of 0.2 to 3.0×10^6 Da. The internal structure is a nucleocapsid core (45 percent RNA) with 10 to 12 spikes. Each RNA molecule is associated with RNA polymerase and becomes transcribed in synchrony. They contain 10 to 12 RNA species of 0.5 to 2.7×10^6 Da as well as many oligonucleotides. Nine structural proteins of 34 to 155×10^6 kDa have been identified. Hosts are mammals, birds, insects, other invertebrates and plants. (*Example:* blue-tongue virus human reovirus type 1) (name from *re*spiratory *en*teric *o*rphan viruses)

Retroviridae: The common feature of this family is the presence of reverse transcriptive that can convert their single-stranded genomic RNA into DNA. Thus, the virion carries reverse transcriptase. The enveloped virions about 90 nm in diameter (550 to 600 S; density in sucrose, 1.16 to 1.21 g/cm^3) are spherical and have a lipid-rich outside membrane with spikes containing two glycoproteins. The RNAs of the nondefective viruses range in molecular weight from 3 to 10×10^6 Da. The RNA occurs in the virion as duplex molecules. The RNAs are capped and have poly(A) on the 3′ end. The virion also contains tRNAs, with one specifically base-paired to a specific site on the virion RNA and serves as primer for the reverse transcriptase. Besides the two envelope proteins and the reverse transcriptase, there are also four nucleocapsid and/or matrix proteins and other accessory gene products depending on the specific virus group. (*Examples:* Rous sarcoma virus, HIV, simian foamy virus, HTLV) (*retro,* "backward" referring to the reverse transcription during replication)

Rhabdoviridae: This RNA virus family includes both animal and plant viruses. The animal rhabdoviruses are generally bullet-shaped, whereas many plant rhabdoviruses are rounded at both ends, and thus, bacilliform. The enveloped virions vary in dimension, from 130 to 380 × 50 to 95 nm (550 to 1000 S; density in CsCl or sucrose, 1.19 g/cm^3). They are unstable below pH 4, at 56°C, and in ether. The nucleocapsid contains a single negative-sense RNA of about 4×10^6 Da within a protein helix (70 percent of total) particle weight, plus the RNA polymerase. The lipid-rich envelope contains a glycoprotein. (*Examples:* rabies virus, vesicular stomatitis virus) (*rhabdos,* "rod")

Tetraviridae: Presently only one genus (Nudaurelia β virus group). Nonenveloped, icosahedral insect virus with a single positive-sense RNA of 1.8 × 10 Da and a single capsid protein. (Name derived from the unusual T-4 particle symmetry).

Togaviridae: The smallest enveloped animal viruses. Particles are spherical, 40 to 70 nm in diameter (density, in CsCl, 1.25 g/cm^3). The icosahedral nucleocapsid is about 30 nm in diameter and consists of a 4.0 to 4.5 $\times$ 10^6 Da positive-sense RNA, plus the capsid protein; the envelope contains two glycoproteins and lipid. (*Examples:* Sindbis virus, Semliki Forest virus) (*toga,* "mantle")

Toroviridae: Viruses with a nucleocapsid of a unique morphology, a hollow tubular helix, 23 nm in diameter bent into a torus, and enveloped with peplomers. Infected cells contain several RNAs, but only the largest (6 $\times$ 10^6) is positive-sense and infectious. Some properties are similar to those of Coronaviridae. Causes diarrhea in mammals. (*Examples:* Bernevirus, Bredavirus) (Name derived from the shape of the nucleocapsid torus, lowest convex molding in the base of a column.)

Plant Virus Taxonomy

As described in Sections 1.6 and 2.6, the basis of plant virus taxonomy has been the **virus group** rather than the family-genus structure adopted by animal and bacterial virologists. A virus group is *not* a recognized taxon but rather a coherent cluster of individual viruses that share certain major characteristics. The fifth report of the International Committee on Taxonomy of Viruses recognized 35 groupings of plant viruses—32 virus groups plus three virus families (i.e., the plant *Rhabdo-, Bunya-, and Reoviridae*).

Even before that report was issued in 1991, however, plant virologists had begun to consider whether or not certain of these virus groups should be combined into families. Impetus for such a philosophical change was provided by an increasing awareness that genomic recombination and reassortment are responsible for much of the genetic diversity observed among RNA viruses. Thus, a recent provisional classification of plant viruses included six new virus families—the *Crypto-, Gemini-, Tombus-, Como-, Poty-,* and *Bromoviridae.* These families would replace 12 of the present 32 virus groups. Additional information about the current status of plant virus taxonomy can be found in a recent article by Martelli (1992) *Plant Disease* 76:436. For the most recent listing of viral taxonomic classes as recommended by the International Committee on Taxonomy of Viruses, see *Archives of virology,* Supplementum 2. Springer-Verlag, Vienna, 1991.

Postscript: The Origin and Evolution of Viruses

Ever since scientist came to understand the true nature of viruses, they have attempted to locate the origin of viruses somewhere on the evolutionary tree between the chemical origins of macromolecules and the appearance of the various cellular organisms. The results of these speculations can be summarized very briefly: We do not know the answer. Three different scenarios have been proposed, but the origins of viruses may well be as complex as the origin of life itself.

Retrograde evolution has frequently been considered as a possible mechanism for generating viruses. This concept presumes that viruses are derived from intracellular parasitic microorganisms that have gradually lost most of their genes responsible for an independent metabolism. Such entities would thus retain only those genes that gave them their identity and replicability. The strongest argument against such a scenario has been the apparent failure to find intermediates between the most degenerate microorganisms (*Chlamydiae*) and viruses. Complex viruses like the poxviruses, which replicate in the cytoplasm independently of host nuclear functions, could represent such intermediate forms, however.

For the emergence of other viruses, especially DNA viruses, **origin from cellular RNA and/or DNA components** is most favored—the idea that viruses are cellular components that have declared their independence and emigrated. As discussed in Chapter 6, several viruses certainly resemble bacterial plasmids or episomes in many respects, and the possibility that some such cellular genomic components, possibly arising through recombination involving parts of more than one genome, may have acquired protein coats cannot be ruled out.

However, this scenario leaves us with the problem of the origin of the many RNA viruses that we encounter in nature. In analogy with the origin of DNA viruses, it has been suggested that they are derived from host cell mRNAs. Messenger RNAs, however, are not replicated in any known organism, and it is difficult to visualize a mechanism by which those RNAs could have acquired at the moment of their emigration the complex capability that would enable them to reenter cells and elicit their own replication. Nevertheless, intriguing sequence similarities have been noted between the promoters for various plant viral genomic and subgenomic RNAs and the **internal control regions** known to be involved in regulating eukaryotic tRNA transcription.

One hypothesis about the origin of retroviruses is also relevant to this proposed origin of RNA viruses in general. In this **protovirus theory,** Temin proposed that portions of cellular RNA were reverse-transcribed by cellular enzymes into DNA products that were integrated into the host cell genome. Over millions of years, these sequences found themselves sufficiently linked to permit production of a replicating virion. This theory predicts the presence of pieces of retroviruses scattered throughout the genomes of both mammals and lower species. This possibility has been substantiated in a number of studies, but it is not known whether the integrated viral genes are the progenitors of retroviruses or merely the results of their evolution. For example, the **Ty elements** of yeast (Chapter 6) contain coding sequences with homology to reverse transcriptase, mobilize via an RNA form, and generate particles that resemble virions, but these particles are not infectious. Integration into the host cell genome would appear to be an almost perfect means for virus perpetuation. Why, then, do so few virus families and groups appear to have retained this survival mechanism?

A third scenario for virus origin proposes that some viruses may be **descendants of primitive precellular life forms.** Viroids, in particular, may represent "living fossils" of a precellular RNA world that assumed an intracellular mode of existence sometime after the evolution of cellular organisms. The inherent stability of viroids and viroidlike RNAs that arises from their small size and circularity would have enhanced the probability of their survival in primitive, error-prone RNA self-replicating systems and assured their complete replication without the need for initiation or termination signals.

Most viroids (but *not* satellite RNAs or random sequences of the same base composition) also display structural periodicities with repeat units of 12, 60, or 80 nucleotides. The high error rate of prebiotic replication systems would be expected to have favored the evolution of polyploid genomes, and the mechanism of viroid replication (i.e., rolling-circle transcription of a circular template) provides an effective means of genome duplication. Viroids and viroidlike satellite RNAs also possess efficient mechanisms for the precise cleavage of the **oligomeric replicative intermediates** to form monomeric progeny. Viroids are not known to code for proteins, which suggests that viroids may be phylogenetically older than introns. Thus, it may be more accurate to consider introns to be "captured" viroids, rather than viroids "escaped" introns. Finally, the organization of the hepatitis delta virus genome into viroidlike and protein-coding domains (see Sec. 9.4) provides further support for the origin of certain components of modern viruses in the precellular RNA world.

Although the origin(s) of viruses may never be established with any degree of certainty, the past several years have witnessed a near-revolution in our concepts regarding **virus evolution.** These changes, due in large part to rapidly increasing amounts of detailed structural information about viral nucleic acids and proteins, are having a profound effect on virus taxonomy. In this regard, we can use the positive-sense RNA viruses to briefly illustrate certain current concepts about virus evolution.

The genetic variation required for virus evolution has several sources. Point mutations, insertion/deletion mutations, DNA or RNA recombination, reassortment of multipartite genomes, and the "capture" of host genes have all made important contributions. Moreover, it has become increasingly clear that **RNA recombination** plays a major role in reshaping viral genomes. Similarities in the order of viral genes as well as in the primary and/or secondary structures of various viral proteins have been shown to traverse many existing taxonomic boundaries such as host group, particle morphology, nature of the viral genomic nucleic acid, and number of genomic segments.

Given the obvious similarities in particle structure and replication strategy, the existence of sequence similarities among several separate taxonomic groupings of plant and animal viruses (e.g., hepadnaviruses and caulimoviruses) is not surprising. Sequence similarities among various plant and animal reo-, rhabdo-, and bunyaviruses also confirm previously suspected evolutionary relationships and support the decision to include the plant viruses within the families *Reo-, Rhabdo-, or Bunyaviridae.* Indeed, both reo- and rhabdoviruses are commonly assumed to have originated within their insect vectors and only later to have become adapted to replication in their animal or plant hosts.

Much more surprising, however, are the sequence and/or structural similarities that have been identified between various structural and nonstructural proteins of plant and animal RNA virus groups that have quite diverse particle morphologies. As shown in the Figure, positive-sense RNA viruses appear to form three **superfamilies** based on comparisons of their RNA-dependent RNA polymerases. Analysis of the relationships among virus-encoded

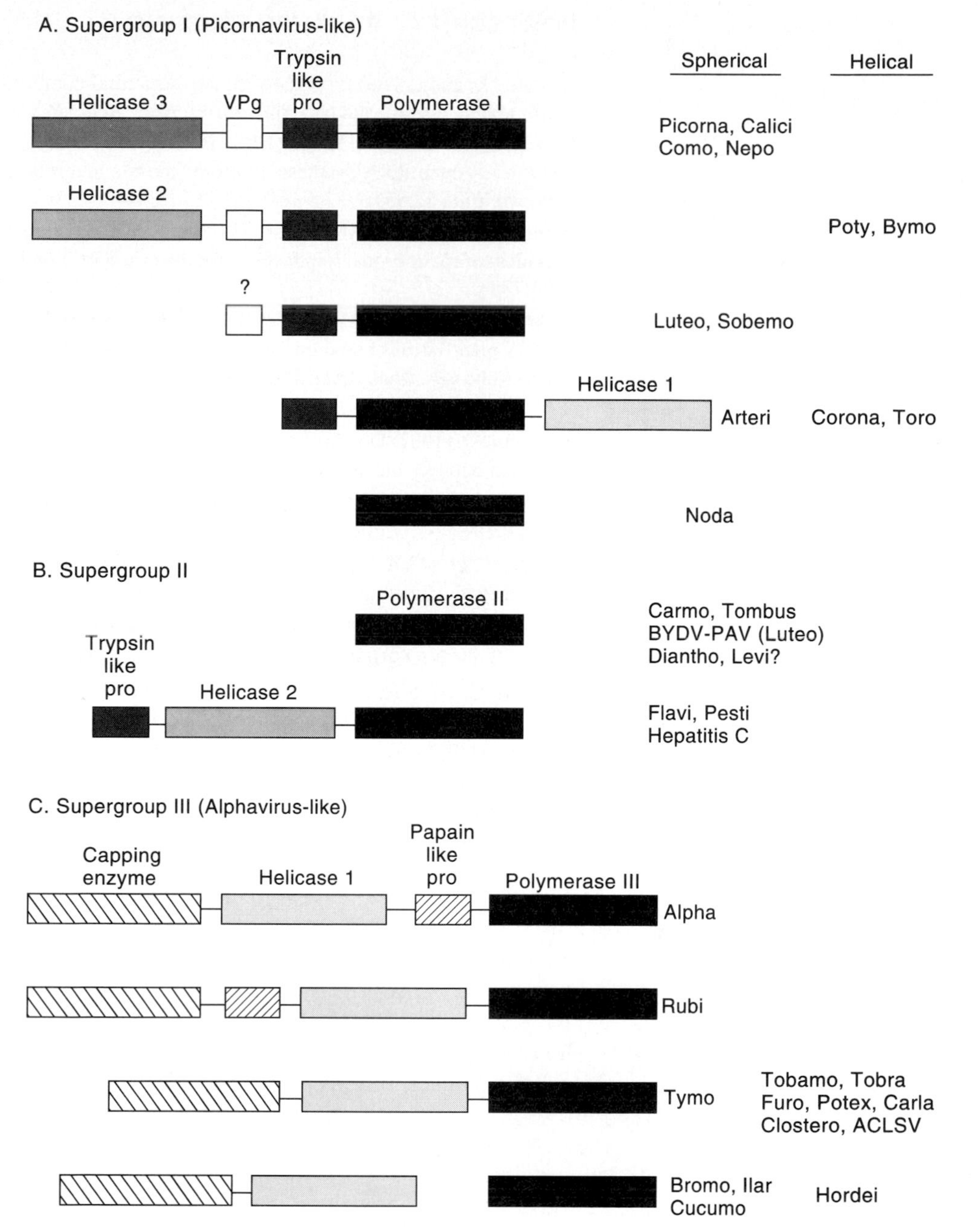

FIGURE 1

Conserved modules within positive-sense RNA virus genomes coding for replication-associated proteins. Virus groups are listed according to their spherical or helical nucleocapsid morphology, and the tentative position of luteo- and sobemovirus VPg is indicated by the question mark. Note that not all modules involved in replication are listed. Pro, proteinase; VPg, genome-linked virus protein; BYMV, barley yellow mosaic virus; ACLSV, apple chlorotic leafspot virus; and BYDV-PAV, barley yellow dwarf virus (PAV isolate). (From Dolja and Carrington. (1992) Evolution of positive-sense RNA viruses. *Seminars in Virol.* 3:315)

proteins shows that viral genomes comprise several modules with independent phylogenies, and only the viral RNA-dependent RNA polymerase can be used to compare all positive-sense RNA viruses. In general, however, virus groupings within a superfamily also exhibit certain other similarities in their genomic organization and replicative strategy.

Where similarities in genome organization parallel sequence similarities, a common evolutionary origin for the viruses within a superfamily seems likely. Where similarities are confined to the active sites of enzymes, the possibility of convergent evolution must be considered. The variety of viruses we know today appear to be the result of three major evolutionary processes—**divergent evolution** from a common ancestor, **convergent evolution** of particular viral proteins, and **modular evolution** involving gene acquisition from other viral or host genomes.

POSTSCRIPT FURTHER READING

Original Papers

KAMER, G., and P. ARGOS. (1984) Primary structural comparisons of RNA-dependent polymerases from plant, animal, and bacterial viruses. *Nucl. Acids Res.* 12:7269.

MARSH, L. E., and T. C. HALL. (1987) Evidence implicating a tRNA heritage for the promoters of positive-strand RNA synthesis in brome mosaic and related viruses. *Cold Spring Harbor Symp. Quant. Biol.* 52:331.

WEINER, A. M., and N. MAIZELS. (1987) tRNA-like structures tag the 3′ ends of genomic RNA molecules for replications: Implications for the origin of protein synthesis. *Proc. Natl. Acad. Sci. USA* 84:7383.

HABILI, N., and R. H. SYMONS. (1989) Evolutionary relationship between luteoviruses and other RNA plant viruses based on the sequence motifs in their putative RNA polymerases and nucleic acid helicases. *Nucl. Acids Res.* 17:9543.

Books and Reviews

ZIMMERN, D. (1982) Do viroids and RNA viruses derive from a system that exchanges genetic information between eukaryotic cells? *Trends Biochem. Sci.* 7:205.

GOLDBACH, R. W. (1987) Genome similarities between plant and animal RNA viruses. *Microbiol. Sci.* 4:197.

DIENER, T. O. (1989) Circular RNAs: Relics of precellular evolution? *Proc. Natl. Acad. Sci. USA* 86:9370.

JOYCE, G. F. (1989) RNA evolution and the origins of life. *Nature* (*London*) 338:217.

LEWIN, B. (1990) *Genes IV.* Chap. 34. Cell Press, Cambridge, Mass.

STRAUSS, E. G., J. H. STRAUSS, and A. J. LEVINE. (1990) Virus evolution. In *Virology,* 2d ed., eds. B. N. Fields and D. M. Knipe. Vol. 1, 167. Raven Press, New York.

MATTHEWS, R. E. F. (1991) *Plant virology,* 3d ed. Chap. 17. Academic Press, New York.

KOONIN, E. V., ed. (1992) Evolution of viral genomes. *Seminars in Virol.* 3:311.

TEMIN, H. (1992) Origin and general nature of retroviruses. In *Retroviridae,* ed. J. A. Levy. Vol. 1, 1. Plenum Press, New York.

Glossary of Technical Terms

A-type inclusion body Cytoplasmic inclusion bodies produced by certain pox viruses late in the infection cycle; similar in some respects to the **polyhedra** produced by insect baculoviruses.

A-type particle A double-shelled spherical particle characteristic of certain retroviruses (65–75 nm. for the outer and 50 nm. for the inner shell) that are found only in the cell cytoplasm.

Abortive infection A condition in which virus infection leads to inefficient replication of virions and often a lack of expression of infectious virus. Used in contrast to productive virus infection.

Acquired immunedeficiency syndrome (AIDS) Disease caused by the human immunodeficiency virus (HIV) and resulting in a wide range of adverse immunological and clinical conditions.

Actinomycin D Antibiotic that inhibits RNA transcription by interacting with the guanine residues of helical DNA; inhibits the replication of DNA-containing viruses and RNA viruses which require DNA-to-RNA transcription.

Activator A DNA-binding protein that binds upstream of a gene and activates its transcription.

Acute leukemia virus A retrovirus that gives rise to leukemia after a short incubation period. It generally carries an oncogene.

Acyclovir A potent and specific antiviral agent against Herpes virus 1 and 2; its inhibitory action results from its selective phosphorylation by the virus-induced thymidine kinase and subsequent inhibition of the virus-encoded DNA polymerase.

Adenine A purine base found in both DNA and RNA.

A (adsorption) protein Minor protein component of RNA bacteriophage that is responsible for specific binding to the host cell.

Adjuvant Substances added to antigens to enhance the immune response.

Adsorption First stage of virus infection often involving the specific, temperature-independent interaction of a receptor on the cell surface with a component of the virus.

Affinity chromatography Chromatographic technique in which *ligands* attached to an insoluble support interact with a molecule of interest, thereby retaining it while unwanted molecules are washed away.

Agarose One of the polysaccharide constituents of *agar* that is widely used as a support medium for electrophoresis.

Aggregation Intramolecular interactions that do not involve covalent linkage.

α-Amanitin A cyclic peptide isolated from the poisonous fungus *Amanita phalloides* that specifically inhibits eukaryotic DNA-dependent RNA polymerase II and III (at higher concentration).

Amber mutation Point mutation producing a UAG termination codon and thereby premature termination of translation.

Amino acid Basic building block of proteins that contains amino and carboxyl groups plus a variable side chain that determines the properties of the individual amino acid. There are 20 amino acids that commonly occur in nature.

Amino acid sequence Linear order of the amino acids in a peptide or protein.

Amino group (-NH_2) A chemical group, characteristically basic because it tends to bind a proton to form –NH_3^+. The *amino-terminus* of a polypeptide is the end with a free α-amino group.

Amphotropic Subclassification of mouse retroviruses that can productively infect cells of the host species as well as cells of heterologous species.

Angstrom 0.1 nm (10^{-10}m).

Antibody See **Antigen.**

Anticodon Group of three bases in a tRNA molecule that recognizes (via base-pairing) a *codon* in a mRNA.

Antigen A substance, usually a protein or carbohydrate, that upon injection into a vertebrate is capable of stimulating the production and release into the bloodstream of **antibodies** (i.e. proteins that bind specifically to the antigen).

Antigenic drift Slight changes in the antigenic structure of a virus following passage in the natural host.

Antigenic shift Sudden major change in the antigenic specificity of viruses having segmented genomes, presumably as a result of genome *reassortment* or *recombination.*

Anti-sense gene Gene in which orientation of the coding sequence with respect to the promoter has been reversed. RNAs transcribed from such a gene can hybridize with mRNAs transcribed from normal gene, thereby inhibiting its ultimate translation into protein.

Antiserum Serum from a vertebrate exposed to an *antigen* and containing specific *antibodies.*

Aphid Phytophagous insect belonging to the family *Aphidae* and a common vector of many plant viruses.

Apoptosis Programmed cell death. A process that regulates the life span of a cell. It is active during normal lymphocyte development. It involves an endonuclease-directed degradation of chromosomal DNA that leads to death of cells. This process can occur as a result of virus infection.

Arbovirus Obsolete term used to describe any virus of vertebrates that is transmitted by an arthopod vector.

Attenuated region DNA sequences that retard or stop RNA transcription.

Attenuator strain Selected strain of a *virulent* virus that does not cause the severe disease associated with the parent virus but still replicates well enough to induce immunity. Attenuated strains have been used to produce many highly successful *vaccines.*

AT/GC ratio DNA base composition, inversely proportional to the stability of base pairing (A = adenine, T= thymine, G = guanine, C = cytosine).

ATP Abbreviation for *adenosine 5′-triphosphate*, the main storage agent of biological energy.

Autogenous regulation Regulation of gene expression by the product of the same gene.

Autoimmunity Production of an immune response against an organism's own tissues; may involve both *humoral* and *cell-meditated* responses.

Autoradiography Technique for the detection of radioactivity in cytological and biochemical preparations by exposure to photographic film.

Avirulent Virus strain lacking in virulence; see **Attenuator strain**.

AZT (azidothymidine) A thymidine analog which blocks DNA replication and is used in the treatment of HIV infection.

B-type particles Extracellular enveloped virus particles found in mouse mammary carcinomas that have an eccentric dense 40–60 nm. core surrounded by a 90–120 nm. envelope.

Bacteriocin Proteins released by some bacteria which are able to kill other bacterial strains. "True" bacteriocins are low molecular weight enzymes, while "particulate" bacteriocins resemble phage particle or subviral particles.

Balanced pathogenicity A state in which an infectious agent and its host coexist without compromising each other. An example is the "carrier" state in hepatitis B virus infection.

Base analogs Compounds which resemble one of the naturally-occurring purine and pyrimidine bases in RNA or DNA. Substitution of an analog for the "normal" base can result in mutation or growth inhibition.

Base pair A pair of nucleotides held together by hydrogen bonding. Double-stranded DNA contains G-C and A-T base pairs, and double-stranded RNA contains G-C and A-U base pairs. Highly-structured, single-stranded RNAs such as tRNA may also contain a variety of additional "non-Watson-Crick" base pairings.

Base plate Structure in some tailed phage to which the tail pins and tail fibers are attached.

B cells Lymphocytes that produce antibodies.

Bicistronic mRNA A messenger RNA containing two ribosome-binding sites.

Bioassay Determination of the amount of virus by measuring its biological activity (infectivity).

Biological control Use of parasites, predators, or pathogens to control pests.

Breakage and rejoining Crossing over by physical breakage and reunion of DNA molecules during recombination or cellular chromosomes during meiosis.

5-Bromouracil Mutagenic thymine analog in which the 5–CH_3 group has been replaced by bromine.

Budding Method of releasing enveloped virus particles from infected cells in which an area of the plasma membrance associates with the viral nucleocapsid as it leaves the cells. During budding, cellular proteins are displaced from the membrane by viral proteins, but the lipid in the viral envelope is host-derived.

Buoyant density Density at which a virus or other macromolecule neither sinks nor floats when suspended in a density gradient (e.g., CsCl or sucrose).

Burst size Yield of infectious virus particles obtained during a lytic one-step growth infection of a host cell.

C-terminus End of the peptide chain with a free α-carboxl group (traditionally written as the right end).

C-type particle Particles having a diameter of 90–110 nm. and containing a central core and lipoprotein envelope covered with knob-like projections. Characteristic of most retroviruses and often seen in leukemic tissues.

^{14}C Radioactive carbon isotope emitting a weak ß particle. The half-life is 5700 years.

Cancer Malignant neoplasm that, if untreated, kills the host.

Cap Sequence of methylated bases joined to the 5′-terminus of a eukaryotic mRNA in the opposite (i.e., 5′ to 5′) orientation and interacting with protein factors involved in the initiation of protein synthesis.

Capsid Regular, shell-like structure composed of aggregated protein subunits which surrounds the virus nucleic acid.

Capsomer Morphological unit from which the virus capsid is built and often consisting of groups of identical protein molecules.

Carboxyl group COOH, a chemical group that is characteristically acidic as a result of the tendency of its proton to dissociate. The carboxy-terminus of a polypeptide is the end that carries a free α-carboxyl group (C-terminus).

Carcinogen Agent (usually chemical) which causes cancer.

Carcinoma Cancer of epithelial tissue.

Carrier state A condition in which the host is persistently infected with a virus, yet often shows no signs of disease. Infectious virus can be isolated from such "carriers".

Catalyst Substance that can increase the rate of a chemical reaction without being consumed (e.g. enzymes are biological catalysts).

CD4 antigen A protein complex first identified on human T helper cells using monoclonal antibodies and subsequently found on other cells of the body.

Cell culture *In vitro* growth and replication of cells in suspension or as monolayer on surfaces of solid matrices.

Cell cycle Sequence of cellular events occurring in the course of mitotic division. The cell cycle consists of metaphase (M phase) and interphase (S phase), and DNA synthesis occurs during S phase. As cells pass through the cell cycle, $G(ap)_1$ phase separates M and S, and G_2 phase separates S and M. Somatic (nondividing) cells are said to be in G_0 phase.

Cell-free extract Extract containing subcellular organelles and soluble molecules that is made by breaking open cells and removing the whole cells, cell walls, and other large particulate components by centrifugation.

Cell fusion Formation of a hybrid cell containing nuclei and cytoplasm from different cells; often induced by treatment of a mixed cell culture with inactivated paramyxoviruses (e.g., Sendai virus). the viral fusion protein, or chemical agents (e.g. polyethylene glycol).

Cell-mediated immunity That part of the immune response involving cell-cell recognition, particularly by macrophages and lymphocytes.

Central dogma of molecular biology The concept that the basic relationship among DNA, RNA, and protein is "one way"; i.e., that DNA serves as a template for both its own duplication and the synthesis of RNA; RNA, in turn, serves as messenger for protein synthesis; and that information cannot be retrieved from the amino acid sequence of a protein.

Chelating agents Chemicals that bind divalent ions (e.g. Mg^{++}) and thus inhibit biological interactions that require such ions as cofactors.

Chimera Organism or molecule containing portions derived from two genetically different "parents".

Chloramphenicol Antibiotic isolated from *Streptomyces venezuelae* that acts as a phenylalanine analog, thereby inhibiting prokaryotic (as well as mitochondrial and chloroplast) protein synthesis.

Chromatin Nucleoprotein fibers composing eukaryotic chromosomes.

Chromosome Thread-like, self-replicating nucleoprotein structures containing a number of genes. In bacteria, the entire genome is contained with one double-stranded circular DNA molecule. Eukaryotic cells have multiple chromosomes, each containing a singular linear DNA molecule.

Chronic infection Infection in which a virus continues to replicate in portions of a host or culture without killing the host or entire culture.

Chronic leukemia virus A retrovirus that causes leukemia after a long incubation period. The mechanism for induction of cancer does not directly involve an oncogene.

***cis-trans* test** A genetic complementation test used to establish whether two mutations are in the same gene. This test was first used in the genetic mapping of bacteriophage T4 and requires that the wild-type allele be dominant.

Cisternae Membranous vesicles that proliferate in the cytoplasm of some virus-infected cells.

Cistron Basic unit of genetic function, usually a gene or protein-coding region.

Clone Term applied to cells, viruses or DNA molecules descended from a single common ancestor.

Cloning Term frequently (and loosely) used for the entire process of isolating, identifying, and manipulating single genes or cells in recombinant DNA research.

Coat protein External structural protein(s) of a virus.

Co-carcinogens Two or more agents that may not be carcinogenic separately, but that together induce cancer. Usually one low dose of a carcinogen *initiates* and a later dose of a non-carcinogen *promotes* tumor formation. See **Promotion**.

Codon Sequence of three adjacent nucleotides coding for either an amino acid or for a chain termination.

Coding sequence That portion of a nucleic acid that directly specifies the amino acid sequence of a protein product. Non-coding sequences may contain various sorts of *control sequences*.

Complement Series of sequentially-acting proteins present in the blood serum that, when activated, lyse foreign cells.

Complementary base sequences Polynucleotide sequences that are able to form perfect hydrogen-bonded duplexes, every G base-paired with a C and every A base-paired with a T or U.

Complementary DNA (cDNA) ssDNA molecule that is complementary to the base sequence of the single-stranded template from which it was transcribed; usually applied to DNA transcribed from RNA using *reverse transcriptase*.

Complementation Process by which one genome provides functions which another lacks. If the mutations occur in different genes, complementation is said to be "intergenic". Mutants defective in the same gene may produce a functional gene product via "intergenetic" complementation.

Concatemer Long molecules containing a number of identical monomers linked in a head-to-tail fashion; such DNA or RNA molecules often occur as intermediates in virus replication.

Conditional lethal mutant Mutant that is able to replicate under "permissive" conditions but is unable to replicate under conditions allowing replication of wild-type virus (e.g.. temperature-sensitive or host-specific mutations).

Conformation Overall shape of macromolecules resulting from multiple weak interactions.

Congenital infection Infection occurring at or before birth.

Connective tissue Cells, primarily fibroblasts, that make up the structure of tissues (opposite = epithelium).

Conservative replication Opposite of semi-conservative replication. By this process, no displacement of single-stranded form from the double-stranded genome takes place during replication.

Contact inhibition Phenomenon in which cells, when in close proximity to each other, stop dividing; also called *topoinhibition*.

Conversion Any change in host cell phenotype due to viral infection, especially cell surface changes.

Core Term used to describe the nucleocapsid and associated proteins remaining upon removal of the virus envelope or outer shell.

Covalent bonds Strong bonds formed by the sharing of electron pairs between atoms; also termed *primary bonds*.

Cross protection Protection against virus infection resulting from previous inoculation of the host with a closely related virus.

Crossing over Exchange of genetic material between homologous chromosomes.

Curing Loss of a lysogenic phage or plasmid from a bacterial culture.

Cyanophages Phages isolated from blue-green algae.

Cycloheximide Antibiotic isolated from *Streptomyces griseus* that reversibly inhibits eukaryotic (but not prokaryotic) protein biosynthesis.

Cytocidal Causing cell death.

Cytokine A general term for substances produced by cells (e.g., *lymphokines*) that play a role in intercellular communication (e.g., interleukin, IL-1, IL-2).

Cytolytic Causing cell lysis.

Cytopathological effect (CPE) Changes in the microscopic appearances of cultured cells following virus infection.

Cytoplasm That part of the cell within the cell membrane but outside the nucleus and other subcellular organelles.

Cytoskeleton Intracellular scaffolding and transport system consisting largely of two proteins, tubulin and actin.

D-type virus particles Morphologically defined group of intra- and extra-cellular enveloped retrovirus particles. The intra-cellular form is ring-shaped (60–90 nm. in diameter), and the extra-cellular form has an eccentric nucleoid and short surface spikes (100–120 nm. in diameter).

Dalton Unit of mass equal to that of a single hydrogen atom.

Defective interfering (DI) particles Virus particles whose nucleic acids lack part of the viral genome; DI particles often interfere with the replication of the standard virus.

Degeneracy of the genetic code The ability of several codons differing in the third base position to encode a given amino acid.

Deletion Loss of a portion of the genetic material. Deletions range in size from a single nucleotide to the entire genes (opposite = **insertion**).

Denaturation Loss of the native conformation of a macromolecule due to breakage of weak bonds (H-bonds, etc.); usually results from heat treatment or exposure to extreme pH or certain chemicals.

Density gradient centrifugation Method for separating macromolecules or organelle based upon differences in their density through use of a density gradient (usually sucrose or cesium chloride).

Deoxyribonucleoside A purine or pyrimidine base attached by an N-C bond to 2-deoxyribose, a five-carbon sugar.

Deoxyribonucleotide A deoxyriboncleoside carrying a phosphate group at the 3′ or 5′ position of sugar.

Differential host Host used to isolate a virus from a mixture of viruses. Such hosts are susceptible to one virus but not to the other(s).

Dimer Structure resulting from covalent bonding or aggregation of two identical subunits.

DNA (deoxyribonucleic acid) Polymer of deoxyribonucleotides and the genetic material of all cells.

DNA-dependent RNA polymerase Enzymes which transcribe RNA from a DNA template. In eukaryotes, DNA-dependent RNA polymerase I synthesizes ribosomal RNA; II, mRNA; and III, tRNA.

DNA polymerase I *E. coli* enzyme able to catalyze the formation of 3′->5′ phosphodiester bonds of DNA which also possesses 3′->5′ "proofreading" and 5′->3′ double-strand exonuclease activities. Sometimes called the "Kornberg enzyme" after its discoverer, it repairs and finishes the replication of DNA.

DNA polymerase III *E. coli* enzyme that catalyzes the formation of 3′->5′ phosphodiester bonds at a very rapid rate. Also possessing a 3′->5′ exonuclease activity for "proofreading", its chief role seems to be DNA replication.

DNA-RNA hybrid Double helix containing one strand of DNA hydrogen-bonded to one strand of RNA.

Early genes Genes transcribed early in the virus replication cycle and often involved in replication of the viral nucleic acid. See **Late genes**.

Eclipse phase Time between virus penetration into a host cell, when it appears to lose its infectiv ity, and the appearance of the newly synthesized infectious progeny.

Ecology Field of science dealing with the mutual relations between organisms (or viruses) and their environment.

Ecotropic Subclassification of retroviruses that show productive infection of cells of their own host species.

eIF-2a Eukaryotic translation initiation factor that can be modified by the interferon system and some viruses.

Electron microscope Instrument which uses a beam of electrons focused by electromagnets to obtain magnified images of specimens. The shorter wavelength of electrons permits greater magnification than is possible with an optical microscope; resolution below 1nm is attainable.

Electrophoresis Separation of charged molecules, either in free solution or through the pores of a matrix, by application of an electrical potential (voltage gradient).

Enations Small outgrowths (often on the underside of leaf veins) caused by certain plant viruses.

Enzyme-linked immunosorbent assay (ELISA) Serological method in which the antigen is immobilized on a solid matrix and is then recognized by means of an antibody-enzyme conjugate. Presence of the enzyme is detected by monitoring the enzymatic conversion of a substrate to a colored product.

Endemic A disease which persists in a given locality or population.

Endocytosis An active process by which a virus is brought into the cytoplasm through the cell membrane without disruption of the virion or rupture of the membrane; also known as "viropexis".

Endogenous virus Virus whose genome integrates into the germ cell DNA and is thus able to spread by *vertical transmission* (e.g., retroviruses).

Endolysin Enzyme that breaks peptide linkages in bacterial cell walls.

Endonuclease Enzyme that cleaves polynucleotides (DNA or RNA) internally.

Endoplasmic reticulum Cytoplasmic membrane system involved in the synthesis and intracellular distribution of proteins.

Enhancer Sequences in viral or genomic DNAs that stimulate the transcription of adjacent genes.

Envelope Membrane surrounding a virus particle and usually containing virus-encoded proteins.

Enzootic Same as *endemic*, but referring to animal populations.

Enzyme Protein (or occasionally RNA) molecules that catalyze chemical reactions.

Epidemic Major increase in disease incidence, affecting either a large number of host organisms or spreading over a large area.

Epidemiology Field of science dealing with the role of various factors that determine the frequency and distribution of infectious diseases.

Episomes See **Plasmids**.

Epithelial tissue Cells making up the exterior and interior surfaces of tissues and organs.

Epitope Small groupings of amino acids that elicit the formation of specific antibodies.

Epizootic Major disease outbreak in animal hosts.

Etiology Cause(s) of a disease.

Eukaryote Organisms that have i) a genome divided among multiple chromosomes contained within a nuclear membrane, ii) a cytoplasm containing other membrane-bound organelles and 80 S ribosomes, and iii) a biochemistry differentiating them from *prokaryotes* (i.e., bacteria and blue-green algae),

Excision Release of an integrated provirus; reversal of integration (also called *induction* or *activation*). Also removal of mismatched or damaged nucleotides from DNA molecules.

Feedback regulation Regulatory mechanism where the *effector molecule* is a product of the reaction/pathway being regulated.

Ferritin Iron-rich liver protein used as a marker in electron microscopy.

Fibroblast Type of connective tissue cell found in fibrous tissue.

Filamentous Composed of long thread-like structures, often used to describe bacterial colonies or virus particles.

Fluorescent antibody Antibody conjugated with a fluorescent dye and used in combination with a fluorescence microscope to detect the presence of viral antigens in cells.

Focus Cluster of morphologically transformed cells in a monolayer culture, initiated by a single transforming virus particle.

Frameshift mutation Mutation caused by the insertion or deletion of one or more nucleotides whose effect is to change the reading frame starting at the affected codon.

Freeze fracture Specialized technique for the preparation of specimens for electron microscopy in which rapidly frozen samples are broken and the exposed surfaces may be etched to reveal further detail.

Fusion Process by which enveloped viruses become incorporated into the cell membrane and thus enter the cell. It is also a mechanism by which such viruses bring cell membranes together to form multinucleated cells or *syncytia*.

G2 phase See **Mitosis**

ß-Galactosidase Enzyme catalyzing the hydrolysis of lactose into glucose plus galactose.

Gamma globulin Fraction of blood serum proteins that contains antibodies.

Gel filtration Type of column chromatography that separates molecules on the basis of size.

Genetic map Graphic representation of the linear arrangement of genes within a genome or chromosome. May be determined by the percentage of recombination in linkage experiments or, in smaller genomes, by direct nucleotide sequence analysis.

Genotype Genetic constitution of an organism, as distinguished from its physical appearance or *phenotype*.

Germ-line transmission Transmission of a virus by egg or sperm.

Glycolipids Lipids that are covalently liked to polysaccharides.

Glycolysis Metabolic pathway utilizing glucose as a source of energy.

Glycoprotein Polypeptide to which at least one carbohydrate residues is covalently attached.

Golgi apparatus Subcellular organelle consisting of flattened, parallel membranes derived from the endoplasmic reticulum and playing a role in intracellular transport.

Granulosis See **Polyhedrosis**

Group-specific antigen Antigen specific to (shared by) a group of viruses.

Guanine A purine base found in both DNA and RNA.

Guarnieri bodies Inclusion bodies in vaccinia virus-infected cells.

Gyrase Enzyme causing double-stranded circular DNA to become negatively supercoiled (a topoisomerase).

^{3}H (tritium) Radioactive isotope of hydrogen, a weak ß emitter with a half-life of 12.43 years.

Hairpin Double-helical regions resulting from the base pairing of contiguous, largely complementary stretches of bases on a single DNA or RNA strand.

HeLa cells Established line of human cervical carcinoma cells often used in studies of the biochemistry and growth of cultured human cells and susceptible to infection by many viruses.

Head Structural component of tailed phages that contains the DNA genome.

Helix Spiral structure with a repeating pattern. It is the natural conformation of many regular biological polymers (e.g., polypeptides and polynucleotides) and virus particles.

Helper virus Virus providing the factor(s) needed by a *defective virus* for replication.

Hemadsorption Binding of erythrocytes to the surface of virus-infected cells due to the presence of virus-encoded gene products in the cell membrane.

Hemagglutination Clumping of erythrocytes (red blood cells) resulting from cell surface binding (e.g. to virus).

Heteroduplex Double-stranded DNA molecule in which the two strands are not completely complementary. This can be the result of mutation, recombination, or biochemical manipulation.

Heterologous Derived from a species with a different genetic constitution.

Hexamer Group of six protein subunits on the triangular faces of an icosahedral virus capsid.

Histones Proteins rich in basic amino acids (e.g., lysine and arginine) and found in the nucleus of eukaryotic cells in association with DNA. See **Chromatin**.

Holoenzyme The active (complete) form of an enzyme.

Homology Degree of relatedness between two nucleic acid or amino acid sequences. Use of the term homology rather than *similarity* implies an evolutionary relationship. See **Similarity**.

Horizontal transmission Transmission of a virus or other pathogen to a host at any age after birth.

Hot spots Sites at which mutations occur with exceptionally high frequency.

Humoral immunity Immunity conferred by antibodies in extracellular fluids (i.e., not cell-mediated).

Hybrid cell Cell created by the fusion of two unrelated cells through the action of certain viruses or other agents.

Hybridization Formation of stable duplexes between complementary RNA or DNA sequences by means of Watson-Crick base-pairing. In genetics and breeding, the formation of a novel diploid organism by normal sexual processes or cell fusion.

Hybridoma Hybrid cell line produced by the fusion of a normal lymphocyte with a myeloma cell and used to produce a *monoclonal antibody*.

Hydrogen bond Weak attractive force between an electronegative atom (e.g., N, O) and a hydrogen atom that is covalently bonded to a second electronegative atom.

Hydrolysis Breakage of a molecule into small molecules by the addition of a water molecule: $A\text{-}B + H_20 - AH + BOH$

Hydrophilic Chemical groups with ability to form hydrogen bonds, thus having an affinity for water and similar solvents (e.g., -OH, = 0, -NH, etc).

Hydrophobic Chemical groups unable to form hydrogen bonds, thus lacking affinity for water (e.g., benzene and other aromatic rings, aliphatic hydrocarbons).

Hydrophobic bonding Association of hydrophobic molecules or parts of molecules with each other in aqueous solution caused by the tendency of water molecules to exclude such nonpolar molecules.
Hyperchromicity Increase in UV absorbance resulting from the denaturation of a macromolecule; provides an indication of the amount of base-pairing in a nucleic acid.
Hypochomicity Decrease in UV absorbance by a molecule due to an increase in secondary structure.

Icosahedron Regular polyhedron composed of 20 equilateral triangular faces.
Idiotype A specific epitope (immunogenic site) on an antibody molecule.
Immunogenicity The ability of a substance or virus to induce an immune response.
Immunoglobulin Set of proteins produced by B (plasma) cells during the immune response. Five classes can be distinguished on the basis of differences in their H (heavy) chains: IgG, IgA, IgM, IgD, and IgE.
Immunology Study of the immune system with its soluble antibodies and cell-mediated responses.
Inclusion bodies Microscopically distinct sites of virus synthesis and/or accumulation of viral products in virus-infected cells; often intracellular virus crystals.
Induction Initiation of virus production in a lysogenic or latently-infected cell.
Infection cycle Time between infection of a cell and production of the first progeny virions.
Infectious mononucleosis A disease caused by Epstein-Barr virus and characterized by enlarged lymph nodes and many atypical lymphocytes (T cells) in the peripheral blood.
Initiation (start) codon Trinucleotide in an mRNA at which ribosomes begin the process of translation, thereby establishing the *reading frame*. Usually AUG (which encodes methionine), but may be GUG (valine) in prokaryotes.
Initiation factors Specific proteins required for ribosome binding and the initiation of mRNA translation.
Insertion sequences Short (300–1000 nucleotide) DNA sequences able to integrate themselves at new positions within the genome without any sequence similarity to the target locus (e.g., bacteriophage Mu). Related to *transposons*.
Integration Insertion of a smaller DNA into a larger DNA. Integration of a viral DNA into the host genome usually requires a virus-encoded enzyme, *integrase*.
Intercalation Insertion of planar molecules (e.g., ethidium bromide) between adjacent base pairs in DNA or RNA.
Interference Ability of one virus to inhibit infection or replication by another (usually similar) virus.
Interferon A large group of species- and cell-specific proteins that are released by animal cells in response to various triggers, including viral infection; they protect other cells from viral infection.
Interleukin-2 (IL-2) A T-cell growth factor that stimulates the replication of T lymphocytes.
In vitro Pertaining to experiments done in a cell-free system (Latin, "in glass").
In vivo Pertaining to experiments done in living cells or organisms (Latin, "in life").
Isoelectric point pH value of a solution in which a macromolecule has no net surface charge and fails to move in an electric field.
Isolate Sample of a virus (or other infectious agent) from a defined source.
Isometric Of equal dimensions.

Kaposi's sarcoma A malignant tumor usually found on the surface of the skin or in lymph nodes. Often associated with AIDS, it is formed by transformation of endothelial cells lining the vascular or lymphatic system.
Kinase Any enzyme catalyzing a phosphorylation reaction.
Kleinschmidt technique Method in which nucleic acid molecules are spread on an air-water interface in a monolayer containing either a small basic protein or other organic molecules. After heavy metal shadowing, the molecules are visualized in the electron microscope.
Koch's postulates Set of criteria used to assess the role of a given microorganism in a disease. For viruses, these include: 1) isolation of the virus from the diseased host; 2) growth of the virus in an experimental host; 3) filterability of the pathogen; 4) production of comparable disease in the original or related host species; 5) reisolation of the virus; and 6) in the case of animal viruses, detection of a specific immune response.

Late genes Gene expressed late in the virus replication cycle, often those encoding capsid proteins.
Latent infection A state of persistent infection in which no symptoms are observed and no infectious virus is produced. In general, most transcriptional and translational processes are blocked.
Latent period Portion of time in the virus infection cycle between the apparent disappearance of the infecting virus and the appearance of newly-synthesized progeny (see **Eclipse phase**).

Leader sequence Untranslated 5′-terminal sequences in mRNAs.

Leaky mutations Mutant retaining some residual "wild-type" activity under *nonpermissive* conditions.

Lectins Plant glycoproteins that bind specifically to sugar residues in cell membrane glycoproteins and, thereby, act as *mitogens.*

Leukemia Cancer involving white blood cells.

Leukocyte White blood cell.

Ligation Process of joining the 5′ and 3′ termini of linear nucleic acid molecule via a phosphodiester bond. May involve either an *inter- or intra*molecular reaction.

Lipid bilayer Model for the structure of cell membranes and viral envelopes based upon the hydrophobic interactions between phospholipids. The polar head groups face the inner and outer surface of the membrane, while the hydrophobic tails are clustered within the body of the membrane.

Lipids Hydrophobic bioorganic molecules. Includes steroids, fats, fatty acids, phospholipids, and water-insoluble vitamins.

Lipopolysaccharides (LPS) Cell wall constituents of bacteria, rich in lipids and sugars.

Long terminal repeats (LTR) Sequences at the ends of the retroviral genome containing promoters and enhancers necessary for efficient virus replication.

Lymphocyte White blood cell involved in the immune response. *B lymphocytes* produce soluble antibodies, and *T lymphocytes* are involved in cell-mediated immunity.

Lymphokine A substance made by lymphocytes that affects the function of other cells.

Lymphoma Solid tumor of lymphatic tissue.

Lysis Cell rupture caused by the destruction of the cell membrane. Often associated with destruction of bacteria by phage-encoded *lysins.*

Lysogenic viruses Viruses (generally bacteriophages) that can become stably established within the host-either by integration into the host's genome or by plasmid formation-and thereby establish immunity in the host cell to reinfection by the same virus.

Lysolecithin Lecithin lacking one fatty acid chain. A very active hemolytic agent, it is also used in making spheroplasts and fusing cells.

Lysosomes Intracellular vesicles containing many hydrolytic and degradative enzymes.

Lysozyme Enzyme that degrades the peptidoglycan present in the cell wall of many bacteria.

Lytic phage Bacterial viruses causing cell lysis and death of their hosts.

Macromolecules Molecules with molecular weights ranging from a few thousand daltons to hundreds of millions.

Macrophage White blood cell that engulfs foreign particles or debris and helps in the recognition of antigens for immunologic response.

Major histocompatibility complex (MHC) Cell-surface antigens involved in cell:cell recognition. They are specific for each individual genotype.

Malignancy A cancer.

Marek's disease A lymphoproliferative disease of birds caused by an avian herpes virus.

Maturation protein Synonymous with the **adsorption (A) protein** of RNA phages.

Mechanical inoculation Inoculation of plants by rubbing their leaves with sap or other viral extracts in the presence of a mild abrasive. Can be a means of transferring wart viruses.

Melting temperature Temperature at which a double-stranded DNA or RNA molecule denatures into separate single strands or the secondary structure of a single-stranded nucleic acid is lost. Dependent upon both the G+C content of the nucleic acid and the salt concentration of the solution.

Mesophyll Photosynthetic parenchyma tissue located between the epidermal layers of a plant leaf.

Messenger RNA (mRNA) RNA that is translated by ribosomes to produce a protein. By definition, mRNA is *positive-sense.*

Metastases Secondary cancers, usually due to malignant cells carried from the primary cancer in the bloodstream or lymphatic system to other tissues.

Microsomes Fraction of a cell homogenate obtained by ultracentrifugation that contains ribosomes and vesicles derived from the rough endoplasmic reticulum.

Mitogen A substance that induces cell proliferation.

Mitomycin C Antibiotic that selectively inhibits DNA replication by cross-linking the strands.

Mitosis (M phase) Portion of the cell cycle when eukaryotic somatic cells divide. DNA synthesis occurs during the preceding S phase. See **Cell cycle**.

Molecular weight Sum of the atomic weights of the constituent atoms in a molecule.

Monocistronic mRNA mRNA carrying the information for the synthesis of only a single protein translation product.

Monoclonal antibody Highly specific antibodies produced by specifically selected cell clones obtained by fusing antibody-producing B lymphocytes with myeloma cells. See **Hybridoma**.

Monolayer cells Animal cells grown in culture while attached to a solid surface, in contrast to cells grown in *suspension culture*.

Monomer Basic subunit from which, by repetition of a single type of reaction, *polymers* are made. For example, amino acids (monomers) condense to yield polypeptides or proteins (polymers).

Mosaic Common symptom of plant virus infections in which the leaves of infected plants show a pattern of dark green and light areas. In the leaves of monocots, these symptoms appear as *stripes*.

mRNA Abbreviation for messenger RNA.

Multicistronic mRNA mRNA carrying the information for the synthesis of two or more proteins.

Multicomponent virus Virus whose genome is divided between two or more particles; especially common in plants.

Multiplicity of infection (m.o.i) Number of infectious virions per cell used to initiate infection.

Mutagens Physical or chemical agents (e.g., radiation, heat, and alkylating or deaminating agents) that raise the frequency of genetic mutation above the spontaneous rate.

Mutant Host containing a gene which has undergone mutation.

Mutation Heritable change in a genome or chromosome.

Myeloma B cell tumor of the immune system.

Nasopharyngeal carcinoma Tumor involving the epithelial cells of the nasopharynx and linked to Epstein-Barr virus infection.

Necrosis Death of cells.

Negative staining Technique used to prepare specimens for electron microscopy in which an electron-dense stain (e.g., uranyl acetate) is allowed to dry down on the grid, thus outlining the particles and revealing surface structure.

Neoplasm Cellular proliferation sometimes leading to malignant tumors.

Neuraminic acid Nine-carbon sugar derivative that is part of the cellular receptor recognized by orthomyxovirus hemaglutinin.

Neutralization Inactivation of a virus by the binding of antibodies to sites required for adsorption and entry into the host cell.

Nitrous acid (HNO_2) Mutagen that acts by converting amino (NH_2) groups on certain bases to keto ($C = O$) groups.

Natural killer (NK) cell Subset of white blood cells that reacts with tumor cells and virus-infected cells without prior exposure to these cells.

NK cell Natural killer cell. A white blood cell that non-specifically kills foreign cells and virus-infected cells in the host.

Non-ionic detergent Detergent with no net surface charge (e.g., Triton X-100 and Nonidet P40).

Nonpermissive cells See **Permissive cells.**

Non-persistent transmission Relationship between certain viruses (e.g., potyviruses) and their insect vectors that is characterized by a short *acquisition period* and a relatively short period during which the vector is able to transmit the virus. Infection involves virus particles associated with mouth parts and not those that have passed into the gut.

Nonsense codon See **Stop codon**.

Nonsense mutation Mutation that produces a *stop codon* and thereby causes protein synthesis to terminate prematurely.

Nonstructural proteins Virus-encoded proteins that are not part of the virus particle, usually functional during replication.

N terminus End of the polypeptide chain which carries a free α-amino group (as traditionally written, the left end.

Nucleases Enzymes that cleave the phosphodiester bonds of nucleic acid chains.

Nucleocapsid Viral nucleic acid enclosed by a protein.

Nucleoprotein Complex containing nucleic acid and protein.

Nucleosomes Basic structural subunit of *chromatin*, consisting of approximately 200 base-pairs of DNA and a histone protein octamer.

Occlusion body Term applied to the large proteinaceous crystals which contain (occlude) the particles of certain viruses (e.g., *nuclear* and *cytoplasmic polyhedrosis virus*).

Okazaki fragments Short (1000–2000 nucleotide) DNA fragments made in the course of discontinuous DNA replication and later joined together to form an intact strand.

2′–5′-Oligo (A) An oligonucleotide containing unusual 2′, 5′-phosphodiester bonds. Plays an important role in controlling interferon production.

Oligonucleotide Short polynucleotide chain (DNA or RNA).

Oligopeptide Short chain of amino acids.

Oligosaccharides Short sugar polymers that, in viruses, are usually attached to asparagine residues in proteins and contain mannose, glucosamine, fucose, sialic acid, etc.

Oncogene Cellular or virus-encoded gene whose products are able to *transform* eukaryotic cells so that they begin to grow like tumor cells.

Operator Site on DNA capable of interacting with a specific *repressor*, thereby controlling the functioning of an adjacent operon.

Operon Group of adjacent genes under the joint control of an operator and a repressor; unit of bacterial gene expression and regulation.

Organelle Particulate subcellular structures in eukaryotic cells such as the nucleus, mitochondria, chloroplasts, Golgi apparatus, and ribosomes.

^{32}P Radioactive isotope of phosphorous that emits a strong ß particle and has a half-life of 14.3 days.

Pactamycin Antibiotic that blocks the initiation of pro- and eukaryotic protein synthesis.

PAGE See **Polyacrylamide gel electrophoresis**.

Palindrome DNA sequence that is the same when one strand is read left to right and the other is read right to left (i.e., sequence consisting of adjacent inverted repeats.)

→
ACTAGT
TGATCA
←

Pancreatic ribonuclease A Endonuclease that cleaves RNA only at the 3′ site of pyrimidine nucleotides.

Papilloma A small neoplasm on the skin. Usually called a wart.

Pathogen An organism or virus that causes a disease.

Pathogenesis related (PR) proteins Proteins which accumulate in plant tissues as part of a *hypersensitive response* to viral or fungal infection or to treatment with certain chemicals (e.g. salicylate).

Penetration Following adsorption, the second step in the initiation of virus infection during which the virus particle enters the cell.

Peplomers Protrusions from the surface of enveloped viruses (also called spikes).

Peptide Two or more amino acids joined by a peptide bond.

Peptidoglycan Glycopeptide constituent of the bacterial cell wall.

Pericular space Region between the two nuclear membranes or close to the outer membrane.

Permissive cells Cells that support complete virus replication. This term is used in contrast to *nonpermisive* cells that show abortive infection.

Persistent infection Continuous presence of virus in infected cells and often associated with the integration of the viral genome into that of the host. Three types of persistence are known: 1) virus persists within the cell but is not released; 2) virus is released sporadically, and 3) virus is continually released without lysis of the host cell.

Persistent transmission Relationship between viruses and their insect vectors characterized by long acquisition times, a latent period before the vector is able to transmit, and an extended period where transmission is possible. The virus must pass through the gut wall, into the hemocoel, and then accumulate in the salivary glands before transmission. Virus may or may not multiply in the vector. Also called *circulative* transmission.

Phage cross Infection of a bacterium by two or more bacteriophage mutants, resulting in the production of recombinant progeny phage that carry genes derived from both parental phage types.

Phage typing Subdivision of bacteria into "phage types" according to their susceptibility to lysis by host-range specific phage.

Phagocytosis Engulfment of particles by macrophages and at times other cells.

Phenotype Apparent properties of an organism or virus resulting from expression of its genotype but modified by environmental factors.

Phenotypic mixing A phenomenon in which viruses exchange their outside envelopes or coats. Thus, the genome of one virus may be encapsidated in the external membrane or coat of the other.

Phosphodiester bond Linkage of phosphoric acid with the 3′–hydroxl group of one ribose or deoxyribos molecule and the 5′–hydroxyl group of the next ribose or deoxyribose molecule in a polynucleotide.

Phosphokinase See **Kinase**

Phospholipids Lipids that contain charged phosphate groups, thus showing both hydrophobic and hydrophilic properties. Phospholipids are a primary component of cell membranes.

Phosphoproteins Proteins containing phosphate groups, most commonly attached to serine, threonine, or tyrosine residues.

Phosphorylation The process of esterifying a compound with phosphoric acid. This event is generally performed by a *kinase* using ATP.

Photosynthate Reduced carbon compounds (especially carbohydrates) produced by photosynthesis.

Pilot protein Protein playing an important role in infection by Micro-and Inoviridae.

Pilus Filamentous non-motile appendage found on many Gram-negative bacteria and necessary for *bacterial conjugation* as well as phage infection involving male *E. coli* strains.

Pinocytosis Uptake of water, solutes, or particles by internalization of fluid-filled vacuoles. Also called *viropexis*.

Pitch Repeat periodicity in a helical structure.

Plaques More or less clear (and usually circular) areas in a bacterial lawn or confluent cell monolayer that result from cell fusion or the killing or lysis of cells by several cycles of virus replication.

Plasma membrane The membrane which makes up the cell surface and encloses the cytoplasm. The plasma membrane is *semi-permeable* and largely composed of phospholipids and proteins.

Plasmids Extrachromosomal elements (usually covalently closed circular DNAs) that replicate autonomously and range in size from ≤1 kbp to ≥300 kpb. Cells (usually bacteria) may contain one to more than 100 copies of a plasmid.

Plasmodesmata Cytoplasmic connections through the cell walls of higher plants. Analogous to the *gap junctions* between certain types of animal cells.

Polar mutation Mutation which has an effect on the expression of adjacent, downstream genes.

Polyacrylamide gel electrophoresis Electrophoresis in a hydrophilic polyacrylamide matrix. Separation usually depends upon molecular size, and PAGE is commonly used to estimate the relative molecular sizes of proteins and nucleic acids. Denaturation of proteins with *sodium dodecyl sulfate (SDS)* provides a uniform charge per unit molecular weight.

Point mutation Mutation resulting from a single base exchange.

Poly(A), poly(U), poly(C), poly(G) Polyribonucleotides containing a single type of base. The 3′ ends of most eukaryotic mRNAs contain up to 300 adenosine nucleotides that are enzymatically added after transcription.

Polyamine Small organic molecules containing two or more amino (NH_2) groups; e.g., spermidine, spermine, and putrescine.

Polycistronic mRNA See **Multicistronic mRNA**.

Polyclonal antibody Preparation containing antibodies against more than one *epitope* of an antigen. Antibodies obtained from whole animals are always polyclonal.

Polyhedrin Matrix protein comprising the major component of *occlusion bodies* produced by *nuclear polyhedrosis virus (NPV)* and *cytoplasmic polyhedrosis virus (CPV)*.

Polyhedrosis Disease caused by invertebrate (e.g., insect) viruses that lead to tissue breakdown and accumulation of virions imbedded in polyhedral crystals in the cytoplasm or nucleus. Similar to *granulosis*.

Polymer Regular covalently bonded chain of subunits (*monomers*).

Polymerase chain reaction (PCR) Selective amplification of DNA by repeated cycles of a) heat denaturation of the DNA, b) annealing of two oligonucleotide *primers* that flank the sequence to be amplified and c) the extension of the annealed primers with a heat-stable DNA polymerase.

Polynucleotide Linear nucleic acid polymer in which the 3′ position of the sugar of a nucleotide is linked through a phosphate group to the 5′ position on the sugar of the adjacent molecule.

Polynucleotide phosphorylase Enzyme catalyzing the polymerization of ribonucleoside diphosphates to yield free phosphate and polynucleotides (e.g., RNA). Its physiological functions remain unknown.

Polypeptide Polymer of amino acids linked together by peptide bonds.

Polyprotein Large primary translation product from which individual functional proteins are subsequently released by proteolytic cleavage.

Poly(ribo)some Complex which contains an mRNA molecule and ribosomes actively engaged in polypeptide synthesis. The number of ribosomes depends on the size of the mRNA.

Polysaccharide Carbohydrate (sugar) polymer; in viruses often mannose, galactose, fucose, sialic acid, and glucosamine.

Polytopic A term used for retroviruses that have both *xenotropic* and *amphotropic* cellular host ranges.

Primary cells Cells directly obtained from multicellular organisms and seeded onto culture plates.

Primary structure Nucleotide sequence of an RNA or DNA molecule or the amino acid sequence of a polypeptide.

Primer Structure that serves as a growing point for polymerization; e.g., a small oligonucleotide with a free 3′–hydroxyl group necessary for the initiation of DNA (and occasionally RNA) synthesis.

Primosome Complex of proteins involved in the priming action that initiates the synthesis of Okazaki fragments during discontinuous DNA replication.

Prion Term applied to the agents responsible for certain neurological diseases of vertebrates (e.g., scrapie). Derived from "proteinaceous infectious particle" and refers to the apparent absence of a genomic nucleic acid.

Prokaryotes Simple unicellular organisms, such as bacteria or blue-green algae, that lack a nuclear membrane and membrane-bound organelles.

Promoter Site on DNA molecule where RNA polymerase binds and initiates transcription.

Promotion (in carcinogenesis) Induction of expression of latent carcinogen-initiated tumors by a noncarcinogen. See **Co-carcinogen**.

Proofreading Mechanism for correcting errors in protein from nucleic acid synthesis that involves examination of individual monomers *after* their incorporation into the growing polymer chain.

Prophage Proviral stage of a lysogenic phage. See **Provirus**.

Proto-oncogene Term used to designate a gene in a normal eukaryotic cell that is believed to have given rise to an *oncogene* in a transforming virus.

Protoplast Plant cell whose thick cellulose walls have been removed by enzymatic treatment.

Provirus A virus genome that has been integrated into either the host genome or into a plasmid. Proviral DNA is passively replicated by the host machinery and thus transmitted from one cell generation to another.

Pseudorecombinant Virus produced by the *in vitro* mixing of nucleic acid segments from two closely related viruses with *multipartite genomes*.

Pseudotype Virus particles containing the genome of one virus in the capsid or envelope of another (see **Phenotypic mixing**).

Pulse-chase experiment Experimental technique in which a radioactively labeled compound is added to living cells or a cell extract (pulse) and then, a short time later, an excess of unlabeled compound is added to dilute out the "hot" compound. Samples are then collected at various times after the puise to follow the course of the label as the compound is metabolized (chase).

Purine Heterocyclic organic compound (or "base") containing fused pyrimidine and imidazole rings. *Adenine* and *guanine* are the normal purine nucleotides found in DNA and RNA.

Puromycin An antibiotic that mimics a charged tRNA. Its use leads to an inhibition of protein synthesis.

Putrescine See **Polyamine**.

Pyknosis Shrinkage and condensation of the cell's nucleus due to viral infection.

Pyrimidine Heterocyclic organic compound (or "base"). *Cytosine* and *thymine (uracil)* are the normal pyrimidine nucleotides present in DNA (RNA).

Quarantine A period of detention for animals, plants, or people coming from a place where a disease is known to exist. Also, the place where individuals or animals are kept for inspection.

Quasi-equivalence theory Theory proposed by Caspar and Klug to explain the fact that the identical chemical subunits in an *icosahedral virus particle* are not arranged in a strictly mathematical equivalent manner.

Quasi-species Term used to describe the genome of an RNA virus which recognizes the presence of one or more sequence differences between every copy of the viral genome, a result of the inherently high error frequency of RNA-dependent RNA polymerase.

Radioimmunoassay Technique that utilizes radioactive labeling of viral antigens or antibodies to increase sensitivity of serological identification.

Reactivation Activation of a virus, (e.g., herpesvirus) from a latent stage.

Readthrough Translation of mRNA through a *stop codon*. See **Suppressor tRNA**.

Reannealing Reassociation of single-stranded nucleic acids after denaturation to restore the H-bonded, double-stranded structure.

Receptor Specific sites or structures on cell surface to which viruses attach.

Recombinant Molecules containing a new combination of nucleic acid or protein sequences; e.g., those made via recombinant DNA techniques.

Recombination Exchange of genetic information from two or more virus genomes to produce *recombinant progeny*. While recombination between two double-stranded DNA molecules usually proceeds through strand "crossover" (breakage and reunion), RNA recombination occurs via a "copy choice" mechanism during replication.

Renaturation Return of a protein or nucleic acid from a denatured state to its "native" conformation.

Replica technique Method for making plastic models of surfaces for study in the electron microscope.

Replicase Enzyme involved in the replication of the viral genomic nucleic acids; e.g., RNA-dependent RNA replicase.

Replication Duplication of the genomic DNA or RNA of a virus.

Replication fork Y-shaped region of a DNA genome that is the growing point during DNA replication.

Replicative form (RF) Structure of a nucleic acid at the time of its replication; the term most frequently used to refer to double-helical intermediates in the replication of single-stranded DNA and RNA viruses.

Replicative intermediate (RI) Structure of a viral nucleic acid during the act of replication. For single-stranded RNA genomes, it is a partially double-stranded molecule formed by the simultaneous synthesis of one or more complementary strands from a single *template* strand.

Repressor Regulatory protein which binds to *operator* sites on a DNA, thereby preventing the transcription of adjacent sequences into RNA.

Restriction endonuclease Enzyme which recognizes short specific sequence in double-stranded DNA and cleaves the duplex (usually at the recognition site, but sometimes elsewhere). Components of the bacterial restriction-modification systems which protect the cell against invasion by foreign nucleic acids.

Reticulocyte Immature red blood cell (or erythrocyte) that is able to synthesize hemoglobin. Cell-free rabbit reticulyte extracts are often used to study protein synthesis *in vitro*.

Reverse transcriptase Enzyme encoded by retroviruses (as well as hepada- and caulimoviruses) that is able to synthesize a complementary DNA copy of a single-stranded RNA template and then to convert this DNA molecule to the double-stranded form.

Ribonuclease Enzyme which hydrolyzes (cleaves) the phosphodiester bonds of RNA.

Ribonucleotide Compound consisting of a purine or pyrimidine base bonded to ribose, which in turn is esterified with phosphoric acid.

Ribosomes Small (approx. 20 nm. in diameter) subcellular ribonucleoprotein particles that are made up of two subparticles and are responsible for the translation of mRNAs into proteins.

Ribozymes RNA molecules having the ability to catalyze a variety of intra- and/or intermolecular reactions.

Rifampicin Synthetic antibiotic which binds to certain prokaryotic RNA polymerases, thereby inhibiting RNA synthesis.

RNA (ribonucleic acid) Polymer of ribonucleotides.

Rolling circle Mechanism for nucleic acid replication in which the template is a circular molecule (either DNA or RNA). The newly synthesized strand may either be released as a single copy or, more commonly, continue on to form *concatameric* progeny molecules. Prominent examples include the replication of bacteriophage lambda (double-stranded DNA) and viroids (single-stranded RNA).

S See **Sedimentation coefficient**.

^{35}S Radioactive isotope of sulfur, a ß emitter with a half-life of 87 days. Useful in both *protein synthesis* studies, because it can be incorporated into proteins via the sulfur-containing amino acids (methionine and cysteine), and *nucleotide sequence analysis*, where it can replace O to give nucleotides containing an α-thiophosphate bond.

SAM (S-adenosylmethionine) Intracellular source of activated methyl groups required for a variety of metabolic processes, including RNA and DNA methylation.

Salting out Precipitation of proteins or nucleic acids in concentrated salt solutions. Ammonium sulphate and lithium chloride are frequently used in the purification of proteins and RNA, respectively.

Sarcoma Cancer of the connective tissue.

Satellite RNA Small RNA molecules dependent upon a *helper virus* for their replication but showing little or no sequence similarity to the genome of their helper virus.

Satellite viruses Defective viruses that are replicated only in the presence of a specific *helper virus*.

SDS (Sodium dodecylsulfate) A strong anionic detergent.

Secondary cells Cells arising from proliferation of cultured *primary cells*. Unless they are transformed (e.g., HeLa cells) secondary cells are able to divide only a finite number of times.

Secondary structure Features of a macromolecular structure that are maintained by hydrogen bonds and other weak interactions. For proteins, such features include the well-known α-helix and ß-pleated sheet; for nucleic acids, the G-C and A-T(U) basepairing interactions.

Sedimentation coefficient Rate at which a macromolecule sediments under a defined gravitational force. Sedimentation coefficients are influenced by both the molecular weight and shape of a macromolecule; the basic unit is the *Svedberg (S)* which is 10^{-13} sec.

Semiconservative replication DNA replication mechanism in which both strands of a double-stranded DNA are used as templates and the two resulting progeny duplexes each contain one parental and one newly-synthesized strand.

Semipersistent transmission Relationship between a plant virus and its vector characterized by short acquisition period and no latent period, but the vector remains able to transmit the virus for hours to days.

Serotype Group of viruses or microorganisms that can be distinguished on the basis of their antigenic properties.

Shadow casting Classical technique used to prepare virus specimens for electron microscopy. Shadows outlining the virions are created by the accumulation of heavy metal atoms (e.g., platinum) deposited in a vacuum.

Sialic acid Neuraminic acid derivative in which free amino acid group is acetylated.

Signal peptide Short segment (usually 15–30 amino acids) at the N-terminus of a protein that allows it to pass through the membrane of the cell or organelle. The signal peptide is usually removed by a specific protease and, in these cases, is not present in the mature protein.

Silent gene (message) Gene that is not expressed because a potential ribosome binding site is blocked/unavailable.

Spermine, spermidine See **Polyamine**.

Spheroplast A bacterium that has lost most or all of its cell wall, usually as the result of enzymatic treatment.

Specific absorbance (A) Absorbance per unit mass of a substance, usually measured at the wavelength where absorbance is maximal. For 0.1% solutions (i.e., 1 mg/ml) the A_{260nm} is 25 for RNA and single-stranded DNA and 20 for double-stranded DNA. The corresponding A_{280nm} value for proteins is approximately 1, but the actual value depends on their aromatic amino acid content.

Splicing Specific cleavage and ligation of mRNA precursors leading to the ultimate removal of *introns* and joining of *exons* to form functional mRNAs.

Spongiform encephalopathy Term describing the degenerative changes that occur in the absence of an inflammatory reaction in the brains of animals (or humans) infected with agents such as *scrapie* or *Creutzfeldt-Jakob disease*.

Start codon See **Initiation codon**.

"Sticky" ends Complementary singlestranded tails projecting form otherwise double-stranded, helical nucleic acid molecules; produced by many Type II restriction endonucleases.

Subgenomic RNA Less-than-genome-length RNA found in infected cells and occasionally encapsidated. Formation of subgenomic RNAs facilitates virus gene expression by circumventing the eukaryotic ribosome's preference to initiate translation at the 5′ proximal open reading frame in a polycistronic mRNA.

Sucrose density gradient centrifugation Technique in which particles having different sedimentation coefficients are separated by sedimentation through a continuous or discontinuous gradient of sucrose solutions. Particles usually do not reach an equilibrium easily.

Supercoiling Coiling of a covalently closed circular, double-stranded DNA molecule such that it crosses over its own axis this action results from the action of *DNA gyrase* and other proteins.

Superinfection Attempt to infect a host with a second virus. May result in *interference*, *synergism*, *recombination*, or *phenotypic mixing*.

Suppressor-sensitive (sus) mutation Point mutation creating a termination codon that tends to be misread by a *suppressor tRNA*.

Suppressor tRNA tRNA molecule containing an altered anticodon that either reads a mutated codon in the same sense as the original or introduces a different amino acid that does not abolish protein function.

Susceptibility Property of a host that allows a virus (or other pathogen) to replicate in it.

Syncytia Multinucleate cells usually resulting from fusion.

Synergism A situation in which the symptoms caused by co-infection with two viruses are more severe than the sum of those caused by the individual viruses.

Systemic infection Infection resulting from the spread of virus from the site of infection to all or most of the cells of the host.

T antigens Proteins associated with tumor production and cell transformation by various viruses (e.g., SV40)

T cells Lymphocytes involved in cell-mediated immunity.

T1 ribonuclease Nuclease from *Aspergillus nidulans* that cleaves RNA only at the 3′ side of guanine residues.

Tail Structure present in the most complex DNA phages that mediates phage attachment to specific receptors on the host cells and acts as a passageway for transfer of the phage genome.

Tautomeric shift Reversible change in the location of hydrogen atom and double bond in a molecule that alters its chemical properties.

Temperate phage Bacterial virus that can exist (replicate) in both a *lysogenic* and *productive* state.

Temperature-sensitive (ts) mutation Mutation leading to gene products that are not functional over as broad a range of temperatures as the wild-type virus products.

Template Nucleic acid molecule from which a complementary nucleic acid molecule is synthesized.

Terminal redundancy Presence of identical sequences at both ends of a linear polynucleotide.

Termination codon One of three trinucleotides (UAG, UAA, and UGA) that cause protein synthesis to stop. Also called *stop codons* or *nonsense codons*.

Thermotherapy Curing of a host or cell line of a virus infection by heat (or rarely cold) treatment.

Theta structure Structure resembling the Greek letter ø that is formed by the movement of a *replication fork* in a circular DNA. DNA replication may be either uni- or bidirectional.

Thymine Pyrimidine base found in DNA.

Topoisomerase Enzyme that can change the *linking number* (i.e., the degree of supercoiling) of a DNA molecule.

Toxin Poisonous substance produced by many bacterial and some other cells, sometimes encoded by a virus. *Exotoxins* are released from the cell, whereas *endotoxins* are part of the cell surface.

Trans Configuration of two sites located on different DNA or RNA molecules; opposite of *cis*.

Transactivation Situation in which the product of one gene increases the expression of other gene(s) located some distance away.

Transcription Synthesis of complementary RNA strand from a DNA or RNA template.

Transduction Transfer of genes from one host cell to another by virus.

Transencapsidation See **Phenotypic mixing**

Transfection Initiation of a virus infection by inoculation with viral nucleic acid; a process similar to *transformation*.

Transfer RNA (tRNA) Small (approx. 75 nucleotide) RNAs responsible for decoding the genetic information in mRNA. Each species of tRNA is able to combine covalently with a specific amino acid and to hydrogen bond with at least one mRNA nucleotide triplet.

Transformation Gene transfer via DNA uptake by "competent" cells. Also used to describe the morphological changes in cells, often associated with malignancy.

Translation Process whereby the genetic information present in an mRNA molecule directs the order of amino acid polymerization during protein synthesis.

Transition Mutation in which one pyrimidine (purine) is substituted for the other (in contrast to *transversion*).

Transmission Transfer of a virus from an infected organism to a non-infected one.

Transposon DNA sequence that is able to insert itself into unrelated DNA sequences within the cell. The ends of a transposon usually contain *inverted repeats*.

Transversion Mutation in which a purine is replaced by a pyrimidine or vice versa.

Trypsin Proteolytic enzyme that is secreted by the pancreas and cleaves only on the carboxyl side of the basic amino acids, arginine and lysine. Its nonenzymatic precursor is *trypsinogen*.

Tubulin Important protein component of the *cytoskeleton*, aggregating in the form of *microtubules* that are essential in many cellular functions (e.g., mitosis).

Tumor promoter Substance or agent that helps the progression of a transformed cell. Croton oil, a source of phorbol esters, is a common tumor promoter.

Tumor virus Virus that induces the formation of a tumor.

Ultracentrifuge High-speed centrifuge that can attain speeds in excess of 70,000 rpm and centrifugal fields up to 500,000 times gravity and thus is capable of rapidly sedimenting macromolecules.
Ultraviolet (UV) light Electromagnetic radiation with wavelengths of 40–400nm. Damages nucleic acids, thus causing mutations and chromosome breaks. Also used in chemical analyses.
Uncoating Removal of the outer layer(s) of a virus particle following infection and leading to the release of the viral nucleic acid.
Unwinding protein Polypeptides that bind to, and thus stabilize, single-stranded DNA. In doing so, they tend to unwind the DNA double helix.
Uracil One of the two principal pyrimidine bases in RNA, the other being *cytosine*.
Uranyl acetate Often used as a *negative strain* in electron microscopy.
Urea Low-molecular-weight organic compound used to denature proteins and nucleic acids. Its formula is $CO(NH_2)_2$.

Vaccination Administration of a vaccine; immunization.
Vacuole Intracytoplasmic liquid-containing vesicle; particularly prominent structural feature of plant cells.
Vector Animals (usually insects) or plants (e.g. fungi) that transmit disease-causing agents, usually without becoming themselves diseased. More recently used to describe a self-replicating DNA molecule into which fragments of DNA can be inserted for **molecular cloning**.
Vertical transmission Passage of a viral genome from one host generation to the next, either as an integrated *provirus* or in close association with the host gametes (e.g., transovarial or seed transmission).
Vesicle Intracellular compartment enclosed by membranes derived from the endoplasmic reticulum.
Vinculin Protein component of the cytoskeleton that readily becomes phosphorylated.
Viremia Presence of virus in the blood.
Virion Virus particle—infectious or non-infectious.
Viropexis Putative special form of a phagocytosis by which some viruses enter animal cells through the plasma membrane.
Viroplasm Amorphous cytoplasmic inclusion body associated with certain plant virus infections; usually not surrounded by a membrane.
Virulence The ability of an infectious agent, such as a virus, to produce a disease.
Virulent virus Virus (especially a bacteriophage) that is unable to establish lysogeny; also known as a *lytic virus*. Sometimes used to denote a virus causing particularly severe disease symptom.
Virus factory Intracellular focus of active virus synthesis and/or maturation.
Virus-like particle Structure resembling a virus particle but which has not been shown to be pathogenic.
VPg Small virus-encoded protein attached through a phosphodiester linkage to the 5′–terminus of the virus nucleic acid. Abbreviation for *v*irion *p*rotein *g*enome-linked.

Western blot Immunological technique in which proteins that have been separated in a polyacrylamide gel are transferred (blotted) to an immobilizing matrix (e.g., nitrocellulose). The proteins can then be analyzed by monitoring their ability to react with suitably labeled antibodies.
Wheat germ extract Cell-free extract prepared from wheat germ (embryos) and commonly used for the *in vitro* translation of eukaryotic mRNAs.
Wobble hypothesis The ability of the third base in a tRNA anticodon to interact with more than a single base in the corresponding position of the mRNA codon. Explains the ability to certain tRNAs to decode more than one codon.

Xenotropic Subclassification of retroviruses denoting the productive infection only of cells of non-host origin.

Zoonosis Disease which is naturally transmitted between vertebrate animals and man.

Definitions of many other commonly-used virological terms can be found in **VIROLOGY—Directory and Dictionary of Animal, Bacterial and Plant Viruses** by Roger Hull, Fred Brown, and Chris Payne (Stockton Press, New York, 1989). 325 pp.

Index